W9-AFO-463

Lectures in Applied Mathematics
Volumes in This Series

Proceedings of the Summer Seminar, Boulder, Colorado, 1960

1 **Lectures in Statistical Mechanics,** *G. E. Uhlenbeck and G .W. Ford with E. W. Montroll*

2 **Mathematical Problems of Relativistic Physics,** *I. E. Segal with G. W. Mackey*

3 **Perturbation of Spectra in Hilbert Space,** *K. O. Friedrichs*

4 **Quantum Mechanics,** *R. Jost*

Proceedings of the Summer Seminar, Ithaca, New York, 1963

5 **Space Mathematics. Part 1,** *J. Barkley Rosser, Editor*

6 **Space Mathematics. Part 2,** *J. Barkley Rosser, Editor*

7 **Space Mathematics. Part 3,** *J. Barkley Rosser, Editor*

Proceedings of the Summer Seminar, Ithaca, New York, 1965

8 **Relativity Theory and Astrophysics. 1. Relativity and Cosmology,** *Jürgen Ehlers, Editor*

9 **Relativity Theory and Astrophysics. 2. Galactic Structure,** *Jürgen Ehlers, Editor*

10 **Relativity Theory and Astrophysics. 3. Stellar Structure,** *Jürgen Ehlers, Editor*

Proceedings of the Summer Seminar, Stanford, California, 1967

11 **Mathematics of the Decision Sciences, Part 1,** *George B. Dantzig and Arthur F. Veinott, Jr., Editors*

12 **Mathematics of the Decision Sciences, Part 2,** *George B. Dantzig and Arthur F. Veinott, Jr., Editors*

Proceedings of the Summer Seminar, Troy, New York, 1970

13 Mathematical Problems in the Geophysical Sciences. 1. Geophysical Fluid Dynamics, *William H. Reid, Editor*

14 Mathematical Problems in the Geophysical Sciences. 2. Inverse Problems, Dynamo Theory, and Tides, *William H. Reid, Editor*

Proceedings of the Summer Seminar, Potsdam, New York, 1972

15 Nonlinear Wave Motion, *Alan C. Newell, Editor*

Proceedings of the Summer Seminar, Troy, New York, 1975

16 Modern Modeling of Continuum Phenomena, *Richard C. DiPrima, Editor*

Proceedings of the Summer Seminar, Salt Lake City, Utah, 1978

17 Nonlinear Oscillations in Biology, *Frank C. Hoppensteadt, Editor*

Proceedings of the Summer Seminar, Cambridge, Massachusetts, 1979

18 Algebraic and Geometric Methods in Linear Systems Theory, *Christopher I. Brynes and Clyde F. Martin, Editors*

Proceedings of the Summer Seminar, Salt Lake City, Utah, 1980

19 Mathematical Aspects of Physiology, *Frank C. Hoppensteadt, Editor*

Proceedings of the Summer Seminar, Chicago, Illinois, 1981

20 Fluid Dynamics in Astrophysics and Geophysics, *Norman R. Lebovitz, Editor*

Proceedings of the Summer Seminar, Chicago, Illinois, 1982

21 Applications of Group Theory in Physics and Mathematical Sciences, *Moshe Flato, Paul Sally and Gregg Zuckerman, Editors*

22 Large-Scale Computations in Fluid Mechanics, Parts 1 and 2, *B. Engquist, S. Osher and R. C. J. Somerville, Editors*

23 Nonlinear Systems of Partial Differential Equations in Applied Mathematics, Parts 1 and 2, *Basil Nicolaenko, Darryl D. Holm and James M. Hyman, Editors*

24 Reacting Flows: Combustion and Chemical Reactors, Parts 1 and 2, *G. S. S. Ludford, Editor*

Proceedings of the Summer Seminar, Minneapolis, Minnesota, 1987

25 Computational Aspects of VLSI Design with an Emphasis on Semiconductor Device Simulation, *Randolph E. Bank, Editor*

Proceedings of the Summer Seminar, Fort Collins, Colorado, 1988

26 Computational Solution of Nonlinear Systems of Equations, *Eugene L. Allgower and Kurt Georg, Editors*

Computational Solution of Nonlinear Systems of Equations

LECTURES IN APPLIED MATHEMATICS

VOLUME 26

Computational Solution of Nonlinear Systems of Equations

Eugene L. Allgower
Kurt Georg
Editors

American Mathematical Society • Providence, Rhode Island

The Proceedings of the 1988 SIAM-AMS Summer Seminar on Computational Solution of Nonlinear Systems of Equations were prepared by the American Mathematical Society with support from the following sources: Air Force Office of Scientific Research, and the National Science Foundation through NSF Grant DMS-8714141.

1980 *Mathematics Subject Classification* (1985 *Revision*). 65H10, 65K05, 65K10, 65N10, 65N20, 65N25, 65N30, 90C30.

Library of Congress Cataloging-in-Publication Data

Computational solution of nonlinear systems of equations/Eugene L. Allgower, Kurt Georg, editors.

p. cm.–(Lectures in applied mathematics; ISSN 0075-8485; v. 26)

Proceedings of the 1988 SIAM-AMS Summer Seminar on Computational Solution of Nonlinear Systems of Equations, which was held July 18–29, 1988 at Colorado State University, Ft. Collins, Colorado with the support of the Air Force Office of Research and the National Science Foundation under grant no. DMS-8714141.

Includes bibliographical references.

ISBN 0-8218-1131-2

1. Differential equations, Nonlinear–Numerical solutions–Data processing–Congresses. I. Allgower, E. L. (Eugene L.) II. Georg, Kurt. III. SIAM-AMS Summer Seminar on Computational Solution of Nonlinear Systems of Equations (1988: Colorado State University) IV. United States. Dept. of the Air Force. Office of Scientific Research. V. National Science Foundation (U.S.) VI. Series: Lectures in applied mathematics (American Mathematical Society); v. 26.

QA372.C6374 1990 90-27

515′.355–dc20 CIP

Printed in the United States of America.

This publication was typeset using $\mathcal{A}\mathcal{M}\mathcal{S}$-TEX, the American Mathematical Society's TEX macro system.

10 9 8 7 6 5 4 3 2 1 94 93 92 91 90

Contents

Foreword

Nonlinear equations arise in essentially every branch of modern science, engineering, and mathematics. However, in only a scant few special cases is it possible to obtain useful solutions to nonlinear equations via analytical calculations by hand. This leads one to resort to computational methods of solving the nonlinear problem with which one is presented. This volume represents the proceedings of an AMS-SIAM Summer Seminar on "Computational Solution of Nonlinear Systems of Equations", which was held July 18–29, 1988 at Colorado State University, Ft. Collins, Colorado with the support of the Air Force Office of Research and the National Science Foundation under grant no. DMS-8714141.

The purpose of the seminar was to provide a wide-ranging survey of current major thrusts in the numerical solution of nonlinear equations. Among the currently active topics treated were: continuation methods, numerical bifurcation, nonlinear optimization, quasi-Newton and nonlinear conjugate gradient methods, complementarity problems, piecewise-linear methods, mesh refinement and multigrid methods, large scale PDE's, complexity of nonlinear methods, fractals. A highly successful evening talk on "The Beauty of Fractals" for a broader academic community was delivered by H.-O. Peitgen. He also kindly provided the illustration for the title page.

A list of speakers and titles of talks is presented below. During the conference the notes of 40 lectures totaling 777 pages were reproduced and distributed to the participants. The papers in this volume are more formal versions of some of these lectures and have undergone a standard refereeing process. During the conference it was suggested that it may be useful to make a collection of some nonlinear test problems for computational methods. Such a collection was undertaken and edited by Jorge Moré, and can be found at the end of the volume.

The proposal for the conference was approved by the AMS-SIAM Committee on Applied Mathematics. Members of the committee were C. M. Dafermos, J. M. Hyman, D. E. McClure, G. C. Papanicolau (Chairman), F. E. Sullivan, and R. F. Warming. The Organizing Committee for the seminar included E. L. Allgower (Chairman), H. B. Keller, H.-O. Peitgen, W. C. Rheinboldt and S. Smale. K. Georg came on board after the proposal was approved and may rightfully be regarded as a co-chairman.

In keeping with the current policy of the AMS, SIAM, and NSF regarding the Summer Seminars, an effort was made to provide assistance and stimulation to young researchers in the mathematical sciences. Thus the seminar involved lectures on both the entry and frontier levels for a number of topics. Consequently, some of the papers in these proceedings are of a survey nature. Furthermore, a significant number of young researchers were given an opportunity to present their work. Finally, participants on the graduate student level were generally excused from paying registration fees.

The conference was international as the participants represented at least 11 countries not in North America. Essentially all neighboring universities in Colorado and the surrounding states were represented as well as several industrial and national laboratories.

It gives us pleasure to express our appreciation to those to whom we are indebted for support and assistance in organizing and carrying the conference through to its conclusion. Robert Gaines, Chairman of the Department of Mathematics at Colorado State University, encouraged the submission of the proposal to hold an AMS-SIAM Summer Seminar, offered the use of department facilities and made available the very capable services of Laurie Feehan, who was the Local Coordinator of the conference. Mrs. Feehan handled this strenuous job with skill and charm. At every stage of the organizing we received skillful help from staff members of the AMS. James Maxwell, Executive Secretary of the AMS, assisted in preparing the proposals which ultimately yielded support for the conference from the Air Force Office of Sponsored Research and the National Science Foundation. Betty Verducci did a fine job as Conference Coordinator for all of the preplanning. Due to the illness of a loved one, she was unable to attend the conference and so Donna Salter and Laurie Feehan served as Conference Coordinators during the seminar. In the preparation of this volume we received excellent help from Donna Harmon of the AMS. Finally we wish to express our appreciation to the mathematics graduate students at Colorado State University who helped with many errands and chores in a very responsible and cheerful manner.

E. Allgower, K. Georg
Fort Collins, 1989

List of Lectures

G. Auchmuty: Duality Methods for Nonconvex Optimization

M. Berger (1): Antidotes to Nonintegrability

M. Berger (2): Bifurcation from Equilibria for Certain Infinite Dimensional Dynamical Systems

K. Böhmer (1): Computation of Bifurcating Manifolds at Higher Singular Points

K. Böhmer (2): Mesh Independence and Defect Correction

P. Chien: Secondary Bifurcations in the Buckling Problem

T. Coleman: On Characterizations of Superlinear Convergence for Constrained Optimization

J. Curry: Methods of Factoring Polynomials

S. K. Dey: A Degenerate Vectorized Implicit Finite Difference Algorithm for Partial Differential Equations

J. Duvallet: Computation of Solutions of Two-Point Boundary Value Problems by a Simplicial Homotopy Algorithm

R. E. Ewing: Nonlinear Systems Arising in Multicomponent and Multiphase Flows

R. Fletcher (1): Recent Methods for Nonlinear Programming

R. Fletcher (2): Low Storage Methods for Unconstrained Optimization

A. Galantai: Block ABS Methods for Nonlinear Systems of Algebraic Equations

K. Georg: Introduction to PL Methods

R. Guenther: Convergence of the Newton-Raphson Method for Boundary Value Problems of Ordinary Differential Equations

P. T. Harker: Recent Results on Finite-Dimensional Variational Inequalities and Nonlinear Complementarity Problems

J. Hunter: Numerical Solutions of Some Nonlinear Dispersive Wave Equations

H. Th. Jongen: Parametric Optimization: Critical Points and Local Minima

R. B. Kearfott: Interval Arithmetic Techniques in the Computational Solution of Nonlinear Systems of Equations

H. B. Keller (1–3): Continuation and Bifurcation in Scientific Computing (1–3)

T. Kelley: Operator Prolongation Methods for Nonlinear Equations

M. M. Kostreva: Computational Experience with Homotopy Continuation for Polynomial Models in Optimization and Noncooperative Game Theory

B. Lundberg: Adams-Bashforth Predictors for Continuation Methods

S. McCormick: Multi-level Adaptive Methods for Nonlinear Partial Differential Equations on Parallel Computers

S. McKay: Application of the Fast Adaptive Composite Grid Method for Solving Nonlinear Partial Differential Equations

R. Mejia: Interactive Program for Continuation of Solutions of Large Systems

H. Mittelmann (1): Continuation and Free Boundary Problems

H. Mittelmann (2): The Augmented Skeleton Method for Parametrized Capillary Surfaces

J. J. Moré (1): Numerical Solution of Nonlinear Equations: Ideas and Algorithms

J. J. Moré (2): Trust Region Methods for Systems of Nonlinear Equations

A. P. Morgan (1): Generically Nonsingular Polynomial Continuation

A. P. Morgan (2): Polynomial Continuation for Mechanism Design Problems

H.-O. Peitgen (1): Continuation of Homoclinic Bifurcation for Discretizations of Nonlinear Boundary Value Problems

H.-O. Peitgen (2): The Beauty of Fractals

A. Poore: Smooth Penalty Functions and Continuation Methods

G. Russo: A Particle Method for Collisional Kinetic Equations

R. Saigal: Computational Solution of Nonlinear Systems of Equations

T. Sauer: Some Remarks on the Numerical Solution of Polynomial Systems by Continuation Methods

J. Scheurle: Splitting of Separatrices and Chaos

P. Schmidt: PL Methods for Constructing a Numerical Implicit Function

K. Schmitt: Oscillatory Perturbations of Linear Problems at Resonance

R. B. Schnabel: Tensor Methods for Nonlinear Equations and Unconstrained Optimization

H. Schwetlick (1): Some Superlinearly Convergent Methods for Singular Nonlinear Equations

H. Schwetlick (2): Nonstandard Scaling Matrices in Trust Region Methods

K. Sikorski: Fast Algorithms for the Computation of Fixed Points

S. Smale: On the Foundations of Numerical Analysis

M. W. Smiley: Numerical Determination of Breathers and Forced Oscillations of Nonlinear Wave Equations

M. Sun: Numerical Solutions of Singular Stochastic Control Problems in Bounded Intervals

G. Tavares: PL Approximation to Manifolds and Its Application to Implicit ODEs

W. C. Thacker: Large Least Squares Problems in Oceanography and Meteorology

R. Vaillancour: Application of Julia-Fatou Iteration Theory in Dielectric Spectroscopy

H. Walker: Newton-like Methods for Underdetermined Systems

Y. Yamamoto: Fixed Point Algorithms for Stationary Point Problems

T. J. Ypma: Approximation of Large Sparse Jacobian Matrices by Finite Differences

Y. Yomdin: Introduction to an Effective Sard's Theorem

D. Zachmann: Modified Newton's Methods for Finite Difference Equations

Lectures in Applied Mathematics
Volume **26** (1990)

Numerically Stable Homotopy Methods without an Extra Dimension

EUGENE L. ALLGOWER[1,3,4,5] AND KURT GEORG[1,2,4]

Abstract. We give new versions of the global Newton method and the Kellogg–Li–Yorke method for calculating zero points and fixed points of nonlinear maps, which are numerically stable, but do not require an extra homotopy dimension. In addition, regularity results are established so that predictor-corrector continuation methods will lead to solutions if appropriate boundary conditions are satisfied.

1. Introduction. The global Newton method for calculating zeros of maps and the Kellogg–Li–Yorke method for calculating fixed points of maps have fallen somewhat out of favor vis-à-vis the global homotopy and linear homotopy methods respectively, because of observed numerical instabilities. The homotopy remedy for overcoming the numerical instability is usually achieved at the mild cost of increasing the dimension of the system by one parameter. Our aim here is to show that by an appropriate reparametrization one may overcome this and some other difficulties without increasing the dimension of the system. In the course of the discussion some striking analogies between the zero point or fixed point problems and aspects of the above-mentioned solution methods will become apparent. Before coming to these aspects, we

---

1980 *Mathematics Subject Classification* (1985 *Revision*). Primary 65H10.

[3]Partially supported by the Office of Naval Research under grant # URI - ONR N 00014 - 86K 0687.

[4]Partially supported by National Science Foundation grant # DMS - 8805682.

[5]Partially supported by the Alexander von Humboldt-Stiftung.

will give brief reviews of the main ideas of predictor-corrector continuation methods, the global Newton method, and the Kellogg–Li–Yorke method.

2. A brief review of predictor-corrector continuation methods. Underlying all of the zero point and fixed point methods we discuss below is the fact that an implicitly defined curve must be numerically traced. Consequently, we briefly review some ideas for doing this task. For readers wishing to learn more of the details we suggest **[3]**, **[14]**, or **[18]**, where also many further references may be found.

Let us make the following

2.1. HYPOTHESES.

(1) *$H : \mathbf{R}^{N+1} \to \mathbf{R}^N$ is a C^∞-map;*
(2) *0 is a regular value of H.*

Under (2.1), it is well known (see e.g. **[16]**) that $H^{-1}(0)$ consists of finitely many smooth curves each of which is homeomorphic either to $\mathbf{R}$ or to the unit circle S^1. The task now is to numerically trace a curve $c \subset H^{-1}(0)$. For simplicity we can assume that c is parametrized according to arclength s. The tangent $\dot{c}(s)$ to $c(s)$ is given by $\tau\big(H'(c(s))\big)$, where H' denotes the Jacobian of H and τ denotes a normalized vector in the kernel of H'.

2.2. DEFINITION. Let A be an $(N+1)\times N$ matrix which has full rank. The *tangent vector* $\tau(A) \in \mathbf{R}^{N+1}$ is uniquely defined by the following three conditions:

(1) $$A\,\tau(A) = 0\,;$$

(2) $$\|\tau(A)\|_2 = 1\,;$$

(3) $$\det\begin{pmatrix} A \\ \tau(A)^* \end{pmatrix} > 0\,.$$

For this special case the *Moore–Penrose inverse* A^+ is defined by the equation

(4) $$\begin{pmatrix} A \\ \tau(A)^* \end{pmatrix}^{-1} = (A^+\,,\ \tau(A))\,.$$

Here * denotes transposition.

Differentiating the equation $H\big(c(s)\big) = 0$ with respect to arclength s, one can view a curve $c(s)$ in $H^{-1}(0)$ as the solution of the following autonomous initial value problem:

$$\text{(2.3)} \qquad \begin{aligned} \dot{x} &= \tau(H'(x)); \\ x(0) &\in H^{-1}(0). \end{aligned}$$

The point $x(0)$ is a given starting point, and we assume positive orientation. One could employ an initial value solver on (2.3) to trace $c(s)$. However, it is preferable to use a predictor-corrector continuation method because, contrary to the situation with the usual initial value solvers, it is possible to precisely correct back *orthogonally* to the curve $c(s)$ via the Newton method:

$$\text{(2.4)} \qquad x_{i+1} = x_i - H'(x_i)^+ H(x_i).$$

This method is locally quadratically convergent; see, e.g., [**5**]. The method consists of the following basic steps:

2.5. GENERIC EULER–NEWTON METHOD. *comment:*

input
  begin
  $x \in \mathbf{R}^{N+1}$ such that $H(x) = 0$; *initial point*
  $h > 0$; *initial steplength*
  end;
repeat
  approximate $A :\approx H'(x)$; *approximate Jacobian*
  $x := x + h\tau(A)$; *predictor step*
  repeat
    approximate $A :\approx H'(x)$; *approximate Jacobian*
    $x := x - A^+ H(x)$; *corrector loop*
  until convergence;
  choose a new steplength $h > 0$; *steplength adaptation*
until traversing is stopped.

Numerous variations of the Euler–Newton method are possible, e.g.,

→ modifications of Newton's method can be used, i.e., the method of obtaining the approximation $A :\approx H'(x)$ may vary;
→ various stepsize selection strategies can be incorporated; see, e.g., [**7**], [**10–11**], [**19**];
→ special points on the curve, e.g., where a homotopy parameter attains the value 1, can be approximated by various techniques; see [**4**, chap. 9] or [**10**].

3. Remarks on the numerical linear algebra. Let us briefly describe how the numerical linear algebra involved in algorithm (2.5) can be performed. Given an $N \times (N+1)$ matrix A which has maximal rank, let us briefly explain how the tangent vector $\tau(A)$ is calculated and how the system

$$\left.\begin{aligned} Aw &= b \\ \tau(A)^* w &= 0 \end{aligned}\right\} \iff w = A^+ b \tag{3.1}$$

can be solved for w whenever a vector $b \in \mathbf{R}^N$ is given. For our first case, assume that a QR factorization of A^* is available:

$$A^* = Q \begin{pmatrix} R \\ 0^* \end{pmatrix},$$

where Q is an $(N+1) \times (N+1)$ orthogonal matrix and R is a nonsingular $N \times N$ upper triangular matrix. Then the last column of Q, viz., $z := Qe_{N+1}$, satisfies (2.2)(1)–(2), and hence $\tau(A) = \pm z$, accordingly as $\det Q \det R$ is positive or negative, since

$$\det \begin{pmatrix} A \\ z^* \end{pmatrix} = \det(A^*, z) = \det Q \det \begin{pmatrix} R & 0 \\ 0^* & 1 \end{pmatrix} = \det Q \det R\,. \tag{3.2}$$

Now, $\det R$ is the product of the diagonal elements of R, and its sign is obvious. Also $\operatorname{sign} \det Q$ is usually easily obtained. For example, if Givens rotations are used, it is equal to unity. If Householder reflections are used, each reflection changes the sign, and so $\operatorname{sign} \det Q = (-1)^p$, where p is the number of reflections which are involved in the factorization of A^* by Householder's method. In any event, the question of determining $\tau(A)$ is easily resolved. Furthermore, the numerical solution of (3.1) is accomplished by a forward solving of $R^* u = b$ and a matrix multiplication

$$w := Q \begin{pmatrix} u \\ 0 \end{pmatrix}.$$

Let us now discuss how in general any linear solver can be incorporated into the continuation methods which we have outlined above; see also [2]. In our situation, a linear solver might be generically described as follows: Given an $N \times (N+1)$ matrix A and some vector $e \in \mathbf{R}^{N+1}$ which is not yet specified, we have a method for obtaining the solution $x \in \mathbf{R}^{N+1}$ for the linear system

$$\begin{aligned} Ax &= y, \\ e^* x &= 0, \end{aligned} \tag{3.3}$$

whenever $y \in \mathbf{R}^N$ is given. Among such methods might be linear conjugate gradient methods, direct factorization methods, etc. The choice of the vector e in (3.3) may be regarded as representing a local parametrization, which usually is changed in the process of numerically traversing a solution curve. Of primary importance in the choice of e is its influence upon the condition of the coefficient matrix in (3.3), viz., we should require that

$$\operatorname{cond}\begin{pmatrix} A \\ e^* \end{pmatrix} \approx \sqrt{\operatorname{cond}(AA^*)} \tag{3.4}$$

are approximately of the same order. Intuitively speaking, the vector e should be as parallel as possible to $\ker A$. A very typical choice for e is the ith coordinate unit vector, where the coordinate i must be carefully chosen. This leads to deleting the corresponding column and coordinate in (3.3). Let us show how the tangent vector $\tau(A)$ and the numerical calculation $w := A^+b$ can be cheaply obtained. For convenience, let us denote by

$$x = By \tag{3.5}$$

the solution operator of (3.3). We emphasize that the $(N+1)\times N$ matrix B is not explicitly given, but instead we have some way of calculating the *result* $x := By$.

The tangent vector $\tau(A)$ is determined as follows. By its definition B satisfies

$$\begin{aligned} AB &= \mathrm{Id}\,; \\ e^*B &= 0^*\,. \end{aligned} \tag{3.6}$$

If we set

$$z := e - BAe\,, \tag{3.7}$$

then it can be seen that

$$\tau(A) = \pm\frac{z}{||z||}\,. \tag{3.8}$$

We note that the cost of calculating $\tau(A)$ requires essentially one calculation of Ae (which is cost free in case $e = e_i$) and one solving of (3.3), i.e., $x := B(Ae)$.

In most applications, the choice of sign in (3.7) will be clear from the context; e.g., we take the tangent which has a small angle with a previously obtained tangent along the curve. Occasionally, it may be desirable to explicitly obtain this sign by calculating

$$\operatorname{sign}\det\begin{pmatrix} A \\ z^* \end{pmatrix} = \operatorname{sign}\det\begin{pmatrix} A \\ e^* \end{pmatrix}, \tag{3.9}$$

which may be cheaply calculated for some linear solvers of (3.3).

Using the tangent vector $\tau(A)$ which we already obtained in the previous step, it is readily checked that

$$A^+ = \left[\mathrm{Id} - \tau(A)\,\tau(A)^*\right] B.$$

Hence, once $\tau(A)$ has been obtained, the cost of calculating $w := A^+y$ amounts to one solving of (3.3), i.e., $x := By$, and then calculating $w = x - \left[\tau(A)^*x\right]\tau(A)$, which is essentially the cost of one scalar product.

4. The global Newton method. Let us begin by stating the problems we want to study and the underlying assumptions. These assumptions could be considerably relaxed in our discussions below. It is not our aim to make the most general hypotheses. Since we want only to convey the essential ideas and have our discussions simple, we make the following

4.1. HYPOTHESES.

(1) *$f : \mathbf{R}^N \to \mathbf{R}^N$ is a C^∞-map;*

(2) *$\Omega \subset \mathbf{R}^N$ is an open bounded set having a smooth connected boundary $\partial\Omega$;*

(3) *0 is a regular value of f.*

The global Newton method calculates a zero point of f in Ω. This method has been promulgated by Branin **[6]** and has found frequent use in scientific applications. Smale **[20]** has studied this method from a theoretical standpoint and given an existence theorem which we will state below. The method consists of the following steps.

4.2. GLOBAL NEWTON METHOD.

(1) *Choose a starting point $p \in \partial\Omega$;*

(2) *Follow the curve defined by the initial value problem*

$$\begin{aligned} \dot{x} &= -\lambda(x) f'(x)^{-1} f(x) \\ x(0) &= p \end{aligned} \tag{4.3}$$

into Ω until a zero point of f is found **(success)** *or until the curve exits from Ω without having found a zero point* **(failure)**.

A number of choices for $\lambda(x)$ are possible. A standard choice is $\lambda(x) = \det f'(x)$. In order to guarantee success of the global Newton method a boundary condition needs to be satisfied. Smale **[20]** has shown that success is assured under the following

4.4. BOUNDARY CONDITION. *For all $p \in \partial\Omega$*

(1) $f(p) \neq 0$;

(2) *$f'(p)$ is nonsingular;*
(3) *the Newton direction $-f'(p)^{-1}f(p)$ is not tangent to $\partial\Omega$ at p.*

We note that the factor $\lambda(x)$ in (4.3) allows the overcoming of simple singular points of f in the flow, but nevertheless the evaluation of the right-hand side remains numerically unstable. Keller **[13]** observed that the solution of (4.3) can be obtained in a numerically stable way using his global homotopy method. Independently, Garcia and Gould **[8]**, **[9]** obtained similar results. The global homotopy method consists of the following steps.

4.5. GLOBAL HOMOTOPY METHOD.
(1) *Choose a starting point $p \in \partial\Omega$;*
(2) *Follow the curve defined by the equation*

$$f(x) - tf(p) = 0 \tag{4.6}$$

into the cylinder $\Omega \times \mathbf{R}$, starting from $(p, 1)$ until the level $t = 0$ is reached **(success)** *or until the curve exits from $\Omega \times \mathbf{R}$ without reaching the level $t = 0$* **(failure)**.

The set $\{t \mid f(x) = tf(p),\ x \in \overline{\Omega}\}$ is bounded for $f(p) \neq 0$. Hence, if 0 is a regular value of the map $(x, t) \mapsto f(x) - tf(p)$, then the curve of (4.5) must exit $\Omega \times \mathbf{R}$. As Keller has observed, the relationship between the global Newton method (4.2) and his global homotopy method (4.5) is easily seen by differentiating (4.6) according to arclength to obtain

$$\dot{x} = \frac{\dot{t}}{t} f'(x)^{-1} f(x).$$

Comparing this with (4.3) we see that the x-component of the solution curve for (4.6) and the solution curve for (4.3) are the same — only their parametrizations are different. However, the global homotopy method (4.5) handles singular points of f in a more natural way. It also enabled Keller to show success for method (4.5), i.e., to give a simple proof of Smale's theorem that the solution curve of (4.2) hits a zero point of f before leaving the domain Ω if the boundary condition (4.4) is satisfied.

Let us define the global homotopy $\widetilde{H} : \mathbf{R}^N \times \mathbf{R} \times \partial\Omega \longrightarrow \mathbf{R}^N$ by

$$\widetilde{H}(x, t, p) := f(x) - tf(p). \tag{4.7}$$

In order to guarantee that the solution curves given by (4.3) and (4.6) are well defined and smooth, we have to assume that the starting point $p \in \partial\Omega$ is chosen in such a way that the restricted map

$$\widetilde{H}(\cdot, \cdot, p) : \mathbf{R}^N \times \mathbf{R} \to \mathbf{R}^N$$

has 0 as a regular value. However, the proof that this is true for almost all $p \in \partial\Omega$ becomes technically complicated, because p only varies over the surface $\partial\Omega$. Percell **[17]** gave such a proof.

We now want to give a numerically stable version of the global Newton method. Using the boundary condition (4.4)(1) we obtain from (4.6)

$$t = \frac{f(p)^* f(x)}{\|f(p)\|_2^2}. \tag{4.8}$$

Thus the map (4.7) induces a new map $H : \mathbf{R}^N \times \partial\Omega \to \mathbf{R}^N$ defined by

$$H(x,p) := f(x) - t(x,p)f(p),$$

where

$$t(x,p) := \frac{f(p)^* f(x)}{\|f(p)\|_2^2}. \tag{4.9}$$

Hence for $p \in \partial\Omega$ and $x \in \mathbf{R}^N$, the value $H(x,p)$ is obtained by orthogonally projecting $f(x)$ onto $\{f(p)\}^\perp$. To show that the homotopy method defined by (4.9) is numerically stable and that for almost all $p \in \partial\Omega$ the solution curve $\mathscr{C}_p \subset H^{-1}(0)$ with starting point $x = p$ consists of regular points of H, we use the following partial derivatives which are routine to check. Note that t_x and t_p are row vectors.

4.10. EQUATIONS.

(1) $$t_x(x,p) = \frac{f(p)^* f'(x)}{\|f(p)\|_2^2};$$

(2) $$H_x(x,p) = f'(x) - f(p)t_x(x,p) = \left(\mathrm{Id} - \frac{f(p)f(p)^*}{\|f(p)\|_2^2}\right) f'(x);$$

(3) $$t_p(x,p) = \frac{1}{\|f(p)\|_2^2}\big(f(x) - 2t(x,p)f(p)\big)^* f'(p);$$

(4) $$t_p(x,p) = -\frac{t(x,p)}{\|f(p)\|_2^2} f(p)^* f'(p) \quad \text{for} \quad H(x,p) = 0;$$

(5) $$H_p(x,p) = -t(x,p)f'(p) - f(p)t_p(x,p);$$

(6) $$H_p(x,p) = -t(x,p)\left(\mathrm{Id} - \frac{f(p)f(p)^*}{\|f(p)\|_2^2}\right) f'(p) \quad \text{for} \quad H(x,p) = 0.$$

From (4.8)–(4.10) and since $\|f(p)\|_2 \neq 0$ we see that the evaluations of $H(x,p)$ and $H_x(x,p)$, which would be required for a predictor-corrector tracing of $H(x,p) = 0$, are numerically stable.

4.11. PROPOSITION. *Let $x \in \mathbf{R}^N$, $p \in \partial\Omega$ be such that $H(x,p) = 0$. Let T_p denote the tangent space of $\partial\Omega$ at p. Then for the Jacobian $H'(x,p) = \big(H_x(x,p), H_p(x,p)\big)$ we have*

$$H'(x,p)\big(\mathbf{R}^N \times T_p\big) = \{f(p)\}^{\perp}.$$

PROOF. We have two cases to consider.

(1) $t(x,p) = 0$. In this case $H(x,p) = 0$ implies $f(x) = 0$. Thus x is a regular point of f by (4.1)(3). Then $H_x(x,p)(\mathbf{R}^N) = \{f(p)\}^{\perp}$ by (4.10)(2).

(2) $t(x,p) \neq 0$. We prove that $H_p(x,p)(T_p) = \{f(p)\}^{\perp}$. As a consequence of (4.10)(6) we need only to show the following:

$$u \in \mathbf{R}^N,\ u \perp f(p),\ u \perp f'(p)(T_p) \quad \Longrightarrow \quad u = 0. \tag{4.12}$$

Since $f'(p)$ is nonsingular by (4.4)(2), we have span

$$\{f(p), f'(p)(T_p)\} = \mathbf{R}^N$$

by (4.4)(3). This proves our claim (4.12).
Hence the assertion of the proposition follows. □

Now we are prepared to show the following regularity result.

4.13. THEOREM. *For almost all $p \in \partial\Omega$, the homotopy map $H(\cdot,p)$: $\mathbf{R}^N \to \{f(p)\}^{\perp}$ defined by* (4.8)–(4.9) *has* 0 *as a regular value.*

PROOF. Let us choose a fixed $q \in \partial\Omega$ and consider the orthogonal projection $P_q : \mathbf{R}^N \to \{f(q)\}^{\perp}$. By the previous proposition there exists a q-neighborhood $U_q \subset \partial\Omega$ such that $P_q H'(x,p)\big(\mathbf{R}^N \times T_p\big) = \{f(q)\}^{\perp}$ for all $p \in U_q$ and $x \in \mathbf{R}^N$ such that $H(x,p) = 0$. By a generalized version of Sard's theorem (see, e.g., **[1]** or **[12]**), we conclude that $P_q H(\cdot,p) : \mathbf{R}^N \to \{f(q)\}^{\perp}$ has 0 as a regular value for almost all $p \in U_q$. Since $H(\mathbf{R}^N, p) = \{f(p)\}^{\perp}$ and P_q is only an orthogonal projection onto $\{f(q)\}^{\perp}$, it follows that also $H(\cdot,p) : \mathbf{R}^N \to \{f(p)\}^{\perp}$ has 0 as a regular value for almost all $p \in U_q$. Since $\partial\Omega \subset \mathbf{R}^N$ is compact, we can cover $\partial\Omega$ with finitely many U_q's, and the theorem follows. □

We conclude this section with some brief remarks about implementing a predictor-corrector method for our version of the global Newton method defined by the homotopy (4.9) which does not have an extra dimension. Given a starting point $p \in \partial\Omega$, we fix $N-1$ basis vectors $u_1, u_2, \ldots, u_{N-1}$ of $\{f(p)\}^{\perp}$, and define $\widehat{H} : \mathbf{R}^N \to \mathbf{R}^{N-1}$ by

$$\widehat{H}(x) := \begin{pmatrix} u_1^* H(x,p) \\ \vdots \\ u_{N-1}^* H(x,p) \end{pmatrix}.$$

Then the equation $\widehat{H}(x) = 0$ is to be used in the general Euler–Newton method outlined in Sections 2 and 3.

5. The Kellogg–Li–Yorke method. The method of Kellogg, Li, and Yorke **[15]** calculates a fixed point of $f : \overline{\Omega} \to \overline{\Omega}$ where $\overline{\Omega}$ is a compact convex set in $\mathbf{R}^N$ and f is a smooth map. Analogously to our sketch of the global Newton method we again simplify our discussion by making the following somewhat stronger

5.1. HYPOTHESES.
(1) $f : \mathbf{R}^N \to \mathbf{R}^N$ *is a* C^∞*-map;*
(2) $\Omega = \{x \in \mathbf{R}^N \mid \|x\|_2 < 1\}$ *is the open unit ball;*
(3) 0 *is a regular value of* $\mathrm{Id} - f$;
(4) $\overline{f(\mathbf{R}^N)} \subset \Omega$.

Let $C := \{x \in \mathbf{R}^N \mid f(x) = x\}$ be the fixed point set of f. Define a map $G : \overline{\Omega} \setminus C \to \partial\Omega$ by $G(x) = f(x) + \mu(x)(x - f(x))$, where $\mu(x) > 0$ is uniquely defined via the condition $G(x) \in \partial\Omega$. The method consists of the following steps.

5.2. KELLOGG–LI–YORKE METHOD.
(1) *Choose a starting point* $p \in \partial\Omega$;
(2) *trace the curve defined by*

$$G(x) := f(x) + \mu(x)(x - f(x)) = p$$

where

$$\mu(x) := \frac{1 - p^* f(x)}{p^*(x - f(x))} \tag{5.3}$$

inward into Ω *starting from* $x = p$, $\mu(p) = 1$ *until a fixed point of* f *is approached* **(success)** *or until the curve exits from* Ω *without having found a fixed point* **(failure)**.

The classical *boundary condition* in this case comes from the Brouwer fixed point theorem, i.e., in our case we assume the somewhat stronger condition (5.1)(4). By Sard's theorem (cf. **[16]**), almost all $p \in \partial\Omega$ are regular values of G, and for such starting points p, Kellogg–Li–Yorke showed the success of method (5.2) under the hypothesis of the Brouwer fixed point theorem.

By examining (5.3) we can see that for x near C the determination of $\mu(x)$ can become numerically unstable and yet this is precisely what one hopes to do, viz., approach C via the curve defined by $G(x) = p$. In order to overcome this difficulty, we replace $\mu(x,p)$ by $(1 - t(x,p))^{-1}$

in the equation (5.3) to obtain a zero point problem for the following homotopy $H : \mathbf{R}^N \times \partial\Omega \to \mathbf{R}^N$ defined by

$$H(x,p) := x - p - t(x,p)\big(f(x) - p\big),$$

where

$$t(x,p) := \frac{p^*(x-p)}{p^*\big(f(x)-p\big)}. \tag{5.4}$$

Clearly, for fixed $p \in \partial\Omega$, the range of the restricted map $H(\cdot,p)$ is contained in $\{p\}^\perp$. Much of the following discussion is now similar to that which was made in Section 4 for the global Newton method. Hence we may abbreviate some of the remarks. Let us list the partial derivatives to be used below. They are routine to check.

5.5. EQUATIONS.

(1) $t_x(x,p) = \dfrac{p^*}{p^*\big(f(x)-p\big)}\big(\mathrm{Id} - t(x,p)f'(x)\big);$

(2) $t_p(x,p) = \dfrac{1}{p^*\big(f(x)-p\big)}\left[(x-2p)^* - t(x,p)\big(f(x)-2p\big)^*\right];$

(3) $t_p(x,p) = \dfrac{t(x,p)-1}{p^*\big(f(x)-p\big)}p^* \quad \text{for} \quad H(x,p) = 0;$

(4) $H_x(x,p) = \left(\mathrm{Id} - \dfrac{\big(f(x)-p\big)p^*}{p^*\big(f(x)-p\big)}\right)\big(\mathrm{Id} - t(x,p)f'(x)\big);$

(5) $H_p(x,p) = \big(t(x,p)-1\big)\,\mathrm{Id} - \big(f(x)-p\big)t_p(x,p);$

(6) $H_p(x,p) = \big(t(x,p)-1\big)\left(\mathrm{Id} - \dfrac{\big(f(x)-p\big)p^*}{p^*\big(f(x)-p\big)}\right) \quad \text{for} \quad H(x,p) = 0.$

Since

$$\inf_{x\in\mathbf{R}^N} \left|p^*\big(f(x)-p\big)\right| > 0$$

by our assumption (5.1)(4), we see from the above equations (5.5) that the evaluations of $H(x,p)$ and $H_x(x,p)$, which would be required for an Euler–Newton tracing of $H(x,p) = 0$ are numerically stable.

5.6. PROPOSITION. *Let $x \in \mathbf{R}^N$, $p \in \partial\Omega$ be such that $H(x,p) = 0$. Let T_p denote the tangent space of $\partial\Omega$ at p, i.e., $T_p = \{p\}^\perp$. Then for the Jacobian $H'(x,p) = \big(H_x(x,p), H_p(x,p)\big)$, we have*

$$H'(x,p)\big(\mathbf{R}^N \times T_p\big) = \{p\}^\perp.$$

PROOF. We have two cases to consider.

(1) $t(x,p) = 1$. In this case $H(x,p) = 0$ implies $f(x) = x$. Thus by (5.1)(3) x is a regular point of $\mathrm{Id} - f$. Then $H_x(x,p)(\mathbf{R}^N) = \{p\}^\perp$ by (5.5)(4).

(2) $t(x,p) \neq 1$. As an immediate consequence of (5.5)(6), we obtain $H_p(x,p)(T_p) = \{p\}^\perp$.

Hence the assertion of the proposition follows. □

In a fashion similar to the discussion for Theorem 4.13, the above proposition gives us the following

5.7. THEOREM. *For almost all $p \in \partial\Omega$, the homotopy map $H(\cdot,p)$: $\mathbf{R}^N \to \{p\}^\perp$ defined by* (5.6)–(5.7) *has* 0 *as a regular value.*

Analogously to the remarks at the end of Section 4, we describe how to implement an Euler–Newton method for our version of the Kellogg–Li–Yorke homotopy defined by (5.4). Given a starting point $p \in \partial\Omega$, we fix $N-1$ basis vectors $u_1, u_2, \ldots, u_{N-1}$ of $\{p\}^\perp$, and define $\widehat{H} : \mathbf{R}^N \to \mathbf{R}^{N-1}$ by

$$\widehat{H}(x) := \begin{pmatrix} u_1^* H(x,p) \\ \vdots \\ u_{N-1}^* H(x,p) \end{pmatrix}.$$

Then the equation $\widehat{H}(x) = 0$ is to be used in the general Euler–Newton method outlined in Sections 2 and 3.

REFERENCES

1. R. Abraham and J. Robbin, *Transversal Mappings and Flows*, W. A. Benjamin, New York, Amsterdam, 1967.

2. E. L. Allgower, C.-S. Chien, and K. Georg, *Large sparse continuation problems*, J. Comput. Appl. Math. **26** (1989), 3–21.

3. E. L. Allgower and K. Georg, *Predictor-corrector and simplicial methods for approximating fixed points and zero points of nonlinear mappings*, Mathematical Programming: The State of the Art, A. Bachem, M. Grötschel, and B. Korte (eds.), Springer-Verlag, Berlin, Heidelberg, New York, 1983, pp. 15–56.

4. ——, *Introduction to numerical continuation methods*, Springer-Verlag, Berlin, Heidelberg, New York, 1990.

5. A. Ben-Israel, *A Newton-Raphson method for the solution of systems of equations*, J. Math. Anal. Appl. **15** (1966), 243–252.

6. F. H. Branin, Jr., *Widely convergent methods for finding multiple solutions of simultaneous nonlinear equations*, IBM J. Res. Develop. **16** (1972), 504–522.

7. C. Den Heijer and W. C. Rheinboldt, *On steplength algorithms for a class of continuation methods*, SIAM J. Numer. Anal **18** (1981), 925–948.

8. C. B. Garcia and F. J. Gould, *A theorem on homotopy paths*, Math. Oper. Res. **3** (1978), 282–289.

9. ——, *Relations between several path following algorithms and local and global Newton methods*, SIAM Review **22** (1980), 263–274.

10. K. Georg, *Zur numerischen Realisierung von Kontinuitätsmethoden mit Prädiktor-Korrektor- oder simplizialen Verfahren*, Habilitationsschrift, University of Bonn, 1982.

11. ——, *A note on stepsize control for numerical curve following*, Homotopy Methods and Global Convergence, B. C. Eaves, F. J. Gould, H.-O. Peitgen, and M. J. Todd (eds.), Plenum Press, New York, pp. 145–154.

12. M. W. Hirsch, *Differential topology*, Springer-Verlag, Berlin, Heidelberg, New York, 1976.

13. H. B. Keller, *Global homotopies and Newton methods*, Recent Advances in Numerical Analysis, C. De Boor and G. H. Golub (eds.), Academic Press, New York, London, 1978, pp. 73–94.

14. ——, *Lectures on numerical methods in bifurcation problems*, Springer-Verlag, Berlin, Heidelberg, New York, 1987.

15. R. B. Kellogg, T. Y. Li, and J. A. Yorke, *A constructive proof of the Brouwer fixed point theorem and computational results*, SIAM J. Numer. Anal. **13** (1976), 473–483.

16. J. W. Milnor, *Topology from the differential viewpoint*, The University Press of Virginia, Charlottesville, Virginia, 1969.

17. P. Percell, *Note on a global homotopy*, Numer. Func. Anal. Optim. **2** (1980), 99–106.

18. W. C. Rheinboldt, *Numerical analysis of parametrized nonlinear equations*, John Wiley and Sons, New York, 1986.

19. H. Schwetlick and J. Cleve, *Higher order predictors and adaptive steplength control in path following algorithms*, SIAM J. Numer. Anal. **14** (1987), 1382–1393.

20. S. Smale, *A convergent process of price adjustment and global Newton method*, J. of Math. Econom. **3** (1976), 1–14.

[1]Department of Mathematics, Colorado State University, Ft. Collins, Colorado 80523

[2]Institut für Angewandte Mathematik, Universität Bonn, 53 Bonn, West Germany

Lectures in Applied Mathematics
Volume 26 (1990)

Duality Algorithms for Smooth Unconstrained Optimization

GILES AUCHMUTY

1. Introduction. In this paper, we shall describe some algorithms for finding critical points of a general smooth function of n real variables.

The algorithms are based on a decomposition theorem of Pommellet **[8]** and the use of nonconvex duality theory as described in **[1]**. A more general study of these algorithms is given in **[2]**; here we concentrate on the special case of unconstrained, smooth optimization.

Each step in these algorithms involves an unconstrained, strictly convex, smooth minimization problem. The successive iterates may be chosen by any of a number of methods involving either an exact minimization, or else a sufficient decrease, at each step. The method does not require any knowledge that the Hessian is nonsingular and one has convergence theorems whenever the iterative sequence remains bounded.

In Section 2, we describe the basic decomposition, and the primal, Lagrangian, and dual, problems associated with this decomposition. The correspondence of critical points of the three problems is described in Section 3, where it is shown that if $(\tilde{\mathbf{x}}, \tilde{\mathbf{y}})$ is a critical point of the Lagrangian, then $\tilde{\mathbf{x}}$ is a critical point of the primal problem, and $\tilde{\mathbf{y}}$ is a critical point of the dual problem. Moreover, the functions have the same value at these points and the corresponding critical points are of the same type (in terms of Morse index).

In Section 4, some algorithms for minimizing the Lagrangian are described. They are based on successively minimizing, or descending, in the primal and the dual variable consecutively. Each of these is a strictly

This research was partially supported by NSF grant DMS 8701886 and the AFOSR.

convex optimization problem and one can show that either the algorithm terminates in a finite number of steps, or else, all limit points of the iterative sequence, provided it remains bounded, are critical points of the Lagrangian.

The algorithms and results described here are related to many recent results on the Morse theory and optimization of difference of convex (d.c.) functions. These include the papers of Bougeard, Penot, and Tuy, amongst others, in [3] and [7]. A more general study of block minimization in different groups of variables is described in [5].

The author would like to thank H. Th. Jongen for bringing Pommellet's result to his attention, and to the organizers of the conference for arranging the pleasant environment that stimulated this paper.

2. Dual problems for unconstrained smooth optimization. A function $f\colon \mathbf{R}^n \to \mathbf{R}$ is said to be coercive if

$$\liminf_{\|\mathbf{x}\|\to\infty} \frac{f(\mathbf{x})}{\|\mathbf{x}\|} = +\infty. \tag{2.1}$$

Throughout this paper we shall assume

(A1) $f\colon \mathbf{R}^n \to \mathbf{R}$ is k-times continuously differentiable (C^k-) with $k \geq 2$ and coercive on $\mathbf{R}^n$.

Our interest is in studying the critical points of f and the problem (P) of minimizing f on $\mathbf{R}^n$. Let

$$\alpha = \inf_{\mathbf{x}\in\mathbf{R}^n} f(\mathbf{x}). \tag{2.2}$$

When f obeys (A1) then α will be finite and it is attained. α is called the value of this problem (P).

The starting point for this analysis is the following modification of a theorem of Pommellet ([8], Theorem 0.16).

THEOREM 2.1. *Suppose $f\colon \mathbf{R}^n \to \mathbf{R}$ is C^k with $k \geq 2$. Then there exist functions $f_1, f_2\colon \mathbf{R}^n \to \mathbf{R}$ which are C^k and strictly convex with*

$$f(\mathbf{x}) = f_1(\mathbf{x}) - f_2(\mathbf{x}). \tag{2.3}$$

When f is coercive so is f_1.

PROOF. Pommellet constructs a function $g\colon \mathbf{R}^n \to \mathbf{R}$ which is real-analytic so that $f + g$ is convex on $\mathbf{R}^n$ and then the result holds with $f_1 = f + g$ and $f_2 = g$.

Obviously, if f is C^k so are f_1 and f_2. If f_1 and f_2 are not strictly convex, then define

$$\tilde{f}_1(\mathbf{x}) = f_1(\mathbf{x}) + \tfrac{\varepsilon}{2}\|\mathbf{x}\|^2 \qquad \text{for some } \varepsilon > 0$$

and

$$\tilde{f}_2(\mathbf{x}) = f_2(\mathbf{x}) + \tfrac{\varepsilon}{2}\|\mathbf{x}\|^2.$$

Then $f(\mathbf{x}) = \tilde{f}_1(\mathbf{x}) - \tilde{f}_2(\mathbf{x})$ with $\tilde{f}_1$, $\tilde{f}_2$ strictly convex.

Since f_2 is convex and continuously differentiable, one has

$$f_2(\mathbf{x}) \geq f_2(0) + \langle \mathbf{x}, \nabla f_2(0)\rangle,$$

so that

$$f_2(\mathbf{x}) \geq f_2(0) - \|\mathbf{x}\|\,\|\nabla f_2(0)\|.$$

Here $\nabla f(\mathbf{x})$ denotes the gradient of f at $\mathbf{x}$. Thus,

$$f(\mathbf{x}) \leq f_1(\mathbf{x}) - f_2(0) + \|\mathbf{x}\|\,\|\nabla f_2(0)\|.$$

Upon dividing by $\|\mathbf{x}\|$ and taking limits, one has

$$\liminf_{\|\mathbf{x}\|\to\infty} \frac{f(\mathbf{x})}{\|\mathbf{x}\|} \leq \liminf_{\|\mathbf{x}\|\to\infty} \frac{f_1(\mathbf{x})}{\|\mathbf{x}\|} + \|\nabla f_2(0)\|.$$

When f obeys (2.1) this implies f_1 does also. □

This decomposition is not unique. For some problems of interest, such as the eigenvalue problems analyzed in Section 7 of **[2]**, there is a natural decomposition of this form. Generally, one tries to make this decomposition in a manner which provides a relatively simple expression for the polar function of f_2.

We shall often require

(A2) f, f_1, f_2 obey (A1); f_1, f_2 are strictly convex; and (2.3) holds.

When $h\colon \mathbf{R}^n \to \mathbf{R}$ is a continuous function, its polar (or convex conjugate) function is $h^*\colon \mathbf{R}^n \to (-\infty, \infty]$ defined by

$$h^*(y) = \sup_{\mathbf{x}\in\mathbf{R}^n} [\langle \mathbf{x}, \mathbf{y}\rangle - h(\mathbf{x})]. \tag{2.4}$$

The basic properties of polar functions are described in Chapter 51 of **[9]**, or other texts on convex analysis. We shall need the following result.

LEMMA 2.2. *Suppose $h\colon \mathbf{R}^n \to \mathbf{R}$ is continuous and coercive. Then h^* is lower semicontinuous, convex, and finite-valued. When h is also strictly convex and C^k with $k \geq 2$, then h^* is strictly convex and C^{k-1}.*

PROOF. For given $\mathbf{y}$ in $\mathbf{R}^n$, define $H_{\mathbf{y}}(\mathbf{x}) = h(\mathbf{x}) - \langle \mathbf{x}, \mathbf{y}\rangle$. When h is continuous and coercive, so is $H_{\mathbf{y}}$ and thus it attains its infimum on $\mathbf{R}^n$. Hence, there exists $\hat{\mathbf{x}}$ in $\mathbf{R}^n$ such that

$$H_{\mathbf{y}}(\hat{\mathbf{x}}) = \inf_{\mathbf{x}\in\mathbf{R}^n} H_{\mathbf{y}}(\mathbf{x}) = -h^*(\mathbf{y}) \tag{2.5}$$

and thus $h^*(\mathbf{y})$ is finite. The other two properties of h^* are standard.

When h is strictly convex and C^k with $k \geq 2$, $\hat{\mathbf{x}}$ is unique and is the solution of

$$\nabla h(\mathbf{x}) = \mathbf{y}. \tag{2.6}$$

Define $\eta\colon \mathbf{R}^n \to \mathbf{R}^n$ by $\eta(\mathbf{y}) = \hat{\mathbf{x}}$ when $\hat{\mathbf{x}}$ is the solution of (2.6). Since h is strictly convex on $\mathbf{R}^n$, its Hessian $D^2h(\mathbf{x})$ is always nonsingular so the solution $\eta(\mathbf{y})$ of (2.6) is C^{k-1} from the implicit function theorem.

Moreover, $h^*(\mathbf{y}) = \langle \eta(\mathbf{y}), \mathbf{y}\rangle - h(\eta(\mathbf{y}))$ from (2.5) and the definition of η. This, and the chain rule, implies that h^* is C^{k-1}.

When h and h^* are convex and differentiable, (2.6) is equivalent to $\eta(\mathbf{y}) = \nabla h^*(\mathbf{y})$ and thus

$$\begin{aligned}\langle \nabla h^*(\mathbf{y}) - \nabla h^*(\mathbf{z}), \mathbf{y} - \mathbf{z}\rangle &= \langle \eta(\mathbf{y}) - \eta(\mathbf{z}), \mathbf{y} - \mathbf{z}\rangle \\ &= \langle \mathbf{x} - \mathbf{w}, \nabla h(\mathbf{x}) - \nabla h(\mathbf{w})\rangle,\end{aligned}$$

where $\mathbf{x} = \eta(\mathbf{y})$, $\mathbf{w} = \eta(\mathbf{z})$. When h is strictly convex this right-hand side is positive whenever $\mathbf{x} \neq \mathbf{w}$ and thus h^* will be strictly convex. □

To construct dual problems and algorithms for finding the critical points of f, we shall use the nonconvex duality theory described in **[1]**. Associated with the decomposition (2.3) is the Lagrangian function $L\colon \mathbf{R}^{2n} \to \mathbf{R}$ defined by

$$L(\mathbf{x}, \mathbf{y}) = f_1(\mathbf{x}) + f_2^*(\mathbf{y}) - \langle \mathbf{x}, \mathbf{y}\rangle. \tag{2.7}$$

The Lagrangian problem (Q) is to minimize L on $\mathbf{R}^{2n}$. The basic properties of L are as follows.

THEOREM 2.3. *Suppose* (A2) *holds and L is defined by* (2.7). *Then*

(i) *$L(\cdot, \mathbf{y})$ is a strictly convex, C^k function for each $\mathbf{y}$ in $\mathbf{R}^n$,*

(ii) *$L(\mathbf{x}, \cdot)$ is a strictly convex, C^{k-1} function and*

$$f(\mathbf{x}) = \inf_{\mathbf{y}\in\mathbf{R}^n} L(\mathbf{x}, \mathbf{y}). \tag{2.8}$$

(iii) *When α is defined by* (2.2), *then*

$$\alpha = \inf_{(\mathbf{x},\mathbf{y})\in\mathbf{R}^{2n}} L(\mathbf{x}, \mathbf{y}). \tag{2.9}$$

(iv) *When f_2^* is coercive on $\mathbf{R}^n$, then there exists $(\hat{\mathbf{x}}, \hat{\mathbf{y}})$ in $\mathbf{R}^{2n}$ such that*

$$L(\hat{\mathbf{x}}, \hat{\mathbf{y}}) = \inf_{\mathbf{R}^{2n}} L(\mathbf{x}, \mathbf{y}). \tag{2.10}$$

PROOF.

(i) When f_1 is strictly convex and C^k, so is $L(\cdot, \mathbf{y})$ for each $\mathbf{y}$ in $\mathbf{R}^n$.

(ii) When f_2 is strictly convex and C^k with $k \geq 2$, then from Lemma 2.2, f_2^* is strictly convex and C^{k-1}. Hence, $L(\mathbf{x}, \cdot)$ is strictly convex and C^{k-1} for each $\mathbf{x}$ in $\mathbf{R}^n$. Also

$$\begin{aligned}\inf_{\mathbf{y}\in\mathbf{R}^n} L(\mathbf{x},\mathbf{y}) &= f_1(\mathbf{x}) + \inf_{\mathbf{y}\in\mathbf{R}^n}(f_2^*(\mathbf{y}) - \langle \mathbf{x},\mathbf{y}\rangle) \\ &= f_1(\mathbf{x}) - f_2^{**}(\mathbf{x}) \\ &= f_1(\mathbf{x}) - f_2(\mathbf{x}) \quad \text{as } f_2 \text{ is convex and continuous} \\ &= f(\mathbf{x}) \qquad\qquad\quad \text{from (2.3).}\end{aligned}$$

(iii) Take infima of both sides of (2.8). Then (2.9) holds as

$$\alpha = \inf_{\mathbf{x}\in\mathbf{R}^n} f(\mathbf{x}) = \inf_{\mathbf{x}\in\mathbf{R}^n}\inf_{\mathbf{y}\in\mathbf{R}^n} L(\mathbf{x},\mathbf{y}).$$

(iv) Since f obeys (A1), α is finite and there is an $\hat{\mathbf{x}}$ such that $f(\hat{\mathbf{x}}) = \alpha$. When f_2^* is coercive, and continuous, there exists a $\hat{\mathbf{y}}$ in $\mathbf{R}^n$ such that

$$\inf_{\mathbf{y}\in\mathbf{R}^n}(f_2^*(\mathbf{y}) - \langle \hat{\mathbf{x}},\mathbf{y}\rangle) = f_2^*(\hat{\mathbf{y}}) - \langle \hat{\mathbf{x}},\hat{\mathbf{y}}\rangle.$$

Thus, $\alpha = f(\hat{\mathbf{x}}) = L(\hat{\mathbf{x}},\hat{\mathbf{y}})$ from (2.8), and (2.10) follows using (iii). □

This theorem shows that the value of this problem (Q) is the same as that of (P).

A point $(\tilde{\mathbf{x}},\tilde{\mathbf{y}})$ is said to be a critical point of L if it obeys

$$D_1 L(\mathbf{x},\mathbf{y}) = \nabla f_1(\mathbf{x}) - \mathbf{y} = 0. \tag{2.11}$$

and

$$D_2 L(\mathbf{x},\mathbf{y}) = \nabla f_2^*(\mathbf{y}) - \mathbf{x} = 0. \tag{2.12}$$

Here D_1 (D_2) indicates differentiation with respect to the first (second) variable, respectively.

The dual problem (P*), associated with this Lagrangian L, is to minimize $g\colon \mathbf{R}^n \to \mathbf{R}$, where

$$g(\mathbf{y}) = f_2^*(\mathbf{y}) - f_1^*(\mathbf{y}) \tag{2.13}$$

$$= \inf_{\mathbf{x}\in\mathbf{R}^n} L(\mathbf{x},\mathbf{y}). \tag{2.14}$$

The essential properties of g may be summarized as follows.

Theorem 2.4. *Assume* (A2) *holds, α is defined by* (2.2), *and g by* (2.13). *Then*

(i) *g is C^{k-1} and* $\inf_{\mathbf{y}\in\mathbf{R}^n} g(\mathbf{y}) = \alpha$; *and*
(ii) *if f_2^* is coercive, then there exists $\hat{\mathbf{y}}$ in $\mathbf{R}^n$ such that $\hat{\mathbf{y}}$ minimizes g on $\mathbf{R}^n$.*

Proof.

(i) When f_1, f_2 are strictly convex and C^k, then from Lemma 2.2, g is C^{k-1}. From (2.9) one has

$$\alpha = \inf_{\mathbf{R}^{2n}} L(\mathbf{x}, \mathbf{y}) = \inf_{\mathbf{y}\in\mathbf{R}^n} g(\mathbf{y})$$

upon taking infima of both sides of (2.14).

(ii) When f_2^* is coercive, then from (iv) of Theorem 2.3, there exists $(\hat{\mathbf{x}}, \hat{\mathbf{y}})$ in $\mathbf{R}^{2n}$ such that

$$L(\hat{\mathbf{x}}, \hat{\mathbf{y}}) = \inf_{\mathbf{R}^{2n}} L(\mathbf{x}, \mathbf{y}) = \inf_{\mathbf{y}\in\mathbf{R}^n} g(\mathbf{y}) = \alpha.$$

Now $g(\hat{\mathbf{y}}) = \alpha$ from (2.14) and the fact that $L(\hat{\mathbf{x}}, \hat{\mathbf{y}}) = \alpha$. Thus, $\hat{\mathbf{y}}$ minimizes g on $\mathbf{R}^n$ as claimed. □

This result shows that the values of (P) and (P*) are the same and that this dual problem (P*) has a minimizer provided that f_2^* is coercive. In the next section, we shall study more closely the relationship between general critical points of (P) and those of (P*).

3. Correspondence of critical points. Our interest is in minimizing, or finding the critical points and local minima, of a smooth function f. In the last section, we showed how, by using a decomposition of f into the difference of two convex functions, we could associate a Lagrangian problem (Q) and a dual optimization problem (P*) with the primal problem (P). From Theorems 2.3 and 2.4, we know that the Lagrangian L, and the dual function g, are C^{k-1} whenever f is C^k. In this section, we shall show that there is a correspondence between the critical points of f and those of g. These corresponding critical points have the same type in terms of Morse theory and their critical values are equal.

When $f\colon \mathbf{R}^n \to \mathbf{R}$ is C^k, $k \geq 2$, then its derivative at a point $\mathbf{x}$ is denoted $\nabla f(\mathbf{x})$ and its second derivative, or Hessian matrix, is denoted $D^2 f(\mathbf{x})$. All derivatives will be taken in the Gâteaux sense and we shall always work with the standard basis in $\mathbf{R}^n$.

A point $\hat{\mathbf{x}}$ in $\mathbf{R}^n$ is said to be a critical point of f if

$$\nabla f(\hat{\mathbf{x}}) = 0. \tag{3.1}$$

$\hat{\mathbf{x}}$ is said to be a nondegenerate critical point if it is a critical point and $D^2 f(\hat{\mathbf{x}})$ is nonsingular. When $\hat{\mathbf{x}}$ is a critical point and $D^2 f(\hat{\mathbf{x}})$ is singular, then $\hat{\mathbf{x}}$ is said to be a degenerate critical point of f.

When $\hat{\mathbf{x}}$ is a nondegenerate critical point of f, the Morse index $i(\hat{\mathbf{x}})$ of $\hat{\mathbf{x}}$ is the number of negative eigenvalues of $D^2 f(\hat{\mathbf{x}})$.

For a detailed study of Morse theory and optimization see the two volumes of Jongen, Jonker, and Twilt [**4**]. The results in this section are specializations, and improvements, of results from Section 5 of [**1**].

The next result relates the critical points of f, g, and L.

THEOREM 3.1. *Suppose* (A2) *holds, L is defined by* (2.7), *and g by* (2.11). *Then*

(i) *if $(\tilde{\mathbf{x}}, \tilde{\mathbf{y}})$ is a critical point of L, then $\tilde{\mathbf{x}}$ is a critical point of f and $\tilde{\mathbf{y}}$ is a critical point of g;*

(ii) *if $\tilde{\mathbf{x}}$ is a critical point of f, $\tilde{\mathbf{y}} = \nabla f_1(\tilde{\mathbf{x}})$, then $(\tilde{\mathbf{x}}, \tilde{\mathbf{y}})$ is a critical point of L and $\tilde{\mathbf{y}}$ is a critical point of g; and*

(iii) *if $\tilde{\mathbf{y}}$ is a critical point of g, $\tilde{\mathbf{x}} = \nabla f_2^*(\tilde{\mathbf{y}})$, then $(\tilde{\mathbf{x}}, \tilde{\mathbf{y}})$ is a critical point of L and $\tilde{\mathbf{x}}$ is a critical point of f.*

PROOF.

(i) When $(\tilde{\mathbf{x}}, \tilde{\mathbf{y}})$ is a critical point of L, (2.11) and (2.12) hold. From (2.12), $\nabla f_2^*(\tilde{\mathbf{y}}) = \tilde{\mathbf{x}}$. Since f_2 and f_2^* are both C^1 and convex, this implies $\nabla f_2(\tilde{\mathbf{x}}) = \tilde{\mathbf{y}}$ (see Proposition 51.5 in [**9**]). From (2.11) one also has $\nabla f_1(\tilde{\mathbf{x}}) = \tilde{\mathbf{y}}$, so $\nabla f(\tilde{\mathbf{x}}) = 0$, or $\tilde{\mathbf{x}}$ is a critical point of f.

Similarly, (2.11) is equivalent to $\tilde{\mathbf{x}} = \nabla f_1^*(\tilde{\mathbf{y}})$ and thus $\nabla g(\tilde{\mathbf{y}}) = 0$ or $\tilde{\mathbf{y}}$ is a critical point of g.

(ii) When $\tilde{\mathbf{x}}$ is a critical point of f one has $\nabla f_1(\tilde{\mathbf{x}}) = \nabla f_2(\tilde{\mathbf{x}})$. Let $\tilde{\mathbf{y}} = \nabla f_1(\tilde{\mathbf{x}})$; then (2.11) holds and $\tilde{\mathbf{y}} = \nabla f_2(\tilde{\mathbf{x}})$. Dualizing (using Proposition 51.5 again), one has $\tilde{\mathbf{x}} = \nabla f_2^*(\tilde{\mathbf{y}})$ and thus (2.12) holds and $(\tilde{\mathbf{x}}, \tilde{\mathbf{y}})$ is a critical point of L. The last part of (ii) now follows from (i).

(iii) is proven by an argument similar to that of (ii). □

Essentially, this theorem depends on the inverse relationship between the gradient of a convex function and its polar. When $(\tilde{\mathbf{x}}, \tilde{\mathbf{y}})$ is a critical point of the Lagrangian we say that the critical points $\tilde{\mathbf{x}}$ of f and $\tilde{\mathbf{y}}$ of g correspond, or are dual critical points.

From (2.8) and (2.14) one sees that both f and g may be obtained from the Lagrangian L by minimizing L in $\mathbf{y}$ or $\mathbf{x}$, respectively. In fact one has the following.

THEOREM 3.2. *Suppose* (A2) *holds, L is defined by* (2.7), *g by* (2.11), *and f_2^* is coercive on $\mathbf{R}^n$. Then there exist C^{k-1} functions $\xi: \mathbf{R}^n \to \mathbf{R}^n$ and $\eta: \mathbf{R}^n \to \mathbf{R}^n$ such that*

$$f(\mathbf{x}) = L(\mathbf{x}, \xi(\mathbf{x})) \qquad \textit{for all } \mathbf{x} \textit{ in } \mathbf{R}^n \tag{3.2}$$

and

$$g(\mathbf{y}) = L(\eta(\mathbf{y}), \mathbf{y}) \qquad \textit{for all } \mathbf{y} \textit{ in } \mathbf{R}^n. \tag{3.3}$$

ξ and η are strictly monotone maps.

PROOF. From (2.8) one has

$$f(\mathbf{x}) = f_1(\mathbf{x}) + \inf_{\mathbf{y}\in\mathbf{R}^n} [f_2^*(\mathbf{y}) - \langle \mathbf{x}, \mathbf{y}\rangle].$$

From Lemma 2.2, f_2^* is strictly convex and at least C^{k-1}. Since it is coercive this implies the infimum here is attained at the unique solution of

$$\nabla f_2^*(\mathbf{y}) = \mathbf{x}.$$

This implies $\mathbf{y} = \nabla f_2(\mathbf{x})$, based on Proposition 51.5 in **[9]** again. Define $\xi(\mathbf{x}) = \nabla f_2(\mathbf{x})$. Then $f(\mathbf{x}) = f_1(\mathbf{x}) + f_2^*(\xi(\mathbf{x})) - \langle \mathbf{x}, \xi(\mathbf{x})\rangle$ or (3.2) holds. ξ is strictly monotone and C^{k-1} as f_2 is strictly convex and C^k.

Similarly, from (2.14) one has

$$g(\mathbf{y}) = f_2^*(\mathbf{y}) + \inf_{\mathbf{x}\in\mathbf{R}^n} [f_1(\mathbf{x}) - \langle \mathbf{x}, \mathbf{y}\rangle].$$

Since f_1 is coercive and strictly convex, this infimum is always attained at the unique solution of

$$\mathbf{y} = \nabla f_1(\mathbf{x}). \tag{3.4}$$

One has $\mathbf{x} = \nabla f_1^*(\mathbf{y})$ by duality, so we take $\eta(\mathbf{y}) = \nabla f_1^*(\mathbf{y})$ this time. The fact that η is actually C^{k-1} comes by using the implicit function theorem on (3.4) knowing that ∇f_1 is C^{k-1} and the Hessian $D^2 f_1(\mathbf{x})$ is always nonsingular. □

The representations (3.2) and (3.3) now allow us to prove the following theorem on the correspondence between critical points and critical values of f and g. A number c is said to be a critical value of f if there is a critical point of f in the set $f^{-1}(c)$.

THEOREM 3.3. *Suppose* (A2) *holds with $k \geq 3$, $\hat{\mathbf{x}}$ is a critical point of f, and $\hat{\mathbf{y}} = \nabla f_1(\hat{\mathbf{x}})$. Then*

(i) $f(\hat{\mathbf{x}}) = L(\hat{\mathbf{x}}, \hat{\mathbf{y}}) = g(\hat{\mathbf{y}})$; (3.5)

(ii) *if* $\hat{\mathbf{x}}$ *is a nondegenerate critical point of* f *with index* $i(\hat{\mathbf{x}})$, *then* $\hat{\mathbf{y}}$ *is a nondegenerate critical point of* g *of the same index; and*

(iii) *if* $\hat{\mathbf{x}}$ *is a degenerate critical point of* f, *then* $\hat{\mathbf{y}}$ *is a degenerate critical point of* g.

PROOF. From Theorem 3.2, one has $f(\hat{\mathbf{x}}) = L(\hat{\mathbf{x}}, \xi(\hat{\mathbf{x}}))$ and $\xi(\hat{\mathbf{x}}) = \nabla f_2(\hat{\mathbf{x}})$.

If $\hat{\mathbf{x}}$ is a critical point of f then $\hat{\mathbf{y}} = \nabla f_1(\hat{\mathbf{x}}) = \nabla f_2(\hat{\mathbf{x}})$ so $f(\hat{\mathbf{x}}) = L(\hat{\mathbf{x}}, \hat{\mathbf{y}})$. Now, if $\hat{\mathbf{y}} = \nabla f_1(\hat{\mathbf{x}})$, then $\hat{\mathbf{x}} = \nabla f_1^*(\hat{\mathbf{y}})$ from duality and $\eta(\mathbf{y}) = \nabla f_1^*(\mathbf{y})$ from the proof of Theorem 3.2. Thus, $L(\hat{\mathbf{x}}, \hat{\mathbf{y}}) = L(\eta(\hat{\mathbf{y}}), \hat{\mathbf{y}}) = g(\hat{\mathbf{y}})$ from (3.3) and (i) holds.

From (3.2) and the chain rule, one has

$$\nabla f(\mathbf{x}) = D_1 L(\mathbf{x}, \xi(\mathbf{x})) + D_2 L(\mathbf{x}, \xi(\mathbf{x})) D_{\mathbf{x}} \xi(\mathbf{x}).$$

Now $D_1 L(\mathbf{x}, \mathbf{y}) = \nabla f_1(\mathbf{x}) - \mathbf{y}$ so $D_1 L(\mathbf{x}, \xi(\mathbf{x})) = \nabla f_1(\mathbf{x}) - \xi(\mathbf{x})$.

Also, $D_2 L(\mathbf{x}, \mathbf{y}) = \nabla f_2^*(\mathbf{y}) - \mathbf{x}$ so $D_2 L(\mathbf{x}, \xi(\mathbf{x})) \equiv 0$ on $\mathbf{R}^n$ as $\xi(\mathbf{x})$ is the solution of

$$\nabla f_2^*(\mathbf{y}) = \mathbf{x}. \tag{3.6}$$

Thus,

$$\nabla f(\mathbf{x}) = \nabla f_1(\mathbf{x}) - \xi(\mathbf{x})$$

and

$$D^2 f(\mathbf{x}) = D^2 f_1(\mathbf{x}) - D_{\mathbf{x}} \xi(\mathbf{x}).$$

Differentiating (3.6), one has that $D^2 f_2^*(\xi(\mathbf{x})) D_{\mathbf{x}} \xi(\mathbf{x}) = \mathbf{I}$, where $\mathbf{I}$ is the $n \times n$ identity matrix. Thus, $D_{\mathbf{x}} \xi(\mathbf{x}) = D^2 f_2^*(\xi(\mathbf{x}))^{-1}$.

Similarly,

$$\nabla g(\mathbf{y}) = \nabla f_2^*(\mathbf{y}) - \eta(\mathbf{y})$$

and

$$D^2 g(\mathbf{y}) = D^2 f_2^*(\mathbf{y}) - D_{\mathbf{y}} \eta(\mathbf{y}).$$

Also, $D^2 f_1(\eta(\mathbf{y})) \cdot D_{\mathbf{y}} \eta(\mathbf{y}) = \mathbf{I}$, so

$$D^2 g_1(\mathbf{y}) = D^2 f_2^*(\mathbf{y}) - D^2 f_1(\eta(\mathbf{y}))^{-1}.$$

Thus, if $\hat{\mathbf{x}}$ is a critical point of f, $D^2 f_1(\hat{\mathbf{x}}) = \mathbf{A}$, $D^2 f_2^*(\xi(\hat{\mathbf{x}})) = \mathbf{B}$, one has $D^2 f(\hat{\mathbf{x}}) = \mathbf{A} - \mathbf{B}^{-1}$.

If $\hat{\mathbf{y}} = \nabla f_1(\hat{\mathbf{x}})$ and $\hat{\mathbf{x}}$ is a critical point of f, then $\hat{\mathbf{y}} = \nabla f_2(\hat{\mathbf{x}}) = \xi(\hat{\mathbf{x}})$ so $D^2 g(\hat{\mathbf{y}}) = \mathbf{B} - \mathbf{A}^{-1}$, as one also has $\hat{\mathbf{x}} = \eta(\mathbf{y})$.

Since f_1 and f_2^* are strictly convex, $\mathbf{A}$ and $\mathbf{B}$ are positive-definite matrices. From elementary arguments, $\mathbf{A} - \mathbf{B}^{-1}$ is singular if and only if $\mathbf{B} - \mathbf{A}^{-1}$ is, so (iii) holds.

From the theory of quadratic forms there is a nonsingular matrix $\mathbf{V}$ such that $\mathbf{V}^*\mathbf{A}\mathbf{V} = \mathbf{I}$ and $\mathbf{V}^*\mathbf{B}^{-1}\mathbf{V} = \mathbf{D}$ with $\mathbf{D}$ being a positive-definite diagonal matrix.

When $\hat{\mathbf{x}}$ is a nondegenerate critical point, the index $i(\hat{\mathbf{x}})$ of $\hat{\mathbf{x}}$ is the number of eigenvalues of $\mathbf{D}$ greater than 1.

Now $\mathbf{A}^{-1} = \mathbf{V}\mathbf{V}^*$, $\mathbf{B} = \mathbf{V}\mathbf{D}^{-1}\mathbf{V}^*$, so $D^2 g(\hat{\mathbf{y}}) = \mathbf{V}(\mathbf{D}^{-1} - \mathbf{I})\mathbf{V}^*$. The index $i(\hat{\mathbf{y}})$ of $\hat{\mathbf{y}}$, from Sylvester's theorem, is the number of negative eigenvalues of $\mathbf{D}^{-1} - \mathbf{I}$, which is again the number of eigenvalues of $\mathbf{D}$ larger than 1. Thus, (ii) holds. □

COROLLARY. *Suppose* (A2) *holds with* $k \geq 3$, $\hat{\mathbf{x}}$ *is a critical point of* f, *and* $\hat{\mathbf{y}} = \nabla f_1(\hat{\mathbf{x}})$. *Then*

(i) *if* $\hat{\mathbf{x}}$ *minimizes* f, *then* $\hat{\mathbf{y}}$ *minimizes* g *and* $(\hat{\mathbf{x}}, \hat{\mathbf{y}})$ *minimizes* L; *and*

(ii) *if* $\hat{\mathbf{x}}$ *is a nondegenerate local minimizer of* f, *then* $\hat{\mathbf{y}}$ *is a nondegenerate local minimizer of* g.

PROOF. Let α be defined by (2.2). Then if $\hat{\mathbf{x}}$ minimizes f one has $f(\hat{\mathbf{x}}) = \alpha$. From (3.5) above one has

$$L(\hat{\mathbf{x}}, \hat{\mathbf{y}}) = g(\hat{\mathbf{y}}) = \alpha.$$

Thus, from (iii) of Theorem 2.3 and (i) of Theorem 2.4, one sees that $(\hat{\mathbf{x}}, \hat{\mathbf{y}})$ minimizes L on $\mathbf{R}^{2n}$ and $\hat{\mathbf{y}}$ minimizes g on $\mathbf{R}^n$, so (i) holds.

If $\hat{\mathbf{x}}$ is a nondegenerate local minimizer of f, then its index $i(\hat{\mathbf{x}}) = 0$. Hence, $i(\hat{\mathbf{y}}) = 0$, from (ii) of the theorem , and thus $\hat{\mathbf{y}}$ is a nondegenerate local minimizer of g, so (ii) of the corollary holds. □

A referee has pointed out that parts (ii) and (iii) of Theorem 3.3 are stated as Theorem 2.14 in Pommellet's thesis **[8]**.

4. Duality algorithms for critical points. In the last two sections we have shown how one can associate a Lagrangian L and a dual function g with the primal problem of minimizing a smooth function f on $\mathbf{R}^n$. Moreover, we have shown how the critical points and critical values of these functions are related.

From Theorem 2.3 one sees that the Lagrangian defined by (2.7) is strictly convex in each of $\mathbf{x}$ and $\mathbf{y}$ separately. Thus, it suggests that one minimize L in each of these variables separately. Each of these problems will have a unique minimizer and there are no critical points that are not global minimizers. By contrast, the functions f and g may have many local minimizers that are not global minimizers of f or g on $\mathbf{R}^n$, as well as critical points that are saddle points or local maximizers. Our assumptions on f have been only that it be C^k for some $k \geq 2$ (or 3) and be coercive. Thus, we shall first look at the following algorithm (ALG 1) for generating a descent sequence for L by minimizing alternately in $\mathbf{x}$ and in $\mathbf{y}$.

(1) Given $(\mathbf{x}^{(0)}, \mathbf{y}^{(0)})$ in $\mathbf{R}^{2n}$, evaluate $\gamma_0 = L(\mathbf{x}^{(0)}, \mathbf{y}^{(0)})$.

(2) For $k \geq 0$, if $\mathbf{y}^{(k)} = \nabla f_1(\mathbf{x}^{(k)})$, put $\mathbf{x}^{(k+1)} = \mathbf{x}^{(k)}$,

$$\text{else let} \quad \mathbf{x}^{(k+1)} = \nabla f_1^*(\mathbf{y}^{(k)}). \tag{4.1}$$

(3) If $\mathbf{x}^{(k+1)} = \nabla f_2^*(\mathbf{y}^{(k)})$, put $\mathbf{y}^{(k+1)} = \mathbf{y}^{(k)}$,

$$\text{else let} \quad \mathbf{y}^{(k+1)} = \nabla f_2(\mathbf{x}^{(k+1)}). \tag{4.2}$$

(4) Evaluate $\gamma_{k+1} = L(\mathbf{x}^{(k+1)}, \mathbf{y}^{(k+1)})$.

(5) If $\gamma_{k+1} = \gamma_k$ stop, else go to (2).

Sometimes ∇f_1^* may not be known explicitly, so instead of (4.1) one requires that $\mathbf{x}^{(k+1)}$ be the solution of

$$\nabla f_1(\mathbf{x}) = \mathbf{y}^{(k)}. \tag{4.3}$$

Let $\Gamma = \{(\mathbf{x}^{(k)}, \mathbf{y}^{(k)}) : k \in \mathscr{K}\}$ be the sequence generated by this algorithm. $\mathscr{K}$ may either be a finite set $\{0, 1, 2, \ldots, K\}$ or be countably infinite. The conditions (4.1) or (4.3) imply that $\mathbf{x}^{(k+1)}$ minimizes $L(\cdot, \mathbf{y}^{(k)})$. Similarly, (4.2) implies that $\mathbf{y}^{(k+1)}$ minimizes $L(\mathbf{x}^{(k+1)}, \cdot)$. The properties of this sequence may be summarized as follows.

THEOREM 4.1. *Suppose f, f_1, f_2 obey* (A2) *and* $\Gamma = \{(\mathbf{x}^{(k)}, \mathbf{y}^{(k)}) : k \in \mathscr{K}\}$ *is defined by* (ALG 1). *Then*

(i) *when $(\mathbf{x}^{(k)}, \mathbf{y}^{(k)})$ is not a critical point of L then $\gamma_{k+1} < \gamma_k$;*

(ii) *when $\mathscr{K}$ is finite, its last element is a critical point of L; and*

(iii) *when Γ is a bounded, infinite set, then Γ has at least one limit point. Any such limit point is a critical point of L.*

PROOF. The stopping criterion in ALG 1 is that

$$\mathbf{y}^{(k)} = \nabla f_1(\mathbf{x}^{(k)}) \quad \text{and} \quad \mathbf{x}^{(k)} = \nabla f_2^*(\mathbf{y}^{(k)}).$$

This is the condition that $(\mathbf{x}^{(k)}, \mathbf{y}^{(k)})$ be a critical point of L. When it does not hold one has $\gamma_{k+1} < \gamma_k$ and thus (i)–(ii) hold.

(iii) follows from Theorem 6.4 in [2] or from the following Theorem 4.2. □

Sometimes one cannot use (4.1) and (4.2), so instead of requiring $\mathbf{x}^{(k+1)}$, $\mathbf{y}^{(k+1)}$ to be the exact minima of $L(\cdot, \mathbf{y}^{(k)})$ and $L(\mathbf{x}^{(k+1)}, \cdot)$, respectively, we require only that there be sufficient decrease at each step.

The following algorithm (ALG 2) implements Goldstein-type tests as described in Luenberger ([**6**], Section 7.5).

(1) Given $(\mathbf{x}^{(0)}, \mathbf{y}^{(0)})$, $0 < c_0 < 1$ and $0 < c_1 < c_0/2$.

(2) For $k \geq 0$, let $\mathbf{r}_1^{(k)} = \nabla f_1(\mathbf{x}^{(k)}) - \mathbf{y}^{(k)}$, $\gamma_k = L(\mathbf{x}^{(k)}, \mathbf{y}^{(k)})$.

(3) If $\mathbf{r}_1^{(k)} = 0$ put $\mathbf{x}^{(k+1)} = \mathbf{x}^{(k)}$ and go to (5), else choose $\mathbf{d}_1^{(k)}$ in $\mathbf{R}^n$ so that $\|\mathbf{d}_1^{(k)}\| = 1$ and

$$-\langle \mathbf{r}_1^{(k)}, \mathbf{d}_1^{(k)} \rangle \geq c_0 \|\mathbf{r}_1^{(k)}\|.$$

(4) Choose $t_{1k} > 0$ so that $\mathbf{x}^{(k+1)} = \mathbf{x}^{(k)} + t_{1k}\mathbf{d}_1^{(k)}$ obeys

$$\begin{aligned} c_1 \langle \mathbf{r}_1^{(k)}, \mathbf{d}_1^{(k)} \rangle &\leq L(\mathbf{x}^{(k+1)}, \mathbf{y}^{(k)}) - L(\mathbf{x}^{(k)}, \mathbf{y}^{(k)}) \\ &\leq (c_0 - c_1) \langle \mathbf{r}_1^{(k)}, \mathbf{d}_1^{(k)} \rangle. \end{aligned} \tag{4.4}$$

(5) Let $\mathbf{r}_2^{(k)} = \nabla f_2^*(\mathbf{y}^{(k)}) - \mathbf{x}^{(k+1)}$.

(6) If $\mathbf{r}_2^{(k)} = 0$, put $\mathbf{y}^{(k+1)} = \mathbf{y}^{(k)}$ and go to (8), else choose $\mathbf{d}_2^{(k)}$ in $\mathbf{R}^n$ so that $\|\mathbf{d}_2^{(k)}\| = 1$ and

$$-\langle \mathbf{r}_2^{(k)}, \mathbf{d}_2^{(k)} \rangle \geq c_0 \|r_2^{(k)}\|.$$

(7) Choose $t_{2k} > 0$ so that $\mathbf{y}^{(k+1)} = \mathbf{y}^{(k)} + t_{2k}\mathbf{d}_2^{(k)}$ obeys

$$\begin{aligned} c_1 \langle \mathbf{r}_2^{(k)}, \mathbf{d}_2^{(k)} \rangle &\leq L(\mathbf{x}^{(k+1)}, \mathbf{y}^{(k+1)}) - L(\mathbf{x}^{(k+1)}, \mathbf{y}^{(k)}) \\ &\leq (c_0 - c_1) \langle \mathbf{r}_2^{(k)}, \mathbf{d}_2^{(k)} \rangle. \end{aligned} \tag{4.5}$$

(8) If $\mathbf{r}_1^{(k)} = \mathbf{r}_2^{(k)} = 0$ stop, else go to (2).

The fact that t_{1k} and t_{2k} in steps 4 and 7 can be so chosen follows from the usual analysis of the Armijo–Goldstein conditions. The convergence results about this theorem may be summarized as follows.

THEOREM 4.2. *Suppose f, f_1, f_2 obey* (A2) *and* $\Gamma = \{(\mathbf{x}^{(k)}, \mathbf{y}^{(k)}) : k \in \mathscr{K}\}$ *is defined by algorithm* ALG 2. *Then*

(i) *when $(\mathbf{x}^{(k)}, \mathbf{y}^{(k)})$ is not a critical point of L, then $\gamma_{k+1} < \gamma_k$;*

(ii) *when $\mathscr{K}$ is finite its last element is a critical point of L; and*

(iii) *when Γ is bounded and infinite, it has at least one limit point. Every limit point of Γ is a critical point of L.*

PROOF. The stopping criterion in ALG 2 is that $(\mathbf{x}^{(k)}, \mathbf{y}^{(k)})$ be a critical point of L, as this is equivalent to $\mathbf{r}_1^{(k)} = \mathbf{r}_2^{(k)} = 0$.

Otherwise, from (4.4)–(4.5), one sees that

$$\begin{aligned} \gamma_{k+1} &\leq \gamma_k + (c_0 - c_1)[\langle \mathbf{r}_2^{(k)}, \mathbf{d}_2^{(k)} \rangle + \langle \mathbf{r}_1^{(k)}, \mathbf{d}_1^{(k)} \rangle] \\ &\leq \gamma_k - c_0(c_0 - c_1)[\|\mathbf{r}_1^{(k)}\| + \|\mathbf{r}_2^{(k)}\|] \end{aligned}$$

so $\gamma_{k+1} < \gamma_k$ whenever $(\mathbf{x}^{(k)}, \mathbf{y}^{(k)})$ is not a critical point of L.

In the terminology of Luenberger ([**6**], Section 6.6), this algorithm is a closed algorithm (use the theorem in Section 7.5) on $\mathbf{R}^{2n}$ with respect to the solution set consisting of the critical points of L. L is a descent function off this set; thus, from the global convergence theorem of Section 6.6 in [**6**], each limit point of Γ will be a critical point of L. □

It is worth noting that this Theorem 4.2 implies Theorem 4.1, as when $\mathbf{x}^{(k+1)}, \mathbf{y}^{(k+1)}$ are chosen as in ALG 1, then they certainly obey the conditions of ALG 2.

5. Examples. A number of functions that arise in optimization problems are described, *ab initio*, as the difference of two convex functions. For example, in studying weighted eigenvalue problems one considers functions of the form $f \colon \mathbf{R}^n \to \mathbf{R}$ defined by

$$f(\mathbf{x}) = \frac{1}{p}\langle \mathbf{Cx}, \mathbf{x}\rangle^{p/2} - \frac{1}{2}\langle \mathbf{Ax}, \mathbf{x}\rangle, \tag{5.1}$$

where $2 < p < \infty$ and $\mathbf{A}$, $\mathbf{C}$ are symmetric, positive-definite, real $n \times n$ matrices (see [**2**], Section 7).

Here we shall look at an example where the representation of f as the difference of two convex functions must be chosen appropriately.

Consider the function $f \colon \mathbf{R}^n \to \mathbf{R}$ defined by

$$f(\mathbf{x}) = \frac{1}{2}\langle \mathbf{Ax}, \mathbf{x}\rangle - \mu \cos(\|\mathbf{x}\|^2) - \langle \mathbf{b}, \mathbf{x}\rangle. \tag{5.2}$$

Here $\mathbf{A}$ is a positive-definite, symmetric, real $n \times n$ matrix, $\mu \geq 0$, and $\mathbf{b}$ is in $\mathbf{R}^n$. When $\mu = 0$ this is the standard functional associated with the equation $\mathbf{Ax} = \mathbf{b}$.

The basic properties of f may be summarized as follows.

THEOREM 5.1. *The function f defined by* (5.2) *is C^∞, bounded below and coercive on $\mathbf{R}^n$. It attains its infimum and the critical points of f obey the equation*

$$\mathbf{Ax} + 2\mu \sin(\|\mathbf{x}\|^2)\mathbf{x} = \mathbf{b}. \tag{5.3}$$

PROOF. From the chain rule, f is C^∞ on $\mathbf{R}^n$ and one has

$$\nabla f(\mathbf{x}) = \mathbf{Ax} + 2\mu \sin(\|\mathbf{x}\|^2)\mathbf{x} - \mathbf{b}$$

so (5.3) follows.

If c_1 (> 0) is the least eigenvalue of $\mathbf{A}$ one has

$$f(\mathbf{x}) \geq \frac{c_1}{2}\|\mathbf{x}\|^2 - \mu - \|\mathbf{b}\|\,\|\mathbf{x}\|$$
$$\geq \frac{c_1}{2}\left(\|\mathbf{x}\| - \frac{\|\mathbf{b}\|}{c_1}\right)^2 - \left(\frac{\|\mathbf{b}\|^2}{2c_1} + \mu\right);$$

so f is coercive and bounded below on $\mathbf{R}^n$. Hence, it attains its infimum. □

To apply the theory developed here to this function, one seeks a decomposition of f into the difference of two strictly convex functions.

The natural decomposition for $\cos(\|\mathbf{x}\|^2)$, from its representation as a Taylor series, is to use the fact that

$$\cos z = e_1(z) - e_2(z)$$

where

$$2e_1(z) = \cosh z + \cos z,$$
$$2e_2(z) = \cosh z - \cos z.$$

The functions e_1 and e_2 are convex and, in fact, $e_j^{(k)}(z) \geq 0$ on $(0, \infty)$ for any integer k and $j = 1$ or 2. Unfortunately, it is difficult to compute f_2^* using these expressions. Instead, let

$$\cos z = h_1(z) - h_2(z)$$

where

$$h_1(z) = e^z \quad \text{and} \quad h_2(z) = e^z - \cos z.$$

These h_1, h_2 are strictly convex on $[0, \infty)$ and one has (2.3) with

$$f_1(\mathbf{x}) = \frac{1}{2}\langle \mathbf{Ax}, \mathbf{x}\rangle + \mu h_2(\|\mathbf{x}\|^2) - \langle \mathbf{b}, \mathbf{x}\rangle,$$
$$f_2(\mathbf{x}) = \mu e^{\|\mathbf{x}\|^2}.$$

Now

$$f_2^*(\mathbf{y}) = \sup_{r\geq 0}\sup_{\mathbf{v}\in S^1}(r\langle \mathbf{v}, \mathbf{y}\rangle - \mu e^{r^2}),$$

where S^1 is the unit ball in $\mathbf{R}^n$. Now $\langle \mathbf{v}, \mathbf{y}\rangle$ is maximized when $\mathbf{v} = \mathbf{y}/\|\mathbf{y}\|$ if $\mathbf{y} \neq 0$ and $f_2^*(0) = -\mu$.

Thus, if $\mathbf{y} \neq 0$, one has

$$f_2^*(\mathbf{y}) = \sup_{r\geq 0}(r\|\mathbf{y}\| - \mu e^{r^2}).$$

This is maximized at the unique positive value of r obeying

$$\|\mathbf{y}\| = 2\mu r e^{r^2}.$$

Let this solution be written $r = E(\|\mathbf{y}\|/(2\mu))$, where E is the inverse function to re^{r^2}. Then

$$f_2^*(\mathbf{y}) = \|\mathbf{y}\| \left[E\left(\frac{\|\mathbf{y}\|}{2\mu}\right) - \frac{1}{2}\left(E\left(\frac{\|\mathbf{y}\|}{2\mu}\right)\right)^{-1} \right].$$

The term in brackets here is an analytic function of $\|\mathbf{y}\|/(2\mu)$ and f_2^* is coercive. Using this expression, one can investigate the Lagrangian defined by (2.7) and implement the algorithms described in Section 4.

The algorithm (ALG 1), in particular, becomes remarkably simple. (4.1) requires that $\mathbf{x}^{(k+1)}$ be the solution of

$$\mathbf{A}\mathbf{x} + 2\mu h_2'(\|\mathbf{x}\|^2)\mathbf{x} = \mathbf{y}^{(k)} + \mathbf{b} \tag{5.4}$$

and then (4.2) becomes, for $k \geq 1$,

$$\mathbf{y}^{(k)} = 2\mu e^{\|\mathbf{x}^{(k)}\|^2}\mathbf{x}^{(k)}. \tag{5.5}$$

Here $h_2'(z) = e^z + \sin z$. The solution $\mathbf{x}^{(k+1)}$ of (5.4) is obtained as the unique minimizer of the strictly convex function

$$F_k(\mathbf{x}) = f_1(\mathbf{x}) - \langle \mathbf{y}^{(k)}, \mathbf{x} \rangle,$$

and (5.5) says that $\mathbf{y}^{(k+1)}$ is a specific multiple of $\mathbf{x}^{(k+1)}$.

Whenever the sequence Γ generated by this algorithm remains bounded, one has from (iii) of Theorem 4.1 that a critical point $(\tilde{\mathbf{x}}, \tilde{\mathbf{y}})$ of L can be extracted. $\tilde{\mathbf{x}}$ will now be a critical point of f from Theorem 3.1.

Even when $n = 1$, one may adjust $\mathbf{A}$, μ in (5.2), so that f has many critical points, including local minima and maxima as well as a global minimum. When $n \geq 2$, the number and type of critical points become even more complicated.

References

1. G. Auchmuty, *Duality for nonconvex variational principles*, J. Diff. Eqs. **50** (1983), 80–145.

2. ——, *Duality algorithm for nonconvex variational principles*, Num. Funct. Anal. Optim. **10** (1989), 211–264.

3. J. B. Hiriart-Urruty (ed.), *Fermat days* '85, North-Holland, New York, 1986.

4. H. Th. Jongen, P. Jonker, and F. Twilt, *Nonlinear optimization in* $\mathbf{R}^n$, vols. 1 (1983) and 2 (1986), Peter Lang Verlag, Frankfurt.

5. H. Th. Jongen, T. Möbert, and K. Tammer, *On iterated minimization in nonconvex optimization*, Math. Oper. Res. **11** (1986), 679–691.

6. D. G. Luenberger, *Linear and nonlinear programming*, 2nd edition, Addison-Wesley, Reading, Massachusetts, 1984.

7. J. P. Penot and M. L. Bougeard, *Approximation and decomposition properties of some classes of locally d.c. functions*, Math. Programming **41** (1988), 195–227.

8. A. Pommellet, *Analyse convexe et théorie de Morse*, Thèse, troisième cycle, Université Paris IX–Dauphine, 1983.

9. E. Zeidler, *Nonlinear functional analysis and its applications*, III. *Variational methods and optimization*, Springer-Verlag, New York, 1985.

UNIVERSITY OF HOUSTON

Lectures in Applied Mathematics
Volume **26** (1990)

Antidotes for Nonintegrability of Nonlinear Systems: Quasi-Periodic Motions

M. S. BERGER

1. Introduction. "A basic question for nonlinear analysis is to find the simplest nonlinear equations and compute their solutions."

The systems that are normally considered the simplest are called integrable systems. These integrable systems for nonlinear dynamics are always Hamiltonian. Their integrability is based on a theorem of Liouville which states that for Hamiltonian systems with n degrees of freedom a system is integrable if it has n conservation laws given by the equation $F_i = 0$, $i = 1, 2, 3, \ldots, n$. Moreover, these conservation laws are required to be independent and such that they are in involution. That is, the Poisson bracket of any two of these conservation laws vanishes.

There is an interesting situation that arises, however, when any integrable system is perturbed preserving the Hamiltonian structure. One can ask if their perturbed system preserves any semblance of its integrability.

In this article we address this question. Our first observation is that such integrable systems, defined by Liouville's theorem, always possess quasi-periodic solutions. In fact, we can say that all finite energy solutions of these integrable systems are quasi-periodic. A good example is the family of quasi-periodic solutions generated by N harmonic oscillations. This system can be written in the form

$$\ddot{\mathbf{x}} + \mathbf{A}\mathbf{x} = \mathbf{0}. \tag{1}$$

1980 *Mathematics Subject Classification* (1985 *Revision*). Primary 94A15.
Research partially supported by an AFOSR grant.

Here $\mathbf{x}(t)$ is an N-vector of functions, and $\mathbf{A}$ is an $N \times N$ matrix that is nonsingular, positive definite, and self-adjoint. Periodic motions for such systems are called normal modes, but general solutions are normally finite linear combinations of these normal modes. Hence, in general, any solution of system (1) will be quasi-periodic. Before we proceed further it is best to display this definition.

DEFINITION.. A continuous function $q(t)$ is called quasi-periodic with frequencies $\omega_1, \omega_2, \dots, \omega_n$ if there is a continuous function Q defined on $\mathbf{R}^n \to \mathbf{R}^1$, denoted $Q(x_1, x_2, \dots, x_n)$, such that the continuous function Q is 2π-periodic in each coordinate x_i and, moreover,

$$q(t) = Q(\omega_1 t, \omega_2 t, \dots, \omega_n t). \tag{2}$$

This definition is important because it shows that to find a quasi-periodic function it is necessary to find functions on $\mathbf{R}^n$ that are periodic in each coordinate. In other words, one considers a torus T^n in n dimensions and seeks real-valued functions on T^n.

At this stage it is important to consider the current situation in nonlinear dynamics. Generally one considers what happens when nonlinear systems are studied with a given initial condition. If these systems are defined by an iteration process and depend on a parameter, say

$$x_{n+1} = \lambda F(x_n), \tag{3}$$

then it is known for a large class of functions, F, even in one dimension, that when the parameter λ gets larger the system exhibits chaotic dynamics; that is, such systems exhibit sensitivity to perturbations to initial data and are unpredictable asymptotically. Such chaotic systems may exhibit bifurcation phenomena in addition to chaos, but in any case it is important to note that the initial value problem is not what is being studied in this article on nonintegrability. Indeed, when one passes to systems of ordinary differential equations one would like to study the initial value problem and predict the outcome of solutions after a long time. For this purpose the notion of an integrable system has been very helpful, since for such systems explicit solutions based on integrability by quadrature are available.

Now we inquire what happens when such systems are perturbed. The first point to note is that the perturbation preserves the Hamiltonian structure. The KAM theory is applicable provided the perturbation is sufficiently small and a certain degeneracy is excluded. Otherwise, there is nothing known about such perturbations.

In this article we propose a new approach to studying quasi-periodic motion of certain Hamiltonian dynamical systems based on optimization techniques. The proposed technique is new, yet, because of the advances in optimization, promises computability as well as insight into the nonlinear process involved. Research discussed here represents joint work with Alex Eydeland.

2. Simple new example to find quasi-periodic motions. Our idea is to write down a standard nonlinear differential second-order equation that may have quasi-periodic solutions. We then find from this nonlinear ordinary differential equation the associated nonlinear partial differential equation whose solution will yield the function of n variables that we seek. Because the system is Hamiltonian we are then able to find a variational principle for this function of n variables. As a final step, we show that in certain cases this function of n variables can be determined by infinite-dimensional minimization techniques. Here is a case in point.

We consider the nonlinear ordinary differential equation

$$\ddot{q} - aq - bq^3 = h(t), \qquad \text{where } a, b \text{ are positive numbers} \tag{1}$$

and $q(t)$ is a given quasi-periodic function of frequencies $(\omega_1, \omega_2, \dots, \omega_n)$. We seek a variational principle that determines a quasi-periodic solution $q(t)$ of frequencies $\omega_1, \omega_2, \dots, \omega_n$ for (1) assuming $h(t)$ is given as above [i.e., we seek a quasi-periodic solution $q(t)$ whose frequencies are identical with those of the forcing term $h(t)$]. In symbols, this means we seek a function $u(t_1, t_2, \dots, t_n)$ of n variables 2π-periodic in each t_i $(i = 1, 2, \dots, n)$ with

$$q(t) = u(\omega_1 t, \omega_2 t, \dots, \omega_n t). \tag{2}$$

We now derive a relationship between the functions $q(t)$ and $u(t_1, t_2, \dots, t_n)$. Differentiating (2) with respect to t, we find

$$\dot{q}(t) = \sum_{i=1}^{n} \omega_i u_i(\omega_1 t, \omega_2 t, \dots, \omega_n t), \tag{3}$$

where $u_i = u_{t_i}$. Differentiating once again, we find

$$\begin{aligned} \ddot{q}(t) &= \sum_{i,j=1}^{n} \omega_i \omega_j u_{t_i t_j} \\ &= Lu. \end{aligned} \tag{4}$$

Here Lu denotes the second-order differential operator defined by the right-hand side of (4). This operator L is positive semidefinite when regarded as a second-order differential operator on $\mathbf{R}^n$. Indeed, L is the square of any operator M, i.e.,

$$Lu = \sum_{i,j=1}^{n} \omega_i \omega_j u_{t_i t_j} = \left(\sum_{i=1}^{n} \omega_i u_{t_i} \right)^2$$

where $Mu = \sum_{i=1}^{n} \omega_i u_{t_i}$. The characteristic form of L, $L_0(\mathbf{k})$, can be written in terms of the vector $\mathbf{k} = (k_1, k_2, \ldots, k_n)$ as

$$L_0(\mathbf{k}) = \left(\sum_{i=1}^{n} \omega_i k_i \right)^2 \geq 0.$$

This form is semidefinite (for $n > 1$), since $L_0(\boldsymbol{\xi}) = 0$ whenever we have $\sum_{i=1}^{n} \omega_i \xi_i = 0$ (i.e., on a hyperplane of codimension 1).

Consider now the following partial differential equation associated with (1):

$$\text{(5)} \qquad Lu - au - bu^3 = H(t_1, t_2, \ldots, t_n) \qquad (a, b \text{ positive constants}),$$

and consider the functional on the Sobolev space H of functions in $W_{1,2}([-\pi, \pi]^n)$

$$\text{(6)} \qquad \Phi(u) = \int_{-\pi}^{\pi} \cdots \int_{-\pi}^{\pi} [(Mu)^2 + au^2 + \tfrac{1}{2} bu^4 + 2Hu] \, dt_1 dt_2 \cdots dt_n.$$

Lemma 1. *If $u(x_1, x_2, \ldots, x_n)$ is a smooth critical point of the functional $\Phi(u)$ on H, then the function*

$$q(t) = u(\omega_1 t, \omega_2 t, \ldots, \omega_n t)$$

is a solution of the problem (1). *Moreover, $\Phi(u)$ is convex on H.*

Proof. Indeed, $u_1(t_1, t_2, \ldots, t_n)$ must satisfy the Euler–Lagrange equation of $\Phi(u)$. Thus, $q(t)$, defined by (2), must satisfy (1). The desired convexity follows immediately from computing the appropriate second derivative of $\Phi(u)$.

Lemma 2. *The functional* (6) *defined above attains its minimum in the space K of odd functions in H.*

Proof. We can prove that for $u \in K$,

$$(\dagger) \qquad \int_{-\pi}^{\pi} \int_{-\pi}^{\pi} \int_{-\pi}^{\pi} \int_{-\pi}^{\pi} (Mu)^2 \geq \alpha \|u\|_H^2 \qquad (\alpha \text{ a positive constant}).$$

As $a, b > 0$, $\inf_K \Phi(u) > -\infty$. Moreover, the coerciveness and lower semicontinuity properties in H prevail by general functional-analytic arguments (see [1]). Thus, general arguments show that $\inf_K \Phi(u)$ is attained.

Thus, we have derived a variational principle for the solution u of (5), assuming the constants $a, b > 0$. Clearly, in order for (1) to have the desired quasi-periodic solution $q(t)$, (5) must possess a smooth solution; however, at the moment this regularity result is not proved by our methods.

At this stage, a fundamental issue appears in computational work:

PROBLEM. Can the infimum of the functional $\Phi(u)$ be used to *compute* quasi-periodic motions of (1) via optimization methods? The resolution of such problems would be important because they yield new approaches *independent* of small divisor and resonance problems.

3. Higher-dimensional cases. The method just discussed is, of course, quite general; it can be applied to higher-dimensional cases equally well. For simplicity we shall limit our discussion here to second-order Hamiltonian systems for the forcing term so that the frequencies of the known oscillators discussed will be clearly defined. Such systems occur naturally in many mechanical situations, but their numerical computations have not yet been carried out. The system we discuss here can be written

$$\ddot{\mathbf{x}} - \nabla U(\mathbf{x}) = \mathbf{h}^{(N)}(t), \tag{7}$$

where $\mathbf{x}(t)$ is an N-vector function of t, $\mathbf{h}^{(N)}(t)$ is an N-vector of quasi-periodic functions with components each of the form (1), and $U(x)$ is a convex even function x with $U(x)/|x| \to \infty$ as $|x| \to \infty$. Associated with (7) we also consider the partial differential equation

$$L_N u - \nabla U(\mathbf{u}) = \mathbf{H}^N(t_1, t_2, \dots, t_n), \tag{8}$$

where $\mathbf{u}$ is an N-vector function of $(t_1, t_2, \dots, t_n)$. Let $\mathbf{M}_N$ be an N-vector operator consisting of N copies of M. Consider also the functional

$$\Phi^{(N)}(\mathbf{u}) = \int_{-\pi}^{\pi} \cdots \int_{-\pi}^{\pi} \left[\frac{1}{2}(\mathbf{M}_N\mathbf{u})^2 + U(\mathbf{u}) + \mathbf{H}^N \cdot \mathbf{u}\right] dt_1 \cdots dt_n. \tag{9}$$

LEMMA 3. *Let* $\mathbf{u}$ *be a smooth critical point of* (9). *Then the* N*-vector function*

$$\mathbf{x}(t) = \mathbf{u}(\omega_1 t, \omega_2 t, \dots, \omega_n t)$$

is a solution of (7).

LEMMA 4. *The functional defined in* (9) *attains its minimum in the space* K_N *of* N*-vector copies of* K.

To prove these results, we simply apply the arguments used in Lemmas 1 and 2, as well as the inequality (†), in the N-vector case.

Once again, there are two theoretical problems: the attainment of the infimum without the parity constraint and the regularity of the infimum so attained.

4. Comparisons with periodic solutions. There is now a well-developed theory of periodic motions of nonlinear Hamiltonian systems of ordinary differential equations (see the surveys **[1]**, **[2]**). I developed this theory initially in the late 1960s, but it has since been greatly developed and expanded in numerous countries. I initially studied second-order systems because the optimization processes involved were easier to isolate. When one discusses quasi-periodic motions, as has been done above, these optimization processes are very crucial for the partial differential equation involved.

Indeed, for first-order nonlinear Hamiltonian systems, the desired periodic solution is *not* obtained as an absolute minimum relative to the constraints. This problem is avoided in our approach to quasi-periodic motion. In the future we hope to discuss

(i) large-amplitude quasi-periodic motions for large forcing, and

(ii) avoiding the small divisors usually associated with quasi-periodic motion.

References

1. M. S. Berger, *Nonlinearity and Functional Analysis*, Academic Press, New York, 1977.

2. M. S. Berger, *Global aspects of nonlinear conservative systems*, Nonlinear Dynamical Systems, Springer Lecture Notes Series in Physics, M. Cawley (ed.), 1986.

University of Massachusetts

Lectures in Applied Mathematics
Volume **26** (1990)

Bifurcation into Folds of Infinite Dimension

M. S. BERGER

1. Introduction. At the conference on computational solutions of nonlinear systems of equations, it was remarkable how many times the notion of computing the behavior of a nonlinear equation near a simple singular point arose. In particular, the talks of Professors H. Schwetlick, Böhmer, and Keller all related to this topic. Thus, I decided to compile a collection of ideas on how infinite-dimensional simple singular points come up in infinite-dimensional analysis and its applications.

The simplest example is in the Riccati equations. The next example is in certain nonlinear Dirichlet equations, and the final example that we discuss is semiconductive device analysis in conjunction with switching phenomena. This idea also occurred in the symposium in the talk of Professor Mittlemann, but I had noticed it a number of years earlier in conjunction with my research on VLSI semiconductor devices.

Throughout this paper we use the technical terms Sobolev space, singular point, singular value, and nonlinear Fredholm operator of a given index. These terms are standard and are defined in **[3]**. For semiconductors the relevant terminology can be found in Sze's book **[3]**. For the ideas of singularity theory, Whitney fold, normal form, and stability, see **[2]**.

2. The periodic Riccati equation. We consider the equation

$$\dot{x} + x^2 = f(t) \qquad \text{where } \dot{x} = dx/dt \tag{1}$$

and $f(t)$ is a given forcing term of fixed period, say T. We seek the solutions $x(t)$ that are also T-periodic. Although equation (1) is known

---

1980 *Mathematics Subject Classification* (1985 *Revision*). Primary 53C30.
Research partially supported by an AFOSR grant.

to be the simplest equation that is not integrable by quadrature, there is a large study of Liouville that goes into great detailed analysis of various functions $f(t)$ for which this equation cannot be solved by quadrature. These cases can all be found in Chapter 4 of Watson's treatise on Bessel functions. Integrable by quadrature means that one searches to find transformations of the variables x and t in (1), so that the equation can be written in new variables Y and S in the form

$$(2) \qquad \dot{Y} = g(S) \qquad \text{where } \dot{Y} = dY/dS.$$

This means essentially that the Riccati operator

$$A(x) = \dot{x} + x^2$$

may be linearized by appropriate change of coordinates. It is important to analyze why this cannot be achieved for equation (1). The answer is very simple:

(1) Every singular point of the periodic Riccati equation is a simple singular point—that is, an infinite-dimensional (Whitney) fold.
(2) There is a large infinite-dimensional hyperplane of singular points for every equation of the type (1) for any periodic forcing term $f(t)$.

Let us prove that each singular point of the periodic Riccati equation is a simple singular point (that is, a fold). This is how we progress. We denote the operator $A(x) = \dot{x} + x^2$. We regard this operator as a mapping

$$A(x) = \frac{dx}{dt} + x^2$$

acting between the Sobolev space $W_{1,2}(0, T)$ and $L_2(0, T)$ with T-periodic boundary conditions.

Our first observation about A is that it is a nonlinear Fredholm operator with index 0. That is, its Fréchet derivative, $A'(x)$, is a linear Fredholm† operator with index 0, at each point $x \in X$. Then we use the following theorem proved by myself and P. Church [**1**].

THEOREM 1. *Let A be a C^2 Fredholm map of index 0 between two Hilbert spaces H_1 and H_2 with a singular point at x. Suppose*

(i) $\dim \ker A'(x) = 1$ *and*
(ii) $(A''(x)(e_0, e_0), h^*) \neq 0$, *where* $e_0 \in \ker A'(x)$ *and* $h^* \in \ker[A'(x)]^*$ *are not both identically zero.*

†L is a linear Fredholm operator of index 0 acting between Banach spaces X and Y if L is continuous, L has a closed range in Y, and $\dim \ker L = \dim \operatorname{coker} L < \infty$.

Then A is a local Whitney fold near x.

REMARK. This theorem implies after a change of coordinates that A can be written (near x) as $(t, v) \to (t^2, v)$. *This means that A has a simple singular point at x if the linearized mapping has a one-dimensional kernel and its second derivative is nondegenerate.*

Now, we compute $A'(x)y$ near a singular point $A'(x)y = dy/dt + 2xy$ with the same T-periodic boundary conditions for y. Thus, via the Sobolev embedding theorems, $A'(x)y$ can be represented as the sum of an invertible linear map plus a compact map. This fact establishes the Fredholm property, our first goal. Then we classify each singular point of A. Now, by Taylor series expansion,

$$A(x+h) = A(x) + [h' + 2xh] + h^2.$$

This shows that the second derivative of A at x, $(A''(x)h, h) = h^2$, is "nondegenerate," so every singular point of A is a fold and so has the desired local normal form.

Indeed, given the above notation, we have the following lemma.

LEMMA.

(i) $\ker A'(x)y = \ker(dy/dt + 2xy)$ *with T-periodic boundary conditions can be at most one-dimensional.*

(ii) $x \in S(A)$ *if and only if* $\int_0^T x(t)\,dt = 0$. *[Here $S(A)$ are the points of the singular set A.]*

(iii) *Every singular point x of A is a Whitney fold.*

PROOF. For $x \in S(A)$, we require $\ker A'(x)$ to be nontrivial. Thus, $A'(x)y = 0$ has a nontrivial solution. Now, this means that

$$A'(x)y = \frac{dy}{dt} + 2xy = 0$$

has a nontrivial nonzero solution $y(t)$ (as is clear by ordinary calculus). By the elementary theory of linear ordinary differential equations, $\dim\ker A'(x)y = 1$.

Here I demonstrate only the necessity of (ii). The sufficiency follows by a simple computation. Assume $x(t)$ is singular. Thus,

$$\left(\frac{1}{x}\right)\frac{dx}{dt} + 2x(t) = 0,$$

so, by integrating over a period T, we obtain

$$\int_0^T \frac{d}{dt}[\log|x(t)|]\,dt + 2\int_0^T x(t)\,dt = 0.$$

Now the periodic boundary conditions on $x(t)$ imply that the first integral in the previous equation vanishes.

(iii) Thus, to verify that $x \in S(A)$ is a fold we note that

$$(A''(x)(e_0, e_0), h^*) = \int_0^T e_0^2(t)h^*(t)\,dt$$

and this integral is necessarily nonzero since h^* is a nonzero solution of the equation $dy/dt - 2xy = 0$, which necessarily is of one sign. Thus, the desired result follows from Theorem 1.

The remarkable fact about our work on the periodic Riccati equation is that this problem, although not integrable by quadrature, can be studied globally. In fact, the mapping A is a global Whitney fold. This is discussed in my papers **[1]** and **[2]** following work of McKean and Scovil. It shows that, in theory at least, the Riccati operator A with periodic boundary conditions can be studied from a global point of view using the ideas of normal forms even though bifurcation in the form of a simple singular point presents itself many times in the problem. This is a new idea but should be useful in computations. Another important feature relevant to numerical computation work is that the arguments used in studying a simple singular point are stable. This view is that if the Riccati equation (1) is perturbed slightly by changing the quadratic term x^2 or the forcing term $f(t)$, then the problem is not altered significantly. The same theoretic structures apply because they are independent of perturbation. Here is the relevant computational issue:

A FUNDAMENTAL[†] PROBLEM FOR COMPUTATION. Develop a stable, efficient computer code for the periodic Riccati equation (1), utilizing the fact that the Riccati operator A is a global Whitney fold.

3. Nonlinear Dirichlet equation. We consider nonlinear partial differential equations on an arbitrary bounded domain $\mathbf{R}^n$ with null boundary conditions imposed on the boundary. This equation can be written

$$(*) \qquad \begin{aligned} \Delta u + f(x, u) &= g(x), \\ u|_{\partial\Omega} &= 0. \end{aligned}$$

Here Δ denotes the Laplace operator. This equation is conditioned by the fact that f has the following asymptotic properties:

$$\lambda_2 > \lim_{t\to\infty} \frac{f(x,t)}{t} > \lambda_1; \qquad \lambda_1 > \lim_{t\to-\infty} \frac{f(x,t)}{t} > 0;$$

[†]Note that stability implies discretization errors, for example, can be minimized because of the nature of the simple (Whitney) fold.

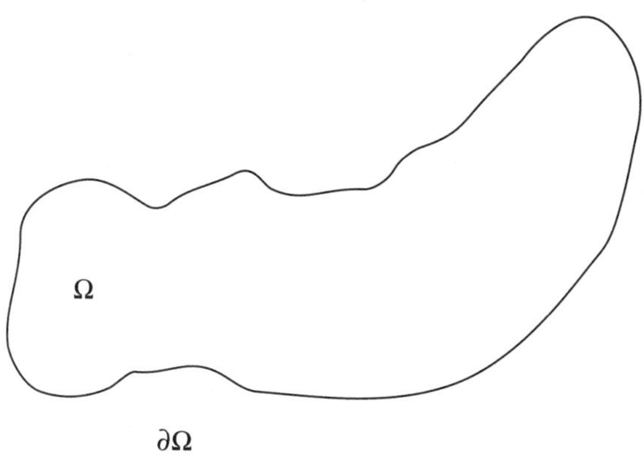

FIGURE 0.

where λ_1 and λ_2 are the lowest two eigenvalues of Δ relative to Ω.

Once again we can prove that if the equation is written $Au = g$, then the associated operator A is a Fredholm operator of index 0 acting between appropriate Sobolev spaces. In fact, one can prove a result analogous to Section 2. In particular, we show the following:

(1) Every singular point of the operator A is a simple singular point (i.e., infinite-dimensional fold of Whitney).
(2) The singular points of A form a hypersurface of codimension 1, in the Sobolev space, $H = \dot{W}_{1,2}(\Omega)$.
(3) The equation $(*)$ has either 2, 0, or 1, solutions, depending on whether g is a singular value of the operator A.
(4) Theoretically, the solution can be written explicitly in terms of computable quantities for equation $(*)$.

The same global properties that applied to the periodic Riccati equation apply for the problem discussed here as well, independent of the shape of the domain Ω or the number of spatial dimensions involved. Thus, it is quite remarkable and will undoubtedly lead to new developments in the numerical study of nonlinear partial differential equations. In particular, the associated mapping, A, is a global Whitney fold; that is, as the structure of a simple singular point, it can be converted after changes of coordinates to a simple quadratic mapping.

REMARK. This idea can also be carried out for nonlinear parabolic equations of the form

$$\frac{\partial u}{\partial t} + \Delta u + f(x,u) = q(t,x),$$
$$u|_{\partial\Omega} = 0,$$

where $q(t,x)$ is T-periodic in t and smooth in x and t; here we require that the function $q(t,x)$ be T-periodic in time.

4. Thyristor: Folds in semiconductive devices. The third example of fold structures involves switching devices for the thyristor. The semiconductor device equations can be written

$$Au(x) = f,$$

where the semiconductor device has an interesting doping profile. It turns out that the I–V curve for this device can be pictured as in Figure 1; when the doping, f, is sufficiently large, then the device has one equilibrium or three depending on the intersections of V with the current axis written above. This gives rise to the switching phenomenon I have in mind, because the doping profile is an S-shaped curve at exactly the points at which the S-shape has maximum curvature on the $J-V$ curve, then at exactly the points a fold structure occurs. These fold structures can be precisely arranged by an external gate.

More explicitly, the simplest semiconductor device exhibiting this behavior is the thyristor. The thyristor exhibits a switching phenomenon that has been explained in physical terms but whose mathematical development up to now has been unclear. In fact, in terms of a J–V curve as drawn in Figure 1, the group of the voltage vs. current exhibits negative resistivity.

The mathematical explanation of the switching process is as a bifurcation phenomenon. The bifurcation equations are a consequence of the mathematical structures inherent in the equation and in the boundary conditions. The important parameter involved is the magnitude of the doping profile (see Fig. 2).

The important issue mathematically is the description of the behavior at the bifurcation point. The nonconvexity of the J–V graph can be made more precise by an analysis of the bifurcation problem posed in mathematical terms. Each bifurcation point is a fold and this leads to a justification of the turning points A and B on the V-axis.

The doping profile generating the J–V curve of Figure 1 is drawn in Figure 2. It consists of dividing the x-axis into four segments and alternating the sign of the doping profile as indicated in the figure. This doping profile coupled with the one-dimensional steady-state van Roosbroeck equations generates the J–V curve of Figure 1.

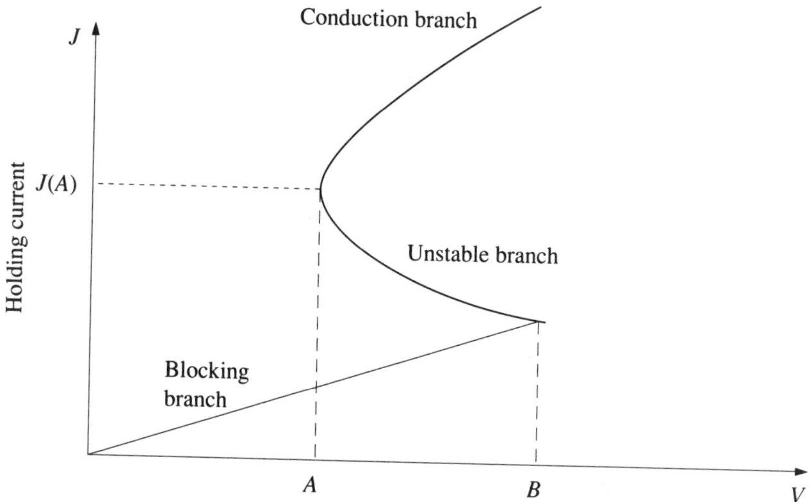

FIGURE 1. The J–V curve for thyristors

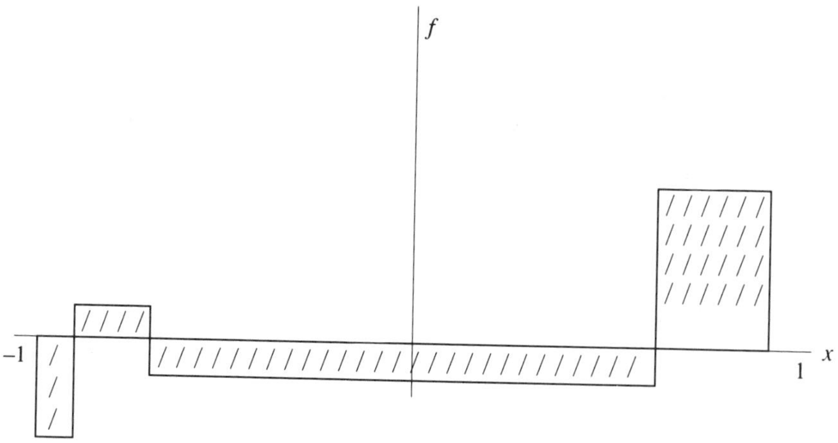

FIGURE 2. Doping profile for the thyristor

The mathematics of bifurcation phenomena. For any physical system described by a system of the form $Au(x) = f$, we can regard the differential operators and boundary conditions A as a mapping between two function spaces X and Y. Generally, one studies the linearization of A, $A'(u)$ at a function u simply by writing the Taylor series

$$A(u + \varepsilon y) = A(u) + \varepsilon A'(u)y + O(|\varepsilon|^2). \tag{3}$$

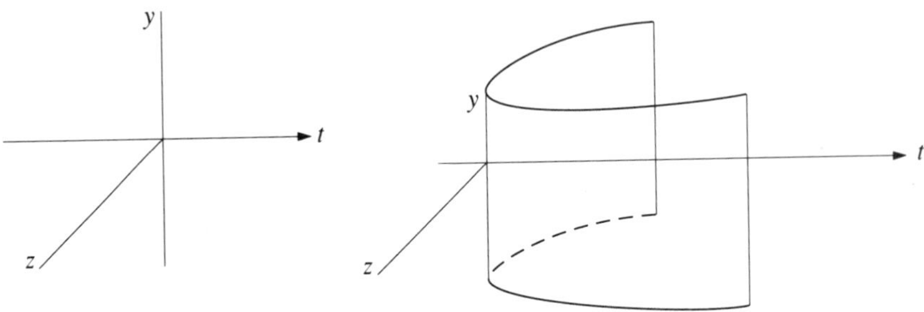

FIGURE 3. The fold $(t, y, z) \to (t^2, y, z)$

Here y is another function, and ε is a real number. In the finite-dimensional case, $A'(u)$ is just the Jacobian matrix evaluated at the vector u and A is simply a vector of functions. For partial differential operators, $A'(u)$ is the linearized operator evaluated at u. For steady-state problems as in the form (3) describing the thyristor, we describe the onset of bifurcation by finding the functions u at which the linear operator $A'(u)$ is not invertible. Using the Fredholm alternative, this means that the equation

$$A'(u)y = 0$$

supplemented by boundary conditions has a nonzero solution y. If the equation $A(v) = f$ has *nonunique solutions* for v near u and f near $A(u)$, we call u a bifurcation point.

If $A'(u)$ is always invertible, no bifurcation exists and for computational purposes the Newton method or a modification of it suffices for computing numerical results.

If bifurcation does occur at u, i.e., u is a bifurcation point, then there is a classification of different kinds of behavior based on the structure of the differential operator A. The simplest case is called a *fold* and can be characterized by the fact that near the bifurcation point u (say) all linear approximations to the behavior of A fail, but (after a change of coordinates) A can be rewritten near u in the form $(t, y) \to (t^2, y)$. This means that after a change of coordinates A has a parabolic shape (see Figure 3).

In terms of the operator A the fold is easily distinguished by virtue of Theorem 1, described above.

The *switching behavior* of the thyristor can be easily explained via the J–V curve. When the voltage V, the biasing, is switched on and is small, the thyristor has small J as shown in the "blocking branch" before A in Figure 1. To get the thyristor into the conduction branch with high J-current without altering V, one utilizes the fact that the "middle branch" joining the blocking and conduction branches is unstable. Thus, introducing a sufficiently large perturbative gate current to the system in the blocking state causes the system to jump "over" the unstable branch into the conducting state. Conversely, once in the conduction state, the system can jump to the blocking state by reducing V below the holding current voltage A (as in Figure 1).

ANOTHER FUNDAMENTAL PROBLEM FOR COMPUTATION. Compute the switching behavior of thyristors in current high technology utilizing their simple singular point behavior.

REFERENCES

1. M. S. Berger and P. T. Church, *Complete integrability for a nonlinear Dirichlet problem*, Indiana Univ. Math. J. (1979), 935–952.

2. M. S. Berger, *The diagonalization of nonlinear differential operators*, Proc. Symp. AMS–SIAM on Nonlinear P.D.E. **23** (1986), 223–239.

3. ——, *Nonlinearity and Functional Analysis*, Academic Press, New York, 1977.

4. S. Sze, *Semiconductor Devices*, 2nd edition, Wiley, New York, 1985.

UNIVERSITY OF MASSACHUSETTS

Lectures in Applied Mathematics
Volume **26** (1990)

On the Geometry of Factorization Algorithms

SHARON L. BLISH AND JAMES H. CURRY

1. Introduction. Possibly one of the best-known algorithms for factoring polynomials was first discovered by Leonard Bairstow in 1916 [**1**]. Bairstow, who was an aeronautical engineer, was motivated by the need to find the roots of polynomials in a single variable and having only real coefficients. Due to the primitive nature of the computational mathematics at the turn of the century, Bairstow, being aware that such polynomials could have complex zeros, wished to avoid complex arithmetic at all cost. Bairstow's idea can be presented as follows: The roots of the real quadratic factor $z^2 + uz + v$ are zeros of

$$P(z) = a_n z^n + \cdots + a_0, \qquad a_n \neq 0, \tag{1.1}$$

if and only if $P(z)$ can be divided by $z^2 + uz + v$ without remainder.

Suppose that

$$P(z) = (z^2 + uz + v)Q(z) + F(u, v)z + G(u, v). \tag{1.2}$$

If an algorithm could be developed that solves the two equations

$$F(u, v) = 0 \quad \text{and} \quad G(u, v) = 0 \tag{1.3}$$

simultaneously, then complex arithmetic could be avoided.

Bairstow's algorithm is the result of applying Newton's method to solve equation (1.3). Specifically,

$$T\begin{pmatrix} u_n \\ v_n \end{pmatrix} = \begin{pmatrix} u_n \\ v_n \end{pmatrix} - \begin{bmatrix} F_u(u_n, v_n) & F_v(u_n, u_n) \\ G_u(u_n, v_n) & G_v(u_n, v_n) \end{bmatrix}^{-1} \begin{bmatrix} F(u_n, v_n) \\ G(u_n, v_n) \end{bmatrix}, \tag{1.4}$$

Research supported, in part, by the National Science Foundation.

where F_u, F_v, G_u, and G_v denote partial derivatives with respect to the obvious variables, and the pair (u_n, v_n) is the result of the composition of T with itself n times.

Several different implementations of Bairstow's algorithm appear in the literature. The version found in reference **[11]** is the concern of this article. (See Appendix B.) The goal of Bairstow's method is to locate pure quadratic factors, that is, pairs of real numbers (u^*, v^*) such that

$$T\begin{pmatrix} u^* \\ v^* \end{pmatrix} = \begin{pmatrix} u^* \\ v^* \end{pmatrix}. \tag{1.5}$$

Section 2 contains known mathematical results related to Bairstow's algorithm. Many of these results have been extended to a broader class of factorization methods [3]. In Section 3 a summary of the results of our computer experiments is presented. To complete this article a proof of the nonconvergence of the "Traub–Bairstow" factorization algorithm is presented in Appendix A and the variant of Bairstow's method that is central to this study is presented in Appendix B.

2. Mathematical results. As we have seen, a polynomial $P(z)$ can be written in the following form:

$$P(z) = (z^2 + uz + v)Q(z) + F(u,v)z + G(u,v).$$

Bairstow's algorithm is Newton's method applied to the nonlinear terms

$$F(u,v) = 0 \quad \text{and} \quad G(u,v) = 0. \tag{2.1}$$

Newton's method is a quadratically converging contraction mapping in a sufficiently small neighborhood of a solution (u^*, v^*); what more can be said?

Let $P(z)$ be a polynomial of degree n having all real simple roots. Then it is clear that the total number of stable fixed points for (1.5) is $\binom{n}{2}$. For more general $P(z)$ we have the following unpublished result due to Stuart Fiedler [7]:

PROPOSITION 1. *Let $P(z)$ be a polynomial having real coefficients. Let $r_1, r_2, \ldots, r_k$ be distinct zeros of $P(z)$. Suppose that $r_1, \ldots, r_j \in \mathbf{R}$ $(j \leq k)$ and that the $k - j$ remaining roots are complex. If l_1 of the real roots and l_2 of the complex roots have multiplicities greater than* 1, *then there are $\frac{1}{2}(j^2 + k) - j + l_1$ quadratic factors or stable fixed points for* (1.4).

PROOF. It is sufficient to observe that quadratic factors can be characterized as members of three disjoint collections: those which are the product of two distinct real roots, a single real root having multiplicity 2, and a complex conjugate pair of roots. The result follows from counting the members of these disjoint classes. □

THEOREM [9]. *Let $z^2 + uz + v$ be a quadratic factor for the polynomial $P(z)$, where $P(z)$ is of degree n, and suppose that the zeros of the quadratic factor are simple and distinct zeros of $P(z)$. Then the Jacobian matrix in (1.4) is nonsingular.*

The next result gives necessary and sufficient conditions for the invertibility of the Jacobian in Bairstow's algorithm and essentially guarantees conditions under which the iteration can continue. Let

$$P(z) = (z^2 + uz + v)Q(z, u, v) + F(u, v)z + G(u, v).$$

THEOREM [6]. *Let u, v be as above. The Jacobian matrix in the Bairstow algorithm is nonsingular if and only if the quadratic factor $(z^2 + uz + v)$ and the polynomial $Q(z, u, v)$ are relatively prime. The rank of the Jacobian is 1 if and only if the quadratic factor and the degree $(n-2)$ factor share one linear factor. The Jacobian is identically zero if and only if $(z^2 + uz + v)$ is a divisor of $Q(z, u, v)$.*

PROOF. For a proof of this result we refer the reader to [6] in the case of simple Bairstow's method and to [8] in the case of more general division algorithms. □

An immediate consequence of this last result is the following:

COROLLARY. *Let $P(z) = (z^2 + u^* z + v^*)Q(z, u^*, v^*)$, that is, the remainder term is zero. Suppose that the quadratic term and the degree $(n - 2)$ have no linear factors in common. Then in a sufficiently small neighborhood of (u^*, v^*), a fixed point of Bairstow's iteration, the iteration converges quadratically.*

The next result, due to D. Boyd [4], is in a strong sense a dynamical systems result. As before, let

$$T\begin{pmatrix} u \\ v \end{pmatrix} = \begin{pmatrix} u' \\ v' \end{pmatrix} = \begin{pmatrix} u \\ v \end{pmatrix} - \begin{bmatrix} F_u & F_v \\ G_u & G_v \end{bmatrix}^{-1} \begin{bmatrix} F(u,v) \\ G(u,v) \end{bmatrix}. \tag{2.2}$$

Following Boyd, let

$$\alpha = \frac{-u + (u^2 - 4v)^{1/2}}{2} \quad \text{and} \quad \beta = -u - \alpha$$

be the roots of $z^2 + uz + v$, and α', β' the roots of $z^2 + u'z + v'$.

THEOREM [4]. *Let ζ be a real zero of $P(z)$. Then the line $A(\zeta) = \zeta^2 + \zeta u + v = 0$, in the (u, v) plane, is invariant under the transformation T. For the pair of points (u, v) on the line $A(\zeta) = 0$, $\alpha' = \alpha = \zeta$ and $\beta' = \beta - P_\zeta(\beta)/P'_\zeta(\beta)$ (Newton's method), where $P_\zeta(z) = P(z)/(z - \zeta)$.*

PROOF. For a proof of this result see [4] or Appendix A, where a generalization of this theorem is given. □

The fact that the Bairstow algorithm reduces to Newton's method on certain invariant lines is very exciting because it allows our research to explore the relationship between behavior in the one-dimensional Newton's method and the associated dynamics of Bairstow's method. For a step in another direction, the reader is referred to Appendix A.

Before proceeding further, let us mention that Bairstow's factorization algorithm is one of many such algorithms [8]. A subclass of these factorization methods is based on representations of the form

$$P(z) = (z^2 + uz + v)Q(z, u, v) + F(u, v)z^m + G(u, v)z^n.$$

One can then introduce, say, a transformation $B\binom{u}{v}$ analogous to the Bairstow transformation $T\binom{u}{v}$. It is then possible to establish the results of Henrici [9], Fiala and Krebsz [6], and Boyd [4] in this broader context. The proofs of these and other results will appear in a forthcoming article by the authors [3].

3. Cubic polynomials: A case study. The simplest case in which to apply Bairstow's algorithm is for $P(z)$ a polynomial of degree 2. In this case it is a simple exercise to show that given any initial condition (u, v) Bairstow's algorithm converges in one step. The next simplest case to explore is the application of factorization algorithms to cubic polynomials.

Let $P_a(z) = z^3 + (a - 1)z - a$, where a is a real parameter and z is a real variable. This family has been used previously to probe the global behavior of other algorithms [5].

For a real and less than $\frac{1}{4}$, $P_a(z)$ has three quadratic factors:

$$\begin{aligned} &z^2 + z + a, \\ &z^2 + z\left(-\frac{1}{2} + \frac{\sqrt{1-4a}}{2}\right) + \left(-\frac{1}{2} - \frac{\sqrt{1-4a}}{2}\right), \\ &z^2 + z\left(-\frac{1}{2} - \frac{\sqrt{1-4a}}{2}\right) + \left(-\frac{1}{2} + \frac{\sqrt{1-4a}}{2}\right), \end{aligned} \tag{3.1}$$

and clearly for $a > \frac{1}{4}$ there is exactly one real quadratic factor.

It follows immediately from the work of Boyd [**4**] (see also [**3**]) that there are three invariant lines determined by the three real roots of $P_a(z)$ when $a < \frac{1}{4}$, and the behavior of Bairstow's method reduces to a one-dimensional Newton's method on these lines. For $P_a(z)$ we can be much more explicit. It is straightforward to show that Bairstow's algorithm applied to the family $P_a(z)$ is equivalent to the following family of endomorphisms of the plane:

$$T_a\begin{pmatrix} u \\ v \end{pmatrix} = \begin{pmatrix} \frac{u^3+u(v-a+1)+a}{2u^2+v} \\ \frac{v(u^2+a-1)+2au}{2u^2+v} \end{pmatrix}, \tag{3.2}$$

which should be contrasted with the family of transformations produced by the synthetic division algorithm in [**10**],

$$S_a\begin{pmatrix} u \\ v \end{pmatrix} = \begin{pmatrix} \frac{2u(v+1-a)+a}{u^2-a+1+2v} \\ \frac{2u^2(v+1-a)+2a(v+u+1)-2v-u^4-a^2-1}{u^2-a+1+2v} \end{pmatrix}. \tag{3.3}$$

Both algorithms that determine the families T_a and S_a are referred to as Bairstow's method in the literature. Apparently, however, only T_a has invariant lines.

In what follows we shall focus on the dynamics of formula (3.2):

$$T_a\begin{pmatrix} u \\ v \end{pmatrix} = \begin{pmatrix} \frac{u^3+u(v-a+1)+a}{2u^2+v} \\ \frac{v(u^2+a-1)+2au}{2u^2+v} \end{pmatrix}.$$

It is clear that the fixed points of T_a are the quadratic factors given above. Let us also note that

$$T_a\begin{pmatrix} 1 \\ v \end{pmatrix} = \begin{pmatrix} 1 \\ a \end{pmatrix}.$$

This last observation is general, as indicated by the following proposition.

PROPOSITION 3.1. *Let (u^*, v^*) be a fixed point for T_a. Then* $T_a\binom{u^*}{v} = \binom{u^*}{v^*}$.

PROOF. Follows immediately from the definitions. □

Proposition 3.1 states that points on vertical lines that pass through fixed points are mapped to the fixed point (u^*, v^*) in one iteration.

It is also clear from the formula which defines T_a that the transformation itself is not defined on the parabola $2u^2 + v = 0$.

3.1. *Contour plots.* A tool that we have found useful for providing insights into the geometry of algorithms is the contour plot. Such plots are more widely used in geology and the atmospheric sciences than in the study of convergence and divergence of iterative methods.

Contour plots result from placing a grid on the plane and constructing lines of constant pressure or height. In the case of iterations, we construct curves of constant increment or, more precisely, the arctangent of the norm of the increment. Why the arctangent is used can be understood as follows. As previously noted, the endomorphisms that are studied here have associated singular sets. When only the increments are plotted, the singular set, $2u^2 + v = 0$, for T_a, dominates the contour plots. The arctangent function is used to bound points in a neighborhood of the singular set.

Figure 1a corresponds to $a = -0.5$ for the mapping T_a. On this plot are superimposed lines corresponding to the invariant lines whose existence was established by Boyd's theorem. Such lines do intersect at fixed points of Bairstow's method. As noted before, the dynamics on the invariant lines become Newton's method in one real variable. In the case of cubic polynomials having all real zeros, Newton's method on the invariant lines becomes an iteration function of degree 2; such iterations were first studied by Cayley. On these lines Bairstow's method must converge for almost all initial conditions. The excluded points are preimages of points on the singular parabola.

Figure 1b is a contour plot associated with S_a with $a = -0.5$. This algorithm is also known as Bairstow's method in the literature but this contour plot differs significantly from Figure 1a.

3.2. *Basin maps.* A further graphics tool that is useful in the analysis of the mappings is the basin map. These illustrations, properly presented in color, provide some insight into the ultimate behavior of initial conditions. Presented in the graphs that follow are basin maps associated with T_a for the parameter values $a = -0.5$ and $a = 1.0$.

For $a = -0.5$ there are three fixed points for equation (3.2); see also expression (3.1). In the corresponding figure, 2a, the black areas are associated with two of the fixed points and the white area is associated with a single stationary solution. Also visible in this plot are various regions indicating that this system can be expected to exhibit sensitivity to initial conditions for this parameter value. Further, this plot strongly

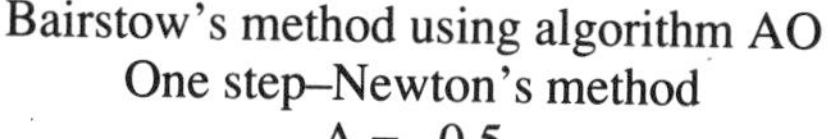

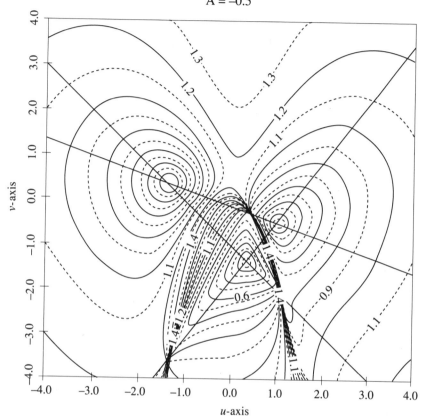

FIGURE 1a. Contour plot showing levels of constant increment for the transformation T_a for $a = -0.5$. Included in this illustration are the invariant lines on which T_a reduces to a one-dimensional Newton's method. Points where the lines intersect correspond to "pure quadratic factors." Also visible in this figure is the singular parabola corresponding to $2u^2 + v = 0$.

suggests that Bairstow's method converges for almost all initial conditions; the proof of this and similar observations is an active area of interest.

Figure 2b, associated with $a = 1.0$, needs some discussion. Here there is only one fixed point, located at $(1.0, 1.0)$, and the invariant line associated with the real root of $P(z) = 0$. That line is clearly visible in Figure 2b. Also visible is the singular parabola and a set of points which, apparently, do not converge after as many as 750 iterations.

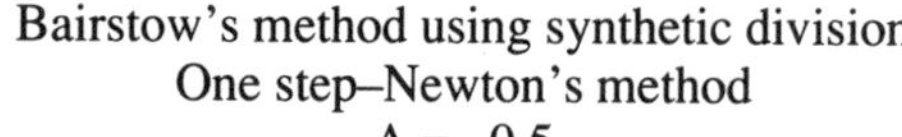

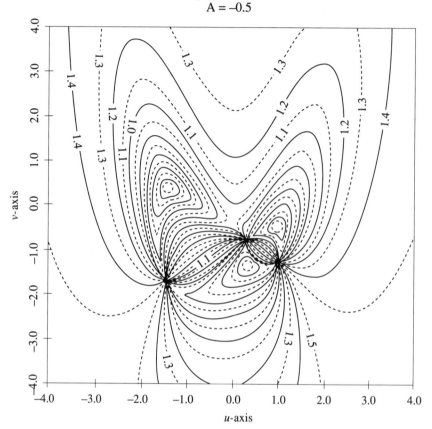

FIGURE 1b. Same as Figure 1a, but for Bairstow's algorithm derived from the synthetic division method. For this variant of the algorithm invariant lines are not present. See equation (3.3) in the text.

The behavior on the invariant line is equivalent to Newton's method applied to $z^2 + 1$ for real initial conditions, which is known to be ergodic. A new feature of the dynamics of this algorithm is the apparent ergodicity of Bairstow's dynamical system for points not on the invariant line.

Finally, in [3] we will present detailed proofs of the materials from Section 2 and also exhibit a family of polynomials for which Bairstow's method displays stable periodic behavior. Unlike the result of Boyd, the existence of a stable periodic attractor is associated with an open

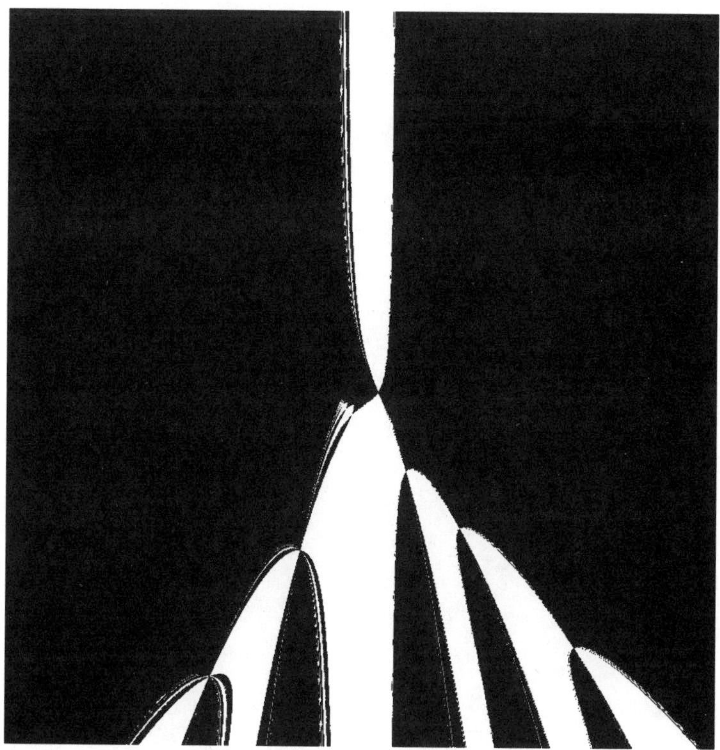

FIGURE 2a. Basin map, in black and white, showing the basin of attractions of the three fixed points in Figure 1a.

set of initial conditions. Such sets have positive planar measure and are therefore significant from the point of view of iteration theory.

Appendix A (Nonconvergence of the Traub–Bairstow method). Assume that

$$P(z) = (z^2 + uz + v)Q(z) + F(u,v)z + G(u,v). \tag{A.1}$$

The goal of factorization algorithms is to solve the two equations

$$F(u,v) = 0 \quad \text{and} \quad G(u,v) = 0 \tag{A.2}$$

simultaneously.

In [12] J. Traub introduced the following multipoint iteration function. Define

$$J_\beta(z) = z - \alpha_1 \frac{P(z)}{P'(z)} - \alpha_2 \frac{P(z + \beta P(z)/P'(z))}{P'(z)} \tag{A.3}$$

FIGURE 2b. Basin mapping associated with T_a for $a = 1.0$. For this value of the parameter the only stationary solution is located at the point $(1.0, 1.0)$. Clearly visible are the invariant line and the singular parabola. Also visible is a strange repeller on which points may, apparently, wander forever. Points in this illustration have been iterated, in the case of the repeller, 120 times and still not converged to the fixed point.

where β is real and $\alpha_1 = (\beta^2 - \beta - 1)/\beta^2$ and $\alpha_2 = 1/\beta^2$. Traub establishes, among other results, that $J_\beta(z)$ is cubically convergent in a sufficiently small neighborhood of a simple zero of $P(z)$.

In this appendix an extension of Boyd's theorem (Section 2) is established for an analogue of $J_\beta(z)$, which may be properly termed the "Traub–Bairstow" method. Let

$$\text{(A.4)}\quad C_\beta\begin{pmatrix} u \\ v \end{pmatrix} = \begin{pmatrix} u \\ v \end{pmatrix} - \begin{bmatrix} F_u(u,v) & F_v(u,v) \\ G_u(u,v) & G_v(u,v) \end{bmatrix}^{-1} \times \left[\alpha_1 \begin{pmatrix} F(u,v) \\ G(u,v) \end{pmatrix} + \alpha_2 \begin{pmatrix} F(u(\beta), v(\beta)) \\ G(u(\beta), v(\beta)) \end{pmatrix}\right]$$

where β, α_1, and α_2 are as above and

$$\begin{pmatrix} u(\beta) \\ v(\beta) \end{pmatrix} = \begin{pmatrix} u \\ v \end{pmatrix} + \beta \begin{bmatrix} F_u(u,v) & F_v(u,v) \\ G_u(u,v) & G_v(u,v) \end{bmatrix}^{-1} \begin{pmatrix} F(u,v) \\ G(u,v) \end{pmatrix}. \tag{A.5}$$

With $(u(\beta), v(\beta))$ as previously defined assume that $\zeta = (-u + (u^2 - 4v)^{1/2})/2$ and $\gamma = -u - \zeta$ are zeros of $z^2 + uz + v$, and let $\zeta(\beta)$ and $\gamma(\beta)$ denote the corresponding zeros of $z^2 + u(\beta)z + v(\beta)$.

LEMMA. *If r is a real and simple zero of $P(z)$ the line $A(r) = r^2 + ru + v = 0$ is invariant under the transformation*

$$\begin{pmatrix} u \\ v \end{pmatrix} \to \begin{pmatrix} u(\beta) \\ v(\beta) \end{pmatrix}.$$

Further, $\zeta = \zeta(\beta) = r$ and $\gamma(\beta) = \gamma + \beta P_r(\gamma)/P_r'(\gamma)$ with $P_r(\gamma) = P(\gamma)/(\gamma - r)$.

PROOF. Differentiating (A.1) with respect to u and v and setting $z = r$, we have the following five equations:

$$rF(u,v) + G(u,v) = 0, \tag{A.6}$$

$$rQ(r) + rF_u(u,v) + G_u(u,v) = 0, \tag{A.7}$$

$$Q(r) + rF_v(u,v) + G(u,v) = 0, \tag{A.8}$$

and

$$(u(\beta) - u)F_u(u,v) + (v(\beta) - v)F_v(u,v) - \beta F(u,v) = 0,$$
$$(u(\beta) - u)G_u(u,v) + (v(\beta) - v)G_v(u,v) - \beta G(u,v) = 0,$$

where the last pair of equations are a consequence of the definition of the transformation

$$\begin{pmatrix} u \\ v \end{pmatrix} \to \begin{pmatrix} u(\beta) \\ v(\beta) \end{pmatrix}. \tag{$*$}$$

Combining the above equations, we have

$$(v(\beta) + u(\beta)r + r^2 - (v + ur + r^2))Q(r) = 0.$$

If $Q(r) = 0$, then the Jacobian in the definition of the mapping $(*)$ is undefined.

To complete the proof we must show that when $\zeta(\beta) = (-u(\beta) + (u(\beta)^2 - 4v(\beta))^{1/2})/2$ and $\gamma(\beta) = -u(\beta) - \zeta(\beta)$, then $\zeta(\beta) = r$ and $\gamma(\beta) = \gamma + \beta P_r(\gamma)/P_r'(\gamma)$, where $P_r(\gamma) = P(\gamma)/(\gamma - r)$.

Since $(u(\beta) - u) = -(\gamma(\beta) - \gamma)$, $(v(\beta) - v) = r(\gamma(\beta) - \gamma)$, and

$$P_r(\gamma) = \frac{P(\gamma) - P(r)}{\gamma - r} = F(-r - \gamma, r\gamma), \tag{A.9}$$

we have

$$\beta F(-r-\gamma, r\gamma) = -(\gamma(\beta)-\gamma)[F_u(-\gamma-r, r\gamma) - rF_v(-\gamma-r, r\gamma)];$$

thus,

$$\gamma(\beta) = \gamma + \beta\frac{P_r(\gamma)}{P_r'(\gamma)}. \quad \square \tag{A.10}$$

Let us note that the proof of the lemma is essentially identical to the one given by Boyd. The only difference is the introduction of the parameter β, which leads to a relaxed Newton's method on the invariant line associated with the root r.

Given the lemma, the central result of the appendix follows easily. More precisely, we have the next theorem.

THEOREM. *Let r be a real and simple zero of $P(z)$. Then the line $A(r) = r^2 + ur + v = 0$ is invariant under the transformation*

$$\begin{pmatrix} u' \\ v' \end{pmatrix} = C_\beta \begin{pmatrix} u \\ v \end{pmatrix}. \tag{A.11}$$

For (u, v) on this line, $\zeta = \zeta' = r$ and

$$\gamma' = \gamma - \alpha_1\frac{P_r(\gamma)}{P_r'(\gamma)} - \alpha_2\frac{P_r(\gamma(\beta))}{P_r'(\gamma)}.$$

PROOF. The proof is a consequence of the previous lemma and is left to the reader. $\square$

Some comments are in order. The theorem shows that the iterative method C_β induces a one-dimensional method, similar to J_β, on the invariant line, $A(r) = 0$. This suggests not only that there is a close relationship between the iterative algorithm used to solve the nonlinear equations

$$F(u, v) = 0 \quad \text{and} \quad G(u, v) = 0$$

in factorization algorithms, but also that a key to the dynamics of such factorization methods could be the one-dimensional dynamics that they induce. This is a theme we plan to explore more fully in a future article.

Finally, a consequence of these results is that for cubic polynomials having a single real zero the Traub–Bairstow method must "fail" unless a quadratic factor lies on the invariant line $A(r) = 0$. This failure may or may not be serious. It may not be serious because such lines have planar Lebesgue measure zero. And the conclusion that "for almost all initial conditions the factorization method works" would be very good. On the other hand, it may prove to be serious because of the apparent close relationship between the dynamics of factorization methods and the induced dynamics on the invariant lines. If the method does not converge for points on an invariant line this may produce a large (in the sense of measure) "hyperbolic set" on which orbits of the dynamical system tend to wander.

Appendix B (Bairstow's algorithm). Let $P(z) = a_n z^n + a_{n-1} z^{n-1} + \cdots + a_0$ be a polynomial having only real coefficients. If $P(z)$ is divided by the quadratic factor $(z^2 + pz + q)$, then

$$\text{(B.1)} \quad P(z) = (z^2 + pz + q)(b_{n-2} z^{n-2} + b_{n-3} z^{n-3} + \cdots + b_0) + uz + v$$

and $(uz + v)$ is the remainder term. If coefficients are now equated, then

$$\text{(B.2)} \qquad b_k = a_{k+2} - pb_{k+1} - qb_{k+2} \qquad (k = n-2, n-3, \ldots, 0);$$

we also require that $b_{n-1} = b_n = 0$.

Continuing in this manner, we can develop equations for $u(p,q)$ and $v(p,q)$:

$$\text{(B.3)} \qquad \begin{aligned} u(p,q) &= a_1 - pb_0 - qb_1 = b_{-1}, \\ v(p,q) &= a_0 - qb_0. \end{aligned}$$

Bairstow's method seeks to determine pure quadratic factors of $P(z)$, i.e., $z^2 + p^* z + q^*$ for which

$$u(p^*, q^*) = v(p^*, q^*) = 0.$$

Indeed, Bairstow's method is simply the two-dimensional Newton–Raphson strategy applied to (B.3).

In order to implement the Newton–Raphson method various partial derivatives must be determined. If we define $d_j = \partial b_j / \partial p$, then

$$\text{(B.4)} \qquad d_j = -b_{j+1} - pd_{j+1} - qd_{j+2} \qquad (j = n-3, \ldots, 0, -1)$$

where $d_{n-2} = d_{n-1} = 0$. It follows that $\partial b_{j-1}/\partial q = \partial b_j / \partial p$ and it is sufficient to use (B.4) to obtain an implementation of the algorithm.

Acknowledgments. Special thanks to Liz Stimmel for her usual excellence in preparing this manuscript. This research was supported in part by a grant from the National Science Foundation's Computational Mathematics Program.

REFERENCES

1. L. Bairstow, *Investigations related to the stability of the aeroplane*, Repts. & Memo., No. 154, Advis. Comm. Aeronaut., 1914.

2. S. L. Blish, *On the geometry of algorithms for factoring polynomials: Bairstow's method*, Master's thesis, University of Colorado, 1988.

3. S. L. Blish and J. H. Curry, *On factorization algorithms*, in preparation.

4. D. Boyd, *Nonconvergence in Bairstow's method*, SIAM J. Numerical Anal. **14** (1977), 571–574.

5. J. H. Curry, L. Garnett, and D. Sullivan, *On the iteration of a rational function: Computer experiments with Newton's method*, Comm. Math. Phys. **91** (1983), 267–277.

6. T. Fiala and A. Krebsz, *On the convergence and divergence of Bairstow's method*, Numerische Mathematik **50** (1987), 477–482.

7. S. L. Fiedler, *Using Bairstow's method to find quadratic factors of polynomials*, Master's thesis, University of Colorado, 1985.

8. A. A. Grau, *A generalization of the Bairstow process*, J. Soc. Indust. Appl. Math. **11** (1963), 508–519.

9. P. Henrici, *Elements of Numerical Analysis*, Wiley, New York, 1964, pp. 108–115.

10. M. J. Maron, *Numerical Analysis: A Practical Approach*, Macmillan, New York, 1982, pp. 75–85.

11. J. Stoer and R. Bulirsch, *Introduction to Numerical Analysis*, Springer-Verlag, New York, 1983.

12. J. F. Traub, *Iterative Methods for the Solution of Equations*, Chelsea Pub. Co., New York, 1982.

UNIVERSITY OF COLORADO-BOULDER

Lectures in Applied Mathematics
Volume **26** (1990)

Defect Corrections and Mesh Independence Principle for Operator Equations and Their Discretizations

KLAUS BÖHMER

1. Introduction. In this introductory article we want to give, on a rather elementary level, the main ideas of defect correction methods and the mesh independence principle (MIP) for operator equations and their discretizations. Section 2 presents operators and their approximations especially in the form of discretization. Section 3 describes the main idea of defect corrections and combines it in Section 4 with discretization methods. The strong connection between an operator and its discretization is reflected in the very similar behaviour of the iterates in Newton's method for the operator and its discretization; see Section 5. This fact is used in Section 6 to develop mesh refinement strategies. A simple demonstration example is used throughout.

2. Operators, approximations, and discretizations. We want to compute, for an operator

$$F\colon D(F) \subset E \to R(F) \subset \hat{E}, \qquad y \in R(F), \tag{2.1}$$

where E, $\hat{E}$ represent Banach spaces, $D(F)$ and $R(F)$ the closed domain and range, resp., the exact solution z^* for

$$Fz^* = y. \tag{2.2}$$

We assume the existence of a unique and isolated solution z^* in the sense that

$$F \in C^1(D(F)) \text{ and } (F'(z^*))^{-1} \text{ exists and is bounded.} \tag{2.3}$$

As an example we use the boundary value problem for an ordinary differential equation

$$(2.4)\quad Fz := \begin{pmatrix} z'(\cdot) - f(\cdot, z(\cdot)) \\ B_0 z(0) + B_1 z(1) \end{pmatrix} = y := \begin{pmatrix} g(\cdot) \\ \alpha \end{pmatrix}, \quad \operatorname{rank}(B_0, B_1) = m.$$

Here $E = C^1([0,1]; \mathbf{R}^m)$, $\hat{E} = C([0,1]; \mathbf{R}^m)$ and $D(F)$, $R(F)$ are appropriate subsets of E and $\hat{E}$, resp., depending upon $D(f)$. For $f \in C^1(D(f))$ and the partial f_z we have to require, see (2.3), that

$$(2.5)\quad F'(z^*)u = \begin{pmatrix} u'(\cdot) - f_z(\cdot, z^*(\cdot))u(\cdot) \\ B_0 u(0) + B_1 u(1) \end{pmatrix} \text{ is boundedly invertible.}$$

Usually (2.1) and (2.4) are not solvable directly. Therefore one tries to define approximate operators

$$(2.6)\qquad \tilde{F}: D(\tilde{F}) \subset E \to R(\tilde{F}) \subset \hat{E},$$

such that in some sense $F \simeq \tilde{F}$, $y \simeq \tilde{y}$, and the problem

$$(2.7)\qquad \tilde{F}\tilde{z}^* = \tilde{y}$$

is again locally uniquely solvable and $\tilde{z}^*$ is "easily" computable. For (2.4) an $\tilde{F}$ may be defined, if an approximation $\tilde{f}$ for f is known, such that

$$Fz := \begin{pmatrix} z'(\cdot) - \tilde{f}(\cdot, z(\cdot)) \\ B_0 z(0) + B_1 z(1) \end{pmatrix}$$

is directly solvable. One might, e.g., use $\tilde{f}(t,u) = a(t) + b(t)u$ if $f(t, z^*(t)) - \tilde{f}(t, z^*(t))$ is small enough for $t \in [0,1]$. This type of transition (2.6), (2.7) from F to $\tilde{F}$ is rather artificial, and therefore more systematic ways are necessary. The original spaces E and $\hat{E}$ are replaced by finite-dimensional counterparts E^h and $\hat{E}^h$, Banach spaces of appropriate dimensions. Here $h \in H$ indicates a sequence of mesh size parameters, appropriate to characterize E^h, with $|h|$ a measure for the fineness of the subdivision of the mesh, and $|h| \to 0$ in H has to be possible. Instead of (2.2) we study, see Stetter **[56]**,

$$(2.8)\qquad F^h z_*^h = y^h$$

with the exact solution z_*^h, $y^h = y^h(y, F^h)$ and

$$(2.9)\qquad F^h: D(F^h) \subset E^h \to R(F^h) \subset \hat{E}^h, \qquad y^h \in R(F^h).$$

The original and the discrete spaces are related by, usually linear, "restriction" operators

$$(2.10)\qquad \Delta^h: E \to E^h, \qquad \hat{\Delta}^h: \hat{E} \to \hat{E}^h$$

satisfying

(2.11) $$\lim_{h\to 0} \|\Delta^h x\| = \|x\|, \qquad \lim_{h\to 0} \|\hat{\Delta}^h y\| = \|\hat{y}\| \quad \text{for } x \in E,\ \hat{y} \in \hat{E}.$$

Now the question arises, under which conditions the locally unique solvability for (2.2) with (2.1), (2.3) is reflected in (2.8). Certainly some type of approximation property of F^h to F has to be required, e.g., the "consistency of order p" in the form

(2.12) $$(F^h \Delta^h z - y^h) - \hat{\Delta}^h (Fz - y) = O(|h|^p)$$

for $\|z - z^*\|$ small enough, or even only for $z = z^*$ as, e.g., for Runge-Kutta methods. We do not discuss $o(|h|)$ behaviour here. Let in addition the discretization be "stable" in the sense that

(2.13) $$\|z_1^h - z_2^h\| \le S \|F^h z_1^h - F^h z_2^h\|,$$

with S independent of h and $\|F^h z_i^h - F^h z_*^h\|$ small enough, $i = 1, 2$. Then, if F^h is continuous in a neighbourhood of $\Delta^h z_*$, (2.8) is locally uniquely solvable and z_*^h "converges of order p" to z^* in the sense, see Stetter **[56]**,

(2.14) $$\|z_*^h - \Delta^h z^*\| = O(|h|^p).$$

For a wide class of discretization methods, the stability (2.13) is a consequence of (2.3) and (2.12); see, e.g., Stummel **[57]**, **[58]**, Grigorieff **[35]**, **[36]**, Beyn **[9]**, Hackbusch **[38]**, **[39]**, Böhmer **[12]**.

We want to give two different discretizations for our example (2.4), both based on the same mesh or grid size vectors $h := (h_0, \dots, h_{n-1})$, $n = n(h)$, $|h| := \max\{h_j, j = 0, \dots, n-1\}$. This h defines a (nonequidistant) grid Γ^h which is used in both discretizations for the spaces E^h, $\hat{E}^h$ and projectors Δ^h and $\hat{\Delta}^h$. For the box-scheme we introduce

(2.15)
$$\begin{cases} \Gamma^h := \{t_0 := 0,\ t_{j+1} := t_j + h_j,\ j = 0, 1, \dots, n-1,\ t_n = 1\}, \\ E^h := \{z^h : \Gamma^h \to \mathbf{R}^m\}, \\ \hat{E}^h := \{(g^h, \alpha) | g^h : (t_j + t_{j+1})/2 \to \mathbf{R}^m,\ j = 0, \dots, n-1,\ \alpha \in \mathbf{R}^m\}, \\ \Delta^h z := z|\Gamma^h,\ \hat{\Delta}^h(g, \alpha) := (g(t_j) + g(t_{j+1}))/2,\ j = 0, 1, \dots, n-1; \alpha. \end{cases}$$

Now the "box-scheme" for (2.4) is defined with $z_j^h := z^j(t_i)$ as

(2.16)
$$F_b^h z^h := \begin{pmatrix} \dfrac{z_{j+1}^h - z_j^h}{h_j} - (f(t_j, z_j^h) + f(t_{j+1}, z_{j+1}^h))/2, j = 0, 1, \dots, n-1 \\ B_0 z_0^h + b_1 z_n^h \end{pmatrix} = \hat{\Delta}^h y.$$

It is easily verified that for $z \in C^3[0,1]$ the consistency condition (2.12) is satisfied for $p = 2$. The stability (2.13) is a consequence of (2.5) and (2.12); see Grigorieff **[36]**, Schild **[50]**.

To define a collocation method for (2.4), see de Boor–Swartz **[20]**, let $\Pi_q[a,b]$ be the polynomials of degree q on $[a,b]$, and

$$\begin{aligned} \Pi_{q,\Gamma^h} &:= \Pi_q[t_0,t_1] \times \Pi_q[t_1,t_2] \times \cdots \times \Pi_q[t_{n-1},t_n], \\ E^h &:= \Pi_{q,\Gamma^h} \cap C[0,1]. \end{aligned}$$

For the special case $-\rho_1 = \rho_q = 1$, see (2.18), we even require $E^h = \Pi_{q,\Gamma^h} \cap C^1[0,1]$. Based upon a fixed choice of

$$-1 \le \rho_1 < \rho_2 < \cdots < \rho_q \le 1, \tag{2.18}$$

we define an extended collocation grid Γ_c^h, see (2.15), as

$$\Gamma_c^h := \{t_{qj+i} := (t_j + t_{j+1} + \rho_i h_j)/2,\ j = 0,\ldots,n-1,\ i = 1,\ldots,q\}. \tag{2.19}$$

The collocation method for (2.4) determines a $z_*^h \in E^h$ from the equation

$$F_c^h z^h := \begin{pmatrix} (z^{h'}(\cdot) - f(\cdot, z^h(\cdot)))|\Gamma_c^h \\ B_0 z^h(0) + B_1 z^h(1) \end{pmatrix} = \begin{pmatrix} g|\Gamma_c^h \\ \alpha \end{pmatrix}. \tag{2.20}$$

In case $\rho_1 = -1$ or $\rho_q = 1$, the derivatives $z^{h'}(t_j)$ have to be interpreted as right or left limits $z^{h'}(t_j+)$ or $z^{h'}(t_j-)$, respectively. Especially we see for $\rho_1 = -1$ and $\rho_q = 1$ that (2.20) implies $z_*^h \in C^1[0,1]$, since $z_*^{h'}(t_j+) = z_*^{h'}(t_j-) = f(t_j, z_*^h(t_j)) + g(t_i)$ and therefore $E^h \subset C^1[0,1]$.

To get back to the general framework, we introduce

$$\begin{cases} \hat{E}^h := \{(g,\alpha) | g\colon \Gamma_c^h \to \mathbf{R}^m,\ \alpha \in \mathbf{R}^m\}, \\ \hat{\Delta}^h(g,\alpha) := (g|\Gamma_c^h, \alpha) \text{ for } (g,\alpha) \in \hat{E}, \\ \Delta^h z := z^h \in E^h \text{ with } (z - z^h)|\Gamma_c^h = 0 \text{ and} \\ B_0(z - z^h)(0) + B_1(z - z^h)(1) = 0; \end{cases} \tag{2.21}$$

here, the first condition determines $z^h = \Delta^h z$ up to a constant and with $\operatorname{rank}(B_0, B_1) = m$, see (2.4), this constant vector is uniquely determined by the second condition. For a function $z \in C^{1+q}[0,1]$ we have

$$(z^h - \Delta^h z)(\tau_i) = 0, \qquad (z' - (\Delta^h z)')(\tau_i) = O(|h|^q); \tag{2.22}$$

hence, the collocation method for (2.4) is consistent of order q. The stability is proved in Schild **[50]**, Böhmer **[12]**. In the gridpoints of the original grid Γ^h superconvergence results up to the order $q^* \le 2q$ are valid, if the ρ_i in (2.18) are chosen as Gaussian points; see de Boor–Swartz **[20]**.

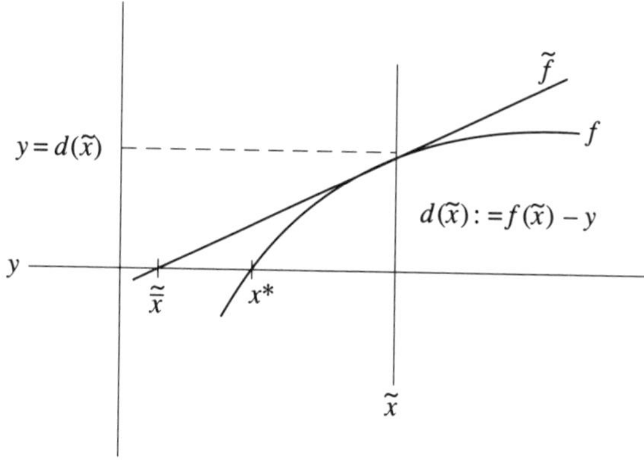

FIGURE 1

3. General defect correction ideas. One of the first examples for defect corrections is Newton's method for the computation of the zeros of a real function. We want to reformulate this method a little bit unusually in order to prepare the general case. For $f\colon [a,b] \to \mathbf{R}$ and $y \in f[a,b]$, assume we want to compute

$$f(x^*) = y \quad \text{with } x^* \in [a,b]. \tag{3.1}$$

Starting with an $\tilde{x} \simeq x^*$ we define an approximate function $\tilde{f}$

$$f(x) \simeq \tilde{f}(x) := m(x - \tilde{x}) + b, \tag{3.2}$$

which is simply invertible, hence, $\tilde{g} := (\tilde{f})^{-1}$ is available. Newton's method is obtained for $m = f'(\tilde{x})$ or an approximate value and with

$$b = f(\tilde{x}) = y + d(\tilde{x}) = \tilde{f}(x), \qquad d(\tilde{x}) := f(\tilde{x}) - y, \tag{3.3}$$

$d(\tilde{x})$ the so-called defect of $\tilde{x}$. Instead of computing x^* from (3.1) we compute a hopefully better approximation $\tilde{\tilde{x}}$ than $\tilde{x}$ from, see Figure 1 and (3.2),

$$\tilde{f}(\tilde{\tilde{x}}) = y. \tag{3.4}$$

A combination of (3.3) and $f(\tilde{x}) = \tilde{f}(\tilde{x})$ shows that

$$\tilde{\tilde{x}} = \tilde{g}(y) = \tilde{x} - (\tilde{g}(f(\tilde{x})) - \tilde{g}(y)). \tag{3.5}$$

This procedure may be immediately generalized to operator equations. For simplicity we assume

$$F\colon D \subset E \to \hat{D} \subset \hat{E}, \qquad D \subset D(F), \quad \hat{D} \subset R(F), \tag{3.6}$$

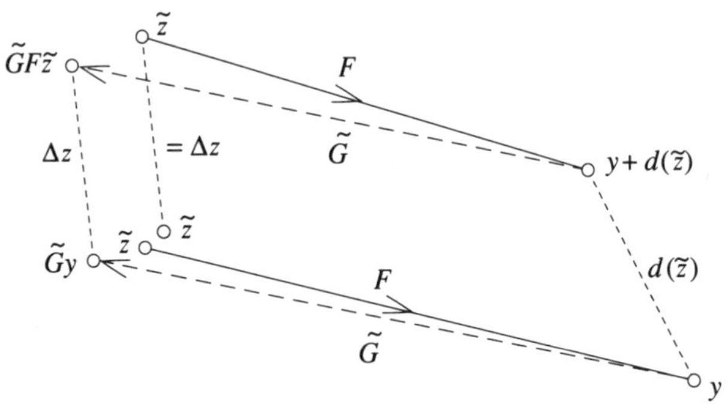

FIGURE 2

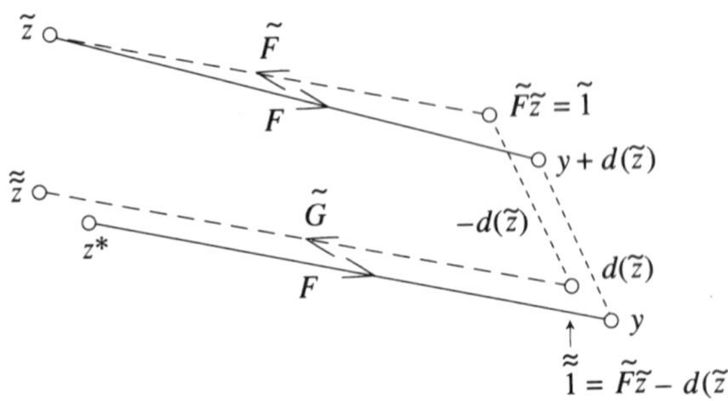

FIGURE 3

to be bijective on the closed sets D, $\hat{D}$ with interior points $z^* \in D^0$, $y \in \hat{D}^0$. Furthermore, we assume the existence of an approximate operator $\tilde{F} \simeq F$, such that its inverse $\tilde{G}$ is computable:

$$\tilde{F}: D \to \hat{D}, \qquad \tilde{G} = (\tilde{F})^{-1}: \hat{D} \to D \tag{3.7}$$

with

$$F z \simeq \tilde{F} z \text{ and } \tilde{G}F z \simeq z \text{ for } z \in D \qquad \text{and } F\tilde{G}y \simeq y \text{ for } y \in \hat{D}. \tag{3.8}$$

Then we may use the defect with respect to (2.2) for an approximation $\tilde{z} \simeq z^*$,

$$d\tilde{z} := F\tilde{z} - y, \tag{3.9}$$

to define a hopefully better approximation $\tilde{\tilde{z}}$ in analogy to (3.5) as

$$\tilde{\tilde{z}} := \tilde{z} - (\tilde{G}F\tilde{z} - \tilde{G}y). \tag{3.10}$$

Based upon (3.8) this procedure may be interpreted as follows: We know y and have computed $y + d(\tilde{z})$.

Instead of the exact change $z^* - \tilde{z}$ we compute with (3.8) the approximation $\tilde{\tilde{z}} - \tilde{z}$, see Figure 2,

$$(3.11)\qquad z^* - \tilde{z} \simeq \tilde{\tilde{z}} - \tilde{z} := \tilde{G}Fz^* - \tilde{G}F\tilde{z} = \tilde{G}y - \tilde{G}F\tilde{z}.$$

In (3.10), (3.11) the knowledge of $d(\tilde{z})$ allowed the computation of an improved approximation $\tilde{\tilde{z}}$. We could as well use $Fz \simeq \tilde{F}z$ in (3.8) to define, again with $d(\tilde{z})$, a better approximation $\tilde{\tilde{y}}$ for the approximate image under $\tilde{F}$, hence $\tilde{y} = \tilde{F}\tilde{\tilde{z}}$ and hopefully $\|\tilde{\tilde{y}} - y\| < \|\tilde{y} - y\|$ and $\|\tilde{\tilde{z}} - z^*\| < \|\tilde{z} - z^*\|$, see Figure 3,

$$\tilde{F}z^* - \tilde{F}\tilde{z} \simeq Fz^* - F\tilde{z} =: \tilde{F}\tilde{\tilde{z}} - \tilde{F}\tilde{z}$$

or

$$\tilde{F}z^* - \tilde{F}\tilde{z} \simeq y - F\tilde{G}\tilde{y} =: \tilde{\tilde{y}} - \tilde{y}$$

and

$$\tilde{F}\tilde{\tilde{z}} = (\tilde{F} - F)\tilde{z} + y \quad \text{or} \quad \tilde{\tilde{y}} = \tilde{y} - F\tilde{G}\tilde{y} + y,$$

so finally

$$(3.12)\qquad \tilde{\tilde{z}} = \tilde{G}((\tilde{F} - F)\tilde{z} + y) \quad \text{or} \quad \tilde{\tilde{y}} = \tilde{y} - F\tilde{G}\tilde{y} + y.$$

In many cases $\tilde{G}$ will be differentiable, such that

$$\tilde{G}(y + \Delta y) - \tilde{G}(y) = \tilde{G}'(y)\Delta y + o(\Delta y).$$

Then we obtain from (3.10) and (3.12), by neglecting the $o(\Delta y)$ term, the approximation

$$(3.13)\qquad \tilde{\tilde{z}} = \tilde{z} - \tilde{G}'(y)\,d(\tilde{z}) = \tilde{z} - \tilde{G}'(y)(F\tilde{z} - y).$$

All these approaches (3.10), (3.12), (3.13) may either be used to estimate the error $z^* - \tilde{z}$ for a given approximation $\tilde{z}$ by $\tilde{\tilde{z}} - \tilde{z}$, or to define an improved approximation $\tilde{\tilde{z}}$. In this case the error estimate gets lost.

This improvement technique may be used, see **[16]**, iteratively to define the Iterated Defect Corrections versions A and B or the Discrete Newton Methods, essentially going back in their most important form to Stetter **[56]**, Pereyra **[44]**, **[45]**, **[46]**, **[47]**, **[48]**, and Böhmer **[10]**. With different intentions Fox **[23]**, **[24]**, **[25]** and Zadunaiski **[60]**, **[61]** had initiated this area of research.

Starting with a $z^0 \in D$ let

$$\text{(IDCA)} \quad z^{\nu+1} := (I - \tilde{G}F)z^\nu + \tilde{G}y,$$

$$\text{(IDCB)} \quad z^{\nu+1} := \tilde{G}((\tilde{F} - F)z^\nu + y) \quad \text{or} \quad y^{\nu+1} := (I - F\tilde{G})y^\nu + y,$$

$$\text{(DNM)} \quad z^{\nu+1} := (I - \tilde{G}'(y)F)z^\nu + \tilde{G}(y)y \quad \text{or}$$

$$\tilde{F}'(z^0)(z^{\nu+1} - z^\nu) = -(Fz^\nu - y) \text{ for } \tilde{F}z^0 = y.$$

The comparison of (3.6) with these iteration methods shows that we have convergence to the same fixed point z^*, if the following mappings are contractions:

$$(I - \tilde{G}F) \text{ for (IDCA)}, \quad (I - F\tilde{G}) \text{ for (IDCB)}, \quad (I - \tilde{G}'(y)F) \text{ for (DNM)}.$$

In many important cases, see §4, it is not necessary to compute the exact defect Fz^ν. Rather, one can use increasingly better defect evaluations for increasing ν. It might be useful in special cases to use different inverse mappings $\tilde{G}_\nu$ as well.

4. Combination of discretization and defect correction. In this section we apply the defect correction from Section 3 to the original problem (2.2). For $\tilde{F}$ we use a low-order discretization $\tilde{F}^h$, e.g., the box-scheme for the boundary value problem (2.4). As target problem we use either a high-order discretization problem F^h, e.g., collocation with superconvergence in the gridpoints $t_j \in \Gamma^h$ for (2.4) or, under suitable conditions, the original problem; see end of this section. Let

$$\tilde{F}^h z^h = \tilde{y}^h \tag{4.1}$$

and

$$F^h z^h = y^h \tag{4.2}$$

represent discretizations of order p and $q > p$ for (2.2) with exact solutions $\tilde{z}^h_*$ and z^h_*. We assume for simplicity that $\tilde{F}^h$ and F^h map the same E^h into $\hat{E}^h$. Then we usually choose as starting value

$$\text{the solution } z^h_0 = \tilde{z}^h \text{ of } \tilde{F}^h z^h_0 = \tilde{y}^h \tag{4.3}$$

and obtain with the inverse $\tilde{G}^h$ for $\tilde{F}^h$

$$\text{(IDCA)}^h \qquad z^h_{\nu+1} := z^h_\nu - \tilde{G}^h F^h z^h_\nu + \tilde{G}^h \tilde{y}^h,$$

$$\text{(IDCB)}^h \qquad z^h_{\nu+1} := \tilde{G}^h((\tilde{F}^h - F^h)z^h_\nu + y^h),$$

$$\text{(DNM)}^h \qquad (\tilde{F}^h)'(z^h_0)(z^h_{\nu+1} - z^h_\nu) = -(F^h z^h_\nu - y^h).$$

It has to be pointed out, that the (IDCA)^h, (IDCB)^h versions require the solution of a system of nonlinear equations for each iteration, whereas

$(\mathrm{DNM})^h$ requires the solution of only one linear system per iteration with the same "matrix" $(\tilde{F}^h)'(z^h)$ for all the iterations. So the computational effort is much smaller than for the first two versions, unless linearization and one iteration is used for their solution. In this case the $(\mathrm{DNM})^h$ is obtained again.

Now it may be shown, that under certain conditions the mappings in the three iteration processes are $O(h^p)$ contractions, and therefore

$$\|z_\nu - \Delta^h z^*\| = O(h^{\min((\nu+1)p,q)}). \tag{4.4}$$

Hence, a relatively small additional computational amount, the evaluation of the defect and the solution of a linear system for $(\mathrm{DNM})^h$ allow one to dramatically increase the $O(h^p)$ accuracy of the original approximation $\tilde{z}^h_*$ to higher and higher orders.

The proof for the $O(h^p)$ contractivity of the mappings in $(\mathrm{IDCA})^h$, $(\mathrm{IDCB})^h$, $(\mathrm{DNM})^h$ requires a very careful analysis, taking into account the different norms in E^h and $\hat{E}^h$. Either the theory of asymptotic expansions may be employed (Böhmer **[10]**, Frank **[26]**, Frank–Hertling–Ueberhuber **[29]**, Frank–Ueberhuber **[33]**, Lindberg **[43]**), or norms in Sobolev spaces of higher smoothness (Hackbusch **[37]**, Skeel **[52]**, **[53]**, Schmitt **[51]**) or direct contractivity arguments may be used (Brakhage **[21]**, **[22]**, Stetter **[56]**, Hemker **[40]**, Schild **[50]**).

For our example (2.4) we want to describe the $(\mathrm{DNM})^h$ case: Frank–Ueberhuber **[33]** observed that a combination of defect corrections and collocation methods produced satisfactory convergence only for the case of equidistant collocation points in (2.18)–(2.21). Therefore it was not possible to obtain the superconvergence results for the especially interesting Gaussian and Lobatto points. The new approach due to Schild **[50]** eliminates this difficulty. Given F^h in (4.2) as the collocation scheme (2.18)–(2.21), he defines $\tilde{F}^h$ as box-scheme on an extended grid Γ^h_e, which is chosen to be equidistant in the subintervals, i.e. (compare (2.15) and (2.19)),

$$\Gamma^h_e := \{t^e_{qj+i} := t_j + h_j i/q,\ j = 0, \ldots, n-1,\ i = 1, \ldots, q\}. \tag{4.5}$$

Now we define $\tilde{F}^h_b$ to be the box-scheme defined on the grid Γ^h_e instead of the grid Γ^h in (2.15), and apply $(\mathrm{DNM})^h$ to this combination of $\tilde{F}^h$ and F^h_c. If the order of superconvergence for F^h is $q^* \le 2q$, we obtain for the z^h_ν in $(\mathrm{DNM})^h$ the result (on the original, nonextended grid Γ^h)

$$\|(z^h_\nu - z^*)|\Gamma^h\| = O(h^{\min(2(\nu+1),q^*)}), \tag{4.6}$$

where $h := \max\{h_j,\ j = 0, \ldots, n-1\}$.

We want to describe the modification when the original problem (2.2) replaces the high-order approximation (4.2) as a target problem for the case of our example (2.4) and the box-scheme. Then

$$Fz = y \text{ is transformed into } F_b^h z^h = \Delta^h y.$$

In this case the "local discretization error", see (2.12), admits an asymptotic expansion in powers of h $(= h_0 = \cdots = h_{n-1})$ for $z \in C_L^{2q+1}[0, 1]$ in the form (with $y_{2\mu}$ independent of h)

$$F_b^h \Delta^h z - \hat{\Delta}^h F z = \sum_{\mu=1}^{q} h^{2\mu} y_{2\mu} + O(h^{2q+1}).$$

This allows us to show that the "global discretization error", see (2.14), admits an asymptotic expansion as well (again $z_{2\mu}$ independent of h)

$$z_*^h - \Delta^h z^* = \sum_{\mu=1}^{q} h^{2\mu} z_{2\mu} + O(h^{2q+1}).$$

If it is possible to smoothly extend (2.4) to a larger interval $[0, 1] \subset [-\sigma, 1+\sigma]$ and hence to compute the $z_*^h(t_j)$ for $j = -k, -k+1, \ldots, 0, \ldots, n, \ldots, n+k$ one may use symmetric approximation formulas for $z'(t_{j+1/2})$ of order $(2\nu + 4)$ in (2.4) based upon the $z(t_\mu)$ to define approximations F_ν^h of order $(2\nu + 4)$

$$F_\nu^h \Delta^h z - \hat{\Delta}^h F z = \sum_{\mu=\nu+2}^{q} h^{2\mu} y_{\mu,\nu} + O(h^{2q+1}).$$

Then $(\mathrm{DNM})^h$ has the form, see (2.15), (2.16) (Böhmer **[15]**),

$$(F_0^h)'(z_0^h)(z_{\nu+1}^h - z_\nu^h) = -(F_\nu^h z_\nu^h - \hat{\Delta}^h y), \qquad \nu = 0, 1, \ldots, \tag{4.7}$$

with z_0^h the solution of (2.16). The $(\mathrm{IDCA})^h$ and $(\mathrm{IDCB})^h$ essentially represent the approaches of Fox **[23]**, **[24]**, **[25]** and Pereyra **[44]**–**[48]** to the problem (2.4); the z_ν^h in (4.7) satisfy

$$\|z_\nu^h - \Delta^h z^*\| = O(h^{\min(2(\nu+1), 2q)}). \tag{4.8}$$

Various types of defect corrections have been applied (no attempt for completeness is made here) to initial value problems in ordinary differential equations for the nonstiff (Zadunaisky **[60]**, **[61]**, Frank **[26]**, **[27]**, Böhmer–Fleischmann **[13]**, Frank–Ueberhuber **[33]**) and the stiff case (Stetter **[55]**, Frank–Hertling–Lehner **[28]**, Frank–Schneid–Ueberhuber **[30]**–**[34]**) to boundary value problems in ordinary differential equations for the nonstiff case (Fox **[23]**, **[24]**, **[25]**, Pereyra **[44]**–**[49]**, Lentini–Pereyra **[41]**, **[42]**, Frank **[26]**, Frank–Ueberhuber **[33]**, Böhmer **[10]**,

Lindberg [**43**], Skeel [**52**], [**53**], Schild [**50**]) and the stiff case (Böhmer–Römer [**17**]) and for singular cases with side conditions in connection with Hartree–Fock equations (Schmitt [**51**], Böhmer–Gross [**14**], Böhmer–Gross–Schmitt–Schwarz [**1**], Böhmer–Schmitt [**18**]), for integral equations (Brakhage [**21**], [**22**]), to elliptic differential equations in Pereyra–Proskurowski–Widlund [**49**] and finally in the context of multigrid methods for integral equations (see, e.g., Hemker–Schippers [**40**]) and differential equations (see, e.g., Hackbusch [**37**] and Wittum [**59**]). We have totally omitted the wide range of multigrid techniques; see, e.g., Böhmer–Hemker–Stetter [**16**] and Hackbusch [**39**].

5. Mesh independence principle (MIP). One of the main problems throughout is the solution of the nonlinear equations $F^h z^h = y^h$, especially for small h. In many cases, Newton-like iteration methods are used. It turns out that the number of Newton iterates for (2.2) and (2.8) are strongly related. Define, for $\|z_0 - z^*\|$ small enough and under the condition (2.3), Newton's method for (2.2) as

$$F'(z_\nu)(z_{\nu+1} - z_\nu) = -(Fz_\nu - y)$$

and let for $\varepsilon > 0$ and z_0

$$M(\varepsilon, z_0) := \min\{\nu\colon \|z_\nu - z^*\| < \varepsilon\}. \tag{5.1}$$

In analogy we define for (2.8), starting with a z_0^h,

$$(F^h)'(z_\nu^h)(z_{\nu+1}^h - z_\nu^h) = -(F^h z_\nu^h - y^h) \tag{5.2}$$

and

$$M^h(\varepsilon, z_0^h) := \min\{\nu\colon \|z_\nu^h - z_*^h\| < \varepsilon\}.$$

Now let

$$\left\{\begin{array}{l} \|z_0 - z^*\| \text{ and } \|z - z^*\| \text{ be small enough, (2.3), (2.11)–} \\ \text{(2.13) be satisfied, } (F^h)' \text{ be Lipschitz-continuous, and, for} \\ \|u\| \le 1,\ (F^h)'(\Delta^h z)\Delta^h u - \hat{\Delta}^h F'(z)u = O(h^p). \end{array}\right. \tag{5.3}$$

Then we have the following Mesh Independence Principle (MIP)

$$M^h(\varepsilon, \Delta^h z_0) = M(\varepsilon, z_0) + O(h^p). \tag{5.4}$$

If in (5.1), $M(\varepsilon, z_0)$ would be defined by $\|z_\nu - z^*\| \le \varepsilon$, one would have to allow ± 1 in (5.4). This result was for the first time derived by Allgower–McCormick **[6]** under much stronger conditions only for $z'' - f(t, z) = 0$, $z(0) = z(1) = 0$, and later on generalized by Allgower–McCormick–Pryor **[7]**, Allgower–Böhmer **[2]**, **[3]** and Allgower–Böhmer–Potra–Rheinboldt **[5]**. Applications to Newton methods and Galerkin type discretizations have been given by Adam **[1]**; Axelsson **[8]** had obtained similar results.

A comparison of (2.5) and the corresponding discretizations

$$(F_b^h)'(\Delta^h z)\Delta^h u \text{ and } (F_c^h)'(\Delta^h z)\Delta^h u$$

shows that especially the last condition in (5.9) is satisfied for F_b^h and F_c^h. For F_b^h this is obvious; for F_c^h we need (2.22).

6. Mesh refinement strategies. For multigrid-techniques the iteration methods for linear and nonlinear equations require very specific refinement strategies in the different grids (often halving of the step sizes). The following discussion is valid only if the solutions of linear equations for different values of h are independent from each other. In this case the MIP is a good tool to develop efficient strategies to solve nonlinear equations. We restrict the discussion to equidistant grids characterized by a positive step length h; for more general cases see Allgower–Böhmer **[3]**. For a given tolerance, tol, we want to compute h and a $z_\nu^h \simeq z_*^h$ such that

$$\|z_*^h - \Delta^h z^*\| \simeq \|z_\nu^h - \Delta z^*\| \lesssim \text{tol}. \tag{6.1}$$

Instead of applying (5.2) to this final grid, we want to determine a sequence of step lengths

$$h_0 > h_1 > \cdots > h_i > h_{i+1} > \cdots > h, \tag{6.2}$$

such that the computation of z_*^h via the $h_0 > \cdots > h$ grids amounts to the equivalence of only two or three iterations on the final grid. The main idea is reflected in the following algorithmic type of procedure.

(6.3)
(i) Choose E^{h_0} and $z_0^{h_0} := \Delta^{h_0} z_0$.
(ii) Perform a full Newton process (5.2) for h_0, until

$$\|z_{n_0+1}^{h_0} - z_{n_0}^{h_0}\| \leq \text{tol}.$$

(iii) Choose E^{h_1} with h_1 slightly smaller than h_0 and compute two or three steps in (5.2) for h_1. If the MIP is well reflected in h_0, h_1, go to (iv), else refine E^{h_0}, e.g., by $h_0 := h_1$, improve z_0, go to (i).
(iv) Based on h_0, h_1 results, choose as few as possible $h_1 > \cdots > h_{m-1} > h$ in (6.2) such that for $z_0^{h_{i+1}} := \Delta_{h_i}^{h_{i+1}} \tilde{z}^{h_i}$ ($\Delta_{h_i}^{h_{i+1}}$ an extension operator from E^{h_i} to $E^{h_{i+1}}$, $\tilde{z}^{h_i}$ the best approximation for $z_*^{h_i}$ available in E^{h_i}) in each E^{h_i} just one and in E^h as few as possible, iterations yield (6.1).

We start with a relatively large h_0 and perform (6.3 i–iii). The (even many) iterations (5.2) for large h_0 (and h_1) cost very little and are necessary for monitoring anyway. In passing from E^{h_i} to $E^{h_{i+1}}$ we have essentially to balance three errors: the discretization error

$$\text{err}_{\text{dis}}^{h_i} := \|z_*^{h_i} - \Delta^{h_i} z^*\| \simeq C_{\text{dis}}(z^*) h_i^p, \tag{6.4}$$

the iteration error in E^{h_i}

$$\text{err}_{\text{it}}^{h_i} := \|z_*^{h_i} - \tilde{z}^{h_i}\| \simeq \|z_n^{h_i} - z_{n+1}^{h_i}\| \quad \text{for } \tilde{z}^{h_i} = z_n^{h_i}, \tag{6.5}$$

and the interpolation, or approximation, error from E^{h_i} to $E^{h_{i+1}}$ (for small $\|z - z^*\|$)

$$\text{err}_{\text{int}}^{h_i} := \|\Delta_{h_i}^{h_{i+1}}(\Delta^{h_i} z) - \Delta^{h_{i+1}} z\| \simeq C_{\text{int}}(z^*) h_i^q, \tag{6.6}$$

if, e.g., piecewise polynomials of degree $q - 1$ are used. Since the maximum of these three errors essentially determines the error of $\|z_0^{h_{i+1}} - z_*^{h_{i+1}}\|$, the best strategy is to choose p, q, and n to equalize the terms in (6.4)–(6.6), especially since the best we can do for z_*^h in $E^{h_i} \neq E^h$ is to come within $\text{err}_{\text{dis}}^{h_i}$.

In [2], [3] we have described ways to estimate $C_{\text{dis}}(z^*)$ and $C_{\text{int}}(z^*)$. The best choice for q would be $q = p$, unless the corresponding $C_{\text{int}}(z^*)$ differs strongly from $C_{\text{dis}}(z^*)$. With an estimation for $C_{\text{dis}}(z^*)$ we may, by comparing (6.1) and (6.4), estimate h from

$$C_{\text{dis}}(z^*) h^p \simeq \text{tol} \quad \text{or} \quad h \simeq \sqrt[p]{\text{tol} / C_{\text{dis}}(z^*)}. \tag{6.7}$$

Now it might be that this h is intolerably small. Then improvement techniques such as defect corrections should be used or equivalently tol should be increased; see Allgower–Böhmer [**3**]. In this case the defect corrections need a much better approximation $\tilde{z}^h$, hence $\|\tilde{z}_*^h - z^h\| \ll \|z_*^h - \Delta^h z^*\|$. In this case, more than one iteration might be necessary on E^h; see [**3**].

We tend to have just one iteration for each E^{h_i}. Starting in $E^{h_{i+1}}$ with

$$\|z_0^{h_{i+1}} - z_n^{h_{i+1}}\| \simeq C_{\text{dis}} h_i^p$$

the quadratically convergent Newton method (5.2) yields

$$\|z_1^{h_{i+1}} - z_*^{h_{i+1}}\| \le K^{h_{i+1}} \|z_0^{h_{i+1}} - z_*^{h_{i+1}}\|^2 \simeq K^{h_{i+1}} (C_{\text{dis}} h_i^p)^2.$$

We want to have

$$\|z_1^{h_{i+1}} - z_*^{h_{i+1}}\| \simeq C_{\text{dis}} h_{i+1}^p$$

and if $K^{h_{i+1}} \simeq K$, see [**4**], we obtain

$$h_{i+1} \simeq \sqrt[p]{K C_{\text{dis}}} h_i^2. \tag{6.8}$$

Even if only very insecure estimates for K and C_{dis} are available and h_{i+1} has to be strongly overestimated, (6.8) means a very strong decay of the step sizes in our refinement process. For details and experimental results see [**3**].

Literature

1. D. Adam, *M.I.P. of Newton's method for Galerkin type discretizations*, Report Nr. 34, Institutul National Pentru Creatic Stiintifica Si Tehnica, Dept. of Mathematics, Bucaresti (1988).

2. E. Allgower and K. Böhmer, *A mesh independence principle for operator equations and their discretizations*, Arbeitspapiere GMD, No. 129 (1985).

3. ———, *Application of the mesh independence principle to mesh refinement strategies*, SIAM J. Numer. Anal. **24** (1987), 1335–1351.

4. E. Allgower, K. Böhmer, and S. F. McCormick, *Discrete defect corrections: The basic ideas*, Zeit. Angew. Math. Mech. **62** (1982), 371–377.

5. E. Allgower, K. Böhmer, F. Potra, and W. C. Rheinboldt, *A mesh independence principle for operator equations and their discretizations*, SIAM J. Numer. Anal. **23** (1986), 160–169.

6. E. Allgower and S. F. McCormick, *Newton's method with mesh refinements for numerical solution of nonlinear two-point boundary value problems*, Numer. Math. **29** (1978), 237–260.

7. E. Allgower, S. F. McCormick, and D. V. Pryor, *A general mesh independence principle for Newton's method applied to second-order boundary value problems*, Computing **23** (1979), 233–246.

8. O. Axelsson, *On global convergence of iterative methods*, in *Iterative Solution of Nonlinear Systems and Equations*, Lecture Notes in Math. **953** (1982) (R. Ansorge, Th. Meis and W. Törnig, eds.), Springer, Berlin-New York.

9. W.-J. Beyn, *Discrete Green's functions and strong stability properties of the finite difference method*, Applicable Anal. **14** (1982), 73–98.

10. K. Böhmer, *Discrete Newton methods and iterated defect corrections*, Numer. Math. **37** (1981), 167–192.

11. ——, *Defect corrections for Hartree-Fock methods in the Schrödinger equation*, in Proc. Symposium on Differential Geometry and Differential Equations—Computation of Partial Differential Equations, Beijing, 1984 (1985), 1–30 (Feng Kang, eds.), Science Press, Beijing, China.

12. ——, *On the stability of consistently differentiable discretization methods*, Report Fachbereich für Mathematik der Universität Marburg (1989).

13. K. Böhmer and H.-J. Fleischmann, *Self-adaptive discrete Newton methods for Runge-Kutta methods*, in Numerical Methods of Approximation **52** (1980), 28–48, International Series in Numerical Mathematics (L. Collatz, G. Meinardus, and H. Werner, eds.), Birkhäuser, Basel.

14. K. Böhmer and W. Gross, *Hartree-Fock methods: A realization of variational methods in computing energy levels in atoms*, in Numerische Behandlung von Eigenwertproblemen, Bd. 3, Internat. Schriftenreihe Num. Math. **69** (1984), 27–40 (J. Albrecht, L. Collatz, and W. Velte, eds.).

15. K. Böhmer, W. Gross, B. Schmitt, and R. Schwarz, *Defect corrections and Hartree-Fock method*, in Comput. Suppl. **5** (1984), 193–210, Springer, Vienna.

16. K. Böhmer, P. Hemker, and H. Stetter, *The defect correction approach*, in Comput. Suppl. **5** (1984), 1–32, Springer, Vienna.

17. K. Böhmer and Th. Römer, *Discrete Newton method and grid strategy for the Kreiss-Kreiss method for stiff boundary value problems*, Computing **39** (1987).

18. K. Böhmer and B. Schmitt, *Highly accurate solutions of the Hartree-Fock equations via defect corrections*, in Numerical Treatment of Differential Equations, Proc. NUMDIFF-4 (1988), 289–294 (K. Strehmel, ed.), Halle, Teubner, Leipzig.

19. K. Böhmer and H. Stetter, eds., *Defect correction methods, theory and applications*, Comput. Suppl. **5** (1984), Springer, Vienna.

20. C. De Boor and B. Swartz, *Collocation at Gaussian points*, SIAM J. Numer. Anal. **10** (1973), 582–606.

21. H. Brakhage, *Über die numerische Behandlung von Integralgleichungen nach der Quadraturformelmethode*, Numer. Math. **2** (1960), 183–196.

22. ——, *Zur Fehlerabschätzung für die numerische Eigenwertbestimmung bei Integralgleichungen*, Numer. Math. **3** (1961), 174–179.

23. L. Fox, *Some improvements in the use of relaxation methods for the solution of ordinary and partial differential equations*, Proc. Roy. Soc. London **A190** (1947), 31–59.

24. ——, *The solution by relaxation methods of ordinary differential equations*, Proc. Cambridge Phil. Soc. **45** (1949), 50–68.

25. ——, *The Numerical Solution of Two-point Boundary Value Problems in Ordinary Differential Equations*, Oxford: University Press, 1957.

26. R. Frank, *The method of iterated defect-correction and its application to two-point boundary value problems*, Numer. Math. **25** (1976), 409–419.

27. ——, *Schätzungen des globalen Diskretisierungsfehlers bei Runge-Kutta Methoden*, ISNM **27** (1975), 45–70.

28. R. Frank, J. Hertling, and H. Lehner, *Defect correction algorithms for stiff ordinary differential equations*, in **[19]**, (1984), 33–41.

29. R. Frank, J. Hertling, and C. W. Ueberhuber, *An extension of the applicability of iterated deferred corrections*, Math. Comp. **31** (1977), 907–915.

30. R. Frank, J. Schneid, and C. W. Ueberhuber, *The concept of B-convergence*, SIAM J. Numer. Anal. **18** (1981), 753–780.

31. ____, *Order results for implicit Runge-Kutta methods applied to stiff systems*, SIAM J. Numer. Anal. (to appear).

32. ____, *Stability properties of implicit Runge-Kutta methods*, SIAM J. Numer. Anal. **22** (1985), 497–534.

33. R. Frank and C. W. Ueberhuber, *Iterated defect correction for differential equations, part I: Theoretical results*, Computing **20** (1978), 207–228.

34. ____, *Iterated defect correction for the efficient solution of stiff systems of ordinary differential equations*, BIT **17** (1977), 46–159.

35. R. D. Grigorieff, *Zur Theorie linearer approximationsregulärer Operatoren*, I und II, Math. Nachr. **55** (1973), 233–249 and 251–263.

36. ____, *On the convergence of stability constants*, manuscript, Technische Universität Berlin.

37. W. Hackbusch, *Bemerkungen zur interierten Defektkorrektur und zu ihrer Kombination mit Mehrgitterverfahren*, Report 79–13, Math. Inst. Universität Köln (1979); Rev. Roum. Math. Pures Appl..

38. ____, *On the regularity of difference schemes—Part II: Regularity estimates for linear and nonlinear problems*, Ark. Mat. **21** (1982), 3–28.

39. ____, *Multigrid methods and applications*, Springer, Berlin-Heidelberg-New York-Tokyo, 1985.

40. P. W. Hemker and H. Schippers, *Multiple grid methods for the solution of Fredholm equations of the second kind*, Math. Comp. **36** (1981), 215–232.

41. M. Lentini and V. Pereyra, *A variable order finite difference method for nonlinear multipoint BVP's*, Math. Comp. **28** (1974), 981–1003.

42. ____, *An adaptive finite difference solver for nonlinear two-point boundary problems with mild boundary layers*, SIAM J. Numer. Anal. **14** (1977), 91–111.

43. B. Lindberg, *Error estimation and iterative improvement for discretization algorithms*, BIT **20** (1980), 486–500.

44. V. Pereyra, *Accelerating the convergence of discretization algorithms*, SIAM J. Numer. Anal. **4** (1967), 508–533.

45. ____, *Iterated deferred correction for nonlinear operator equations*, Numer. Math. **10** (1967), 316–323.

46. ____, *Iterated deferred corrections for nonlinear boundary value problems*, Numer. Math. **11** (1968), 111–125.

47. ____, *Highly accurate numerical solution of quasilinear elliptic boundary-value problems in n dimensions*, Math. Comp. **24** (1970), 771–783.

48. ____, *Deferred corrections software*, in **[19]**, (1984), 211–226.

49. V. Pereyra, W. Proskurowski, and O. Widlund, *High order fast Laplace solvers for the Dirichlet problem on general regions*, Math. Comp. **31** (1977), 1–16.

50. K. H. Schild, *Gaussian collocation via defect correction*, Report Philipps-Universität Marburg, Fachbereich Mathematik (1989).

51. B. Schmitt, *Defektkorrekturen für Hartree-Fock Gleichungen*, Habilitationsschrift, Univ. of Marburg, West Germany (1986).

52. R. D. Skeel, *A theoretical framework for proving accuracy results for deferred corrections*, SIAM J. Numer. Anal. **19** (1982), 171–196.

53. ____, *The order of accuracy for deferred corrections using uncentered formulas*, SIAM J. Numer. Anal. **23** (1986), 393–402.

54. H. J. Stetter, *Analysis of discretization methods for ordinary differential equations*, Springer, Berlin-Heidelberg-New York, 1973.

55. ____, *Economical global error estimation*, Stiff Differential Systems (Willoughby, R. A., ed.), Plenum Press, New York-London, 1974, pp. 245–258.

56. ——, *The defect correction principle and discretization methods*, Numer. Math. **29** (1978), 425–443.

57. F. Stummel, *Diskrete Konvergenz linearer Operatoren*, I. Math. Ann. **190** (1970), 45–92; II, Math. Z. **120** (1971), 231–264, Ii. Proc. Oberwolfach 1971, ISNM **20** (1972), 196–216.

58. ——, *Stability and discrete convergence of differentiable mappings*, Rev. Roum. Math. Pures e. Appl. **21** (1976), 63–96.

59. G. Wittum, *Distributive Iterationen für indefinite Systeme*, SFB-Preprint Nr. 454, Heidelberg, 1988.

60. P. E. Zadunaisky, *A method for the estimation of errors propagated in the numerical solution of a system of ordinary differential equations*, In: The Theory of Orbits in the Solar System and in Stellar Systems, Proc. Intern. Astronomical Union, Symp. 25, Thessaloniki (1964) (G. Contopoulos, ed.).

61. ——, *On the estimation of errors propagated in the numerical integration of ordinary differential equations*, Numer. Math. **27** (1976), 21–40.

Lectures in Applied Mathematics
Volume **26** (1990)

On a Numerical Lyapunov-Schmidt Method

KLAUS BÖHMER AND MEI ZHEN

1. Introduction. Recently, in **[1]** we discussed equations of the form

$$G(z,\lambda)=0, \qquad G\colon E:=E_0\times\mathbf{R}^q\to\hat{E}, \tag{1.1}$$
$$E_0, \hat{E} \text{ Hilbert spaces}, \quad q\geq 0.$$

To simultaneously determine a singular zero point $x_0:=(z_0,\lambda_0)\in E$ and the kernel $N(G_0')$ of its derivative, i.e.,

$$G(x_0)=0, \qquad \dim N(G_0')=m+q>0, \qquad G_0':=G'(x_0), \tag{1.2}$$

we defined an inflated mapping H. x_0 is called a higher-order singular point if $\dim N(G_0')>q+1$.

There are two essentially different approaches to determine the solution manifold of an equation (1.1). In a more or less global computation of the manifolds, candidates for bifurcation points are avoided by introducing some type of perturbations. This idea has for the first time been realized in Keener–Keller **[18]**, and has been applied again in Keller **[19]**, Georg **[13]**, **[14]**, Peitgen **[23]**, and Allgower–Chien **[2]**. Since for more complicated manifolds the first approach yields submanifolds rather randomly, a systematic study of the local properties of the manifolds bifurcating at a singular point will yield much insight into the structure of the whole solution. Therefore our aim here is the study of the local behaviour at a singular point.

Based on the (approximate) knowledge of x_0 and $N(G_0')$, we want to compute numerically the manifolds bifurcating at x_0 and discuss the problem, how a discretization of G has to be modified to yield these manifolds. The main tool is the Lyapunov-Schmidt method, which we present in Section 2. Section 3 is devoted to the introduction of discretization methods and to the transformation of the Lyapunov-Schmidt equations into their discrete counterparts. These equations are separated into a fixed-point equation studied in Section 4, and the nonlinear bifurcation equations studied in Sections 5 and 6. To simplify the determination of a good enough starting value for these nonlinear bifurcation equations we give a slight modification in Section 6. The final Section 7 includes numerical experiments with the proposed method applied to an elliptic boundary value problem. This paper is confined to the presentation of the main ideas; technical details are given in [5].

2. Lyapunov-Schmidt method. For the problem (1.1), (1.2) we assume (let $[v_1, \dots, v_n] := \operatorname{span}\{v_1, \dots, v_n\}$ and see [1])

$$(2.1)\qquad \begin{cases} \text{(i)} & G \in C^2(D(G)),\ D(G) \text{ a convex neighbourhood of } x_0,\ G(x_0) = 0; \\ \text{(ii)} & N(G_0') = [\varphi_1, \dots, \varphi_{m+q}] \subset E_0 \times \mathbf{R}^q = E \text{ with } G_0' := G'(x_0); \\ \text{(iii)} & N(G_0'^*) = [\hat{\psi}_1, \dots, \hat{\psi}_m] \subset \hat{E}; \\ \text{(iv)} & G_0' \text{ and } G''(x_0)w, \text{ for every } w \in N(G_0'), \text{ are linear (bounded) operators with closed range.} \end{cases}$$

Now let, see e.g. Stakgold [27] or Golubitsky–Schaeffer [15],

$$(2.2)\qquad \begin{cases} N := N(G_0'),\ \hat{N} := N(G_0'^*) \text{ and} \\ E = N + M, \text{ with orthogonal subspaces } N \text{ and } M \text{ and let} \\ P\colon E \to N,\ Q\colon E \to M,\ I = P + Q \end{cases}$$

be the corresponding orthogonal projections. Then any $u \in E$ allows the representation

$$(2.3)\qquad u := x - x_0 = (z - z_0, \lambda - \lambda_0) = v + w = Pu + Qu = \sum_{i=1}^{m+p} c_i \varphi_i + Qu$$

and the Taylor formula yields

$$(2.4)\qquad \begin{cases} 0 = G(x) = G_0'u + R_0(u) \text{ or } G_0'u = -R_0(u) \text{ with} \\ R_0(u) = \int_0^1 (G'(x_0 + tu) - G_0')u\,dt = O(\|u\|^2). \end{cases}$$

The conditions (2.1) imply, with respect to the inner products $\langle \cdot, \cdot \rangle$ in E and $\hat{E}$, see Yosida [33],

$$(2.5)\qquad \mathbf{R}(G_0') = N(G_0'^*)^{\perp} = [\hat{\psi}_1, \dots, \hat{\psi}_m]^{\perp}.$$

Now (2.3)–(2.5) imply

$$(2.6)\qquad \langle R_0(v + w), \hat{\psi}_j \rangle = 0, \quad j = 1, \dots, m,$$

(2.7)
$$G_0'w = -R_0(v+w), \qquad v = \sum_{i=1}^{m+q} c_i\varphi_i, \qquad \langle w, \varphi_i\rangle = 0, \quad i = 1, \dots, m+q.$$

In particular, (1.7) is solvable if and only if (2.6) and (2.7) are solvable.

Introducing the generalized inverse

$$T := G_0'^{+} := (G_0' \mid M)^{-1}\hat{Q}, \tag{2.8}$$

with the projection $\hat{Q}\colon \hat{E} \to \mathbf{R}(G_0')$, we thus obtain the following equations, which are equivalent to (2.6), (2.7), or (1.7) for arbitrary u:

(2.9)
$$w = -TR_0(v+w) \text{ with } v = \sum_{i=1}^{m+q} c_i\varphi_i, \quad \langle w, \varphi_i\rangle = 0, \quad i = 1, \dots, m+q,$$

$$\left\langle R_0\left(\sum_{i=1}^{m+q} c_i\varphi_i + w\right), \hat{\psi}_j\right\rangle = 0, \quad j = 1, \dots, m. \tag{2.10}$$

Our main goal, however, is the computation of the branching manifolds in a neighbourhood of x_0, i.e., for small u. Let

$$v = \sum_{i=1}^{m+q} c_i\varphi_i \in N, \quad \|v\| \text{ small enough.} \tag{2.11}$$

We show that, under suitable conditions, (2.9) is uniquely solvable for w if v is given.

THEOREM 1. *Let G satisfy* (2.1). *Then there exists an $\varepsilon > 0$ such that for any v in* (2.11) *with $\|v\| \le \varepsilon$, Equation* (2.9) *is uniquely solvable, $w = w(v)$ depends continuously upon v and satisfies $\|w(v)\| = O(\|v\|^2)$.*

For the proof we apply Banach's fixed-point theorem and study the iteration

$$w^\nu = Bw^{\nu-1} := -TR_0(v + w^{\nu-1}), \quad \text{with e.g. } w^0 := 0. \tag{2.12}$$

We have seen that (1.1) is equivalent to (2.6), (2.7) or (2.9), (2.10). For a chosen v in (2.11) the equation (2.9) is uniquely solvable by a $w = w(v) = w(c_1, \dots, c_{m+q})$. To compute the solution manifolds for (1.1), bifurcating at x_0, we thus have to solve (2.10) for the $c_1, \dots, c_{m+q}$ in terms of $w(c_1, \dots, c_{m+q})$. By admitting perturbation terms in $N(G_0'^{*})$ in (2.6), (2.7) we obtain

COROLLARY 2. *Under the conditions of Theorem* 1 *there is an* $\varepsilon > 0$ *such that for every* $v \in N$ *with* $\|v\| \le \varepsilon$ *there is a unique solution* $(w, d_1, \ldots, d_m) = O(\|v\|^2)$ *for the nonlinear system*

$$\begin{cases} G_0'w + \sum_{j=1}^m d_j \hat{\psi}_j = -R_0(v+w) & \text{for } v = \sum_{i=1}^{m+q} c_i \varphi_i \\ \langle w, \varphi_i \rangle = 0, \quad i = 1, \ldots, m+q. \end{cases} \tag{2.13}$$

Furthermore, the solution of (2.13) *satisfies* $d_1 = \cdots = d_m = 0$ *iff* $R_0(v+w) \in R(G_0')$.

Studying discretization methods in Section 3, especially with respect to the necessary consistency, we need dense subspaces E_s of E (and $\hat{E}_s$ of $\hat{E}$) with increased smoothness properties. For the example in Section 7 and the second-order method used there, $z \in W^4(\Omega)$ is appropriate. This increased smoothness is guaranteed, whenever G is smooth enough, e.g. for $g \in C^2(\Omega, \mathbf{R})$ and boundary conditions, which are sufficiently consistent, implying $G \in C^2(D(G))$. With these smooth dense subspaces indicated by an index s, we therefore require

$$\begin{cases} G\colon E_s = E_{0s} \times \mathbf{R}^q \subset E \to \hat{E}_s \subset \hat{E}, \; x_0 \in E_s, \; N(G_0') \subset E_s, \\ \text{furthermore, every solution } x \text{ of } G(x) = 0 \text{ is in } E_s. \end{cases} \tag{2.14}$$

Usually, we still regard E_s and $\hat{E}_s$ as (dense) subspaces of E and $\hat{E}$, equipped with the original norms in E and $\hat{E}$, respectively. The conditions in (2.14) concerning G_0' usually impose one additional smoothness condition compared with those sufficient for G.

In the situation of (2.14) the equivalences of (2.4) with (2.6), (2.7) or (2.9), (2.10) should imply $x_0, x, u \in E_s$. This is indeed the case, since for the solutions x_0, x of (2.4) the relations (2.6), (2.7), (2.9), (2.10) are satisfied. If (2.6), (2.7) or (2.9), (2.10) is satisfied, $x = x_0 + v + w$ is, due to the equivalence, a solution of $G(x) = 0$, and by (2.14) an element of E_s.

In a more formal way we may introduce smooth operators $P_s, Q_s, \hat{Q}_s$ and subsets M_s. In addition to (2.14) we need the condition

$$\begin{cases} G_0'\colon E_s \to \hat{E}_s \text{ and } G_0'(E \setminus E_s) \subset \hat{E} \setminus \hat{E}_s, \\ \text{there exists a constant } C > 0 \text{ such that} \\ \|x\|_E \le C\|x\|_{E_s}, \; \forall x \in E_s, \; \|\hat{y}\|_{\hat{E}} \le C\|\hat{y}\|_{\hat{E}_s}, \; \forall \hat{y} \in \hat{E}_s. \end{cases} \tag{2.15}$$

In this situation, (2.6) had to be modified and we are able to show that

$$\begin{cases} E_s = N + M_s, \qquad M_s := M \cap E_s \perp N \subset E_s, \\ P_s := P|E_s\colon E_s \to N, \qquad Q_s := Q|E_s\colon E_s \to M_s \\ \text{are bounded with respect to } \|\cdot\|_E \text{ and to } \|\cdot\|_{E_s}. \end{cases} \tag{2.16}$$

If we replace in (2.8) and (2.9) the $Q, \hat{M}, T$ by the corresponding smooth $\hat{Q}_s : \hat{E}_s \to \hat{M}_s := \hat{E}_s \cap \hat{M}, M_s, T_s$, we obtain the equivalence of (2.4) with (2.9), (2.10) in the form, see **[5]**,

$$w = -T_s R_0(v + w) \text{ with } T_s := (G'_0|M_s)^{-1}\hat{Q}_s. \tag{2.17}$$

Therefore all the computations as described in Theorem 1 and Corollary 2 may be performed for the smooth setting as well and yield smooth solutions. For the case $q = 1$, many analytical results are proved including higher bifurcations ($m > 1$). Usually, see Berger **[3]**, **[4]** or Knightly–Sather **[20]** and Sather **[25]**, under suitable assumptions on the linear and nonlinear terms in G, the existence of solutions $w = w(v)$ for (2.9) allows us, via (2.10) to show existence of "at least one" or "at least two" additional, and hence bifurcating, solutions of (1.1).

3. Discretization of the main equation. Now we want to solve the equations (2.6), (2.7) or (2.9), (2.10) numerically. For the finite-dimensional case, because of the unavoidable round-off errors, and for the infinite-dimensional case by discretization, it is not possible to evaluate the exact G and G', but only approximations $\widetilde{G}$ and $\widetilde{G}'$ with, e.g.,

$$\begin{cases} \|G(x) - \widetilde{G}(x)\| \leq \varepsilon \text{ and } \|G'(x) - \widetilde{G}'(x)\| \leq \varepsilon \\ \text{for } \|x - x_0\| \text{ small enough.} \end{cases} \tag{3.1}$$

To simplify matters we only discuss discretization errors (indicated by $O(h^p)$ below) and assume that they dominate the round-off errors, which will be neglected from now on. In the following we outline only the main ideas and suppress some of the tedious details.

To define a discretization for (1.1), see e.g. Stetter **[28]**, we need some sequences, indicated by $h \in H$. Let (finite-dimensional) spaces E^h and $\hat{E}^h$ be given such that

$$\Delta^h : E = E_0 \times \mathbf{R}^q \to E^h = E_0^h \times \mathbf{R}^q, \qquad \hat{\Delta}^h : \hat{E} \to \hat{E}^h, \tag{3.2}$$

usually with $\Delta^h E_0 := \Delta^h(E_0 \times \{0\})$, are (linear) bounded operators satisfying

$$\begin{cases} \|\Delta^h x\| = \|x\| + O(h^p), \qquad \|\hat{\Delta}^h \hat{y}\| = \|\hat{y}\| + O(h^p) \text{ for fixed} \\ x, \hat{y} \text{ in appropriate (smooth) subspaces } E_s, \hat{E}_s. \end{cases} \tag{3.3}$$

Then the discretization for (1.1) is given by an operator $G^h : D(G) \subset E \to \hat{E}^h$, which is assumed to be "consistent", i.e.,

$$G^h(\Delta^h x) = \hat{\Delta}^h G x + O(h^p) \text{ for fixed } x \in E_s, \tag{3.4}$$

and "consistently differentiable", i.e.,

$$G^{h'}((\Delta^h x)\Delta^h u) = \hat{\Delta}^h G'(x)u + O(h^p) \text{ for fixed } x, u \in E_s \tag{3.5}$$

with $\|x - x_0\|$ and $\|u\|$ sufficiently small (w.r.t. $\|\cdot\|_E$)—compare the concept of "approximating sequences" G^h for G in Brezzi–Rappaz–Raviart [7]. We assume the necessary stability properties of G^h (or $G^{h'}$) in the form

$$\|x_1^h - x_2^h\|_{E^h} \leq S\|G^h(x_1^h) - G^h(x_2^h)\|_{\hat{E}^h}, \tag{3.6}$$

for $\|x_i^h - x_0^h\|_{E^h}$ small enough, $i = 1, 2$. If the defining functions in G are smooth enough, e.g. $g(\cdot, x)$ in (7.2), and the boundary conditions fit well, differencing of $G^h(x^h) = 0$ combined with induction and (3.6) may be used to yield estimates where $\|\cdot\|_{E^h}$ and $\|\cdot\|_{\hat{E}^h}$ in (3.6) are replaced by $\|\cdot\|_{E_s^h}$ and $\|\cdot\|_{\hat{E}_s^h}$, respectively. Finally, we require that

$$\begin{cases} G^h(D(G^h) \cap E_s^h) \subset \hat{E}_s^h \text{ and} \\ G^{h'}(x^h)E_s^h \subset \hat{E}_s^h \text{ for } x^h \in E_s^h. \end{cases} \tag{3.7}$$

With the solution $x_0^h \simeq \Delta^h x_0$ for $G^h(x_0^h) = 0$, we assume G^h (and $G^{h'}$) to be Lipschitz continuous in a neighbourhood of x_0^h. We try to compute the bifurcating manifolds and define R_0^h from, see (2.4),

(3.8)

$$0 = G^h(x^h) = (G^{h'})(x^h - x_0^h) + R_0^h(x^h - x_0^h), \qquad (G^{h'})_0 := (G^{h'})(x_0^h).$$

The $O(h^p)$ are used here and below in a slightly loose way, since we want to avoid a careful book-keeping of boundedly many elementary algebraic operations with bounded constants, which preserve the size of $O(h^p)$ terms for $h \in H$ and h tending to zero.

The degeneracy of G' in x_0 violates the usual stability results for G^h (or $G^{h'}$) in a neighbourhood of $\Delta^h x_0$. So we have to approach the problems in an unusual way. That the introduction of the discretization errors usually will unfold the original problem is reflected in the fact that, in general,

$$q \leq \dim N((G^{h'})_0) \leq m + q \quad \text{and} \quad 0 \leq \dim N((G^{h'})_0^*) \leq m,$$

and hence, a direct transformation of (2.2)–(2.10) into the discrete counterparts will not be possible. In **[1]** we have presented a method for computing the approximations $x_0^h, \varphi_i^h, \hat{\psi}_j^h$ and defining subspaces N^h

and $\hat{N}^h$ of the form

$$(3.9)\quad \begin{cases} x_0^h = \Delta^h x_0 + O(h^p), \\ \varphi_i^h = \Delta^h \varphi_i + O(h^p), \quad i = 1, \dots, m+q, \\ \hat{\psi}_j^h = \hat{\Delta}^h \hat{\psi}_j + O(h^p), \quad j = 1, \dots, m, \\ N^h := [\varphi_1^h, \dots, \varphi_{m+q}^h] \neq N((G^{h'})(x_0^h)) = N((G^{h'})_0), \\ \hat{N}^h := [\hat{\psi}_1^h, \dots, \hat{\psi}_m^h] \neq N((G^{h'})_0^*). \end{cases}$$

N^h and $\hat{N}^h$ usually do not represent the respective kernels of $(G^{h'})_0$ and $(G^{h'})_0^*$, although these operators will be "nearly singular" and will have "approximate" rank deficiency m.

Besides the restriction operators in (3.2) we need linear extension operators usually realized by interpolation or approximation (see **[2]** and Swartz–Varga **[31]**, Schultz **[26]**, Dahmen **[10]**, Dahmen–Micchelli **[11]**, Micchelli **[22]**, Chui **[8]**, Chui–Schumaker–Ward **[9]**) such that

$$(3.10)\quad \begin{cases} \Delta_h \colon E^h \to E, \quad \hat{\Delta}_h \colon \hat{E}^h \text{ with } \Delta_h(E_s^h) \subset E_s, \quad \hat{\Delta}_h(\hat{E}_s^h) \subset \hat{E}_s, \\ \Delta_h \Delta^h x = x + O(h^p), \quad \Delta^h \Delta_h x^h = x^h + O(h^p), \\ \hat{\Delta}_h \hat{\Delta}^h \hat{y} = \hat{y} + O(h^p), \quad \hat{\Delta}^h \hat{\Delta}_h y^h = \hat{y}^h + O(h^p), \end{cases}$$

for fixed $x \in E_s$, $\hat{y} \in E_s$ with small $\|x - x_0\|$, $\|\hat{y}\|$ and sequences $x^h \in E_s^h$, $\hat{y}^h \in \hat{E}_s^h$ with small $\|x^h - x_0^h\|$ and bounded $\|x^h\|_{E_s^h}$, $\|\hat{y}^h\|_{\hat{E}_s^h}$. Here the $O(h^p)$ terms in (3.10) mean, as in (3.9), e.g.,

$$\|\Delta^h \Delta_h x^h - x^h\|_{E^h} \leq C \cdot h^p \quad \text{for } x^h \in E_s^h \text{ and } \|x^h\|_{E_s^h} \leq K.$$

Unless in E or $\hat{E}$ the sup-norm is chosen, a reconstruction in the sense of (3.10) requires the incorporation of higher divided differences, especially close to the boundary, into the definition of Δ_h. Hence Δ_h is only bounded for bounded $\|x^h\|_{E_s^h}$; the same is true for $\hat{\Delta}_h$.

A transformation of (2.6), (2.7) or (2.9), (2.10) into their discrete counterparts requires discrete inner products $\langle \cdot, \cdot \rangle^h$. We assume, for $\|u\|, \|v\|, \|u^h\|_{E_s^h}, \|v^h\|_{E_s^h} \leq C$, see (3.10),

$$(3.11)\quad \begin{cases} \langle u, v \rangle = \langle \Delta^h u, \Delta^h v \rangle^h + O(h^p), \\ \langle u^h, v^h \rangle^h = \langle \Delta_h u^h, \Delta_h v^h \rangle + O(h^p). \end{cases}$$

Corresponding relations hold for Δ^h and Δ_h replaced by $\hat{\Delta}^h$ and $\hat{\Delta}_h$, respectively.

To numerically solve (2.4) we may either proceed along the lines of Theorem 1 or of Corollary 2 employing $\hat{Q}$ and T or the $\sum_{j=1}^m d_j \hat{\psi}_j$ and their discrete counterparts, respectively. With the N^h and $\hat{N}^h$ in (3.9) we decompose, analogously to (2.2) and (2.6), and introduce projectors and operators as

(3.12)
$$\begin{cases} E^h = N^h + M^h \text{ with } M^h := (N^h)^{\perp_h}, \\ P^h\colon E^h \to N^h, \quad Q^h\colon E^h \to M^h = (N^h)^{\perp_h}, \quad I^h = P^h + Q^h, \\ \hat{E}^h = \hat{N}^h + (\hat{N}^h)^{\perp_h}, \quad P^{\perp_h}\colon \hat{E}^h \to \hat{N}^h, \quad Q^{\perp_h}\colon \hat{E}^h \to (\hat{N}^h)^{\perp_h}, \\ \qquad\qquad \hat{I}^h = P^{\perp_h} + Q^{\perp_h}, \end{cases}$$

where $\perp_h$ indicates orthogonality w.r.t. the scalar products $\langle\cdot,\cdot\rangle^h$, and $I^h, \hat{I}^h$ represent the identities in E^h and $\hat{E}^h$.

Similarly to (2.14) and (2.15) we may, for the spaces E_s^h and $\hat{E}_s^h$, see (3.3), and their subspaces

$$N^h \subset E_s^h, \quad M_s^h := M^h \cap E_s^h, \quad \hat{N}^h \subset \hat{E}_s^h, \quad (\hat{N}^h)_s^{\perp_h} := (\hat{N}^h)^{\perp_h} \cap \hat{E}_s^h,$$

introduce "smooth" projectors

(3.13)
$$\begin{cases} P_s^h, Q_s^h, P_s^{\perp_h}, Q_s^{\perp_h} \text{ by restriction of } P^h, Q^h, P^{\perp_h}, Q^{\perp_h} \text{ to } E_s^h \text{ and } \hat{E}_s^h, \\ \text{with } P_s^h x^h = P^h x^h \text{ for } x^h \in E_s^h, \text{ a.s.o.} \end{cases}$$

Then we may even split $\hat{E}_s^h$ in a new way, see Lemma 3, as

$$\begin{cases} \hat{E}_s^h = \hat{N}^h + \hat{M}_s^h \text{ with } \hat{M}_s^h := (G^h)_0' M_s^h \neq (\hat{N}^h)_s^{\perp_h}, \\ \hat{P}_s^h\colon \hat{E}_s^h \to \hat{N}^h, \quad \hat{Q}_s^h\colon \hat{E}_s^h \to \hat{M}_s^h, \quad \hat{I}^h = \hat{P}_s^h + \hat{Q}_s^h. \end{cases} \tag{3.14}$$

The usual stability has to be modified as

$$\begin{cases} (G_0^h)' \mid M^h\colon M^h \to \hat{M}^h = (G_0^h)'(M^h) \\ \text{is boundedly invertible with a bound independent of } h. \end{cases} \tag{3.15}$$

Stability results of this type follow for many discretization methods from results of, e.g., Böhmer **[6]**, Grigorieff **[16]**, Hackbusch **[17]**, Stummel **[29]**, **[30]**, Vainikko **[32]**.

In a similar way as for (3.6) differencing may be used to replace M^h and $\hat{M}^h$ in (3.18) by M_s^h and $\hat{M}_s^h$.

The following lemma is basic for our further discussion:

LEMMA 3. *Let the stability condition* (3.15) *be satisfied for* M_s^h *and* $\hat{M}_s^h$ *and* h *be small enough. Then the decompositions* (3.12) *for* E^h *and* $\hat{E}^h$ *are correct (the complements are closed subsets), and we have for a sequence* $\hat{y}^h \in \hat{E}_s^h$:

(3.16)
$$\hat{Q}^h \hat{y}^h = Q^{\perp_h} \hat{y}^h + O(h^p), \qquad \hat{P}^h \hat{y}^h = P^{\perp_h} \hat{y}^h + O(h^p) \quad \textit{for } \|\hat{y}^h\|_{\hat{E}_s^h} \le c.$$

Furthermore, the projectors $L \in \{P, Q\}$, $\tilde{L} \in \{\hat{P}, \hat{Q}\}$ *and the corresponding discrete* $R^h \in \{P^h, Q^h\}$, $\hat{R}^h \in \{P^{\perp_h}, Q^{\perp_h}, \hat{P}^h, \hat{Q}^h\}$ *satisfy the relations*

$$(3.17) \qquad \begin{cases} L^h\Delta^h u = \Delta^h Lu + O(h^p) \text{ for } u \in E \text{ and } \|u\|_{E_s} \le c, \\ \tilde{L}^h\hat{\Delta}^h \hat{u} = \hat{\Delta}^h \tilde{L}\hat{u} + O(h^p) \text{ for } \hat{u} \in \hat{E} \text{ and } \|\hat{u}\|_{\hat{E}_s} \le c. \end{cases}$$

The discretization of the equations (2.6), (2.7), *and* (2.12) *yields,*

$$(3.18) \qquad \begin{cases} \text{(a) } (G^{h'})_0 w^h = -R_0^h(v^h + w^h) \quad \text{with } v^h = \sum_{i=1}^{m+q} c_i^h \varphi_i^h, \\ \text{(b) } \langle w^h, \varphi_i^h \rangle^h = 0, \quad i = 1, \dots, m+q, \\ \text{(c) } \langle R_0^h(v^h + w^h), \hat{\psi}_j^h \rangle^h = O(h^p), \quad j = 1, \dots, m, \end{cases}$$

and

$$(3.19) \quad \begin{cases} \text{(a) } (G^{h'})_0 w^h + \sum_{j=1}^m d_j^h \hat{\psi}_j^h = -R_0^h(v^h + w^h) \quad \text{with } v^h = \sum_{i=1}^{m+q} c_i^h \varphi_i^h, \\ \text{(b) } \langle w^h, \varphi_i^h \rangle^h = 0, \quad i = 1, \dots, m+q, \\ \text{(c) } \langle R_0^h(v^h + w^h), \hat{\psi}_j^h \rangle^h = O(h^p) = d_j^h, \quad j = 1, \dots, m. \end{cases}$$

The properties (3.3)–(3.5) and (3.8)–(3.11) show that (3.18) and (3.19) are satisfied up to an error term $O(h^p)$ if we define $c_i^h := c_i$, $v^h := \Delta^h v$, $w^h := \Delta^h w(v)$, $d_j^h := d_j$ for the solutions $c_i, v, w(v), d_j$ of (2.6), (2.7), and (2.12). The $O(h^p)$-error terms in (3.18c) could have been assigned as well to all three lines. For the iteration process described in the following sections, yielding $O(h^p)$ approximations to the solutions, the version given in (3.18) and (3.19) is more appropriate.

To solve these systems we decompose them into the fixed-point equation consisting of the first two lines (3.18a,b) or (3.19a,b) and the remaining bifurcation equations in the last line, (3.18c) or (3.19c). We have to impose appropriate conditions to guarantee that the numerical solutions approximate the bifurcating manifolds.

4. Numerical solution of the fixed-point equation. Using (3.12) and (3.18) and choosing an arbitrary $v^h \in N^h$ we obtain w^h from (3.18a,b) as

$$(4.1) \qquad (G_0^{h'} \mid M^h) w^h = -\hat{Q}^h R_0^h(v^h + w^h).$$

Applying the bounded operator $\hat{Q}^h$ first, we obtain from (3.6) the "modified stability condition"

$$(4.2) \qquad \begin{cases} T^h := (G_0^{h'} \mid M^h)^{-1} \cdot \hat{Q}^h \text{ exists and} \\ \|T^h\| \text{ is bounded independent of } h. \end{cases}$$

The existence of T^h, and results for smooth Sobolev-norms by differencing, may be obtained similarly as above.

If the discretization yields an $E^h = \mathbf{R}^{n+q}$, then $(G_0^h)': \mathbf{R}^{n+q} \to \mathbf{R}^n$, and $\langle w^h, \varphi_i^h \rangle^h = 0$, $i = 1, \ldots, m+q$, impose $m+q$ additional equations, then e.g. Householder matrices may be used to transform

(4.3)

$$\begin{matrix} n \\ m+q \end{matrix} \begin{pmatrix} (G^h)_0' \\ \hline \langle \cdot, \varphi_i^h \rangle, i = 1, \ldots, m+q \end{pmatrix} w^h = \begin{pmatrix} -R_0^h(v^h + w^h) \\ \hline 0 \\ \vdots \\ 0 \end{pmatrix} \begin{matrix} n \\ \\ m+q \\ \end{matrix}$$

$n+q$ columns

into the following form

$$(4.4) \qquad \begin{matrix} n-m \\ m \\ m+q \end{matrix} \begin{pmatrix} \ddots & & \\ 0 & & \\ \cdots\cdots & \cdot & \cdots\cdots \\ \simeq O(h^p) & \cdot & \overline{} \\ \hline 0 & & \ddots \end{pmatrix} w^h = \begin{pmatrix} * \\ \vdots \\ * \\ \cdots\cdots \\ \underline{\text{defect}} \\ * \\ \vdots \\ * \end{pmatrix} \begin{matrix} n-m \\ m \\ m+q \end{matrix}$$

Neglecting the m lines between $\underline{\cdots\cdots}$, marked by $O(h^p)$ and defect on the left and right-hand side, respectively, we obtain the solution w^h of (4.1), which we formally compute with the "modified generalized inverse" T^h in (4.2) as

$$w^h = -T^h R_0^h(v^h + w^h) \tag{4.5}$$

compared to the exact solution w of (2.7) in the form (2.9). The existence and uniqueness of a solution w^h for (4.5) for a given v^h may be proven along the lines of the proof of Theorem 1.

To obtain a convergence result of the form $w^h - \Delta^h w = O(h^p)$ we want to compare the exact sequence in (2.12) for the operator $-TR_0(v + \cdot)$ with an approximate sequence:

$$w^{h,\nu} := B^h w^{h,\nu-1} := -T^h R_0^h(w^{h,\nu-1} + v^h). \tag{4.6}$$

To this end we recall that for perturbed sequences, e.g.,

$$\widetilde{w}^\nu := \widetilde{B}\widetilde{w}^{\nu-1} \text{ with } \widetilde{w}^0 = w^0 \text{ and } \|\widetilde{B}x - Bx\| \le \varepsilon \tag{4.7}$$

we have, with the Lipschitz constant $q \in (0,1)$ for B,

$$\|\widetilde{w}^\nu - w\| \le \frac{1}{1-q}(q\|\widetilde{w}^\nu - \widetilde{w}^{\nu-1}\| + \|\widetilde{B}\widetilde{w}^{\nu-1} - B\widetilde{w}^{\nu-1}\|),$$

where $w = Bw$ is the fixed point we are interested in.

With the extension operators $\Delta_h, \hat{\Delta}_h$ in (3.10) and

$$\widetilde{B} := \Delta_h B^h \Delta^h, \qquad \widetilde{w}^\nu := \Delta_h w^{h,\nu}, \qquad w^{h,0} := \Delta^h w^0$$

we have the conditions in (3.18) satisfied with $\varepsilon = O(h^p)$ and obtain with an appropriate $q \in (0,1)$, see Theorem 1,

$$\|w^{h,\nu} - \Delta^h w\| \leq \frac{1}{1-q}(q\|w^{h,\nu} - w^{h,\nu-1}\| + O(h^p)), \tag{4.8}$$

with the generally unique least squares solution $w^{h,\nu} = -T^h R_0^h(w^{h,\nu-1} + v^h)$ of

$$\begin{aligned} \langle w^{h,\nu}, \varphi_i^h \rangle^h &= 0, \quad i = 1, \dots, m+q, \\ G_0^{h'} w^{h,\nu} &= -R_0^h(w^{h,\nu-1} + v^h). \end{aligned} \tag{4.9}$$

Now, we iterate in (4.9) until for the first $\nu = \mu$

$$\|w^{h,\mu} - w^{h,\mu-1}\| = O(h^p) \text{ and define } \widetilde{w}^h := w^{h,\mu}. \tag{4.10}$$

Then (4.8) implies

$$\|\Delta_h \widetilde{w}^h - w\| = O(h^p) \quad \text{for} \quad \|\widetilde{w}^h - \Delta^h w\| = O(h^p). \tag{4.11}$$

From Theorem 1 we know that $w = Bw = O(\|v\|^2)$ and (2.9) has exactly one solution. Then (4.11) with (3.5), (3.8)–(3.11) imply that (3.18a,b) is satisfied by $\widetilde{w}^h$ up to an error term $O(h^p)$ in the sense of (4.11). Since we cannot expect more than (4.11) from a discretization method of order p, we will accept $\widetilde{w}^h$ in (4.10) as discrete approximation for w solving (2.9).

THEOREM 4. *Let for the problem* (1.1) *and its discretization* (3.8) *the conditions* (2.1), (2.11), (3.4), (3.5), *and* (3.8)–(3.11) *be satisfied. Then there exists a unique solution w^h to* (4.1). *The $\widetilde{w}^h := w^{h,\mu}$ in* (4.10) *solves* (4.1) *up to a $O(h^p)$ error and is an approximation for the solution w of* (2.9) *of order p; see* (4.11).

To obtain significant data from (4.10) we have to choose $\|v\|$ or $\|v^h\|$ large enough in order to guarantee $\|R_0(v+w)\|$ or $\|R_0^h(v^h + w^h)\|$ and $\widetilde{w}^h$ are distinctly larger than $O(h^p)$. This requires with $w = O(\|v\|^2)$ that we have to start with $\|v^h\| \gg O(h^p)$.

5. Numerical solution of the bifurcation equations. In Sections 2 to 4 we are able to prove the unique existence of solutions $w = w(v)$ for (2.9) and $w^h = w^h(v^h)$ for (4.1), w^h a discrete $O(h^p)$ approximation for w, satisfying (4.1) or (3.18a, b) in the least squares sense. Both results were possible under rather natural and unrestrictive conditions such as (2.1), (3.4), (3.5), (3.8)–(3.11). The situation changes considerably if we proceed to (2.10). Even for the case of "ordinary singular" points, hence $\dim N(G_0') = q+1$, the analytic results, see e.g. **[3]**, **[4]**, **[20]**, **[25]**, **[27]**, require rather complicated and incisive conditions to guarantee the existence of a solution v_0 for (2.6) such that v_0, $w(v_0)$ satisfy (2.6), (2.7), or (2.9), (2.10). With the solution $w = w(v)$ of (2.9) the nonlinear system (2.10) represents m equations for $m+q$ unknowns $c_1, \dots, c_{m+q}$. So for every up to q arbitrarily chosen parameters among the $c_1, \dots, c_{m+q}$, one has to determine (if they exist) the remaining m or more parameters, i.e., the defining manifolds of dimension q or less, bifurcating in $x_0 = (z_0, \lambda_0)$.

There are two different, but essentially equivalent ways, to compute the solutions of (2.10). The discrete $O(h^p)$ approximations $x_0^h, \varphi_i^h, \hat{\psi}_j^h$, $\widetilde{w}^h$ and $(G^h)_0'$, R_0^h, see (3.9) and (4.10), may be used directly or one defines, with the extension operators Δ_h and $\hat{\Delta}_h$ in (3.10), continuous $O(h^p)$ approximations $\tilde{x}, \tilde{\varphi}_i, \tilde{\psi}_j, \widetilde{w}$ for the exact x_0, φ_i, $\hat{\psi}_j$, w respectively. With the continuity of the inner product $\langle \cdot, \cdot \rangle$, $R_0(u) = O(\|u\|^2)$ and the solution $w = w(v)$ of (2.9), we obtain the following perturbed system instead of (2.10):

(5.1)
$$\left\langle R_0 \left(\sum_{i=1}^{m+q} \tilde{c}_i \tilde{\varphi}_i + w \left(\sum_{i=1}^{m+q} \tilde{c}_i \tilde{\varphi}_i \right) \right), \tilde{\psi}_j \right\rangle = S_j(\tilde{c}_1, \dots, \tilde{c}_{m+q}) + O(h^p) = 0,$$
$$j = 1, \dots, m,$$

with

$$S_j(c_1, \dots, c_{m+q}) := \left\langle R_0 \left(\sum_{i=1}^{m+q} c_i \varphi_i + w \left(\sum_{i=1}^{m+q} c_i \varphi_i \right) \right), \hat{\psi}_j \right\rangle. \tag{5.2}$$

If the inner products $\langle \cdot, \cdot \rangle$ in (5.1) can be computed exactly, this nonlinear system may be solved to yield $\tilde{c}_i \simeq c_i + O(h^p)$ whenever an $O(h^p)$-perturbation of the original problem (2.10) correspondingly perturbs the solution only by $O(h^p)$ terms. This "stability" property has to be verified for every specific problem separately. More precisely, it may be stated as follows. Let for $C^\mu := (c_{1,\mu}, \dots, c_{m+q,\mu})$, $\Delta^\mu := (\delta_{1,\mu}, \dots, \delta_{m,\mu})$, $\mu = 1, 2$, the systems

(5.3) $(S_j(C^\mu))_{j=1}^m = \Delta^\mu$, $\mu = 1, 2$, be solvable simultaneously and let the solutions C^1, C^2 satisfy $\|C^1 - C^2\| \le L \|\Delta^1 - \Delta^2\|$ for $\Delta^\mu = O(h^p)$, $\mu = 1, 2$.

Usually, the inner products in (5.1) cannot be computed exactly and quadrature formulas of order p, see (3.11), have to be used and again $O(h^p)$-perturbations are introduced. In this case it would not make sense to compute the extended $\tilde{x}_0, \tilde{\varphi}_j, \tilde{\psi}_j, \tilde{w}$; instead one would prefer to work directly with the discrete R_0^h, introduced in (3.8). As a consequence of (3.4), (3.5), (3.8)–(3.11) we would obtain

(5.4)

$$\left\langle R_0^h \left(\sum_{i=1}^{m+q} c_i^h \varphi_i^h + w^{h,\nu-1} \left(\sum_{i=1}^{m+q} c_i^h \varphi_i^h \right) \right), \hat{\psi}_j^h \right\rangle^h = 0, \quad j = 1, \ldots, m.$$

Again we need the above "stability" property (5.3) to obtain

THEOREM 5. *Let the system* (2.10) *satisfy the "stability" condition* (5.3). *Then for every solution* $C = (c_1, \ldots, c_{m+q})$ *of* (3.18c) *there are corresponding solutions* $\widetilde{C} = (\tilde{c}_1, \ldots, \tilde{c}_{m+q})$ *and* $C^h = (c_1^h, \ldots, c_{m+q}^h)$ *of* (5.1) *and* (5.4), *and vice versa, such that*

$$\|C - C^h\| = O(h^p), \qquad \|C - \widetilde{C}\| = O(h^p), \tag{5.5}$$

that is, the bifurcating manifolds may be computed via (5.1) *or* (5.4).

To obtain appropriate results we assume for simplicity

$$G \in C_L^2(K_\rho(x_0)), \tag{5.6}$$

where C_L^2 indicates twice Lipschitz continuously differentiable operators and $K_\rho(x_0)$ the usual ball with center x_0. Then we obtain

$$R_0(u) = \frac{1}{2} G_0'' u^2 + O(\|u\|^3) \tag{5.7}$$

and as linearization (for $\|s\| \ll \|u\| \ll 1$)

$$R_0(u + s) - R_0(u) = G_0'' us + O(\|u\|^2 \|s\| + \|s\|^2). \tag{5.8}$$

To linearize $R_0(v + w(v))$ with respect to v we need a linearization for $w(v)$.

With (2.9), (5.8) and $\|w\| = O(\|v\|^2)$ we obtain

$$\Delta w := \Delta w(s) := w(v + s) - w(v) = -TG_0'' v(s + \Delta w) + \text{smaller terms},$$

ending up with an equation for the correction Δw of the form

$$\Delta w = -TG_0'' v(s + \Delta w), \qquad \Delta w = \Delta w(s), \tag{5.9}$$

or, re-interpreted in iteration form, as

(5.10)

$$G_0' \Delta w^\nu = -Q \hat{G}_0'' v(s + \Delta w^{\nu-1}), \qquad \Delta w^0 := 0, \qquad \langle \Delta w^\nu, \varphi_i \rangle = 0,$$
$$i = 1, \ldots, m + q.$$

This system may be solved cheaply, since we may use the information obtained during the computations described in Section 4.

Given a starting value v^0, see Section 6, we want to compute the correction s by linearizing (5.4). A combination of (5.8) and (5.9) and their linearity yields a system

(5.11)
$$\sum_{\mu=1}^{m+q} \alpha_\mu \langle G_0''(v+w(v))(\varphi_\mu + \Delta w(\varphi_\mu)), \hat{\psi}_j\rangle = -\langle R_0(v+w(v)), \hat{\psi}_j\rangle, \qquad j=1,\dots,m,$$

of m equations for the $m+q$ unknowns α_μ in

$$s = \sum_{\mu=1}^{m+q} \alpha_\mu \varphi_\mu. \tag{5.12}$$

The coefficients in (5.11) require knowledge of the $\Delta w(\varphi_\mu)$, which are obtained as $m+q$ solutions of the systems (5.10) with $s = \varphi_1, \dots, \varphi_{m+q}$.

We have confined our discussion to a study of the original reduced problem (2.10), rather than to its discretization (5.4). As a consequence of the stability property (5.3) the difference in the results would only be an $O(h^p)$ term.

In all our discussion a careful monitoring of the $O(h^p)$ behaviour is essential. This can only be achieved by careful comparisons of the results for different and appropriate discretization parameters h. If the $O(h^p)$ behaviour is reflected well enough, $x_0^h + v^h + w^h(v^h)$ represent good approximation for usually regular points on the bifurcating manifolds. Then the usual numerical methods may be applied to increase the accuracy, if necessary, or to verify, via stability and consistency properties of the regular problem, the fact that the computed $x_0^h + v^h + w^h(v^h)$ approximate the solution $x_0 + v + w(v)$ of (1.1).

6. Modifications for the starting value for v. One of the difficulties in this approach is the unknown starting value for the nonlinear system (5.1) or (5.4) for v or v^h, respectively. To this end we use the nullspaces N and $\hat{N}$ or N^h and $\hat{N}^h$, to slightly modify the computation of the bifurcating manifolds. We confine the presentation again to the case of the continuous equation and omit the obvious $O(h^p)$ modifications for the discrete equations. With the notation in (2.3) we represent $u = v+w$ in the form

$$\begin{cases} u = \|\tau\| v + \|\tau\|^2 w, & \text{with } v \in N,\ w \in N^\perp \text{ and} \\ \tau = (\tau_1, \dots, \tau_q) \in \mathbf{R}^q, & \|\tau\|^2 = \tau_1^2 + \cdots + \tau_q^2. \end{cases} \tag{6.1}$$

This representation permits a good discussion for $\tau \simeq 0$, and hence for the bifurcating manifolds near the singular point. We obtain for G:

(6.2) $G(x) = G(x_0+\|\tau\|v+\|\tau\|^2 w) = \|\tau\|^2 G_0' w + R_0(\|\tau\|v+\|\tau\|^2 w) = 0.$

By (2.5) the R_0-term is $O(\|\tau\|^2)$ and thus we may divide (6.1) by $\|\tau\|^2$ to obtain the equations for v and w in (6.1):

$$(6.3) \qquad \begin{cases} G_0' w + \dfrac{R_0(\|\tau\|v + \|\tau\|^2 w)}{\|\tau\|^2} = 0, \quad v, w \text{ as in } (6.1), \\ \langle w, \varphi_i \rangle = 0, \quad i = 1, \dots, m+q. \end{cases}$$

As discussed in Sections 2 and 4, (6.3) allows the computation of w and w^h for a given v and v^h.

A good enough starting value for v may now be computed via (6.3), if we study the remainder term $R_0/\|\tau\|^2$ in (6.3) more closely. For bounded v and, hence, w in (6.1) we have by (5.7)

$$(6.4) \qquad \begin{cases} \dfrac{1}{\|\tau\|^2} R_0(\|\tau\|v + \|\tau\|^2 w) \\ = \frac{1}{2} G_0''(v + \|\tau\| w)^2 + O(\|\tau\|) = \frac{1}{2} G'' v^2 + O(\|\tau\|). \end{cases}$$

A combination of (6.3), (6.4) and $\langle G_0' w, \hat{\psi}_j \rangle = 0$, $j = 1, \dots, m$, yields for $\tau = 0$ the relatively simple nonlinear system for a $v = \sum c_i \varphi_i$

$$(6.5) \qquad \langle G_0'' v^2, \hat{\psi}_j \rangle = \left\langle G_0'' \left(\sum_{i=1}^{m+q} c_i \varphi_i \right)^2, \hat{\psi}_j \right\rangle = 0, \quad j = 1, \dots, m.$$

This system may either be solved directly or by linearization. A linearization of (6.5) at a v^0 for a correction $u = \sum c_i \varphi_i$ has the form, see (5.8) and (5.11),

$$(6.6) \qquad 2 \cdot \sum_{i=1}^{m+q} c_i \langle G_0'' v^0 \varphi_i, \hat{\psi}_j \rangle = -\langle G_0''(v^0)^2, \hat{\psi}_j \rangle, \quad j = 1, \dots, m.$$

A comparison with (5.11) shows that the linear operators on the left in (6.6) and (5.11) coincide if $w(v) = 0$ is introduced in (5.11). This change of (5.11) is very natural, since for small v in (2.11) and (5.11), we have the extremely small $w = O(\|v\|^2)$. Therefore the linearized operators in (5.11) and (6.6) are regular simultaneously if only $\|w\|$ (and hence $\|v\|$) is small enough in (5.11).

Other modifications are possible as well. Instead of using $\|\tau\|$, simple components τ_j, $j = 1, \dots, q$, in τ may be used in (6.1)–(6.6), especially if a continuation with respect to the parameter τ_j is interesting. We do not want to present the obvious modifications here.

There are several benefits to this approach. The relatively small quadratic system (6.5) is much easier to solve than the systems (5.4) or (6.3) which still involve v and $w(v)$. With all solutions $v(0)$ of (6.5) known, we may use the combination of (6.4) and (6.3) to compute a $w(v(0))$. Now we choose a sequence of τ^l with $\|\tau^0\| = 0 < \|\tau^1\| < \|\tau^2\| < \cdots$, $\tau^l = \tau^{l-1} + \sigma^l$. The known $v(0)$ and $w(v(0))$ then may be used in

$$\left\langle \frac{R_0(\|\tau^1\|v + \|\tau^1\|^2 w(v(0)))}{\|\tau^1\|}, \hat{\psi}_j \right\rangle = 0, \quad j = 1, \ldots, m, \tag{6.7}$$

to compute an improved approximation $v^1(\tau^1)$ starting with the known $v(0)$, similarly to (6.5), (6.6). This improved $v^1(\tau^1)$ and $w(v(0))$ may be used to compute a better $w^1(\tau^1)$ from (6.3). The same procedure may be repeated until the approximations for $v(\tau^1), w(\tau^1)$ are good enough. Then one proceeds from τ^1 to τ^2, etc. By choosing different values for the τ^l, e.g., only one nontrivial component, a systematic discussion of the solution manifolds bifurcating at $x_0 = (z_0, \lambda_0)$ is possible.

7. A numerical example. For numerical experiments we choose the following example; see Allgower–Chien **[2]**, where $g \in C^2(\mathbf{R})$ and $g(0) = 0$, $g'(0) = 1$,

$$G(z, \lambda) := \Delta z + \lambda g(z) = 0 \text{ in } \Omega = [0, 1]^2 \subset \mathbf{R}^2; \tag{7.1}$$

this $G(z, \lambda)$ is defined on $E = H_0^2(\Omega) \times \mathbf{R} \to L^2(\Omega)$, hence $(z, \lambda) \in D(G)$ implies $z = 0$ on $\partial\Omega$. Furthermore, G is a densely defined operator with closed range in $L^2(\Omega)$. From (7.1) we find for the special case $(z, \lambda) = x_0 = (z_0, \lambda_0)$, using $g(0) = 0$, $g'(0) = 1$,

$$\begin{cases} G_0'(s, \tau) = \Delta s + \lambda s \\ G_0''(s, \tau)^2 = \lambda g''(0) s^2 + 2\tau s \\ R_0(s, \tau) = \frac{\lambda}{2} g''(\delta_2 s) s^2 + \tau g'(\delta_1 s) s, \ 0 \le \delta_1, \delta_2 \le 1. \end{cases} \tag{7.2}$$

This implies

$$\begin{cases} N(G_0') = [(\sigma_i, 0), i = 1, \ldots, m, (0, 1)] \\ \text{with } N(G_{u0}') = [\sigma_i, i = 1, \ldots, m] \text{ and } (G_{u0}', G_{\lambda 0}') := G_0'. \end{cases} \tag{7.3}$$

Now it is well known, Kuttler–Sigillito **[21]**, that for $\lambda = \lambda(\mu, \kappa) := (\mu^2 + \kappa^2)\pi^2$ and $\mu \ne \kappa$, $\mu^2 + \kappa^2 \ne \nu^2 + \rho^2$ for all $\nu, \rho \in \mathbf{N}$, there are two independent solutions of $G'_{u0}\sigma = 0$,

$$(7.4)\qquad \begin{cases} \sigma_1(x,y) := \sin(\mu\pi x)\sin(\kappa\pi y) = \hat{\psi}_1 \text{ and} \\ \sigma_2(x,y) := \sin(\kappa\pi x)\sin(\mu\pi y) = \hat{\psi}_2. \\ N(G'_0) = [(\sigma_1, 0), (\sigma_2, 0), (0, 1)] =: [\varphi_1, \varphi_2, \varphi_3]. \end{cases}$$

For our computations we choose in (7.1)

$$(7.5)\qquad \begin{cases} \mu = 1,\ \kappa = 2,\ \lambda_0 := 5\pi^2,\ z = z_0 = 0, \\ g(z) := z + fz^2 \text{ with } f(x,y) := \frac{1}{10}(\sin(2\pi x) + \frac{1}{10}x)y^2, \\ G(0) = 0,\ G'(0) = 1,\ G''(0) = 2f. \end{cases}$$

Corresponding to (6.1), (6.5) we choose

$$v = \tau(x_1\varphi_1 + x_2\varphi_2 + \varphi_3)$$

to obtain

$$(7.6)\qquad \begin{cases} G''_0 v^2/2 = (\lambda_0 + \tau)(\tau x_1\sigma_1 + \tau x_2\sigma_2)^2 f + \tau(\tau x_1\sigma_1 + \tau x_2\sigma_2) \\ \qquad\qquad = \tau^2[(\lambda_0 + \tau)(x_1\sigma_1 + x_2\sigma_2)^2 f + (x_1\sigma_1 + x_2\sigma_2)]. \end{cases}$$

Dividing by τ^2 and with $\sigma_i = \hat{\psi}_i$ we use (6.5) in the form

$$\frac{1}{2\tau^2}\langle G''_0 v^2, \sigma_i\rangle = x_i + (\lambda_0 + \tau)\langle(x_1\sigma_1 + x_2\sigma_2)^2 f, \sigma_i\rangle = 0, \quad i = 1,2,$$

with the abbreviations

$$a_{ijk} := \langle\sigma_i, \sigma_j\cdot\sigma_k\cdot f\rangle;$$

we obtain for the special value $\tau = 0$ the system

$$g(x_1, x_2) := (x_i + \lambda_0(a_{i11}x^2 + 2a_{i12}x_1x_2 + a_{i22}x^2))_{i=1}^2 = 0.$$

This system admits the trivial solution $x_1 = x_2 = 0$, corresponding to the trivial solution of the original problem and two nontrivial solutions. Since

$$Dg(x_1, x_2) = \begin{pmatrix} 1 + 2\lambda_0(a_{111}x_1 + a_{112}x_2) & 2\lambda_0(a_{112}x_1 + a_{122}x_2) \\ 2\lambda_0(a_{211}x_1 + a_{212}x_2) & 1 + 2\lambda_0(a_{212}x_1 + a_{222}x_2) \end{pmatrix}$$

is regular for small enough x_1, x_2, the stability condition (5.2) is satisfied and the behaviour of the solutions of (6.2) will represent the behaviour of the solution manifolds, bifurcating in $(0, \lambda_0)$. We find the following solutions (for $\tau = 0$)

$$(x_{10}, x_{20}) = (0, 0),$$
$$(x_{11}, x_{21}) = (.3012210, .4768256),$$
$$(x_{12}, x_{22}) = (-.2670384, -.5385926).$$

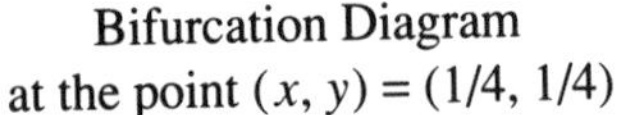

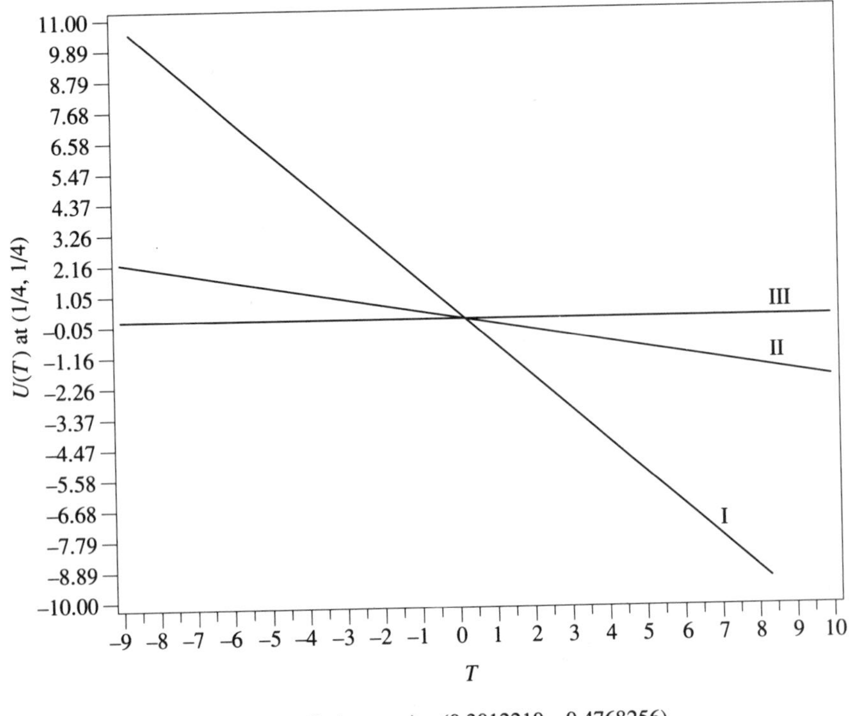

I. $(x_{10}, x_{20}) = (0.3012210, -0.4768256)$
II. $(x_{10}, x_{20}) = (-0.2670384, -0.5385926)$
III. $(x_{10}, x_{20}) = (0, 0)$

FIGURE 1

The behaviour of the corresponding solution branches for a variable $\tau = T$, represented by the solution value at the point $(\frac{1}{4}, \frac{1}{4})$, is plotted in Figure 1; Figure 2 shows a solution for $T = 0.1$ on branch I. For simplicity we have formulated the computation in this section for the original problem. A transition to an appropriate discretization essentially reproduces the results due to (5.3).

ACKNOWLEDGMENTS. We want to thank Prof. Allgower for stimulating discussions and the Friedrich Naumann Foundation, W. Germany, for the support for the second author.

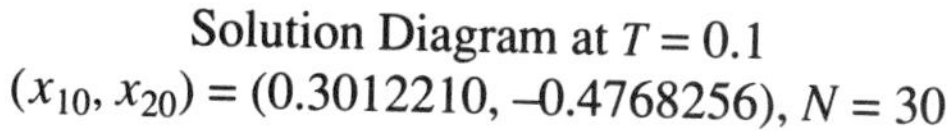

Solution Diagram at $T = 0.1$
$(x_{10}, x_{20}) = (0.3012210, -0.4768256)$, $N = 30$

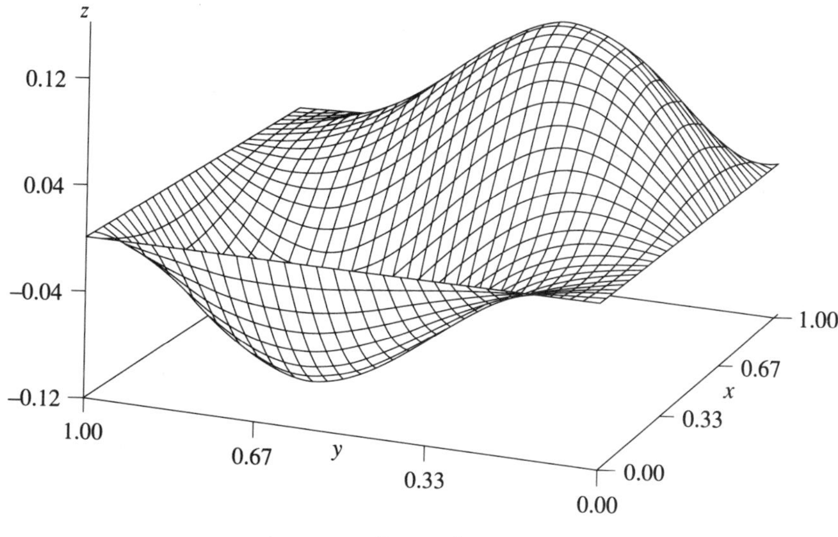

FIGURE 2

LITERATURE

1. E. Allgower and K. Böhmer, *Resolving highly singular equations and bifurcation*, Rocky Mountain Journal of Mathematics **18** (1988), 225–268.

2. E. Allgower and C.-S. Chien, *Continuation and local perturbation for multiple bifurcations*, SIAM J. Sci. Statist. Comput **7** (1986), 1265–1281.

3. M. S. Berger, *A Sturm-Liouville theorem for nonlinear elliptic partial differential equations*, Ann. Scuola di Pisa **20** (1966), 543–582.

4. ———, *On nonlinear perturbations of the eigenvalues of a compact self-adjoint operator*, Bull. Amer. Math. Soc. **73** (1967), 704–708.

5. K. Böhmer, *On the stability of consistently differentiable discretization methods*, Report Fachbereich für Mathematik der Universität Marburg (1989).

6. ———, *On a numerical Lyapunov-Schmidt Method for operator equations*, Report Fachbereich für Math. der Universität Marburg (1989).

7. F. Brezzi, J. Rappaz, and P.-A. Raviart, *Finite-dimensional approximation of nonlinear problems, Part I: Branches of nonsingular solutions*, Numer. Math. **36** (1980), 1–25.

8. C. K. Chui, *Multivariate splines*, CBMS-NSF Regional Conference Series in Applied Mathematics, SIAM, 1988.

9. C. K. Chui, L. L. Schumaker, and J. D. Ward, *Approximation Theory*, IV, Proceedings International Symposium at Texas A and M University, Academic Press (1983).

10. W. Dahmen, *Konstruktion mehrdimensionaler B-Splines und ihre Anwendung auf Approximationsprobleme*, ISNM **52** (1980), 84–100.

11. W. Dahmen and C. A. Micchelli, *Recent progress in multivariate splines*, in **[9]**, 21–121.

12. H. Esser, *Stabilitätsungleichungen für Diskretisierungen von Randwertaufgaben gewöhnlicher Differentialgleichungen*, Numer. Math **28** (1977), 69–100.

13. K. Georg, *A numerically stable update for simplicial algorithms*, In: Numerical solution of nonlinear equations, E. Allgower, K. Glashoff, and H.-O. Peitgen, editors, Lecture Notes in Math. **878**, Springer-Verlag, Berlin, Heidelberg, New York, 1981, pp. 117–127.

14. ———, *On tracing an implicitly defined curve by quasi-Newton steps and calculating bifurcation by local perturbation*, SIAM J. Sci. Statist. Comput. **2** (1981), 35–50.

15. M. Golubitsky and D. G. Schaeffer, *Singularities and Group Theory in Bifurcation Theory*, Vol. I, Springer-Verlag, New York, Berlin, Heidelberg, 1985.

16. R. D. Grigorieff, *Zur Theorie linearer approximationsregulärer Operatoren*, I und II, Math. Nachr. **55** (1973), 233–249 and 251–263.

17. W. Hackbusch, *Multigrid Methods and Applications*, Springer-Verlag, Berlin, New York, Heidelberg, 1985.

18. J. P. Keener and H. B. Keller, *Perturbed bifurcation theory*, Arch . Rat. Mech. Anal. **50** (1974), 159–175.

19. H. B. Keller, *The bordering algorithm and path following near singular points of higher nullity*, SIAM J. Sci. Statist. Comput **4** (1983), 573–584.

20. G. Knightly and D. Sather, *On nonuniqueness of solutions of the von Karman equations*, Arch. Rat. Mech. Anal. **36** (1970), 65–78.

21. J. Kuttler and V. G. Sigillito, *Eigenvalues of the Laplacian in two dimensions*, SIAM Review **26** (1984), 163–183.

22. C. A. Micchelli, *Smooth multivariate piecewise polynomials: a method for computing multivariate B-splines*, Pubblicazioni Serie III, 187, Istituto per le Applicazioni del Calcolo "Mauro Picone" (IAC), Rome, 1979.

23. H.-O. Peitgen, *Topologische Perturbationen beim globalen numerischen Studium nichtlinearer Eigenwert- und Verzweigungsprobleme*, In: Jahresbericht des Deutschen Mathematischen Vereins **84** (1982), 107–162.

24. H. J. Reinhard, *Analysis of approximation methods for differential and integral equations*, Springer-Verlag, Berlin, Heidelberg, Tokyo, New York, 1985.

25. D. Sather, *Branching of solutions of an equation in Hilbert space*, Arch. Rat. Mech. Anal. **36** (1970), 47–64.

26. H. M. Schultz, *Spline Analysis*, Prentice Hall, Englewood Cliffs, 1973.

27. I. Stakgold, *Branching of solutions of nonlinear equations*, SIAM Review **13** (1971), 289–332.

28. H. Stetter, *Analysis of discretization methods for ordinary differential equations*, Springer, Berlin-Heidelberg-New York, 1973.

29. F. Stummel, *Diskrete Konvergenz linearer Operatoren, I*, Math. Ann. **190** (1970), 45–92; *II*, Math. Z. **120** (1971), 231–264; *III*, Proc. Oberfach 1971 ISNM **20** (1972), 196–216.

30. ———, *Stability and discrete convergence of differentiable mappings*, Rev. Roum. Math. Pures e. Appl. **21** (1976), 63–96.

31. B. K. Swartz and R. S. Varga, *Error bounds for spline and L-spline interpolation*, J. Approx. Theory **6** (1972), 6–49.

32. G. Vainikko, *Funktionsanalysis der Diskretisierungsmethoden*, Teubner-Texte zur Mathematik, Teubner, Leipzig, 1976.

33. K. Yosida, *Functional Analysis*, Fourth Edition, Springer-Verlag, Berlin, Heidelberg, New York, 1974.

Fachbereich Mathematik der Universität Marburg, Lahnberge, 3550 Marburg/Lahn, Federal Republic of Germany

Department of Mathematics, Xi'an Jiaotong University, Xi'an, Shaanxi Province, People's Republic of China

Lectures in Applied Mathematics
Volume **26** (1990)

PL Approximation to Manifolds and Its Application to Implicit ODEs

A. CASTELO, S. DE FREITAS, AND G. TAVARES

0. Introduction. Computational methods to study nonlinear equations began with Scarf [**Sc**] for fixed points and continued with Eaves' homotopy methods [**E1**]; for a survey on these methods up to 1980 see [**AGe**].

Recently some algorithms have been devised in higher codimensions. Allgower–Schmidt [**AS**] and Allgower–Gnutzman [**AGn**] presented algorithms based on triangulations and a piecewise-linear version of Sard's theorem. Rheinboldt [**R**] introduced a moving-frame algorithm that blends simplicial and continuation methods to solve locally nonlinear equations.

In [**CFT**] a simplicial algorithm based on triangulation and transversality was developed. We present here an overview of the algorithm and illustrate it with some computer graphics. We also present an algorithm to solve implicit ordinary differential equations based on simplicial methods.

The last-named author would like to thank B. Plohr from the Computer Sciences Department at the University of Wisconsin-Madison for the lively discussions on the subject and its potential applications as well as for the hospitality extended to him while visiting that department where part of this work was done.

1. The topology and geometry framework. The main problem is then to represent computationally all the solutions of a system of nonlinear equations:

$$F_1(x_1, \ldots, x_m) = 0$$
$$\vdots$$
$$F_n(x_1, \ldots, x_m) = 0$$

in a given domain. This problem cannot be solved in all this generality; therefore, we make the following *topological assumption*:

Zero is a regular value of the C^1 map $\mathbf{F}\colon \mathbf{R}^m \to \mathbf{R}^n$, $\mathbf{F} = (F_1, \ldots, F_n)$. (Zero a regular value for $\mathbf{F}$ means that at every point $x \in \mathbf{F}^{-1}(0)$, the Jacobian matrix of $\mathbf{F}$ is full-rank.)

This assumption implies, via the implicit function theorem, that $\mathscr{S} = \mathbf{F}^{-1}(0)$ is a smooth $(m-n)$-dimensional manifold in $\mathbf{R}^m$.

As for the geometry setting we adopt the following strategy:

We triangularize the unit cube in $\mathbf{R}^m$ by m-dimensional simplices in such a way that the triangulation of each of its faces is invariant under the unit translation along the coordinate axis. By scaling and translating the unit cube we obtain a triangulation of $\mathbf{R}^m$ with given $\delta_1, \ldots, \delta_m$ cube edge sizes.

For the unit cube in $\mathbf{R}^3$ we have the following picture:

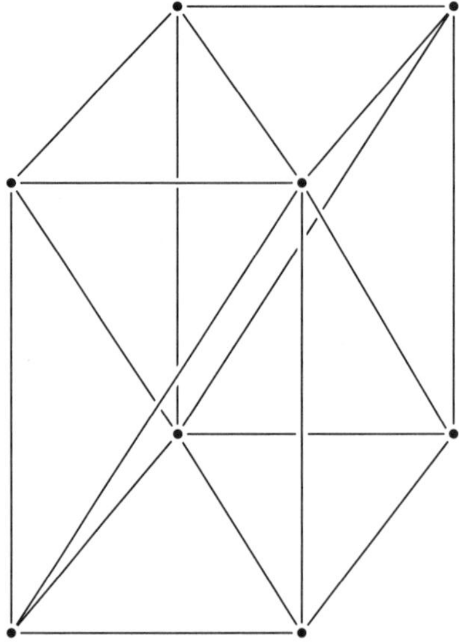

FIGURE 1

This triangulation of $\mathbf{R}^m$ is invariant under δ_i-translations, $i = 1, \ldots, m$, along the respective coordinate axes. It is called the CFK-triangulation of $\mathbf{R}^m$ (CFK for Coxeter, Freudenthal, Kuhn, v. **[C]**, **[Fr]**, **[K]**) and it will be denoted by T. In Section 2 we will describe this triangulation combinatorially.

2. Combinatorics. The vertices of the unit cube are indexed by associating to each vertex the number for which it is a binary representation. These numbers range from 0 to $2^m - 1$. For the unit cube in $\mathbf{R}^3$ we have the following table:

$$\begin{array}{lcl}
(0,\ 0,\ 0) & \to & 0.2^0 + 0.2^1 + 0.2^2 = 0 \\
(1,\ 0,\ 0) & \to & 1.2^0 + 0.2^1 + 0.2^2 = 1 \\
(0,\ 1,\ 0) & \to & 0.2^0 + 1.2^1 + 0.2^2 = 2 \\
(1,\ 1,\ 0) & \to & 1.2^0 + 1.2^1 + 0.2^2 = 3 \\
(0,\ 0,\ 1) & \to & 0.2^0 + 0.2^1 + 1.2^2 = 4 \\
(1,\ 0,\ 1) & \to & 1.2^0 + 0.2^1 + 1.2^2 = 5 \\
(0,\ 1,\ 1) & \to & 0.2^0 + 1.2^1 + 1.2^2 = 6 \\
(1,\ 1,\ 1) & \to & 1.2^0 + 1.2^1 + 1.2^2 = 7
\end{array}$$

To generate all m-dimensional simplices in the unit cube we apply the following *pivoting rule* to the vertex indices of each simplex except 0 and $2^m - 1$:

Starting with the simplex $\langle v_0, v_1, \ldots, v_i, \ldots, v_m \rangle$ with vertex indices $2^i - 1$, $i = 0, \ldots, m$, replace the vertex v_i with $\bar{v}_i = v_{i+1} + v_{i-1} - v_i$ to build a new simplex $\langle v_0, v_1, \ldots, \bar{v}_i, \ldots, v_m \rangle$.

Applied to the simplices of the unit cube in $\mathbf{R}^3$, this pivoting rule gives the following table:

$$\begin{array}{lcccccc}
v_0: & 0 & 0 & 0 & 0 & 0 & 0 \\
v_1: & \to 1 & 2 & \to 2 & 4 & \to 4 & 1 \\
v_2: & 3 & \to 3 & 6 & \to 6 & 5 & 5 \\
v_3: & 7 & 7 & 7 & 7 & 7 & 7
\end{array}$$

where the arrow indicates the vertex index being pivoted.

For the unit cube in $\mathbf{R}^3$ we have the following vertex indexing picture:

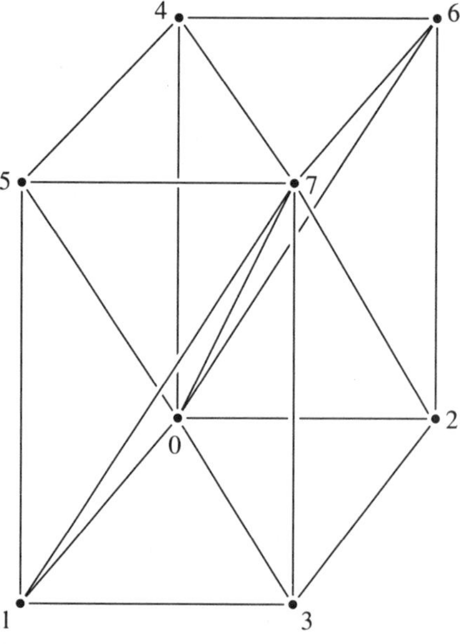

FIGURE 2

The n-dimensional faces of each m-simplex are generated by considering the $\binom{m+1}{n+1}$ combinations of the vertex indices of the simplex.

3. The approach to solve the equation $F(x) = 0$. A point $\mathbf{x}$ in a simplex σ generated by $v_0, \ldots, v_m$ can be written $\mathbf{x} = \sum_{i=0}^{m} \lambda_i v_i$ with $\sum_{i=0}^{m} \lambda_i = 1$, $\lambda_i \geq 0$, $i = 0, \ldots, m$. The λ's are called the barycentric coordinates of $\mathbf{x}$.

The piecewise-linear approximation of $\mathbf{F}$ is the one which in the simplex σ of the triangulation is defined by

$$\mathbf{F}_\sigma \left(\sum_{i=0}^{m} \lambda_i v_i \right) = \sum_{i=0}^{m} \lambda_i \mathbf{F}(v_i).$$

Let us denote by $\mathbf{F}_T$ the piecewise-linear approximation of $\mathbf{F}$, such that on each simplex σ, $\mathbf{F}_T = \mathbf{F}_\sigma$.

We take D to be a finite union of cubes used to build the CFK-triangulation. Instead of solving the original equation

$$\mathbf{F}(\mathbf{x}) = \mathbf{0}$$

in D we solve the linearized equation

$$\mathbf{F}_T(\mathbf{x}) = \mathbf{0}.$$

To look for the solution of $\mathbf{F}_\sigma(\mathbf{x}) = \mathbf{0}$, $\mathbf{x} = \sum_{i=0}^m \lambda_i v_i$, we have then to solve the system of equations

$$\lambda_i \geq 0,$$
$$\sum_{i=0}^{m} \lambda_i = 1,$$
$$\sum_{i=0}^{m} \lambda_i \mathbf{F}(v_i) = 0.$$

In $n \times m$-matrix form:

$$\begin{pmatrix} 1 & \cdots & 1 \\ F(v_0) & \cdots & F(v_m) \end{pmatrix} \begin{pmatrix} \lambda_0 \\ \vdots \\ \lambda_m \end{pmatrix} = \begin{pmatrix} 1 \\ \vdots \\ 0 \end{pmatrix}$$

under the condition

$$\lambda_i \geq 0, \qquad i = 0, \ldots, m.$$

To determine each face of $\mathbf{F}_\sigma^{-1}(0) \cap \sigma$, it is enough to find its vertices, that is, its intersection with the n-dimensional faces of σ. For this we need the following lemma.

LEMMA (THE DOOR-IN/DOOR-OUT PRINCIPLE [E2]). *Generically, any $(n+1)$-dimensional simplex σ has either two n-simplices with vertices of $\mathbf{F}_\sigma^{-1}(0)$ or none.*

We have a vertex of $\mathbf{F}_\sigma^{-1}(0)$ in the interior of an n-dimensional face $\sigma^n = \langle v_{i_1}, \ldots, v_{i_n} \rangle$ of σ if the linear system

$$\begin{pmatrix} 1 & \cdots & 1 \\ F(v_{i_1}) & \cdots & F(v_{i_n}) \end{pmatrix} \begin{pmatrix} \lambda_{i_1} \\ \vdots \\ \lambda_{i_n} \end{pmatrix} = \begin{pmatrix} 1 \\ \vdots \\ 0 \end{pmatrix}$$

has a solution $\lambda_{i_1}, \ldots, \lambda_{i_n}$ with

$$\lambda_i > 0, \qquad i = i_1, \ldots, i_n.$$

Given a vertex v in the cube decomposition we take a small $\varepsilon(v) > 0$ as the radius of the ball $B(v, \varepsilon(v))$ centered at v. Assume that for each v, $\varepsilon(v)$ is small enough to guarantee that by taking a point in each ball $B(v, \varepsilon(v))$ we build a new m-dimensional cube-like decomposition in the following way:

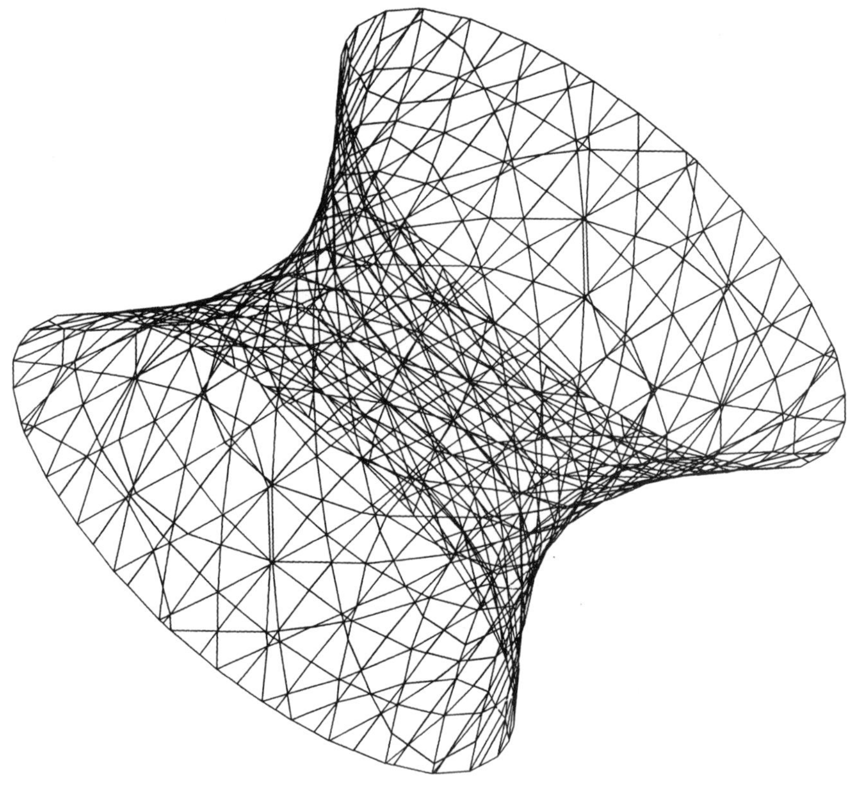

FIGURE 3

If $Q_i = \bigcup_{j=1}^{m!} \sigma_i^j$ is a cube in the CFK triangulation T of $\mathbf{R}^m$ then $Q_i(\varepsilon) = \bigcup_{j=1}^{m!} \sigma_i^j(\varepsilon)$ is cube-like in the new triangulation $T_\varepsilon = \{\sigma_i^j(\varepsilon);\ i \in \mathbf{Z},\ 1 \le j \le m!\}$ of $\mathbf{R}^m$.

The triangulation $T_\varepsilon(v)$ is still called a CFK triangulation of $\mathbf{R}^m$. Let $B = \bigcup_{v \in T^0} B(v, \varepsilon(v))$, the disjoint union of the balls constructed above, T^0 the set of vertices of T, and $\operatorname{diam}(T)$ the diameter of any of the cubes in the decomposition of T.

THEOREM [CFT]. *If* $\operatorname{diam}(T)$ *and* $\varepsilon(v)$, $v \in T^0$, *are sufficiently small, then for a dense set of vertices in* B, $\mathbf{F}_{T_\varepsilon}^{-1}(0)$ *is an* $(m-n)$*-dimensional PL manifold.*

This theorem is very much associated with its computer implementation. Its applicability comes from the strong connection between the data structure of the cube-like and simplicial decomposition of the underlying space together with the PL structure of the manifold $\mathbf{F}_{T_\varepsilon}^{-1}(0)$.

In this work we address two applications of the above theorem: computer graphics and implicit ordinary differential equations. In Section 7 we discuss some potential applications of this theorem.

4. The algorithm in a cube.

Apply the pivoting rule successively to the simplices σ^m starting with the one with vertex indices $0, \ldots, 2^m - 1$;

Find an n-simplex σ^n of σ^m with a vertex of $\mathbf{F}_{\sigma^m}$;

Apply the door-in/door-out principle on each $(n+1)$-simplex containing σ^n.

5. Computer graphics. Next we give several examples which demonstrate how the algorithm performs in low dimensions.

Hyperboloid of one sheet.
$HO := z^2 - x^2 - y^2 + 0.5$;
Domain $D = [-0.7,\ 0.7] \times [-0.7,\ 0.7] \times [-0.7,\ 0.7]$;
Number of cubes in D: $1{,}000 = 10 \times 10 \times 10$ (x, y, and z-directions).

Hyperboloid of two sheets.
$HT := z^2 - x^2 - y^2 - 0.02$;
Domain $D = [-0.8,\ 0.8] \times [-0.8,\ 0.8] \times [-0.7,\ 0.7]$;
Number of cubes in D: $1{,}000 = 10 \times 10 \times 10$ (x, y, and z directions).

Sliced torus.
$ST := (\sqrt{x^2 + y^2} - 2)^2 + z^2 - 1.0$;
Domain $D = [-3.05,\ 3.05] \times [-3.05,\ 3.05] \times [-1.05,\ 0.0]$;
Number of cubes in D: $2{,}304 = 24 \times 24 \times 4$ (x, y, and z directions).

Sliced linked tori.
$SLT := [(\sqrt{y^2 + (x+1)^2} - 2)^2 + z^2 - 0.44]$;
$\qquad [(\sqrt{z^2 + (x-1)^2} - 2)^2 + y^2 - 0.44]$.
Domain $D = [-4.1,\ 4.1] \times [-3.05,\ 0.00] \times [-3.05,\ 3.05]$;
Number of cubes in D: $3{,}888 = 24 \times 18 \times 9$ (x, y, and z directions).

The three-dimensional sphere. Figure 7 represents the orthogonal plane projection of one-fourth of the three-dimensional sphere.
$S = x^2 + y^2 + z^2 + w^2 - 1$;
Domain $D = [-1.0,\ 1.0] \times [-1.0,\ 1.0] \times [0.0,\ 1.0] \times [0.0,\ 1.0]$;
Number of cubes in D: $375 = 5 \times 5 \times 5 \times 3$ (x, y, z, and w directions).

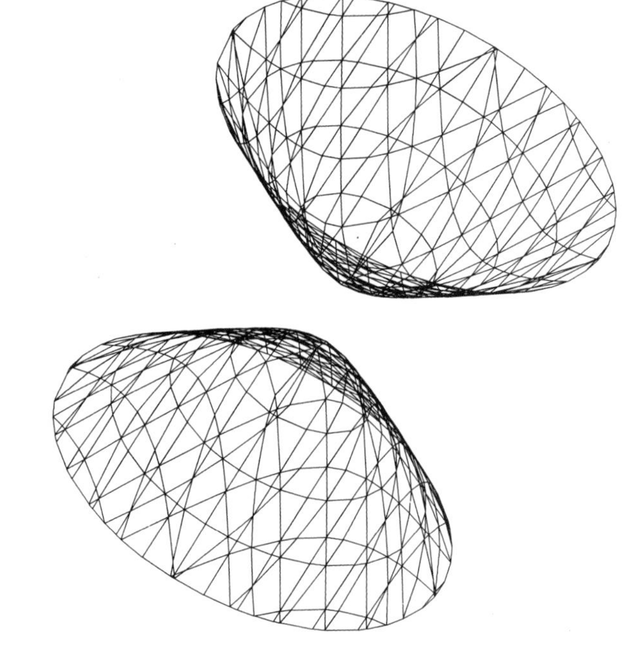

FIGURE 4

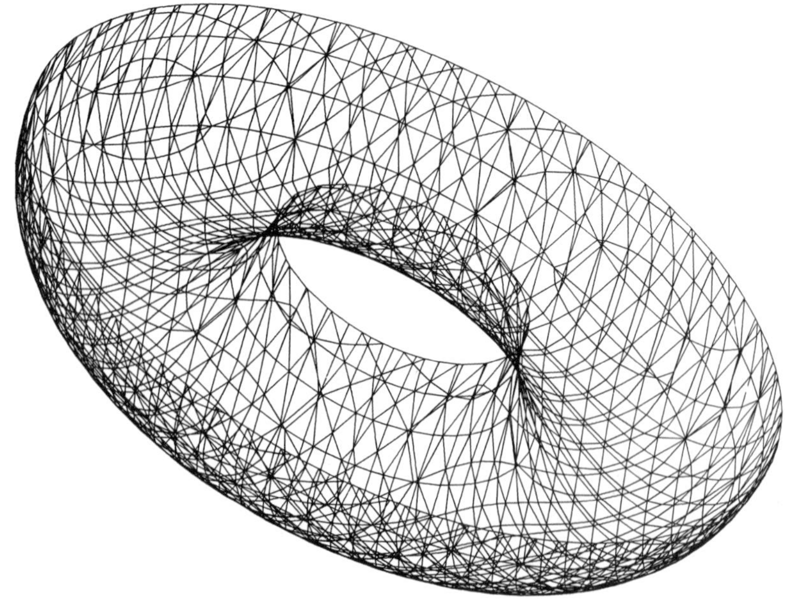

FIGURE 5

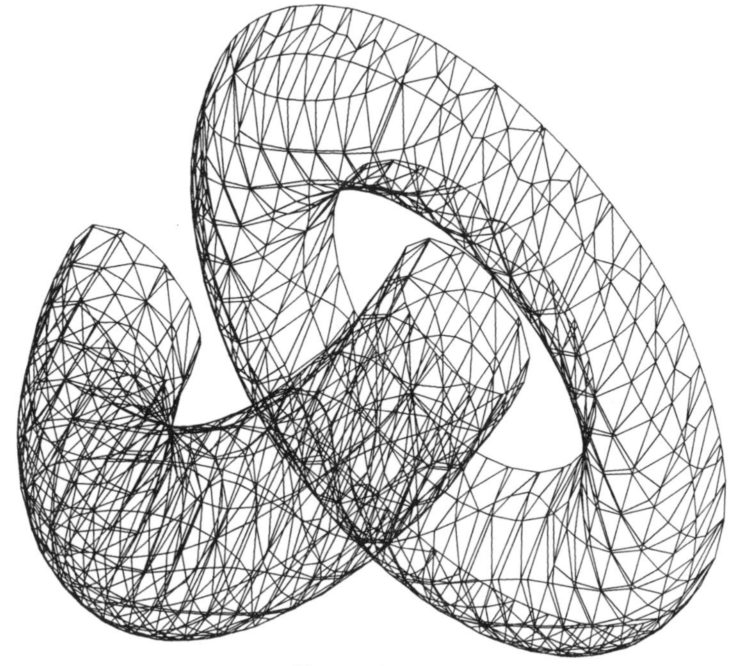

FIGURE 6

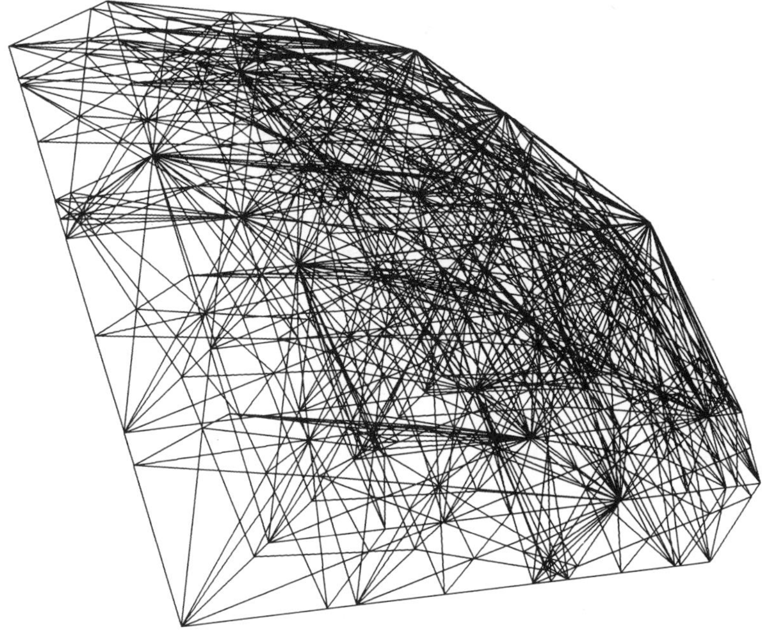

FIGURE 7

6. Solving $\mathbf{F(x,y,y')} = \mathbf{0}$. Given the ordinary implicit differential equation $F(x,y,y') = 0$ we call $\mathcal{S} = \{(x,y,p);\ F(x,y,p) = 0\}$ its *associated surface.*

We can desingularize the equation $F(x,y,y') = 0$ by considering the following line-field, defined at each point $(x,y,z) \in S$,

$$F(x,\ y,\ z) = 0,$$
$$z\,dx - dy = 0.$$

Projecting this line-field back to the plane by $\pi: \mathcal{S} \to \mathbf{R}^2$, $(x,\ y,\ z) \to (x,y)$, gives the solutions of the implicit differential equation. Thus, given a point $(x_0,\ y_0,\ z_0) \in S$, what we have to solve is the following system of equations:

$$F(x_0,\ y_0, z_0) = 0,$$
$$F_{x^\bullet}(x - x_0) + F_{y^\bullet}(y - y_0) + F_{z^\bullet}(z - z_0) = 0,$$
$$z_0(x - x_0) - (y - y_0) = 0.$$

The families of planes $z_0(x - x_0) - (y - y_0) = 0$ with $(x_0,\ y_0,\ z_0) \in \mathcal{S}$ are called *differential planes.* A fold point for the projection $\pi: S \to \mathbf{R}^2$ is the one where $F = \partial F/\partial z = 0$ and $\partial^2 F/\partial z^2 \neq 0$. A cusp point satisfies $F = \partial F/\partial z = \partial^2 F/\partial z^2 = 0$ and $\partial^3 F/\partial z^3 \neq 0$. We will denote the set of fold points by FP and the set of cusp points by CP.

MAIN ASSUMPTION FOR π. The critical set C is the union of FP and CP.

The projection via π of the solutions of the line-field defined on $\mathcal{S}$ gives the solutions of the equation $F(x,\ y,\ y') = 0$. Now let's place the above theoretical framework in an algorithmic context.

The critical set C_T of the piecewise-linear approximation $\mathcal{S}_T$ of $\mathcal{S}$ is the critical set of the projection $\pi_T: \mathcal{S}_T \to \mathbf{R}^2$, $(x,\ y,\ z) \to (x,\ y)$, which is defined in the following way.

First, we orient each connected component of $\mathcal{S}_T$ by the normal of its faces. $(x,\ y,\ z) \in C_T$ if it belongs to an edge $e \subset \mathcal{S}_T$ such that $e = \sigma \cap \omega$, σ and ω faces of $\mathcal{S}_T$ whose normals have opposite z-coordinate signs. The next picture illustrates this definition.

To follow an approximate solution of the line-field in the triangularized surface $\mathcal{S}_T$ we will assume that our initial condition $(x_0,\ y_0,\ z_0) \notin C_T$. We take the plane containing the face $\sigma \ni (x_0,\ y_0,\ z_0)$ and intersect it with the differential plane through the same point to determine two points $(x_1,\ y_1,\ z_1)$, $(x_1',\ y_1',\ z_1')$. Choose either of these points and repeat the intersection procedure.

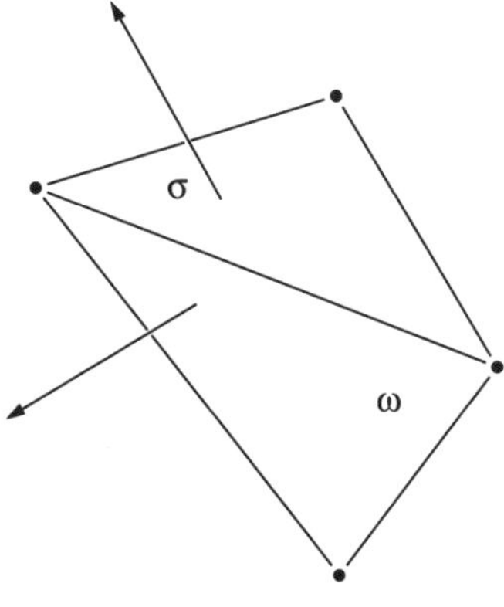

FIGURE 8

An approximation for the multiple solution of the ordinary implicit differential equation through (x_0, y_0) is obtained by projecting on $\mathbf{R}^2$ via π_T the solution through $(x_0, y_0, z_0) \in \mathscr{S}_T$. For more details see **[FT]**.

7. Potential applications for the algorithm. The algorithm described in this paper has many applications which require an understanding of how the chosen triangulation is related to the particular mappings involved in a specific problem. We point out some of the areas to which our algorithm seems to be tailored. We intend to address some of these areas in our future work.

7.1. *Singularities of mappings.*

Shadows. If we take the surfaces in 3-space to be opaque geometrical objects, their shadows on a plane are characterized by the boundary of its projection (the light source being the center of projection). This boundary is a subset of the critical set of this projection **[A]**.

Transparencies. If a surface is completely transparent (no refraction of light rays when illuminated from a light source), then there is the formation of caustics (where light concentrates). These caustics form the critical set of the projection from the light source **[A]**.

Surgery on manifolds. This problem can be characterized by a one-parameter family of manifolds which goes through bifurcation **[M]**. In the context of implicitly defined manifolds it means we have a family of functions $\mathbf{F}_t\colon \mathbf{R}^n \to \mathbf{R}$ such that $\mathbf{F}_t^{-1}(0)$ is a manifold except at $t = 0$, where $\mathbf{F}_0^{-1}(0)$ has only one singular point for $\mathbf{F}_0$ which is a Morse point (the Hessian of $\mathbf{F}_0$ is nonsingular at this point). The problem here is to devise an algorithm to represent computationally $\mathbf{F}_0^{-1}(0)$.

7.2. *Geometric modeling.*

Generation of forms in classical computer graphics. In classical computer graphics parametrized surfaces together with interpolation methods play an essential role in the generation of forms. The simplicial method for implicitly defined manifolds suggests composing functions in three variables via a smoothing procedure combined with adaptive techniques of the domains of the functions involved.

Off-set geometry procedures. Given an implicitly defined surface $\mathscr{S} = \mathbf{F}^{-1}(0)$, for a small displacement along the normal to $\mathscr{S}$ we still have a surface, that is, 0 is a regular value for the map $\mathbf{G}\colon \mathbf{R}^3 \to \mathbf{R}$ given by $\mathbf{G}_\varepsilon = \mathbf{F} + \varepsilon \nabla \mathbf{F}$. The surface $\mathbf{G}_\varepsilon^{-1}(0)$ is called an off-set surface of $\mathscr{S}$ in the geometric modeling framework **[Fa]** and called a wave front set in the singularities of mappings literature **[A]**. If we increase ε beyond a certain value, 0 is no longer a regular value of $\mathbf{G}_\varepsilon$. An interesting problem is to determine how computationally critical points for the map $\mathbf{G}_\varepsilon$ develop.

Physically-based models. Recently computer graphics animation models became more involved with techniques already familiar to the engineering and physical sciences. They are a combination of finite element methods with solid modeling techniques (e.g., elasticity **[PTFB]**); they are called physically-based models. The idea here is to apply finite element techniques to the PL manifold associated with some kind of solid modeling technique to produce computer graphics animation.

References

[A] V. I. Arnold, *Catastrophe Theory*, Springer-Verlag, New York, 1984.

[AGe] E. Allgower and K. Georg, *Simplicial and continuation methods for approximating fixed points and solutions to systems of equations*, SIAM Review **22** (1) (1980), 28–85.

[AGn] E. Allgower and S. Gnutzman, *An algorithm for piecewise-linear approximation of implicitly defined two-dimensional surfaces*, SIAM J. Numer. Anal. **24** (2) (1987), 452–469.

[AS] E. Allgower and P. H. Schmidt, *An algorithm for piecewise-linear approximation of an implicitly defined manifold*, SIAM J. Numer. Anal. **22** (2) (1985), 322–346.

[BFK] W. Böhm, G. Farin, and J. Kahmann, *A survey of curve and surface methods in CAGD*, Comp. Aided Geom. Design **1** (1984), 1–60.

[C] H. S. M. Coxeter, *Discrete groups generated by reflections*, Ann. Math. **35** (1934), 588–621.

[CFT] A. Castelo, S. R. Freitas, and G. Tavares, *Simplicial approximation of implicitly-defined manifolds*, preprint, Department of Mathematics, Catholic University, Rio de Janeiro-Brazil.

[DLTW] D. P. Dobkin, S. V. F. Levy, W. P. Thurston, and A. B. Wilks, preprint, *Robust contour tracing*, Department of Mathematics, Princeton University, Princeton, New Jersey, 1987.

[E1] B. C. Eaves, *A course in triangulation for solving equations with deformations*, Springer-Verlag Lecture Notes in Economics and Math. Systems #**234** (1987).

[E2] ——, *Properly labeled simplexes*, Studies in Optimization (G. B. Dantzig and B. C. Eaves, eds.), vol. **10**, 1974, pp. 71–93.

[Fa] R. T. Farouki, *The approximation of nondegenerate off-set surfaces*, Comp. Aided Geom. Design **3** (1985), 15–43.

[Fr] H. Freudenthal, *Simplizialzerlegungen von beschränkter Flachheit*, Ann. Math. **43** (1942), 580–582.

[FT] S. R. Freitas and G. Tavares, *Solving implicit differential equations by simplicial methods*, preprint, Department of Mathematics, Catholic University, Rio de Janeiro-Brazil.

[K] H. W. Kuhn, *Simplicial approximation of fixed points*, Proc. Nat. Acad. Sci. USA **61** (1968), 1238–1242.

[M] J. Milnor, *Morse theory*, Princeton University Press, Princeton, New Jersey, 1963.

[PTFB] J. Platt, D. Terzopoulos, K. Fleicher, and A. Barr, *Elastically deformable models*, SIGGRAPH 87 course notes.

[R] W. C. Rheinboldt, *On the computation of multidimensional solution manifolds of parametrized equations*, Numerische Mathematik **53** (1988), 165–181.

[Sa] R. Saigal, *On piecewise linear approximation to smooth mappings*, Math. Oper. Res. **4** (2) (1979), 153–161.

[Sc] H. Scarf, *The approximation of fixed points of a continuous mapping*, SIAM J. Appl. Math. **15** (5) (1967), 1328–1343.

[T] M. J. Todd, *The computation of fixed points and applications*, Springer-Verlag Lecture Notes in Economics and Math. Systems #**124** (1976).

CATHOLIC UNIVERSITY, RIO DE JANEIRO

Lectures in Applied Mathematics
Volume **26** (1990)

On Characterizations of Superlinear Convergence for Constrained Optimization

THOMAS F. COLEMAN

Abstract. We show how the Dennis–Moré characterization of superlinear convergence for unconstrained optimization can be applied, and usefully restricted, for use in the constrained setting.

1. Introduction. The main purpose of this paper is to illustrate how the Dennis–Moré [7] characterization of superlinear convergence can be adapted to the (equality) constrained optimization setting. In particular, we follow Goodman [**11**] and replace the constrained minimization problem with a smooth zero-finding problem (valid in a neighborhood of the solution). It is then possible to apply the Dennis–Moré characterization directly. However, there is a subtle point: the function in the zero-finding problem is smooth but is not computable because its definition depends on information computable at the solution only. Nevertheless, we demonstrate the usefulness of this viewpoint.

In Section 2 we first consider the case where exact second derivatives are known; we provide a new and easy proof of quadratic convergence of algorithms in the sequential quadratic programming (SQP)

1980 *Mathematics Subject Classification* (1985 *Revision*). Primary 65H10, 65K05, 65K10.

Key words and phrases. Constrained optimization, equality constraints, numerical optimization, quasi-Newton method, secant method, superlinear characterization, sequential quadratic programming, SQP-method.

Research partially supported by the Applied Mathematical Sciences Research Program (KC-04-02) of the Office of Energy Research of the U.S. Department of Energy under grant DE-FG02-86ER25013.A000 and by the U. S. Army Research Office through the Mathematical Sciences Institute, Cornell University.

class. Section 3 follows with a discussion of characterizations of superlinear convergence and various restrictions. The Boggs–Tolle–Wang [**1**] "characterization" is shown to be a restriction of the general characterization (i.e., it is applicable to a subclass of methods only). Finally, in Section 4, we provide a new proof of the superlinear convergence of the Coleman–Conn [**4**] method, using the results developed in the previous section.

2. Quadratic convergence of sequential quadratic programming (SQP). We are interested in the nonlinear equality constrained problem,

$$\text{minimize } \{f(\mathbf{x}) : \mathbf{c}(\mathbf{x}) = \mathbf{0}\}$$

where $f : \mathbf{R}^n \to \mathbf{R}$, $\mathbf{c} : \mathbf{R}^n \to \mathbf{R}^t$, $t \le n$, and $\mathbf{c}(\mathbf{x}) = (c_1(\mathbf{x}), \ldots, c_t(\mathbf{x}))^{\mathrm{T}}$. In this section we provide a simple proof of the quadratic convergence of the SQP method for problem (1). The proof is new though it is inspired by the viewpoint developed by Goodman [**11**]; however, in our approach we use a smooth zero-finding problem valid in a neighborhood of the solution, whereas Goodman defines a sequence of smooth zero-finding problems, each valid around the current point.

The technique and terminology introduced in this section will be useful in the remainder of the paper. Let $\mathbf{A}(\mathbf{x})$ denote the matrix of constraint gradients $\mathbf{A}(\mathbf{x}) = (\nabla c_1(\mathbf{x}), \ldots, \nabla c_t(\mathbf{x}))$. The following assumptions are frequently used in this paper:

(A_1) f and $c_i, i = 1, t$, are twice continuously differentiable on an open convex set D;

(A_2) $\mathbf{x}_* \in D$, where x_* is a local solution to (1); $\mathbf{A}_* \overset{\text{def}}{=} \mathbf{A}(\mathbf{x}_*)$ is of full column rank t;

(A_3) $\nabla^2 f_* + \sum \lambda_i^* \nabla^2 c_i^*$ is positive definite on null($\mathbf{A}_*^{\mathrm{T}}$), where $\nabla f_* + \mathbf{A}_* \lambda_* = 0$;

(A_4) $\lambda(\mathbf{x})$ is a continuous function on D satisfying a Lipschitz condition at $\mathbf{x}_*$.

We say that $\{\mathbf{x}_k\}$ is generated by the SQP method if $\mathbf{x}_{k+1} = \mathbf{x}_k + \mathbf{s}_k^{\mathrm{SQP}}$ and $\mathbf{x}_k^{\mathrm{SQP}}$ solves

$$\min\left\{\mathbf{s}^{\mathrm{T}} \nabla L_k + \tfrac{1}{2}\mathbf{s}^{\mathrm{T}} \nabla^2 L_k \mathbf{s} : \mathbf{A}_k^{\mathrm{T}} \mathbf{s} = -\mathbf{c}_k\right\}, \tag{2}$$

where $L_k(\mathbf{x}) = f(\mathbf{x}) + \lambda_k^{\mathrm{T}} \mathbf{c}(\mathbf{x})$, $\lambda_k = \lambda(\mathbf{x}_k)$, and λ satisfies $\mathbf{A}_4$. (For example, the least-squares multiplier function is given by $\lambda(\mathbf{x}) = -\mathbf{R}(\mathbf{x})^{-1}\mathbf{Y}(\mathbf{x})^{\mathrm{T}} \nabla f(\mathbf{x})$ where $\mathbf{A}(\mathbf{x}) = \mathbf{Y}(\mathbf{x})\mathbf{R}(\mathbf{x})$; $\mathbf{Y}$ is n-by-t with orthonormal columns, $\mathbf{R}$ is t-by-t and upper triangular.)

Notational note. For simplicity we will write L_k (i.e., without argument) to mean $L_k(\mathbf{x}_k)$. In addition, we write ∇L_k and $\nabla^2 L_k$ to refer to the gradient and Hessian of $L_k(\mathbf{x}_k)$, respectively, with λ_k treated as a constant.

If $\mathbf{A}_k$ is of full column rank t and the two-sided projection of $\nabla^2 L_k$ onto null$(\mathbf{A}_k^{\mathrm{T}})$ is positive definite, then (2) has a unique solution. Specifically, let $\mathbf{Z}_k$ be *any basis* for null$(\mathbf{A}_k^{\mathrm{T}})$ ($\mathbf{Z}_k$ is not necessarily an orthogonal basis). We can write $\mathbf{s} = \mathbf{Y}_k\mathbf{v}_k + \mathbf{Z}_k\mathbf{h}$, where $\mathbf{v}_k = -\mathbf{R}_k^{-\mathrm{T}}\mathbf{c}_k$. Hence, (2) is equivalent to

$$\text{(3)} \qquad \operatorname{minimize} \mathbf{h}^{\mathrm{T}}\mathbf{Z}_k^{\mathrm{T}}(\nabla f_k + \nabla^2 L_k \mathbf{Y}_k\mathbf{v}_k) + \tfrac{1}{2}h^{\mathrm{T}}\mathbf{Z}_k^{\mathrm{T}}\nabla^2 L_k\mathbf{Z}_k\mathbf{h}.$$

But we have assumed that $\mathbf{Z}_k^{\mathrm{T}}\nabla^2 L_k\mathbf{Z}_k > 0$ and therefore the solution to (3) is given by

$$\text{(4)} \qquad (\mathbf{Z}_k^{\mathrm{T}}\nabla^2 L_k\mathbf{Z}_k)\mathbf{h}_k = -\mathbf{Z}_k^{\mathrm{T}}(\nabla f_k + \nabla^2 L_k\mathbf{Y}_k\mathbf{v}_k)$$

and the solution to (2) is

$$\text{(5)} \qquad \mathbf{s}_k^{\mathrm{SQP}} = \mathbf{Z}_k\mathbf{h}_k + \mathbf{Y}_k\mathbf{v}_k.$$

But (4) and (5) can be combined to give

$$\begin{pmatrix} \mathbf{Z}_k^{\mathrm{T}}\nabla^2 L_k \\ \mathbf{A}_k^{\mathrm{T}} \end{pmatrix}(\mathbf{Z}_k\mathbf{h}_k + \mathbf{Y}_k\mathbf{v}_k) = \begin{pmatrix} -\mathbf{Z}_k^{\mathrm{T}}\nabla f_k \\ -\mathbf{c}_k \end{pmatrix}$$

which is equivalent to

$$\text{(6)} \qquad \begin{pmatrix} \mathbf{Z}_k^{\mathrm{T}}\nabla^2 L_k \\ A_k^{\mathrm{T}} \end{pmatrix}\mathbf{s}_k^{\mathrm{SQP}} = \begin{pmatrix} -\mathbf{Z}_k^{\mathrm{T}}\nabla f_k \\ -\mathbf{c}_k \end{pmatrix}.$$

Note that $\mathbf{s}_k^{\mathrm{SQP}}$ is independent of the particular basis $\mathbf{Z}_k$.

To be more precise, we should include a dependence on λ when we refer to an SQP direction, e.g., $\mathbf{s}_k^{\mathrm{SQP}_\lambda}$, because the SQP direction will differ with different λ-rules. However, the differences are low-order: All results presented in this section are valid for any continuous λ-function satisfying a Lipschitz condition at $\mathbf{x}_*$, and therefore we suppress the λ subscript. We do assume that the rule for choosing λ is consistently applied so that $\lambda(\mathbf{x})$ can be viewed as a specific continuous function, Lipschitz continuous at $\mathbf{x}_*$.

In order to analyze the convergence behavior of $\{\mathbf{x}_k\}$, where $\mathbf{x}_{k+1} = \mathbf{x}_k + \mathbf{s}_k^{\mathrm{SQP}}$, we need the following result (Theorem 3.4 of **[8]**).

LEMMA 1. *Let* $\mathbf{F}: \mathbf{R}^n \to \mathbf{R}^n$ *satisfy*

(a) $\mathbf{F}$ *is continuously differentiable on an open convex set* D;
(b) *there exists* $\mathbf{x}_* \in D$ *such that* $\mathbf{F}(\mathbf{x}_*) = 0$ *and* $\mathbf{F}'(\mathbf{x}_*)$ *is nonsingular;*
(c) $\|\mathbf{F}'(\mathbf{x}) - \mathbf{F}'(\mathbf{x}_*)\| \le \eta(\|\mathbf{x} - \mathbf{x}_*\|)$ *for all* $\mathbf{x} \in D$ *and some* $\eta \ge 0$.

Assume $\{\mathbf{x}_k\} \in D$ *and* $\{\mathbf{x}_k\} \to \mathbf{x}_*$. *If there exists a sequence of nonsingular matrices* $\{\mathbf{B}_k\}$ *such that* $\mathbf{x}_{k+1} = \mathbf{x}_k + \mathbf{s}_k$, $\mathbf{B}_k\mathbf{s}_k = -\mathbf{F}(\mathbf{x}_k)$, *and* $\|\mathbf{B}_k - \mathbf{F}'_*\| = O(\|\mathbf{x}_k - \mathbf{x}_*\|)$ *then* $\{\mathbf{x}_k\}$ *converges quadratically to* $\mathbf{x}_*$.[1]

In order to apply Lemma 1 we must phrase (1) as a continuously differentiable system of equations (at least in a neighborhood of a solution $\mathbf{x}_*$). System (6) provides a strong hint as to a possible form; however, a difficulty arises because the standard method of computing $\mathbf{Z}(\mathbf{x})$ does not yield a continuously differentiable basis representation [5]. Moreover, Byrd and Schnabel [3] have shown that a smooth representation of a basis of the null space does not exist, in general. Nevertheless, various (theoretical) forms can be used [3], [11] to yield a continuously differentiable representation in the neighborhood of any given point. Below we derive and use a specific form; other forms are possible. Indeed, it is possible to proceed without using an explicit form; however, in the interest of clarity, we prefer to be concrete. Specifically, let $\overline{\mathbf{x}} \in \mathbf{R}^n$ and suppose that $\overline{\mathbf{A}} \stackrel{\text{def}}{=} \mathbf{A}(\overline{\mathbf{x}})$ is of full rank t. Let $\overline{\mathbf{Z}}$ be any basis for null$(\overline{\mathbf{A}}^{\mathrm{T}})$ and define

$$\mathbf{Z}(\mathbf{x}) = (\mathbf{I} - \overline{\mathbf{A}}(\mathbf{A}^{\mathrm{T}}\overline{\mathbf{A}})^{-1}\mathbf{A}^{\mathrm{T}})\overline{\mathbf{Z}}. \tag{7}$$

In a neighborhood of $\overline{\mathbf{x}}$, $\mathbf{Z}(\mathbf{x})$ is continuously differentiable with

$$\dot{\mathbf{Z}} = \overline{\mathbf{A}}(\mathbf{A}^{\mathrm{T}}\overline{\mathbf{A}})^{-1}\dot{\mathbf{A}}^{\mathrm{T}}\overline{\mathbf{A}}(\mathbf{A}^{\mathrm{T}}\overline{\mathbf{A}})^{-1}\mathbf{A}^{\mathrm{T}}\overline{\mathbf{Z}} - \overline{\mathbf{A}}(\mathbf{A}^{\mathrm{T}}\overline{\mathbf{A}})^{-1}\dot{\mathbf{A}}^{\mathrm{T}}\overline{\mathbf{Z}}. \tag{8}$$

Therefore,

$$\dot{\mathbf{Z}}(\overline{\mathbf{x}}) = -\overline{\mathbf{A}}(\overline{\mathbf{A}}^{\mathrm{T}}\overline{\mathbf{A}})^{-1}\dot{\mathbf{A}}(\overline{\mathbf{x}})^{\mathrm{T}}\overline{\mathbf{Z}} = -\overline{\mathbf{Y}}\,\overline{\mathbf{R}}^{-\mathrm{T}}\dot{\mathbf{A}}(\overline{\mathbf{x}})\overline{\mathbf{Z}}, \tag{9}$$

where $\overline{\mathbf{A}} = \overline{\mathbf{Y}}\,\overline{\mathbf{R}}$. (Note that $\dot{\mathbf{Z}}(\overline{\mathbf{x}})^{\mathrm{T}}\nabla f(\overline{\mathbf{x}}) = \overline{\mathbf{Z}}^{\mathrm{T}}\sum\overline{\lambda}_i\nabla^2 c_i(\overline{\mathbf{x}})$, where $\overline{\lambda} = -\overline{\mathbf{R}}^{-1}\overline{\mathbf{Y}}^{\mathrm{T}}\nabla f(\overline{\mathbf{x}})$.)

Using this definition of $\mathbf{Z}$ defined around $\overline{\mathbf{x}} = \mathbf{x}_*$, consider the nonlinear system $\mathbf{F}(\mathbf{x}) = \mathbf{0}$, where

$$\mathbf{F}(\mathbf{x}) \stackrel{\text{def}}{=} \begin{pmatrix} \mathbf{Z}(\mathbf{x})^{\mathrm{T}}\nabla f(\mathbf{x}) \\ \mathbf{c}(\mathbf{x}) \end{pmatrix}. \tag{10}$$

Clearly, under assumptions (A_1)–(A_3), $\mathbf{x}_*$ is an isolated zero of (10).

Note that

$$\mathbf{J}(\mathbf{x}) \stackrel{\text{def}}{=} \mathbf{F}'(\mathbf{x}) = \begin{pmatrix} \dot{\mathbf{Z}}(\mathbf{x})^{\mathrm{T}}\nabla f(\mathbf{x}) + \mathbf{Z}(\mathbf{x})^{\mathrm{T}}\nabla^2 f(\mathbf{x}) \\ \mathbf{A}^{\mathrm{T}}(\mathbf{x}) \end{pmatrix} \tag{11}$$

---

[1] We make extensive use of the "O" and "o" notation: $\varphi_k = O(\psi_k)$ means that the ratio φ_k/ψ_k remains bounded as $k \to \infty$ and $\varphi_k = o(\psi_k)$ means that the ratio $\varphi_k/\psi_k \to 0$ as $k \to \infty$.

and hence

$$\mathbf{J}_* = \begin{pmatrix} \mathbf{Z}_*^{\mathrm{T}}(\nabla^2 f_* + \sum \lambda_i^* \nabla^2 c_i^*) \\ \mathbf{A}_*^{\mathrm{T}} \end{pmatrix}. \tag{12}$$

Assumptions (A_2) and (A_3) imply that $\mathbf{J}_*$ is nonsingular.

Unfortunately, the SQP step is not a Newton step for (10) and therefore quadratic convergence is not automatic. Indeed, the construction of a Newton process for (10) is unclear since

$$\mathbf{Z}(\mathbf{x}) = (\mathbf{I} - \mathbf{A}_*(\mathbf{A}(\mathbf{x})^{\mathrm{T}}\mathbf{A}_*)^{-1}\mathbf{A}(\mathbf{x})^{\mathrm{T}})\mathbf{Z}_* \tag{13}$$

and typically $\mathbf{Z}_*$ and $\mathbf{A}_*$ are unknown, except at $\mathbf{x}_*$.

Our approach to this problem is to show that the solution to (6) (which is independent of the choice of $\mathbf{Z}_k$) is actually a solution to an *approximate* Newton system, $\mathbf{B}_k\mathbf{s}_k^{\mathrm{SQP}} = -\mathbf{F}(\mathbf{x}_k)$, where $\|\mathbf{B}_k - \mathbf{J}_*\| = O(\|\mathbf{x}_k - \mathbf{x}_*\|)$. In particular, since $\mathbf{s}_k^{\mathrm{SQP}}$ is independent of the choice of basis $\mathbf{Z}_k$ in (6), $\mathbf{s}_k^{\mathrm{SQP}}$ satisfies

$$\begin{pmatrix} \mathbf{Z}(\mathbf{x}_k)^{\mathrm{T}}\nabla^2 L_k \\ \mathbf{A}_k^{\mathrm{T}} \end{pmatrix} \mathbf{s}_k^{\mathrm{SQP}} = \begin{pmatrix} -\mathbf{Z}(\mathbf{x}_k)^{\mathrm{T}}\nabla f_k \\ -c_k \end{pmatrix} \tag{14}$$

where $\mathbf{Z}(\mathbf{x}_k)$ is defined by (13). Define

$$\mathbf{B}_k = \begin{pmatrix} \mathbf{Z}(\mathbf{x}_k)^{\mathrm{T}}\nabla^2 L_k \\ \mathbf{A}_k^{\mathrm{T}} \end{pmatrix}. \tag{15}$$

Hence, $\mathbf{s}_k^{\mathrm{SQP}}$ satisfies $\mathbf{B}_k\mathbf{s}_k^{\mathrm{SQP}} = -\mathbf{F}(\mathbf{x}_k)$.

LEMMA 2. *Let assumptions* (A_1)–(A_4) *hold. Further, assume that* $\nabla^2 f$, $\nabla^2 c_i$ ($i = 1, \dots, t$) *are Lipschitz continuous at* $\mathbf{x}_*$. *Let* $\mathbf{x}_c$ *be an arbitrary point and define* $L_c(\mathbf{x}) = f(\mathbf{x}) + \lambda_c^{\mathrm{T}}\mathbf{c}(\mathbf{x})$, *where* $\lambda_c = \lambda(\mathbf{x}_c)$; *define*

$$\mathbf{B}_c(\mathbf{x}) = \begin{pmatrix} \mathbf{Z}(\mathbf{x})^{\mathrm{T}}\nabla^2 L_c(\mathbf{x}) \\ \mathbf{A}(\mathbf{x})^{\mathrm{T}} \end{pmatrix}.$$

Then, for all $\mathbf{x}_c$ *sufficiently close to* $\mathbf{x}_*$, $\|\mathbf{B}_c(\mathbf{x}_c) - \mathbf{J}_*\| = O(\|\mathbf{x}_c - \mathbf{x}_*\|)$.

PROOF. First consider rows $1, \dots, n-t$:

$$\begin{aligned}(\mathbf{I}_{n-t}, \mathbf{0})(\mathbf{B}_c(\mathbf{x}_c) - \mathbf{J}_*) &= \mathbf{Z}(\mathbf{x}_c)^{\mathrm{T}}\nabla^2 L_c(\mathbf{x}_c) - \mathbf{Z}_*^{\mathrm{T}}(\nabla^2 L_*(\mathbf{x}_*)) \\ &= (\mathbf{Z}(\mathbf{x}_c)^{\mathrm{T}} - \mathbf{Z}_*^{\mathrm{T}})(\nabla^2 L_*(\mathbf{x}_*)) \\ &\quad + \mathbf{Z}(\mathbf{x}_c)^{\mathrm{T}}(\nabla^2 L_c(\mathbf{x}_c) - \nabla^2 L_*(\mathbf{x}_*)).\end{aligned}$$

But $\dot{\mathbf{Z}}$ is bounded in a neighborhood of $\mathbf{x}_*$ and therefore, using Taylor's theorem, we have

$$\|\mathbf{Z}(\mathbf{x}_c) - \mathbf{Z}_*\| = O(\|\mathbf{x}_c - \mathbf{x}_*\|) \tag{16}$$

for $\mathbf{x}_c$ in a neighborhood of $\mathbf{x}_*$; using the Lipschitz continuity of $\nabla^2 f$, $\nabla^2 c_i$, and λ, it follows that

$$\|\nabla^2 L_c(\mathbf{x}_c) - \nabla^2 L_*(\mathbf{x}_*)\| = O(\|\mathbf{x}_c - \mathbf{x}_*\|) \tag{17}$$

in a neighborhood of $\mathbf{x}_*$. Equations (16) and (17) yield

$$\|(\mathbf{I}_{n-t}, \mathbf{0})(\mathbf{B}_c - \mathbf{J}_*)\| = O(\|\mathbf{x}_c - \mathbf{x}_*\|), \tag{18}$$

for all $\mathbf{x}_c$ in a neighborhood of $\mathbf{x}_*$. Next consider the last t rows:

$$(\mathbf{0}, \mathbf{I}_t)(\mathbf{B}_c - \mathbf{J}_*) = (\mathbf{A}_c^{\mathrm{T}} - \mathbf{A}_*^{\mathrm{T}}).$$

But $\dot{\mathbf{A}}(\mathbf{x})$ is bounded in a neighborhood of $\mathbf{x}_*$ and so by Taylor's theorem we get

$$\|\mathbf{A}_c - \mathbf{A}_*\| = O(\|\mathbf{x}_c - \mathbf{x}_*\|) \tag{19}$$

for all $\mathbf{x}_c$ sufficiently close to $\mathbf{x}_*$.

The result now follows from (18) and (19). ■

We are now poised to use Lemma 1 to prove quadratic convergence of the SQP method; we need only establish assumption (c) of Lemma 1 with respect to our definition of $\mathbf{F}$. However, the fact that $\dot{\mathbf{A}}(\mathbf{x})$ is bounded in a neighborhood of $\mathbf{x}_*$ and definition (8) yield

$$\|\dot{\mathbf{Z}}(\mathbf{x}) - \dot{\mathbf{Z}}_*\| = O(\|\mathbf{x} - \mathbf{x}_*\|) \tag{20}$$

for all $\mathbf{x}$ sufficiently close to $\mathbf{x}_*$. Next, using (20), the Lipschitz continuity of $\nabla^2 f$, and the boundedness of $\dot{\mathbf{A}}(\mathbf{x})$, we obtain the required result:

$$\|\mathbf{J}(\mathbf{x}) - \mathbf{J}_*\| = O(\|\mathbf{x} - \mathbf{x}_*\|) \tag{21}$$

for all $\mathbf{x}$ sufficiently close to $\mathbf{x}_*$.

We now have all the necessary prerequisites to establish quadratic convergence of the SQP method.

THEOREM 3. *Let assumptions* (A_1)–(A_4) *hold; assume that* $\nabla^2 f$, $\nabla^2 c_i$, $i = 1, \ldots, t$, *are Lipschitz continuous at* $\mathbf{x}_*$. *Assume* $\{\mathbf{x}_k\} \in D$ *and* $\{\mathbf{x}_k\} \to \mathbf{x}_*$, *where* $\{\mathbf{x}_k\}$ *is generated by the* SQP *method. Then* $\{\mathbf{x}_k\}$ *converges quadratically.*

PROOF. Clearly, assumptions (a), (b), and (c) of Lemma 1 hold. Moreover, by Lemma 2 there exists a $\mathbf{B}_k$ such that $\mathbf{B}_k \mathbf{s}_k^{\mathrm{SQP}} = -\mathbf{F}(\mathbf{x}_k)$ and $\|\mathbf{B}_k - \mathbf{J}_*\| = O(\|\mathbf{x}_k - \mathbf{x}_*\|)$, provided D is small enough. The result follows.

The local convergence of the SQP method is easily established by Theorem 5.1 in **[8]**.

REMARK. Goodman's [**11**] proof of quadratic convergence differs from ours in the following respect. Goodman's technique involves the fact that the SQP steps—with the least-squares multipliers—form a sequence of *exact* Newton steps for a *sequence* of functions, $\mathbf{F}_i$. In contrast, we use the fact that the SQP step—with any Lipschitz continuous multiplier function—is an *approximate* Newton step for a *fixed* function $\mathbf{F}$. Our proof is also closely related to the approach in Tapia [**14**] in which an approximate Newton point of view is also taken. Tapia's development differs in that a projection-based function is used instead of defining a smooth basis $\mathbf{Z}$.

3. Superlinear convergence for constrained optimization algorithms. For unconstrained minimization the characterization of Dennis and Moré ([**7**], Theorem 2.2 or [**8**], Theorem 3.1) has proven to be very useful. It can also be used for constrained optimization, using $\mathbf{F}(\mathbf{x})$ and $\mathbf{J}_*$ defined in (10) and (12), respectively. For completeness we reproduce the result here, reworded to reflect the constrained optimization setting.

THEOREM 4. *Let assumptions* (A_1)–(A_3) *hold. Let* $\{\mathbf{M}_k\}$ *be a sequence of nonsingular matrices. Define* $\mathbf{F}$ *and* $\mathbf{J}_*$ *by* (7), (10), *and* (12). *Suppose that for some* $\mathbf{x}_0$ *in* D *the sequence*

$$\mathbf{x}_{k+1} = \mathbf{x}_k - \mathbf{M}_k^{-1}\mathbf{F}(\mathbf{x}_k), \qquad k = 0, 1, \ldots, \tag{22}$$

remains in D, $\mathbf{x}_k \neq \mathbf{x}_*$ *for* $k \geq 0$, *and converges to* $\mathbf{x}_*$. *Then* $\{\mathbf{x}_k\}$ *converges superlinearly in* $\mathbf{x}_*$ *if and only if*

$$\lim_{k\to+\infty} \frac{\|[\mathbf{M}_k - \mathbf{J}_*](\mathbf{x}_{k+1} - \mathbf{x}_k)\|}{\|\mathbf{x}_{k+1} - \mathbf{x}_k\|} = 0. \tag{23}$$

As noted in [**8**], if we define $\mathbf{s}_k = \mathbf{x}_{k+1} - \mathbf{x}_k$ and $\mathbf{s}_k^N = -\mathbf{J}_k^{-1}\mathbf{F}(\mathbf{x}_k)$ then it is possible to rephrase (23) without explicitly referring to an iteration matrix $\mathbf{M}_k$.

THEOREM 5. *Let assumptions* (A_1)–(A_3) *hold. Let* $\{\mathbf{x}_k\}$ *be a sequence of points that remains in* D, $\mathbf{x}_{k+1} \neq \mathbf{x}_k$ *and* $\mathbf{x}_k \neq \mathbf{x}_*$, *for* $k \geq 0$, *and converges to* $\mathbf{x}_*$. *Then* $\{\mathbf{x}_k\}$ *converges superlinearly to* $\mathbf{x}_*$ *if and only if*

$$\lim_{k\to+\infty} \frac{\|\mathbf{s}_k - \mathbf{s}_k^N\|}{\|\mathbf{s}_k\|} = 0. \tag{24}$$

PROOF. It is easy to see that the assumptions of Theorem 4 imply the assumptions of Theorem 5. To see that the converse is true, assume that the assumptions of Theorem 5 hold. But we can define a nonsingular matrix mapping the vector $-\mathbf{F}(\mathbf{x}_k)$ to the vector $\mathbf{x}_{k+1}-\mathbf{x}_k$: Define $\mathbf{W}_k$ to be the matrix $\alpha \times \mathbf{Q}$ where $\mathbf{Q}$ is the orthogonal rotator that brings the vector $-\mathbf{F}(\mathbf{x}_k)$ onto the ray $(\mathbf{x}_{k+1}-\mathbf{x}_k)/\|\mathbf{x}_{k+1}-\mathbf{x}_k\|$ and define $\alpha = \|\mathbf{x}_{k+1}-\mathbf{x}_k\|/\|\mathbf{F}(\mathbf{x}_k)\|$. Obviously, $\mathbf{W}_k$ is well defined and nonsingular; let $\mathbf{M}_k = \mathbf{W}_k^{-1}$. The equivalence of the assumptions is established.

To complete the proof we must merely establish the equivalence of (23) and (24). But

$$\mathbf{s}_k - \mathbf{s}_k^N = \mathbf{J}_k^{-1}[\mathbf{J}_k - \mathbf{M}_k]\mathbf{s}_k$$

and the result follows. ∎

The application of (23) or (24) to constrained optimization is subtle: $\mathbf{F}$, as defined in (10), is a theoretical construction and is not, in general, computable at $\mathbf{x}_k$. A possible alternative is to derive a characterization that does not depend explicitly on $\mathbf{F}$. We do this next.

First we show that $\mathbf{J}_*$ can be replaced in (23).

LEMMA 6. *Let the assumptions of Theorem 4 hold. Further, let $\{\mathbf{B}_k\}$ be any sequence of matrices satisfying*

$$\lim_{k\to+\infty} \|\mathbf{B}_k - \mathbf{J}_*\| = 0. \tag{25}$$

Then $\{\mathbf{x}_k\}$ converges superlinearly to $\mathbf{x}_*$ if and only if

$$\lim_{k\to+\infty} \frac{\|[\mathbf{M}_k - \mathbf{B}_k](\mathbf{x}_{k+1}-\mathbf{x}_k)\|}{\|\mathbf{x}_{k+1}-\mathbf{x}_k\|} = 0. \tag{26}$$

PROOF. We need only show that condition (26) is equivalent to (23). However,

$$\frac{[\mathbf{M}_k - \mathbf{B}_k](\mathbf{x}_{k+1}-\mathbf{x}_k)}{(\mathbf{x}_{k+1}-\mathbf{x}_k)} = \frac{[\mathbf{M}_k - \mathbf{J}_*](\mathbf{x}_{k+1}-\mathbf{x}_k)}{(\mathbf{x}_{k+1}-\mathbf{x}_k)} + \frac{[\mathbf{J}_* - \mathbf{B}_k](\mathbf{x}_{k+1}-\mathbf{x}_k)}{(\mathbf{x}_{k+1}-\mathbf{x}_k)} \tag{27}$$

from which the result follows directly. ∎

THEOREM 7. *Let the assumptions of Theorem 5 hold. Let $\mathbf{s}_k = \mathbf{x}_{k+1} - \mathbf{x}_k$ and let $\mathbf{s}_k^{\mathrm{SQP}}$ be the* SQP *step (i.e.,* (3) *or* (6) *or* (14)). *Then $\{\mathbf{x}_k\}$ converges superlinearly to $\mathbf{x}_*$ if and only if*

$$\lim_{k\to+\infty} \frac{\|\mathbf{s}_k - \mathbf{s}_k^{\mathrm{SQP}}\|}{\|\mathbf{s}_k\|} = 0. \tag{28}$$

PROOF. The SQP step is the solution to a system $\mathbf{B}_k\mathbf{s}_k^{\mathrm{SQP}} = -\mathbf{F}(\mathbf{x}_k)$, where $\mathbf{B}_k$ and $\mathbf{F}$ are defined in (15) and (10), respectively; hence, by

Lemma 2, $||\mathbf{B}_k - \mathbf{J}_*|| = O(||\mathbf{x}_k - \mathbf{x}_*||)$. But our assumptions imply the existence of a nonsingular matrix $\mathbf{M}_k$ such that $\mathbf{M}_k\mathbf{s}_k = -\mathbf{F}(\mathbf{x}_k)$ and therefore,

$$\mathbf{s}_k - \mathbf{s}_k^{\mathrm{SQP}} = \mathbf{B}_k^{-1}[\mathbf{B}_k - \mathbf{M}_k]\mathbf{s}_k \tag{29}$$

and the result follows using Lemma 6. ■

Hence, a superlinear rate is achieved if and only if the steps approach the SQP steps, asymptotically, both in size and direction.

For completeness, we remark that the superlinearity characterization given in **[6]** can be expressed in the context of problem (1) using **F** given by (10).

THEOREM 8. *Let the assumptions of Theorem 5 hold. Then* $\mathbf{x}_k$ *converges to* $\mathbf{x}_*$ *superlinearly if and only if*

$$\lim_{k\to\infty} \frac{\left\| \begin{pmatrix} \mathbf{Z}(\mathbf{x}_k)^{\mathrm{T}}\nabla^2 L_k \\ \mathbf{A}_k^{\mathrm{T}} \end{pmatrix} \mathbf{s}_k + \begin{pmatrix} \mathbf{Z}(\mathbf{x}_k)^{\mathrm{T}}\nabla f_k \\ \mathbf{c}_k \end{pmatrix} \right\|}{\left\| \begin{pmatrix} \mathbf{Z}(\mathbf{x}_k)^{\mathrm{T}}\nabla f_k \\ \mathbf{c}_k \end{pmatrix} \right\|} = 0. \tag{30}$$

The proof is straightforward; we omit it. This type of characterization has proven useful in the unconstrained setting in the context of an iterative technique for solving the current linearized approximation inexactly; (30) is not immediately useful in the constrained setting because $\mathbf{Z}$ is not computable at $\mathbf{x}_k$, in general. However, if we define $\mathbf{P}_k^Z$ to be the orthogonal projector onto null$(\mathbf{A}_k^{\mathrm{T}})$, then (30) is equivalent to

$$\lim_{k\to\infty} \frac{\left\| \begin{pmatrix} \mathbf{P}_k^Z(\nabla^2 L_k \mathbf{s}_k + \nabla f_k) \\ \mathbf{A}_k^{\mathrm{T}}\mathbf{s}_k + \mathbf{c}_k \end{pmatrix} \right\|}{\left\| \begin{pmatrix} \mathbf{P}_k^Z \nabla f_k \\ \mathbf{c}_k \end{pmatrix} \right\|} = 0 \tag{31}$$

which is computable using local information only. Note that (31) is independent of the choice of basis $\mathbf{Z}_k$.

3.1. *Quasi-Newton* SQP *methods.* A characterization of superlinear convergence for constrained optimization, in the special case when a quasi-Newton SQP algorithm is used, was first given in **[1]**. Theorem 4 is more general because it does not presuppose an algorithm class (also, linear convergence is assumed in **[1]**, whereas convergence only is assumed above). However, the application of Theorem 4 is not obvious since **F**—given by (7), (10)—is a theoretical device and is not, in general, computable at $\mathbf{x}_k$. Nevertheless, the restricted characterization in

[1] is a direct consequence of the following result which, in turn, is an easy consequence of Theorem 4.

LEMMA 9. *Let the assumptions of Theorem* 4 *hold. Further, let* $\{\mathbf{C}_k\}$ *be a sequence of matrices such that* $\{\|\mathbf{C}_k\|\}$ *is bounded above and*

$$\lim_{k\to\infty} \frac{\|(\mathbf{I}-\mathbf{C}_k)(\mathbf{M}_k-\mathbf{J}_*)\mathbf{s}_k\|}{\|\mathbf{s}_k\|} = 0. \tag{32}$$

Then $\{\mathbf{x}_k\}$ *converges to* $\mathbf{x}_*$ *superlinearly if and only if*

$$\lim_{k\to\infty} \frac{\|\mathbf{C}_k(\mathbf{M}_k-\mathbf{J}_*)\mathbf{s}_k\|}{\|\mathbf{s}_k\|} = 0. \tag{33}$$

PROOF. First suppose $\{\mathbf{x}_k\}$ converges superlinearly; therefore, by Theorem 4, (23) holds. But,

$$\|\mathbf{C}_k(\mathbf{M}_k-\mathbf{J}_*)\mathbf{s}_k\| \le \|\mathbf{C}_k\|\cdot\|(\mathbf{M}_k-\mathbf{J}_*)\mathbf{s}_k\| = O(\|(\mathbf{M}_k-\mathbf{J}_*)\mathbf{s}_k\|)$$

and therefore (33) holds. On the other hand, suppose (33) is true. But,

$$\|(\mathbf{M}_k-\mathbf{J}_*)\mathbf{s}_k\| \le \|(\mathbf{I}-\mathbf{C}_k)(\mathbf{M}_k-\mathbf{J}_*)\mathbf{s}_k\| + \|\mathbf{C}_k(\mathbf{M}_k-\mathbf{J}_*)\mathbf{s}_k\|,$$

which implies (23) and hence superlinear convergence. ■

It is worthwhile noting at this point that our application will involve the use of a specific choice for $\mathbf{C}_k$, constant for all k: $\mathbf{C}_k$ will be the orthogonal projector onto $\langle \mathbf{e}_1,\ldots,\mathbf{e}_{n-t}\rangle$, denoted by $\mathbf{P}$.

The following result was originally established by Boggs, Tolle, and Wang [1]. Since then alternative proofs have been provided in [9], [12], [13].

COROLLARY 10. *Let assumptions* (A_1)–(A_3) *hold. Let* $\mathbf{H}_k$ *be a symmetric matrix with the restriction of* $\mathbf{H}_k$ *onto* $\mathrm{null}(\mathbf{A}_*^{\mathrm{T}})$ *positive definite (by restriction we mean the two-sided projection). Let* $\{\mathbf{x}_k\}$ *be defined by* $\mathbf{x}_{k+1}\leftarrow \mathbf{x}_k+\mathbf{s}_k$, *where* $\mathbf{s}_k$ *solves*

$$\text{minimize}\left\{\mathbf{s}^{\mathrm{T}}\nabla f_k + \tfrac{1}{2}\mathbf{s}^{\mathrm{T}}\mathbf{H}_k\mathbf{s}\colon \mathbf{A}_k^{\mathrm{T}}\mathbf{s} = -\mathbf{c}_k\right\}.$$

Assume $\{\mathbf{x}_k\}\to\mathbf{x}_*$; *let* $\mathbf{P}_k^Z$ *be the orthogonal projector onto* $\mathrm{null}(\mathbf{A}_k^{\mathrm{T}})$. *Then* $\{\mathbf{x}_k\}$ *converges superlinearly to* $\mathbf{x}_*$ *if and only if*

$$\lim_{k\to+\infty} \frac{\|\mathbf{P}_k^Z(\mathbf{H}_k-\nabla^2 L_*)\mathbf{s}_k\|}{\|\mathbf{s}_k\|} = 0. \tag{35}$$

PROOF. For sufficiently small D the restriction of $\mathbf{H}_k$ onto $\mathrm{null}(\mathbf{A}_k^{\mathrm{T}})$ is positive definite and therefore $\mathbf{s}_k$ is the solution to the system

$$\mathbf{M}_k\mathbf{s}_k = \begin{pmatrix} -\mathbf{Z}(\mathbf{x}_k)^{\mathrm{T}}\nabla f_k \\ -\mathbf{c}_k \end{pmatrix} = -\mathbf{F}(\mathbf{x}_k), \tag{36}$$

where

$$\mathbf{M}_k = \begin{pmatrix} \mathbf{Z}(\mathbf{x}_k)^{\mathrm{T}}\mathbf{H}_k \\ \mathbf{A}_k^{\mathrm{T}} \end{pmatrix} \overset{\text{def}}{=} \begin{pmatrix} \mathbf{M}_k^1 \\ \mathbf{M}_k^2 \end{pmatrix}, \tag{37}$$

$\mathbf{Z}(\mathbf{x}_k)$ is defined by (7), and $\mathbf{F}$ is given by (10). By (12),

$$\mathbf{J}_* = \begin{pmatrix} \mathbf{Z}_*^{\mathrm{T}}\nabla^2 L_* \\ \mathbf{A}_*^{\mathrm{T}} \end{pmatrix} \overset{\text{def}}{=} \begin{pmatrix} \mathbf{J}_*^1 \\ \mathbf{J}_*^2 \end{pmatrix}. \tag{38}$$

Let $\mathbf{P}$ be the orthogonal projector onto $\langle \mathbf{e}_1, \dots, \mathbf{e}_{n-t} \rangle$; hence,

$$(\mathbf{I} - \mathbf{P})(\mathbf{M}_k - \mathbf{J}_*)\mathbf{s}_k = \begin{pmatrix} \mathbf{0} \\ (\mathbf{M}_k^2 - \mathbf{J}_*^2)\mathbf{s}_k \end{pmatrix}$$

and

$$\lim_{k\to\infty} \frac{\|(\mathbf{M}_k^2 - \mathbf{J}_*^2)\mathbf{s}_k\|}{\|\mathbf{s}_k\|} = \lim_{k\to\infty} \frac{\|(\mathbf{A}_k^{\mathrm{T}} - \mathbf{A}_*^{\mathrm{T}})\mathbf{s}_k\|}{\|\mathbf{s}_k\|} = 0 \tag{39}$$

and therefore assumption (32) of Lemma 9 is satisfied (with $\mathbf{C}_k = \mathbf{P}$ for all k). To complete the proof we must merely show the equivalence of (35) and (33) in this case where we restrict the algorithm to class (34). But,

$$\mathbf{P}(\mathbf{M}_k - \mathbf{J}_*)\mathbf{s}_k = \begin{pmatrix} (\mathbf{M}_k^1 - \mathbf{J}_*^1)\mathbf{s}_k \\ \mathbf{0} \end{pmatrix} \tag{40}$$

and

$$\begin{aligned} \|(\mathbf{M}_k^1 - \mathbf{J}_*^1)\mathbf{s}_k\| &= \|(\mathbf{Z}(\mathbf{x}_k)^{\mathrm{T}}\mathbf{H}_k - \mathbf{Z}_*^{\mathrm{T}}\nabla^2 L_*)\mathbf{s}_k\| \\ &\le \|\mathbf{Z}(\mathbf{x}_k)^{\mathrm{T}}(\mathbf{H}_k - \nabla^2 L_*)\mathbf{s}_k\| + \|(\mathbf{Z}(\mathbf{x}_k) - \mathbf{Z}_*)^{\mathrm{T}}\nabla^2 L_*\mathbf{s}_k\|. \end{aligned}$$

But,

$$\begin{aligned} &\|\mathbf{Z}(\mathbf{x}_k)^{\mathrm{T}}(\mathbf{H}_k - \nabla^2 L_*)\mathbf{s}_k\| \\ &\quad = \|[\mathbf{Z}(\mathbf{x}_k)^{\mathrm{T}}\mathbf{Z}(\mathbf{x}_k)][\mathbf{Z}(\mathbf{x}_k)^{\mathrm{T}}\mathbf{Z}(\mathbf{x}_k)]^{-1}\mathbf{Z}(\mathbf{x}_k)^{\mathrm{T}}(\mathbf{H}_k - \nabla^2 L_*)\mathbf{s}_k\| \\ &\quad \le \|\mathbf{Z}(\mathbf{x}_k)\| \cdot \|\mathbf{P}_k^Z(\mathbf{H}_k - \nabla^2 L_*)\mathbf{s}_k\|. \end{aligned}$$

Therefore (35) implies (33). To see that (33) implies (35) use (40) and note that

$$\begin{aligned} \|(\mathbf{M}_k^1 - \mathbf{J}_*^1)\mathbf{s}_k\| &= \|(\mathbf{Z}(\mathbf{x}_k)^{\mathrm{T}}\mathbf{H}_k - \mathbf{Z}_*^{\mathrm{T}}\nabla^2 L_*)\mathbf{s}_k\| \\ &= \|\mathbf{Z}(\mathbf{x}_k)^{\mathrm{T}}(\mathbf{H}_k - \nabla^2 L_*)\mathbf{s}_k + (\mathbf{Z}(\mathbf{x}_k) - \mathbf{Z}_*)^{\mathrm{T}}\nabla^2 L_*\mathbf{s}_k\| \\ &\ge \|\mathbf{Z}(\mathbf{x}_k)^{\mathrm{T}}(\mathbf{H}_k - \nabla^2 L_*)\mathbf{s}_k\| - \|(\mathbf{Z}(\mathbf{x}_k) - \mathbf{Z}_*)^{\mathrm{T}}\nabla^2 L_*\mathbf{s}_k\|. \end{aligned}$$

But

$$\begin{aligned} \|\mathbf{Z}(\mathbf{x}_k)^{\mathrm{T}}(\mathbf{H}_k - \nabla^2 L_*)\mathbf{s}_k\| &= \frac{\|\mathbf{Z}(\mathbf{x}_k)[\mathbf{Z}(\mathbf{x}_k)^{\mathrm{T}}\mathbf{Z}(\mathbf{x}_k)]^{-1}\|}{\|\mathbf{Z}(\mathbf{x}_k)[\mathbf{Z}(\mathbf{x}_k)^{\mathrm{T}}\mathbf{Z}(\mathbf{x}_k)]^{-1}\|} \\ &\qquad \cdot \|\mathbf{Z}(\mathbf{x}_k)^{\mathrm{T}}(\mathbf{H}_k - \nabla^2 L_*)\mathbf{s}_k\| \\ &\ge \frac{\|\mathbf{P}_k^Z(\mathbf{H}_k - \nabla^2 L_*)\mathbf{s}_k\|}{\|\mathbf{Z}(\mathbf{x}_k)[\mathbf{Z}(\mathbf{x}_k)^{\mathrm{T}}\mathbf{Z}(\mathbf{x}_k)]^{-1}\|}. \end{aligned}$$

Hence, (33) implies (35) and the proof is complete. ■

Next we make additional restrictions on the approximating matrix $\mathbf{M}_k$.

LEMMA 11. *Let the assumptions of Theorem 4 hold; assume that* $\{\mathbf{C}_k^L\}$ *and* $\{\mathbf{C}_k^R\}$ *are sequences of matrices such that* $\{||\mathbf{C}_k^L||\}$ *is bounded from above. Further, assume*

$$\lim_{k\to+\infty} \frac{||(\mathbf{I}-\mathbf{C}_k^L)(\mathbf{M}_k-\mathbf{J}_*)\mathbf{s}_k||}{||\mathbf{s}_k||} = 0 \tag{41}$$

and

$$\lim_{k\to+\infty} \frac{||\mathbf{C}_k^L(\mathbf{M}_k-\mathbf{J}_*)(\mathbf{I}-\mathbf{C}_k^R)\mathbf{s}_k||}{||\mathbf{s}_k||} = 0. \tag{42}$$

Then $\{\mathbf{x}_k\}$ *converges to* $\mathbf{x}_*$ *superlinearly if and only if*

$$\lim_{k\to+\infty} \frac{||\mathbf{C}_k^L(\mathbf{M}_k-J_*)\mathbf{C}_k^R\mathbf{s}_k||}{||\mathbf{s}_k||} = 0. \tag{43}$$

PROOF. First suppose $\{\mathbf{x}_k\}$ converges superlinearly to $\mathbf{x}_*$. But

$$\begin{aligned}||\mathbf{C}_k^L(\mathbf{M}_k-\mathbf{J}_*)\mathbf{C}_k^R\mathbf{s}_k|| &\le ||\mathbf{C}_k^L(\mathbf{M}_k-\mathbf{J}_*)\mathbf{s}_k|| + ||\mathbf{C}_k^L(\mathbf{M}_k-\mathbf{J}_*)(\mathbf{I}-\mathbf{C}_k^R)\mathbf{s}_k|| \\ &\le ||\mathbf{C}_k^L||\cdot||(\mathbf{M}_k-\mathbf{J}_*)\mathbf{s}_k|| \\ &\qquad + ||\mathbf{C}_k^L(\mathbf{M}_k-\mathbf{J}_*)(\mathbf{I}-\mathbf{C}_k^R)\mathbf{s}_k||\end{aligned}$$

and therefore by Theorem 4 and assumption (42), (43) follows. Next assume (43). But

$$\begin{aligned}(\mathbf{M}_k-\mathbf{J}_*)\mathbf{s}_k &= (\mathbf{I}-\mathbf{C}_k^L)(\mathbf{M}_k-\mathbf{J}_*)\mathbf{s}_k \\ &\quad + \mathbf{C}_k^L(\mathbf{M}_k-\mathbf{J}_*)\mathbf{C}_k^R\mathbf{s}_k + \mathbf{C}_k^L(\mathbf{M}_k-\mathbf{J}_*)(\mathbf{I}-\mathbf{C}_k^R)\mathbf{s}_k\end{aligned}$$

and therefore by Theorem 4 superlinear convergence follows. ■

We can apply Lemma 11 to a quasi-Newton SQP-algorithm by making specific choices for matrices $\mathbf{C}_k^L$ and $\mathbf{C}_k^R$.

THEOREM 12. *Let the assumptions of Corollary 10 hold; further, assume*

$$\lim_{k\to\infty} \frac{||\mathbf{P}_k^Z(\mathbf{H}_k-\nabla^2 L_*)(\mathbf{I}-\mathbf{P}_k^Z)\mathbf{s}_k||}{||\mathbf{s}_k||} = 0. \tag{44}$$

Then $\{\mathbf{x}_k\}$ *converges superlinearly to* $\mathbf{x}_*$ *if and only if*

$$\lim_{k\to\infty} \frac{||\mathbf{P}_k^Z(\mathbf{H}_k-\nabla^2 L_*)\mathbf{P}_k^Z\mathbf{s}_k||}{||\mathbf{s}_k||} = 0. \tag{45}$$

PROOF. From Corollary 10 it follows that superlinear convergence is achieved if and only if

$$(46) \qquad \lim_{k\to+\infty} \frac{\|\mathbf{P}_k^Z(\mathbf{H}_k - \nabla^2 L_*)\mathbf{s}_k\|}{\|\mathbf{s}_k\|} = 0.$$

But,

$$(47) \quad \mathbf{P}_k^Z(\mathbf{H}_k - \nabla^2 L_*)\mathbf{s}_k = \mathbf{P}_k^Z(\mathbf{H}_k - \nabla^2 L_*)(\mathbf{I} - \mathbf{P}_k^Z)\mathbf{s}_k + \mathbf{P}_k^Z(\mathbf{H}_k - \nabla^2 L_*)\mathbf{P}_k^Z\mathbf{s}_k.$$

The result now follows trivially. ■

Theorem 12 has interesting algorithmic implications. Specifically, it turns out that it is possible to satisfy assumption (44) while only maintaining an explicit approximation to $\mathbf{P}_k^Z\nabla^2 L_k\mathbf{P}_k^Z$. To see this consider the following (two-step) algorithm, where $\mathbf{H}_k$ is a symmetric matrix, of order n, and positive definite on $\text{null}(\mathbf{A}_k^{\text{T}})$.

Solve

$$(48) \qquad \begin{pmatrix} \mathbf{Z}_k^{\text{T}}\mathbf{H}_k\mathbf{Z}_k & \mathbf{0} \\ \mathbf{0} & \mathbf{R}_k^{\text{T}} \end{pmatrix} \begin{pmatrix} \mathbf{h}_k \\ \mathbf{v}_k \end{pmatrix} = \begin{pmatrix} -\mathbf{Z}_k^{\text{T}}(\nabla L_k(\mathbf{x}_k + \mathbf{Y}_k\mathbf{v}_k)) \\ -\mathbf{c}_k \end{pmatrix}$$

where $\mathbf{Z}_k$ is any basis for $\text{null}(\mathbf{A}_k^{\text{T}})$, and then set

$$(49) \qquad \mathbf{s}_k \leftarrow \mathbf{Z}_k\mathbf{h}_k + \mathbf{Y}_k\mathbf{v}_k, \quad \mathbf{x}_{k+1} \leftarrow \mathbf{x}_k + \mathbf{s}_k.$$

COROLLARY 13. *Let assumptions* (A_1)–(A_3) *hold. Let* $\mathbf{H}_k$ *be a symmetric matrix with the restriction of* $\mathbf{H}_k$ *onto* $\text{null}(\mathbf{A}_*^{\text{T}})$ *positive definite (by restriction we mean the two-sided projection). Let* $\{\mathbf{x}_k\}$ *be defined by* (48)–(49). *Assume that* $\{\lambda_k\} \to \lambda_*$; $\{\mathbf{x}_k\} \to \mathbf{x}_*$. *Let* $\mathbf{P}_k^Z$ *be the orthogonal projector onto* $\text{null}(\mathbf{A}_k^{\text{T}})$. *Then* $\{\mathbf{x}_k\}$ *converges superlinearly to* $\mathbf{x}_*$ *if and only if* (45) *holds.*

PROOF. By Taylor's theorem, for any vector $\mathbf{u}_k$,

$$\nabla L_k(\mathbf{x}_k + \mathbf{u}_k) = \nabla L_k(\mathbf{x}_k) + \nabla^2 L_k(\mathbf{x}_k)\mathbf{u}_k + \mathbf{w}_k$$

where $\|\mathbf{w}_k\| = o(\|\mathbf{u}_k\|)$. Equivalently,

$$(50) \qquad (\nabla^2 L_k(\mathbf{x}_k) + \mathbf{E}_k)\mathbf{u}_k = \nabla L_k(\mathbf{x}_k + \mathbf{u}_k) - \nabla L_k(\mathbf{x}_k)$$

where $\mathbf{E}_k = \mathbf{w}_k(\mathbf{u}_k^{\text{T}}\mathbf{u}_k)^+\mathbf{u}_k^{\text{T}}$ and $(\cdot)^+$ denotes the pseudo-inverse (i.e., $\alpha^+ = \alpha^{-1}$ if $\alpha \neq 0$; otherwise, $\alpha^+ = 0$). But

$$\|\mathbf{E}_k\| \leq \|\mathbf{w}_k\| \cdot \|(\mathbf{u}_k^{\text{T}}\mathbf{u}_k)^+\mathbf{u}_k^{\text{T}}\|$$

and therefore, if $\mathbf{u}_k = 0$ then $\mathbf{E}_k = \mathbf{0}$; otherwise,

$$\|\mathbf{E}_k\| \leq \frac{o(\|\mathbf{u}_k\|)}{\|\mathbf{u}_k\|}$$

and so

$$\text{if } \{\|\mathbf{u}_k\|\} \to 0 \quad \text{then } \{\|\mathbf{E}_k\|\} \to 0. \tag{51}$$

If $\mathbf{u}_k = \mathbf{Y}_k\mathbf{v}_k$ then by (50) we can rewrite (48)–(49) as

$$\begin{pmatrix} \mathbf{Z}_k^{\mathrm{T}}\mathbf{H}_k\mathbf{Z}_k & \mathbf{Z}_k^{\mathrm{T}}(\nabla^2 L_k(\mathbf{x}_k) + \mathbf{E}_k)\mathbf{Y}_k \\ \mathbf{0} & \mathbf{R}_k^{\mathrm{T}} \end{pmatrix} \begin{pmatrix} \mathbf{h}_k \\ \mathbf{v}_k \end{pmatrix} = \begin{pmatrix} -\mathbf{Z}_k^{\mathrm{T}}\nabla f_k \\ -\mathbf{c}_k \end{pmatrix} \tag{52}$$

and then set

$$\mathbf{s}_k \leftarrow \mathbf{Z}_k\mathbf{h}_k + \mathbf{Y}_k\mathbf{v}_k, \quad \mathbf{x}_{k+1} \leftarrow \mathbf{x}_k + \mathbf{s}_k. \tag{53}$$

But (52)–(53) can be expressed in the form (34) using matrix $\widehat{\mathbf{H}}_k$, where

$$\widehat{\mathbf{H}}_k = \mathbf{P}_k^Z\mathbf{H}_k\mathbf{P}_k^Z + \mathbf{P}_k^Z(\nabla^2 L_k(\mathbf{x}_k) + \mathbf{E}_k)(\mathbf{I} - \mathbf{P}_k^Z). \tag{54}$$

However,

$$\mathbf{P}_k^Z(\widehat{\mathbf{H}}_k - \nabla^2 L_*)(\mathbf{I} - \mathbf{P}_k^Z) = \mathbf{P}_k^Z(\nabla^2 L_k(\mathbf{x}_k) + \mathbf{E}_k - \nabla^2 L_*)(\mathbf{I} - \mathbf{P}_k^Z)$$

and therefore, using (51) and $\{\lambda_k\} \to \lambda_*$, assumption (44) of Theorem 12 is satisfied and the result follows. ∎

This last method is an *approximate* quasi-Newton SQP method: $\mathbf{s}_k$ is not computed by solving a problem of the form (34). However, we include it in this subsection because theoretically it is easily expressed in this form (as we have seen). This is not the case for the class of methods described next.

3.2. *Approximate quasi-Newton* SQP *methods.* Unfortunately, the characterizations of superlinear convergence for quasi-Newton SQP methods are not as useful as one would hope: existing quasi-Newton methods do not always fit precisely into the SQP mode. However, it is possible to view such methods as *approximate* quasi-Newton SQP procedures; the (unrestricted) superlinear characterizations, discussed at the beginninig of this section, can be applied to establish a superlinear characterization for a broad class of *approximate* quasi-Newton SQP methods. We do this next.

The following algorithm uses, at each step, matrices $\widehat{\mathbf{Z}}_k$ and $\widehat{\mathbf{A}}_k$; note that, in general, $\langle\widehat{\mathbf{Z}}_k\rangle \neq \langle\mathbf{Z}(\mathbf{x}_k)\rangle$, $\langle\widehat{\mathbf{A}}_k\rangle \neq \langle\mathbf{A}(\mathbf{x}_k)\rangle$, and $\widehat{\mathbf{A}}_k^{\mathrm{T}}\widehat{\mathbf{Z}}_k \neq 0$.[2] Let $\widehat{\mathbf{A}}_k = \widehat{\mathbf{Y}}_k\widehat{\mathbf{R}}_k$.

[2]If $\mathbf{M}$ is a matrix then $\langle\mathbf{M}\rangle$ refers to the space spanned by the columns of $\mathbf{M}$.

Solve

$$\begin{pmatrix} \widehat{\mathbf{Z}}_k^{\mathrm{T}}\mathbf{H}_k\widehat{\mathbf{Z}}_k & \widehat{\mathbf{Z}}_k^{\mathrm{T}}\mathbf{H}_k\widehat{\mathbf{Y}}_k \\ \mathbf{0} & \widehat{\mathbf{R}}_k^{\mathrm{T}} \end{pmatrix} \begin{pmatrix} \mathbf{h}_k \\ \mathbf{v}_k \end{pmatrix} = \begin{pmatrix} -\widehat{\mathbf{Z}}_k^{\mathrm{T}}\nabla L_k(\mathbf{x}_k) \\ -\mathbf{c}_k \end{pmatrix}, \tag{55}$$

$$\mathbf{s}_k \leftarrow \widehat{\mathbf{Z}}_k\mathbf{h}_k + \widehat{\mathbf{Y}}_k\mathbf{v}_k, \tag{56}$$

$$\mathbf{x}_{k+1} \leftarrow \mathbf{x}_k + \mathbf{s}_k. \tag{57}$$

THEOREM 14. *Let assumptions* (A_1)–(A_3) *hold. Let* $\{\mathbf{s}_k\}$ *and* $\{\mathbf{x}_k\}$ *be generated as above. Assume* $\{\mathbf{x}_k\} \to \mathbf{x}_*$. *Further, assume*

(i) *there exist matrices* $\mathbf{S}_k$ *and* $\mathbf{T}_k$ *such that*

$$\widehat{\mathbf{Z}}_k = \mathbf{Z}(\mathbf{x}_k)\mathbf{T}_k + \mathbf{S}_k$$

where the singular values of $\mathbf{T}_k$ *are bounded below away from zero and bounded above, and*

$$||\mathbf{S}_k|| = O(||\mathbf{s}_k||);$$

(ii) $\lim_{k\to\infty} \widehat{\mathbf{A}}_k = \mathbf{A}_*$;

(iii) $\{\mathbf{H}_k\}$ *is a sequence of matrices such that the restriction of* $\mathbf{H}_k$ *onto* $\mathrm{null}(\mathbf{A}_*^{\mathrm{T}})$ *is positive definite;* $\{||\mathbf{H}_k||\}$ *is bounded above;*

(iv) $\lim_{k\to\infty}\{\lambda_k\} = \lambda_*$.

Then, $\{\mathbf{x}_k\}$ *converges superlinearly to* $\mathbf{x}_*$ *if and only if*

$$\lim_{k\to+\infty} \frac{||\mathbf{P}_k^Z(\mathbf{H}_k - \nabla^2 L_*)\mathbf{s}_k||}{||\mathbf{s}_k||} = 0. \tag{58}$$

PROOF. Our proof technique is to establish that $\mathbf{s}_k$ solves a system

$$\begin{pmatrix} \mathbf{Z}(\mathbf{x}_k)^{\mathrm{T}}\mathbf{H}_k + \mathbf{E}_k^1 \\ \widehat{\mathbf{A}}_k^{\mathrm{T}} + \mathbf{E}_k^2 \end{pmatrix} \mathbf{s}_k = \begin{pmatrix} -\mathbf{Z}(\mathbf{x}_k)^{\mathrm{T}}\nabla f_k \\ -\mathbf{c}_k \end{pmatrix} \tag{59}$$

such that $\lim_{k\to\infty} ||\mathbf{E}_k^1|| = \lim_{k\to\infty} ||\mathbf{E}_k^2|| = 0$, and $\mathbf{Z}(\mathbf{x}_k)$ is defined by (7). The result then follows directly from Lemma 9 (with $\mathbf{C}_k = \mathbf{P}$, the orthogonal projector onto $\langle \mathbf{e}_1, \dots, \mathbf{e}_{n-t}\rangle$).

But (55)–(57) can be written

$$\begin{pmatrix} \widehat{\mathbf{Z}}_k^{\mathrm{T}}\mathbf{H}_k \\ \widehat{\mathbf{A}}_k^{\mathrm{T}} + \mathbf{E}_k^2 \end{pmatrix} \mathbf{s}_k = \begin{pmatrix} -\widehat{\mathbf{Z}}_k^{\mathrm{T}}\nabla L_k \\ -\mathbf{c}_k \end{pmatrix} \tag{60}$$

where $\mathbf{E}_k^2 = -\mathbf{w}_k^2\mathbf{s}_k^{\mathrm{T}}/(\mathbf{s}_k^{\mathrm{T}}\mathbf{s}_k)$ and $\mathbf{w}_k^2 = \widehat{\mathbf{A}}_k^{\mathrm{T}}\widehat{\mathbf{Z}}_k\mathbf{h}_k$. Note that

$$||\mathbf{E}_k^2|| \le \frac{||\mathbf{w}_k^2||}{||\mathbf{s}_k||} = O(||\widehat{\mathbf{A}}_k^{\mathrm{T}}\widehat{\mathbf{Z}}_k||)$$

and therefore, by convergence and assumptions (i) and (ii), $\lim_{k\to\infty} ||\mathbf{E}_k^2|| = 0$.

Next, using (i), we write

$$\widehat{\mathbf{Z}}_k = \mathbf{Z}(\mathbf{x}_k)\mathbf{T}_k + \mathbf{S}_k$$

where $\{\mathbf{T}_k\}$ is uniformly nonsingular and $\|\mathbf{S}_k\| = O(\|\mathbf{s}_k\|)$. Hence,

$$\widehat{\mathbf{Z}}_k^{\mathrm{T}}\nabla L_k = \mathbf{T}_k^{\mathrm{T}}\mathbf{Z}(\mathbf{x}_k)^{\mathrm{T}}\nabla L_k + \mathbf{w}_k^1$$

where $\mathbf{w}_k^1 \stackrel{\text{def}}{=} \mathbf{S}_k^{\mathrm{T}}\nabla L_k$. If we define $\widetilde{\mathbf{E}}_k^1 = \mathbf{w}_k^1\mathbf{s}_k^{\mathrm{T}}/(\mathbf{s}_k^{\mathrm{T}}\mathbf{s}_k)$ then (60) can be written

$$\begin{pmatrix} \widehat{\mathbf{Z}}_k^{\mathrm{T}}\mathbf{H}_k + \widetilde{\mathbf{E}}_k^1 \\ \widehat{\mathbf{A}}_k^{\mathrm{T}} + \mathbf{E}_k^2 \end{pmatrix} \mathbf{s}_k = \begin{pmatrix} -\mathbf{T}_k^{\mathrm{T}}\mathbf{Z}(\mathbf{x}_k)^{\mathrm{T}}\nabla L_k \\ -\mathbf{c}_k \end{pmatrix}. \tag{61}$$

Note that $\|\widetilde{\mathbf{E}}_k^1\| \le \|\mathbf{w}_k^1\|/\|\mathbf{s}_k\|$ and

$$\|\mathbf{w}_k^1\| \le \|\mathbf{S}_k\| \cdot \|\nabla L_k\|. \tag{62}$$

But, by assumption (i), $\|\mathbf{S}_k\| = O(\|\mathbf{s}_k\|)$ and $\nabla L_k \to \nabla L_* = 0$, using (iv); therefore, $\lim_{k\to\infty} \|\widetilde{\mathbf{E}}_k^1\| = 0$.

Finally,

$$\begin{aligned} \widehat{\mathbf{Z}}_k^{\mathrm{T}}\mathbf{H}_k &= \mathbf{T}_k^{\mathrm{T}}\mathbf{Z}(\mathbf{x}_k)^{\mathrm{T}}\mathbf{H}_k + \mathbf{S}_k^{\mathrm{T}}\mathbf{H}_k \\ &= \mathbf{T}_k^{\mathrm{T}}\mathbf{Z}(\mathbf{x}_k)^{\mathrm{T}}\mathbf{H}_k + \mathbf{E}_k^3 \end{aligned}$$

where $\mathbf{E}_k^3 \stackrel{\text{def}}{=} \mathbf{S}_k^{\mathrm{T}}\mathbf{H}_k$ and therefore, using assumptions (iii) and (i), $\lim_{k\to\infty} \|\mathbf{E}_k^3\| = 0$. System (59) is now obtained with $\mathbf{E}_k^1 = \mathbf{T}_k^{-\mathrm{T}}(\widetilde{\mathbf{E}}_k^1 + \mathbf{E}_k^3)$. ∎

Note. The assumption that $\|\mathbf{S}_k\| = O(\|\mathbf{s}_k\|)$ is practical since it is satisfied by most (if not all) known superlinearly convergent updating schemes. However, it is possible to replace this assumption with conditions on $\{\mathbf{x}_k\}$ and $\{\lambda_k\}$. Specifically, we can replace assumption (i) with (i′):

(i′) There exist matrices $\mathbf{S}_k$ and $\mathbf{T}_k$ such that

$$\widehat{\mathbf{Z}}_k = \mathbf{Z}(\mathbf{x}_k)\mathbf{T}_k + \mathbf{S}_k$$

where the singular values of $\mathbf{T}_k$ are bounded below away from zero and bounded above, and

$$\|\mathbf{S}_k\| \to 0.$$

Moreover, we assume that $\{\mathbf{x}_k\}$ converges to $\mathbf{x}_*$ *linearly* and $\|\lambda_k - \lambda_*\| = O(\|\mathbf{x}_k - \mathbf{x}_*\|)$.

Then (62) still holds but, using Taylor's theorem,

(63)
$$||\nabla L_k|| = O(||\mathbf{x}_k - \mathbf{x}_*||) + O(||\lambda_k - \lambda_*||)$$
(64) $$= O(||\mathbf{x}_k - \mathbf{x}_*||) \quad (by\ (\mathrm{i}'))$$
(65) $$= O(||\mathbf{s}_k||) \quad (by\ the\ linear\ convergence\ assumption\ in\ (\mathrm{i}')).$$

Consequently, $\lim_{k\to\infty} ||\widehat{\mathbf{E}}_k^1|| = 0$ and the rest of the argument follows unchanged.

If, in addition, we assume

(66) $$\lim_{k\to\infty} \frac{||\mathbf{P}_k^Z(\mathbf{H}_k - \nabla^2\mathbf{L}_*)(\mathbf{I} - \mathbf{P}_k^Z)\mathbf{s}_k||}{||\mathbf{s}_k||} = 0,$$

then the remaining requirement for a superlinear characterization is on $\mathbf{P}_k^Z\mathbf{H}_k\mathbf{P}_k^Z$ alone.

THEOREM 15. *Let the assumptions of Theorem* 14 *hold. Further, assume* (66). *Then* $\{\mathbf{x}_k\}$ *converges to* $\mathbf{x}_*$ *superlinearly if and only if*

(67) $$\lim_{k\to\infty} \frac{||\mathbf{P}_k^Z(\mathbf{H}_k - \nabla^2 L_*)\mathbf{P}_k^Z\mathbf{s}_k||}{||\mathbf{s}_k||} = 0.$$

PROOF. Assume superlinear convergence to $\mathbf{x}_*$. But,

$$||\mathbf{P}_k^Z(\mathbf{H}_k - \nabla^2 L_k)\mathbf{P}_k^Z\mathbf{s}_k|| \le ||\mathbf{P}_k^Z(\mathbf{H}_k - \nabla^2 L_k)\mathbf{s}_k|| + ||\mathbf{P}_k^Z(\mathbf{H}_k - \nabla^2 L_k)(\mathbf{I} - \mathbf{P}_k^Z)\mathbf{s}_k||$$

and so by (58) and assumption (66), (67) follows. On the other hand, assume (67). But,

$$\mathbf{P}_k^Z(\mathbf{H}_k - \nabla^2 L_k)\mathbf{s}_k = \mathbf{P}_k^Z(\mathbf{H}_k - \nabla^2 L_k)(\mathbf{I} - \mathbf{P}_k^Z)\mathbf{s}_k + \mathbf{P}_k^Z(\mathbf{H}_k - \nabla^2 L_k)\mathbf{P}_k^Z\mathbf{s}_k.$$

Hence, (66) and (67) yield (58) and by Theorem 14 superlinear convergence is established. ∎

It is possible to satisfy assumption (66) using an extra gradient evaluation. Specifically, let us replace algorithm (55)–(57) with

Solve

(68) $$\begin{pmatrix} \widehat{\mathbf{Z}}_k^{\mathrm{T}}\mathbf{H}_k\widehat{\mathbf{Z}}_k & \mathbf{0} \\ \mathbf{0} & \widehat{\mathbf{R}}_k^{\mathrm{T}} \end{pmatrix} \begin{pmatrix} \mathbf{h}_k \\ \mathbf{v}_k \end{pmatrix} = \begin{pmatrix} -\widehat{\mathbf{Z}}_k^{\mathrm{T}}\nabla L_k(\mathbf{x}_k + \widehat{\mathbf{Y}}_k\mathbf{v}_k) \\ -\mathbf{c}_k \end{pmatrix},$$

(69) $$\mathbf{s}_k \leftarrow \widehat{\mathbf{Z}}_k\mathbf{h}_k + \widehat{\mathbf{Y}}_k\mathbf{v}_k,$$

(70) $$\mathbf{x}_{k+1} \leftarrow \mathbf{x}_k + \mathbf{s}_k.$$

THEOREM 16. *Let the assumptions of Theorem* (14) *hold with the exception that algorithm* (55)–(57) *is replaced with* (68)–(70). *Then,* $\{\mathbf{x}_k\}$ *converges to* $\mathbf{x}_*$ *superlinearly if and only if*

$$\lim_{k\to\infty} \frac{\|\mathbf{P}_k^Z(\mathbf{H}_k - \nabla^2 L_*)\mathbf{P}_k^Z \mathbf{s}_k\|}{\|\mathbf{s}_k\|} = 0. \tag{71}$$

PROOF. Repeating the technique used in the proof of Corollary 13, we obtain a matrix $\mathbf{G}_k$ such that

$$\|\mathbf{G}_k\| \le \frac{o(\|\widehat{\mathbf{Y}}\mathbf{v}_k\|)}{\|\widehat{\mathbf{Y}}\mathbf{v}_k\|}$$

and (68)–(70) is equivalent to

Solve

$$\begin{pmatrix} \widehat{\mathbf{Z}}_k^{\mathrm{T}}\mathbf{H}_k\widehat{\mathbf{Z}}_k & \widehat{\mathbf{Z}}_k^{\mathrm{T}}(\nabla^2 L_k(\mathbf{x}_k) + \mathbf{G}_k)\widehat{\mathbf{Y}}_k \\ \mathbf{0} & \widehat{\mathbf{R}}_k^{\mathrm{T}} \end{pmatrix} \begin{pmatrix} \mathbf{h}_k \\ \mathbf{v}_k \end{pmatrix} = \begin{pmatrix} -\widehat{\mathbf{Z}}_k^{\mathrm{T}}\nabla L_k \\ -\mathbf{c}_k \end{pmatrix}. \tag{72}$$

Therefore, if we define

$$\widetilde{\mathbf{H}}_k = \mathbf{P}_k^Z\mathbf{H}_k\mathbf{P}_k^Z + \mathbf{P}_k^Z(\nabla^2 L_k(\mathbf{x}_k) + \mathbf{G}_k)(\mathbf{I} - \mathbf{P}_k^Z),$$

then (68)–(70) can be expressed in the form of algorithm (55)–(57) using $\widetilde{\mathbf{H}}_k$ to play the role of the matrix $\mathbf{H}_k$ in (55)–(57). The result now follows easily. ■

4. An application. In this section we illustrate the usefulness of the viewpoint developed in the previous sections by providing a new proof of superlinear convergence of the constrained quasi-Newton method due to Coleman and Conn [4]. Coleman and Conn [4] established two-step superlinear convergence; subsequently, Byrd [2] strengthened the result to (one-step) superlinear convergence. Byrd's proof is quite different from the proof we present here.

4.1. *The Coleman–Conn algorithm.* The algorithm recurs a positive definite matrix of order $n-t$, $\overline{\mathbf{H}}_k$. Let $\mathbf{x}_{k\_}$ be a previous point with constraint matrix $\mathbf{A}_{k\_} = \mathbf{Y}_{k\_}\mathbf{R}_{k\_}$. The columns of the matrix $\overline{\mathbf{Z}}(\mathbf{x})$ form an orthonormal basis for the null space of $\mathbf{A}(\mathbf{x})^{\mathrm{T}}$. The function UPDATE($\mathbf{M}, \mathbf{s}, \mathbf{y}$) refers to either of the well-known positive definite secant updates, BFGS or DFP: the matrix $\mathbf{M}$ is a positive definite matrix, $\mathbf{s}$ is the current "step," $\mathbf{y}$ is the difference in "gradients," and $\mathbf{s}^{\mathrm{T}}\mathbf{y} > 0$.

THE ALGORITHM.

(1) *Solve* $\mathbf{R}_{k_-}^{\mathrm{T}} \mathbf{v}_k = -\mathbf{c}_k$.
(2) $\mathbf{x}_{k_+} \leftarrow \mathbf{x}_k + \mathbf{Y}_{k_-}\mathbf{v}_k$
(3) *Compute* $\mathbf{A}_{k_+} = \mathbf{Y}_{k_+}\mathbf{R}_{k_+}$; *compute* $\overline{\mathbf{Z}}_{k_+}$, *an orthonormal basis for* $\mathrm{null}(\mathbf{A}_{k_+}^{\mathrm{T}})$.
(4) *Compute* $\nabla f_{k_+} \stackrel{\mathrm{def}}{=} \nabla f(\mathbf{x}_{k_+})$.
(5) *Solve* $\mathbf{R}_{k_+}\lambda_{k_+} = \mathbf{Y}_{k_+}^{\mathrm{T}} \nabla f_{k_+}$.
(6) *Solve* $\overline{\mathbf{H}}_k \mathbf{h}_k = -\overline{\mathbf{Z}}_{k_+}^{\mathrm{T}} \nabla f_{k_+}$.
(7) $\mathbf{x}_{k+1} \leftarrow \mathbf{x}_{k_+} + \overline{\mathbf{Z}}_{k_+}\mathbf{h}_k$
(8) $\overline{\mathbf{H}}_{k+1} \leftarrow \mathrm{UPDATE}\{\overline{\mathbf{H}}_k, \mathbf{h}_k, \overline{\mathbf{Z}}_{k_+}^{\mathrm{T}}(\nabla f_{k+1} - \mathbf{A}_{k+1}\lambda_{k_+}) - \overline{\mathbf{Z}}_{k_+}^{\mathrm{T}} \nabla f_{k_+}\}$
(9) $k \leftarrow k+1,\ k_- \leftarrow k_+$

The key to the proof of superlinearity is that the correction to $\mathbf{x}_k$ can be expressed in the form (68)–(70). To see this note that $\mathbf{x}_{k+1}$ can be obtained via:

Solve

$$\begin{pmatrix} \overline{\mathbf{H}}_k & \mathbf{0} \\ \mathbf{0} & \mathbf{R}_{k_-} \end{pmatrix} \begin{pmatrix} \mathbf{h}_k \\ \mathbf{v}_k \end{pmatrix} = \begin{pmatrix} -\overline{\mathbf{Z}}_{k_+}^{\mathrm{T}} \nabla L_k(\mathbf{x}_k + \mathbf{Y}_{k_-}\mathbf{v}_k) \\ -\mathbf{c}_k \end{pmatrix}, \tag{73}$$

$$\mathbf{s}_k \leftarrow \overline{\mathbf{Z}}_{k_+}\mathbf{h}_k + \mathbf{Y}_{k_-}\mathbf{v}_k, \tag{74}$$

$$\mathbf{x}_{k+1} \leftarrow \mathbf{x}_k + \mathbf{s}_k. \tag{75}$$

Therefore, if we let $\mathbf{H}_k = \overline{\mathbf{Z}}_{k_+}\overline{\mathbf{H}}_k\overline{\mathbf{Z}}_{k_+}^{\mathrm{T}}$ we see that the Coleman-Conn algorithm is in the form (68)–(70).

THEOREM 17. *Let assumptions* (A_1)–(A_3) *hold. Let* $\{\mathbf{x}_k\}$ *and* $\{\mathbf{H}_k\}$ *be generated as described above with* $\{\mathbf{x}_0, \overline{\mathbf{H}}_0\}$ *the starting pair,* $\mathbf{x}_{0_-} = \mathbf{x}_0$, *and* $\overline{\mathbf{H}}_0$ *is symmetric positive definite. Further, assume that* $\overline{\mathbf{Z}}(\mathbf{x})$ *is a Lipschitz continuous function on D; the Hessian matrices* $\nabla^2 f$ *and* $\nabla^2 c_i$, $i = 1, \ldots, t$, *are Lipschitz continuous on D.*

Then, there exist positive scalars ε *and* Δ *such that if* $\|\mathbf{x}_0 - \mathbf{x}_*\| \le \varepsilon$ *and* $\|\mathbf{H}_0 - \mathbf{M}_*\| \le \Delta$, *where* $\mathbf{M}_* = \overline{\mathbf{Z}}_*^{\mathrm{T}}(\nabla^2 f_* + \sum \lambda_i^* \nabla^2 c_i^*)\overline{\mathbf{Z}}_*$ *and* $\overline{\mathbf{Z}}_*$ *is the limit point of* $\{\overline{\mathbf{Z}}_k\}$, *then* $\{\mathbf{x}_k\}$ *converges at a superlinear rate.*

PROOF. Since Coleman and Conn **[4]** have established convergence, we can prove superlinearity by applying Theorem 16 provided we establish that the assumptions of Theorem 14 hold. But assumptions (ii)–(iv) of Theorem 14 follow straightforwardly (with the boundedness of $\{\|\overline{\mathbf{H}}_k\|\}$ given by Coleman and Conn **[4]**). To establish that assumption

(i) of Theorem 14 holds, note that we can write

$$\mathbf{Z}(\mathbf{x}_{k_+}) = \mathbf{Z}(\mathbf{x}_k) + \left(\int_0^1 \dot{\mathbf{Z}}(\mathbf{x}_k + \tau(\mathbf{x}_{k_+} - \mathbf{x}_k))\partial\tau \right) \cdot (\mathbf{x}_{k_+} - \mathbf{x}_k). \tag{76}$$

But $||\dot{\mathbf{Z}}(\mathbf{x})||$ is bounded above for sufficiently small D, and therefore

$$\mathbf{Z}(\mathbf{x}_{k_+}) = \mathbf{Z}(\mathbf{x}_k) + \widehat{\mathbf{S}}_k \tag{77}$$

where $\widehat{\mathbf{S}}_k = O(||\mathbf{s}_k||)$. Since both $\overline{\mathbf{Z}}_{k_+}$ and $\mathbf{Z}(x_{k_+})$ are bases for null$(\mathbf{A}_{k_+}^{\mathrm{T}})$, there exists a nonsingular matrix $\mathbf{T}_k$ such that

$$\overline{\mathbf{Z}}_{k_+} = \mathbf{Z}(\mathbf{x}_{k_+})\mathbf{T}_k. \tag{78}$$

In fact, continuity of $\overline{\mathbf{Z}}_k$ and $\mathbf{Z}(\mathbf{x}_k)$ and convergence yield that the columns of $\mathbf{T}_k$ are uniformly linearly independent and $\{||\mathbf{T}_k||\}$ is bounded above, for sufficiently small D. Using (77) and (78), we have

$$\overline{\mathbf{Z}}_{k_+} = \mathbf{Z}(\mathbf{x}_k)\mathbf{T}_k + \mathbf{S}_k \tag{79}$$

where $\mathbf{S}_k = \widehat{\mathbf{S}}_k\mathbf{T}_k$ and $||\mathbf{S}_k|| = O(||\mathbf{s}_k||)$. Therefore, assumption (i) of Theorem 14 is established.

Next an argument similar to that used to establish Theorem 3.6 and Corollary 3.11 of Coleman and Conn **[4]** can be used to give

$$\frac{||(\overline{\mathbf{H}}_k - \mathbf{M}_*)\mathbf{h}_k||}{||\mathbf{s}_k||} \to 0. \tag{80}$$

But (80) implies (71); the theorem is established.

4.2. *Concluding remarks.* (1) The two-step superlinear convergence result of Coleman and Conn **[4]** was actually with respect to the sequence $\{\mathbf{x}_{k_+}\}$ as defined above. It was Richard Byrd who first suggested—and later proved **[2]**—that the $\{\mathbf{x}_k\}$ sequence might be (one-step) superlinear.

(2) The proof of (80) does depend on the use of a smooth $\overline{\mathbf{Z}}(\mathbf{x})$; as we have mentioned, this can be a thorny issue. Practical constructions have been suggested in **[5, 10]**; however, Byrd and Schnabel **[3]** proposed a modification to the Coleman–Conn algorithm that is *not* dependent on the choice of basis. The Byrd–Schnabel algorithm can be expressed in the same form as above with an additional "adjustment" to $\overline{\mathbf{H}}_k$ to reflect the change in $\mathbf{Z}$. Specifically, precede step (6) with

$$\mathbf{H}_k \leftarrow \mathbf{T}_k^{\mathrm{T}}\mathbf{H}_k\mathbf{T}_k + \mathbf{Z}_{k_+}^{\mathrm{T}}\mathbf{Y}_{k_-}\overline{\mathbf{C}}_k\mathbf{Y}_{k_-}^{\mathrm{T}}\mathbf{Z}_{k_+} \tag{6$^-$}$$

where $\mathbf{T}_k = \mathbf{Z}_{k_-}^{\mathrm{T}}\mathbf{Z}_{k_+}$. This is actually a slight generalization of the Byrd–Schnabel suggestion: they proposed what amounts to a specific choice for $\overline{\mathbf{C}}_k$, $\overline{\mathbf{C}}_k = \beta_k \mathbf{I}_t$, where β_k is a scale factor. Here, $\overline{\mathbf{C}}_k$ is symmetric but otherwise arbitrary. (The idea is that the matrix $\mathbf{Z}_k\mathbf{H}_k\mathbf{Z}_k^{\mathrm{T}} + \mathbf{Y}_k\mathbf{C}_k\mathbf{Y}_k^{\mathrm{T}}$ represents an approximation to the Hessian of the Lagrangian.)

Convergence properties for this algorithm are unknown; our development here suggests a possible way to proceed. Specifically, because the algorithm can be expressed in the form (68)–(70), a major step toward a proof is to show that (71) holds. It appears that a bounded deterioration result could be used, similar to the manner suggested by Coleman and Conn [**4**], provided the perturbation introduced by (6^-) is bounded by $O(\max\{||\mathbf{x}_k - \mathbf{x}_*||, ||\mathbf{x}_{k+1} - \mathbf{x}_*||\})$.

REFERENCES

1. P. T. Boggs, J. W. Tolle, and P. Wang, *On the local convergence of quasi-Newton methods for constrained optimization*, SIAM Journal on Control and Optimization **20** (1982), 161–171.

2. R. H. Byrd, *On the convergence of constrained optimization methods with accurate Hessian information on a subspace*, Tech. Rep. CU-CS-270-84, Department of Computer Science, University of Colorado (1984).

3. R. H. Byrd and R. B. Schnabel, *Continuity of the null space basis and constrained optimization*, Mathematical Programming **35** (1986), 32–41.

4. T. F. Coleman and A. R. Conn, *On the local convergence of a quasi-Newton method for the nonlinear programming problem*, SIAM J. Num. Anal. **21** (1984), 755–769.

5. T. F. Coleman and D. C. Sorensen, *A note on the computation of an orthonormal basis for the null space of a matrix*, Mathematical Programming **29** (1984), 234–242.

6. R. S. Dembo, S. C. Eisenstat, and T. Steihaug, *Inexact Newton methods*, SIAM J. Num. Anal. **19** (1982), 400–408.

7. J. E. Dennis and J. J. Moré, *A characterization of superlinear convergence and its application to quasi-Newton methods*, Mathematics of Computation **28** (1974), 549–560.

8. ———, *Quasi-Newton methods, motivation and theory*, SIAM Review **19** (1977), 46–89.

9. R. Fontecilla, T. Steihaug, and R. A. Tapia, *A convergence theory for a class of quasi-Newton methods for constrained optimization*, SIAM J. Num. Anal. **24** (1987), 1133–1151.

10. P. Gill, W. Murray, M. Saunders, G. Stewart, and M. Wright, *Properties of a representation of a basis for the null space*, Mathematical Programming **33** (1985), 172–186.

11. J. Goodman, *Newton's method for constrained optimization*, Mathematical Programming **33** (1985), 162–171.

12. J. Nocedal and M. Overton, *Projected Hessian updating algorithms for nonlinearly constrained optimization*, SIAM J. Num. Anal. **22** (1985), 821–850.

13. J. Stoer and R. Tapia, *On the characterization of Q-superlinear convergence of quasi-Newton methods for constrained optimization*, Tech. Rep. 84-2, Dept. of Mathematical Sciences, Rice University (1984, revised 1986).

14. R. Tapia, *A stable approach to Newton's method for optimization problems with equality constraints*, Journal of Optimization Theory and Applications **14** (1974), 453–476.

CORNELL UNIVERSITY

Lectures in Applied Mathematics
Volume **26** (1990)

Computation of Solutions of Two-Point Boundary Value Problems by a Simplicial Homotopy Algorithm

JEANNE DUVALLET

Abstract. Our aim is to compute an approximation of the solution of two-point boundary value problems (TPBVPs) by a simplicial homotopy method such as the one proposed by Allgower and Georg (1980) and Saigal (1983). Thanks to the particular structure of the discrete nonlinear systems associated with the problem, the method can be applied successfully. First, we give theoretical results about the algorithm convergence and precision. Then we develop these results in the case of discretized TPBVPs.

1. Introduction. Let F be a continuous mapping from $\mathbf{R}^N$ to $\mathbf{R}^N$. To solve the equation $F(\mathbf{X}) = 0$, the homotopy method consists of introducing a continuous mapping R from $\mathbf{R}^N$ to $\mathbf{R}^N$ such that a zero point $\mathbf{X}^0$ of R is known. A homotopy H is a continuous mapping from $[0, 1] \times \mathbf{R}^N$ to $\mathbf{R}^N$ such that:

$$\begin{aligned} H(0,\mathbf{X}) &= R(\mathbf{X}), \\ H(1,\mathbf{X}) &= F(\mathbf{X}), \qquad \forall\, \mathbf{X} \in \mathbf{R}^N. \end{aligned}$$

The aim is to follow the connected component of $H^{-1}(0) \cap ([0, 1] \times \mathbf{R}^N)$ that contains $(0, \mathbf{X}^0)$. We denote it by $\Gamma(\mathbf{X}^0)$. From a numerical point of view, a piecewise linear approximation of $\Gamma(\mathbf{X}^0)$ is obtained using a simplicial algorithm (see, e.g., **[1]**, **[2]**, **[10]**, **[12]**, **[13]**, **[14]**, **[15]**). The one we consider here is described in **[2]**, **[14]**.

In Section 2, we first recall the main steps of the algorithm. Our purpose is to study how this algorithm can be applied when F is strongly monotone.

DEFINITION 1.1. A mapping F from $\mathbf{R}^N$ to $\mathbf{R}^N$ is said to be strongly monotone iff

$$\exists\, \alpha > 0, \quad \forall\, \mathbf{X}, \mathbf{Y} \in \mathbf{R}^N, \qquad (F(\mathbf{X}) - F(\mathbf{Y}), \mathbf{X} - \mathbf{Y}) \geq \alpha \|\mathbf{X} - \mathbf{Y}\|_2^2.$$

Here $(\cdot,\cdot)$ denotes the Euclidean inner product and $\|\cdot\|_2$ the associated norm in any space $\mathbf{R}^q$, $q \in \mathbf{N}^*$. It may be specified that F is α-strongly monotone.

We suppose that the homotopy H preserves this property in the following sense:

DEFINITION 1.2. A homotopy H is said to be strongly monotone iff $\exists\, \alpha > 0$, $\forall\, \lambda \in [0, 1]$, $H(\lambda, \cdot)$ is α-strongly monotone.

Then we prove that $H^{-1}(0)$ is the graph of a function φ continuous on $[0, 1]$ (Theorem 2.1). Moreover, if the initial mapping R is linear, the algorithm converges (Theorem 2.8). With a stronger hypothesis of regularity on F, we obtain an evaluation of the computed solution (Proposition 2.13), and we propose an upper bound of the distance between $\Gamma(\mathbf{X}^0)$ and its piecewise linear approximation (Theorem 2.14).

The property (1.1) appears quite naturally in the resolution of two-point boundary value problems. These problems are treated in Section 3. Let f be a continuous mapping from $[0, 1] \times \mathbf{R}^p$ to $\mathbf{R}^p$, $p \geq 1$. We consider the following differential system:

$$\text{(P)} \qquad \begin{cases} \ddot{x}(t) = f(t, x(t)), & \forall t \in]0, 1[, \\ x(0) = \boldsymbol{\alpha}, \quad x(1) = \boldsymbol{\beta}, \end{cases}$$

and its discrete approximation (cf. [**8**]),

$$(\mathrm{P}_n) \qquad \begin{cases} -\mathbf{x}_{i-1} + 2\mathbf{x}_i - \mathbf{x}_{i+1} + h^2 f(t_i, \mathbf{x}_i) = 0, & i = 1, \dots, n, \\ \mathbf{x}_0 = \boldsymbol{\alpha}, \quad \mathbf{x}_{n+1} = \boldsymbol{\beta}, \quad h = 1/(n+1). \end{cases}$$

Thus, a nonlinear equation $F_n(\mathbf{X}) = 0$ is to be solved, where F_n is a continuous mapping from $\mathbf{R}^{np}$ to $\mathbf{R}^{np}$. We recall some usual hypotheses ((3.2) or (3.3)) that ensure existence and uniqueness of solutions to (P). Under these assumptions, F_n is strongly monotone (Proposition 3.4) and the results of Section 2 can be applied. In particular, if H is strongly monotone, the algorithm converges and we get an evaluation of the convergence rate (Theorems 3.6 and 3.8). To shorten the path length, we choose the initial mapping as the discretized form of a linear TPBVP (with the same boundary conditions). We propose different ways of selecting such a linear problem and give numerical results.

Many efficient numerical methods are known for solving two-point boundary value problems. So a simplicial method can be recommended only in cases where the classical methods encounter difficulties. For example, a first approximation of the solution can be obtained and then induce an initialization for a quicker method.

2. Convergence and precision of the simplicial algorithm. In this section, we study the applicability of the homotopy method, and the convergence of the simplicial algorithm, in the case of a strongly monotone homotopy H.

THEOREM 2.1. *If the homotopy H is strongly monotone then there exists a mapping φ continuous on* $[0,1]$ *such that*

$$H(\lambda, \mathbf{X}) = 0 \Leftrightarrow \mathbf{X} = \varphi(\lambda), \qquad \forall \lambda \in [0,1].$$

To prove this theorem, we need the following important property of strongly monotone mappings (cf. [**5**]):

LEMMA 2.2. *If a continuous mapping F is strongly monotone on* $\mathbf{R}^N$, *then F is one-to-one.*

PROOF OF THEOREM 2.1. By Lemma 2.2, we get the existence of the mapping φ. Its continuity is the consequence of the following inequalities:

$$\begin{aligned}
&\forall \lambda, \lambda_0 \in [0,1]\\
&\alpha\|\varphi(\lambda) - \varphi(\lambda_0)\|_2^2 \leq (H(\lambda, \varphi(\lambda)) - H(\lambda, \varphi(\lambda_0)), \varphi(\lambda) - \varphi(\lambda_0))\\
&\qquad \leq \|H(\lambda_0, \varphi(\lambda_0)) - H(\lambda, \varphi(\lambda_0))\|_2 \|\varphi(\lambda) - \varphi(\lambda_0)\|_2. \quad \square
\end{aligned}$$

The principle of the algorithm is to construct a sequence of simplices that imbed the continuous path φ. For any positive integer m, if $\mathbf{V}^0, \dots, \mathbf{V}^m$ are affinely independent points of $\mathbf{R}^{N+1}$, the (m-)simplex of vertices $\mathbf{V}^0, \dots, \mathbf{V}^m$ is the convex hull $[\mathbf{V}^0, \dots, \mathbf{V}^m]$. An N-simplex is called a face. All the simplices considered here belong to a standard triangulation $K(\boldsymbol{\delta})$ of $\mathbf{R}^{N+1}$. Let $\boldsymbol{\delta} = (\delta_i)$ be an element of $(\mathbf{R}^{+*})^{N+1}$, and $\mathbf{e}_1, \dots, \mathbf{e}_{N+1}$ be the unit base of $\mathbf{R}^{N+1}$. If $\mathbf{V}^0$ is a vector such that for any $i \in \{1, \dots, N+1\}$ the ith coordinate is an integer multiple of δ_i, for any permutation π of $\{1, \dots, N+1\}$ we denote by $[\mathbf{V}^0, \pi]$ the simplex whose vertices $\mathbf{V}^0, \dots, \mathbf{V}^{N+1}$ are defined by

$$\mathbf{V}^i = \mathbf{V}^{i-1} + \delta_{\pi(i)}\mathbf{e}_{\pi(i)}, \qquad i = 1, \dots, N+1. \tag{2.3}$$

The triangulation $K(\boldsymbol{\delta})$ consists of all such simplices $[\mathbf{V}^0, \pi]$. The diameter of any simplex σ of $K(\boldsymbol{\delta})$ is defined by

$$\sup_{\mathbf{x},\mathbf{y}\in\sigma} \|\mathbf{x} - \mathbf{y}\| = \|\boldsymbol{\delta}\|,$$

where $\|\cdot\|$ denotes either the Euclidean norm $\|\cdot\|_2$ or the supremum norm $\|\cdot\|_\infty$. Thus, we say that $\|\boldsymbol{\delta}\|$ is the size of the triangulation $K(\boldsymbol{\delta})$. There exist exactly two simplices that have a common face and when the simplex σ is given, the adjacent one is built by a simple pivoting rule (cf. [2]).

For any simplex σ, we denote by H_σ the linear mapping that coincides with H at each vertex of σ. The sequence generated by the algorithm is such that each simplex σ contains a segment included in the straight line $H_\sigma^{-1}(0)$. More precisely, it is usual to introduce the following notion (cf. [2]):

DEFINITION 2.4. A face $\tau = [\mathbf{V}^0, \dots, \mathbf{V}^N]$ is called completely labeled iff the $N+1$ points $H(\mathbf{V}^0), \dots, H(\mathbf{V}^N)$ are affinely independent and if there exists $\varepsilon_0 > 0$ such that all the vectors $(\varepsilon, \varepsilon^2, \dots, \varepsilon^N)^t$, $0 \le \varepsilon \le \varepsilon_0$, are included in the simplex $[H(\mathbf{V}^0), \dots, H(\mathbf{V}^N)]$.

In particular, 0 belongs to $[H(\mathbf{V}^0), \dots, H(\mathbf{V}^N)]$ and τ contains a point $\mathbf{X}$ such that $\exists \beta_0, \dots, \beta_N$, $\beta_i \in \mathbf{R}^+$, $\forall i = 0, \dots, N$,

$$X = \sum_{i=0}^{N} \beta_i \mathbf{V}^i, \qquad \sum_{i=0}^{N} \beta_i = 1, \tag{2.5a}$$

$$0 = \sum_{i=0}^{N} \beta_i H(\mathbf{V}^i). \tag{2.5b}$$

The interest of the notion (2.4) is the following result (cf. [2], Lemma 3.11):

PROPOSITION 2.6. *Let* $\sigma = [\mathbf{V}^0, \dots, \mathbf{V}^{N+1}]$ *be a simplex of* $\mathbf{R}^{N+1}$. *If* $[\mathbf{V}^0, \dots, \mathbf{V}^N]$ *is a completely labeled face, there exists a unique index* $k \in \{0, \dots, N\}$ *such that* $[\mathbf{V}^0, \dots, \mathbf{V}^{k-1}, \mathbf{V}^{k+1}, \dots, \mathbf{V}^{N+1}]$ *is a completely labeled face.*

Now the simplicial algorithm may be informally described as follows. A sequence

$$\tau_0, \quad \sigma_0, \quad \tau_1, \quad \sigma_1, \quad \dots, \quad \tau_i, \quad \sigma_i, \quad \tau_{i+1}, \quad \dots$$

is generated in such a way that each simplex σ_i has exactly two completely labeled faces, τ_i and τ_{i+1}. The set $\sigma_i \cap H_{\sigma_i}^{-1}(0)$ is a segment of line that "enters" in the simplex by τ_i and "goes out" by τ_{i+1}. The union of these segments is a piecewise linear path that is an approximation of $H^{-1}(0)$. We denote it by C^δ. To initialize the algorithm, we define the first simplex σ_0 in $K(\boldsymbol{\delta})$ such that one of its faces, τ_0, is a completely labeled face included in $\{0\} \times \mathbf{R}^N$ and admitting $(0, \mathbf{X}^0)$ as its isobarycenter. By Proposition 2.6, the exit face τ_1 is found and the adjacent simplex σ_1 is built by the pivoting rule. This continues until the exit face is in $\{0\} \times \mathbf{R}^N$ or in $\{1\} \times \mathbf{R}^N$. The algorithm is said to converge if a completely labeled face is included in $\{1\} \times \mathbf{R}^N$. In this case, we denote by $(1, \mathbf{X}^\delta)$ the intersection of C^δ with $\{1\} \times \mathbf{R}^N$.

If there exists a compact K such that each simplex σ_i meets K, then the sequence is finite ([2], Lemma 3.18) and either C^δ meets $\{1\} \times \mathbf{R}^N$ or comes back to $\{0\} \times \mathbf{R}^N$. Moreover, the following lemma may be proved (see [13]):

LEMMA 2.7. *Let K be a compact set in $[0,1] \times \mathbf{R}^N$ such that $K \cap H^{-1}(0)$ is empty. There exists $\varepsilon_0 > 0$ such that for any triangulation $K(\boldsymbol{\delta})$, if $\|\boldsymbol{\delta}\|_2 < \varepsilon_0$ then C^δ does not meet K.*

So under the assumption of strong monotonicity, we get the following convergence result:

THEOREM 2.8. *Let H be a strongly monotone homotopy. If the initial mapping R is linear, then for any triangulation $K(\boldsymbol{\delta})$, if $\|\boldsymbol{\delta}\|_2$ is small enough, C^δ crosses $\{1\} \times \mathbf{R}^N$.*

PROOF. By Theorem 2.1 and Lemma 2.7, if $\|\boldsymbol{\delta}\|_2$ is small enough, either C^δ meets $\{1\} \times \mathbf{R}^N$ or it comes back to $\{0\} \times \mathbf{R}^N$ at a point different from $(0, \mathbf{X}^0)$. This last alternative is impossible as $\mathbf{X}^0$ is the only zero point of the strongly monotone mapping R. The complete proof and a more general result may be found in [6]. □

From now on, we suppose that the algorithm converges. Now we address the problem of the precision—how to evaluate the distance between the exact solution $\mathbf{X}^*$ and the computed one $\mathbf{X}^\delta$. There exist $\mathbf{W}^0, \ldots, \mathbf{W}^N$ vectors of $\mathbf{R}^N$ and $\beta_0, \ldots, \beta_N$ nonnegative reals such that

$$\left\{ \begin{array}{l} \mathbf{X}^\delta = \sum_{i=0}^N \beta_i \mathbf{W}^i, \quad \sum_{i=0}^N \beta_i = 1, \\ 0 = \sum_{i=0}^N \beta_i F(\mathbf{W}^i). \end{array} \right. \tag{2.9}$$

Thanks to the definition of the triangulation $K(\boldsymbol{\delta})$, the diameter of any face included in $\{1\} \times \mathbf{R}^N$ is $\|\bar{\boldsymbol{\delta}}\|$, where $\bar{\boldsymbol{\delta}}$ is defined by

$$\bar{\boldsymbol{\delta}} = (\delta_2, \ldots, \delta_{N+1}), \tag{2.10}$$

and $\|\cdot\|$ is either the Euclidean norm or the supremum norm. So we have

$$\|\mathbf{X}^\delta - \mathbf{W}^i\| \leq \|\bar{\boldsymbol{\delta}}\|, \qquad \forall i = 1, \ldots, N.$$

To get an upper bound of the error we must make the regularity assumption on F more precise.

CONDITION 2.11. *Let g be a differentiable mapping from $\mathbf{R}^{n+p}$ to $\mathbf{R}^n$, $p \geq 0$. We denote by E_g the function from $(\mathbf{R}^{n+p})^2$ to $\mathbf{R}^n$ defined by*

$$E_g(\mathbf{u}, \mathbf{v}) = g(\mathbf{v}) - g(\mathbf{u}) - Dg(\mathbf{u})(\mathbf{v} - \mathbf{u}).$$

We suppose that there exists a function θ_g from $\mathbf{R}^+$ to $\mathbf{R}^+$ such that

$$\|\mathbf{u} - \mathbf{v}\|_\infty \leq \eta \Rightarrow \|E_g(\mathbf{u}, \mathbf{v})\|_\infty \leq \theta_g(\eta).$$

In particular, if the mapping g satisfies

$$\exists\, L \geq 0, \quad \forall\, \mathbf{u}, \mathbf{v}, \qquad \|Dg(\mathbf{u}) - Dg(\mathbf{v})\|_\infty \leq L\|\mathbf{u} - \mathbf{v}\|_\infty,$$

then we have (cf. **[11]**)

$$\|E_g(\mathbf{u}, \mathbf{v})\|_\infty \leq \tfrac{1}{2}L\|\mathbf{u} - \mathbf{v}\|_\infty^2.$$

So we get an upper bound of $\|F(\mathbf{X}^\delta)\|_\infty$:

LEMMA 2.12. *If the mapping F satisfies Condition* 2.11, *we have*

$$\|F(\mathbf{X}^\delta)\|_\infty \leq \theta_F(\|\bar{\boldsymbol{\delta}}\|_\infty).$$

PROOF. With the notation (2.9), (2.10), for any $i \in \{0, \dots, N\}$, we have

$$F(\mathbf{W}^i) = F(\mathbf{X}^\delta) + DF(\mathbf{X}^\delta)(\mathbf{W}^i - \mathbf{X}^\delta) + E_F(\mathbf{X}^\delta, \mathbf{W}^i).$$

By linear combination, we get (cf. (2.9)):

$$0 = F(\mathbf{X}^\delta) + \sum_{i=0}^{N} \beta_i E_F(\mathbf{X}^\delta, \mathbf{W}^i).$$

Hence the result follows. □

Moreover, the strong monotonicity gives an upper bound for the distance between $\mathbf{X}^*$ and $\mathbf{X}^\delta$:

PROPOSITION 2.13. *Let F be an α-strongly monotone mapping satisfying* (2.11). *We get*

$$\|\mathbf{X}^* - \mathbf{X}^\delta\|_2 \leq (1/\alpha)\|F(\mathbf{X}^\delta)\|_2 \leq (\sqrt{N}/\alpha)\theta_F(\|\bar{\boldsymbol{\delta}}\|_\infty).$$

The same method leads to an evaluation of the distance between any point $(\lambda, \mathbf{X})$ of C^δ and the corresponding point $(\lambda, \varphi(\lambda))$ of $H^{-1}(0)$:

PROPOSITION 2.14. *If the strongly monotone homotopy H satisfies Condition* 2.11, *for any triangulation $K(\delta)$, we obtain*

$$\forall (\lambda, \mathbf{X}) \in C^\delta, \qquad \|\mathbf{X} - \varphi(\lambda)\|_2 \leq (\sqrt{N}/\alpha)\theta_H(\|\boldsymbol{\delta}\|_\infty).$$

3. Solution of TPBVPs. In this section we adapt the preceding results to solve two-point boundary value problems such as

$$\text{(P)} \qquad \begin{cases} \ddot{x}(t) = f(t, x(t)), & \forall t \in]0,1[, \\ x(0) = \boldsymbol{\alpha}, \quad x(1) = \boldsymbol{\beta}, \end{cases}$$

where f is a continuous mapping from $[0,1] \times \mathbf{R}^p$ to $\mathbf{R}^p$. Without restricting the generality of the problem we can suppose that

$$\boldsymbol{\alpha} = 0, \qquad \boldsymbol{\beta} = 0.$$

For any $n \in \mathbf{N}^*$, a discretized approximation is obtained by a classical finite difference scheme (cf. [**8**]):

$$(\mathrm{P}_n) \qquad \begin{cases} -\mathbf{x}_{i-1} + 2\mathbf{x}_i - \mathbf{x}_{i+1} + h^2 f(t_i, \mathbf{x}_i) = 0, & i = 1, \dots, n, \\ \mathbf{x}_0 = \mathbf{x}_{n+1} = 0, \quad h = 1/(n+1), \\ t_i = ih, \quad \mathbf{x}_i = (x_{i,1}, \dots, x_{i,p}), \quad i = 0, \dots, n+1. \end{cases}$$

Then (P_n) is equivalent to a nonlinear equation $F_n(\mathbf{X}) = 0$ which is solved by the simplicial homotopy algorithm described in Section 2. We isolate the linear part of the mapping F_n which is continuous from $\mathbf{R}^N$ to $\mathbf{R}^N$, $N = np$. Let $\mathbf{A}_n$ be the $n \times n$ symmetric positive definite matrix defined by

$$\begin{aligned} (A_n)_{i,i} &= 2, \quad i = 1, \dots, n, \\ (A_n)_{i,i+1} &= -1, \quad i = 1, \dots, n-1, \\ (A_n)_{i,j} &= 0, \quad i = 1, \dots, n-2, \ j \geq i+2. \end{aligned}$$

We denote by $(\mathbf{I}_p \otimes \mathbf{A}_n)$ the $np \times np$ block diagonal matrix such that each $n \times n$ diagonal block is $\mathbf{A}_n$. For any $i \in \{1, \dots, n\}$, we can consider $\mathbf{x}_i$ as an approximation of the solution at $t_i = ih$. We have organized the unknown $\mathbf{X}$ as follows:

$$\begin{aligned} \mathbf{X} &= (\mathbf{X}_1, \dots, \mathbf{X}_p) \in (\mathbf{R}^n)^p, \\ \mathbf{X}_j &= (x_{1,j}, x_{2,j}, \dots, x_{n,j}) \in \mathbf{R}^n, \qquad j = 1, \dots, p. \end{aligned}$$

So $\mathbf{X}_j$ approximates the jth coordinate of the solution. Thus, F_n is defined by

$$F_n(\mathbf{X}) = (\mathbf{I}_p \otimes \mathbf{A}_n)\mathbf{X} + h^2 \Phi_n(\mathbf{X}), \tag{3.1}$$

where

$$\Phi_n(\mathbf{X}) = (f_1(t_1, \mathbf{x}_1), \dots, f_1(t_n, \mathbf{x}_n), f_2(t_1, \mathbf{x}_1), \dots, f_p(t_n, \mathbf{x}_n))$$

and f_i is the ith coordinate of f.

Two classical hypotheses induce existence and uniqueness of the solution to (P). The first one is a Lipschitz condition:

$$\exists\, K \in\,]0, 3\sqrt{10}[, \quad \forall\, t \in [0,1], \quad \forall\, \mathbf{x}, \mathbf{y} \in \mathbf{R}^p, \qquad \|f(t,\mathbf{x}) - f(t,\mathbf{y})\|_2 \le K\|\mathbf{x}-\mathbf{y}\|_2. \tag{3.2}$$

It was introduced in 1893 by Picard (see [3]). The second one is a monotonicity property:

$$\forall\, t \in [0,1], \quad \forall\, \mathbf{x}, \mathbf{y} \in \mathbf{R}^p, \qquad (f(t,\mathbf{x}) - f(t,\mathbf{y}), \mathbf{x}-\mathbf{y}) \ge 0. \tag{3.3}$$

This last hypothesis is widely used in the scalar case (cf. [8], [9]); it has more recently been extended to the vector case (cf. [7]).

The two hypotheses (3.2) and (3.3) are not only adapted to the theoretical treatment of TPBVPs, but they also guarantee the strong monotonicity of the discretized versions of f.

Proposition 3.4. *If the mapping f satisfies either* (3.2) *or* (3.3), *then for any n sufficiently large, the mapping F_n is strongly monotone.*

Proof. For any $\mathbf{X}$, $\mathbf{Y} \in \mathbf{R}^N$, we get

$$\begin{aligned}(F_n(\mathbf{X}) &- F_n(\mathbf{Y}), \mathbf{X}-\mathbf{Y}) \\ &= \sum_{j=1}^{p} (\mathbf{X}_j - \mathbf{Y}_j)^{t}\mathbf{A}_n(\mathbf{X}_j - \mathbf{Y}_j) + h^2 \sum_{i=1}^{n} (f(t_i,\mathbf{x}_i) - f(t_i,\mathbf{y}_i), \mathbf{x}_i - \mathbf{y}_i).\end{aligned}$$

The smallest eigenvalue of $\mathbf{A}_n$ is

$$\lambda_n = 4\sin^2(\pi h/2) > 0.$$

So under the hypothesis (3.3), we have for any $n \in \mathbf{N}^*$,

$$(F_n(\mathbf{X}) - F_n(\mathbf{Y}), \mathbf{X}-\mathbf{Y}) \ge \lambda_n \|\mathbf{X}-\mathbf{Y}\|_2^2.$$

With (3.2), we get

$$\begin{aligned}(F_n(\mathbf{X}) - F_n(\mathbf{Y}), \mathbf{X}-\mathbf{Y}) &\ge \lambda_n \|\mathbf{X}-\mathbf{Y}\|_2^2 \\ &\quad - h^2 \sum_{i=1}^{n} \|f(t_i,\mathbf{x}_i) - f(t_i,\mathbf{y}_i)\|_2 \|\mathbf{x}_i - \mathbf{y}_i\|_2 \\ &\ge (\lambda_n - Kh^2)\|\mathbf{X}-\mathbf{Y}\|_2^2.\end{aligned}$$

As K is less than π^2, the mapping F_n is strongly monotone for n sufficiently large. □

Under the assumption of Proposition 3.4, F_n is α_n-strongly monotone, where α_n is defined by

$$\alpha_n = \begin{cases} \lambda_n = 4\sin^2(\pi h/2), & \text{if } f \text{ satisfies (3.2)}, \\ \lambda_n - Kh^2, & \text{if } f \text{ satisfies (3.3)}. \end{cases} \tag{3.5}$$

So, by Lemma 2.2, the problem (P_n) has a unique solution $\mathbf{X}^*_{(n)}$ and we denote by $\mathbf{X}^\delta_{(n)}$ the computed result when it exists. Before discussing the choice of the homotopy, we give two upper bounds of the distance between $\mathbf{X}^*_{(n)}$ and $\mathbf{X}^\delta_{(n)}$. The first one (Theorem 3.6) is obtained when the mapping f satisfies the Lipschitz condition (3.2). The second one (Theorem 3.8) supposes that the regularity of f can be made precise (Condition 3.7). In both cases, the error depends on $\|\bar{\boldsymbol{\delta}}\|$, the diameter of the "last" face (cf. (2.10)).

THEOREM 3.6. *If the mapping f satisfies* (3.2) *then for any n large enough, we have*

$$\|\mathbf{X}^*_{(n)} - \mathbf{X}^\delta_{(n)}\|_2 \le (K/(3\sqrt{10} - K))\|\bar{\boldsymbol{\delta}}\|_2.$$

PROOF. With the same notation as (2.9), we have

$$\begin{cases} \mathbf{X}^\delta_{(n)} = \sum_{i=0}^N \beta_i \mathbf{W}^i, \quad \sum_{i=0}^N \beta_i = 1, \quad \beta_i \ge 0, \quad i = 0, \dots, N, \\ 0 = \sum_{i=0}^N \beta_i F(\mathbf{W}^i). \end{cases}$$

From the expression for F_n, we get

$$0 = (\mathbf{I}_p \otimes \mathbf{A}_n)\mathbf{X}^\delta_{(n)} + h^2 \sum_{j=0}^N \beta_j \Phi_n(\mathbf{W}^j),$$

and as $F_n(\mathbf{X}^*_{(n)}) = 0$, we obtain

$$0 = (\mathbf{I}_p \otimes \mathbf{A}_n)(\mathbf{X}^*_{(n)} - \mathbf{X}^\delta_{(n)}) + h^2 \sum_{j=0}^N \beta_j (\Phi_n(\mathbf{X}^*_{(n)}) - \Phi_n(\mathbf{W}^j)).$$

Then

$$\mathbf{X}^*_{(n)} - \mathbf{X}^\delta_{(n)} = -h^2(\mathbf{I}_p \otimes \mathbf{A}_n^{-1}) \sum_{j=0}^N \beta_j (\Phi_n(\mathbf{X}^*_{(n)}) - \Phi_n(\mathbf{W}^j)).$$

The matrix $\mathbf{A}_n$ is symmetric and positive definite. Its smallest eigenvalue is $\lambda_n = 4\sin^2(\pi h/2)$. Hence, the matrix norm $\|h^2 \mathbf{A}_n^{-1}\|_2$ equals h^2/λ_n and the ratio h^2/λ_n converges toward π^{-2}. Let $Q' = 1/(3\sqrt{10})$. For n sufficiently large, we have

$$\|h^2 \mathbf{A}_n^{-1}\|_2 \le Q'.$$

So

$$\begin{aligned}\|\mathbf{X}^*_{(n)} - \mathbf{X}^\delta_{(n)}\|_2 &\le Q' \sum_{j=0}^{N} \beta_j \|\Phi_n(\mathbf{X}^*_{(n)}) - \Phi_n(\mathbf{W}^j)\|_2 \\ &\le Q' \sum_{j=0}^{N} \beta_j \left(\sum_{i=1}^{n} \|f(t_i, \mathbf{x}^*_i) - f(t_i, \mathbf{W}^j_i)\|_2^2 \right)^{1/2} \\ &\le Q' \sum_{j=0}^{N} \beta_j K \|\mathbf{X}^*_{(n)} - \mathbf{W}^j\|_2 \\ &\le Q'K \left(\|\mathbf{X}^*_{(n)} - \mathbf{X}^\delta_{(n)}\|_2 + \sum_{j=0}^{N} \beta_j \|\mathbf{X}^\delta_{(n)} - \mathbf{W}^j\|_2 \right).\end{aligned}$$

Then

$$\|\mathbf{X}^*_{(n)} - \mathbf{X}^\delta_{(n)}\|_2 \le Q'K(\|\mathbf{X}^*_{(n)} - \mathbf{X}^\delta_{(n)}\|_2 + \|\bar{\boldsymbol{\delta}}\|_2).$$

Hence the result follows. □

REMARK. It may be more convenient to have an upper bound for the supremum norm. As (3.2) is verified, there exists K_∞ such that

$$\forall\, t \in [0,1], \quad \forall\, \mathbf{x}, \mathbf{y} \in \mathbf{R}^p, \qquad \|f(t,\mathbf{x}) - f(t,\mathbf{y})\|_\infty \le K_\infty \|\mathbf{x} - \mathbf{y}\|_\infty.$$

If K_∞ is less than 8, we can prove by a similar method that

$$\|\mathbf{X}^*_{(n)} - \mathbf{X}^\delta_{(n)}\|_\infty \le (K_\infty/(8 - K_\infty))\|\bar{\boldsymbol{\delta}}\|_\infty.$$

As we did in Section 2, we can make precise the regularity of f by the following condition (cf. Condition 2.11):

CONDITION 3.7. *Let f be a mapping from $[0,1] \times \mathbf{R}^p$ to $\mathbf{R}^p$ such that for any $t \in [0,1]$ the functions $f(t,\cdot)$ are differentiable. We denote by e_f the mapping from $[0,1] \times \mathbf{R}^p \times \mathbf{R}^p$ to $\mathbf{R}^p$ defined by*

$$e_f(t,\mathbf{x},\mathbf{y}) = f(t,\mathbf{y}) - f(t,\mathbf{x}) - D_x f(t,\mathbf{x})(\mathbf{y} - \mathbf{x}).$$

We suppose that there exist $C \ge 0$ and $\alpha \ge 1$ such that

$$\|\mathbf{x} - \mathbf{y}\|_\infty \le \eta \Rightarrow \|e_f(t,\mathbf{x},\mathbf{y})\|_\infty \le C\eta^\alpha, \qquad \forall\, t \in [0,1].$$

We deduce from Proposition 2.13 the following result:

THEOREM 3.8. *If the mapping f satisfies Condition 3.7 and either (3.2) or (3.3), there exists a constant Q (independent of n and δ) such that*

$$\|\mathbf{X}^*_{(n)} - \mathbf{X}^\delta_{(n)}\|_2 \le Q\sqrt{n}\|\bar{\boldsymbol{\delta}}\|_\infty^\alpha.$$

PROOF. The first point is to prove that F_n satisfies (2.11). The mapping F_n is differentiable and

$$DF_n(\mathbf{X}) = \mathbf{I}_p \otimes \mathbf{A}_n + h^2 D\Phi_n(\mathbf{X}).$$

For any subscript $i \in \{1, \dots, np\}$, there exist $j \in \{1, \dots, n\}$ and $k \in \{1, \dots, p\}$ such that $i = n(k-1) + j$ and by a direct calculation, we get

$$(E_{F_n}(\mathbf{X}, \mathbf{Y}))_i = h^2 (e_f(t_j, \mathbf{x}_j, \mathbf{y}_j))_k.$$

So by (3.11), if $\|\mathbf{X} - \mathbf{Y}\|_\infty < \eta$, we have

$$\|e_f(t_j, \mathbf{x}_j, \mathbf{y}_j)\|_\infty \le C\eta^\alpha, \qquad \forall j \in \{1, \dots, n\},$$

and

$$\|E_{F_n}(\mathbf{X}, \mathbf{Y})\|_\infty \le h^2 C \eta^\alpha.$$

Then we can apply Proposition 2.13 and we obtain

$$\|\mathbf{X}^*_{(n)} - \mathbf{X}^\delta_{(n)}\|_2 \le (\sqrt{np}/\alpha_n) h^2 C \|\bar{\boldsymbol{\delta}}\|^\alpha_\infty,$$

where α_n is defined in (3.5). As the ratio h^2/α_n is bounded, we have the result. □

The last point is to make precise how to choose homotopies and initial mappings that ensure convergence of the algorithm. To solve the equation $F(\mathbf{X}) = 0$, most authors **[4]**, **[10]**, **[12]**, **[16]** consider the following "linear" homotopy:

$$H(\lambda, \mathbf{X}) = (1-\lambda) R(\mathbf{X}) + \lambda F(\mathbf{X}), \qquad \forall (\lambda, \mathbf{X}) \in [0,1] \times \mathbf{R}^N. \tag{3.9}$$

To study large systems (and discretized problems such as (P_n)), Allgower **[1]** and Saigal **[14]** introduce a homotopy that acts coordinate by coordinate. The ith coordinate of this "piecewise linear" homotopy is defined as follows: For any $\lambda \in [0,1]$ there exists $k \in \{1, \dots, N\}$ such that $\lambda \in [(k-1)/N, k/N]$; then

$$H_i(\lambda, \mathbf{X}) = \begin{cases} F_i(\mathbf{X}), & \text{if } i < k, \\ (N\lambda - k + 1) F_k(\mathbf{X}) + (k - N\lambda) R_k(\mathbf{X}), & \text{if } i = k, \\ R_i(\mathbf{X}), & \text{if } i > k. \end{cases} \tag{3.10}$$

The next lemma may be immediately verified.

LEMMA 3.11.

(a) *The linear homotopy* (3.9) *is strongly monotone iff R and F are strongly monotone.*

(b) *The piecewise linear homotopy* (3.10) *is strongly monotone iff for any $k \in \{0, \dots, N\}$ the mapping $(F_1, \dots, F_k, R_{k+1}, \dots, R_N)$ is strongly monotone.*

We can notice that the mapping $R(\mathbf{X}) = \mathbf{X} - \mathbf{X}^0$ is strongly monotone for any $\mathbf{X}^0 \in \mathbf{R}^N$. Thus, the problem is not to find an initial mapping that ensures the convergence but, better, to choose an initialization that could reduce computing cost.

To solve the discretized equation $F_n(\mathbf{X}) = 0$, we propose to define the initial mapping as the discretized problem associated with a linear TPBVP. Let r be a continuous mapping from $[0,1] \times \mathbf{R}^p$ to $\mathbf{R}^p$ such that

$$r(t, \mathbf{x}) = \mathbf{u}(t)\mathbf{x} + \mathbf{v}(t),$$

where $\mathbf{u}(t)$ is a $p \times p$ matrix and $\mathbf{v}(t)$ a vector of $\mathbf{R}^p$. We consider

$$\text{(L)} \qquad \begin{cases} \ddot{x}(t) = r(t, x(t)), & \forall t \in]0,1[, \\ x(0) = \boldsymbol{\alpha}, \quad x(1) = \boldsymbol{\beta}. \end{cases}$$

This problem is discretized by the same finite difference scheme as the one used for (P). Thus we get a linear equation $R_n(\mathbf{X}) = 0$. This mapping R_n is strongly monotone if one of the following conditions is satisfied (cf. Proposition 3.4):

$$\text{(3.12)} \qquad \exists\, L \in]0, 3\sqrt{10}[, \quad \forall\, t \in [0,1], \qquad \|\mathbf{u}(t)\|_2 \leq L,$$

or

$$\text{(3.13)} \qquad \forall\, t \in [0,1], \quad \forall\, \mathbf{x} \in \mathbf{R}^p, \qquad (\mathbf{u}(t)\mathbf{x}, \mathbf{x}) \geq 0.$$

By Lemma 3.11 and Theorem 2.8, we conclude immediately that:

PROPOSITION 3.14. *We suppose that f (resp. r) satisfies either* (3.2) *or* (3.3) (*resp. either* (3.12) *or* (3.13)).

(a) *If the homotopy is defined by*

$$H(\lambda, \mathbf{X}) = (1-\lambda)R_n(\mathbf{X}) + \lambda F_n(\mathbf{X}), \qquad \forall (\lambda, \mathbf{X}) \in [0,1] \times \mathbf{R}^N,$$

then the simplicial algorithm converges.

(b) *If the homotopy is defined as in* (3.10) *then the simplicial algorithm converges in the scalar case* ($p = 1$).

With such an initialization, we can evaluate the distance between any point $(\lambda, \mathbf{X})$ of C^δ and the corresponding point $(\lambda, \varphi(\lambda))$ of $H^{-1}(0)$ (cf. Proposition 2.14). We give the result in the case of the linear homotopy (3.9); a similar result can be obtained in the case of the homotopy (3.10).

THEOREM 3.15. *We suppose that the mapping f (resp. r) satisfies Condition* 3.7 *and either* (3.2) *or* (3.3) (*resp. either* (3.12) *or* (3.13)). *If, moreover,*

$$\exists\, M \geq 0, \qquad \sup_{\mathbf{x} \in \mathbf{R}^p} \|D_x f(t, \mathbf{x}) - \mathbf{u}(t)\|_\infty \leq M, \qquad \forall\, t \in [0,1],$$

then there exists a constant Q such that

$$\forall(\lambda, \mathbf{X}) \in C^{\delta}, \qquad \|\mathbf{X} - \varphi(\lambda)\|_2 \leq Q\sqrt{n}\|\boldsymbol{\delta}\|_{\infty}^{\beta},$$

where $\beta = \inf(2, \alpha)$.

PROOF. The only point is to verify that H satisfies (2.11) to apply Proposition 2.14. For any $(\lambda, \mathbf{X})$, $(\mu, \mathbf{Y})$ in $[0,1] \times \mathbf{R}^N$, by a direct calculation we get

$$\begin{aligned} E_H((\lambda, \mathbf{X}),\ (\mu, \mathbf{Y})) = {} & (1-\mu)E_{R_n}(\mathbf{X}, \mathbf{Y}) + \mu E_{F_n}(\mathbf{X}, \mathbf{Y}) \\ & + (\mu - \lambda)(DF_n(\mathbf{X}) - DR_n(\mathbf{X}))(\mathbf{Y} - \mathbf{X}). \end{aligned}$$

So the ith coordinate of E_H depends on the kth coordinate of e_f and e_r at the instant t_j where k and j are such that $i = n(k-1) + j$ and $k \in \{1, \ldots, p\}$, $j \in \{1, \ldots, n\}$:

$$\begin{aligned} (E_H((\lambda, \mathbf{X}),\ (\mu, \mathbf{Y})))_i = {} & h^2[(1-\mu)e_f(t_j, \mathbf{x}_j, \mathbf{y}_j) + \mu e_r(t_j, \mathbf{x}_j, \mathbf{y}_j) \\ & + (\mu - \lambda)(D_x f(t_j, \mathbf{x}_j) - D_x r(t_j, \mathbf{x}_j))(\mathbf{x}_j - \mathbf{y}_j)]_k. \end{aligned}$$

Hence, if $\|(\lambda, \mathbf{X}) - (\mu, \mathbf{Y})\|_{\infty} \leq \eta$ we get

$$\|E_H((\lambda, \mathbf{X}), (\mu, \mathbf{Y}))\|_{\infty} \leq h^2(2C\eta^{\alpha} + M\eta^2). \quad \square$$

On a numerical point, we have tested different initializations by discretized linear systems. The simplest linear differential problem (L) we consider is defined by

$$r^0(t, \mathbf{x}) = 0, \qquad \forall\, t, \mathbf{x}.$$

We denote by $\mathbf{y}_0$ the linear mapping such that $\mathbf{y}_0(0) = \boldsymbol{\alpha}$ and $\mathbf{y}_0(1) = \boldsymbol{\beta}$, where $\boldsymbol{\alpha}$ and $\boldsymbol{\beta}$ are the boundary conditions of (P). We consider then the first iterate of the Picard method (see [3]):

$$r^1(t, \mathbf{x}) = f(t, \mathbf{y}_0(t)), \qquad \forall\, t, \mathbf{x}.$$

When the mapping f is differentiable, a quite natural way to approach it by a linear mapping is to define

$$r^2(t, \mathbf{x}) = D_x f(t, \mathbf{y}_0(t))(\mathbf{x} - \mathbf{y}_0(t)) + f(t, \mathbf{y}_0(t)).$$

We can simplify this expression to have the coefficient $\mathbf{u}(t)$ independent of t:

$$\begin{aligned} r^3(t, \mathbf{x}) &= D_x f(0, \boldsymbol{\alpha})(\mathbf{x} - \boldsymbol{\alpha}) + f(0, \boldsymbol{\alpha}), \\ r^4(t, \mathbf{x}) &= D_x f(1, \boldsymbol{\beta})(\mathbf{x} - \boldsymbol{\beta}) + f(1, \boldsymbol{\beta}), \\ r^5(t, \mathbf{x}) &= \left(\int_0^1 D_x f(s, \mathbf{y}_0(s))\, ds \right) (\mathbf{x} - \mathbf{y}_0(t)) + f(t, \mathbf{y}_0(t)). \end{aligned}$$

If f satisfies (3.2) (resp. (3.3)), for any $i = 0, \ldots, 5$, the mapping r^i verifies (3.12) (resp. (3.13)).

If it is not possible to compute $D_x f$, the following family of mappings may be helpful:

$$r^6(t, \mathbf{x}) = (f(1, \boldsymbol{\beta}) - f(0, \boldsymbol{\alpha}))g(t, \mathbf{x}) + f(0, \boldsymbol{\alpha}),$$

where the ith coordinate of g is defined by

$$g_i(t, \mathbf{x}) = \mu_i(x_i - \alpha_i) + (1 - \mu_i(\beta_i - \alpha_i))t,$$

and the reals μ_i are chosen to satisfy (3.12) or (3.13). In the vector case ($p > 1$), these last mappings are quite interesting: the matrix $\mathbf{u}(t)$ is constant and diagonal. Moreover, they give rather good results (cf. [**5**]).

NUMERICAL EXAMPLE. We have tested these different initializations to compare the number of simplices built by the algorithm. We consider the following problem:

$$\begin{aligned} \ddot{x}(t) &= 2(x(t) - t + 1)^3, \\ x(0) &= 1, \quad x(1) = \tfrac{2}{3}. \end{aligned}$$

The exact solution is $x(t) = t - 1 + 2/(2t + 1)$.

For $i = 0, \ldots, 6$, the linear system (L_i) is defined by

$$(\mathrm{L}_i) \qquad \begin{cases} \ddot{x}(t) = r^i(t, x(t)), & \forall t \in]0, 1[, \\ x(0) = 1, \quad x(1) = \tfrac{2}{3}, \end{cases}$$

where r^i was defined previously. In the case r^6, we choose

$$\mu = (\beta - \alpha)^{-1}.$$

Each problem is discretized by the same finite difference scheme as the one used for (P). So we get a linear equation $R^i_n(\mathbf{X}) = 0$ and let $\mathbf{X}^{0,i}_n$ be its solution.

We work with the piecewise linear homotopy (3.10). A triangulation well adapted to this homotopy is such that each vertex belongs to $\{k/N\} \times \mathbf{R}^N$, $k \in \{0, \ldots, N\}$. To get a regular triangulation we defined δ by $\delta_i = 1/N$, $i = 1, \ldots, N + 1$. We have discretized with $n = 15$ points, so the step in time is $h = \frac{1}{16}$.

Let N_s be the number of simplices built by the algorithm. The results presented in Table 1 are obtained when the initial mapping is a discretized system R^i_n, $i = 0, \ldots, 6$. It is interesting to notice that N_s may be quite different according to the initialization. For $i = 0, \ldots, 6$,

$n = 15$	r^0	r^1	r^2	r^3	r^4	r^5	r^6
N_s	633	471	301	335	535	319	483

TABLE 1

$n = 15$	$i = 0$	$i = 1$	$i = 2$	$i = 3$	$i = 4$	$i = 5$	$i = 6$
N_s	895	713	477	439	691	437	547

TABLE 2

if the initial mapping is $\bar{R}_n^i(\mathbf{X}) = \mathbf{X} - \mathbf{X}_n^{0,i}$, the path has the same initial point $\mathbf{X}_n^{0,i}$ but it is quite longer: see Table 2.

If we denote by $\mathbf{X}^{\delta,i}$ the computed result in each case, we have

$$\|F(\mathbf{X}^{\delta,i})\|_\infty = 0.4 \times 10^{-5}, \qquad i = 0, \dots, 6.$$

Let $\mathbf{X}$ be the vector $\mathbf{X}_i = x(ih)$, $i = 1, \dots, 15$, where x is the exact solution. The error is

$$\|\mathbf{X}^{\delta,i} - \mathbf{X}\|_\infty = 1.2 \times 10^{-3}, \qquad i = 0, \dots, 6,$$

and

$$\|\mathbf{X}^{\delta,i} - \mathbf{X}\|_2 = 3.1 \times 10^{-3}, \qquad i = 0, \dots, 6;$$

it is independent of the homotopy (cf. Proposition 2.13).

REFERENCES

1. E. L. Allgower, *A survey of homotopy methods for smooth mappings*, Numerical Solution of Nonlinear Equations (E. L. Allgower, K. Galshoff, and H.-O. Peitgen, eds.), Lecture Notes in Mathematics, **878**, Springer-Verlag, New York, 1981, pp. 1–29.

2. E. L. Allgower and K. Georg, *Simplicial and continuation methods for approximating fixed points and solutions to systems of equations*, SIAM Review **22** (1980), 28–85.

3. P. B. Bailey, L. F. Shampine, and P. E. Waltman, *Nonlinear Two-point Boundary Value Problems*, Academic Press, New York, 1968.

4. S. N. Chow, J. Mallet-Paret, and J. A. Yorke, *Finding zeros of maps: Homotopy methods that are constructive with probability one*, Math. Comp. **32** (1978), 887–899.

5. J. Duvallet, *Étude de systèmes differentiels du second ordre avec conditions aux deux bornes et résolution par la méthode homotopique simpliciale*, Thèse de l'Université de Pau, 1986.

6. ——, *Resolution de systèmes non linéaires par une méthode homotopique simpliciale*, preprint U.A. 1204, No. 87/09, 1987.

7. C. P. Gupta, *Existence and uniqueness theorems for boundary value problems involving reflection of the argument*, Nonlinear Analysis, Theory, Methods and Applications **11** (1987), 1075–1083.

8. P. Henrici, *Discrete Variable Methods in Ordinary Differential Equations*, Wiley, New York, 1962.

9. M. Lees, *A boundary value problem for nonlinear ordinary differential equations*, J. Math. Mech. **10** (1961), 423–430.

10. O. H. Merill, *Applications and extensions of an algorithm that computes fixed points of a certain upper semi-continuous point to set mapping*, Ph.D. thesis, University of Michigan, 1972.

11. J. M. Ortega and W. C. Rheinboldt, *Iterative Solutions of Nonlinear Equations in Several Variables*, Academic Press, New York, 1970.

12. R. Saigal, *On paths generated by fixed-point algorithms*, Math. Oper. Res. **1** (1976), 359–380.

13. ——, *On the convergence rate of algorithms for solving equations that are based on methods of complementary pivoting*, Math. Oper. Res., **2**, 1977, pp. 108–124.

14. ——, *A homotopy for solving large, sparse and structured fixed-point problems*, Math. Oper. Res., **8**, 1983, pp. 557–578.

15. G. Van Der Laan, *Simplicial fixed-point algorithms*, Mathematisch Centrum, Amsterdam, 1980.

16. L. T. Watson, *An algorithm that is globally convergent with probability one for a class of nonlinear two-point boundary value problems*, SIAM J. Numer. Anal. **16** (1979), 394–401.

LABORATOIRE DE MATHÉMATIQUES APPLIQUÉES, U.A. C.N.R.S. 1204, FACULTÉ DES SCIENCES, 64000 PAU, FRANCE

Lectures in Applied Mathematics
Volume **26** (1990)

Computational Methods for Nonlinear Systems of Partial Differential Equations Arising in Contaminant Transport in Porous Media

RICHARD E. EWING

Abstract. In large-scale applications involving coupled systems of nonlinear partial differential equations, the method of linearization can cause serious effects on the accuracy of the solution process. Often the systems are discretized directly, with little regard to the properties of the equations or the solutions. This can result in extremely large, ill-conditioned, nonlinear systems; the accurate solution of these equations can be extremely difficult and expensive. If properties of the solution are used in the discretization process, often the resulting nonlinear system can be much better behaved. A modified method of characteristics is presented to follow the physical flow in advection-dominated applications. Then, full linearization and linear solution techniques are discussed for the resulting systems.

1. Introduction. Large coupled systems of nonlinear partial differential equations are frequently used to model complex, dynamic physical processes. Usually the systems are discretized and linearized without regard to important mathematical and physical properties of the equations, leading to badly conditioned and computationally difficult problems. We will indicate that, by using a knowledge of the physical process to guide our discretization, we can often greatly reduce the computational complexity arising from the nonlinearities.

This paper illustrates techniques for effectively solving nonlinear systems via a two-dimensional model for the simulation of flow and transport processes in geologic media. The model was developed for use

---

1980 *Mathematics Subject Classification* (1985 *Revision*). Primary 65N20, 65N30.

by the Nuclear Regulatory Commission in the analysis of deep geologic nuclear waste disposal facilities. The mathematical model is a fully transient, incompressible, two-dimensional model that solves the coupled equations for transport in geologic media **[1]**. The processes described via the model are: (1) fluid flow, (2) heat transport, (3) dominant-species miscible displacement (brine), and (4) trace-species miscible displacement (radionuclide).

The transport equations used here are obtained by combining the appropriate continuity and constitutive relations and have been derived by several authors **[2–4]**. The resulting relations for the total fluid, heat, brine, and the ith component of radionuclides may be stated as follows:

$$-\nabla\cdot\mathbf{u}-q+R_s'=\nabla\cdot\left(\frac{k}{\mu}\right)\left(\nabla p-\rho_0\frac{g}{g_c}\nabla z\right)-q+R_s'=0, \tag{1.1}$$

$$-\nabla(H\mathbf{u})+\nabla\cdot(\mathbf{E}_H\nabla T)-q_L-qH-q_H=[\varphi c_p+(1-\varphi)\bar{\rho}_R c_{pR}]\frac{\partial T}{\partial t}, \tag{1.2}$$

$$-\nabla\cdot(\hat{c}\mathbf{u})+\nabla\cdot(\mathbf{E}_c\nabla\hat{c})-q\hat{c}-q_c+R_s=\varphi\frac{\partial\hat{c}}{\partial t}, \tag{1.3}$$

$$\begin{aligned}-\nabla\cdot(c_i\mathbf{u})+\nabla\cdot(\mathbf{E}_c\nabla c_i)-qc_i-q_{c_i}+q_{0_i}+\sum_{j=1}^{N}k_{ij}\lambda_jK_j\varphi c_j-\lambda_iK_i\varphi c_i\\ =\varphi K_i\frac{\partial c_i}{\partial t},\qquad i=1,2,\ldots,N.\end{aligned} \tag{1.4}$$

The quantity $q=q(x,t)$ is a production term, and

$$R_s'=R_s'(\hat{c})=\frac{c_s\varphi k_sf_s}{1+c_s}(1-\hat{c})$$

is a salt-dissolution term for equation (1.1). $q_L=q_L(T)$ is a term for heat loss to the under/overburden, qH is an injected enthalpy term, and $q_H=q_H(T)$ is a produced enthalpy term for equation (1.2). $q\hat{c}$ is an injected brine term, $q_c=q_c(\hat{c})$ is a produced brine term, and $R_s=R_s(\hat{c})=\varphi k_sf_s(1-\hat{c})$ is a salt-dissolution term for equation (1.3). qc_i are injected component terms, $q_{c_i}=q_{c_i}(c_i)$ are produced component terms, $q_{0_i}=q_{0_i}(c_i)$ are waste leach terms, $\sum_{j=1}^{N}k_{ij}\lambda_jK_j\varphi c_j$ describe generation of component i by decay of j, and $\lambda_iK_i\varphi c_i$ is the decay term for component i.

Several quantities in equations (1.1)–(1.4) require further definition in terms of the basic parameters. The tensors in equations (1.2), (1.3), and (1.4) are defined as sums of dispersion and molecular terms:

$$\mathbf{E}_c = \mathbf{D} + D_m \mathbf{I} \tag{1.5}$$

and

$$\mathbf{E}_H = \mathbf{D} c_{pw} + K_m \mathbf{I}, \tag{1.6}$$

where

$$D_{ij} = \alpha_T |\mathbf{u}| \delta_{ij} + (\alpha_L - \alpha_T) u_i u_j / |\mathbf{u}| \tag{1.7}$$

in a Cartesian system. The nonlinearities in the dispersion terms greatly complicate both the analysis and the related computational schemes. The tensor form requires a nine-point communication star in two space dimensions, for both finite differences and finite elements.

Also, absorption of radionuclides is included via an assumption of a linear equilibrium isotherm. This yields the distribution coefficient k_{di} and the retardation factor, for each component i,

$$K_i = 1 + \rho_R k_{di} (1 - \varphi)/\varphi. \tag{1.8}$$

Equations (1.1)–(1.4) are coupled by a mixing rule for viscosity, $\mu = \mu(\hat{c})$, and three auxiliary relations for Darcy velocity, fluid enthalpy, and fluid internal energy given, respectively, by

$$\mathbf{u} = -\left(\frac{\mathbf{k}}{\mu}\right)\left(\nabla p - \rho_0 \frac{g}{g_c} \nabla z\right), \tag{1.9}$$

$$H = U_0 + U + p/\rho_0, \tag{1.10}$$

$$U = c_p (T - T_0), \tag{1.11}$$

where $U_0 = U_0(x)$ and $T_0 = T(x, 0)$.

We shall assume that no fluid flow occurs across the boundary and shall assume a zero Dirichlet boundary condition for T. In addition, the initial conditions must be given. We need the compatibility condition:

$$(q - R_s', 1) = \int_\Omega [q(x,t) - R_s'(\hat{c})]\, dx = 0. \tag{1.12}$$

It is the combination of strong nonlinearities and close couplings between the equations that causes the difficulties in solving these systems. The dominance of advection and the accompanying presence of very sharp, moving fluid interfaces also greatly increases the computational complexity. If we do not do something special to treat the advection accurately, the errors involved in the linearization can destroy the accuracy of the resulting approximations. Also, due to the enormous size of the applications, the equations cannot be solved in a fully coupled, fully implicit fashion. Choices of implicitness and decoupling of the equations are forms of linearization and must be analyzed and treated with great care for these difficult problems.

In Section 2, we treat the advection in a special way via an operator splitting to treat the advection directly and accurately. Since very accurate fluid velocities are needed for this splitting, mixed finite element methods are utilized for velocity estimation. The spatial discretization, the full description of the modified method of characteristics, and the basic time-stepping schemes are presented in Section 3. The superiority of the modified method of characteristics over standard implicit or explicit treatment of the advection has been established in similar petroleum-related applications **[5–9]**. Although detailed analysis does not appear in this paper, a typical asymptotic convergence theorem is presented in Section 3 for completeness. Finally, in Section 4, we address the decoupling, linearization, and linear solution techniques which are applied after the advection has been treated via characteristic methods.

2. Operator-splitting techniques. Due to the advection-dominance of equations (1.2)–(1.4), we will discuss a modified method of characteristics, which is an operator-splitting technique, for time-stepping these equations. This method is motivated by a desire to address the advection directly and accurately and try to follow the physics of the flow process. For simplicity of exposition, we will describe the method for equation (1.3). The full technique has been described in detail and analyzed by Ewing, Yuan, and Li **[5]**.

In this method, the first and last terms in equation (1.3) are combined to form a directional derivative along what would be the characteristics for the equation if the tensor $\mathbf{E}_c$ were zero. The resulting equation is

$$\nabla \cdot (\mathbf{E}_c \nabla \hat{c}) = q\hat{c} - q_c + R_s = \varphi \frac{\partial \hat{c}}{\partial t} + \nabla \cdot (\hat{c}\mathbf{u}) \equiv \varphi \frac{\partial \hat{c}}{\partial s}. \tag{2.1}$$

The system obtained by modifying equations (1.2)–(1.4) via similar directional derivatives is solved sequentially. An approximation for $\mathbf{u}$ is first obtained at time level $t = t^n$ from a solution of equation (1.1) with the fluid viscosity μ evaluated via some mixing rule at time level t^{n-1}. Equation (1.1) can be solved as an elliptic equation for the pressure p or, via a mixed finite element method presented by Ewing et al. **[5]**, for a more accurate fluid velocity. Let $\hat{C}^n(x)$ and $\mathbf{U}^n(x)$ denote the approximations of $\hat{c}(x,t)$ and $\mathbf{u}(x,t)$, respectively, at time level $t = t^n$. The directional derivative is then discretized along the "characteristic" mentioned above as

$$\varphi \frac{\partial \hat{c}}{\partial s}(x, t^n) \approx \varphi \frac{\hat{C}^n(x) - \hat{C}^{n-1}(\bar{x}^{n-1})}{\Delta t}, \tag{2.2}$$

where $\bar{x}^{n-1}$ is defined for any x as

$$\bar{x}^{n-1} = x - \frac{\mathbf{U}^n(x)\Delta t}{\varphi}. \tag{2.3}$$

This technique, described by Russell **[6]** for petroleum applications, is a discretization back along the "characteristic" generated by the first-order derivatives from equation (2.1). Since there is no reduction in spatial or temporal dimensions, this is not a full method of characteristics, but a modified method. Equations (1.2) and (1.4) are treated in a similar fashion. Although the advection-dominance in the original equations (1.2)–(1.4) makes them non-selfadjoint, the form with directional derivatives is selfadjoint, and discretization techniques for selfadjoint equations can be utilized. Similarly, this modified method of characteristics can be combined with either finite difference or finite element Galerkin spatial discretizations. These methods have been described and analyzed by Ewing, Yuan, and Li **[5]**, **[7]**.

In applications where the contaminant forms an additional phase, the constitutive equations become more complex. For example, due to nonlinear relative permeability effects, the first or transport term in equation (1.3) becomes a nonlinear, nonconvex function of the saturation $\hat{c}$ of the separate phase. The nonconvexity tends to sharpen fluid interfaces, forming narrow, moving interior layers. This nonlinearity complicates the formulation of the modified method of characteristics considerably.

For purposes of exposition, we will discuss the differences with the modified equation

$$-\nabla \cdot f(\hat{c}, \mathbf{u}) + \nabla \cdot (\mathbf{E}_c \nabla \hat{c}) - q\hat{c} - q_c + R_s = \varphi \frac{\partial \hat{c}}{\partial t}, \tag{2.4}$$

where $f(\hat{c}, \mathbf{u})$ is a nonlinear, nonconvex function of $\hat{c}$. In order to use the same concepts as we did before, we will consider a splitting of the flux term f as follows:

$$f(\hat{c}) = \bar{f}^n(\hat{c}) + b^n(\hat{c})\hat{c}, \tag{2.5}$$

where

$$\bar{f}^n(\hat{c}) = \begin{cases} \dfrac{1}{\hat{c}_a^n} f(\hat{c}_a^n)\hat{c}, & 0 \le \hat{c} \le \hat{c}_a^n, \\ f(\hat{c}), & \hat{c}_a^n \le \hat{c} \le 1. \end{cases} \tag{2.6}$$

$\hat{c}_a^n$ is the top saturation of a shock at $t = t^n$, $0 \le c_a^n \le c_b < 1$; this means that a shock could develop, depending on the initial saturation, for $\hat{c} < \hat{c}_b$, where $\hat{c}_b$ is the saturation that determines the tangent of the line from the origin to the $f(\hat{c})$ curve (see Espedal and Ewing [8]). Further, $b^n(\hat{c}_a^n) = 0$; $b^n(\hat{c}) < 0$, for $0 \le \hat{c} \le \hat{c}_a^n$. This splitting, defined by Espedal and Ewing [8], yields the following operator splitting for equation (2.4):

$$\varphi \frac{\partial \tilde{c}}{\partial t} + \frac{d}{d\tilde{c}} \bar{f}^n(\tilde{c}) \nabla \tilde{c} \equiv \varphi \frac{\partial}{\partial s} \tilde{c} = 0, \tag{2.7}$$

$$-\nabla \cdot (b^n(\hat{c})\hat{c}) + \nabla \cdot (\mathbf{E}_c \nabla \hat{c}) - q\hat{c} - q_c + R_s = \varphi \frac{\partial \hat{c}}{\partial s}, \tag{2.8}$$

where, as in the previous section, we have defined a directional derivative along the characteristic curves.

Integrating equation (2.7) backward along the characteristics from a fixed grid-point x, we obtain, as in Dahle, Espedal, and Ewing [9],

$$\varphi \frac{\partial \tilde{c}}{ds} \approx \varphi \frac{\tilde{C}(x) - \tilde{C}^n(\bar{x}^{n-1})}{\Delta t}, \tag{2.9}$$

where, as in equation (2.3),

$$\bar{x}^{n-1} = x - \frac{1}{\varphi} \frac{d}{d\hat{c}} (\bar{f}(\hat{c}^n)(x)) \Delta t. \tag{2.10}$$

The other modified equations in the system would be treated analogously.

Although equation (2.3) is selfadjoint, equation (2.8) is not, due to the b^n term, which tends to sharpen fronts until it balances the diffusion-dispersion term. Therefore, some type of upwind stabilization is still required for the discretization of equation (2.8). The analogue of upstream weighting for finite differences is Petrov–Galerkin weighting techniques in a finite element setting. The choice of test functions that are optimal or "quasi-optimal" for similar petroleum-related applications has been described by Dahle et al. [9]. Numerical results for a single spatial variable were also presented in that paper.

3. Spatial discretization procedures. In the previous section, we used a modified method of characteristics to treat the advection portions of our equations by following the flow lines. In this section we will describe finite element and mixed finite element methods to discretize the remaining spatial operators.

Let $W^{k,q}(\Omega)$, $1 \le q \le \infty$, be the standard Sobolev spaces with k finite distributional derivatives in $L^q(\Omega)$. Also denote the norm in $W^{k,q}(\Omega)$

by $\|\cdot\|_{k,q}$ and the norm in $H^k(\Omega)$ by $\|\cdot\|_k \equiv \|\cdot\|_{k,2}$. Denote by $H(\text{div};\Omega)$ the set of vector functions $\mathbf{v}$ in $(L^2(\Omega))^2$ such that the components of $\mathbf{v}$ and $\operatorname{div}\mathbf{v}$ are in $L^2(\Omega)$. Let $V = \{\mathbf{v} \in H(\text{div};\Omega)\colon \mathbf{v}\cdot\nu = 0 \text{ on } \partial\Omega\}$ and $W = L^2(\Omega)$. Then a saddle-point weak form of (1.1) is given by the system

$$(3.1)\qquad \begin{aligned} &\text{(a)}\quad (\nabla\cdot\mathbf{u}, w) = (-q + R_s'(\hat{c}), w), && w \in W,\\ &\text{(b)}\quad (a(\hat{c})^{-1}\mathbf{u}, \mathbf{v}) - (\nabla\cdot\mathbf{v}, p) = 0, && \mathbf{v} \in V,\end{aligned}$$

for $0 < t \le \overline{T}$, where $a(\hat{c}) = k(x)/\mu(\hat{c})$.

The standard finite element method for the problem (1.2)–(1.4) is based on the weak form given by

$$(3.2)\quad \left(d_2\frac{\partial T}{\partial t}, z\right) + c_p(\mathbf{u}\cdot\nabla T, z) + (\widetilde{\mathbf{E}}_H\nabla T, \nabla z) = (Q(\mathbf{u}, T, \hat{c}, p), z),$$

$$(3.3)\quad \left(\varphi\frac{\partial \hat{c}}{\partial t}, z\right) + (\mathbf{u}\cdot\nabla\hat{c}, z) + (\mathbf{E}_c\nabla\hat{c}, \nabla z) = (g(\hat{c}), z),$$

$$(3.4)\quad \left(\varphi K_i\frac{\partial c_i}{\partial t}, z\right) + (\mathbf{u}\cdot\nabla c_i, z) + (\mathbf{E}_c\nabla c_i, \nabla z) = (f_i(\hat{c}, c_1, c_2, \dots, c_N), z),$$

$i = 1, 2, \dots, N$, for $z \in H^1(\Omega)$ and $0 < t \le \overline{T}$.

Let $h = (h_c, h_T, h_p)$, where h_c, h_T, and h_p, all positive, are grid spacings for the concentrations, the temperature, and the pressure, respectively. Let $V_h \times W_h$ be a Raviart–Thomas **[10–13]** space of index at least k associated with a quasi-triangulation of Ω such that the elements have diameters bounded by h_p.

As was described in the last section, due to the advection-dominance of equations (3.2)–(3.4), we will use the modified method of characteristics for time-stepping these equations, as described in Section 2. In this method, the first two terms of each of equations (3.2)–(3.4) are combined to form a directional derivative along what would be the characteristics for the equation if the diffusion-dispersion tensors $\widetilde{\mathbf{E}}_H$ and $\mathbf{E}_c$ were zero. The directional derivative is then differenced back along the characteristic directions to perform the time-stepping. Since the directional derivatives and hence the characteristics depend strongly on the Darcy velocity $\mathbf{u}$, mixed finite element methods are used to obtain a more accurate approximation to $\mathbf{u}$ at time level t^n from the system (3.1) and some mixing rule for the fluid velocity. Once this approximation $\mathbf{U}_h^n$ for $\mathbf{u}(\cdot, t^n)$ is obtained, the approximate characteristic direction can be obtained and equations (3.2)–(3.4) can be discretized along this approximate characteristic. This modified method of characteristics has been described in more detail by Douglas and Russell in **[6]**, **[14–16]**. The discretization and sequential solution of equations (3.1)–(3.4) are described below.

Characteristics—Mixed finite element scheme. When $t = t_{n-1}$, if the approximate solution $\{P_h^{n-1}, \mathbf{U}_h^{n-1}, \widehat{C}_h^{n-1}, C_{hi}^{n-1}\ (i = 1, \ldots, N),\ T_h^{n-1}\} \in V_h \times W_h \times M_h \times M_h^N \times R_h$ is known, we find the approximate solution $\{P_h^n, \mathbf{U}_h^n, \widehat{C}_h^n, C_{hi}^n\ (i = 1, \ldots, N),\ T_h^n\} \in V_h \times W_h \times M_h \times M_h^N \times R_h$ at $t = t_n$ by

(3.5)

(a) $(\nabla \cdot \mathbf{U}_h^n, w) = (-q + R_s'(\widehat{C}_h^{n-1}), w), \qquad w \in W_h,$

(b) $(a^{-1}(\widehat{C}_h^{n-1})\mathbf{U}_h^n, \mathbf{v}) - (\nabla \cdot \mathbf{v}, P_h^n) = 0, \qquad \mathbf{v} \in V_h,$

(c) $$\left(\varphi \frac{\widehat{C}_h^n - \overline{\widehat{C}}_h^{n-1}}{\Delta t}, z\right) + (\mathbf{E}_c \nabla \widehat{C}_h^n, \nabla z) = (g(\widehat{C}_h^{n-1}), z), \qquad z \in M_h,$$

(d) $$\left(\varphi K_i \frac{C_{hi}^n - \overline{C}_{hi}^{n-1}}{\Delta t}, z\right) + (\mathbf{E}_c \nabla C_{hi}^n, \nabla z) = (f_i(\widehat{C}_h^n, C_{h1}^{n-1}, \ldots, C_{hN}^{n-1}), z),$$
$$z \in M_h, \quad i = 1, 2, \ldots, N,$$

(e) $$\left(d_2 \frac{T_h^n - \overline{T}_h^{n-1}}{\Delta t}, z\right) + (\widetilde{\mathbf{E}}_H \nabla T_h^n, \nabla z) = (Q(\mathbf{U}_h^n, T_h^{n-1}, \widehat{C}_h^n, P_h^n), z),$$
$$z \in R_h.$$

In addition, the initial approximations $\widehat{C}_h(0)$, $C_{hi}(0)$ $(i = 1, 2, \ldots, N)$, $T_h(0)$, and $P_h(0)$ must be determined. From (3.5a,b) we obtain P_h^n and $\mathbf{U}_h^n$, and from (c) we obtain $\widehat{C}_h^n$, where $\overline{\widehat{C}}_h^{n-1} = \widehat{C}_h^{n-1}(\bar{x}^{n-1})$, and $\bar{x}^{n-1} = x - \mathbf{U}_h^n \Delta t/\varphi$ denotes the projection of x back along the approximate characteristic direction. If $\bar{x}^{n-1}$ is located outside Ω, we can join $\bar{x}^{n-1}$ with $Y \in \partial\Omega$ so that $(\bar{x}^{n-1} - Y)/\|\bar{x}^{n-1} - Y\|$ is the outer-normal direction to the boundary $\partial\Omega$ at Y. Take $x^* \in \Omega$ so that $Y - x^{*n-1} = \bar{x}^{n-1} - Y$. Now, we can define $\widehat{C}_h^{n-1}(\bar{x}^{n-1}) = \widehat{C}_h^{n-1}(x^{*n-1})$. Note that boundary condition $(\partial c/\partial \nu|_{\partial\Omega} = \nabla c \cdot \nu|_{\partial\Omega} = 0)$ makes this choice reasonable. Similarly, by (d) and (e), we obtain C_{hi}^n $(i = 1, 2, \ldots, N)$ and T_h^n, where $\overline{C}_{hi}^{n-1} = C_{hi}^{n-1}(\bar{x}_i^{n-1})$, $\bar{x}_i^{n-1} = x - \mathbf{U}_h^n \Delta t/(\varphi K_i)$, $\overline{T}_h^{n-1} = T_h^{n-1}(\bar{x}_T^{n-1})$, and $\bar{x}_T^{n-1} = x - c_p \mathbf{U}_h^n \Delta t/d_2$.

The decoupling of the equations described above still does not linearize the equations in all situations. However, the use of the method of characteristics linearizes the advection except in the case of the multiphase flow with nonlinear and nonconvex flux function f from equation (2.4). The full linearization of the remaining nonlinearities will be treated in the next section.

We shall suppose that the coefficients and data of (3.1)–(3.4) are locally bounded and locally Lipschitz continuous. Using these assumptions and others (see **[5]**), we can obtain asymptotic error estimates of optimal order. A typical theorem follows.

THEOREM 1 **[5]**. *Let $p(x,t)$, $\hat{c}(x,t)$, $c_i(x,t)$ $(i = 1,2,\dots,N)$, and $T(x,t)$ be the solution of problem (1.1)–(1.12), and let P_h^n, $\widehat{C}_h^n$, C_{hi}^n $(i = 1,2,\dots,N)$, and T_h^n be the approximate solution obtained from equation (3.5). Suppose $k \geq 1$, and the spatial and time discretizations satisfy the relations*

$$\text{(a)}\quad h_p^{-1}\{h_c^{l+1} + h_T^{r+1} + \Delta t_c + (\Delta t_p)^2 + (\Delta t_p^0)^{3/2}\} \to 0,$$

$$\text{(b)}\quad (\Delta t_p + h_p^{k+1})(\log h_c^{-1})^{1/2} \to 0,$$

$$\text{(c)}\quad (\Delta t_p + h_p^{k+1})(\log h_T^{-1})^{1/2} \to 0.$$

Then the following error estimates hold:

(3.6)

$$\text{(a)}\quad \|\mathbf{u}^m - \mathbf{U}_h^m\|_{\widetilde{L}^\infty(0,\overline{T};L^2(\Omega)^2)} + \|p^m - P_h^m\|_{\widetilde{L}^\infty(0,\overline{T};L^2(\Omega))} \leq M\{h_c^{l+1} + h_p^{k+1} + (\Delta t_p)^2 + (\Delta t_p^0)^{3/2} + \Delta t_c\},$$

$$\text{(b)}\quad \|\hat{c}^n - \widehat{C}_h^n\|_{\widetilde{L}^\infty(0,\overline{T};L^2(\Omega))} + h_c\|\hat{c}^n - \widehat{C}_h^n\|_{\widetilde{L}^2(0,\overline{T};H^1(\Omega))} \leq M\{h_c^{l+1} + h_p^{k+1} + (\Delta t_p)^2 + (\Delta t_p^0)^{3/2} + \Delta t_c\},$$

$$\text{(c)}\quad \sum_{i=1}^N \|c_i^n - C_{hi}^n\|_{\widetilde{L}^\infty(0,\overline{T};L^2(\Omega))} + h_c \sum_{i=1}^N \|c_i^n - C_{hi}^n\|_{\widetilde{L}^2(0,\overline{T};H^1(\Omega))} \leq M\{h_c^{l+1} + h_p^{k+1} + (\Delta t_p)^2 + (\Delta t_p^0)^{3/2} + \Delta t_c\},$$

$$\text{(d)}\quad \|T^n - T_h^n\|_{\widetilde{L}^\infty(0,\overline{T};L^2(\Omega))} + h_T\|T^n - T_h^n\|_{\widetilde{L}^2(0,\overline{T};H^1(\Omega))} \leq M\{h_c^{l+1} + h_p^{k+1} + h_T^{r+1} + (\Delta t_p)^2 + (\Delta t_p^0)^{3/2} + \Delta t_c\},$$

where

$$\|\varphi^k\|_{\widetilde{L}^\infty(0,\overline{T};X)} = \sup_{0<t^k\leq\overline{T}} \|\varphi^k\|_X, \quad \|\varphi^k\|_{\widetilde{L}^2(0,\overline{T};X)} = \left(\sum_{0<t^k\leq\overline{T}} \|\varphi^k\|_X^2 \Delta t\right)^{1/2},$$

and the constants M depend on the smoothness of the solutions.

Although asymptotic error estimates such as those presented here are very important for obtaining stability concepts and limitations to optimal order accuracy, they are of limited value in assessing the true accuracy for large problems on coarse grids. In addition, we must perform extensive computations on simpler, model problems, using the same computational techniques proposed for the large systems. In testing the methods on model problems, one should still try to incorporate some of the key nonlinearities that are present in the full applications. Samples of these computations on simpler problems have been presented in **[9]**, **[17–19]**.

4. Linearization and linear solution. The discretization methods presented above are designed to partially linearize and formally decouple the equations via a sequential solution process. In the cases where the nonlinearities in our partial differential equations are strong or the fractional flow terms are nonlinear, this linearization process is not sufficiently accurate for the desired application. In those cases, a better linearization and linear solution process must be considered. The discretization methods could be thought of as the first step in a full Newton linearization of the coupled nonlinear system. In this context, the convergence analysis presented in **[5]** will indicate the convergence properties of the nonlinear system and could be extended to a fully rigorous analysis, including effects of iteration tolerances. Since many simulators would require the accuracy of a full Newton–Raphson type of treatment, in this section we will consider the full nonlinear systems generated for each time level t^n by a standard discretization process that does not include linearization or decoupling. The enormous size of these time-dependent, nonlinear systems motivates iterative solution techniques. We will discuss the different levels in this overall iterative process. The ultimate goal of this research is to determine criteria for picking time-step size and iteration parameters to optimize the solution process.

Standard fully-discrete schemes for equations (1.1)–(1.12) would generate large, nonlinear systems for each time level t^n. The transport present in our physical problem means that the differential operators are non-selfadjoint, and thus the resulting nonlinear and linear systems are nonsymmetric. The nonsymmetries cause serious difficulties for many iterative processes, such as conjugate gradient methods. We will discuss other iterative procedures that have worked well for nonsymmetric systems.

Although the discretization methods produce a different system of nonlinear equations at each time step, these systems are closely related. Since the different systems arise from an evolution process, the approximate solution at time level t^n will be a good initial guess for the nonlinear system produced at time level t^{n+1}. Clearly, the smaller we take the time steps Δt_c and Δt_p, the better these initial guesses are. With good initial guesses, a Newton–Raphson linearization of the nonlinear systems will converge quadratically. We see our first efficiency tradeoff: the smaller the time steps, the faster the Newton convergence, but the larger the total number of nonlinear systems that must be solved to reach a specified time level.

The next point to note is that the construction of the Jacobian and its evaluation for each iteration is a very large and time-consuming process. We are thus led to consideration of inexact or quasi-Newton linearizations that allow cheaper updates at the expense of possibly slower convergence rates. The study of partial or more efficient updates for Newton-like methods is a major area of current research.

In any of the linearization methods mentioned, a new large, nonsymmetric linear system must be solved at each iteration and for each time step. The fill-in that would result from direct solution of each linear system for large, three-dimensional applications would swamp the computational effort. Therefore, iterative procedures for these large nonsymmetric systems must be considered.

Since a nonsymmetric system does not generate an inner product that defines a norm, the notion of conjugacy on which gradient methods are based is not so useful. Techniques such as ORTHOMIN **[20]** and ORTHODIR are useful in this context, but have severe storage requirements. An effective iterative procedure is the generalized conjugate residual (GCR) algorithm described by Elman **[21]**, which is a generalization of the conjugate gradient method for nonsymmetric systems. The main requirement for the GCR to be applicable is that the symmetric part, $\operatorname{symm}\mathbf{J} \equiv (\mathbf{J} + \mathbf{J}^{\mathrm{T}})/2$ (where $\mathbf{J}$ is the Jacobian of our nonlinear system), be positive definite. This requirement is generally satisfied for the matrices arising in these applications.

Theoretically, the algorithm must terminate in d stages, since it will have minimized the error over all of $\mathbf{R}^d$. In practice, however, the finite precision of machine arithmetic causes a loss of strict orthogonality in the search directions, and d must be considered as enormous, so GCR behaves computationally like a true iterative scheme.

Since this linear solution is a part of the Newton iteration, it is natural to consider how to choose the tolerances for the linear, inner iteration and the Newton outer iteration to minimize computational effort. Again, the size of the time-step should also be considered, since an even larger outer time loop is in operation.

Of course, the convergence rates for the iterative processes described are heavily dependent on the conditioning of the matrices involved. In general, the matrices arising from these partial differential equation systems are highly ill-conditioned, with the condition number growing as the reciprocal of the square of the spatial discretization grid size. Therefore, efficient preconditioners are essential for these applications. Many types of incomplete Gaussian elimination (ILU) preconditioners have been used successfully on these problems **[22]**. One very efficient preconditioner, based on a nested factorization procedure, was introduced by Appleyard, Cheshire, and Pollard **[23]** and has been applied successfully **[24]** to these nonsymmetric problems in conjunction with the GCR algorithm described above.

Finally, we note that changing the preconditioner at each iteration of each time-step would be very inefficient. The same preconditioner can be fixed and used for all iterations in the linear and nonlinear solution algorithms. Similarly, due to the evolution nature of our problems, we have shown in **[25]**, **[26]** that the same preconditioner can actually be used very effectively for several time-steps.

ACKNOWLEDGMENTS. This research was supported in part by National Science Foundation Grant No. DMS-8504360, by Office of Naval Research Contract No. N00014-88-K-0370, and by the Institute for Scientific Computation at the University of Wyoming through NSF Grant No. RII-8610680.

REFERENCES

1. M. Reeves and R. M. Cranwell, *User's manual for the Sandia waste-isolation flow and transport model (SWIFT), release* 4.81, Sandia Report Nureg/CR-2334, SAND 81-2516, GF, 1981.

2. K. Aziz and A. Settari, *Petroleum reservoir simulation*, Applied Science Publishers, London and New York, 1979.

3. J. Bear, *Hydraulics of groundwater*, McGraw-Hill, New York, 1979.

4. H. Cooper, *The equation of ground-water flow in fixed and deforming coordinates*, J. Geophys. Res. **71** (1966), 4783–4790.

5. R. E. Ewing, Y. Yuan, and G. Li, *Time-stepping along characteristics for finite element approximations for contamination by nuclear waste-disposal in porous media*, SIAM J. Numer. Anal. (to appear).

6. T. F. Russell, *Finite elements with characteristics for two-component incompressible miscible displacement*, SPE 10500, Proc. Sixth SPE Symp. on Reservoir Simulation, New Orleans, 1982.

7. R. E. Ewing, Y. Yuan, and G. Li, *Finite element methods for contamination by nuclear-waste disposal in porous media*, Numerical Analysis (D. F. Griffiths and G. A. Watson, eds.), Longman Scientific and Technical, Essex, U. K., 1988, pp. 53–66.

8. M. S. Espedal and R. E. Ewing, *Characteristic Petrov–Galerkin subdomain methods for two-phase immiscible flow*, Comp. Meth. in Appl. Mech. and Eng. **64** (1987), 113–135.

9. H. K. Dahle, M. S. Espedal, and R. E. Ewing, *Characteristic Petrov–Galerkin subdomain methods for convection-diffusion problems*, IMA Volume 11, Numerical Simulation in Oil Recovery (M. F. Wheeler, ed.), Springer-Verlag, Berlin, 1988, pp. 77–88.

10. J. Douglas, Jr., R. E. Ewing, and M. F. Wheeler, *The approximation of the pressure by a mixed method in the simulation of miscible displacement*, RAIRO Anal. Numér. **17** (1983), 17–23.

11. ———, *A time-discretization procedure for a mixed finite element approximation of miscible displacement in porous media*, RAIRO Anal. Numér. **17** (1983), 249–265.

12. R. E. Ewing, T. F. Russell, and M. F. Wheeler, *Convergence analysis of an approximation of miscible displacement in porous media by mixed finite elements and a modified method of characteristics*, Comp. Meth. in Appl. Mech. and Eng. **47** (1984), 73–92.

13. P. A. Raviart and J. M. Thomas, *A mixed finite element method for second-order elliptic problems: Mathematical aspects of the finite element method*, Lecture Notes in Mathematics, vol. 606, Springer-Verlag, Berlin and New York, 1977.

14. J. Douglas, Jr., *Simulation of miscible displacement in porous media by a modified method of characteristics procedure* (Numerical Analysis, Proc., Dundee, 1981), Lecture Notes in Mathematics, vol. 912, Springer-Verlag, Berlin and New York, 1982.

15. J. Douglas, Jr. and T. F. Russell, *Numerical methods for convection dominated diffusion problems based on combining the method of characteristics with finite element or finite difference procedures*, SIAM J. Numer. Anal. **19** (1982), 871–885.

16. T. F. Russell, *An incomplete iterated characteristic finite element method for a miscible displacement problem*, Ph.D. thesis, University of Chicago, 1980.

17. R. E. Ewing and M. F. Wheeler, *Computational aspects of mixed finite element methods*, Numerical Methods for Scientific Computing, R. S. Stepleman (ed.), North-Holland, New York, 1983, pp. 163–172.

18. R. E. Ewing, T. F. Russell, and M. F. Wheeler, *Simulation of miscible displacement using mixed methods and a modified method of characteristics*, Proc. Seventh SPE Symp. on Reservoir Simulation, SPE No. 12241, San Francisco, 1983, pp. 71–82.

19. H. Dahle, M. S. Espedal, R. E. Ewing, and O. Sævareid, *Characteristic adaptive subdomain methods for reservoir flow problems*, Numerical Solutions of Partial Differential Equations (to appear).

20. P. K. W. Vinsome, *Orthomin, an iterative method for solving sparse sets of simultaneous linear equations*, SPE 5729, Proc. Fourth SPE Symp. on Numerical Simulation of Reservoir Performance, Los Angeles, California, 1976.

21. H. C. Elman, *Preconditioned conjugate-gradient methods for nonsymmetric systems of linear equations*, Research Report 203, Yale University Department of Computer Science, New Haven, Connecticut, 1981.

22. J. A. Meijerink, *Iterative methods for the solution of linear equations based on incomplete block factorizations of the matrix*, SPE 12262, Proc. Seventh SPE Symp. on Reservoir Simulation, San Francisco, 1983.

23. J. R. Appleyard, I. M. Cheshire, and R. K. Pollard, *Special techniques for fully-implicit simulators*, presented at the European Symp. on Enhanced Oil Recovery, Bournemouth, U. K., 1981.

24. U. Obeysekare, M. Allen, R. E. Ewing, and J. H. George, *Application of conjugate gradient-like methods to a hyperbolic problem in porous media flow*, Int. J. for Numerical Methods in Fluids **7** (1987), 551–566.

25. J. Douglas, Jr., T. Dupont, and R. E. Ewing, *Incomplete iteration for time-stepping a Galerkin method for a quasilinear parabolic problem*, SIAM J. Numer. Anal. **16** (1979), 503–522.

26. R. E. Ewing, *Time-stepping Galerkin methods for nonlinear Sobolev partial differential equations*, SIAM J. Numer. Anal. **15** (1978), 1125–1150.

Institute for Scientific Computation, University of Wyoming, Laramie, Wyoming 82071

Lectures in Applied Mathematics
Volume **26** (1990)

Low Storage Methods for Unconstrained Optimization

R. FLETCHER

Abstract. Current low storage methods use $2m$ vectors to store second-order information. A new approach is described in which only $m+1$ vectors are used to store an equivalent amount of information. Numerical results are described when this information structure is used in a method which requires periodic restarts.

The problem of using this structure without periodic restarts, and hence of deleting second-order information, is also considered. Canonical **QL** factors are defined which provide an equivalent Lanczos sequence of conjugate gradient vectors. It is then shown how a Lanczos sequence containing fewer vectors can be determined.

An alternative way of reusing information is also considered. It is argued that the merit of the steepest descent method of Barzilai and Borwein is due to the use of step lengths based on Rayleigh quotient estimates of eigenvalues. Eigenvalue estimates are also available in a low storage method and can be used to define a sequence of step lengths. Some features of this idea are discussed.

1. Introduction. This paper considers the calculation of a local minimizer, $\mathbf{x}^*$ say, for the problem

$$\text{minimize } f(\mathbf{x}), \qquad \mathbf{x} \in \mathbf{R}^n, \tag{1.1}$$

where $f \in \mathbf{C}^2$. It is assumed that expressions for the elements of the gradient vector $\mathbf{g}(x) = \nabla f(\mathbf{x})$ are available, but that the Hessian matrix $\mathbf{G}(\mathbf{x}) = \nabla^2 f(\mathbf{x})$ is not available. It is also assumed that n is large and the problem contains no structure in the Hessian (e.g., sparsity or partial separability) which could usefully be used. Standard methods for solving this problem (e.g., **[4]**) include the *conjugate gradient* (*CG*) *method*, which requires $4n$ locations of computer storage to implement, and the *BFGS method,* which requires $n^2/2 + O(n)$ locations. If this latter amount of storage were available, then the BFGS method would be the usual choice of method. However, this paper considers the case that less than $n^2/2$ but more than $4n$ locations are available, say $10n$ or $16n$ locations. Low storage methods attempt to use this additional storage to improve on the performance of the CG method.

All methods fall into the class of *line search descent methods*. On iteration k a current iterate $\mathbf{x}^{(k)}$ is available and the corresponding gradient vector is $\mathbf{g}^{(k)}$ ($\mathbf{g}^{(k)} = \nabla g(\mathbf{x}^{(k)})$, etc.). A search direction $\mathbf{s}^{(k)}$ ($\mathbf{s}^{(k)\mathrm{T}}\mathbf{g}^{(k)} < 0$) is calculated and a sequence of iterates $\{\mathbf{x}^{(k)}\}$ is generated from given $\mathbf{x}^{(1)}$ by

$$\mathbf{x}^{(k+1)} = \mathbf{x}^{(k)} + \alpha^{(k)}\mathbf{s}^{(k)}. \tag{1.2}$$

In an *exact line search,* $\alpha^{(k)}$ is determined by

$$\alpha^{(k)} = \operatorname{argmin}_\alpha f(\mathbf{x}^{(k)} + \alpha\mathbf{x}^{(k)}), \tag{1.3}$$

although in practice an approximate solution of (1.3) would be accepted (e.g., [**4**]).

In the conjugate gradient method, $\mathbf{s}^{(k)}$ is defined by

$$\begin{aligned} \mathbf{s}^{(1)} &= -\mathbf{g}^{(1)}, \\ \mathbf{s}^{(k)} &= -\mathbf{g}^{(k)} + \beta^{(k)}\mathbf{s}^{(k-1)}, \qquad k > 1, \end{aligned} \tag{1.4}$$

where $\beta^{(k)}$ is given either by

$$\beta^{(k)} = \mathbf{g}^{(k)\mathrm{T}}\mathbf{g}^{(k)}/\mathbf{g}^{(k-1)\mathrm{T}}\mathbf{g}^{(k-1)} \tag{1.5}$$

in the *Fletcher–Reeves method,* or by

$$\beta^{(k)} = \mathbf{g}^{(k)\mathrm{T}}(\mathbf{g}^{(k)} - \mathbf{g}^{(k-1)}/\mathbf{g}^{(k-1)\mathrm{T}}\mathbf{g}^{(k-1)} \tag{1.6}$$

in the *Polak–Ribiere method.* The latter is usually preferred in practice. In the BFGS method, $\mathbf{s}^{(k)}$ is defined by

$$\mathbf{s}^{(k)} = -\mathbf{H}^{(k)}\mathbf{g}^{(k)}, \tag{1.7}$$

where $\mathbf{H}^{(k)} \in \mathbf{R}^{n\times n}$ is a symmetric positive definite matrix. $\mathbf{H}^{(1)}$ is given, and for $k > 1$, $\mathbf{H}^{(k)}$ is defined by the updating formula

$$\mathbf{H}^{(k)} = \mathrm{bfgs}(\mathbf{H}^{(k-1)}, \boldsymbol{\delta}^{(k-1)}, \boldsymbol{\gamma}^{(k-1)}), \tag{1.8}$$

where

$$\boldsymbol{\delta}^{(k-1)} = \mathbf{x}^{(k)} - \mathbf{x}^{(k-1)}, \qquad \boldsymbol{\gamma}^{(k-1)} = \mathbf{g}^{(k)} - \mathbf{g}^{(k-1)}, \tag{1.9}$$

and

$$\mathrm{bfgs}(\mathbf{H}, \boldsymbol{\delta}, \boldsymbol{\gamma}) = \left(\mathbf{I} - \frac{\boldsymbol{\delta}\boldsymbol{\gamma}^{\mathrm{T}}}{\boldsymbol{\delta}^{\mathrm{T}}\boldsymbol{\gamma}}\right)\mathbf{H}\left(\mathbf{I} - \frac{\boldsymbol{\gamma}\boldsymbol{\delta}^{\mathrm{T}}}{\boldsymbol{\delta}^{\mathrm{T}}\boldsymbol{\gamma}}\right) + \frac{\boldsymbol{\delta}\boldsymbol{\delta}^{\mathrm{T}}}{\boldsymbol{\delta}^{\mathrm{T}}\boldsymbol{\gamma}}. \tag{1.10}$$

Another related method with aspects of both the above is the *preconditioned conjugate gradient* (*PCG*) *method*, defined by

$$\begin{aligned} \mathbf{s}^{(1)} &= -\mathbf{H}\mathbf{g}^{(1)}, \\ \mathbf{s}^{(k)} &= -\mathbf{H}\mathbf{g}^{(k)} + \beta^{(k)}\mathbf{s}^{(k-1)}, \qquad k > 1, \end{aligned} \tag{1.11}$$

where $\beta^{(k)}$ is given either by

$$\beta^{(k)} = \mathbf{g}^{(k)\mathrm{T}}\mathbf{H}\mathbf{g}^{(k)}/\mathbf{g}^{(k-1)\mathrm{T}}\mathbf{H}\mathbf{g}^{(k-1)} \tag{1.12}$$

or by a similar formula derived from (1.6). Here $\mathbf{H}$ is a *fixed* symmetric positive definite matrix, and the method is equivalent to the CG method applied in a transformed coordinate system ($\mathbf{H} = \mathbf{I}$ gives the CG method). For quadratic functions, the BFGS method with $\mathbf{H}^{(1)} = \mathbf{H}$ is equivalent to the PCG method in that they generate the same sequences $\{\mathbf{s}^{(k)}\}$ and $\{\mathbf{x}^{(k)}\}$.

Various low storage methods have been suggested, but there are two particular methods of note. In the method of Nocedal **[8]**, the m most recent difference pairs $\boldsymbol{\delta}^{(i)}, \boldsymbol{\gamma}^{(i)}$, $i = k-1, k-2, \dots, k-m$ are stored. The search direction is calculated as in (1.7) but $\mathbf{H}^{(k)}$ is defined as the outcome of updating a given diagonal matrix, $\mathbf{D}$ say, by using the BFGS formula and the above difference pairs. That is (for $k > m$),

$$\begin{aligned} \mathbf{H}^{(k)} = \mathrm{bfgs}(\cdots(\mathrm{bfgs}(\mathrm{bfgs}(\mathbf{D}, \boldsymbol{\delta}^{(k-m)}, \boldsymbol{\gamma}^{(k-m)}), \boldsymbol{\delta}^{(k-m+1)}, \boldsymbol{\gamma}^{(k-m+1)}), \\ \dots, \boldsymbol{\delta}^{(k-1)}, \boldsymbol{\gamma}^{(k-1)}). \end{aligned} \tag{1.13}$$

It is shown in **[8]** that the product $\mathbf{H}^{(k)}\mathbf{g}^{(k)}$ can be computed in $4nm + O(n)$ flops, without explicitly calculating $\mathbf{H}^{(k)}$. Usually $\mathbf{D}$ is chosen as a multiple of the unit matrix. After each iteration (once $k > m$) the oldest difference pair is deleted from storage and a new difference pair is introduced.

The method of Buckley and LeNir **[2]** is a two-phase method, starting with the *quasi-Newton* (*QN*) *phase*, in which for $k = 1, 2, \dots, m$, $\mathbf{H}^{(k)}$ is built up as in Nocedal's method. The *CG phase* follows, in which for $k = m+1, m+2, \dots$, $H^{(m)}$ is used as a fixed preconditioner in the PCG method. The CG phase is terminated when it is judged that a sufficient deviation from quadratic behaviour is detected; in particular, use of the test

$$|\mathbf{g}^{(k-1)\mathrm{T}}\mathbf{g}^{(k)}| \geq 0.2\,\mathbf{g}^{(k-1)\mathrm{T}}\mathbf{g}^{(k-1)} \tag{1.14}$$

is suggested. The QN phase of the algorithm is then reentered, but with $\mathbf{H}^{(1)} = \mathrm{bfgs}(\mathbf{D}, \boldsymbol{\delta}^{(k-1)}, \boldsymbol{\gamma}^{(k-1)})$. Both methods are claimed to be effective, at least insofar as increasing the value of m gives an increasing improvement over the conjugate gradient method. Two recent papers ([6],[7]) claim a preference for Nocedal's method and this may be an indication that restarting the Buckley and LeNir method by deleting most of the current second-order information is a source of inefficiency. Moreover the BFGS method is still superior if sufficient storage is available. Other work on low storage methods is also referenced in [7].

In Section 2 of this paper a new approach to low storage minimization is described, related to the BFGS method, in which the information content of m difference pairs is stored in only $m+1$ vectors, rather than $2m$. Some numerical results are given when the information structure is used in the manner of Buckley and LeNir. The problem of avoiding periodic restarts is considered in Section 3. The new approach is shown to be related to the Lanczos tridiagonalization of a symmetric positive definite matrix, and certain canonical QL factors are seen to be significant. It is also shown how this information structure can be compressed so that meaningful QL factors of lower dimension can be determined. Finally in Section 4, an alternative way of reusing information is considered. The recent steepest descent method of Barzilai and Borwein is described, and it is argued that it is best regarded as using step lengths based on Rayleigh quotient estimates of eigenvalues. It is pointed out that good eigenvalue estimates are also available from the low storage method of Section 2, which can be used to define a sequence of step lengths. Some possibilities for using this idea are discussed.

Before proceeding with this material, two general points are made. The effectiveness of an iterative method is correlated with the extent to which it achieves rapid local convergence. For example, the BFGS method is known to exhibit Q-superlinear convergence under mild conditions. The Dennis–Moré characterization of Q-superlinear convergence is that the correction must be asymptotic to that of Newton's method. This usually requires asymptotically exact estimates of the entire Hessian matrix to be available, and hence it may be unrealistic to expect any low storage method to supersede the BFGS method as an effective general-purpose routine. On the other hand, there do exist problems which possess some implicit structure, such as near-multiple eigenvalues in the Hessian or near-zero eigenvector components in $g^{(1)}$, in which case it is not necessary to estimate the entire Hessian matrix. For such problems we can expect rapid convergence from a low storage method.

Another point refers to the common practice of using double precision storage when coding minimization methods. If some low storage method requires say $2m$ double precision vectors, then it should not be forgotten that this is the equivalent of $4m$ single precision vectors, and it might be more effective to use the storage in this way. It is also my experience that the BFGS method performs satisfactorily in single precision, and it is important to recognize this when comparing storage requirements.

2. A new approach. In this section a new scheme is described in which the information content of m difference pairs is stored in only $m+1$ vectors, rather than in $2m$. It is hoped that this will enable an improved rate of convergence to be obtained from a given amount of storage. The idea devolves from the BFGS method, and in particular to a form in which Choleski factors

$$\mathbf{H}^{-1} = \mathbf{B} = \mathbf{L}\mathbf{L}^{\mathrm{T}} \tag{2.1}$$

are updated (deleting superscript (k)). The BFGS formula (1.10) can be expressed in terms of updating $\mathbf{B}$ as

$$\mathbf{B}^{(k+1)} = \mathbf{B} - \frac{\mathbf{B}\boldsymbol{\delta}\boldsymbol{\delta}^{\mathrm{T}}\mathbf{B}}{\boldsymbol{\delta}^{\mathrm{T}}\mathbf{B}\boldsymbol{\delta}} + \frac{\gamma\gamma^{\mathrm{T}}}{\boldsymbol{\delta}^{\mathrm{T}}\gamma}, \tag{2.2}$$

and this can be written as

$$\mathbf{B}^{(k+1)} = \mathbf{L}\left(\mathbf{I} - \frac{\mathbf{L}^{\mathrm{T}}\boldsymbol{\delta}\boldsymbol{\delta}^{\mathrm{T}}\mathbf{L}}{\boldsymbol{\delta}^{\mathrm{T}}\mathbf{L}\mathbf{L}^{\mathrm{T}}\boldsymbol{\delta}}\right)\mathbf{L}^{\mathrm{T}} + \frac{\gamma\gamma^{\mathrm{T}}}{\boldsymbol{\delta}^{\mathrm{T}}\gamma}. \tag{2.3}$$

Note that the vector $\mathbf{L}^{\mathrm{T}}\boldsymbol{\delta} = -\alpha\mathbf{L}^{-1}\mathbf{g}$ is a by-product of the computation of $\mathbf{s}^{(k)}$, since we assume that $\mathbf{s}^{(k)}$ is computed using (1.7). To update $\mathbf{L}$, an orthogonal matrix $\mathbf{P}$ is determined such that

$$\mathbf{P}^{T}\mathbf{L}^{T}\boldsymbol{\delta} = \pm\mathbf{e}_1\|\mathbf{L}^{\mathrm{T}}\boldsymbol{\delta}\| \tag{2.4}$$

($\mathbf{e}_1$ denotes the first coordinate vector). $\mathbf{P}^{\mathrm{T}}$ is calculated by applying a backward sequence of plane rotation operations to $\mathbf{L}^{\mathrm{T}}\boldsymbol{\delta}$ in positions $(n, n-1), (n-1, n-2), \ldots, (2, 1)$ in turn. This gives

$$\mathbf{B}^{(k+1)} = \mathbf{L}\mathbf{P}(\mathbf{I} - \mathbf{e}_1\mathbf{e}_1^{\mathrm{T}})\mathbf{P}^{\mathrm{T}}\mathbf{L}^{\mathrm{T}} + \frac{\gamma\gamma^{\mathrm{T}}}{\boldsymbol{\delta}^{\mathrm{T}}\gamma}. \tag{2.5}$$

Calculating $\mathbf{LP}$ by applying the plane rotations in the forward direction gives rise to a lower Hessenberg matrix, and the $\mathbf{I} - \mathbf{e}_1\mathbf{e}_1^{\mathrm{T}}$ operation deletes column 1 of $\mathbf{LP}$, which can be overwritten by the vector $\gamma/\sqrt{\boldsymbol{\delta}^{\mathrm{T}}\gamma}$. Finally the resulting lower Hessenberg matrix can be restored to lower triangular form by another forward sequence of plane rotations, giving $\mathbf{L}^{(k+1)}$.

Omitting the last step in the above process gives another form of the BFGS method in which the matrix L ultimately becomes full. In fact we can replace $\mathbf{L}$ by any matrix $\mathbf{LX}$ where $\mathbf{X}$ is orthogonal, and yet continue to use the above updating scheme. The following is an equivalent form of the BFGS method ($\mathbf{B}^{(1)} = \mathbf{I}$) using factors

$$\mathbf{B}^{(k)} = \left[\mathbf{Z} \middle| \mathbf{Z}^{\perp}\right] \left[\frac{\mathbf{Z}^{\mathrm{T}}}{\mathbf{Z}^{\perp\mathrm{T}}}\right] \tag{2.6}$$

where $\mathbf{Z} \in \mathbf{R}^{n\times k}$, $\mathbf{Z}^{\perp} \in \mathbf{R}^{n\times(n-k)}$, $k < n$, and where $\mathbf{Z}^{\perp\mathrm{T}}\mathbf{Z}^{\perp} = \mathbf{I}$ and $\mathbf{Z}^{\mathrm{T}}\mathbf{Z}^{\perp} = \mathbf{0}$. The columns of $\mathbf{Z}$ span $\mathbf{g}^{(1)}, \mathbf{g}^{(2)}, \dots, \mathbf{g}^{(k)}$ and contain the nontrivial information in $\mathbf{B}^{(k)}$, while columns of $\mathbf{Z}^{\perp}$ contain only unit matrix-like information. These properties enable the computation of $\mathbf{s} = -\mathbf{B}^{-1}\mathbf{g}$ to be simplified (see (i) below). Initially $\mathbf{Z}^{(1)} = \mathbf{g}^{(1)}/\|\mathbf{g}^{(1)}\|$ and for $k = 1, 2, \dots$ the following steps are repeated:

$$\begin{array}{ll} \text{(i)} & s := -(\mathbf{Z}\mathbf{Z}^{\mathrm{T}})^{+}\mathbf{g} \\ \text{(ii)} & \text{line search} \quad \rightarrow \quad \mathbf{x}^{(k+1)}, \mathbf{g}^{(k+1)} \\ \text{(iii)} & \mathbf{g}^{\perp} := \mathbf{Z}^{\perp}\mathbf{Z}^{\perp\mathrm{T}}\mathbf{g}^{(k+1)} \\ \text{(iv)} & \text{if } \mathbf{g}^{\perp} \neq \mathbf{0} \text{ extend } \left[\mathbf{Z} \middle| \mathbf{g}^{\perp}/\|\mathbf{g}^{\perp}\|\right] =: \widehat{\mathbf{Z}} \quad \text{else} \quad \widehat{\mathbf{Z}} := \mathbf{Z} \\ \text{(v)} & \text{update } \widehat{\mathbf{Z}} \text{ to give } \mathbf{Z}^{(k+1)} \text{ as for } \mathbf{L} \text{ in the BFGS method.} \end{array} \tag{2.7}$$

For $k = n, n+1, \dots$, $\mathbf{g}^{\perp}$ is usually zero and need not be calculated in step (iii). The vectors $\mathbf{s}$ and $\mathbf{g}^{\perp}$ are readily computed if, for example, QR factors of $\mathbf{Z}$ are available. (Alternatively it is possible to use a conjugate gradient iteration based on the matrix $\mathbf{A} = \mathbf{Z}\mathbf{Z}^{\mathrm{T}}$.) The motivation behind this approach is that the matrix $\mathbf{Z}^{\perp}$ *is not stored*. In step (v) the update could be carried out on the entire matrix $\left[\mathbf{Z} \middle| \mathbf{Z}^{\perp}\right]$. However, we may regard $\mathbf{g}^{\perp}/\|\mathbf{g}^{\perp}\|$ as being the first column of $\mathbf{Z}^{\perp}$, in which case it is possible to neglect the remaining columns and only update the matrix $\widehat{\mathbf{Z}}$ in step (v). This motivates the bordering operation in step (iv).

It is now possible to describe a BFGS-like low storage method based on this information structure. Let storage for m vectors in $\mathbf{Z}$ be available. The method in (2.7) can be followed for $m-1$ iterations, after which it is possible to carry out PCG steps using $\mathbf{H}^{(k)} = (\mathbf{B}^{(m)})^{-1}$ as the preconditioner for $k > m$. These steps are continued as in the Buckley–LeNir method until some test like (1.14) is satisfied. Then $\mathbf{Z}$ is reset to $\mathbf{g}^{(k)}/\|\mathbf{g}^{(k)}\|$ and the whole process is restarted. For a quadratic function, the method is equivalent to the BFGS and CG methods. For general functions, the method is equivalent to the BFGS method for the first $m-1$ iterations. In theory, for a problem with some symmetry, and if m is sufficiently large, it is possible to get $\mathbf{g}^{\perp} = \mathbf{0}$ occurring on successive iterations, and hence complete equivalence to the BFGS method is obtained. (This occurs with the Chebyquad test problem when $m > n/2$.) However, this neglects the effects of round-off, and to see the effect in practice some tolerance on zero must be set. Finally a disadvantage of the approach is that restarting the iteration discards all the information in $\mathbf{Z}$ that has been built up. We return to this point in Section 3.

Some numerical tests have been carried out with the low storage method outlined above. These tests are on an $n = 10$ realization of the trigonometric problem of Fletcher and Powell [**5**]. Of course this is only a small problem whereas ultimately these methods must be proved on large problems. On the other hand, if the method does not show an improvement over the CG method on difficult small problems, then it can scarcely be expected to do so on difficult large problems. Even small-dimensional trigonometric function test problems are difficult for CG methods, as the results in [**4**] show. This can possibly be ascribed to a lack of any symmetry and the fact that $\nabla^2 f$ is much larger than $\mathbf{I}$. The results are obtained in single precision ($\varepsilon = 2^{-27}$) on a SUN 3/50 workstation. A tolerance of $f^{(k)} - f^* \leq 10^{-7}$ is requested but any accuracy of better than 10^{-5} is regarded as being adequate.

For this test problem the best CG method (Polak–Ribiere) takes at least 102 line searches, and the Fletcher–Reeves method is generally worse. The BFGS method takes 24 line searches. The number of line searches taken by the new low storage method is indicated in the following table (σ denotes the reduction in the modulus of the direction derivative that is required in the line search, PR denotes the Polak–Ribiere method, and (r) indicates the use of restarts).

Table 1. Relative performance of a low storage method

m	2	3	4	5	6	7	8	9	10	PR	PR(r)
$\sigma = 0.1$	108	128	(63)	89	(59)	60	65	52	54	154	157
$\sigma = 0.05$	98	97	101	79	91	82	51	54	49	172	102

$(\cdot)$ = *premature termination*

Allowing for some random behaviour in the results, these figures nonetheless suggest an improvement in performance over the CG methods, which increases as m is increased. However, the rate of improvement as m increases is rather slow, particularly for small values of m, which is disappointing. In comparison with the BFGS method the break-even point for storage occurs when $m = \frac{1}{2}n$, at which point the low storage method is significantly inferior. Even when $m = n$, the low storage method does not perform as well as BFGS, which can be ascribed to the inefficiency of the Buckley–LeNir type of approach, in which all second-order information is periodically rejected. In the next section of this paper a mechanism is described whereby the data structure can be used without periodic restarts.

Nocedal's method has also been applied to this problem and the results are shown in Table 2, in which m denotes the number of difference *pairs* that are stored.

Table 2. Performance of Nocedal's method

m	1	2	3	4	5	6	7	8	9	10
$\sigma = 0.1$	88	120	84	85	64	49	39	35	34	32
$\sigma = 0.05$	92	136	74	72	58	44	43	35	32	31

Allowing for the fact that $2m$ vectors are required by Nocedal's method, it can be seen that there is little to choose between the methods on this test problem. We hope to conduct a more extensive comparison and follow up some of the ideas in a later paper.

3. Deleting information from Z. In this section the problem is considered of how to delete information from the $\mathbf{Z}$ matrix that is built up by algorithm (2.7). It is important to realize that it is unsatisfactory merely to delete a column of $\mathbf{Z}$ at any stage. This is because $\mathbf{Z}$ contains information about differences in both $\mathbf{x}$ and $\mathbf{g}$, and deleting a column of $\mathbf{Z}$ leaves a matrix with no meaningful information. This explanation is clarified in what follows by introducing *canonical factors* for $\mathbf{Z}$ which relate the method of (2.7) to the CG method. A mechanism is then described whereby information may be deleted from these factors in a meaningful way.

It has been pointed out in Section 2 that $\mathbf{Z}$ may be replaced by a matrix $\mathbf{ZP}$, where $\mathbf{P}$ is orthogonal, without changing the Hessian approximation $\mathbf{B}$. Allowing such a transformation, the canonical factors are

$$\mathbf{Z} = \mathbf{QL} \tag{3.1}$$

where $\mathbf{Q} \in \mathbf{R}^{n\times k}$, $\mathbf{Q}^{\mathrm{T}}\mathbf{Q} = \mathbf{I}$ with $\mathbf{q}_k = \mathbf{g}^{(k)}/\|\mathbf{g}^{(k)}\|$, and where

$$\mathbf{L} = \begin{bmatrix} a_1 & & & & \\ b_1 & a_1 & & & \\ & b_2 & \ddots & & \\ & & \ddots & a_{k-1} & \\ & & & b_{k-1} & a_k \end{bmatrix} \tag{3.2}$$

is lower bidiagonal with $b_i \leq 0$. (Note $\mathbf{q}_k$ denotes column k of Q.)

It is possible to give a method for calculating canonical factors when $\mathbf{Z}$ is built up by algorithm (2.7). Clearly the initial matrix $\mathbf{Z}^{(1)} = [\mathbf{g}^{(1)}/\|\mathbf{g}^{(1)}\|]$ has canonical factors $\mathbf{Q} = \mathbf{Z}$ and $\mathbf{L} = [\,1\,]$. Now consider updating $\mathbf{Z}$ after the kth line search. A consequence of the property $\mathbf{q}_k = \mathbf{g}^{(k)}/\|\mathbf{g}^{(k)}\|$ is that $\widehat{\mathbf{Z}}^{\mathrm{T}}\boldsymbol{\delta}$ is a multiple of $\mathbf{e}_k$. Thus step (v) of (2.7) is simply effected by overwriting column k of $\widehat{\mathbf{Z}}$ by the vector $\gamma/\sqrt{\boldsymbol{\delta}^{\mathrm{T}}\gamma}$. At this stage we can typically express $\widehat{\mathbf{Z}}$ in the form

$$\widehat{Z} = \begin{bmatrix} \mathbf{Q} & \Big| & \mathbf{g}^{(k+1)} \end{bmatrix} \begin{bmatrix} \times & & & & \circ \\ \times & \times & & & \circ \\ & \times & \times & & \circ \\ & & \times & * & \circ \\ & & & * & \circ \end{bmatrix} \tag{3.3}$$

where the elements marked $\times$ are what remains after deleting column k of $\mathbf{L}$, those marked $*$ are the contribution of $\gamma/\sqrt{\boldsymbol{\delta}^{\mathrm{T}}\gamma}$, and those marked $\circ$ are known coefficients arising in the computation of $\mathbf{g}^{\perp}$.

It now remains to calculate new canonical factors

$$\widehat{\mathbf{Z}}\mathbf{P} = \mathbf{QL}. \tag{3.4}$$

First a vector v such that

$$\widehat{\mathbf{Z}}\mathbf{v} = \mathbf{g}^{(k+1)} \tag{3.5}$$

is required. This is readily calculated to within a multiplicative factor by a forward substitution process derived from $[\cdot]\mathbf{v} = \mathbf{e}_{k+1}$ where $[\cdot]$ denotes the rightmost matrix in (3.3). Now a forward sequence of plane rotations is applied to reduce $\mathbf{v}$ to a multiple of $\mathbf{e}_{k+1}$. Denoting this operation by $\mathbf{P}_{k+1}^{\mathrm{T}}\mathbf{v} = \pm\mathbf{e}_{k+1}\|\mathbf{v}\|$, we can apply the same sequence of operations to $\widehat{\mathbf{Z}}$, denoted by $\mathbf{Z} := \widehat{\mathbf{Z}}\mathbf{P}_{k+1}$. The effect of this calculation is that column $k+1$ of this new $\mathbf{Z}$ matrix has been set to $\mathbf{g}^{(k+1)}/\|\mathbf{g}^{(k+1)}\|$. Next there follows what is essentially the standard calculation of $\mathbf{QL}$ factors by applying the modified Gram–Schmidt process to the columns of $\mathbf{Z}$ in backwards order, replacing them by columns of $\mathbf{Q}$. Initially $\mathbf{q}_{k+1} = \mathbf{z}_{k+1}$ and on stage i of this process $(i = k, \dots, 2, 1)$, the components of $\mathbf{q}_{i+1}$ are removed from the columns $\mathbf{z}_j$ for $j = 1, 2, \dots, i$. This involves calculating the scalars $u_j = \mathbf{q}_{i+1}^{\mathrm{T}}\mathbf{z}_j, j \le i$, and these would normally be the subdiagonal elements in row $i+1$ of the resulting $\mathbf{L}$ matrix. However we are free on stage i to apply a sequence of forward rotations, $\mathbf{P}_i^{\mathrm{T}}\mathbf{u} = \pm\mathbf{e}_i\|\mathbf{u}\|$ ($\mathbf{u}$ has $u_j = 0,\ j > i$), and $\mathbf{Z} := \mathbf{Z}\mathbf{P}_i$. It will then follow that the matrix $\mathbf{L}$ which results is lower bidiagonal. (The diagonal elements of L are determined by the normalizing constants in the modified Gram–Schmidt process.) It is also possible, by careful choice of sign in the plane rotation matrices, to ensure that $b_i \le 0$ in (3.2). The matrix $\mathbf{P}$ in (3.4) is defined by $\mathbf{P} = \mathbf{P}_{k+1}\mathbf{P}_k \cdots \mathbf{P}_2$, but we do not need to store either $\mathbf{P}$ or the individual matrices $\mathbf{P}_i$. The whole updating calculation takes $O(nm^2)$ flops, which is worse asymptotically than the $O(nm)$ flops required by Nocedal's method. However, for small m the difference will not be significant.

When this algorithm is applied to a quadratic function, the columns of $\mathbf{Q}$ on iteration k satisfy

$$\mathbf{q}_i = \mathbf{g}^{(i)}/\|\mathbf{g}^{(i)}\|, \quad i = 1, 2, \dots, k, \tag{3.6}$$

and the elements of the $k \times k$ tridiagonal matrix $\mathbf{L}\mathbf{L}^{\mathrm{T}}$ are submatrices of the conjugate gradient (Lanczos) tridiagonalization of the Hessian (e.g., [3]), except for the $(\mathbf{L}\mathbf{L}^{\mathrm{T}})_{kk}$ element. Also the plane rotation operations are trivially given by $\mathbf{P} = \mathbf{I}$. Consequently the elements of $\mathbf{L}$ are related to the parameters $\alpha^{(i)}$ and $\beta^{(i)}$ in the conjugate gradient method by

$$a_i = \alpha^{(i)-\frac{1}{2}} \quad \text{and} \quad b_i = -\sqrt{\beta^{(i+1)}/\alpha^{(i)}}, \quad \text{for } i < k. \tag{3.7}$$

Herein therefore lies the significance of the canonical factors in that, for the BFGS method applied to a nonquadratic function, they provide an equivalent CG sequence, albeit one which changes from iteration to iteration. Thus the information about differences that is contained in the BFGS matrix is equivalent to the differences in x and g that would be defined by the equivalent CG sequence. It can now be seen that to delete information by making an arbitrary change to $\mathbf{Z}$ would give canonical factors (if they exist), and hence implied difference vectors $\boldsymbol{\delta}^{(i)}$ and $\boldsymbol{\gamma}^{(i)}$, which no longer satisfy $\nabla^2 f \boldsymbol{\delta}^{(i)} \approx \boldsymbol{\gamma}^{(i)}$.

An important aspect of the canonical factors is the property that $\mathbf{q}_k = \mathbf{g}^{(k)}/\|\mathbf{g}^{(k)}\|$. The problem of how to partially remove information from $\mathbf{Z} = \mathbf{QL}$ in a meaningful way can be solved by answering the question: Does there exist a (shorter) CG sequence for the *same* Hessian for which $\mathbf{g}^{(k)}$ is the $(k-1)$-th gradient vector and not the kth? In the rest of this section it is shown how this question can be answered.

Assume then that $k-1$ iterations of algorithm (2.7) are applied to a quadratic function with positive definite Hessian $\mathbf{G}$, let $\mathbf{g}^{(k)} \neq \mathbf{0}$ be the current gradient vector, and let $\mathbf{Z} = \mathbf{QL}$ be the canonical factors. Denote $\mathbf{T} = \mathbf{LL}^{\mathrm{T}}$ (tridiagonal), and let $\widetilde{\mathbf{T}}$ and $\widetilde{\mathbf{Q}}$ be the matrices that result from deleting column k of $\mathbf{T}$ and $\mathbf{Q}$, respectively. Also let $\widehat{\mathbf{T}}$ be the matrix that results from deleting row k of $\widetilde{\mathbf{T}}$ ($\widehat{\mathbf{T}}$ is the leading $(k-1)\times(k-1)$ submatrix of $\mathbf{T}$). The equivalence of the BFGS and CG methods (e.g., **[4]**) and the tridiagonalization property of the CG method (e.g., **[3]**) imply that

$$\mathbf{G}\widetilde{\mathbf{Q}} = \mathbf{Q}\widetilde{\mathbf{T}}. \tag{3.8}$$

Let $\widehat{\mathbf{T}}$ have an eigenvalue λ (preferably the closest eigenvalue to t_{11} since this would correspond to deleting the oldest information). By using the QR method with shift λ, a forward sequence of $k-2$ rotations, denoted by the orthogonal matrix $\mathbf{P}^{\mathrm{T}}$, can be determined for which

$$\mathbf{P}^{\mathrm{T}}\widehat{\mathbf{T}}\mathbf{P} = \begin{bmatrix} \overline{\mathbf{T}} & \mathbf{0} \\ & \\ \mathbf{0}^{\mathrm{T}} & \lambda \end{bmatrix} \tag{3.9}$$

where $\overline{\mathbf{T}} \in \mathbf{R}^{(k-2)\times(k-2)}$. Equation (3.8) can be written

$$\mathbf{G}\widetilde{\mathbf{Q}}\mathbf{P} = \mathbf{Q}\begin{bmatrix} \mathbf{P} & \\ & \mathbf{1} \end{bmatrix}\begin{bmatrix} \mathbf{P}^{\mathrm{T}} & \\ & \mathbf{1} \end{bmatrix}\widetilde{\mathbf{T}}\mathbf{P}. \tag{3.10}$$

Denote $\mathbf{Q}' = \mathbf{Q}\left[\begin{smallmatrix}\mathbf{P} & \\ & \mathbf{1}\end{smallmatrix}\right]$ and let $\widetilde{\mathbf{Q}}'$ be obtained by deleting column k of $\mathbf{Q}'$. Then (3.10) becomes

$$\mathbf{G}\widetilde{\mathbf{Q}}' = \mathbf{Q}'\begin{bmatrix} \overline{\mathbf{T}} & \\ & \lambda \\ \mu\mathbf{e}_{k-2}^{\mathrm{T}} & \nu \end{bmatrix} \tag{3.11}$$

where μ and ν are the rightmost entries in the last row and represent the effect of the last plane rotation on $\tilde{\mathbf{t}}_{k,k-1}$. Now deleting column $k-1$ of $\mathbf{Q}'$ (and $\widetilde{\mathbf{Q}}'$) gives a system like (3.8), with $\mathbf{g}^{(k)}/\|\mathbf{g}^{(k)}\|$ as the last column of $\mathbf{Q}'$, but having one less dimension. Hence this determines a CG sequence of one less dimension which retains $\mathbf{g}^{(k)}$ as the last gradient vector. The computational requirements are modest, being $O(nm)$ flops for the plane rotations, and the calculation of a single eigenvalue of an $(m-2)\times(m-2)$ tridiagonal matrix.

The simplest way to apply this idea is to use algorithm (2.7) for the first $m-1$ iterations, until the available storage is filled. On subsequent iterations, a BFGS (or CG) step is first taken, followed by deletion of a vector of information using the above process. No periodic restarts are necessary, so in this respect it is similar to Nocedal's method. Preliminary numerical experience with this idea has been disappointing, one possible reason being that, used in this way, the property of quadratic termination is lost. However, the idea is thought to be of intrinsic interest, and we hope to make use of it in future work.

4. Steepest descent steps. An alternative possibility also exists for using information contained in the canonical factors. Barzilai and Borwein [1] describe a steepest descent method

$$\mathbf{x}^{(k+1)} = \mathbf{x}^{(k)} - \alpha^{(k)}\mathbf{g}^{(k)} \tag{4.1}$$

in which the step $\alpha^{(k)}$ is not the optimum value given by (1.3), but is given by either

$$\alpha^{(k)} = \frac{\boldsymbol{\delta}^{(k-1)\mathrm{T}}\boldsymbol{\gamma}^{(k-1)}}{\boldsymbol{\gamma}^{(k-1)\mathrm{T}}\boldsymbol{\gamma}^{(k-1)}} \quad \text{or} \quad \alpha^{(k)} = \frac{\boldsymbol{\delta}^{(k-1)\mathrm{T}}\boldsymbol{\delta}^{(k-1)}}{\boldsymbol{\delta}^{(k-1)\mathrm{T}}\boldsymbol{\gamma}^{(k-1)}}. \tag{4.2}$$

The method requires no line search, and is invariant to rotations like the CG method. However, $f(\mathbf{x}^{(k)})$ is *not* monotonically decreasing, so there are doubts about reliability. An interesting feature is that Barzilai and Borwein give an example of a quadratic function of two variables for which R-superlinear convergence occurs in exact arithmetic from almost any $\mathbf{x}^{(1)}$. Numerical experiments with other quadratic functions also indicate that convergence is much more rapid than is commonly experienced with the optimum steepest descent method.

Barzilai and Borwein base their analysis on a study of the underlying recurrence relation, but it seems to me that there is a much simpler explanation of why rapid convergence can occur. Consider applying the method to a quadratic function with positive definite Hessian $\mathbf{G}$. Because of the invariance property we may assume that $\mathbf{G} = \operatorname{diag}(\lambda_i)$. Then it follows from (4.1) that $\mathbf{g}^{(k)}$ recurs like

$$g_i^{(k+1)} = g_i^{(k)} - \alpha^{(k)}(\mathbf{G}\mathbf{g}^{(k)})_i = g_i^{(k)} - \alpha^{(k)}\lambda_i g_i^{(k)}. \tag{4.3}$$

Choosing $\alpha^{(k)} = 1/\lambda_i$ zeroes element $g_i^{(k+1)}$. Hence choosing

$$\alpha^{(k)} = 1/\lambda_k, \quad k = 1, 2, \dots, n, \tag{4.4}$$

would give quadratic termination. Of course the eigenvalues of $\mathbf{G}$ are not usually available to take advantage of this result. One way of interpreting the Barzilai and Borwein method is that the $\alpha^{(k)}$ are Rayleigh quotient estimates of some value $1/\lambda_i$ (respectively, $\boldsymbol{\delta}^T\mathbf{G}\boldsymbol{\delta}/\boldsymbol{\delta}^T\mathbf{G}^2\boldsymbol{\delta}$ or $\boldsymbol{\delta}^T\boldsymbol{\delta}/\boldsymbol{\delta}^T\mathbf{G}\boldsymbol{\delta}$ in (3.2)). Detailed examination of the numerical progress of the method supports this viewpoint.

This interpretation does not however provide any expectation that R-superlinear convergence might be possible. When round-off errors are present it can be argued that R-superlinear convergence cannot be expected, as follows. Consider applying the method with the step lengths (4.4) that give termination. This can be expected to be better than using either of the step lengths given by (4.2). Because of rounding error, exact termination does not occur, and we repeat this sequence of steps indefinitely. Each n iterations therefore reduce $\|\mathbf{g}\|$ by a factor of about the relative precision ε. Hence the method is R-linear convergent with rate constant $\varepsilon^{1/n}$. For single precision on the SUN 3/50 and $n = 50$, $\varepsilon \approx 10^{-7}$ and the asymptotic linear rate is 0.72. In practice this is the best that could be expected from the Barzilai and Borwein method.

A more significant consequence of (4.4) is that any optimization process which provides eigenvalue estimates might usefully use steepest descent steps based on eigenvalue estimates. In particular if algorithm (2.7) is used, then eigenvalue estimates are available from those of the tridiagonal matrix $\mathbf{T} = \mathbf{L}\mathbf{L}^T$, where $\mathbf{L}$ is the bidiagonal matrix in the canonical factors. Moreover, if algorithm (2.7) is followed by a PCG phase which terminates when a significant deviation from quadratic behaviour is detected, then $\mathbf{T}$ can be extended in an obvious way and more accurate eigenvalue estimates can be obtained. In the quadratic case, in exact arithmetic, exact eigenvalues are obtained when the process terminates. In general then the idea would be to follow the PCG phase by calculating the eigenvalues of $\mathbf{T}$, followed by a sequence of steepest descent steps akin to (4.1) and (4.4). Let the eigenvalue estimates be exact and ordered as

$$\lambda_1 \geq \lambda_2 \geq \cdots \geq \lambda_t > 0 \tag{4.5}$$

where t is the total number of eigenvalue estimates available. Renumbering the steepest descent steps as $k = 1, 2, \ldots, t$, then it is readily proved in the quadratic case that

$$\|\mathbf{g}^{(k)}\|^2 - \|\mathbf{g}^{(k+1)}\|^2 = \sum_{i>k} \frac{\lambda_i}{\lambda_k}(2 - \frac{\lambda_i}{\lambda_k}) g_i^{(k)2} \geq \frac{\lambda_{k+1}}{\lambda_k}(2 - \frac{\lambda_{k+1}}{\lambda_k})\|\mathbf{g}^{(k)}\|^2 \tag{4.6}$$

and

$$(4.7)\qquad f^{(k)} - f^{(k+1)} = \sum_{i>k} \frac{1}{2\lambda_k}(2 - \frac{\lambda_i}{\lambda_k}) g_i^{(k)2} \geq \frac{1}{2\lambda_k}(2 - \frac{\lambda_{k+1}}{\lambda_k})\|\mathbf{g}^{(k)}\|^2.$$

Hence *strict descent* in both $\|\mathbf{g}\|$ and f is predicted. In practice these conditions give criteria for detecting nonquadratic behaviour in the sequence of steepest descent steps and hence for terminating the sequence when a sufficient decrease in $\|\mathbf{g}\|$ and f is not obtained.

The question of how to continue the algorithm after the sequence of steepest descent steps now arises. One option would merely be to reject all information and restart algorithm (2.7). However, it is also possible to recover useful second-order information from the sequence of steepest descent steps. Assuming that a quadratic model is valid, and because of the steepest descent property, the gradient vectors $\mathbf{g}^{(k)}, \mathbf{g}^{(k+1)}, \ldots$ are in the span of a Krylov sequence $\mathbf{g}^{(k)}, \mathbf{G}\mathbf{g}^{(k)}, \mathbf{G}^2\mathbf{g}^{(k)}, \ldots,$ starting from *any* k. Thus an equivalent CG sequence can be reconstructed, starting from any $\mathbf{g}^{(k)}$ that occurs during the steepest descent phase. Hence one possibility would be to store as many back gradient vectors as the storage would allow, and to use this information to reconstruct the equivalent CG sequence and hence **QL** factors. These could then be used to restart a PCG phase, for example.

Preliminary numerical experiments with these ideas show a similar performance to that described in Table 1. These results are encouraging in that they support the theoretical derivation of the methods. However, although a number of potentially interesting approaches have been indicated here, no single clear procedure which uses them has yet become apparent. Thus research on these ideas will continue, in particular more extensive numerical testing will enable the methods to be evaluated and compared with the methods of Nocedal and of Buckley and LeNir.

References

1. J. Barzilai and J. M. Borwein, *Two-point step size gradient methods*, IMA J. Numer. Anal. **8** (1988), 141–148.
2. A. Buckley and A. LeNir, *QN-like variable storage conjugate gradients*, Math. Programming **27** (1983), 155–175.
3. R. Fletcher, *Conjugate gradient methods for indefinite systems*, Numerical Analysis, Dundee 1975 (G. A. Watson, ed.), Springer-Verlag, Berlin, 1976.
4. R. Fletcher, *Practical Methods of Optimization*, 2nd edition, Wiley, Chichester 1987.
5. R. Fletcher and M. J. D. Powell, *A rapidly convergent descent method for minimization*, Comput. J. **6** (1963), 163–168.

6. J. C. Gilbert and C. Lemarechal, *Some experiments with variable storage quasi-Newton algorithms*, IIASA working paper WP-88 (1988).

7. D. C. Liu and J. Nocedal, *On the limited memory BFGS method for large scale optimization*, Northwestern Univ., Tech. report NAM 03 (1988).

8. J. Nocedal, *Updating quasi-Newton matrices with limited storage*, Math. Comput. **35** (1980), 773–782.

Lectures in Applied Mathematics
Volume **26** (1990)

Block ABS Methods for Nonlinear Systems of Algebraic Equations

A. GALANTAI

A block generalization of the ABS class of methods for the solution of nonlinear systems of algebraic equations is investigated. It covers subclasses of Brent–Brown-type methods as well as the conjugate direction methods suggested by Stewart [**16**].

1. Introduction. The ABS class of methods for the solution of linear equations was introduced by Abaffy, Broyden, and Spedicato [**1**] and Abaffy and Spedicato [**3**]. An extension of the algorithm described in [**1**] to solve nonlinear systems was given by Abaffy, Galantai, and Spedicato [**2**]. The block generalization of the ABS class of methods was developed by Abaffy and Galantai ([**4**], [**5**]).

Consider first the linear systems of the form

$$\mathbf{A}^{\mathrm{T}}\mathbf{x} = \mathbf{b}. \tag{1.1}$$

The next definition is given by Stewart [**16**].

DEFINITION 1. Let $\mathbf{A}^{\mathrm{T}}, \mathbf{U}, \mathbf{V} \in \mathbf{C}^{m\times m}$ be nonsingular and assume that the matrices $\mathbf{U}$ and $\mathbf{V}$ are partitioned in the form $\mathbf{U} = [\mathbf{U}_1, \mathbf{U}_2, \dots, \mathbf{U}_r]$ and $\mathbf{V} = [\mathbf{V}_1, \mathbf{V}_2, \dots, \mathbf{V}_r]$, where $\mathbf{U}_k$, $\mathbf{V}_k \in \mathbf{C}^{m\times m_k} (k = 1, 2, \dots, r)$. The pair $(\mathbf{U}, \mathbf{V})$ is said to be block $\mathbf{A}^{\mathrm{T}}$-conjugate (with respect to the partition $\{m_1, m_2, \dots, m_r\}$) if $\mathbf{V}_k^{\mathrm{H}}\mathbf{A}^{\mathrm{T}}\mathbf{U}_j = 0$ holds for $k < j$, $k = 1, \dots, r-1$, and $j = k+1, \dots, r$.

The $\mathbf{A}^{\mathrm{T}}$-conjugacy is a special case of block $\mathbf{A}^{\mathrm{T}}$-conjugacy when $r = m$. The pair $(\mathbf{U}, \mathbf{V})$ is (block) biorthogonal if the matrix $\mathbf{L} = \mathbf{V}^{\mathrm{H}}\mathbf{A}^{\mathrm{T}}\mathbf{U}$ is (block) diagonal. Assume that $\mathbf{A}^{\mathrm{T}}$ is nonsingular. Let $\mathbf{V} \in \mathbf{R}^{m\times m}$

1980 *Mathematics Subject Classification* (1985 *Revision*). Primary 65H.

be a given regular matrix and let it be partitioned in the form $\mathbf{V} = [\mathbf{V}_1, \dots, \mathbf{V}_r], \mathbf{V}_k \in \mathbf{R}^{m \times m_k}$ $(k = 1, \dots, r;\ 1 \le r \le m)$. The block ABS methods for linear systems may be formulated as follows.

ALGORITHM 1. *Let $\mathbf{x}_0 \in \mathbf{R}^m$ be arbitrary and set $\mathbf{H}_1 = \mathbf{I}_m$. For $k = 1$ to r, set*

$$\mathbf{P}_k = \mathbf{H}_k^{\mathrm{T}} \mathbf{Z}_k \tag{1.2}$$

with an arbitrary matrix $\mathbf{Z}_k \in \mathbf{R}^{m \times m_k}$ for which $[\mathbf{P}_k^{\mathrm{T}} \mathbf{A} V_k]^{-1}$ exists.

Set

$$\mathbf{r}_{k-1} = \mathbf{A}^{\mathrm{T}} \mathbf{x}_{k-1} - \mathbf{b}, \tag{1.3}$$

$$\mathbf{d}_k = [\mathbf{V}_k^{\mathrm{T}} \mathbf{A}^{\mathrm{T}} \mathbf{P}_k]^{-1} \mathbf{V}_k^{\mathrm{T}} \mathbf{r}_{k-1}, \tag{1.4}$$

$$\mathbf{x}_k = \mathbf{x}_{k-1} - \mathbf{P}_k \mathbf{d}_k. \tag{1.5}$$

Choose a matrix $\mathbf{W}_k \in \mathbf{R}^{m \times m_k}$ such that $\mathbf{W}_k^{\mathrm{T}} \mathbf{H}_k \mathbf{A} \mathbf{V}_k = \mathbf{I}_{m_k}$ and set

$$\mathbf{H}_{k+1} = \mathbf{H}_k - \mathbf{H}_k \mathbf{A} \mathbf{V}_k \mathbf{W}_k^{\mathrm{T}} \mathbf{H}_k. \tag{1.6}$$

It can be proven ([**4**], [**5**]) that Algorithm 1 finitely terminates; that is, $\mathbf{A}^{\mathrm{T}} \mathbf{x}_r = \mathbf{b}$. The following properties of the algorithm should be mentioned. The update matrix $\mathbf{H}_k$ is a projector of rank $m - (m_1 + \cdots + m_{k-1})$ with $N(\mathbf{H}_k) = R([\mathbf{A}\mathbf{V}_1, \dots, \mathbf{A}\mathbf{V}_{k-1}])$ and $R(\mathbf{H}_k) = R^{\perp}([\mathbf{W}_1, \dots, \mathbf{W}_{k-1}])$. $\mathbf{H}_k$ is symmetric if and only if $R([\mathbf{W}_1, \dots, \mathbf{W}_{k-1}]) = R([\mathbf{A}\mathbf{V}_1, \dots, \mathbf{A}\mathbf{V}_{k-1}])$. The pair $(\mathbf{P}, \mathbf{V})$ is block $\mathbf{A}^{\mathrm{T}}$-conjugate. Block $\mathbf{A}^{\mathrm{T}}$-biorthogonal pairs $(\mathbf{P}, \mathbf{V})$ can be generated by (1.2) and (1.6) if $\mathbf{Z}_k$'s are chosen such that

$$R(\mathbf{Z}_k) \subset R^{\perp}(\mathbf{H}_k \mathbf{A} \mathbf{V}_{k+1}, \dots, \mathbf{H}_k \mathbf{A} \mathbf{V}_r) \qquad (k = 1, \dots, r-1).$$

The matrix $\mathbf{H}_k$ can be written in the form

$$\mathbf{H}_{k+1} = \mathbf{I} - \widetilde{\mathbf{A}}_k (\widetilde{\mathbf{W}}_k^{\mathrm{T}} \widetilde{\mathbf{A}}_k)^{-1} \widetilde{\mathbf{W}}_k^{\mathrm{T}}$$

where $\widetilde{\mathbf{A}}_k = [\mathbf{A}\mathbf{V}_1, \dots, \mathbf{A}\mathbf{V}_k], \widetilde{\mathbf{W}}_k = [\mathbf{W}_1, \dots, \mathbf{W}_k]$, and $k \ge 1$. Using this, we can prove that Algorithm 1 produces all of the possible block $\mathbf{A}^{\mathrm{T}}$-conjugate pairs $(\mathbf{P}, \mathbf{V})$ with respect to the given partition ([**5**]). Consequently, Algorithm 1 and Stewart's class are equivalent. The ideas of Algorithm 1 can be traced back to Huang [**14**], Egervary [**11**], and others including the Purcell–Motzkin method. For details see [**4**], [**5**], and Abaffy and Spedicato [**7**].

2. The ABS class of nonlinear systems. We formulate a joint block generalization of the nonlinear ABS class ([**21**]) and the nonlinear conjugate direction methods suggested by Stewart [**16**]. The system of nonlinear equations is given by

$$\mathbf{F}(\mathbf{x}) = \mathbf{0} \qquad (\mathbf{F} \in C^1(\mathbf{R}^m, \mathbf{R}^m)) \tag{2.1}$$

and in componentwise form by $f_k(\mathbf{x}) = 0$ $(k = 1, \dots, m)$. The Jacobian matrix is denoted by $\mathbf{J}^{\mathrm{T}}(\mathbf{x}) = \mathbf{F}'(\mathbf{x})$. The rows of $\mathbf{J}^{\mathrm{T}}(\mathbf{x})$ are denoted by $\mathbf{a}_k^{\mathrm{T}}(\mathbf{x})$ $(k = 1, \dots, m)$. Let the vector $\mathbf{x}^* \in \mathbf{R}^m$ be an isolated solution of (4.1) and assume that $\mathbf{J}(\mathbf{x}^*)$ is regular.

ALGORITHM 2. *Let $\mathbf{x}_0 \in \mathbf{R}^m$ be given and sufficiently close to $\mathbf{x}^*$. For $n = 1, \dots,$ perform the following calculations.*

Set

$$\mathbf{y}_0 = \mathbf{x}_{n-1}, \qquad \mathbf{H}_{n1} = \mathbf{I}, \tag{2.2}$$

and let $\mathbf{V}_n = (\mathbf{V}_{n1}, \dots, \mathbf{V}_{n,r(n)})$ be invertible and partitioned arbitrarily $(\operatorname{rank}(\mathbf{V}_{nk}) = m_{nk},\ k = 1, \dots, r(n),\ 1 \le r(n) \le m)$.

For $k = 1$ to $r(n)$, set

$$\mathbf{t}_{n,k-1} = \sum_{j=0}^{k-1} t_{nkj}\mathbf{y}_j \tag{2.3}$$

with $t_{nkj} \ge 0, \sum_{j=0}^{k-1} t_{nkj} = 1$.

Set

$$\mathbf{P}_{nk} = \mathbf{H}_{nk}^{\mathrm{T}}\mathbf{Z}_{nk} \tag{2.4}$$

with an arbitrary matrix $\mathbf{Z}_{nk} \in \mathbf{R}^{m \times m_{nk}}$ for which $[\mathbf{P}_{nk}^{\mathrm{T}}\mathbf{J}(\mathbf{t}_{n,k-1})\mathbf{V}_{nk}]^{-1}$ exists.

Set

$$\mathbf{d}_k = [\mathbf{V}_{nk}^{\mathrm{T}}\mathbf{J}^{\mathrm{T}}(\mathbf{t}_{n,k-1})\mathbf{P}_{nk}]^{-1}\mathbf{V}_{nk}^{\mathrm{T}}\mathbf{F}(\mathbf{y}_{k-1}), \tag{2.5}$$

$$\mathbf{y}_k = \mathbf{y}_{k-1} - \mathbf{P}_{nk}\mathbf{d}_k. \tag{2.6}$$

Choose a matrix $\mathbf{W}_{nk} \in \mathbf{R}^{m \times m_{nk}}$ such that $\mathbf{W}_{nk}^{\mathrm{T}}\mathbf{H}_{nk}\mathbf{J}(\mathbf{t}_{n,k-1})\mathbf{V}_{nk} = \mathbf{I}_{m_k}$ and let

$$\mathbf{H}_{n,k+1} = \mathbf{H}_{nk} - \mathbf{H}_{nk}\mathbf{J}(\mathbf{t}_{n,k-1})\mathbf{V}_{nk}\mathbf{W}_{nk}^{\mathrm{T}}\mathbf{H}_{nk}, \tag{2.7}$$

$$x_n = y_{r(n)}. \tag{2.8}$$

In the sequel we omit n without any danger of confusion. Algorithm 2 coincides with Algorithm 1 on linear systems of the form $\mathbf{F}(\mathbf{x}) = \mathbf{A}^{\mathrm{T}}\mathbf{x} - \mathbf{b} = 0$. In general, $\mathbf{t}_{k-1}$ is a convex linear combination of the previous minor approximations $\mathbf{y}_0, \dots, \mathbf{y}_{k-1}$. Two particular choices should be mentioned. The choice

$$\mathbf{t}_{k-1} = \mathbf{y}_0 \qquad (k = 1, \dots, r(n);\ n \ge 1) \tag{2.9}$$

results in Stewart's **[16]** class. Note that $\mathbf{J}(\mathbf{x})$ is fixed and only $\mathbf{F}(\mathbf{x})$ is evaluated in the minor iterations. The second case is defined by

$$\mathbf{t}_{k-1} = \mathbf{y}_{k-1} \qquad (k = 1, \dots, r(n);\ n \ge 1). \tag{2.10}$$

The latter is a generalization of the nonlinear ABS class [2] and includes continuous versions of the algorithms due to Brown, Brent, and Gay ([9], [8], [13]) (for details, see [2]). Here $\mathbf{J}(\mathbf{x})$ is evaluated at the last minor approximation $\mathbf{y}_{k-1}$ satisfying a formal Seidel relaxation principle.

3. Local convergence of the method. Algorithm 2 has local convergence under the following assumptions:

(a) There exist two numbers $\rho_0, K_0 > 0$ such that $\mathbf{F} \in C^1(B(\mathbf{x}^*, \rho_0))$ and

$$\|\mathbf{F}(\mathbf{x}) - \mathbf{F}(\mathbf{y})\| \leq K_0\|\mathbf{x} - \mathbf{y}\| \qquad (\mathbf{x}, \mathbf{y} \in B(\mathbf{x}^*, \rho_0)). \tag{3.1}$$

(b) There are two constants $0 < \alpha \leq 1, K_1 > 0$ such that

$$\|\mathbf{J}(\mathbf{x}) - \mathbf{J}(\mathbf{y})\| \leq K_1\|\mathbf{x} - \mathbf{y}\|^{\alpha} \qquad (\mathbf{x}, \mathbf{y} \in B(\mathbf{x}^*, \rho_0)). \tag{3.2}$$

THEOREM 1. *Assume that the pairs* $(\mathbf{P}_n, \mathbf{V}_n)$ *generated by Algorithm 2 satisfy the inequality*

$$\|\mathbf{P}_{nk}[\mathbf{V}_{nk}^{\mathrm{T}}\mathbf{J}^{\mathrm{T}}(\mathbf{t}_{n,k-1})\mathbf{P}_{nk}]^{-1}\mathbf{V}_{nk}^{\mathrm{T}}\| \leq K_2 \tag{3.3}$$

$(k = 1, \ldots, r(n),\ n \geq 1)$ *for some* $K_2 > 0$. *It is also assumed that* $k(\mathbf{V}_n) \leq K_3$ $(n \geq 1)$. *Then there exists a* $\rho^* > 0$ $(\rho^* \leq \rho_0)$ *such that for all* $\mathbf{x}_0 \in B(\mathbf{x}^*, \rho^*)$ *the sequence* $\mathbf{x}_n$ *of the major iterates converges to* $\mathbf{x}^*$ *with Q-order of no less than* $1 + \alpha$.

The proof of Theorem 1 can be found in [4], [5]. The most crucial point of the proof of convergence is condition (3.3). If $r(n) = 1$ is chosen for all n, then Algorithm 2 changes to the Newton method and (3.3) takes the form $\|J^{\mathrm{T}}(\mathbf{x}_{n-1})^{-1}\| \leq K_2$. For the choice $r(n) > 1$ condition (3.3) has been proven only in special cases.

If $r(n) = m$ for all n, then the $\mathbf{V}_k$'s and $\mathbf{P}_k$'s are vectors and condition (3.3) has the form

$$|\mathbf{p}_k\mathbf{v}_k^{\mathrm{T}}/(\mathbf{v}_k^{\mathrm{T}}\mathbf{J}(\mathbf{t}_{k-1})\mathbf{p}_k)| \leq K_2 \qquad (k = 1, \ldots, m). \tag{3.4}$$

Using the techniques of [2] and [4], we can formulate three simple statements which ensure (3.4).

THEOREM 2. *Let* $0 \leq \beta < \pi/2$ *be arbitrary but fixed. If*

$$|(\mathbf{p}_k,\ \mathbf{J}(\mathbf{t}_{k-1})\mathbf{v}_k)_{\sphericalangle}| \leq \beta \qquad (k = 1, \ldots, m) \tag{3.5}$$

is satisfied, then there exists a $K_2 > 0$ *such that* (3.4) *holds.*

PROOF. By elementary calculations one obtains

$$|\mathbf{p}_k\mathbf{v}_k^{\mathrm{T}}/(\mathbf{v}_k^{\mathrm{T}}\mathbf{J}(\mathbf{t}_{k-1})\mathbf{p}_k)| \leq \|\mathbf{p}_k\|\|\mathbf{v}_k\|/(\|\mathbf{p}_k\|\|\mathbf{J}(\mathbf{t}_{k-1})\mathbf{v}_k\| \cos\beta_k),$$

where $\beta_k = (\mathbf{p}_k, \mathbf{J}(\mathbf{t}_{k-1})\mathbf{v}_k)_{\sphericalangle}$. There is a neighborhood $B(\mathbf{x}^*, \rho')$ of $\mathbf{x}^*$ in which the Jacobian $\mathbf{J}(\mathbf{x})$ is invertible and the inverse is bounded by a constant $\Gamma_1 > 0$. Inequality (3.5) and

$$\|\mathbf{J}(\mathbf{t}_{k-1})\mathbf{v}_k\| \geq \|\mathbf{v}_k\|/\|\mathbf{J}^{-1}(\mathbf{t}_{k-1})\|$$

imply that $\|\mathbf{p}_k\mathbf{v}_k^{\mathrm{T}}/(\mathbf{v}_k^{\mathrm{T}}\mathbf{J}(\mathbf{t}_{k-1})\mathbf{p}_k)\| \leq \Gamma_1/\cos\beta = K_2$.

THEOREM 3. *If the update matrices $\mathbf{H}_k$ are uniformly bounded by a constant $\Gamma_2 > 1$ and the directions $\mathbf{z}_k$ satisfy*

$$|(\mathbf{z}_k, \mathbf{H}_k\mathbf{J}(\mathbf{t}_{k-1})\mathbf{v}_k)_{\sphericalangle}| \leq \beta \qquad (k = 1, \ldots, m) \tag{3.6}$$

then (3.4) *holds with $K_2 = \Gamma_2/\cos\beta$.*

Condition (3.5) requires a posteriori control of the angle between $\mathbf{p}_k$ and $\mathbf{J}(\mathbf{t}_{k-1})\mathbf{v}_k$, whereas condition (3.6) means a priori control of the $\mathbf{z}_k$'s. Using the results of **[4]** and **[5]**, one can prove the next theorem.

THEOREM 4. *The update matrices $\mathbf{H}_k$ are bounded if the $\mathbf{w}_k$'s are chosen such that $\mathbf{H}_k$ is symmetric or $|(\mathbf{w}_k, \mathbf{H}_k\mathbf{J}(\mathbf{t}_{k-1})\mathbf{v}_k)_{\sphericalangle}| \leq \beta$ or $|(\mathbf{H}_k^{\mathrm{T}}\mathbf{w}_k, \mathbf{J}(\mathbf{t}_{k-1})\mathbf{v}_k)_{\sphericalangle}| \leq \beta$ is satisfied for $k = 1, \ldots, m$.*

We recall that $\mathbf{H}_k$ is symmetric ($k > 1$) if and only if $R(\mathbf{W}_{k-1}) = R([\mathbf{J}(\mathbf{t}_0)\mathbf{v}_1, \ldots, \mathbf{J}(\mathbf{t}_{k-2})\mathbf{v}_{k-1}])$.

Let us consider now the block case when $\mathbf{J}(\mathbf{x})$ is fixed at the point $\mathbf{x} = \mathbf{y}_0$. Consider Algorithm 2 for any partitions but restrict $\mathbf{t}_{k-1}$ to $\mathbf{y}_0$ ($k = 1, \ldots, r(n)$) and the choice of $\mathbf{Z}_k$'s such that ($\mathbf{P}$, $\mathbf{V}$) is block biorthogonal; that is,

$$\mathbf{P}_i^{\mathrm{T}}\mathbf{J}(\mathbf{y}_0)\mathbf{V}_j = 0 \qquad (i \neq j) \tag{3.7}$$

and

$$\mathbf{P}_k^{\mathrm{T}}\mathbf{J}(\mathbf{y}_0)\mathbf{V}_k = \mathbf{C}_k \qquad (k = 1, \ldots, r(n)) \tag{3.8}$$

with $k(\mathbf{C}) \leq \gamma$, where $\mathbf{C} = \mathrm{diag}(\mathbf{C}_1, \ldots, \mathbf{C}_r)$. Then

$$\|\mathbf{P}_k(\mathbf{V}_k^{\mathrm{T}}\mathbf{J}(\mathbf{y}_0)^{\mathrm{T}}\mathbf{P}_k)^{-1}\mathbf{V}_k^{\mathrm{T}}\| \leq \gamma k(\mathbf{V})\|\mathbf{J}(\mathbf{y}_0)^{-1}\| \tag{3.9}$$

holds for $k = 1, \ldots, r(n)$, from which (3.3) follows if $k(\mathbf{V}_n)$ is bounded. It is easy to see that (3.7) and (3.8) imply $\mathbf{P}^{\mathrm{T}}\mathbf{J}(\mathbf{y}_0)\mathbf{V} = \mathbf{C}$ and

$$\begin{aligned}\|\mathbf{P}_k(\mathbf{V}_k^{\mathrm{T}}\mathbf{J}(\mathbf{y}_0)^{\mathrm{T}}\mathbf{P}_k)^{-1}\mathbf{V}_k^{\mathrm{T}}\| &\leq \|\mathbf{P}\|\|\mathbf{C}_k^{-1}\|\|\mathbf{V}\| \\ &\leq \|\mathbf{C}\mathbf{V}^{-1}\mathbf{J}(\mathbf{y}_0)^{-1}\|\|\mathbf{C}^{-1}\|\|\mathbf{V}\| \leq k(\mathbf{C})k(\mathbf{V})\|\mathbf{J}(\mathbf{y}_0)^{-1}\|.\end{aligned}$$

We note that such $\mathbf{C}_k$'s always exist; e.g., $\mathbf{C}_k = \mathbf{I}$ is a possible choice. The requirement of biorthogonality can be fulfilled by choosing $\mathbf{Z}_k$'s such that

$$R(\mathbf{Z}_k) \subset R^{\perp}\{\mathbf{H}_k\mathbf{J}(\mathbf{y}_0)\mathbf{V}_{k+1}, \ldots, \mathbf{H}_k\mathbf{J}(\mathbf{y}_0)\mathbf{V}_r\}.$$

The latter construction to ensure condition (3.3) can be extended to a more general case still assuming that $\mathbf{t}_{k-1} = \mathbf{y}_0$ holds for $k = 1, \ldots, r(n)$. Using the fact that $\mathbf{V}^T\mathbf{J}^T(\mathbf{y}_0)\mathbf{P}$ is block lower-triangular and the fact that its inverse is also block lower-triangular with the entries $(\mathbf{V}_k^T\mathbf{J}^T(\mathbf{y}_0)\mathbf{P}_k)^{-1}$ in the main diagonal ($k = 1, \ldots, r(n)$), we have the estimation

$$\|(\mathbf{V}_k^T\mathbf{J}^T(\mathbf{y}_0)\mathbf{P}_k)^{-1}\| \le \|(\mathbf{V}^T\mathbf{J}^T(\mathbf{y}_0)\mathbf{P})^{-1}\|.$$

Hence,

$$\|\mathbf{P}_k(\mathbf{V}_k^T\mathbf{J}^T(\mathbf{y}_0)\mathbf{P}_k)^{-1}\mathbf{V}_k^T\| \le k(\mathbf{P})k(\mathbf{V})\|\mathbf{J}(\mathbf{y}_0)^{-1}\|$$

gives a bound K_2.

Finally, a condition due to Abaffy [6] is mentioned. He observed that condition (3.3) can be satisfied by choosing $\mathbf{P}_{nk}$'s such that $\mathbf{P}_{nk} = \mathbf{J}(\mathbf{t}_{n,k-1})\mathbf{V}_{nk}$ or $\mathbf{V}_{n,k} = \mathbf{J}^T(\mathbf{t}_{n,k-1})\mathbf{P}_{nk}$ holds.

4. Remarks. The numerical stability of the nonlinear class has not yet been studied in detail. However, the stability on linear systems has been studied by Galantai [12] using the backward error analysis technique due to Broyden [10] and the projection technique due to Stewart [16]. Consider again the linear system

$$\mathbf{A}^T\mathbf{x} = \mathbf{b} \qquad (\mathbf{A}\ m \times m) \tag{4.1}$$

and assume that $r = m$. Assume that $\mathbf{x}_{k-1}$ is computed with error and that no other error occurs during the subsequent steps. Denote by $\mathbf{x}'_j$ the result of step j for $j = k-1, \ldots, m$. It can then be shown ([12]) that

$$\mathbf{x}_n - \mathbf{x}'_n = \mathbf{Q}_{nk}(\mathbf{x}_{k-1} - \mathbf{x}'_{k-1}), \tag{4.2}$$

where $\mathbf{Q}_{nk}$ is a projector onto $R^{\perp}(\mathbf{AV}^{n-k+1|})$ along $R^{\perp}(\mathbf{AV}^{|k-1})$. The Frobenius norm of $\mathbf{Q}_{nk}$ is minimal if and only if it is symmetric. $\mathbf{Q}_{nk}$ is symmetric if and only if $\mathbf{V}^T\mathbf{A}^T\mathbf{AV}$ is diagonal. We note that the methods satisfying this are called optimal by Broyden [10]. In the general case we have the bound $\|\mathbf{Q}_{nk}\| \le k(\mathbf{AV}) \le k(\mathbf{A})k(\mathbf{V})$ (see [12] , [7]). The residual perturbation is defined by $r'_k = \mathbf{A}^T(\mathbf{x}_k - \mathbf{x}'_k)$. It can be shown that the residual perturbation r'_s is minimal for all s if and only if $\mathbf{V}^T\mathbf{V}$ is diagonal.

BIBLIOGRAPHY

1. J. Abaffy, C. G. Broyden, and E. Spedicato, *A class of direct methods for linear systems*, Numerische Mathematik **45** (1984), 361–376.

2. J. Abaffy, A. Galantai, and E. Spedicato, *The local convergence of* ABS *methods for nonlinear algebraic equations*, Numerische Mathematik **51** (1987), 429–439.

3. J. Abaffy and E. Spedicato, *A generalization of the* ABS *algorithm for linear systems*, Quaderni DMSIA 1985/4, Universita di Bergamo, 1985.

4. J. Abaffy and A. Galantai, *Conjugate direction methods for linear and nonlinear systems of algebraic equations*, Quaderni DMSIA 1986/7, Universita di Bergamo, 1986.

5. ——, *Conjugate direction methods for linear and nonlinear systems of algebraic equations*. Colloquia Mathematica Soc. J. Bolyai, 50. Numerical methods, Miskolc, 1986 (G. Stoyan, ed.), North-Holland, Amsterdam, 1987, pp. 481–502.

6. J. Abaffy, *A superlinear convergency theorem in the* ABSg *class for nonlinear algebraic equations*, JOTA, to appear.

7. J. Abaffy and E. Spedicato, ABS *projection algorithms: Mathematical techniques for linear and nonlinear algebraic equations*, Ellis Horwood, Chichester, England, in preparation.

8. R. P. Brent, *Some efficient algorithms for solving systems of nonlinear equations*, SIAM J. Numer. Anal. **10** (1973), 327–344.

9. K. M. Brown, *A quadratically convergent Newton-like method based upon Gaussian elimination*, SIAM J. Numer. Anal. **6** (1969), 560–569.

10. C. G. Broyden, *On the numerical stability of Huang's and related methods*, JOTA **47** (1985), 401–412.

11. E. Egervary, *On rank-diminishing operations and their applications to the solution of linear equations*, ZAMP **11** (1960), 376–386.

12. A. Galantai, *A study of error propagation for the* ABS *method*, Quaderni DMSIA 1987/17, Universita di Bergamo, 1987.

13. D. M. Gay, *Brown's method and some generalizations, with applications to minimization problems*, Comput. Sci. Tech. Rep. 75–225, Cornell University, 1975.

14. H. Y. Huang, *A direct method for the general solution of a system of linear equations*, JOTA **16** (1975), 429–445.

15. J. M. Ortega and W. C. Rheinboldt, *Iterative Solution of Nonlinear Equations in Several Variables*, Academic Press, New York, 1970.

16. G. W. Stewart, *Conjugate direction methods for solving systems of linear equations*, Numerische Mathematik **21** (1973), 285–297.

Institute of Mathematics, University of Agriculture, Godollo, Hungary

Lectures in Applied Mathematics
Volume 26 (1990)

Application of Julia–Fatou Iteration Theory in Dielectric Spectroscopy

SYLVIE GÉLINAS
AND
RÉMI VAILLANCOURT

Abstract. In using a shielded open-circuited coaxial line of general length for broad-band measurements to obtain the dielectric permittivity of a given sample, one needs to solve the transcendental equation $z \tan z = c$, where the complex number c is the negative of the measured normalized admittance and iz is the normalized propagation constant. The iteration theory of Julia–Fatou for rational and entire functions is applied, in a slightly extended form, to the meromorphic function $c \cot z$ and its inverse in order to obtain any root of the given equation in a prescribed order for any given value of c. The equation $z \cot z = c$ for the short-circuited line is treated in a similar way.

1. Introduction. The coaxial transmission line has been used as a sample cell in dielectric measurements for many years. Reflection measurements with the sample terminating in an open circuit [3], [19] lead to solving the elementary transcendental equation

$$z \tan z = c \tag{1.1}$$

1980 *Mathematics Subject Classification*. Primary 30D05, 35-04, 65E05, 65H05; Secondary 78A40.

This work was supported in part by the Natural Sciences and Engineering Research Council of Canada under grant A 7691. The authors wish to acknowledge many fruitful discussions with Dr. V. N. Tran on the engineering aspects of the problem while he visited the University of Ottawa.

for the unknown normalized propagation constant iz, for a set of experimentally obtained complex values c of the normalized admittance $-c$. Similarly, a short circuit termination [**17**, p. 505], cf. [**13**, p. 1522] leads to the equation

$$z \cot z = c. \tag{1.2}$$

The length of the sample, its position, and the impedance terminating the line were chosen, in the past, to provide the best accuracy at each frequency being used. However, over the past twenty years, commercially available automated network analyzers have been able to measure impedance over an increasing range of frequencies. For optimal use of this instrumentation, it is not practical to adjust the length of the sample or the termination to obtain the best performance at each frequency. Thus equations (1.1) and (1.2) are to be solved over a wide range of values of c, for some of which the roots z may come close to double roots of these equations. To choose the correct root near such a double root, the Lagrange series of the multiple-valued inverse function, $z = H(c)$, associated with (1.1) is used in [**15**], [**16**] to investigate the mapping of each sheet of the Riemann surface of H onto a fundamental region of the z-plane; however, in [**15**] Muller's method is used to solve (1.1) numerically. The fundamental regions of (1.2) are drawn in [**13**, p. 405] in polar coordinates.

It may be noticed here that Henrici, in [**8**], indicated a possible method for the solution of equations like (1.1) and (1.2); that method is closely related to the qd-algorithm as applied to meromorphic functions to find their zeros and poles.

The purpose of this presentation is to establish a numerical strategy to find all solutions of (1.1) and (1.2). Rough graphs of the images of each of the four quadrants of the c-plane in the z-plane by the multiple-valued functions $c \to \{z; z \tan z = c\}$, respectively $c \to \{z; z \cot z = c\}$, are given. These images establish an order on the roots of (1.1), respectively (1.2). According to the Julia–Fatou theory for the iteration of meromorphic functions [**9**], each root of (1.1) is reached by iterating $F(z) = c \cot z$,

$$z_{n+1} = c \cot z_n, \tag{1.3}$$

or a properly chosen branch of its multiple-valued inverse $G(z) = F^{-1}(z)$,

$$z_{n+1} = \operatorname{arccot}(\frac{1}{c} z_n). \tag{1.4}$$

The main idea used here is that F has very few attractive fixed points, and the basins of attraction of all but a few attractive fixed points of appropriate determinations of G, for any given value of c, are relatively vast. A second idea is to apply Newton's method to (1.1) rewritten in the form

$$z \sin z - c \cos z = 0, \tag{1.5}$$

with appropriate starting values to effectively find roots that are known to be located in definite bounded convex regions of the plane. Similar iteration techniques can be used to solve (1.2).

Equation (1.1) will be treated in detail. Fundamental regions for (1.3) and (1.4) are described in Section 2. Results from the Julia–Fatou iteration theory are quoted in Section 3. Starting values for Newton's method are given in Section 4. Two families of classical plane curves, which are used in the convergence theorems, are recalled in Section 5. Sections 6 and 7 deal respectively with the iteration functions $F(z)$ and $G(z)$. Application of the analytical results to dielectric spectroscopy is mentioned in Section 8. Finally, similar results for equation (1.2) are summarized in Section 9.

2. The images of the multiple-valued function $c \to \{z; z \tan z = c\}$. The equation

$$f(z) := z \tan z - c = 0 \tag{2.1}$$

has infinitely many solutions for each value of c. The roots of (2.1) are simple except for an infinity of values of c, $c = c_j$, $j = 0, 1, 2, \ldots$; for each of these c_j, $f(z)$ has only two double roots, $z = p_j$ and $z = -p_j$. The $\pm p_j$ are simultaneously the roots of (2.1) and

$$f'(z) := z \sec^2 z + \tan z = 0.$$

The last equation is independent of c and can be rewritten in the form

$$z + \sin z \cos z = 0. \tag{2.2}$$

The solutions $\pm p_j$ of (2.2) and the corresponding values of c,

$$c_j = p_j \tan p_j, \tag{2.3}$$

are obtained numerically (see **[9]** and **[15**, p. 14**]**). The first seven double roots p_j and the corresponding values of c_j are listed in Table 1.

j	p_j	c_j
0	0	0
1	$2.106 + 1.125i$	$-1.651 + 2.060i$
2	$5.356 + 1.552i$	$-2.058 + 5.335i$
3	$8.537 + 1.776i$	$-2.278 + 8.523i$
4	$11.699 + 1.929i$	$-2.431 + 11.689i$
5	$14.854 + 2.047i$	$-2.548 + 14.846i$
6	$18.005 + 2.142i$	$-2.643 + 17.998i$

Table 1. The first seven double roots of $z \tan z = c$ and the corresponding values of c

It is enough to consider the first quadrant I_z of the z-plane and the upper-half plane $\mathrm{I} \cup \mathrm{II}$ of the c-plane since

$$(-z)\tan(-z) = c$$

and

$$\bar{z} \tan \bar{z} = \bar{c}.$$

To obtain a rough graph of the images of I and II into I_z by the multiple-valued function

$$c \to \{z; z \tan z - c = 0\}, \tag{2.4}$$

one draws the images of the positive imaginary c-axis and for each c_j the image of a properly chosen branch cut joining c_j and $\bar{c}_j$ through ∞. The images of I and II, denoted respectively by $\mathrm{I}_0, \mathrm{II}_0, \mathrm{I}_1, \mathrm{II}_1$, etc., according to an obvious ordering, are shown in Fig. 1. It is noted that the region II_0 is unbounded.

In [**16**, Fig. 3] the branch cuts were taken as the vertical lines $\Re c = \Re c_j$ joining c_j to $\bar{c}_j$ through ∞ and produced fundamental regions that extend slightly to the right of the lines $\Re z = (2j+1)\pi/2$. It would be nice, although not essential for our purpose, to have our regions II_j, $j = 1, 2, \ldots,$ lying completely inside vertical strips of width π bounded by the lines $\Re z = (2j-1)\pi/2$ and $\Re z = (2j+1)\pi/2$. To this effect, one considers the values taken by $c = z \tan z$ along such lines, namely, with $x(j) = (2j+1)\pi/2$ and $z = x + iy$,

$$c(j) = -\frac{y}{\tanh y} + i\frac{x(j)}{\tanh y}.$$

By choosing a branch cut that lies between the curves $c(j-1)$ and $c(j)$ from some point on, the region II_j will be inside the desired strips.

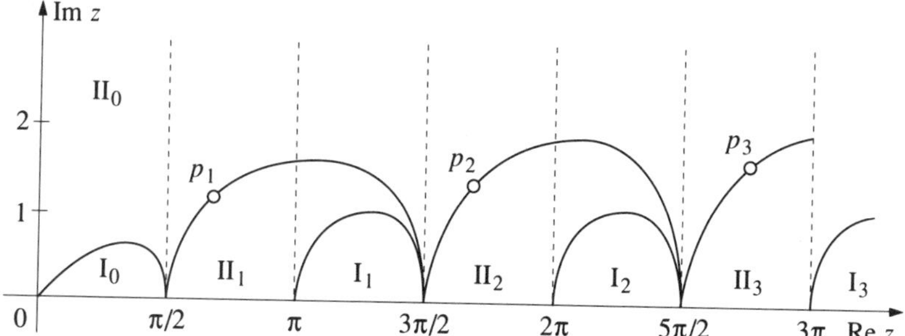

FIGURE 1. The figure shows the regions I_k and II_k, $k \geq 0$, which are images of the upper-half of the c-plane into the first quadrant of the z-plane by the multiple-valued function $c \to \{z; z \tan z = c\}$. The points $p_1, p_2, \ldots$ are the double roots of the equation $z \tan z = c$. The region II_0 is unbounded.

3. The Julia–Fatou iteration theory for meromorphic functions. A few results of the Julia–Fatou theory, as extended to the iteration of meromorphic functions **[9]** and their inverses, will be listed. A general presentation of the Julia–Fatou theory for rational functions, notation, and references are found in the recent survey by Blanchard **[2]**.

Let φ be a transcendental meromorphic function,

$$\varphi: \mathbf{C} \to \overline{\mathbf{C}},$$

and consider the iteration

$$z_{n+1} = \varphi(z_n), \qquad n = 0, 1, 2, \ldots.$$

A fixed point s of φ,

$$s = \varphi(s),$$

is attractive, repulsive, or indifferent, as the absolute value of its multiplier $\varphi'(s)$ satisfies, respectively,

$$|\varphi'(s)| < 1, \qquad > 1, \qquad \text{or} \quad = 1.$$

The zeros of $\varphi'(z)$ are the algebraic critical points of φ, and the exceptional values of φ are the transcendental critical points of φ.

Let

$$\varphi^n(z) = \varphi[\varphi^{n-1}(z)], \qquad \varphi^0(z) = z,$$

denote the nth iterate of z by φ. The Julia set of φ, $\mathcal{J}(\varphi)$, is the set of nonnormality of φ defined as follows:

$$\mathcal{J}(\varphi) = \{z; \{\varphi^n(z)\}_{n=1}^{\infty} \text{ is not a normal family}\}.$$

The Fatou set of φ, $\mathscr{F}(\varphi)$, is the set of normality of φ. $\mathscr{F}$ is open. The Rådtröm [14] set of φ, $\mathscr{R}(\varphi)$, is the set of predecessors of the essential singularity of φ, namely,

$$\mathscr{R}(\varphi) = \{z; \varphi^n(z) \text{ is not defined for some } n \in \mathbf{N}\}.$$

Then

$$\mathscr{F}(\varphi) = \overline{\mathbf{C}} \setminus (\mathscr{J}(\varphi) \cup \mathscr{R}(\varphi)).$$

For the meromorphic functions considered here, $\mathscr{J}$ and $\mathscr{R}$ are nonempty sets without isolated points. Moreover, $\mathscr{J}$ and $\mathscr{F}$ are completely invariant with respect to φ, i.e., invariant under φ and φ^{-1}.

A k-cycle of φ is a set of k distinct points, $s_0, s_1, \dots, s_{k-1}$, satisfying the relations

$$s_1 = \varphi(s_0), \quad s_2 = \varphi^2(s_0), \quad \dots, \quad s_{k-1} = \varphi^{k-1}(s_0), \quad s_0 = \varphi^k(s_0).$$

The multiplier of a k-cycle,

$$(\varphi^k)'(s_j) = \varphi'(s_{k-1}) \cdots \varphi'(s_1)\varphi'(s_0), \qquad j = 0, 1, \dots, k-1,$$

is seen to be the same at every point of the cycle. A k-cycle is attractive, repulsive, or indifferent, respectively, as

$$|(\varphi^k)'(s_j)| < 1, \qquad > 1, \qquad \text{or} \quad = 1.$$

A fixed point is a 1-cycle. Any element s_j of a k-cycle is a fixed point of φ^k. The attractive cycles of φ are the repulsive cycles of φ^{-1} and conversely, since $\varphi'(z)(\varphi^{-1})'(\varphi(z)) = 1$. The immediate basin of attraction of an attractive fixed point s is the largest connected open set Ω such that

$$z_n = \varphi^n(z_0) \to s \quad \text{as} \quad n \to \infty, \qquad \forall z_0 \in \Omega.$$

The immediate basin of attraction of a k-cycle is the union of the immediate basins of attraction of the elements s_j considered as fixed points of φ^k. Attractive fixed points and attractive cycles are in the Fatou set of φ.

The following result (cf. [1]), which is derived in [9], will be needed.

THEOREM 1. *The immediate basin of attraction of every attractive fixed point or cycle of $\varphi = c\cot z$, respectively, $\varphi = c\tan z$, contains at least one critical point of φ.*

In studying dynamical systems (see [5]), one is interested in the Julia set of φ. In solving equations iteratively (see [6], [7], [9]), however, one is mainly interested in finding fixed points of the iteration function φ while avoiding being trapped by the attractive k-cycles, $k > 1$.

4. Newton's method. Newton's method for (1.5) is given by

$$z_{n+1} = z_n - \frac{z_n \sin z_n - c \cos z_n}{(1+c)\sin z_n + z_n \cos z_n}. \tag{4.1}$$

Since each of the regions $I_0, I_1, \ldots$ is bounded and convex, the following starting values obtained in [**4**] by truncated continued fraction expansions for the solutions of (1.1):

$$z_0 = \pi \sqrt{\frac{c}{\pi^2 + 4c}}, \qquad \text{for } s^{(0)} \in I_0, \tag{4.2a}$$

$$z_0 = k\pi + \frac{\pi}{2} \frac{c}{c + 2k + 1}, \qquad \text{for } s^{(k)} \in I_k, \quad k > 0, \tag{4.2b}$$

provide convergence to the desired roots $s^{(0)}$ and $s^{(k)}$ to an accuracy of 10^{-10} in very few iterations. It is to be noted that applying Newton's method to (1.5) rather than (2.1) (see [**9**], [**18**]), with the above initial values, is at times faster and somewhat better .

5. The ovals of Cassini and the level curves of $\sin z$ **and** $\cos z$. The regions of convergence of the iteration functions (1.3) and (1.4) used to solve (1.1) will involve two families of plane curves, namely, the ovals of Cassini and the level curves of $\sin z$. Similarly, in solving (1.2) the same ovals and the level curves of $\cos z$ will play an important role. These curves are described in the present section.

DEFINITION 1. *Given two fixed points f and g, called foci, the locus of the point z in the plane such that the product of the distances r_1 and r_2 of z to the foci is constant,*

$$|z - f||z - g| = k^2,$$

is called an oval of Cassini.

After a translation and a rotation, f can be taken to be real and positive and $g = -f$. Thus ovals are given in standard form in [**12**, pp. 153–155], [**20**, pp. 218–219] as

$$|z - f||z + f| = k^2. \tag{5.1}$$

As is shown in Fig. 2, when $f > k$, the ovals are made of two disjoint simple closed curves or loops; when $f = k$, they describe a lemniscate of Bernoulli; and when $f < k$, they reduce to one simple closed curve or loop.

It may be noticed that an oval of Cassini, with foci $\pm f$ and $r_1 r_2 = k^2$, is a section of a torus with great, resp. small, generating circles of radii

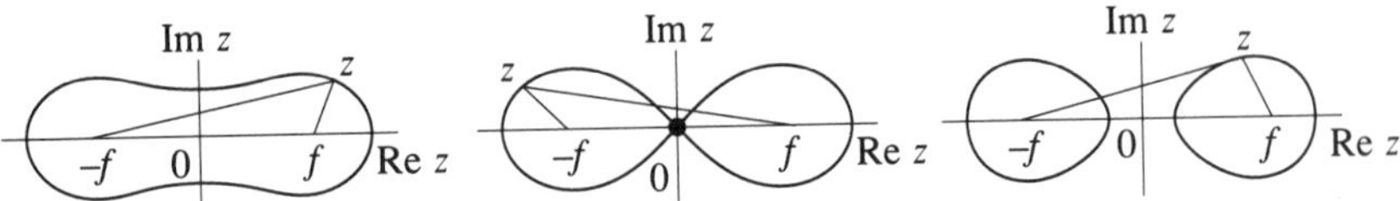

FIGURE 2. The figure shows ovals of Cassini $\{z; |z-f||z+f| = k^2\}$ with foci $f > 0$ and $-f$; the product of the distances r_1 and r_2 from the point z to the foci is constant: $r_1 r_2 = k^2$. Left, $f < k$; center, $f = k$; right, $f > k$.

$R = f$, resp. $r = k^2/(2f)$, on a plane parallel to the axis of the torus and at a distance $d = r$ from this axis.

The farthest and closest points of an oval of Cassini to the midpoint of the segment joining the two foci, f and $-f$, satisfy, respectively,

$$|z|_{\max} = \sqrt{|k|^2 + |f|^2}, \qquad |z|_{\min} = \sqrt{||k|^2 - |f|^2|}. \tag{5.2}$$

It will be useful later to notice that the normal direction to an oval at the point z is simply expressed as the gradient of $r_1 r_2$,

$$r_2 \frac{z-f}{r_1} + r_1 \frac{z+f}{r_2}, \tag{5.3}$$

where $r_1 = |z-f|$ and $r_2 = |z+f|$.

The level curves $|\sin z| = k$, resp. $|\cos z| = k$, of the functions $\sin z$, resp. $\cos z$, are given in [**11**, Addenda, p. 66, Fig. 25], and are shown in Fig. 3. The normal directions to these curves are, respectively,

$$\sin 2x + i \sinh 2y, \qquad -\sin 2x + i \sinh 2y. \tag{5.4}$$

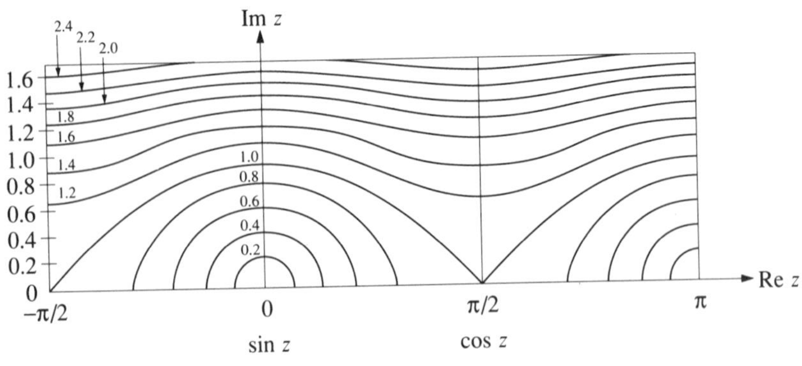

FIGURE 3. The figure shows the level curves $|\sin z| = k$ of $\sin z$, resp. $|\cos z| = k$ of $\cos z$, in the upper half-plane, by taking the origin at $0 + 0i$, resp. at $\pi/2 + 0i$.

The ovals of Cassini and the level curves of $\sin z$, resp. $\cos z$, play a similar role in the solution of the equations $\cos z = cz$, resp. $\sin z = cz$; see [7]. In the solution of $\exp(z) = cz$, the ovals of Cassini reduce to the unit circle with center at the origin (see [6]).

6. The iteration function $F(z)$. The numerical scheme (1.3) is given by the iteration function

$$F(z) = c \cot z. \tag{6.1}$$

The convergence properties of $F(z)$ will be established in Theorems 2 and 3 of this section.

THEOREM 2. *For every given value $c \neq 0$, the iteration function $F(z) = c \cot z$ has either no attractive fixed point or exactly two attractive fixed points, $\pm s(c)$. These can be reached from the transcendental critical points $z_0 = \pm ic$ by iterating F.*

PROOF. The function (6.1) has no algebraic critical points since its derivative,

$$F'(z) = -c \csc^2 z, \tag{6.2}$$

never vanishes for $c \neq 0$; it has, however, two transcendental critical points, namely,

$$c_{\pm} = \pm ic. \tag{6.3}$$

It then follows from Theorem 1 that $F(z)$ has at most two attractive fixed points, and these can be reached from $z_0 = c_{\pm}$. It has exactly two such points, if any, since it is an odd function. □

By (6.2), (2.3), and (2.2) the double roots p_j of (1.1) are indifferent fixed points with multiplier

$$F'(p_j) = 1;$$

hence they belong to the Julia set of F. Theoretically, they can be reached by (1.3) from $z_0 = \pm ic_j$; but, in practice, with 15 decimal digit arithmetic, only three significant digits can be obtained without recourse to Steffensen's acceleration [**10**, pp. 103–108].

The general position of the two attractive fixed points of F, if they exist, can be located by geometric considerations. To this end, it will be useful to define the region in the z plane where attractive fixed points of a general iteration function can lie.

DEFINITION 2. *For any iteration function F, the region*

$$\mathscr{A} := \{z; |F'(z)| < 1\}$$

is said to be the region of attractivity of F.

The region of attractivity of the iteration function (6.1) is defined by the inequality

$$|F'(z)| = |c \csc^2 z| < 1, \tag{6.4}$$

i.e.,

$$\mathscr{A} = \{z; |\sin z| > \sqrt{|c|}\}; \tag{6.5}$$

thus it consists of one or two infinite regions bounded by the level curves

$$|\sin z| = \sqrt{|c|} \tag{6.6}$$

shown in Fig. 3 (see [**11**, Addenda, p. 66, Fig. 25]).

Since attractive fixed points are, by definition, invariants under F, in an attempt to locate them, it is natural to consider the image $\mathscr{B}$ of the region of attractivity $\mathscr{A}$ under the mapping F, i.e.,

$$\mathscr{B} = F(\mathscr{A}). \tag{6.7}$$

In the next lemma $\mathscr{B}$ is characterized geometrically.

LEMMA 1. *For every given value c, the iteration function $F(z) = c \cot z$ maps its region of attractivity, $\mathscr{A} = \{z; |\sin z| > \sqrt{|c|}\}$, onto the bounded region $\mathscr{B} = \{z; |z + ic||z - ic| < |c|\}$.*

PROOF. Given any $z \in \mathscr{A}$, by adding, resp. subtracting, ic to both sides of $w = c \cot z$, one has

$$\begin{aligned} w + ic &= c \cot z + ic, \\ w - ic &= c \cot z - ic; \end{aligned}$$

now, multiplying the left-hand sides, resp. the right-hand sides, and taking absolute values, one obtains, by (6.4),

$$\begin{aligned} |w + ic||w - ic| &= |c^2(1 + \cot^2 z)| \\ &= |c^2 \csc^2 z| \\ &< |c|, \qquad \forall z \in \mathscr{A}. \end{aligned}$$

Therefore the set $\mathscr{B} = F(\mathscr{A})$ is given geometrically by the inequality

$$\mathscr{B} = \{z; |z + ic||z - ic| < |c|\}, \tag{6.8}$$

and it is immediately seen that it is bounded by the oval of Cassini

$$\mathscr{O} = \{z; |z + ic||z - ic| = |c|\} \tag{6.9}$$

with foci $\pm ic$ and distance product $r_1 r_2 = |c|$ from z to the foci. □

Referring to Section 5, it is seen that when $|c| = 1$, the oval (6.9) describes a lemniscate of Bernoulli; when $|c| > 1$, it is made of two disjoint simple closed curves or loops; and when $|c| < 1$, it reduces to one simple closed curve or loop. The farthest and closest points of the oval to the origin satisfy, respectively,

$$|z|_{\max} = \sqrt{|c| + |c|^2}, \qquad |z|_{\min} = \sqrt{||c| - |c|^2|}.$$

For large $|c|$, the diameter of each loop of the oval $\mathscr{O}$ is nearly one; thus, the area of each component of $\mathscr{B}$ is roughly $\pi/4$, hence rather small.

Necessary and sufficient conditions for the existence of attractive fixed points of F are now derived from the previous lemma.

THEOREM 3. *For any given value $c \neq 0$, the iteration function $F(z) = c \cot z$ has two attractive fixed points, $\pm s(c)$, if $\mathscr{B} \subset \mathscr{A}$, and only if $\mathscr{B} \cap \mathscr{A} \neq \varnothing$, where $\mathscr{A} = \{z; |\sin z| > \sqrt{|c|}\}$ is the region of attractivity of F and the bounded region $\mathscr{B} = \{z; |z + ic||z - ic| < |c|\} = F(\mathscr{A})$ is the image of $\mathscr{A}$ by F.*

PROOF. Suppose $\mathscr{B} \subset \mathscr{A}$. Then $F(\mathscr{B}) \subset \mathscr{B}$ by the above lemma. Since $\mathscr{B}$ is bounded, and $|F'(z)| < 1$ in $\mathscr{B}$ by (6.4), the mapping is contracting and the iterations started respectively at $\pm ic$ converge to the fixed points $\pm s(c)$. Now suppose F has two attractive fixed points. Since these belong to the region $\mathscr{A}$ of attractivity of F defined by the inequality (6.5) and also to $\mathscr{B}$ by the above lemma, then $\mathscr{B} \cap \mathscr{A} \neq \varnothing$. □

It is to be noted that, for $c = c_j \in \text{II}$, $j = 1, 2, \ldots$, the ovals go through the double roots $\pm p_j$ of (1.1), and the two loops are tangent to the boundary of the region of attractivity $\mathscr{A}$ of F, i.e., to the level curves $|\sin z| = \sqrt{|c_j|}$, as can be computed by (5.3) and (5.4), and as shown for the case $c = c_1$ in Fig. 4.

The loop of the oval in the first quadrant lies inside the infinite region II_0, as can be seen from Table 1. When the oval is moved slightly away from II_j by moving the focus, $-ic$, in a direction parallel to the perpendicular to the boundary of $\mathscr{A}$ through p_j, the double root bifurcates along this perpendicular; one root enters the loop in II_0, while the other root goes inside II_j. On the other hand, when the oval is moved

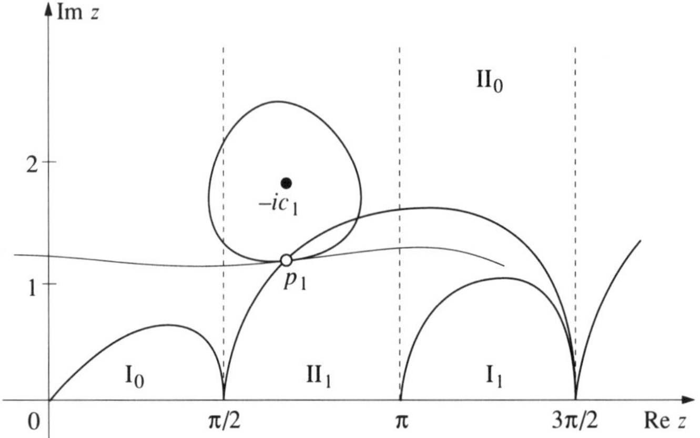

FIGURE 4. The figure shows the right-hand loop of the oval of Cassini $\{z; |z - ic_1||z + ic_1| = |c_1|\}$ and the level curve $|\sin z| = \sqrt{|c_1|}$. These two curves are mutually tangent at the double root p_1 of the equation $z \tan z = c_1$.

slightly inside II_j along a direction parallel to the same perpendicular, the bifurcation occurs in a direction orthogonal to this perpendicular and both roots go outside the loop, one in II_j and the other in II_0, both outside $\mathscr{A}$, thus getting out of the reach of F.

7. The inverse iteration function $G(z)$. The roots of (1.1) which do not lie inside, or on, the ovals of Cassini (6.9) are repulsive fixed points of F, but attractive fixed points of certain determinations of G.

Let G be the inverse of F,

$$G(z) = \frac{1}{2i}\log\left(\frac{z+ic}{z-ic}\right) + k\pi, \qquad k \in \mathbf{Z}. \tag{7.1}$$

To obtain proper determinations of G, set

$$t = \frac{z+ic}{z-ic}$$

and

$$\log t = \log|t| + i\arg t.$$

Firstly, if $-\pi < \arg t \le \pi$, the branch cut of $\log t$ is taken along the negative real axis in the t-plane and correspondingly from $-ic$ to ic through the origin in the z-plane. In this case, $G(z)$ will be denoted by $G_0(z,;k)$; thus

$$-\frac{\pi}{2} + k\pi < \Re G_0(z;k) \le \frac{\pi}{2} + k\pi. \tag{7.2}$$

Secondly, if $0 < \arg t \leq 2\pi$, the branch cut of $\log t$ is taken along the positive real axis in the t-plane and from $-ic$ to ic through ∞ in the z-plane. In this case $G(z)$ will be denoted by $G_\infty(z;k)$; thus

$$k\pi < \Re G_\infty(z;k) \leq \pi + k\pi. \tag{7.3}$$

Let $B_0(k)$ and $B_\infty(k)$ be vertical strips of width π defined, respectively, by

$$B_0(k) = \{z; -\frac{\pi}{2} + k\pi < \Re z \leq \frac{\pi}{2} + k\pi\}, \tag{7.4}$$

$$B_\infty(k) = \{z; k\pi < \Re z \leq (k+1)\pi\}. \tag{7.5}$$

The convergence properties of the iteration functions $G_0(z_n;k)$ and $G_\infty(z_n;k)$ will now be derived in the following lemma and two theorems.

LEMMA 2. *The iterates*

$$z_{n+1} = G_0(z_n;k), \qquad \textit{resp.} \qquad z_{n+1} = G_\infty(z_n;k),$$

converge to an attractive fixed point of G provided the segments $[z_n, z_{n+1}]$ lie outside the oval of Cassini $\{z; |z+ic||z-ic| = |c|\}$ for all $n > n_0$.

PROOF. Since $\mathscr{B} = F(\mathscr{A})$, then, from the assumption, $[z_n, z_{n+1}] \notin \mathscr{A} \cup \mathscr{B}$ for all $n > n_0$. For G_0, with the integral below taken along the segment $[z_n, z_{n+1}]$, one has

$$\begin{aligned}|z_{n+2} - z_{n+1}| &= |G_0(z_{n+1};k) - G_0(z_n;k)| \\ &= |\int_{z_n}^{z_{n+1}} G'(z)\,dz| \\ &\leq d|z_{n+1} - z_n|, \qquad d < 1,\end{aligned}$$

since $|G'| < 1$ on the segment $[z_n, z_{n+1}]$. Similarly for G_∞. □

It is important to remark that $\mathscr{B}/\mathscr{A}$ does not contain any fixed points of F since $\mathscr{B} = F(\mathscr{A})$.

THEOREM 4. *If s is the unique root of $z\tan z = c$ lying in the first quadrant and in a strip $B_0(k) = \{z; -\frac{\pi}{2} + k\pi < \Re z \leq \frac{\pi}{2} + k\pi\}$ to the right of the oval of Cassini $\mathscr{O} = \{z; |z+ic||z-ic| = |c|\}$, then the iteration*

$$z_{n+1} = G_0(z_n;k), \tag{7.6}$$

where $-\frac{\pi}{2} + k\pi < \Re G_0(z;k) \leq \frac{\pi}{2} + k\pi$, converges to s for any initial value $z_0 \in B_0(k)\backslash\mathscr{A}$.

THEOREM 5. *If s is the unique root of $z\tan z = c$ lying in the first quadrant and in a strip $B_\infty(k) = \{z; k\pi < \Re z \leq (k+1)\pi\}$ to the left of* *the right-hand loop of the oval of Cassini* $\mathscr{O} = \{z; |z+ic||z-ic| = |c|\}$, *then the iteration*

$$z_{n+1} = G_\infty(z_n; k), \tag{7.7}$$

where $k\pi < \Re G_\infty(z;k) \leq \pi + k\pi$, *converges to* s *for any initial value* $z_0 \in B_\infty(k)\backslash\mathscr{A}$.

PROOF. The theorems follow from Lemma 2 above and the assumed unicity of the root. □

It is to be noted that the iterations (7.6), resp. (7.7), in the above theorems, do not cross over the chosen branch cuts of log, and all the iterates lie on the same sheet of the Riemann surface of G. This explains why the sequence $\{G(z_n; k)\}$ is normal; indeed, it converges (see [**1**, Lemma 1, p. 4]).

The presence of two roots, namely, $s^{(0)}$ and $s^{(k)}, k > 0$, in the same strip $B_0(k)$ or $B_\infty(k)$, when $c \in \text{II}$, and/or the nonempty intersection of the ovals of Cassini (6.9) with the strip containing one or two roots restrict the basins of attraction of the roots. In such cases the iteration may converge to a root or may end up in an attractive cycle, depending upon the starting value z_0. At most two such situations can occur for any given value of c.

It is easy to show by means of Rouché's theorem (see [**9**]) that the root $s^{(0)} \in \text{II}_0$ lies in the square bounded by the real and imaginary axes and the lines $\Re z = (2m+1)\pi/2$ and $\Im z = (2m+1)\pi/2$, if $m \geq 0.35|c| - 0.5$. Once $s^{(0)}$ is located, the other roots are easily obtained by means of F or G. Near the boundary of the region $\mathscr{A}$ of attractivity of F given by (6.5), Steffensen's method can be used to accelerate a slowly but already converging sequence; for, Steffensen's acceleration will turn any root of (1.1) into an attractive fixed point of F and G and thus may divert convergence to another root. Finally, if $c \in \text{I}$, Newton's method (4.1) with starting values (4.2) can be used effectively to find the desired roots.

8. Application to dielectric spectroscopy. In dielectric spectroscopy with an open-circuited coaxial line, typical measurements start at a low frequency ω which gives a measured normalized admittance $-c(\omega)$ with $c(\omega) \in \text{I}$ and a normalized propagation constant $iz(\omega)$, where $z(\omega) \in \text{I}_0$. Since in practice $10^{-2} < |c(\omega)| < 10^2$, for $c \in \text{I}$ Newton's method (4.1) with Cohen's starting values (4.2) will converge to the root to an accuracy of 10^{-10} in double precision arithmetic in not more than five iterations. As the frequency increases, the solution $z(\omega)$ of (1.1) will describe a smooth curve that may pass near the double roots p_1 and p_2 (see Fig. 1). When $z(\omega)$ moves into II_0, F, G_0 or G_∞ is used to continue the solution as the curve lies in the region $\mathscr{A}$ of attractivity of F given by (6.5), or not. Near a double root, the experimental error in the measurement of $c(\omega)$ increases due to resonance **[3]**. The interpolation technique developed in **[15]**, **[16]** can be used with F, G_0, and G_∞ to determine whether the solution curve goes above the double root in Fig. 1 and hence in $\mathscr{A}$, or below and hence outside $\mathscr{A}$. Then F, G_0 or G_∞ are used accordingly, eventually with Steffensen's acceleration method. Also in **[15]**, **[16]** a linear extrapolation is used to obtain the next starting value, i.e., $z_{(0)\mathrm{new}} = z_{\mathrm{old}}\omega_{\mathrm{new}}/\omega_{\mathrm{old}}$.

9. The equation $z \cot z = c$. The treatment of (1.2) is almost similar to that of (1.1). The double roots q_j of (1.2) are solutions of the equation

$$z - \sin z \cos z = 0.$$

The first few q_j and corresponding values of c_j are listed in Table 2.

j	q_j	c_j
0	0	1
1	$3.749 + 1.384i$	$1.895 - 3.719i$
2	$6.950 + 1.676i$	$2.180 - 6.933i$
3	$10.119 + 1.858i$	$2.361 - 10.107i$
4	$13.277 + 1.992i$	$2.493 - 13.268i$
5	$16.430 + 2.097i$	$2.598 - 16.422i$
6	$19.579 + 2.183i$	$2.684 - 19.573i$

Table 2. The first seven double roots of $z \cot z = c$ and the corresponding values of c

A rough graph of the images of the third and fourth quadrants, III and IV, of the c-plane into the first quadrant, I_z, of the z-plane by the multiple-valued function

$$c \to \{z; z \cot z = c\}$$

is given in Fig. 5. The region IV_0 is unbounded.

When $c \in \mathrm{III}$, Newton's method applied to (1.2) in the form

$$z \cos z - c \sin z = 0,$$

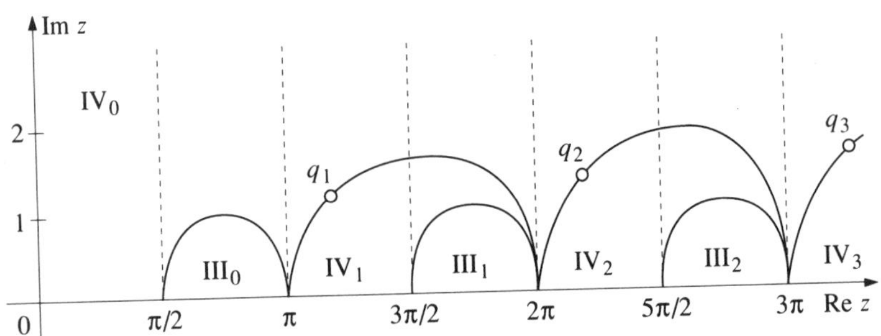

FIGURE 5. The figure shows the regions III_k and IV_k, $k \geq 0$, which are images of the lower-half of the c-plane into the first quadrant of the z-plane by the multiple-valued function $c \to \{z; z\cot z = c\}$. The points $q_1, q_2, \dots$ are the double roots of the equation $z \cot z = c$. The region IV_0 is unbounded.

namely,

$$z_{n+1} = z_n - \frac{z_n \cos z_n - c \sin z_n}{(1-c)\cos z_n - z_n \sin z_n}, \tag{9.1}$$

with the initial values

$$z_0 = \pi\sqrt{\frac{1-c}{4-c}}, \qquad \text{for } s^{(1)} \in \mathrm{III}_1, \tag{9.2a}$$

$$z_0 = (k+\frac{1}{2})\pi + \frac{\pi}{2}\frac{c}{c-2k-1}, \qquad \text{for } s^{(k)} \in \mathrm{III}_k, \quad k > 1, \tag{9.2b}$$

converges quickly to the desired roots $s^{(1)}$ or $s^{(k)}$.

The iteration function

$$F(z) = c \tan z \tag{9.3}$$

has transcendental critical points $\pm ic$. Its region of attractivity is given by

$$\mathscr{A} = \{z; |\cos z| > \sqrt{|c|}\}. \tag{9.4}$$

The image of $\mathscr{A}$, $\mathscr{B} = F(\mathscr{A})$, is given geometrically by the inequality

$$\mathscr{B} = \{z; |z+ic||z-ic| < |c|\} \tag{9.5}$$

and is bounded by the oval of Cassini

$$\mathscr{O} = \{z; |z+ic||z-ic| = |c|\} \tag{9.6}$$

with foci $\pm ic$ and distance product $r_1 r_2 = |c|$. The inverse iteration function

$$G(z) = \frac{1}{2i}\log\left(\frac{ic-z}{ic+z}\right) + k\pi \tag{9.7}$$

has determination $G_\infty(z;k)$ when the branch cut of $\log t$ is taken along the negative real axis; in this case $-\pi < \arg t \leq \pi$. For the other determination, $G_0(z;k)$, the branch cut of $\log t$ is taken along the positive real axis, and $0 < \arg t \leq 2\pi$. Theorems similar to those established for (1.1) also hold for (1.2). By Rouché's theorem, one can show that the root $s^{(0)} \in \mathrm{IV}_0$, for a given value of $c \in \mathrm{IV}$, lies in the square bounded by the real and imaginary axes and the lines $\Re z = m\pi$ and $\Im z = m\pi$ provided $m > 0.32|c|$.

References

1. I. N. Baker, *Limit functions and sets of non-normality in iteration theory*, Ann. Acad. Sci. Fenn. Ser. A I **467** (1970), 11 pp.

2. P. Blanchard, *Complex analytic dynamics on the Riemann sphere*, Bull. Amer. Math. Soc. (N.S.) **11** (1984), 85–141.

3. H. E. Bussey, *Dielectric measurements in a shielded open circuit coaxial line*, IEEE Trans. Instrum. Meas. **IM-29** (1980), 120–124.

4. E. R. Cohen, Science Center, Rockwell International, P.O. Box 1085, Thousand Oaks, CA 91360, private correspondence, 18 April 1985.

5. R. L. Devaney, *Chaotic burst in nonlinear dynamical systems*, Science **237** (1987), 342–345.

6. N. Doual, J. L. Howland, and Rémi Vaillancourt, *Global iterative solutions of elementary transcendental equations*, Proc. of Intern. Conf. on Numer. Math., Singapore, Intern. Series of Numer. Math., Vol. 86, pp. 127–136, Birkhäuser Verlag, Basel, Boston, Stuttgart, 1988.

7. Sylvie Gélinas and Rémi Vaillancourt, *Recherche globale des zéros d'une fonction entière par la théorie de Fatou-Baker*, Ann. Sci. Math. Québec (1990) (to appear).

8. P. Henrici, *Some applications of the quotient-difference algorithm*, Proc. Sympos. Appl. Math., Vol. XV, pp. 159–183, Amer. Math. Soc., Providence, R.I, 1963.

9. J. L. Howland and Rémi Vaillancourt, *Attractive cycles in the iteration of meromorphic functions*, Numer. Math. **46** (1985), 323–337.

10. E. Isaacson and H. B. Keller, *Analysis of Numerical Methods*, Wiley, New York, London, Sydney, 1966, pp. 103–108.

11. E. Jahnke and F. Emde, *Tables of Functions with Formulae and Curves*, Dover, New York, 1945, Addenda, p. 66.

12. J. D. Lawrence, *A Catalog of Special Plane Curves*, Dover, New York, 1972, pp. 153–155.

13. P. H. Morse and H. Feshbach, *Methods of Theoretical Physics*, McGraw-Hill, New York, Toronto, London, 1953, pp. 405 and 1522.

14. H. Rådström, *On the iteration of analytic functions*, Math. Scand. **1** (1953), 85–92.

15. W. R. Scott, Jr., *Dielectric Spectroscopy Using Shielded Open-Circuited Coaxial Lines and Monopole Antennas of General Length*, Georgia Institute of Technology, Atlanta, GA 30332, October, 1985, Ph.D. Thesis, pp. xviii–230.

16. W. R. Scott, Jr. and G. S. Smith, *Error analysis for dielectric spectroscopy using shielded open-circuited coaxial lines of general length*, IEEE Trans. Instrum. Meas. **IM-35** (1986), 130–137.

17. M. Sucher and J. Fox, eds., *Handbook of Microwave Measurements*, Polytechnic Press, distributed by Wiley, New York, London, Third edition, Vol. II, 1963, p. 505.

18. Lucie Tanguay and Rémi Vaillancourt, *Numerical solution of the dielectric equation for a coaxial line*, IEEE Trans. Instrum. Meas. **IM-33** (1984), 88–90.

19. V. N. Tran, S. Stuckly, and M. Stuckly, *Measurement of dielectric properties of biological or agricultural materials*, Proc. 19th Intern. Elect. Convent., Inst. Radio Elec. Eng., Sydney, Australia, 1983, pp. 595–597.

20. *CRC Standard Mathematical Tables*, 27th Edition, W. H. Beyer (ed.), CRC Press, Boca Raton, FL, 1984, pp. 218–219.

Université du Québec à Trois-Rivières

Université d'Ottawa

Lectures in Applied Mathematics
Volume **26** (1990)

An Introduction to PL Algorithms

KURT GEORG

Abstract. We give a brief introduction to piecewise linear (PL) algorithms, also called complementary pivot or fixed point algorithms. Our approach is based on the fundamental presentation of Eaves [**14**]; hence we describe the algorithms in the general setting of PL manifolds. In particular, we introduce the PL homotopy method of Eaves and Saigal [**16**]. The recently established class of variable dimension algorithms will be presented. We use a particular cone construction for handling the homotopy parameter. Special attention is given to convergence results. Numerical details of the algorithms can only be sketched. For a more detailed presentation of such algorithms and bibliographical remarks, we refer to [**4**].

I. Introduction.

1. The first and most prominent example of a PL algorithm was designed by Lemke and Howson [**33**] and Lemke [**30**] to calculate a solution of the linear complementarity problem. This algorithm played a crucial role in the development of subsequent PL algorithms. Linear complementarity problems arise in quadratic programming, bimatrix games, variational inequalities and economic equilibria problems, and numerical methods for their solution have been of considerable interest. Scarf [**39**] gave a numerically implementable proof of the Brouwer fixed point theorem, based upon Lemke's algorithm. Eaves [**13**] observed that a related class of algorithms can be obtained by considering PL approximations of homotopy maps. Thus the PL continuation methods began to emerge as a parallel to the classical imbedding or predictor-corrector methods.

1980 *Mathematics Subject Classification* (1985 *Revision*). Primary 65H10, 90C30.
Partially supported by the National Science Foundation under grant # DMS-8805682.

2. The PL methods require no smoothness of the underlying equations and hence have, at least in theory, a more general range of applicability than classical imbedding methods. In fact, they can be used to calculate fixed points of set-valued maps. They are more combinatorial in nature and are closely related to the topological degree; see Peitgen and Siegberg [**37**]. PL continuation methods are usually considered to be less efficient than the predictor-corrector methods when the latter are applicable, especially in higher dimensions. The reasons for this lie in the fact that steplength adaptation and exploitation of special structure are more difficult to implement for PL methods.

II. PL manifolds.

3. In order to give an idea of the flexibility which is possible and to describe the construction of PL algorithms for all kinds of special purposes, Eaves [**14**] has given a very elegant geometric approach to general PL methods, which has strongly influenced the writing of this introduction; see also Eaves and Scarf [**17**]. Hence, we are going to cast the notion of PL algorithms into the general setting of subdivided manifolds which we will call *PL manifolds*.

4. Throughout this paper, we assume that $\mathscr{E}$ denotes some ambient finite-dimensional Euclidean space which contains all points arising in the sequel. We denote the transposition of a column $x \in \mathscr{E}$ by x^*, and hence the scalar product in $\mathscr{E}$ is written as x^*y for $x, y \in \mathscr{E}$. A *half-space* η and the corresponding *hyperplane* $\partial\eta$ are defined by $\eta = \{y \in \mathscr{E} : x^*y \leq \xi\}$ and $\partial\eta = \{y \in \mathscr{E} : x^*y = \xi\}$, respectively, for some $x \in \mathscr{E}$ with $x \neq \mathbf{0}$ and some $\xi \in \mathbf{R}$. A finite intersection of half-spaces is called a *cell*. If σ is a cell and $\tau \subset \sigma$ is convex, then we call τ a *face* of σ if the condition

$$(4.1) \qquad x, y \in \sigma,\ 0 < \lambda < 1,\ \lambda x + (1 - \lambda)y \in \tau \quad \Longrightarrow \quad x, y \in \tau$$

holds for all $x, y \in \sigma$. Trivially, σ is a face of itself. For notational reasons, we also consider the empty set to be a face of σ. In the theory of convex sets the above definition of a face coincides with that of an *extremal set*. By using separation theorems for convex sets (see, e.g., [**12**] or [**38**]), it can be shown that a subset $\tau \subset \sigma$, $\tau \neq \sigma$, is a face of σ if and only if there is a half-space ξ such that $\sigma \subset \xi$ and $\tau = \sigma \cap \partial\xi$. From this characterization it follows immediately that a face of a cell is again a cell. The *dimension* of a cell is the dimension of its affine hull, i.e., $\dim \sigma := \dim \operatorname{aff} \sigma$. In particular, the dimension of a singleton is

0 and the dimension of the empty set is -1. If the singleton $\{v\}$ is a face of σ, then v is called a *vertex* of σ. If τ is a face of σ such that $\dim \tau = \dim \sigma - 1$, then τ is called a *facet* of σ.

5. A *PL manifold* of dimension n is a system $\mathscr{M} \neq \varnothing$ of cells of dimension n such that the following conditions hold:

(5.1) If $\sigma_1, \sigma_2 \in \mathscr{M}$, then $\sigma_1 \cap \sigma_2$ is a common face of σ_1 and σ_2.

(5.2) A cell τ of dimension $n-1$ can be a facet of at most two cells in $\mathscr{M}$.

(5.3) The family $\mathscr{M}$ is locally finite, i.e., any relatively compact subset of

(5.4) $|\mathscr{M}| := \bigcup_{\sigma \in \mathscr{M}} \sigma$

meets only finitely many cells $\sigma \in \mathscr{M}$.

We say that the PL manifold $\mathscr{M}$ *subdivides* the set $|\mathscr{M}|$. Let us further introduce the notation

$$\mathscr{M}^k := \{\tau \colon \tau \text{ is a face of some } \sigma \in \mathscr{M} \text{ such that } \dim \tau = k\}, \tag{5.5}$$

for $k = -1, 0, 1, 2, 3, \ldots$, and

$$\mathscr{M}^\infty := \bigcup_{k=-1}^{\infty} \mathscr{M}^k. \tag{5.6}$$

We see that $\mathscr{M} = \mathscr{M}^n$ if n is the dimension of $\mathscr{M}$. From (5.2) it follows that a facet $\tau \in \mathscr{M}^{n-1}$ is common to at most two cells in $\mathscr{M}$. We therefore introduce the *boundary* $\Delta\mathscr{M} \subset \mathscr{M}^{n-1}$ of $\mathscr{M}$ as the system of facets $\tau \in \mathscr{M}^{n-1}$ which are common to exactly one cell of $\mathscr{M}$. Generally, we cannot expect $\Delta\mathscr{M}$ to again be a PL manifold. However, this is true for subdivisions of convex sets.

6. Theorem. *Let $\mathscr{M}$ be a PL manifold of dimension $n > 0$ such that $|\mathscr{M}|$ is convex. Then $\Delta\mathscr{M}$ is a PL manifold of dimension $n-1$ without boundary.*

7. Proof. The proof is rather technical, and we give only a sketch of the main ideas. The theorem is trivial in case $n = 1$, and we hence assume $n > 1$. If $\sigma \subset \mathscr{E}$ is a set, we denote by $\operatorname{int} \sigma$ the *interior* of σ relative to the affine hull $\operatorname{aff} \sigma$. Note that the relative interior of a singleton is the singleton itself. From properties (5.1)–(5.3) it is evident that the system

$$\{\operatorname{int} \tau \colon \tau \in \mathscr{M}^\infty\} \tag{7.1}$$

is a pairwise disjoint subdivision of $|\mathcal{M}|$. Therefore we only have to show that a cell $\xi \in (\Delta\mathcal{M})^{n-2}$ is a common facet to exactly two cells in $\Delta\mathcal{M}$. Hence, let $\tau_1 \in \Delta\mathcal{M}$, and let $\xi \in \mathcal{M}^{n-2}$ be a facet of τ_1. We choose points $y_1 \in \operatorname{int}\tau_1$ and $z \in \operatorname{int}\xi$ and consider the path

$$b\colon \varepsilon \in \mathbf{R} \longmapsto (1-\varepsilon)y_1 + \varepsilon z \in |\mathcal{M}|.$$

Let $\sigma \in \mathcal{M}$ be the unique cell which contains τ_1, and let $x_0 \in \operatorname{int}\sigma$ be a chosen point. If x is any point in $|\mathcal{M}|$, then $(1-\lambda)x_0 + \lambda x \in |\mathcal{M}|$ for $0 \le \lambda \le 1$, and by the local finiteness (5.3) it follows that $(1-\lambda)x_0 + \lambda x \in \sigma$ for $\lambda > 0$ being sufficiently small. Hence, $\operatorname{aff}\sigma = \operatorname{aff}|\mathcal{M}|$, and $x_0 \in \operatorname{int}|\mathcal{M}|$. The function $\rho\colon \operatorname{aff}\sigma \to (0,\infty]$ defined by

$$\rho(x) := \sup\{\lambda > 0\colon (1-\lambda)x_0 + \lambda x \in |\mathcal{M}|\}$$

is continuous. Such functions play a fundamental role in the theory of convex sets; see, e.g., **[12]** or **[38]**. Actually, we need only the continuity of $\varepsilon \mapsto \rho\big(b(\varepsilon)\big)$ for ε in a neighborhood of 1. It is clear that $\rho\big(b(\varepsilon)\big) = 1$ for $0 \le \varepsilon \le 1$. Furthermore, since τ_1 is a boundary facet, a straightforward convexity argument shows that $\rho\big(b(\varepsilon)\big) \le 1$ for all $\varepsilon \in \mathbf{R}$. We now consider the projection of the path $b(\varepsilon)$ onto the boundary of $\mathcal{M}$:

$$c(\varepsilon) := \Big(1 - \rho\big(b(\varepsilon)\big)\Big)x_0 + \rho\big(b(\varepsilon)\big)b(\varepsilon).$$

From the property of the above subdivision (7.1) we see that the following two assertions hold:

(7.2) For $\varepsilon < 1$ and ε sufficiently near 1 we have $c(\varepsilon) \in \operatorname{int}\tau_1$.

(7.3) There is a unique $\tau_2 \in \Delta\mathcal{M}$ such that for $\varepsilon > 1$ and ε sufficiently near 1 we have $c(\varepsilon) \in \operatorname{int}\tau_2$.

We conclude that ξ is a common facet to exactly two cells in $\Delta\mathcal{M}$, namely, τ_1 and τ_2. This proves the assertion. □

8. As we will see, PL algorithms perform steps from one cell into an adjacent cell across a common facet. Such steps are possible in a unique way because of property (5.2). More precisely, let us consider a PL manifold $\mathcal{M}$ of dimension $n > 0$, a cell $\sigma \in \mathcal{M}$ and a facet $\tau \in \mathcal{M}^{n-1}$ such that $\tau \subset \sigma$. Then we have two possible cases:

(8.1) If $\tau \notin \Delta\mathcal{M}$, then there exists a unique cell $\tilde{\sigma} \in \mathcal{M}$ such that $\tilde{\sigma} \ne \sigma$ and $\tau \subset \tilde{\sigma}$. We say that σ is *pivoted into* $\tilde{\sigma}$ *across* the facet τ.

(8.2) If $\tau \in \Delta\mathcal{M}$, then the cell $\sigma \in \mathcal{M}$ containing τ is unique, and a pivoting of σ across τ is not possible.

9. EXAMPLE. For the implementation of PL algorithms, it is important to note that a PL manifold is "entered" into a computer by storing some data which describes a current cell σ and a current facet τ of σ, and by writing subroutines which handle the above pivoting steps. We illustrate this for the simple example where $\mathbf{R}^n$ is subdivided into cubes. Let $\mathbf{Z}$ denote the set of integers. For $p \in \mathbf{Z}^n$, $k \in \{1, 2, \dots, n\}$, and $s \in \{+1, -1\}$ we define the cell

$$\sigma_p := \left\{ x \in \mathbf{R}^n : ||x - p||_\infty \leq \frac{1}{2} \right\}$$

and its facet

$$\tau_{p,k,s} := \left\{ x \in \sigma_p : x[k] = p[k] + \frac{s}{2} \right\},$$

where $p[k]$ denotes the kth *coordinate* of p. Clearly, $\mathscr{M} := \{\sigma_p\}_{p \in \mathbf{Z}^n}$ is a PL manifold of dimension n which subdivides $\mathbf{R}^n$. It is evident that σ_p is pivoted into $\sigma_{\tilde{p}}$ across $\tau_{p,k,s}$ if we take $\tilde{p} = p + se_k$. Here e_k denotes the kth unit base vector of $\mathbf{R}^n$. The following pseudo code enables us to run through this manifold by pivoting across facets. It is assumed at each step that a decision has been made for determining across which facet the current cell has to be pivoted next. The code is very trivial and is given here only in order to familiarize the reader with the approach which will be adopted in this paper.

10. PIVOTING CUBES. *comment:*

input $p \in \mathbf{Z}^n$ *data characterizing initial cell*

repeat

  enter $s \in \{+1, -1\}$, $k \in \{1, \dots, n\}$; *data characterizing facet*

  $p[k] \leftarrow p[k] + \frac{s}{2}$ *pivoting*

until pivoting is stopped.

11. A cell of particular interest is a *simplex* $\sigma = [v_1, v_2, \dots, v_{n+1}]$ of dimension n which is defined as the convex hull of $n+1$ affinely independent points $v_1, v_2, \dots, v_{n+1} \in \mathscr{E}$. These points are the vertices of σ. If a PL manifold $\mathscr{M}$ of dimension n consists only of simplices, then $\mathscr{M}$ is called a *pseudo manifold* of dimension n. Such manifolds are of special importance, since PL methods simplify considerably; see, e.g., **[2]**, **[3]**, **[4]**, **[25]**, **[40]**, **[42]**. In this case, a pivoting step can also be described in a different way. Let $\sigma = [v_1, v_2, \dots, v_{n+1}] \in \mathscr{M}$.

Then a facet τ of σ is obtained by deleting one vertex, say the vertex v_i for some $i \in \{1, 2, \dots, n+1\}$. Thus we obtain the facet $\tau = [v_1, \dots, v_{i-1}, v_{i+1}, \dots, v_{n+1}]$ *lying opposite the vertex* v_i of σ. Again we consider two cases.

(11.1) If $\tau \notin \Delta\mathscr{M}$, then σ is pivoted across τ into

$$\tilde{\sigma} = [v_1, \dots, v_{i-1}, \tilde{v}_i, v_{i+1}, \dots, v_{n+1}],$$

where this one vertex v_i is replaced by a new vertex $\tilde{v}_i$. We say that the *vertex* v_i *of* σ *is pivoted into* $\tilde{v}_i$.

(11.2) If $\tau \in \Delta\mathscr{M}$, then the cell $\sigma \in \mathscr{M}$ containing τ is unique, and a pivoting of v_i is not possible.

If a pseudo manifold $\mathscr{T}$ subdivides a set M, then we also say that $\mathscr{T}$ *triangulates* M or that $\mathscr{T}$ is a *triangulation* of M. Some triangulations of $\mathbf{R}^n$ of practical importance were already considered by Coxeter **[9]** and Freudenthal **[18]**; see also Todd **[42]**. Eaves **[15]** gave an overview of standard triangulations. An affine image of the simplest such triangulation can be generated by the following pivoting rules (see **[1]** or **[10]**):

12. STANDARD TRIANGULATION OF $\mathbf{R}^n$. *comment:*

input $[v_1, v_2, \dots, v_{n+1}] \subset \mathbf{R}^n$; *starting simplex*
for $k = 1, \dots, n$ **do** $\rho(k) \leftarrow k + 1$; *cyclic right shift*
$\rho(n+1) \leftarrow 1$;
repeat
  enter $i \in \{1, 2, \dots, n+1\}$; *index of vertex to be pivoted next*
  $v_i \leftarrow v_{\rho^{-1}(i)} - v_i + v_{\rho(i)}$ *pivoting by reflection*
until pivoting is stopped.

A triangulation of $\mathbf{R}^n$ which is preferable to 12 from various viewpoints has been introduced by Todd **[42]** but can also be obtained via certain reflection rules of Coxeter **[9]**:

13. UNION JACK TRIANGULATION OF $\mathbf{R}^n$. *comment:*

input $[v_1, v_2, \ldots, v_{n+1}] \subset \mathbf{R}^n$; *starting simplex*

repeat

enter $i \in \{1, 2, \ldots, n+1\}$; *index of vertex to be pivoted next*

$$v_i \leftarrow \begin{cases} 2v_2 - v_1 & \text{for } i = 1, \\ 2v_n - v_{n+1} & \text{for } i = n+1, \\ v_{i-1} - v_i + v_{i+1} & \text{else;} \end{cases}$$ *pivoting by reflection*

until pivoting is stopped.

III. PL algorithms.

14. Let $\mathscr{M}$ be a PL manifold of dimension $n+1$. We call $H: \mathscr{M} \to \mathbf{R}^n$ a *PL map* if

(14.1) $H: |\mathscr{M}| \to \mathbf{R}^n$ is a map;

(14.2) the restriction $H_\sigma: \sigma \to \mathbf{R}^n$ of H to σ is an affine map for all $\sigma \in \mathscr{M}$.

In this case, H_σ can be uniquely extended to an affine map $H_\sigma:$ aff $\sigma \to \mathbf{R}^n$. If we introduce the $(n+1)$-dimensional linear space $\operatorname{tng} \sigma := \{x - y : x, y \in \operatorname{aff} \sigma\}$ as the *tangent space* of σ, then the *Jacobian* $H'_\sigma: \operatorname{tng} \sigma \to \mathbf{R}^n$ is defined as the linear map which has the property $H'_\sigma(x - y) = H_\sigma(x) - H_\sigma(y)$ for $x, y \in \operatorname{aff} \sigma$. Note that H'_σ corresponds to an $n \times (n+1)$ matrix which has a one-dimensional kernel in case of nondegeneracy, i.e., if its rank is maximal.

15. Let $\mathscr{M}$ be a PL manifold of dimension $n+1$ and $H: \mathscr{M} \to \mathbf{R}^n$ a PL map. A PL algorithm is essentially a method for following a polygonal path in $H^{-1}(0)$. To avoid degeneracies, we introduce a concept of regularity; see **[14]**. A point $x \in |\mathscr{M}|$ is called a *regular point* of H if

(15.1) x is not contained in any lower-dimensional face $\tau \in \mathscr{M}^k$ for $k < n$;

(15.2) H'_σ has maximal rank n for all $\sigma \in \mathscr{M}^n \cup \mathscr{M}^{n+1}$ such that $x \in \sigma$.

A value $y \in \mathbf{R}^N$ is a *regular value* of H if all points in $H^{-1}(y)$ are regular. By definition, y is vacuously a regular value if it is not contained in the range of H. If a point is not regular it is called *singular*. Analogously, if a value is not regular it is called singular. A standard perturbation technique (see, e.g., **[11]**, **[14]**, **[37]**) is used in order to prove a Sard type theorem:

16. Perturbation Lemma. *Let $H: \mathcal{M} \to \mathbf{R}^n$ be a PL map where $\mathcal{M}$ is a PL manifold of dimension $n+1$. Then for any relatively compact subset $C \subset |\mathcal{M}|$ there are at most finitely many $\varepsilon > 0$ such that $C \cap H^{-1}(\vec{\boldsymbol{\varepsilon}})$ contains a singular point of H. Consequently, $\vec{\boldsymbol{\varepsilon}}$ is a regular value of H for almost all $\varepsilon > 0$. Here we use the notation*

$$\vec{\boldsymbol{\varepsilon}} := \begin{pmatrix} \varepsilon \\ \varepsilon^2 \\ \vdots \\ \varepsilon^n \end{pmatrix}.$$

17. Proof. The proof will be given by contradiction. Let us assume that there is a strictly decreasing sequence $\{\varepsilon_i\}_{i\in\mathbf{N}}$ of positive numbers, converging to zero, for which a bounded sequence $\{x_i\}_{i\in\mathbf{N}} \subset |\mathcal{M}|$ of singular points can be found such that the equations

$$H(x_i) = \vec{\boldsymbol{\varepsilon}}_i, \tag{17.1}$$

for $i \in \mathbf{N}$, are satisfied. For any subset $I \subset \mathbf{N}$ of cardinality $n+1$ we see that the $\{\vec{\boldsymbol{\varepsilon}}_i\}_{i\in I}$ are affinely independent, and by (17.1) and the piecewise linearity (14.2) of H the $\{x_i\}_{i\in I}$ cannot all be contained in the same lower-dimensional face $\tau \in \mathcal{M}^k$ for $k < n$. Since this holds for all index sets I, we use this argument repeatedly, and the local finiteness (5.3) of $\mathcal{M}$ permits us to find a strictly increasing function $\nu: \mathbf{N} \to \mathbf{N}$ (to generate a subsequence), and to find a face $\sigma \in \mathcal{M}^{n+1} \cup \mathcal{M}^n$ such that the subsequence $\{x_{\nu(i)}\}_{i\in\mathbf{N}}$ is contained in $\operatorname{int}\sigma$. But now we can again use the above argument: for an index set $I \subset \nu(\mathbf{N})$ of cardinality $n+1$ the $\{\vec{\boldsymbol{\varepsilon}}_i\}_{i\in I}$ are affinely independent, and we conclude that H'_σ has maximal rank n. However, this means that all points $\{x_{\nu(i)}\}_{i\in\mathbf{N}}$ are regular, a contradiction to the choice of $\{x_i\}_{i\in\mathbf{N}}$. The last assertion of the perturbation lemma follows since $|\mathcal{M}|$ can be written as a countable union of relatively compact subsets. □

18. The last lemma enables us to confine ourselves to regular values. We now show that inverse images of regular values consist of polygonal paths whose vertices are always in the interior of some facet. More precisely, we have

19. Lemma. *Let $H: \mathcal{M} \to \mathbf{R}^n$ be a PL map where $\mathcal{M}$ is a PL manifold of dimension $n+1$, and let $\varepsilon > 0$ be such that $\vec{\boldsymbol{\varepsilon}}$ is a regular value of* **H**. *We consider a facet $\tau \in \mathcal{M}^n$ and a cell $\sigma \in \mathcal{M}$. Then the following is*

true:

(19.1) *If* $\tau \cap H^{-1}(\vec{\varepsilon}) \neq \emptyset$, *then* $H^{-1}(\vec{\varepsilon})$ *meets* τ *in exactly one point* $a \in \operatorname{int}\tau$. *Hence, facets are intersected* **transversely** *by* $H^{-1}(\vec{\varepsilon})$.

If $\sigma \cap \mathbf{H}^{-1}(\vec{\varepsilon}) \neq \emptyset$, *then one of the following three cases holds*:

Segment: $H^{-1}(\vec{\varepsilon})$ *meets exactly two facets* τ_1, τ_2 *of* σ, *and we have*

$$\tau_1 \cap H^{-1}(\vec{\varepsilon}) = \{a_1\}, \quad \tau_2 \cap H^{-1}(\vec{\varepsilon}) = \{a_2\}\},$$
$$(\operatorname{int}\sigma) \cap H^{-1}(\vec{\varepsilon}) = \{(1-\lambda)a_1 + \lambda a_2 \colon 0 < \lambda < 1\}\}.$$

Ray: $H^{-1}(\vec{\varepsilon})$ *meets exactly one facet* τ *of* σ, *and we have*

$$\tau \cap H^{-1}(\vec{\varepsilon}) = \{a_1\}, \quad a_2 \in \operatorname{int}\sigma,$$
$$(\operatorname{int}\sigma) \cap H^{-1}(\vec{\varepsilon}) = \{(1-\lambda)a_1 + \lambda a_2 \colon 0 < \lambda\}\}.$$

Line: $H^{-1}(\vec{\varepsilon})$ *meets no facet of* σ, *and we have*

$$a_1, a_2 \in \operatorname{int}\sigma, \quad a_1 \neq a_2,$$
$$(\operatorname{int}\sigma) \cap H^{-1}(\vec{\varepsilon}) = \sigma \cap H^{-1}(\vec{\varepsilon}) = \{(1-\lambda)a_1 + \lambda a_2 \colon \lambda \in \mathbf{R}\}.$$

20. PROOF. "Ad (19.1)": Since $\vec{\varepsilon}$ is a regular value, (15.1) implies $\tau \cap H^{-1}(\vec{\varepsilon}) \subset \operatorname{int}\tau$. If $x_1, x_2 \in \tau \cap H^{-1}(\vec{\varepsilon})$, then $H'_\tau(x_1 - x_2) = 0$, and (15.2) implies $x_1 = x_2$. This proves assertion (19.1). "Ad (19.2)": By (19.1), $H^{-1}(\vec{\varepsilon})$ cannot intersect σ only in a boundary point. Hence $\sigma \cap H^{-1}(\vec{\varepsilon})$ is a 1-dimensional cell which has its boundary points in the boundary of σ. The assertion is now straightforward. □

21. Let us consider a connected component $\mathscr{C}$ of $H^{-1}(\vec{\varepsilon})$. From the above discussion it is clear that we can introduce a direction of traversing as orientation, and we can parametrize $\mathscr{C}$ with respect to, say, arclength, so that we obtain a polygonal path

$$c \colon I \longrightarrow H^{-1}(\vec{\varepsilon})$$

which is defined on some suitable interval $I \subset \mathbf{R}$. We call a parameter $s \in I$ a *node* of c if $c(s)$ is in some facet $\tau \in \mathscr{M}^n$, we call s a *boundary node* if $c(s)$ is in some boundary facet $\tau \in \partial\mathscr{M}$, and we call s an *inner node* if $c(s)$ is in some inner facet $\tau \in \mathscr{M}^n \setminus \partial\mathscr{M}$. We say that the path c has a *boundary start* if it has a smallest node which is a boundary node, a *ray start* if it has a smallest node which is an inner node, and *no start* if it does not have a smallest node. Analogously, c has a *boundary termination* if it has a largest node which is a boundary node, a *ray termination* if it has a largest node which is an inner node, and *no termination* if it does not have a largest node. From the regularity of the value $\vec{\boldsymbol{\varepsilon}}$ it is clear that the path c is locally injective, and hence it is either injective and has one of the above starts and terminations (which makes 9 cases), or it is cyclic. In the latter case, $I = \mathbf{R}$, and there is a number $p > 0$ such that $c(s) = c(\tilde{s})$ if and only if $s - \tilde{s}$ is an integer multiple of the period p. Furthermore, a cyclic path has no start and no termination and infinitely many nodes. Let us also note that there is a pathological case where the path c does not have any node at all. This corresponds to the case "line" of (19.2) and is of no interest for PL algorithms.

22. *Completely labeled facets and transverse cells.* Let $H\colon \mathscr{M} \to \mathbf{R}^n$ be a PL map where $\mathscr{M}$ is a PL manifold of dimension $n+1$. Roughly speaking, a *PL algorithm* traces the polygonal path c discussed in the previous section from node to node. In order to also handle the case where zero is a singular value of H, we choose $\varepsilon > 0$ such that $\vec{\boldsymbol{\varepsilon}}$ is a regular value, and then we let ε tend to zero. This is possible via lemma 16. We are thus led to the following definition:

(22.1) A facet $\tau \in \mathscr{M}^n$ is called *completely labeled* with respect to H if $\tau \cap H^{-1}(\vec{\boldsymbol{\varepsilon}}) \neq \varnothing$ for all sufficiently small $\varepsilon > 0$.

(22.2) A cell $\sigma \in \mathscr{M}$ is called *transverse* with respect to H, if $\sigma \cap H^{-1}(\vec{\boldsymbol{\varepsilon}}) \neq \varnothing$ for all sufficiently small $\varepsilon > 0$.

Since $\sigma \cap H^{-1}(\vec{\boldsymbol{\varepsilon}})$ cannot contain any singular point for all sufficiently small $\varepsilon > 0$ (see lemma 16), the three distinct cases of lemma 19 do not vary for $\varepsilon > 0$ tending to zero, provided that $\varepsilon > 0$ is sufficiently small. Hence, a transverse cell σ contains

(22.3) no completely labeled facet in case of a line,

(22.4) exactly one completely labeled facet in case of a ray,

(22.5) exactly two completely labeled facets in case of a segment.

For $\varepsilon > 0$ sufficiently small, a node of the above polygonal path c corresponds to a completely labeled facet which is intersected by c, and hence the PL algorithm traces such completely labeled facets. It is usually started either on the boundary, i.e., in a completely labeled facet $\tau \in \partial\mathscr{M}$, or on a ray, i.e., in a transverse cell $\sigma \in \mathscr{M}$ which has only one completely labeled facet. We are thus led to the following two generic versions of a PL algorithm.

23. GENERIC PL ALGORITHM WITH BOUNDARY START. *comment:*

input $\tau_1 \in \partial\mathcal{M}$ completely labeled; *starting facet*
find the unique $\sigma_1 \in \mathcal{M}$ such that $\tau_1 \subset \sigma_1$; *starting cell*
for $i = 1, 2, 3, \ldots$ **do**
  begin
  if τ_i is the only completely labeled
    facet of σ_i **then** stop *ray termination*
  else find the unique completely labeled facet *PL step*
    τ_{i+1} of σ_i such that $\tau_{i+1} \neq \tau_i$;
  if $\tau_{i+1} \in \partial\mathcal{M}$ **then** stop *boundary termination*
  else pivot σ_i across τ_{i+1} into σ_{i+1} *pivoting step*
  end.

24. GENERIC PL ALGORITHM WITH RAY START. *comment:*

input $\sigma_1 \in \mathcal{M}$ transverse *starting cell*
  which possesses exactly one completely labeled facet;
find the unique facet τ_2 of σ_1 *initial PL step*
  which is completely labeled;
for $i = 2, 3, \ldots$ **do**
  begin
  if $\tau_i \in \partial\mathcal{M}$ **then** stop *boundary termination*
  else pivot σ_{i-1} across τ_i into σ_i; *pivoting step*
  if τ_i is the only completely labeled
    facet of σ_i **then** stop *ray termination*
  else find the unique completely labeled facet *PL step*
    τ_{i+1} of σ_i such that $\tau_{i+1} \neq \tau_i$
end.

IV. Numerical considerations.

25. From a numerical point of view, two steps have to be efficiently implemented. The *pivoting step* finds a new cell which is adjacent to a given cell across a given facet. Usually, the current cell is stored in the computer via some characteristic data, and a pivoting step is implemented by describing how this data is changed. In order to do

this, a facet has to be specified at each step. Usually, this facet will be completely labeled. The pseudo codes 10 and 12 are simple examples. See **[15]** and **[42]** for further implementations of pivoting rules.

26. The *PL step* finds the second completely labeled facet of a transverse cell or decides that a second such facet does not exist. This is usually a much more time-consuming step than the pivoting step since it involves some numerical linear algebra as is typical for linear programming methods. Let us consider an example. We assume that a cell of dimension $n+1$ is given by

$$\sigma := \{x \in \mathbf{R}^{n+1} : Lx \geq c\}, \tag{26.1}$$

where $L: \mathbf{R}^{n+1} \to \mathbf{R}^m$ is a linear map and $c \in \mathbf{R}^m$ a given value. Furthermore, let us assume that

$$\tau_i := \{x \in \mathbf{R}^{n+1} : Lx \geq c,\ e_i^* Lx = e_i^* c\}, \tag{26.2}$$

for $i = 1, 2, \ldots, m$, characterizes all the facets of σ. Here e_i denotes the ith *unit base vector* in $\mathbf{R}^m$ and '$*$' denotes *transposition*. On the cell σ, the PL map $H: \mathscr{M} \to \mathbf{R}^n$ reduces to an affine map, and hence there is a linear map $A: \mathbf{R}^{n+1} \to \mathbf{R}^n$ and a vector $b \in \mathbf{R}^n$ such that

$$\sigma \cap H^{-1}(0) = \{x \in \mathbf{R}^{n+1} : Ax = b,\ Lx \geq c\}. \tag{26.3}$$

Let τ_i be completely labeled. If we exclude degeneracies, then $\tau_i \cap H^{-1}(0) = \{x_0\}$ is a singleton, and there is a unique vector t in the one-dimensional kernel $A^{-1}(0)$ such that $e_i^* t = -1$. Furthermore, we have $e_j^* Lx_0 > e_j^* c$ for $j = 1, \ldots, m$, $j \neq i$, and hence $x_0 - \lambda t \in \operatorname{int} \sigma$ for small $\lambda > 0$. If (26.3) is a ray (cf. (19.2)), then $e_j^* L(x_0 - \lambda t) > e_j^* c$ for all $\lambda > 0$. Otherwise we have $e_j^* t > 0$ for at least one index j, and since we are excluding degeneracies, the minimization

$$k := \arg\min \left\{ \frac{e_j^*(Lx_0 - c)}{e_j^* t} : j = 1, \ldots, m,\ e_j^* t > 0 \right\} \tag{26.4}$$

yields the unique completely labeled facet τ_k of σ with $k \neq i$. For

$$\lambda_0 := \frac{e_k^*(Lx_0 - c)}{e_k^* t} > 0$$

we obtain $\sigma \cap H^{-1}(0) = \{x_0 - \lambda t : 0 \leq \lambda \leq \lambda_0\}$. Minimizations such as (26.4) are typical for linear programming, and the numerical linear algebra can be efficiently handled by standard routines. Successive LP steps can often make use of previous matrix factorizations via update methods **[22]**. In case of a pseudo manifold $\mathscr{M}$ where the cell σ is a simplex, it is convenient to handle the numerical linear algebra with respect to the barycentric coordinates based on the vertices of σ, and the equations become particularly simple; see, e.g., [2], [4], [42] for details.

V. PL homotopy algorithms.

27. Let us show how the above ideas can be used to approximate zero points of a map $G\colon \mathbf{R}^n \to \mathbf{R}^n$ by applying PL methods to an appropriate homotopy map. Eaves [**13**] presented the first such method. A restart method based on somewhat similar ideas was developed by Merrill [**34**]. As an example of a PL homotopy method we will take the algorithm of Eaves and Saigal [**16**]. To insure success of the algorithms, we will follow a presentation [**21**] which uses a quite general boundary condition extending somewhat that of [**34**].

28. Let us first introduce some notation. For $x \in \mathbf{R}^n$ we denote by $\mathscr{U}(x)$ the *system of neighborhoods* of x. By $\overline{\mathrm{co}}(X)$ we denote the closed convex hull of a set $X \subset \mathbf{R}^n$. By $\mathbf{R}^n_\Sigma$ we denote the system of compact convex nonempty subsets of $\mathbf{R}^n$. We call the map $G\colon \mathbf{R}^n \to \mathbf{R}^n$ *asymptotically linear* if the following three conditions hold:

(28.1) G is *locally bounded*, i.e., each point $x \in \mathbf{R}^n$ has a neighborhood $U \in \mathscr{U}(x)$ such that $G(U)$ is a bounded set.

(28.2) G is *differentiable at* ∞, i.e., there exists a linear map $G'_\infty\colon \mathbf{R}^n \to \mathbf{R}^n$ such that $\|x\|^{-1}\|G(x) - G'_\infty x\| \to 0$ for $\|x\| \to \infty$.

(28.3) G'_∞ is nonsingular.

If a map $G\colon \mathbf{R}^n \to \mathbf{R}^n$ is locally bounded, then we can define its *set-valued hull* $G_\Sigma\colon \mathbf{R}^n \to \mathbf{R}^n_\Sigma$ by setting

$$G_\Sigma(x) := \bigcap_{U \in \mathscr{U}(x)} \overline{\mathrm{co}}\Big(G(U)\Big). \tag{28.4}$$

It is not difficult to see that G_Σ is upper semicontinuous, and that G is continuous at x if and only if $G_\Sigma(x)$ is a singleton. By using a degree argument [**24**] on the set-valued homotopy

$$H_\Sigma(x, \lambda) := (1 - \lambda)G'_\infty x + \lambda G_\Sigma(x), \tag{28.5}$$

it can be seen that G_Σ has at least one zero point, i.e., a point $\bar{x}$ such that $0 \in G_\Sigma(\bar{x})$.

29. It is known [**36**], [**37**] that degree arguments in nonlinear analysis are essentially constructive. We describe how the above situation can be implemented by a PL homotopy method. Let us introduce the following canonical projections $\pi_n: \mathbf{R}^n \times \mathbf{R} \to \mathbf{R}^n$ and $\pi_1: \mathbf{R}^n \times \mathbf{R} \to \mathbf{R}$ by setting $\pi_n(x,\lambda) = x$ and $\pi_1(x,\lambda) = \lambda$. It is possible to construct a pseudo manifold $\mathscr{M}$ which triangulates $\mathbf{R}^n \times [0,\infty)$ and has the following *refining properties*:

(29.1) For each $\sigma \in \mathscr{M}$ there is an integer $i \geq 0$ which we call the *level* of σ such that $\pi_1(\sigma) \subset [i, i+1]$.

(29.2) There is a number $0 < \rho < 1$ which we call the *rate of refinement* of $\mathscr{M}$ such that for each $\sigma \in \mathscr{M}$ we have $\operatorname{diam} \pi_n(\sigma) \leq \delta\rho^i$, where i is the level of σ and $\delta > 0$ is a constant which characterizes the initial mesh of the triangulation.

The first such triangulation was proposed by Eaves [**13**]. Todd [**42**] gave a triangulation with refining factor $\frac{1}{2}$. Subsequently, many triangulations with arbitrary refining factors were developed (see [**15**]).

30. We now construct a PL homotopy for an asymptotically linear map $G: \mathbf{R}^n \to \mathbf{R}^n$. First we define $h: \mathbf{R}^n \times [0,\infty) \to \mathbf{R}^n$ by setting

$$h(x,\lambda) := \begin{cases} G'_\infty(x - x_1) & \text{for } \lambda = 0, \\ G(x) & \text{for } \lambda > 0. \end{cases} \tag{30.1}$$

Then we choose a pseudo manifold $\mathscr{M}$ which triangulates $\mathbf{R}^n \times [0,\infty)$ and has the refining properties (29.1)–(29.2). Since each cell in $\mathscr{M}$ is a simplex, there is a unique PL map $H: \mathscr{M} \to \mathbf{R}^n$ which coincides with h on all vertices: if $\sigma = [v_1, \ldots, v_{n+2}] \in \mathscr{M}$, see Section 11, and if a point $u \in \sigma$ is expanded into its *barycentric coordinates* $u = \sum_{i=1}^{n+2} c_i v_i$, then $H(u) = \sum_{i=1}^{n+2} c_i h(v_i)$.

31. *Eaves–Saigal algorithm.* We continue to consider the above refining pseudo manifold $\mathscr{M}$ and the PL map H generated by the asymptotically linear map G. According to theorem 6 it is clear that the boundary $\Delta\mathscr{M}$ is a pseudo manifold which triangulates the sheet $\mathbf{R}^n \times \{0\}$. If we assume that the starting point $u_1 := (x_1, 0)$ is in the interior of a facet $\tau_1 \in \Delta\mathscr{M}$, then it is immediately clear that τ_1 is the only completely labeled facet of $\Delta\mathscr{M}$. Hence, the PL algorithm started in τ_1 cannot terminate in the boundary, and since all cells of $\mathscr{M}$ are compact, it cannot terminate in a ray. Hence, it has no termination; see program 23.

Thus the PL algorithm generates a sequence $\tau_1, \tau_2, \ldots$ of completely labeled facets in $\mathscr{M}^n$. Let us also consider the polygonal path generated by the PL algorithm. This path is characterized by the nodes $(x_1, \lambda_1), (x_2, \lambda_2), \ldots$ such that (x_i, λ_i) is the unique zero point of the PL map H in τ_i for $i = 1, 2, \ldots$. We call $\bar{x} \in \mathbf{R}^n$ an *accumulation point* of the algorithm if some subsequence of $x_1, x_2, \ldots$ converges to $\bar{x}$.

32. LEMMA. *The set A of accumulation points of algorithm* 31 *is compact, connected, and nonempty. Each point $\bar{x} \in A$ is a zero point of G_Σ, i.e., we have $0 \in G_\Sigma(\bar{x})$.*

33. PROOF. From the construction of the PL map H it follows that

$$\lim_{||x|| \to \infty} ||x||^{-1} ||H(x, \lambda) - G'_\infty x|| = 0$$

uniformly for $\lambda \in [0, \infty)$. Since G'_∞ is nonsingular, $H(x_i, \lambda_i) = 0$ implies that the sequence x_i is bounded. Hence the set A is nonempty and compact.

Let us assume that A can be written as a disjoint union of two nonempty compact sets A_1 and A_2. Then $\operatorname{dist}(A_1, A_2) > 0$, and

$$\liminf_{i \to \infty} \operatorname{dist}(x_i, A_j) = 0 \text{ for } j = 1, 2.$$

On the other hand, $\lim_{i \to \infty} \operatorname{dist}(x_i, A) = 0$, and the refining property 29 implies that $\lim_{i \to \infty} ||x_i - x_{i+1}|| = 0$. This leads to a contradiction, and hence A is connected.

Since a PL manifold is locally bounded and the projections $\pi_n(\tau_i)$ stay in a bounded set, it follows that the level of the τ_i tends to ∞ for $i \to \infty$. Hence, for i sufficiently large, the definition 30 of H and the fact that the facets τ_i are completely labeled imply that $0 \in \operatorname{co} G(\pi_n(\tau_i))$. Since a point $\bar{x} \in A$ is an accumulation point of the sequence x_i, and since $\lim_{i \to \infty} \operatorname{diam} \pi_n(\tau_i) = 0$, we have that for each neighborhood $U \in \mathscr{U}(\bar{x})$ there is an i such that $\operatorname{co} G(\pi_n(\tau_i)) \subset G(U)$. Intersecting over all $U \in \mathscr{U}(\bar{x})$ and applying (28.4) gives $0 \in G_\Sigma(\bar{x})$. □

34. COROLLARY. *If the set-valued hull G_Σ has only isolated zero points, then the sequence x_i generated by algorithm* 31 *converges to a zero point of G_Σ.*

35. EXAMPLE. As a simple example, we consider the situation of the celebrated Brouwer fixed point theorem [5]. Let $F: C \to C$ be a continuous map on a convex, compact, nonempty subset $C \subset \mathbf{R}^n$ with nonempty interior. We define an asymptotically linear map $G: \mathbf{R}^n \to \mathbf{R}^n$

by setting

$$(35.1) \qquad G(x) := \begin{cases} x - F(x) & \text{for } x \in C, \\ x - x_1 & \text{for } x \notin C. \end{cases}$$

Here, a point x_1 in the interior of C is used as a starting point. The above PL algorithm generates a point $\bar{x} \in \mathbf{R}^n$ such that $0 \in G_\Sigma(\bar{x})$. If $\bar{x} \notin C$, then $G_\Sigma(\bar{x}) = \{\bar{x} - x_1\}$, but $\bar{x} \neq x_1$ implies that this case is impossible. If $\bar{x}$ is an interior point of C, then $G_\Sigma(\bar{x}) = \{\bar{x} - F(\bar{x})\}$, and hence $\bar{x}$ is a fixed point of F. If $\bar{x}$ is in the boundary ∂C, then $G_\Sigma(\bar{x})$ is the convex hull of $\bar{x} - x_1$ and $\bar{x} - F(\bar{x})$, and hence $\bar{x} = (1-\lambda)x_1 + \lambda F(\bar{x})$ for some $0 \leq \lambda \leq 1$. But $\lambda < 1$ would imply that $\bar{x}$ is an interior point of C, and hence we have $\lambda = 1$, and again $\bar{x}$ is a fixed point of F. Hence, the above PL homotopy algorithm generates a fixed point of F in either case. Many similar asymptotically linear maps can be constructed which correspond to important nonlinear problems; (see, e.g., **[4]**).

VI. Index and orientation.

36. Nearly all PL manifolds $\mathcal{M}$ which are of importance for practical implementations are orientable. If $\mathcal{M}$ is orientable and of dimension $n+1$, and if $H: \mathcal{M} \to \mathbf{R}^n$ is a PL map, then it is possible to introduce an index for the PL solution manifold $H^{-1}(0)$ which has important invariance properties and occasionally yields some useful information; see **[14]**, **[17]**, **[23]**, **[32]**, **[41]**, **[43]**. It should be noted that this index is closely related to the topological index which is a standard tool in topology and nonlinear analysis; see **[35]**, **[36]**, **[37]**. There are many ways to introduce the index for the present case. Our discussion follows **[2]**, **[4]**.

37. We begin with some basic definitions. Let F be a linear space of dimension k. An *orientation* of F is a function $\text{or}: F^k \to \{-1, 0, 1\}$ such that the following conditions hold:

(37.1) $\text{or}(b_1, \ldots, b_k) \neq 0$ if and only if $b_1, \ldots, b_k$ are linearly independent.

(37.2) $\text{or}(b_1, \ldots, b_k) = \text{or}(c_1, \ldots, c_k) \neq 0$ if and only if the transformation matrix between $b_1, \ldots, b_k$ and $c_1, \ldots, c_k$ has positive determinant.

It is clear from the basic facts of linear algebra that any finite dimensional linear space permits exactly two orientations.

A cell σ of dimension k is oriented by orienting its tangent space $\text{tng}(\sigma)$. Such an orientation or_σ of σ *induces an orientation* $\text{or}_{\tau,\sigma}$ on a facet τ of σ by the following convention:

$$\text{or}_{\tau,\sigma}(b_1,\ldots,b_{k-1}) := \text{or}_\sigma(b_1,\ldots,b_k) \tag{37.3}$$

whenever b_k points from τ into the cell σ. It is routine to check that the above definition of $\text{or}_{\tau,\sigma}$ verifies the conditions (37.1)–(37.2).

If $\mathcal{M}$ is a PL manifold of dimension $n+1$, then an *orientation of* $\mathcal{M}$ is a choice of orientations $\{\text{or}_\sigma\}_{\sigma\in\mathcal{M}}$ such that

$$\text{or}_{\tau,\sigma_1} = -\,\text{or}_{\tau,\sigma_2} \tag{37.4}$$

for each facet $\tau \in \mathcal{M}^N$ which is adjacent to two different cells $\sigma_1, \sigma_2 \in \mathcal{M}$. By making use of the standard orientation

$$\text{or}(b_1,\ldots,b_{n+1}) := \text{sign}\det(b_1,\ldots,b_{n+1}) \tag{37.5}$$

of $\mathbf{R}^{n+1}$, it is clear that any PL manifold of dimension $n+1$ which subdivides a subset of $\mathbf{R}^{n+1}$ is oriented in a natural way. But many oriented PL manifolds are known which are less trivial.

38. If $H: \mathcal{M} \to \mathbf{R}^n$ is a PL map on a PL manifold of dimension $n+1$ such that zero is a regular value of H, then it is clear that the system

$$\ker H := \{\sigma \cap H^{-1}(0)\}_{\sigma\in\mathcal{M}}$$

is a 1-dimensional PL manifold that subdivides the solution set $H^{-1}(0)$. For the case that $\mathcal{M}$ is oriented, the orientation of $\mathcal{M}$ and the natural orientation of $\mathbf{R}^n$ induce a *natural orientation of* $\ker H$. Namely, for $\xi \in \ker H$, $v \in \text{tng}(\xi)$ and $\sigma \in \mathcal{M}$ such that $\xi \subset \sigma$, the definition

$$\text{or}_\xi(v) := \text{or}_\sigma(b_1,\ldots,b_n,v)\ \text{or}(H'_\sigma b_1,\ldots,H'_\sigma b_n) \tag{38.1}$$

is independent of the special choice of $b_1,\ldots,b_n \in \text{tng}(\sigma)$, provided the $b_1,\ldots,b_n$ are linearly independent. Clearly, an orientation of the 1-dimensional manifold $\ker H$ is just a rule which indicates a direction for traversing each connected component of $\ker H$. Keeping this in mind, we now briefly indicate why the above definition indeed yields an orientation for $\ker H$. Let $\tau \in \mathcal{M}^n$ be a facet which meets $H^{-1}(0)$ and does not belong to the boundary $\Delta\mathcal{M}$, let $\sigma_1, \sigma_2 \in \mathcal{M}$ be the two cells adjacent to τ, and let $\xi_j := H^{-1}(0) \cap \sigma_j \in \ker H$ for $j = 1, 2$. If $b_1,\ldots,b_n$ is a basis of $\text{tng}(\tau)$, and if $v_j \in \text{tng}(\xi_j)$ points from τ into σ_j, then from condition (37.4) it follows that

$$\text{or}_{\sigma_1}(b_1,\ldots,b_n,v_1) = -\,\text{or}_{\sigma_2}(b_1,\ldots,b_n,v_2),$$

and hence (38.1) implies that

$$\operatorname{or}_{\xi_1}(v_1) = -\operatorname{or}_{\xi_2}(v_2),$$

which is exactly the right condition in the sense of (37.4) to ensure that the manifold $\ker H$ is oriented.

39. Let $H: \mathcal{M} \to \mathbf{R}^n$ be a PL map on an oriented PL manifold of dimension $n+1$. Given a facet τ of a cell $\sigma \in \mathcal{M}$, we can define an index by setting

$$\text{(39.1)} \qquad \operatorname{index}_{\tau,\sigma}(H) := \operatorname{or}_{\tau,\sigma}(b_1, \ldots, b_n) \operatorname{or}(H'_\sigma b_1, \ldots, H'_\sigma b_n)$$

if τ is completely labeled with respect to H, and $\operatorname{index}_{\tau,\sigma}(H) := 0$ else. It is clear that this definition is independent of the special choice of the basis $b_1, \ldots, b_n$ of $\operatorname{tng}(\tau)$. Furthermore, if zero is a regular value of H, τ is completely labeled and $\xi := H^{-1}(0) \cap \sigma \in \ker H$; then (38.1) implies that $\operatorname{index}_{\tau,\sigma}(H) = 1$ if and only if a positively oriented vector in $\operatorname{tng}\xi$ points from τ into σ. By possibly using the perturbation technique 22, the results of Section 38 immediately yield

$$\text{(39.2)} \qquad \operatorname{index}_{\tau_1,\sigma}(H) = -\operatorname{index}_{\tau_2,\sigma}(H)$$

for the case that τ_1 and τ_2 are two different completely labeled facets of a cell σ (see the PL step in algorithms 23–24), and

$$\text{(39.3)} \qquad \operatorname{index}_{\tau,\sigma_1}(H) = -\operatorname{index}_{\tau,\sigma_2}(H)$$

for the case that τ is a completely labeled facet of two different cells σ_1 and σ_2 (see the pivoting step in algorithms 23–24). A case of special importance is a facet τ in the boundary $\Delta\mathcal{M}$, since then we do not have to specify the cell σ which contains τ since σ is unique. If the PL algorithm 23 starts on the boundary in a completely labeled facet τ_1 and stops again in the boundary in a completely labeled facet τ_k, then the above formulae imply that

$$\text{(39.4)} \qquad \operatorname{index}_{\tau_1}(H) = -\operatorname{index}_{\tau_k}(H)$$

holds. Hence, for a compact PL manifold (i.e., if $|\mathcal{M}|$ is compact) where only a boundary start or termination is possible, we obtain the following celebrated index formula

$$\text{(39.5)} \qquad \sum_{\tau \in \Delta\mathcal{M}} \operatorname{index}_\tau(H) = 0.$$

VII. Lemke's algorithm.

40. The first and most prominent example of a PL algorithm was designed by Lemke **[30]**, **[33]** to calculate a solution of the linear complementarity problem. We present the Lemke algorithm as an example of a PL algorithm since it played a crucial role in the development of subsequent PL algorithms. Let us consider the following *linear complementarity problem*: Given an affine map $g\colon \mathbf{R}^n \to \mathbf{R}^n$, find an $x \in \mathbf{R}^n$ such that

$$x \in \mathbf{R}^n_+; \quad g(x) \in \mathbf{R}^n_+; \quad x^* g(x) = 0. \tag{40.1}$$

Here $\mathbf{R}_+$ denotes the set of nonnegative real numbers, and in the sequel we also denote the set of positive real numbers by $\mathbf{R}_{++}$. If $g(0) \in \mathbf{R}^n_+$, then $x = 0$ is a trivial solution to the problem. Hence this trivial case is always excluded and the additional assumption

$$g(0) \notin \mathbf{R}^n_+ \tag{40.2}$$

is made. Linear complementarity problems arise in quadratic programming, bimatrix games, variational inequalities, and economic equilibria problems, and numerical methods for their solution have been of considerable interest **[6]**, **[7]**, **[8]**, **[31]**. For $x \in \mathbf{R}^n$ we introduce the positive part $x_+ \in \mathbf{R}^n_+$ by setting $e_i^* x_+ := \max\{e_i^* x, 0\}$, $i = 1, \dots, n$ and the negative part $x_- \in \mathbf{R}^n_+$ by $x_- := (-x)_+$. The following formulae are then obvious: $x = x_+ - x_-$, $(x_+)^*(x_-) = 0$. The next proposition is not difficult to prove and reduces the linear complementarity problem (40.1) to a zero point problem in a simple way.

41. Proposition. *Under the assumptions of Section* 40, *let us define* $f\colon \mathbf{R}^n \to \mathbf{R}^n$ *by* $f(z) := g(z_+) - z_-$. *If x is a solution of the linear complementarity problem* (40.1), *then $z := x - g(x)$ is a zero point of f. Conversely, if z is a zero point of f, then $x := z_+$ solves* (40.1).

42. The advantage which f provides is that it is obviously a PL map if we subdivide $\mathbf{R}^n$ into orthants. This is the basis for our description of Lemke's algorithm. For a fixed $d \in \mathbf{R}^n_{++}$ we define the homotopy $H\colon \mathbf{R}^n \times [0, \infty) \to \mathbf{R}^n$ by

$$H(x, \lambda) := f(x) + \lambda d\,. \tag{42.1}$$

For a given subset $I \subset \{1, 2, \dots, n\}$ we introduce the complement $I' := \{1, 2, \dots, n\} \setminus I$. Furthermore we introduce the power set

$$\mathscr{P}_n := \Big\{I : I \subset \{1, 2, \dots, n\}\Big\}. \tag{42.2}$$

Then an orthant in $\mathbf{R}^n \times [0,\infty)$ can be written in the form

(42.3) $\sigma_I := \{ (x,\lambda) : \lambda \geq 0,\ e_i^* x \geq 0 \text{ for } i \in I,\ e_i^* x \leq 0 \text{ for } i \in I' \}$,

and the family

(42.4) $$\mathscr{M} := \{ \sigma_I \}_{I \in \mathscr{P}_n}$$

is a PL manifold (of dimension $n+1$) which subdivides $\mathbf{R}^n \times [0,\infty)$. Furthermore it is clear that $H: \mathscr{M} \to \mathbf{R}^n$ is a PL map since $x \mapsto x_+$ switches its linearity character only at the hyperplanes $\{x \in \mathbf{R}^n : e_i^* x = 0\}_{i=1,2,\ldots,n}$.

43. Let us assume for simplicity that zero is a regular value of H. We note, however, that the case of a singular value is treated in the same way by using the perturbation techniques described in Section 22. Lemke's algorithm is started on a ray: if $\lambda > 0$ is sufficiently large, then

$$\big(-g(0) - \lambda d\big)_+ = 0 \quad \text{and} \quad \big(-g(0) - \lambda d\big)_- = g(0) + \lambda d \in \mathbf{R}^N_{++},$$

and consequently

$$H\big(-g(0)-\lambda d\,,\lambda\big) = f\big(-g(0)-\lambda d\big)+\lambda d = g(0)-\big(g(0)+\lambda d\big)+\lambda d = 0.$$

Hence, the ray $\xi \in \ker H$ defined by

$$\lambda \in [\lambda_0,\infty) \longmapsto -g(0) - \lambda d \in \sigma_\varnothing$$

$$\text{for} \quad \lambda_0 := \max_{i=1,\ldots,N} \frac{-g(0)[i]}{d[i]}$$

is used (for decreasing λ-values) for the ray start. This ray is usually called the *primary ray*, and all other rays in $\ker H$ are called *secondary rays*. Note that $\lambda_0 > 0$ by assumption (40.2). Since the PL manifold $\mathscr{M}$ consists of the orthants of $\mathbf{R}^n \times [0,\infty)$, it is finite, and there are only two possibilities:

(43.1) The algorithm terminates on the boundary $|\partial\mathscr{M}| = \mathbf{R}^n \times \{0\}$ at a point $(z,0)$. Then z is a zero point of f, and proposition 41 implies that z_+ solves the linear complementarity problem (40.1).

(43.2) The algorithm terminates on a secondary ray. Then it can be shown **[6]** that (40.1) has no solution if the Jacobian g' belongs to a certain class of matrices.

44. Let us illustrate the use of index and orientation by showing that the algorithm generates a solution in the sense of (43.1) under the

assumption that all principle minors of the Jacobian g' are positive. Note that the Jacobian g' is a constant matrix since g is affine.

For $\sigma_I \in \mathcal{M}$ (see (42.3)–(42.4)), we immediately calculate the Jacobian

$$H'_{\sigma_I} = (f'_{\sigma_I}, d),$$

$$\text{(44.1)} \qquad \text{where} \quad f'_{\sigma_I} e_i = \begin{cases} g' e_i & \text{for } i \in I, \\ e_i & \text{for } i \in I'. \end{cases}$$

If $\xi \in \ker H$ is a solution path in σ_I, then formula (38.1) yields

$$\operatorname{or}_\xi(v) = \operatorname{sign} \det f'_{\sigma_I} \ \operatorname{or}_{\sigma_I}(e_1, \dots, e_n, v),$$

and since $\operatorname{or}_{\sigma_I}(e_1, \dots, e_n, v) = \operatorname{sign} v^* e_{n+1}$ by the standard orientation in $\mathbf{R}^{n+1}$, we have that $\det f'_{\sigma_I}$ is positive or negative if and only if the λ-direction is increasing or decreasing, respectively, while ξ is traversed according to its orientation. It is immediately seen from (44.1) that $\det f'_{\sigma_I}$ is obtained as a *principle minor of* g', i.e., by deleting all columns and rows of g' with index $i \in I'$ and taking the determinant of the resulting matrix (where the determinant of the "empty matrix" is assumed to be 1). Since we start in the negative orthant $\sigma_\varnothing$ where the principle minor is 1, we see that the algorithm traverses the primary ray against its orientation, because the λ-values are initially decreased. Hence, the algorithm continues to traverse $\ker H$ against its orientation. For the important case that all principle minors of g' are positive, the algorithm must continue to decrease the λ-values and thus stops in the boundary $|\Delta\mathcal{M}| = \mathbf{R}^n \times \{0\}$. Hence, in this case the algorithm finds a solution. Furthermore, it is clear that this solution is unique, since $\ker H$ can contain no other ray than the primary ray.

VIII. Variable dimension algorithms.

45. In recent years, a new class of PL algorithms has attracted considerable attention. They are called *variable dimension algorithms* since they all start from a single point, a zero-dimensional simplex, and successively generate simplices of varying dimension, until a completely labeled simplex is found. Numerical results [27] indicate that these algorithms improve the computational efficiency of PL homotopy methods. The first variable dimension algorithm is due to Kuhn [**28**]. However, this algorithm had the disadvantage that it could only be started from a vertex of a large triangulated standard simplex S, and therefore PL homotopy algorithms were preferred. By increasing the sophistication

of Kuhn's algorithm considerably, van der Laan and Talman [**29**] developed an algorithm which could start from any point inside S. It soon became clear (see Todd [**44**]), that this algorithm could be interpreted as a homotopy algorithm. Numerous other variable dimension algorithms were developed. Two unifying approaches have been given, one due to Kojima and Yamamoto [**26**], the other due to Freund [**19**], [**20**]. We present here a modified version of the first approach. The modification consists of introducing a cone construction for dealing with the homotopy parameter. In a special case, this construction was also used by Kojima and Yamamoto; see their lemma 5.13.

46. Before we can give a description of these algorithms, we introduce the notion of a primal-dual pair of PL manifolds due to Kojima and Yamamoto [**26**]. In fact, we only need a special case. Let $\mathscr{P}$ and $\mathscr{D}$ be two PL manifolds of dimension n. We call $(\mathscr{P}, \mathscr{D})$ a *primal-dual pair* if there is a bijective map

$$\tau \in \mathscr{P}^k \longmapsto \tau^d \in \mathscr{D}^{n-k}, \quad k = 0, 1, \dots, n,$$

such that

$$\tau_1 \subset \tau_2 \quad \Leftrightarrow \quad \tau_2^d \subset \tau_1^d \tag{46.1}$$

holds for all $\tau_1 \in \mathscr{P}^{k_1}$ and $\tau_2 \in \mathscr{P}^{k_2}$.

47. We will deal with a homotopy parameter via the following cone construction. Throughout the rest of this paper, ω denotes a point which is affinely independent from all cells under consideration. The introduction of ω is only formal and may be obtained, e.g., by increasing the dimension of the ambient finite-dimensional Euclidean space $\mathscr{E}$ introduced in Section 4. If σ is a cell, then

$$\langle \omega, \sigma \rangle := \left\{ (1-\lambda)\omega + \lambda x \colon x \in \sigma,\ \lambda \geq 0 \right\} \tag{47.1}$$

denotes the cone containing σ with vertex ω. Clearly, $\langle \omega, \sigma \rangle$ is again a cell and $\dim\langle \omega\sigma \rangle = \dim \sigma + 1$. If $H \colon \sigma \to \mathbf{R}^k$ is an affine map, then the affine extension $\langle \omega, H \rangle \colon \langle \omega, \sigma \rangle \to \mathbf{R}^k$ is defined by

$$\langle \omega, H \rangle \big((1-\lambda)\omega + \lambda x \big) := \lambda H(x) \tag{47.2}$$

for $x \in \sigma$ and $\lambda \geq 0$. If $\mathscr{M}$ is a PL manifold of dimension n, then

$$\langle \omega, \mathscr{M} \rangle := \left\{ \langle \omega, \sigma \rangle \right\}_{\sigma \in \mathscr{M}} \tag{47.3}$$

is a PL manifold of dimension $n+1$, and a PL map $H \colon \mathscr{M} \to \mathbf{R}^k$ is extended to a PL map $\langle \omega, H \rangle \colon \langle \omega, \mathscr{M} \rangle \to \mathbf{R}^k$.

48. We will be interested below in rays traversing a cone $\langle\omega,\sigma\rangle$, and we therefore collect some formulae. A ray in $\langle\omega,\sigma\rangle$ is given as

$$\left\{(1-\varepsilon)z_1+\varepsilon z_2\colon \varepsilon\geq 0\right\}\subset\langle\omega,\sigma\rangle,$$
$$\text{where}\quad z_j=(1-\lambda_j)\omega+\lambda_j x_j,\quad j=1,2,$$

for some suitable $\lambda_1,\lambda_2\geq 0$ and $x_1,x_2\in\sigma$. A simple calculation using the affine independence of ω yields

$$(1-\varepsilon)z_1+\varepsilon z_2=(1-\lambda_\varepsilon)\omega+\lambda_\varepsilon x_\varepsilon,$$
$$\text{where}\quad \lambda_\varepsilon=(1-\varepsilon)\lambda_1+\varepsilon\lambda_2$$
$$\text{and}\quad x_\varepsilon=\frac{(1-\varepsilon)\lambda_1x_1+\varepsilon\lambda_2x_2}{\lambda_\varepsilon}.$$

Since $\lambda_\varepsilon\geq 0$ for all $\varepsilon\geq 0$, it follows that $\lambda_2\geq\lambda_1$. This leaves two cases to consider:

(48.1) $\lambda_2>\lambda_1\geq 0 \quad\Rightarrow\quad \lim_{\varepsilon\to\infty}x_\varepsilon=\dfrac{\lambda_2x_2-\lambda_1x_1}{\lambda_2-\lambda_1}\in\sigma,$

(48.2) $\lambda_2=\lambda_1>0 \quad\Rightarrow\quad x_1\neq x_2,$

$x_\varepsilon=(1-\varepsilon)x_1+\varepsilon x_2\in\sigma \quad\text{for}\quad \varepsilon\geq 0.$

The second case is only possible if the cell σ is unbounded.

49. Let $\mathcal{T}$ and $\mathcal{M}$ be manifolds of dimension n. We call $\mathcal{T}$ a *refinement* of $\mathcal{M}$ if for all $\sigma\in\mathcal{M}$ the restricted PL manifold $\mathcal{T}_\sigma:=\{\xi\colon\xi\in\mathcal{T},\ \xi\subset\sigma\}$ subdivides σ.

50. We are now in a position to introduce primal-dual manifolds. Let $(\mathcal{P},\mathcal{D})$ be a primal-dual pair of n-dimensional PL manifolds, and let $\mathcal{T}$ be a refinement $\mathcal{P}$. Then

(50.1) $\mathcal{T}\otimes\mathcal{D}:=\{\xi\times\tau^d\colon k\in\{0,1,\ldots,N\},\ \xi\in\mathcal{T}^k,\ \tau\in\mathcal{M}^k,\ \xi\subset\tau\}$

is an n-dimensional PL manifold with empty boundary. A proof of this and related results was given by Kojima and Yamamoto [**26**]. We call $\mathcal{T}\otimes\mathcal{D}$ the *primal-dual manifold* generated by $\mathcal{T}$ and $\mathcal{D}$. An essential part of the proof consists of discussing the possible pivoting steps. Let $\xi\times\tau^d\in\mathcal{T}\otimes\mathcal{D}$ with $k=\dim\xi$ as above, and let κ be a facet of $\xi\times\tau^d$. We now describe the pivoting of $\xi\times\tau^d$ across the facet κ (see (8.1)), i.e., we have to find a cell $\eta\in\mathcal{T}\otimes\mathcal{D}$ such that $\eta\neq\xi\times\tau^d$ and $\kappa\subset\eta$.

There are three possible cases:

(50.2) *Increasing the dimension.* Let $\kappa = \xi \times \sigma^d$ such that $\sigma \in \mathscr{M}^{k+1}$ contains τ. Then there is exactly one $\rho \in \mathscr{T}^{k+1}$ such that $\xi \subset \rho$ and $\rho \subset \sigma$. This is a consequence of the fact that $\mathscr{T}$ refines $\mathscr{P}$ and is not difficult to prove. Then $\eta := \rho \times \sigma$ is the desired second cell. In this case the dimension k of the primal cell ξ is increased when performing the pivoting step.

(50.3) *Decreasing the dimension.* Let $\kappa = \delta \times \tau^d$ such that $\delta \in \mathscr{T}^{k-1}$ is a facet of ξ. If $\delta \subset \partial\tau$, then there exists exactly one facet $\nu \in \mathscr{M}^{k-1}$ of τ such that $\delta \subset \nu$, and $\eta := \delta \times \nu^d$ is the desired second cell. In this case the dimension k of the primal cell ξ is decreased when performing the pivoting step.

(50.4) *Keeping the dimension.* Let $\kappa = \delta \times \tau^d$ such that $\delta \in \mathscr{T}^{k-1}$ is a facet of ξ. If $\delta \not\subset \partial\tau$, then there exists exactly one cell $\xi' \in \mathscr{T}^k$ such that $\xi' \neq \xi$, $\xi' \subset \tau$, and $\delta \subset \xi'$. This is again a consequence of the fact that $\mathscr{T}$ refines $\mathscr{P}$ and is not difficult to prove. Now $\eta := \xi' \times \tau$ is the desired second cell. In this case the dimension k of the primal cell ξ is left invariant when performing the pivoting step.

The main point for practical purposes is that the above three different kinds of pivoting steps must be easy to implement on a computer. This is of course mainly a question of choosing a simple primal-dual pair $(\mathscr{P}, \mathscr{D})$ and either $\mathscr{T} = \mathscr{P}$ or some standard refinement $\mathscr{T}$ of $\mathscr{P}$ which can be handled well.

51. We now slightly modify the construction of primal-dual manifolds to include cones for the refinement $\mathscr{T}$ of the primal manifold:

(51.1)
$$\langle \omega, \mathscr{T} \rangle \otimes \mathscr{D} := \{\langle \omega, \xi \rangle \times \tau^d \colon k = 0, 1, \ldots, n,\ \xi \in \mathscr{T}^k,\ \tau \in \mathscr{M}^k,\ \xi \subset \tau\}.$$

If $\dim \xi = k > 0$, then the facets of $\langle \omega, \xi \rangle$ are simply the $\langle \omega, \rho \rangle$ where $\rho \in \mathscr{T}^{k-1}$ is a facet of ξ, and it is readily seen that the pivoting steps (50.2)–(50.4) apply. The only exception is the case $\dim \xi = k = 0$. In this case it follows that $\xi = \tau$, and ξ is a vertex of the primal manifold $\mathscr{P}$, but $\langle \omega, \xi \rangle$ is a ray which has one vertex, namely $\{\omega\}$. Hence, we now have a boundary

(51.2) $$\Delta(\langle\omega,\mathscr{T}\rangle\otimes\mathscr{D}) = \{\,\{\omega\}\times\{v\}^d : \{v\}\in\mathscr{P}^0\,\}.$$

Clearly, such a boundary facet $\{\omega\}\times\{v\}^d$ belongs to the $(n+1)$-dimensional cell $\langle\omega,\{v\}\rangle\times\{v\}^d\in\langle\omega,\mathscr{T}\rangle\otimes\mathscr{D}$. We will later see that such boundary facets are used for starting a PL algorithm. This corresponds to starting a homotopy method on the trivial level $\lambda=0$ at the point v. We will now apply the above concept of primal-dual manifolds in order to describe some PL algorithms.

52. *Lemke's algorithm revisited.* We consider again the linear complementarity problem (40.1) and introduce a primal-dual pair $(\mathscr{P},\mathscr{D})$ by defining for $I\subset\{1,2,\dots,n\}$ and $I':=\{1,2,\dots,n\}\setminus I$ the primal and dual faces

(52.1) $$\begin{aligned}\alpha_I &:= \{x\in\mathbf{R}^n : e_i^*x\ge 0 \text{ for } i\in I,\ e_i^*x=0 \text{ for } i\in I'\},\\ \alpha_I^d &:= \alpha_{I'}.\end{aligned}$$

The primal and dual manifolds consist of just one cell: $\mathscr{P}=\mathscr{D}=\{\mathbf{R}^n_+\}$. We now define a PL map $H:\mathscr{P}\otimes\mathscr{D}\times[0,\infty)\longrightarrow\mathbf{R}^n$ by $H(x,y,\lambda):=y-g(x)-\lambda d$, where $d\in\mathbf{R}^N_{++}$ is fixed. Note that the variables x and y are placed into complementarity with each other by the construction of $\mathscr{P}\otimes\mathscr{D}$, and hence a more complex definition of H as in (42.1) is not necessary. For sufficiently large $\lambda>0$ the solutions of $H(x,y,\lambda)=0$ are given by the primary ray $(x,y,\lambda)=(0,\ g(0)+\lambda d\,,\ \lambda)$. Here the PL algorithm following $H^{-1}(0)$ is started in the negative λ-direction. If the level $\lambda=0$ is reached, a solution $H(x,y,0)=0$ solves the LCP since the complementarity $x\in\mathbf{R}^n_+$, $y=g(x)\in\mathbf{R}^n_+$, $x^*y=0$ holds by the construction of $\mathscr{P}\otimes\mathscr{D}$.

53. As a typical representative of the class of variable dimension algorithms we choose the *octahedral algorithm* of Wright **[45]**, since numerical experiments indicate that it performs favorably **[27]**, and since it can be described in a reasonably simple way. Let us point out, however, that similar arguments hold for many other algorithms where the refinement $\mathscr{T}$ of the primal manifold $\mathscr{P}$ is a pseudo manifold which triangulates $\mathbf{R}^n$, and where the dual manifold $\mathscr{D}$ subdivides a compact subset of $\mathbf{R}^n$; see **[4]**, **[19]**, **[20]**, **[26]**, **[27]**, **[29]**.

We denote by $\Sigma:=\{+1,0,-1\}^n\setminus\{0\}$ the set of all nonzero sign vectors. For two vectors $s,t\in\Sigma$ we introduce the relation

(53.1) $$s\prec p \quad:\iff\quad \forall_{i=1,\dots,n}\Big(e_i^*s\ne 0\Rightarrow e_i^*s=e_i^*p\Big).$$

Then we define a primal-dual pair $(\mathscr{P},\mathscr{D})$ of n-dimensional manifolds by introducing the following duality:

$$\alpha_0 := \{0\}, \quad \alpha_0^d := \{\, y \in \mathbf{R}^n : \|y\|_1 \leq 1 \,\},$$

and for $s \in \Sigma$ we consider

$$\alpha_s := \left\{ \sum_{\substack{p\in\Sigma \\ s\prec p}} \lambda_p p \ : \ \lambda_p \geq 0 \right\}, \tag{53.2}$$

$$\alpha_s^d := \{\, y \in \mathbf{R}^n : \|y\|_1 \leq 1,\ s^* y = 1 \,\}.$$

Hence, the primal manifold $\mathscr{P}$ subdivides $\mathbf{R}^n$ into $2n$ cones centered around the unit base vectors $\pm e_i$ for $i = 1, 2, \ldots, n$, and the dual manifold $\mathscr{D}$ just consists of the unit ball with respect to the $\|\cdot\|_1$-norm. We easily check that

$$y \in \alpha_s^d,\ s \prec p \quad \Rightarrow \quad y^* p \geq 0$$

and hence

$$(x, y) \in \mathscr{P} \otimes \mathscr{D} \quad \Rightarrow \quad x^* y \geq 0\,. \tag{53.3}$$

We now consider a pseudo manifold $\mathscr{T}$ which is a refinement of $\mathscr{P}$; for example, it is easy to see that the Union Jack triangulation 13 of $\mathbf{R}^n$ has this property.

54. Our aim is to find an approximate zero point of an asymptotically linear map $G\colon \mathbf{R}^n \to \mathbf{R}^n$. To do this, we first need to introduce the *PL approximation* $G_{\mathscr{T}}$ of G with respect to the pseudo manifold $\mathscr{T}$; see also Section 30. In fact, there is a unique PL map $G_{\mathscr{T}} : \mathscr{T} \to \mathscr{T}$ such that $G(v) = G_{\mathscr{T}}(v)$ holds for all vertices $v \in \mathscr{T}^0$. If $\sigma = [v_1, \ldots, v_{n+1}] \in \mathscr{T}$, and if a point $u \in \sigma$ is expanded into its barycentric coordinates $u = \sum_{i=1}^{n+1} c_i v_i$, then $G_{\mathscr{T}}(u) = \sum_{i=1}^{n+1} c_i G(v_i)$. It is clear that $G_{\mathscr{T}}$ is also asymptotically linear and $G'_{\mathscr{T}}(\infty) = G'(\infty)$. A homotopy $\widetilde{H} : \mathscr{T} \otimes \mathscr{D} \times [0, \infty) \to \mathbf{R}^N$ is introduced by setting

$$\widetilde{H}(x, y, \lambda) := G'(\infty) y + \lambda G_{\mathscr{T}}(x)\,. \tag{54.1}$$

Here, for simplicity, $y = 0$ plays the role of a starting point. Unfortunately, $\widetilde{H}$ is not PL. Hence, we use the cone construction to identify $\widetilde{H}$ with a PL map $H : \langle \omega, \mathscr{P} \rangle \otimes \mathscr{D} \longrightarrow \mathbf{R}^n$ by collecting the variables in a different way:

$$H(z, y) := G'(\infty) y + \langle \omega, G_{\mathscr{T}} \rangle (z)\,. \tag{54.2}$$

For $z = \omega$, which corresponds to $\lambda = 0$, there is exactly one solution of $H(z,y) = 0$, namely $(z,y) = (\omega,0)$. Hence $H^{-1}(0)$ intersects the boundary $\Delta(\langle\omega,\mathscr{P}\rangle \otimes \mathscr{D})$ in just one point. This is the starting point for our PL algorithm which traces $H^{-1}(0)$.

55. Let us first demonstrate that there is a constant $C > 0$ such that $\widetilde{H}(x,y,\lambda) = 0$ implies $||x|| < C$. Indeed, otherwise we could find a sequence $\{(x_k,y_k,\lambda_k)\}_{k=1,2,\ldots} \subset H^{-1}(0)$ such that $\lim_{k\to\infty} ||x_k|| = \infty$. It follows from $\widetilde{H}(x_k,y_k,\lambda_k) = 0$ and (54.1) that

$$\lambda_k^{-1} y_k + G'(\infty)^{-1} G_{\mathscr{T}}(x_k) = 0 . \tag{55.1}$$

If we multiply this equation from the left with x_k^* and divide by $||x_k||^2$, the asymptotic linearity of $G_{\mathscr{T}}$ yields

$$\lim_{k\to\infty} ||x_k||^{-2} x_k^* G'(\infty)^{-1} G_{\mathscr{T}}(x_k) = 1 , \tag{55.2}$$

and the boundedness $||y_k|| \le 1$ implies that

$$x_k^* G'(\infty)^{-1} G_{\mathscr{T}}(x_k) > 0 \tag{55.3}$$

for all sufficiently large k, and by (55.1), (55.3) we have that $x_k^* y_k > 0$ for all sufficiently large k, which is a contradiction to (53.3).

56. Now Section 55 implies that the algorithm can only traverse finitely many cells, and since the solution on the boundary $\partial(\langle\omega,\mathscr{T}\rangle \otimes \mathscr{D})$ is unique, it can only terminate in a ray

$$\{((1-\varepsilon)z_1 + \varepsilon z_2 , (1-\varepsilon)y_1 + \varepsilon y_2) : \varepsilon \ge 0\} \subset \langle\omega,\tau\rangle \times \alpha_I^d \in \langle\omega,\mathscr{T}\rangle \otimes \mathscr{D} ,$$

where $\tau \in \mathscr{T}^k$ such that $\tau \subset \alpha_I$ and $k = \#I$. We use the notation and remarks of Section 48. It follows from

$$H((1-\varepsilon)z_1 + \varepsilon z_2 , (1-\varepsilon)y_1 + \varepsilon y_2) = 0 \tag{56.1}$$

and (54.1)–(54.2) that

$$(1-\varepsilon)y_1 + \varepsilon y_2 + \lambda_\varepsilon G'(\infty)^{-1} G_{\mathscr{T}}(x_\varepsilon) = 0 \quad \text{for} \quad \varepsilon \ge 0 . \tag{56.2}$$

Since the k-cell τ is bounded, we only have to consider the case $\lambda_2 > \lambda_1 \ge 0$; see (48.1). Dividing equation (56.1) by $\varepsilon > 0$ and letting $\varepsilon \to \infty$ yields

$$G_{\mathscr{T}}(x) = 0 , \quad \text{where} \quad x := \frac{\lambda_2 x_2 - \lambda_1 x_1}{\lambda_2 - \lambda_1} \in \tau$$

is the desired approximate zero point of G.

57. *Concluding remarks.* By the above discussions we wanted to illustrate that the concept of primal-dual manifolds $\mathscr{P} \otimes \mathscr{D}$ enables a unifying description of many variable dimension algorithms. One class of such algorithms is the homotopy methods where the homotopy parameter caused our cone construction. An important feature of primal-dual manifolds is that a complementarity property of the variables (x, y) may be incorporated into the construction of $\mathscr{P} \otimes \mathscr{D}$ so that this property need not be assumed by extra conditions or constructions. This is a very convenient trick for dealing with complementarity problems or related questions, and was illustrated here for the case of the linear complementarity problem in Section 52, but many more applications have been considered; see the literature cited in [**4**].

References

1. E. L. Allgower and K. Georg, *Generation of triangulations by reflections*, Utilitas Math. **16** (1979), 123–129.

2. ——, *Simplicial and continuation methods for approximating fixed points and solutions to systems of equations*, SIAM Rev. **22** (1980), 28–85.

3. ——, *Predictor-corrector and simplicial methods for approximating fixed points and zero points of nonlinear mappings*, Mathematical Programming: The State of the Art, A. Bachem, M. Grötschel, and B. Korte (eds.), Springer-Verlag, Berlin, Heidelberg, New York, 1983, pp. 15–56.

4. ——, *Introduction to Numerical Continuation Methods*, to appear in Springer-Verlag (1990).

5. L. E. J. Brouwer, *Über Abbildung von Mannigfaltigkeiten*, Math. Ann. **71** (1912), 97–115.

6. R. W. Cottle, *Solution rays for a class of complementarity problems*, Math. Programming Stud. **1** (1974), 58–70.

7. R. W. Cottle and G. B. Dantzig, *Complementary pivot theory of mathematical programming*, Linear Algebra and its Applications **1** (1968), 103–125.

8. R. W. Cottle, G. H. Golub, and R. S. Sacher, *On the solution of large structured linear complementarity problems: The block partitioned case*, Appl. Math. Optim. **4** (1978), 347–363.

9. H. S. M. Coxeter, *Discrete groups generated by reflections*, Ann. of Math. **6** (1934), 13–29.

10. ——, *Regular Polytopes*, third edition, Dover Publ., New York.

11. G. B. Dantzig, *Linear Programming and Extensions*, Princeton Univ. Press, Princeton, NJ, 1963.

12. N. Dunford and J. T. Schwartz, *Linear Operators. Part II: Spectral Theory*, Interscience Publ., New York, 1963.

13. B. C. Eaves, *Homotopies for the computation of fixed points*, Math. Programming **3** (1972), 1–22.

14. ——, *A short course in solving equations with PL homotopies*, Nonlinear Programming, SIAM-AMS Proc. **9**, R. W. Cottle and C. E. Lemke (eds.), AMS, Providence, RI, 1976, pp. 73–143.

15. ———, *A course in triangulations for solving equations with deformations*, Lecture Notes in Economics and Mathematical Systems **234**, Springer-Verlag, Berlin, Heidelberg, New York, 1984.

16. B. C. Eaves and R. Saigal, *Homotopies for computation of fixed points on unbounded regions*, Math. Programming **3** (1972), 225–237.

17. B. C. Eaves and H. Scarf, *The solution of systems of piecewise linear equations*, Math. Oper. Res. **1** (1976), 1–27.

18. H. Freudenthal, *Simplizialzerlegungen von beschränkter Flachheit*, Ann. of Math. **43** (1942), 580–582.

19. R. M. Freund, *Variable dimension complexes. Part I: Basic theory*, Math. Oper. Res. **9** (1984), 479–497.

20. ———, *Variable dimension complexes. Part II: A unified approach to some combinatorial lemmas in topology*, Math. Oper. Res. **9** (1984), 498–509.

21. K. Georg, *Zur numerischen Realisierung von Kontinuitätsmethoden mit Prädiktor-Korrektor- oder simplizialen Verfahren*, Habilitationsschrift, University of Bonn, 1982.

22. P. E. Gill, G. H. Golub, W. Murray, and M. A. Saunders, *Methods for modifying matrix factorizations*, Math. Comp. **28** (1974), 505–535.

23. S. Gnutzmann, *Stückweise lineare Approximation implizit definierter Mannigfaltigkeiten*, Ph.D. thesis, University of Hamburg, 1987.

24. L. Górniewicz, *Homological methods in fixed point theory of multivalued maps*, Diss. Math. **129** (1976).

25. F. J. Gould and J. W. Tolle, *Complementary pivoting on a pseudomanifold structure with applications on the decision sciences*, Sigma Series in Applied Mathematics **2**, Heldermann Verlag, Berlin, 1983.

26. M. Kojima and Y. Yamamoto, *Variable dimension algorithms: basic theory, interpretation, and extensions of some existing methods*, Math. Programming **24** (1982), 177–215.

27. ———, *A unified approach to the implementation of several restart fixed point algorithms and a new variable dimension algorithm*, Math. Programming **28** (1984), 288–328.

28. H. W. Kuhn, *Approximate search for fixed points*, Computing Methods in Optimization Problems 2, L. A. Zadek, L. W. Neustat, and A. V. Balakrishnan (eds.), Academic Press, New York, London, 1969, pp. 199–211.

29. G. van der Laan and A. J. J. Talman, *A restart algorithm for computing fixed points without an extra dimension*, Math. Programming **17** (1979), 74–84.

30. C. E. Lemke, *Bimatrix equilibrium points and mathematical programming*, Management Sci. **11** (1965), 681–689.

31. ———, *A survey of complementarity theory*, Variational Inequalities and Complementarity Problems, R. W. Cottle, F. Gianessi, and J. L. Lions (eds.), John Wiley and Sons, London, 1980.

32. C. E. Lemke and S. J. Grotzinger, *On generalizing Shapley's index theory to labelled pseudo manifolds*, Math. Programming **10** (1976), 245–262.

33. C. E. Lemke and J. T. Howson, *Equilibrium points of bimatrix games*, SIAM J. Appl. Math. **12** (1964), 413–423.

34. O. Merrill, *Applications and extensions of an algorithm that computes fixed points of a certain upper semi-continuous point to set mapping*, Ph. D. Thesis, Univ. of Michigan, Ann Arbor, MI, 1972.

35. H.-O. Peitgen, *Topologische Perturbationen beim globalen numerischen Studium nichtlinearer Eigenwert- und Verzweigungsprobleme*, Jahresbericht des Deutschen Mathematischen Vereins **84** (1982), 107–162.

36. H.-O. Peitgen and M. Prüfer, *The Leray-Schauder continuation method is a constructive element in the numerical study of nonlinear eigenvalue and bifurcation problems*, Functional Differential Equations and Approximation of Fixed Points, Lecture Notes in Math. **730**, H.-O. Peitgen and H.-O. Walther (eds.), Springer-Verlag, Berlin, Heidelberg, New York, 1979, pp. 326–409.

37. H.-O. Peitgen and H. W. Siegberg, *An $\bar{\varepsilon}$-perturbation of Brouwer's definition of degree*, Fixed Point Theory, E. Fadell and G. Fournier (eds.), Lecture Notes in Math. **886**, Springer-Verlag, Berlin, Heidelberg, New York, 1981, pp. 331–366.

38. R. T. Rockafellar, *Convex Analysis*, Princeton University Press, Princeton, NJ, 1970.

39. H. E. Scarf, *The approximation of fixed points of a continuous mapping*, SIAM J. Appl. Math. **15** (1967), 1328–1343.

40. K. Schilling, *Simpliziale Algorithmen zur Berechnung von Fixpunkten mengenwertiger Operatoren*, WVT Wissenschaftlicher Verlag, Trier, West Germany, 1986.

41. L. S. Shapley, *A note on the Lemke-Howson algorithm*, Pivoting and extensions: in honor of A. W. Tucker, Math. Programming Stud. **1**, M. L. Balinski (ed.), North-Holland, Amsterdam, New York, 1974, pp. 175–189.

42. M. J. Todd, *The computation of fixed points and applications*, Lecture Notes in Economics and Mathematical Systems **124**, Springer-Verlag, Berlin, Heidelberg, New York, 1976.

43. ——, *Orientation in complementary pivot algorithms*, Math. Oper. Res. **1** (1976), 54–66.

44. ——, *Fixed-point algorithms that allow restarting without extra dimension*, Preprint, Cornell University, Ithaca, NY.

45. A. H. Wright, *The octahedral algorithm, a new simplicial fixed point algorithm*, Math. Programming **21** (1981), 47–69.

Department of Mathematics, Colorado State University

Institut für Angewandte Mathematik, Universität Bonn

Lectures in Applied Mathematics
Volume **26** (1990)

Nonlinear Convection Diffusion Equations and Newton-like Methods

KURT GEORG AND DAVID ZACHMANN

1. Introduction. A number of physical applications call for solving the nonlinear diffusion convection equation

$$s(u)\frac{\partial u}{\partial t} = \nabla \cdot [a(u)\nabla u + \mathbf{v}(u)], \quad \mathbf{x} \text{ in } \Omega, \quad t > 0, \tag{1.1}$$

subject to an initial condition

$$u(\mathbf{x}, 0) = f(\mathbf{x}), \quad \mathbf{x} \text{ in } \Omega, \tag{1.2}$$

and a boundary condition of the form

$$u(\mathbf{x}, t) = g(\mathbf{x}, t) \tag{1.3a}$$

or

$$\frac{\partial u}{\partial \mathbf{n}}(\mathbf{x}, t) = g(\mathbf{x}, t) \tag{1.3b}$$

or

$$\alpha u(\mathbf{x}, t) + \beta \frac{\partial u}{\partial \mathbf{n}}(\mathbf{x}, t) = g(\mathbf{x}, t) \tag{1.3c}$$

specified at each point of S, the boundary of the region Ω.

For example, for the case of partially saturated flow in a homogeneous porous medium occupying a region Ω, let $u(\mathbf{x}, t)$ represent the capillary pressure in the fluid, $\theta = \theta(u)$ represent the volumetric fluid content, and assume that the Darcy law for volumetric flux,

$$\mathbf{q} = -K(u)\ \nabla(\mathbf{u} + \mathbf{z}), \tag{1.4}$$

1980 *Mathematics Subject Classification* (1985 *Revision*). Primary 65N10, 65H10.

0082-0717/90 $1.00 + $.25 per page

holds, with the z-axis oriented downward. Then from a material balance argument it follows that, in the absence of sources or sinks, the continuity equation

$$\frac{\partial \theta}{\partial t} = -\nabla \cdot \mathbf{q} \tag{1.5}$$

must hold. Now, if we define the capacity function, or storage coefficient,

$$s(u) = \frac{d\theta}{du},$$

and use the chain rule to write

$$\frac{\partial \theta}{\partial t} = \frac{d\theta}{du}\frac{\partial u}{\partial t} = s(u)\frac{\partial u}{\partial t},$$

then Equations (1.4) and (1.5) combine to yield

$$s(u)\frac{\partial u}{\partial t} = \nabla \cdot [K(u)\nabla u + \mathbf{K}(u)], \quad \mathbf{x} \text{ in } \Omega, \quad t > 0. \tag{1.6}$$

Equation (1.6) is seen to have the form of (1.1). The hydraulic conductivity function $K(u)$ plays the role of both the diffusion coefficient $a(u)$ and the convection coefficient $\mathbf{v}(u)$ in (1.1).

To illustrate another occurrence of the diffusion convection equation, continue to assume that fluid is moving through a region Ω according to the model equations (1.4)–(1.6). Further assume that a nonreactive contaminant is also present in the flow system. Let $c = c(\mathbf{x}, t)$ represent the concentration of the contaminant. The contaminant moves by diffusion from regions of high concentration to regions of lower concentration, moves by dispersion due to the tortuous flow path through the porous medium and moves by convection as it is swept along with the fluid flow. If the diffusive and dispersive effects are accounted for in an empirically determined dispersion coefficient tensor $\mathbf{D}(\mathbf{q})$, then the flux of contaminant can be assumed to be given by

$$\mathbf{Q} = -\mathbf{D}(\mathbf{q})\ \nabla \cdot c\mathbf{u} + c\mathbf{q}. \tag{1.7}$$

Provided no contaminant sources or sinks are present in Ω, a material balance argument shows

$$\frac{\partial}{\partial t}(\theta c) = -\nabla \cdot \mathbf{Q}. \tag{1.8}$$

Using the product rule to expand the derivatives in Equation (1.8) and recalling Equation (1.5) lead to

$$\theta \frac{\partial c}{\partial t} = \nabla \cdot \mathbf{D}(\mathbf{q})\nabla c + \mathbf{q} \cdot \nabla c, \quad \mathbf{x} \text{ in } \Omega, \quad t > 0. \tag{1.9}$$

Equation (1.9) contains the effects of diffusion/dispersion in the term $\nabla \cdot \mathbf{D}(\mathbf{q})\nabla c$ and the effects of convection in the term $\mathbf{q} \cdot \nabla c$.

If the concentration of the contaminant is sufficiently low that the presence of the contaminant does not affect the fluid flow properties, then Equation (1.9) can be viewed as a linear equation in c provided the Darcy flow field $\mathbf{q}$ and the fluid content θ are known at all points of Ω. However, at high concentration levels both $\mathbf{q}$ and θ must be viewed as c-dependent, and (1.9) again represents a nonlinear diffusion convection equation.

When dealing with an equation of the form (1.1), one must recognize that when $s(u)$ and $a(u)$ are both positive, then Equation (1.1) is of parabolic type. However, when s vanishes and a remains positive, the equation is elliptic. Moreover, when s is positive and $\mathbf{v}$ dominates a, then Equation (1.1) is formally parabolic, but for computational purposes must be treated as being of hyperbolic type. These observations will impact the following discussion of numerical methods for solving Equation (1.1).

2. Finite difference equations. The nonlinearity introduced by the u-dependence of the coefficients $s(u)$, $a(u)$, and $\mathbf{v}(u)$ require that, in general, the solution to Equation (1.1) be approximated by numerical methods. The method of finite differences is a popular numerical method for estimating the solution to a problem of the form (1.1)–(1.3). Two approaches for obtaining finite difference equations that correspond to Equation (1.1) are the following:

(i) use a truncated Taylor series to represent the derivatives in (1.1);
(ii) use a material balance argument to obtain a discrete version of (1.1).

Each of these approaches has its advantages. The Taylor series approach often provides more insight into the nature of the truncation error that arises when the continuous model equation (1.1) is replaced by a discrete set of finite difference equations. However, the material balance approach often yields a more physically based system of difference equations that, when solved, ensure that material is conserved, at least on the balance elements used in the derivation. See the book of DuChateau and Zachmann (1989) for more details on difference equations.

To illustrate a material balance approach to developing difference equations, consider the one-dimensional version of Equation (1.1) in which, for convenience, the convection term has been set to zero. Namely, consider

$$s(u)u_t - (a(u)u_x)_x = 0, \quad x \text{ in } (0,1), \tag{2.1}$$

where it is assumed that Equation (2.1) represents a conservation law in the sense that a density ρ and a flux q can be related to u with equations of the form

$$\rho_t = s(u)u_t \qquad \text{and} \qquad q = -a(u)u_x,$$

so that Equation (2.1) is equivalent to the balance equation

$$\rho_t + q_x = 0. \tag{2.2}$$

Define a uniform x-grid,

$$x_n = nh, \qquad n = 0, 1, 2, \dots, N+1,$$

with $x_0 = 0$ and $x_{N+1} = 1$. Also define a set of points on the x-axis by

$$\xi_n = -\tfrac{1}{2}h + nh, \qquad n = 0, 1, 2, \dots, N+2,$$

so that x_n is the center of the *finite difference block* (ξ_n, ξ_{n+1}). See Figure 2.1.

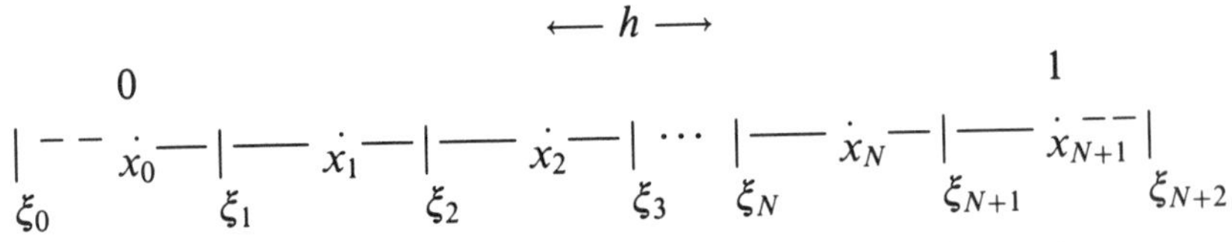

$$\longleftarrow h \longrightarrow$$

$$\begin{array}{ccccccccccccc} & 0 & & & & & & & & & 1 & \\ | \; - - & \dot{x}_0 & - | - & \dot{x}_1 & - | - & \dot{x}_2 & - | \;\cdots\; | - & \dot{x}_N & - | - & \dot{x}_{N+1} & - - \; | \\ \xi_0 & & \xi_1 & & \xi_2 & & \xi_3 \quad \xi_N & & \xi_{N+1} & & \xi_{N+2} \end{array}$$

FIGURE 2.1. Block centered finite difference grid on $0 < x < 1$ with uniform grid spacing h.

Finally, define a uniform time grid by

$$t_j = jk, \quad j = 0, 1, \dots .$$

Now, consider the region in the xt-plane defined by

$$\xi_n < x < \xi_{n+1}, \quad t_j < t < t_{j+1}.$$

If we integrate the differential equation conservation law (2.2) over this region, we can recover the integral equation conservation law.

$$\begin{aligned} \int_{t_j}^{t_{j+1}} \int_{\xi_n}^{\xi_{n+1}} (\rho_t + q_x) dx\, dt &= \int_{\xi_n}^{\xi_{n+1}} \int_{t_j}^{t_{j+1}} \rho_t dt\, dx + \int_{t_j}^{t_{j+1}} \int_{\xi_n}^{\xi_{n+1}} q_x dx\, dt \\ &= \int_{\xi_n}^{\xi_{n+1}} [\rho(x, t_{j+1}) - \rho(x, t_j)]\, dx \\ &\quad + \int_{t_j}^{t_{j+1}} [q(\xi_{n+1}, t) - q(\xi_n, t)]\, dt = 0 \end{aligned}$$

or

$$\int_{\xi_n}^{\xi_{n+1}} \left[\rho(x,t_{j+1}) - \rho(x,t_j)\right] dx = \int_{t_j}^{t_{j+1}} \left[q(\xi_n,t) - q(\xi_{n+1},t)\right] dt. \tag{2.3}$$

In physical terms, Equation (2.3) states that the difference in the amount of material M in the block (ξ_n, ξ_{n+1}) from time t_j to time t_{j+1} is equal to the flow of M into the block across $x = \xi_n$ minus the flow of M out of the block across $x = \xi_{n+1}$ during the time interval $t_j < t < t_{j+1}$. See Figure 2.2.

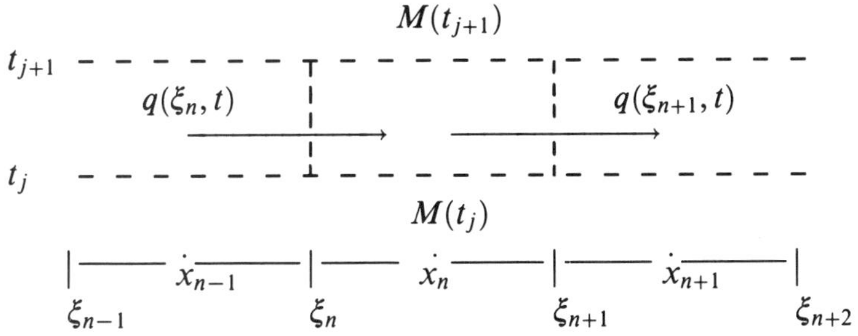

FIGURE 2.2. Material balance diagram showing that the change in $M(t) = \int_{\xi_n}^{\xi_{n+1}} \rho(x,t)dx$ equals flow in minus flow out.

Using the integral equation conservation law (2.2), we can derive a number of finite difference methods for Equation (2.1). If we use the midpoint quadrature rule to approximate the integral of the densities in (2.3), then

$$\int_{\xi_n}^{\xi_{n+1}} \left[\rho\left(x,t_{j+1}\right) - \rho\left(x,t_j\right)\right] dx \cong \left[\rho\left(x_n,t_{j+1}\right) - \rho\left(x_n,t_j\right)\right] h.$$

The choice of the midpoint rule for the last integral relates the densities to the x-grid points.

To approximate the integral of the fluxes in (2.3) in a manner consistent with the implicit, backward-in-time method, we choose the right endpoint quadrature rule to obtain

$$\int_{t_j}^{t_{j+1}} \left[q\left(\xi_n,t\right) - q\left(\xi_{n+1},t\right)\right] dt = \left[q\left(\xi_n,t_{j+1}\right) - q\left(\xi_{n+1},t_{j+1}\right)\right] k. \tag{2.4}$$

The use of the left endpoint rule for the the integral (2.4) results in the explicit, forward-in-time difference method. The trapezoidal rule approximation of (2.4) leads to the well-known Crank–Nicolson method.

Suppose the implicit, backward-in-time difference method has been selected. Namely,

$$[\rho(x_n,t_{j+1})-\rho(x_n,t_j)]\,h=[q(\xi_n,t_{j+1})-q(\xi_{n+1},t_{j+1})]\,k. \tag{2.5}$$

Equation (2.5) represents a single difference equation for the two quantities ρ and q. To derive a useful difference equation for dealing with Equation (2.1), we need to relate ρ and q to u. From (2.2) and the Mean Value Theorem we obtain

$$\rho_n^{j+1}-\rho_n^j=s\left(x_n,V_n^j\right)\left[U_n^{j+1}-U_n^j\right],$$

where V_n^j is between U_n^j and U_n^{j+1}. If we make the approximation $V_n^j = U_n^j$, then it is said that the nonlinearity is being lagged. The result of lagging this nonlinearity is that the resulting term in the difference equation is linear in the unknown quantity U_n^{j+1}. In some circumstances, to maintain accuracy, it is necessary to estimate V_n^j by U_n^{j+1} (or by the average of U_n^j and U_n^{j+1}), which leads to a nonlinear expression of the form

$$\left[\rho_n^{j+1}-\rho_n^j\right]h=s\left(x_n,U_n^j\right)\left[U_n^{j+1}-U_n^j\right]h.$$

It remains to find a replacement for $q(\xi_n,t_{j+1})$ and $q(\xi_{n+1},t_{j+1})$ in terms of U_n^j.

Let us now consider the flux $q(\xi_n,t_{j+1})$ in the case that a is u-dependent. From $q=-a(u)u_x$ we have

$$q(x,t_{j+1})=-a\sigma(u(x,t_{j+1}))\,u_x(x,t_{j+1}),$$

which implies

$$u_x(x,t_{j+1})=-\frac{q(x,t_{j+1})}{a(u(x,t_{j+1}))}.$$

Integrating this last equation from x_{n-1} to x_n yields

$$\begin{aligned}
u_n^{j+1}-u_{n-1}^{j+1}&=\int_{x_{n-1}}^{x_n}u_x(x,t_{j+1})\,dx=-\int_{x_{n-1}}^{x_n}\frac{q(x,t_{j+1})}{a(x,u(x,t_{j+1}))}\,dx\\
&\cong -q(\xi_n,t_{j+1})\int_{x_{n-1}}^{x_n}\frac{1}{a(x,u(x,t_{j+1}))}\,dx\\
&=-q(\xi_n,t_{j+1})\left\{\int_{x_{n-1}}^{\xi_n}\frac{1}{a(x,u(x,t_{j+1}))}\,dx+\int_{\xi_n}^{x_n}\frac{1}{a(x,u(x,t_{j+1}))}\,dx\right\}\\
&\cong -q(\xi_n,t_{j+1})\left\{\frac{1}{a(x_{n-1},u(x_{n-1},t_{j+1}))}\frac{h}{2}+\frac{1}{a(x_n,u(x_n,t_{j+1}))}\frac{h}{2}\right\}\\
&=-q(\xi_n,t_{j+1})\left\{\frac{1}{a_{n-1}^{j+1}}\frac{h}{2}+\frac{1}{a_n^{j+1}}\frac{h}{2}\right\}=-q(\xi_n,t_{j+1})\left\{\frac{ha_n^{j+1}+ha_{n-1}^{j+1}}{2a_n^{j+1}a_{n-1}^{j+1}}\right\}.
\end{aligned}$$

Now we can write

$$q(\xi_n, t_{j+1}) \cong -\frac{2a_n^{j+1}a_{n-1}^{j+1}}{\left(a_n^{j+1}+a_{n-1}^{j+1}\right)}\frac{U_n^{j+1}-U_{n-1}^{j+1}}{h}. \tag{2.6}$$

Similarly, we have

$$q(\xi_{n+1}, t_{j+1}) \cong -\frac{2a_n^{j+1}a_{n+1}^{j+1}}{\left(a_n^{j+1}+a_{n+1}^{j+1}\right)}\frac{U_{n+1}^{j+1}-U_{n}^{j+1}}{h}. \tag{2.7}$$

In Equations (2.6) and (2.7), the coefficient involving a is called a *harmonic average* or *harmonic mean.* For a more compact notation, we will use

$$\mu(a_m, a_n) = \frac{2a_m a_n}{a_m + a_n}$$

to denote the harmonic mean of a_m and a_n. Also, to avoid a division by zero, we define $\mu(0,0) = 0$. For example, in terms of this harmonic mean notation, Equations (2.6) and (2.7) take the form

$$q(\xi_n, t_{j+1}) = -\mu\left(a_n^{j+1}, a_{n-1}^{j+1}\right)\left[U_n^{j+1} - U_{n-1}^{j+1}\right] / h$$

and

$$q(\xi_{n+1}, t_{j+1}) = -\mu\left(a_n^{j+1}, a_{n+1}^{j+1}\right)\left[U_{n+1}^{j+1} - U_{n}^{j+1}\right] / h.$$

Now, if we put Equations (2.6)–(2.7) into (2.5), then the difference equation for s, $u_t - (a(u)u_x)_x = 0$, that results is

$$s\left(x_n, U_n^j\right)\left[U_n^{j+1} - U_n^j\right]h = \Big\{-\mu\left(a_n^{j+1}, a_{n-1}^{j+1}\right)\left[U_n^{j+1} - U_{n-1}^{j+1}\right] / h + \mu\left(a_n^{j+1}, a_{n+1}^{j+1}\right)\left[U_{n+1}^{j+1} - U_n^{j+1}\right] / h\Big\}k.$$

Introducing the notation

$$s_n^j = s\left(x_n, U_n^j\right);$$
$$\mu_{n-1}^{j+1} = \mu\left(a_n^{j+1}, a_{n-1}^{j+1}\right); \qquad \mu_{n+1}^{j+1} = \mu\left(a_n^{j+1}, a_{n+1}^{j+1}\right)$$

allows the finite difference system for Equation 2.1 to be expressed in the form

$$s_n^j\left[U_n^{j+1} - U_n^j\right] = r\left[\mu_{n-1}^{j+1}U_{n-1}^{j+1} - \left(\mu_{n-1}^{j+1} + \mu_{n+1}^{j+1}\right)U_n^{j+1} + \mu_{n+1}^{j+1}U_{n+1}^{j+1}\right], \quad r = k/h^2, \tag{2.8}$$

or, more compactly,

$$F\left(U_{n-1}^{j+1}, U_n^{j+1}, U_{n+1}^{j+1}\right) = 0, \qquad n = 1, 2, \ldots, N.$$

One important point to be made regarding the preceding development of the material balance difference equation (2.8) is that, even if the diffusion coefficient $a(u)$ introduces only a mild nonlinearity into Equation (2.1), the corresponding difference equation (2.8) is found to be nonlinear in a quite complicated way. For example, in the case that $a(u)$ is simply equal to u, the harmonic averages have the following more complicated appearance:

$$\mu_{n-1}^{j+1} = \mu\left(a_n^{j+1}, a_{n-1}^{j+1}\right) = \frac{2u_n^{j+1} u_{n-1}^{j+1}}{u_n^{j+1} + u_{n-1}^{j+1}}.$$

This property has significant computational implications when one applies Newton's method to find the solution of a nonlinear difference equation of the form (2.8). Recall that Newton's method requires that all derivatives of the difference equation expression

$$F\left(U_{n-1}^{j+1}, U_n^{j+1}, U_{n+1}^{j+1}\right)$$

with respect to all arguments U_{n-1}^{j+1}, U_n^{j+1}, and U_{n+1}^{j+1} be calculated.

In some treatments of problems of this type the investigators attempt to avoid having to deal with harmonic averages in the calculations of the coefficients in the difference equation. For example, the difference equation (2.8) is somewhat more tractable if arithmetic means are used in the calculation of the coefficients. Unfortunately, for problems that originate from a conservation law, such an approach is likely to lead to physically unrealistic difference equations.

To close this discussion of the formulation of difference equations for parabolic conservation laws, we point out that only minor modifications of the preceding one-dimensional development are required if the problem is set in two or three space dimensions. On a two-dimensional grid (x_m, y_n) with x and y grid spacings of hx and hy, respectively, the analogue of Equation (2.8) is

$$\begin{aligned} s_{mn}^j \left[U_{mn}^{j+1} - U_{mn}^j\right] = rx \Big[&\mu_{m-1,n}^{j+1} U_{m-1,n}^{j+1} - \left(\mu_{m-1,n}^{j+1} + \mu_{m+1,n}^{j+1}\right) U_{mn}^{j+1} \\ &+ \mu_{m+1,n}^{j+1} U_{m+1,n}^{j+1}\Big] \\ + ry \Big[&\mu_{m,n-1}^{j+1} U_{m,n-1}^{j+1} - \left(\mu_{m,n-1}^{j+1} + \mu_{m,n+1}^{j+1}\right) U_{mn}^{j+1} \\ &+ \mu_{m,n+1}^{j+1} U_{m,n+1}^{j+1}\Big] \end{aligned}$$

where $rx = k/(hx)^2$ and $ry = k/(hy)^2$, which can be expressed in compact notation as

$$F\left(U^{j+1}_{m,n-1}, U^{j+1}_{m,n+1}, U^{j+1}_{m-1,n}, U^{j+1}_{m+1,n}, U^{j+1}_{mn}\right) = 0,$$
$$m = 1, 2, \ldots, M, \qquad n = 1, 2, \ldots, N.$$

3. Iterative solution of difference equations. As shown in Section 2, a system of finite difference equations for an initial-boundary value problem of the form (1.1)–(1.3) leads to a nonlinear set of algebraic equations of the form

$$F_n(U_1, U_2, \ldots, U_N) = 0\,, \quad n = 1, 2, \ldots, N, \tag{3.1}$$

or, in more compact notation,

$$F(\mathbf{U}) = 0. \tag{3.2}$$

The set of equations represented by (3.1)–(3.2) must be solved each time the solution of the difference system is to be advanced from time t to time $t + k$. The number of equations and unknowns in (3.1) is equal to the number of finite difference grid points. The vector $\mathbf{U}$ represents the U^{j+1}-values at the grid points arranged in a one-dimensional array. Having originated from a finite difference scheme, the system of equations represented by (3.1) possesses a good deal of structure. For example, in the case of one space dimension, each of the component expressions F_n contains at most three of the U^{j+1}-values and moreover, in most problems, the U^{j+1}-array can be labeled so that those three U^{j+1}-values have consecutive subscripts. Similar structure exists when difference equations are formulated in two or three space dimensions.

The usual first approximation to the solution to the nonlinear difference equations (3.1) or (3.2) is obtained by "lagging the nonlinearities." That is, the coefficients in the difference equation are set to their time t values, rather than their time $t + k$ values, called for by the backward-in-time method. Specifically, if the nonlinearities in Equation (2.8) are lagged, the result is the linear system

$$s^j_n\left[U^{j+1}_n - U^j_n\right] = r\left[\mu^j_{n-1}U^{j+1}_{n-1} - \left(\mu^j_{n-1} + \mu^j_{n+1}\right)U^{j+1}_n + \mu^j_{n+1}U^{j+1}_{n+1}\right].$$

Let such a linearized system be denoted by

$$\mathbf{AU} = \mathbf{b}, \tag{3.3}$$

where $\mathbf{A}$ is an N by N matrix, $\mathbf{U}$ represents the U^{j+1}-array and $\mathbf{b}$ contains the U^{j+1}-independent terms in the linearized difference equation. Let

$\mathbf{U}^1$ denote the solution of (3.3). Notice that the superscript on $\mathbf{U}$ refers to the first approximation to the solution of the nonlinear system of difference equations (3.1).

In some applications $\mathbf{U}^1$ is a satisfactory estimate of the solution to the system (3.2). However, in many cases, such as when sharp fronts cause the solution of the finite difference system to vary substantially across the grid, an iterative procedure, within each time step, is required to obtain a more accurate estimate of the solution to (3.2).

Newton's method is known to be a robust and rapidly convergent method for dealing with the inner iteration in problems of this type. The method calls for producing a sequence of corrections defined by

$$\mathbf{U}^{i+1} = \mathbf{U}^i + \mathbf{c}^i \tag{3.4}$$

$$i = 1, 2, \ldots,$$

$$\mathbf{c}^i = -\mathbf{J}^i\mathbf{F}\left(\mathbf{U}^i\right) \tag{3.5}$$

where, in Equation (3.5), $\mathbf{J}$ denotes the Jacobian matrix defined by

$$\mathbf{J} = [j_{mn}], \quad j_{mn} = \frac{\partial F_m}{\partial U_n} \tag{3.6}$$

and $\mathbf{J}^i$ denotes the Jacobian evaluated at $\mathbf{U}^i$. Even though Newton's method is robust and rapidly convergent, it has a number of practical shortcomings for difference systems of the forms shown in Section 2. First, the form of the Jacobian requires that not only must computer code be provided for the harmonic average expression μ, but also code must be supplied for the derivatives of μ at each grid point with respect to each of the $\mathbf{U}$-values that enter the finite difference stencil at that grid point. This shortcoming is particularly striking in applications in which the form of the coefficients is subject to frequent change. One is very reluctant to change the form of $a(u)$ if that means that a number of derivatives of $\mu(a_n^{j+1}, a_{n-1}^{j+1})$ must also be changed. Even if one is willing to implement Newton's method, the resulting code often has an extremely high operation count, especially when harmonic averages are used. The computing expense of a high operation count is often compounded by numerous calls to relatively slow functions, such as arctan or exp, which sometimes appear in the expressions for coefficients such as $a(u)$. Of course, using difference quotients to approximate the derivative that appear in the Jacobian will avoid having to produce code for the derivative expressions. However, this approach requires repeated evaluations of the often-complicated harmonic mean expressions $\mu(a_n^{j+1}, a_{n-1}^{j+1})$.

A number of modifications of Newton's method have been developed to avoid the complications cited above. For a reference to Newton-like methods and their convergence theory, we refer to the book of Ortega and Rheinboldt (1970). In particular, we want to mention the class of quasi-Newton methods as an alternative to the methods we describe below. The book of Dennis and Schnabel (1983) is a general reference. However, in the present context, quasi-Newton methods can only be suggested if the special structure (sparsity) of the equations is respected. Some recent papers of Toint deal with this problem. We cite only one of his early discussions. It seems that the quasi-Newton methods lose some of their attractiveness when applied to discretizations of partial differential equations. The modification that we pursue in this presentation is based on the following approach.

Let a Newton-like method be defined by

$$\mathbf{U}^{i+1} = \mathbf{U}^{i} + \mathbf{c}^{i} \tag{3.6}$$

$$i = 1, 2, \ldots,$$

$$\mathbf{c}^{i} = -\mathbf{Q}^{i}\mathbf{F}\left(\mathbf{U}^{i}\right) \tag{3.7}$$

where $\mathbf{Q}$ denotes an approximation to the Jacobian $\mathbf{J}$ in Equation (3.5). We explore methods of the form (3.6)–(3.7) with the properties:

(i) The calculation of $\mathbf{Q}$ requires only those expressions that were required in defining the linearized problem defined by (3.3).

(ii) $\mathbf{Q}$ is a sufficiently good approximation to $\mathbf{J}$ so that the Newton-like iteration (3.6)–(3.7) and the Newton iteration (3.4)–(3.5) have similar convergence properties.

(iii) $\mathbf{Q}$ has a sufficiently simple structure so that the operation count associated with the Newton-like iteration (3.6)–(3.7) is significantly lower than that associated with the Newton iteration (3.4)–(3.5).

4. Newton-like methods. In this section we describe some Newton-like methods for dealing with the nonlinear difference equations (3.2). To simplify the presentation, we consider difference equations for Equation (1.1) when only one space variable is present. In that case, the difference equations represented by (2.8) can be expressed in the form

$$F_n\left(U_{n-1}^{j+1}, U_n^{j+1}, U_{n+1}^{j+1}\right) = 0$$

or

$$-r\mu_{n-1}^{j+1}U_{n-1}^{j+1} + \left(s_n^j + r\mu_{n-1}^{j+1} + r\mu_{n+1}^{j+1}\right)U_n^{j+1} - r\mu_{n+1}^{j+1}U_{n+1}^{j+1} - s_n^jU_n^j = 0. \tag{4.1}$$

A typical row of the linearization of (4.1), $\mathbf{AU} = \mathbf{b}$, reads

$$(4.2)\quad -r\mu_{n-1}^{j}U_{n-1}^{j+1} + \left(s_n^j + r\mu_{n-1}^{j} + r\mu_{n+1}^{j}\right)U_n^{j+1} - r\mu_{n+1}^{j}U_{n+1}^{j+1} = s_n^j U_n^j.$$

For the case of one space dimension, the Jacobian matrix is tridiagonal and can be represented in terms of its subdiagonal $\overrightarrow{\alpha}$, its diagonal $\overrightarrow{\beta}$, and its superdiagonal $\overrightarrow{\gamma}$.

$$\mathbf{J} = \left[\overrightarrow{\alpha}; \overrightarrow{\beta}; \overrightarrow{\gamma}\right],$$

with the mth entries in the respective arrays given as follows:

$$(4.3)\quad \beta_m = \frac{\partial F_m}{\partial U_n} = (s_n + r\mu_{n-1} + r\mu_{n+1}) + r\frac{\partial \mu_{n-1}}{\partial U_n}[U_n - U_{n-1}] + r\frac{\partial \mu_{n+1}}{\partial U_n}[U_n - U_{n+1}],$$

$$(4.4)\quad \alpha_m = \frac{\partial F_m}{\partial U_{n-1}} = -r\mu_{n-1} + r\frac{\partial \mu_{n-1}}{\partial U_{n-1}}[U_n - U_{n-1}],$$

and

$$(4.5)\quad \gamma_m = \frac{\partial F_m}{\partial U_{n+1}} = -r\mu_{n+1} + r\frac{\partial \mu_{n+1}}{\partial U_{n+1}}[U_n - U_{n+1}],$$

where, for notational convenience, the time superscript has been omitted.

Since it is the calculation and coding of the various derivatives of μ that lead to practical difficulties in implementing Newton's method, one of the first Newton-like methods that comes to mind consists of choosing $\mathbf{Q}$ in Equation (3.7) to be that approximation to $\mathbf{J}$ that is obtained by just ignoring the μ-derivatives in Equations (4.3)–(4.5). For later reference, this Newton-like method is indicated by

METHOD NL1.

$$(4.6)\quad \mathbf{U}^{i+1} = \mathbf{U}^i + \mathbf{c}^i$$

$$i = 1, 2, \ldots,$$

$$(4.7)\quad \mathbf{c}^i = -\mathbf{Q1}^i\mathbf{F}\left(\mathbf{U}^i\right)$$

where

$$\mathbf{Q1} = \left[\overrightarrow{\alpha 1}; \overrightarrow{\beta 1}; \overrightarrow{\gamma 1}\right],$$

(4.8) $$\beta 1_m = (s_n + r\mu_{n-1} + r\mu_{n+1}),$$

(4.9) $$\alpha 1_m = \frac{\partial F_M}{\partial U_{n-1}} = -r\mu_{n-1},$$

and

(4.10) $$\gamma 1_m = \frac{\partial F_m}{\partial U_{n+1}} = -r\mu_{n+1}.$$

The NL1 method requires that a tridiagonal system of equations be solved each time Equation (4.7) is updated. A Newton-like method that does not require any systems of equations to be solved can be obtained by keeping only the lower triangular part of **Q** from Method NL1. This leads to

METHOD NL2.

(4.11) $$\mathbf{U}^{i+1} = \mathbf{U}^i + \mathbf{c}^i$$

$$i = 1, 2, \ldots,$$

(4.12) $$\mathbf{c}^i = -\mathbf{Q2}^i \mathbf{F}\left(\mathbf{U}^i\right)$$

where

$$\mathbf{Q2} = \left[\overrightarrow{\alpha 1}; \overrightarrow{\beta 1}; \overrightarrow{0}\right].$$

An examination of Method NL2 shows it to be equivalent to the nonlinear Gauss–Seidel method. This observation suggests a third Newton-like method for dealing with the nonlinear difference equations. Namely, add an overrelaxation step to Method NL2 to obtain

METHOD NL3.

For $n = 1, 2, \ldots, N$ perform the calculation

(4.13) $$U_n^{i+1} = \omega V_n^{i+1} + (1 - \omega) U_n^i,$$

(4.14) $$V_n^{i+1} = U_n^i + c_n^i, \quad i = 1, 2, \ldots,$$

where

(4.15) $$c_n^i = -\alpha 1_n F_{n-1}\left(\mathbf{U}^i\right) - \beta 1_n F_n\left(\mathbf{U}^i\right),$$

and iterate on i until numerical convergence is attained.

Each of the Newton-like methods NL1 through NL3 is recursive. In some computing environments, such as when vector processors are used, it is highly desirable to have available a nonrecursive method. Such a nonrecursive Newton-like method can be obtained from Method NL1 by dropping both the sub and superdiagonal arrays to obtain

METHOD NL4.

$$\mathbf{U}^{i+1} = \mathbf{U}^i + \mathbf{c}^i \tag{4.16}$$

$$i = 1, 2, \ldots,$$

$$\mathbf{c}^i = -\mathbf{Q3}^i F\left(\mathbf{U}^i\right) \tag{4.17}$$

where

$$\mathbf{Q3} = \left[\vec{0}; \beta\vec{1}; \vec{0}\right].$$

5. Convergence results. When Equation (1.1) is everywhere of parabolic type, with $s(u)$ strictly positive, it is not difficult to show that the Jacobian matrix $\mathbf{J}$ is strictly diagonally dominant. Moreover, it is easy to see that the off-diagonal terms in $\mathbf{J}$ all have a factor of k, the time step used in the discretization. These two observations indicate that, at least for small time steps, the essence of the Jacobian is contained in its diagonal terms. Thus, one would expect a Newton-like method that well represents the diagonal of $\mathbf{J}$ to produce a convergent iterative scheme. Note that each of the methods NL1–NL4 agrees with Newton's method in the diagonal terms, at least to order k.

The following development puts the preceding remarks on convergence into a somewhat more formal setting. Suppose that $\mathbf{U}^*$ is a solution to the nonlinear system of finite difference equations, so

$$\mathbf{F}(\mathbf{U}) = \mathbf{0}. \tag{5.1}$$

Further, let one of the Newton-like methods be denoted by

$$\mathbf{U}^{i+1} = \mathbf{U}^i - \mathbf{Q}_i^{-1}\mathbf{F}\left(\mathbf{U}^i\right). \tag{5.2}$$

For notational convenience the iteration index i on $\mathbf{Q}$ is displayed here as a subscript, rather than the superscript position used above. Subtracting $\mathbf{U}^*$ from both sides of Equation (5.2) and using (5.1), we conclude

$$\mathbf{U}^{i+1} - \mathbf{U}^* = \mathbf{U}^i - \mathbf{U}^* - \mathbf{Q}_i^{-1}\left[\mathbf{F}\left(\mathbf{U}^i\right) - \mathbf{F}\left(\mathbf{U}^*\right)\right].$$

Using Taylor's Theorem, we can write

$$\mathbf{F}\left(\mathbf{U}^i\right) - \mathbf{F}\left(\mathbf{U}^*\right) = \mathbf{J}_i\left(\mathbf{U}^i - \mathbf{U}^*\right) + \mathbf{R}_i,$$

where $\mathbf{R}$ denotes the remainder which consists of higher-order terms having the property that

$$\delta_i : = \|\mathbf{R}_i\| / \|\mathbf{U}^i - \mathbf{U}^*\| \to 0 \text{ as } \|\mathbf{U}^i - \mathbf{U}^*\| \to 0.$$

Thus, we conclude that

$$\begin{aligned}\mathbf{U}^{i+1} - \mathbf{U}^* &= \mathbf{U}^i - \mathbf{U}^* - \mathbf{Q}_i^{-1}\left[\mathbf{J}_i\left(\mathbf{U}^i - \mathbf{U}^*\right)\right] - \mathbf{Q}_i^{-1}\mathbf{R}_i \\ &= \left[\mathbf{I} - \mathbf{Q}_i^{-1}\mathbf{J}_i\right]\left(\mathbf{U}^i - \mathbf{U}^*\right) - \mathbf{Q}_i^{-1}\mathbf{R}_i \\ &= \mathbf{Q}_i^{-1}\left[\mathbf{Q}_i - \mathbf{J}_i\right]\left(\mathbf{U}^i - \mathbf{U}^*\right) - \mathbf{Q}_i^{-1}\mathbf{R}_i,\end{aligned}$$

so

$$\begin{aligned}\left(\mathbf{U}^{i+1} - \mathbf{U}^*\right) \le \mathbf{Q}_i^{-1} \bullet \mathbf{Q}_i - \mathbf{J}_i \bullet \left(\mathbf{U}^i - \mathbf{U}^*\right) \\ + \delta_i \mathbf{Q}_i^{-1} \bullet \left(\mathbf{U}^i - \mathbf{U}^*\right).\end{aligned} \tag{5.3}$$

For each of the methods NL1–NL4, it can be shown that

$$\mathbf{Q}_i^{-1} \le \text{ constant } \bullet \max_n \left[\frac{1}{s(U_n) + O(k)}\right] \tag{5.4}$$

and

$$\mathbf{Q}_i - \mathbf{J}_i \le \text{ constant } \bullet k. \tag{5.5}$$

Now, using the inequalities (5.3)–(5.5) it is easy to conclude that the Newton-like method (5.2) converges to the solution of (5.1) provided

(i) $\mathbf{U}^i$ is sufficiently close to $\mathbf{U}^*$;
(ii) $s(U_n)$ is bounded away from zero;
(iii) the time step k is sufficiently small.

In practice, condition (iii) ensures that condition (i) is satisfied.

Experience has shown the Newton-like methods NL1–NL4 to be very robust alternatives to the full Newton iteration within the time step provided the underlying partial differential equation (PDE) is of parabolic type. However, in applications that involve an equation of parabolic/elliptic type ($s(U_n)$ is zero at some grid points), difficulties are sometimes encountered with such Newton-like methods. In light of the above convergence analysis, these difficulties should come as no surprise. It is the positivity of $s(u)$ that ensures the strict diagonal dominance on which the methods depend.

6. Example problem. To illustrate the preceding methods, we consider the following initial-boundary value problem:

$$\frac{\partial u}{\partial t} = \frac{\partial}{\partial x}\left[2u^2 \frac{\partial u}{\partial x}\right], \quad 0 < x < 1, \quad t > 0, \tag{6.1}$$

$$u(x,0) = \sqrt{x+1}, \quad 1 < x < 2, \tag{6.2}$$

$$u(1,t) = \sqrt{t+1}, \quad u(2,t) = \sqrt{t+3}, \quad t > 0, \tag{6.3}$$

whose exact solution is

$$u^*(x,t) = \sqrt{x+t+1}.$$

Using the methods of Section 2 and the grid $x_n = n \cdot h$, $h = 0.1$, $k = 0.2$, we obtain the following difference equations for (6.1)–(6.3):

$$\left[U_n^{j+1} - U_n^j\right] = r\left[\mu_{n-1}^{j+1} U_{n-1}^{j+1} - \left(\mu_{n-1}^{j+1} + \mu_{n+1}^{j+1}\right) U_n^{j+1} + \mu_{n+1}^{j+1} U_{n+1}^{j+1}\right], \quad r = k/h^2, \tag{6.4}$$

$$U_n^0 = \sqrt{x_n + 1}, \quad n = 1, 2, \ldots, N, \tag{6.5}$$

$$U_0^j = \sqrt{t_j + 1}, \quad U_N^j = \sqrt{t_j + 2}, \quad j = 1, 2, \ldots, \tag{6.6}$$

where

$$\mu_{n-1}^{j+1} = \left[\frac{2U_n^2 U_{n-1}^2}{U_n^2 + U_{n-1}^2}\right]^{j+1} \quad \text{and} \quad \mu_{n+1}^{j+1} = \left[\frac{2U_n^2 U_{n+1}^2}{U_n^2 + U_{n+1}^2}\right]^{j+1}.$$

Having incorporated the boundary and initial conditions, the difference system (6.4)–(6.6) can be expressed in the form

$$F_n\left(U_{n-1}^{j+1}, U_n^{j+1}, U_{n+1}^{j+1}\right) = 0, \quad n = 1, 2, \ldots, N, \tag{6.7}$$

or, more compactly, as

$$\mathbf{F}(\mathbf{U}) = \mathbf{0}. \tag{6.8}$$

To illustrate how Newton's method and the various Newton-like methods NL1–NL4 behave, the convergence properties of each approach are displayed in Figures 6.1–6.4. In each figure the horizontal axis displays the iteration counter i for the number of passes through the iterative procedure for solving Equation (6.7) to advance the solution one time step using a time step of $k = 0.1$. The vertical axis displays the L^1 norm of the left side of Equation (6.8) and, thus, represents the residual associated with the numerical solution after i-iterations during the second time step. The behavior indicated in Figures 6.1–6.4 is very nearly duplicated within each time step. When a time step is chosen sufficiently small so that Newton's method and the Newton-like methods NL1–NL4 all converge, the behavior of the various methods is qualitatively as illustrated in Figures 6.1–6.4.

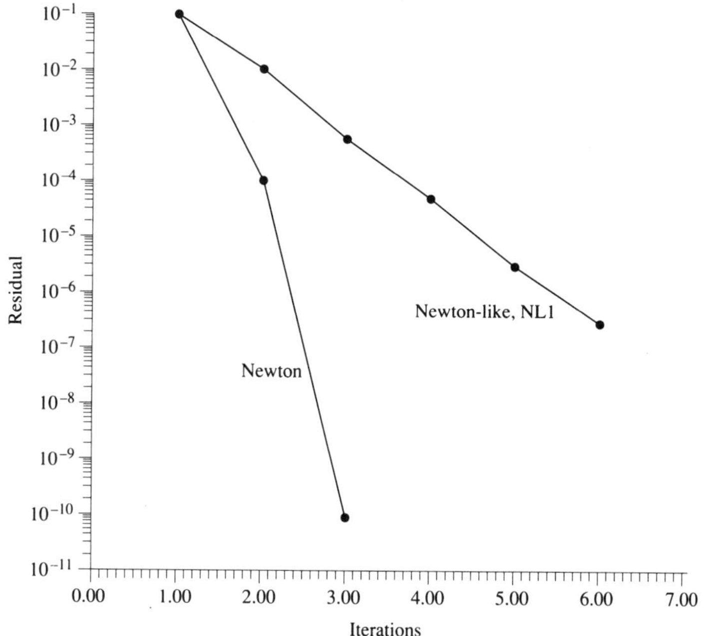

FIGURE 6.1

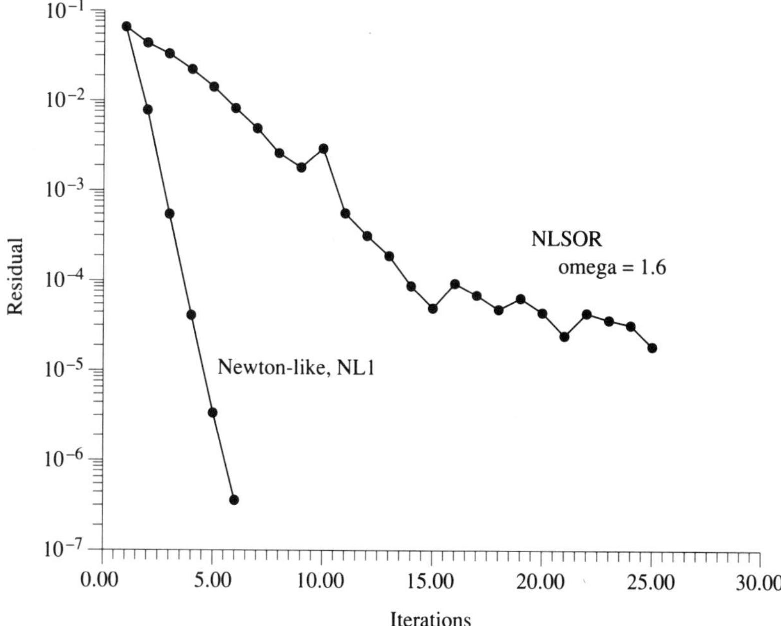

FIGURE 6.2

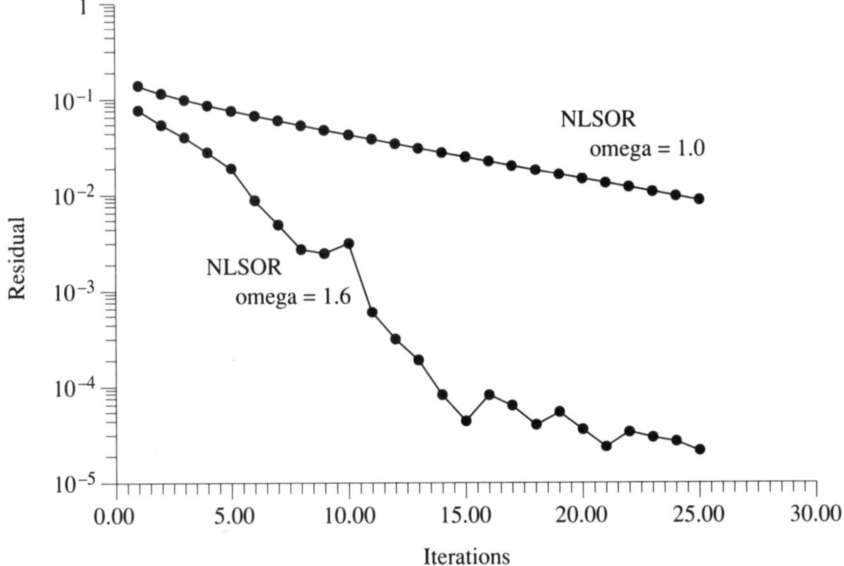

FIGURE 6.3

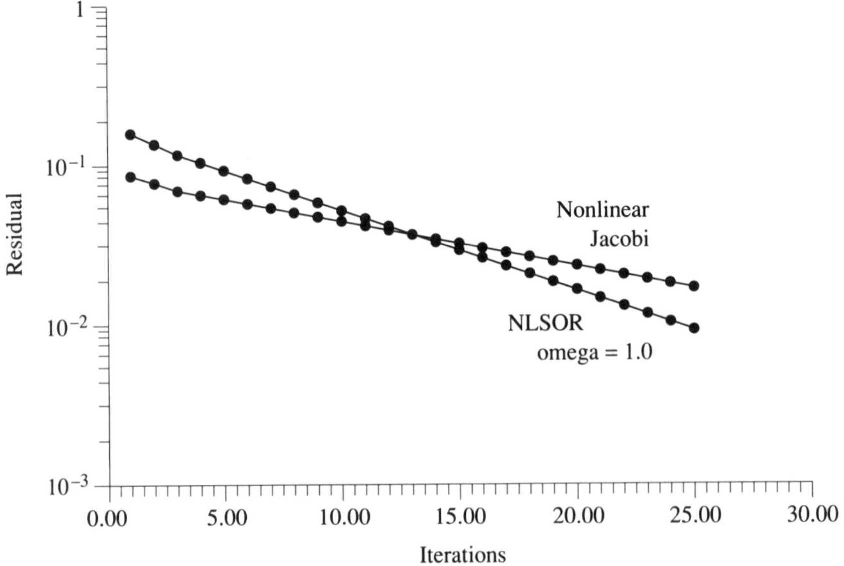

FIGURE 6.4

We refer to the first of the Newton-like methods, NL1, as simply Newton-like. Methods NL2 and NL3 are both versions of nonlinear successive overrelaxation. We refer to the Method NL2 as NLSOR, omega=1 to indicate that the overrelaxation parameter ω has been put to one to produce the nonlinear Gauss–Seidel method. To illustrate Method NL3, we have selected $\omega = 1.6$ as a demonstration value of the overrelaxation parameter. We have not attempted to optimize the convergence rate of Method NL3. Finally, Method NL4 is recognized as a nonlinear version of the usual Jacobi iterative procedure from linear algebra. Thus, in Figure 6.4, Jacobi refers to method NL4.

As Figure 6.1 shows, Newton's method is very rapidly convergent. If one does not plan on changing the coefficients in the underlying PDE, it probably pays to invest the time and effort in providing the code with the necessary functions to implement Newton's method. However, the calculation and coding of these derivatives can be very time consuming and can lead to the introduction of errors. Figure 6.1 also indicates that the Newton-like method NL1 might serve as a substitute for Newton's method if one does not want to confront the problem of differentiating the harmonic means of the coefficients.

Figures 6.2 and 6.3 show a substantial degradation of the convergence rate for methods NL2 and NL3 as compared with Newton's method and Method NL1. The only real simplification afforded by Method NL2 and NL3 is that, in those iterative methods, a tridiagonal system of equations is avoided. As Figure 6.3 shows, one should consider overrelaxing even when the optimal overrelaxation parameter is not known. One caveat in this area involves problems in which the solution is not monotone. Experience has shown that, in such cases, overrelaxation can have a destabilizing effect on the iterative procedure.

Finally, Figure 6.4 shows that, even in the nonlinear case, the convergence rate associated with Jacobi's method is about half the convergence rate for the Gauss–Seidel method. Probably, the only reason for implementing Method NL4 in such problems would be to take advantage of some special computer architecture, such as a vector or parallel processor. Both experience and analysis show that Jacobi's method cannot be effectively accelerated by overrelaxation.

References

1. Dennis, J. E., Jr. and Schnabel, R. B., *Numerical Methods for Unconstrained Optimization and Nonlinear Equations*, Prentice-Hall, Englewood Cliffs, NJ, 1983.

2. DuChateau, P. C. and Zachmann, D. W., *Applied Partial Differential Equations*, Harper and Row, New York, NY, 1989.

3. Ortega, J. M. and Rheinboldt, W. C., *Iterative Solution of Nonlinear Equations in Several Variables*, Academic Press, New York, London, 1970.

4. Toint, Ph., *On the Superlinear Convergence of an Algorithm for Solving a Sparse Minimization Problem*, SIAM J. Numer. Anal. **16** (1979), 1036–1045.

Colorado State University

Lectures in Applied Mathematics
Volume **26** (1990)

Convergence of the Newton–Raphson Method for Boundary Value Problems of Ordinary Differential Equations

RONALD B. GUENTHER AND JOHN W. LEE

1. Introduction. The mathematical treatment of boundary value problems for systems of ordinary differential equations of the second order

$$\mathbf{y}'' = f(x, \mathbf{y}, \mathbf{y}') \tag{1.1}$$

where $f \in C[I \times \mathbf{R}^n \times \mathbf{R}^n,\ \mathbf{R}^n]$, $I = [\alpha, \beta] \subset \mathbf{R}$ is a bounded interval, and where $\mathbf{y}$ is assumed to satisfy prescribed conditions at the endpoints, α and β, of I, centers around questions of existence, uniqueness, and the numerical solution of (1.1). Typically, the discussion of the numerical approximation is carried out under much stronger regularity assumptions than are required to prove existence and uniqueness theorems.

We shall show here how the convergence of the approximating algorithms can be obtained under conditions analogous to those needed in proving the existence and uniqueness. Moreover, the analysis turns out to be simpler than is normal since no special assumptions are needed to guarantee the existence of families of solutions over the entire interval.

The analysis and the requisite discussion make use of certain topological ideas in a Banach space and we first give the definitions and theorems needed in the sequel. See [2] and [4] for additional results.

2. Topological background. Let K be a convex subset of a Banach space E, X a metric space, and $F: X \to K$ a continuous map. We say

1980 *Mathematics Subject Classification* (1985 *Revision*). Primary 34A34, 34B15.
Patially supported by URI-ONR N00014-86K0687.

that F is *compact* if $F(X)$ is contained in a compact subset of K. F is completely continuous if it maps bounded subsets of X into compact subsets of K. A homotopy $\{H_t: X \to K\}_{0 \le t \le 1}$ is said to be compact provided the map $H: X \times [0,1] \to K$ given by $H(x,t) = H_t(x)$ for (x,t) in $X \times [0,1]$ is compact.

The Schauder fixed-point theorem can be formulated as follows.

THEOREM 2.1. *Let K be a convex subset of E and $F: K \to K$ be a compact map. Then F has a fixed point in K.*

Let $U \subset K$ be open in K and $\overline{U}$ and ∂U denote, respectively, the closure and boundary of U in K. A compact map $F: \overline{U} \to K$ is called admissible if it is fixed-point-free on ∂U. The set of all such maps will be denoted $\mathscr{K}_{\partial U}(\overline{U}, K)$.

DEFINITION 2.2. A map F in $\mathscr{K}_{\partial U}(\overline{U}, K)$ is *inessential* if there is a fixed-point-free, compact map $G: \overline{U} \to K$ such that the restrictions of F and G to ∂U agree. The map F is *essential* if it is not inessential.

THEOREM 2.3. *Let p be an arbitrary point in U and $F \in \mathscr{K}_{\partial U}(\overline{U}, K)$ be the constant map $F(x) = p$. Then F is essential.*

PROOF. Let $G: \overline{U} \to K$ be a compact map which agrees with F on ∂U. Define $H: K \to K$ by $H(x) = p$ for x in $K \backslash \overline{U}$ and $H(x) = G(x)$ for x in $\overline{U}$. It is evident that $H: K \to K$ is compact and so has a fixed point z by Schauder's theorem. Thus, $H(z) = z$. By definition of H, z must lie in U where H equals G, and we conclude that $G(z) = z$. Therefore, F is essential.

DEFINITION 2.4. Two maps F and G in $\mathscr{K}_{\partial U}(\overline{U}, K)$ are called *homotopic* ($F \sim G$) if there is a compact homotopy $H_t: \overline{U} \to K$ for which $F = H_0$, $G = H_1$, and H_t is admissible for each t in $[0,1]$.

Inessential maps have the following important characterization.

LEMMA 2.5. *A map F in $\mathscr{K}_{\partial U}(\overline{U}, K)$ is inessential if and only if it is homotopic to a fixed-point-free, compact map.*

PROOF. If F is inessential and G is a fixed-point-free, compact map which agrees with F on ∂U, then a compact homotopy joining F and G is given by $H_t(x) = tF(x) + (1-t)G(x)$.

Conversely, suppose $H_0: \overline{U} \to K$ is a fixed-point-free, compact map and that $H_t: \overline{U} \to K$ is an admissible homotopy joining H_0 to F. We claim that each H_t, in particular, H_1, is an inessential map. To see this, consider the map $H: \overline{U} \times [0,1] \to K$ and define $B = \{u \in \overline{U}: H(u,t) = u$

for some t in $[0,1]\}$. If B is empty, $H_1 = F$ has no fixed point and so is inessential. Thus, we may assume without loss of generality that B is nonempty. Note that B is closed and disjoint from the ∂U. Take a Urysohn function $\lambda\colon \overline{U} \to [0,1]$ with $\lambda(a) = 1$ for $a \in \partial U$ and $\lambda(b) = 0$ for $b \in B$. Put $H_t^*(u) = H(u, \lambda(u)t)$ for $(x,t) \in \overline{U} \times [0,1]$. It is now easily seen that H_t^* is a fixed-point-free, compact homotopy such that H_t^* agrees with H_t on ∂U for each t in $[0,1]$. Consequently, each H_t is inessential and the proof is complete.

The topological transversality theorem follows at once from the lemma.

THEOREM 2.6 (A. GRANAS [2]). *Let F and G in $\mathscr{K}_{\partial U}(\overline{U}, K)$ be homotopic maps. Then one of these maps is essential if and only if the other is essential.*

We use the topological transversality theorem in a slightly different but equivalent form. The idea is this. Suppose we wish to solve the equation $u = F(u)$ in some Banach space; however, it is not obvious that a solution exists. Assume that we can find a related, simpler problem $u = G(u)$ with G essential (for example, $G(u) = 0$ for all u) and that there is a compact homotopy $H(u,t)$ connecting the map G with F. Then the problem $u = F(u)$ will be solvable. In this context, the homotopy used is the usual linear homotopy $H(u,t) = tF(u) + (1-t)G(u)$, and Theorem 2.6 takes the following form.

THEOREM 2.7. *Let K be a convex subset of a Banach space E and let $U \subset K$ be open. Suppose*

(i) *F and G are compact maps from $\overline{U}$ into K;*
(ii) *G is essential in $\mathscr{K}_{\partial U}(\overline{U}, K)$;*
(iii) *$H(u,t)$ is a compact homotopy joining G and F, i.e., $H(u,0) = G(u)$, $H(u,1) = F(u)$;*
(iv) *$H(u,t)$ is fixed-point-free on ∂U for each t in $[0,1]$.*

Then there exists at least one fixed point $u \in U$ for $H(u,t)$, $0 \le t \le 1$; in particular, there is a $u \in U$ such that $u = F(u)$.

In typical applications of Theorem 2.7 to boundary value problems for differential equations, U will be some large ball of radius R centered at the origin. The radius R is chosen so that the known fixed point of G lies in the ball and so that the linear homotopy above has all its fixed points in U. That is, if u_t denotes a fixed point of H_t, we choose R so large that $\|u_t\| < R$. This guarantees that the usual linear homotopy is fixed-point-free on ∂U. To determine such an R amounts to finding an a priori bound on the solutions to a family of differential equations related to the original problem. Thus, the role of a priori bounds in differential equations is to guarantee that the underlying family of operators H_t is fixed-point-free on ∂U.

3. Existence and uniqueness theorems. We shall restrict ourselves to Dirichlet-type boundary conditions, that is, we assume that y is given at the endpoints of the interval $[\alpha, \beta]$. Although more general theorems are available than the ones we give below, the general flavor of the analysis will be apparent if we restrict ourselves to the simplest result for systems. More general theorems can be proven but the basic ideas are unchanged and the details are more complicated.

THEOREM 3.1. *Suppose $f \in C[I \times \mathbf{R}^n, \mathbf{R}^n]$ and satisfies the monotonicity condition*

$$\langle f(x,\mathbf{u}) - f(x,\mathbf{v}), \mathbf{u}-\mathbf{v}\rangle \geq 0 \quad \textit{with } \langle f(x,\mathbf{u}),\mathbf{u}\rangle > 0 \textit{ for } |\mathbf{u}| > M,$$

where $\langle \mathbf{u},\mathbf{u}\rangle = \sum_{i=1}^n u_i v_i$ is the inner product in $\mathbf{R}^n$, $|\mathbf{u}|^2 = \langle \mathbf{u},\mathbf{u}\rangle$, and M is a constant. Then there exists precisely one solution to the Dirichlet problem

$$y'' = f(x,\mathbf{y}), \tag{3.1}$$

$$\mathbf{y}(\alpha) = \mathbf{a}, \quad \mathbf{y}(\beta) = \mathbf{b}, \tag{3.2}$$

where $\mathbf{a}$, $\mathbf{b} \in \mathbf{R}^n$ *are given.*

PROOF. The uniqueness is immediate. Suppose there were two solutions, $\mathbf{u}$ and $\mathbf{v}$, to (3.1) and (3.2). Then integration by parts yields

$$\begin{aligned} -\int_\alpha^\beta |\mathbf{u}'-\mathbf{v}'|^2\,dx &= \int_\alpha^\beta \langle \mathbf{u}-\mathbf{v}, \mathbf{u}''-\mathbf{v}''\rangle\,dx \\ &= \int_\alpha^\beta \langle \mathbf{u}-\mathbf{v},\ f(x,\mathbf{u}) - f(x,\mathbf{v})\rangle\,dx \geq 0, \end{aligned}$$

whence, $|\mathbf{u}'-\mathbf{v}'| = 0$ and from the boundary conditions $\mathbf{u} = \mathbf{v}$.

To prove the existence we must consider the family of problems

$$\begin{gathered} \mathbf{y}'' = \lambda f(x,\mathbf{y}), \quad 0 \leq \lambda \leq 1, \\ \mathbf{y}(\alpha) = \mathbf{a}, \quad \mathbf{y}(\beta) = \mathbf{b}. \end{gathered}$$

To prove existence, we need a priori bounds on $\mathbf{y}$, $\mathbf{y}'$, and $\mathbf{y}''$ which are independent of λ. Let us suppose that $|\mathbf{y}(x)|$ takes on values greater than $\max(|\mathbf{a}|, |\mathbf{b}|)$. Then $|\mathbf{y}(x)|^2/2$ has a maximum at some point $\overline{x}$:

$$0 \geq \frac{d^2}{dx^2}\frac{|\mathbf{y}(\overline{x})|^2}{2} = \langle \mathbf{y}(\overline{x}), f(\overline{x}, \mathbf{y}(\overline{x}))\rangle > 0 \quad \text{if } |\mathbf{y}(\overline{x})| > M.$$

Hence, $|\mathbf{y}(x)| \leq M_0 = \max(|\mathbf{a}|, |\mathbf{b}|, M)$. The differential equation now yields

$$|\mathbf{y}''| \leq \max_{x\in I, |\mathbf{y}|\leq M_0} |f(x, \mathbf{y})| \equiv M_2.$$

Finally, from the differential equation, we find

$$\int_\alpha^\beta |\mathbf{y}'|^2\, dx \leq M_0 M_2(\beta - \alpha).$$

By the mean value theorem there is a point $x^* \in I$ such that

$$|\mathbf{y}'(x^*)| \leq \sqrt{M_0 M_2}$$

and consequently

$$|\mathbf{y}'(x)| \leq |\mathbf{y}'(x^*)| + M_2(\beta - \alpha) \leq \sqrt{M_0 M_2} + M_2(\beta - \alpha) \equiv M_1.$$

Theorem 2.6 now implies the existence of at least one solution.

There are additional existence and uniqueness theorems, the proofs of which are based on the topological results of Section 2 (see [3] and the references cited there) but instead of pursuing these topics further, we shall take up the problem of the numerical approximation of the solutions.

4. Convergence of Newton–Raphson algorithms. In this section we shall assume that $\mathbf{f}: I \times \mathbf{R}^n \to \mathbf{R}^n$ is continuous and the derivatives $\partial f_i(\mathbf{x}, \mathbf{y})/\partial y_j$ $(i,\ j = 1, \ldots, n)$ are continuous. Moreover, we assume that the $n \times n$ matrix $(\partial f_i/\partial y_j)$ is nonnegative definite and $\mathbf{f}$ satisfies $\langle \mathbf{y}, \mathbf{f}(x, \mathbf{y})\rangle > 0$ for $|\mathbf{y}| > M$. Then we know on the basis of Theorem 3.1 that there is a unique solution to the Dirichlet problem

$$\mathbf{y}'' = \mathbf{f}(x, \mathbf{y}), \tag{4.1}$$

$$\mathbf{y}(\alpha) = \mathbf{a}, \quad \mathbf{y}(\beta) = \mathbf{b}. \tag{4.2}$$

The first method to be described is the so-called shooting method. It consists of the following. Replace the system (4.1), (4.2) by the initial value problem

$$\mathbf{u}'' = \mathbf{f}(x, \mathbf{u}), \tag{4.3}$$

$$\mathbf{u}(\alpha) = \mathbf{a}, \quad \mathbf{u}'(\alpha) = \mathbf{c}. \tag{4.4}$$

The solution $\mathbf{u}$ to (4.3), (4.4) depends on both x and $\mathbf{c}$. The idea is to try to choose $\mathbf{c}$ so that

$$\mathbf{u}(\beta, \mathbf{c}) = \mathbf{b}. \tag{4.5}$$

(4.5) represents a nonlinear system. Newton's method for solving it is:

$$\mathbf{c}^{(0)} \text{ is an initial guess.} \tag{4.6}$$

$\mathbf{c}^{(l+1)}$ is constructed recursively by solving

$$\mathbf{A}^{(l)}(\mathbf{c}^{(l+1)} - \mathbf{c}^{(l)}) = -\mathbf{u}(\beta, \mathbf{c}^{(l)}) + \mathbf{b}, \tag{4.7}$$

where the elements of the matrix $\mathbf{A}^{(l)}$ are given by $\partial u_i(\beta, \mathbf{c}^{(l)})/\partial c_j$ and the $u_i(\beta, \mathbf{c})$ are differentiable functions of $\mathbf{c}$ by standard theorems on the differentiability of solutions with respect to parameters.

To show the convergence of (4.6), (4.7), it is simplest to consider (4.1), (4.2) with $\mathbf{y}(\beta) = \mathbf{p}$, where $\mathbf{p} \in \mathbf{R}^n$ is arbitrary. The solution $\mathbf{y}$ which we know exists is a function of x and $\mathbf{p}$ so $\mathbf{y} = \mathbf{y}(x, \mathbf{p})$ and as such has an initial slope $\mathbf{y}'(\alpha, \mathbf{p}) \equiv \mathbf{g}(\mathbf{p})$. $\mathbf{g}(\mathbf{p})$ is a function defined on $\mathbf{R}^n$ and takes values in $\mathbf{R}^n$. It is one-to-one by the uniqueness theorem for the initial value problem and is differentiable. To show that the inverse mapping is differentiable, we must show that the matrix $(\partial g_i(\mathbf{p})/\partial p_j)$ is nonsingular. To prove that, let $\mathbf{A} = (\partial f_i(x, \mathbf{y})/\partial y_j)$, $\mathbf{B} = (\partial y_i/\partial p_j)$, and $\mathbf{I}$ be the identity matrix. Then $\mathbf{B}$ satisfies the differential equation

$$\mathbf{B}'' = \mathbf{AB},$$
$$\mathbf{B}(\alpha) = \mathbf{0}, \quad \mathbf{B}(\beta) = \mathbf{I}.$$

Multiply by the transpose, $\mathbf{B}^*$, of $\mathbf{B}$ and integrate from α to β to obtain

$$\mathbf{B}'(\beta) = \int_\alpha^\beta \mathbf{B}'^*\mathbf{B}'\,dx + \int_\alpha^\beta \mathbf{B}^*\mathbf{AB}\,dx$$

since $\mathbf{B}^*(\beta) = \mathbf{I}^* = \mathbf{I}$. Let $\xi \in \mathbf{R}^n$ be arbitrary. Then

$$\langle \mathbf{B}'(\beta)\boldsymbol{\xi}, \boldsymbol{\xi}\rangle = \int_\alpha^\beta |\mathbf{B}'\boldsymbol{\xi}|^2\,dx + \int_\alpha^\beta \langle \mathbf{AB}\boldsymbol{\xi}, \mathbf{B}\boldsymbol{\xi}\rangle\,dx \geq 0$$

for $\boldsymbol{\xi} \neq \mathbf{0}$. The possibility of equality holding is ruled out as follows. Let $\mathbf{w}(x) = \mathbf{B}(x)\boldsymbol{\xi}$ with $\boldsymbol{\xi} \neq \mathbf{0}$. Then $\mathbf{w}$ satisfies $\mathbf{w}'' = \mathbf{A}\mathbf{w}$ with $\mathbf{w}(\alpha) = \mathbf{0}$, $\mathbf{w}(\beta) = \boldsymbol{\xi}$. If equality were to hold, the computation just given would imply that $\mathbf{B}'(x)\boldsymbol{\xi} = (\mathbf{B}(x)\boldsymbol{\xi})' = \mathbf{w}'(x) = \mathbf{0}$ for all x and since $\mathbf{w}(\alpha) = \mathbf{0}$, $\mathbf{w}(x) = \mathbf{0}$. But this would lead to $\boldsymbol{\xi} = \mathbf{w}(\beta) = \mathbf{0}$, a contradiction. Thus, the matrix $\mathbf{B}'(\beta) = (\partial y_i'(\beta, \mathbf{p})/\partial p_j)$ is invertible. The function $\mathbf{g}$ is a one-to-one, continuously differentiable function defined on $\mathbf{R}^n$ with range an open set $\mathbf{J}$ in $\mathbf{R}^n$. The inverse is also continuously differentiable.

The rest of the discussion now follows word-for-word the discussion in [3]. We give only the final theorem.

THEOREM 4.1. *Let* $\mathbf{y}$ *be the solution to the Dirichlet problem* (4.1), (4.2). *Then there is a* $\delta > 0$ *such that if* $|\mathbf{c}^{(0)} - \mathbf{y}'(\alpha)| < \delta$, *the Newton iterates* $\mathbf{c}^{(1)}, \mathbf{c}^{(2)}, \ldots$ *defined by* (4.6), (4.7) *all exist and the initial value problem* (4.3), (4.4) *has a unique solution,* $\mathbf{u}(x, \mathbf{c}^{(l)})$ *which converges uniformly to* $\mathbf{y}(x)$ *for* $x \in [\alpha, \beta]$.

THEOREM 4.2. *If the equation* (4.1) *is linear, more precisely, if* $\mathbf{f}$ *satisfies the assumptions made at the beginning of this section, and if*

$$\mathbf{f}(\mathbf{x}, \mathbf{y}) = \mathbf{A}(x)\mathbf{y} + \phi(x),$$

then the shooting method converges in one step.

An alternative procedure, the so-called method of quasilinearization, is also often useful. To describe this method, let us define the operator

$$F(\mathbf{y}) = (\mathbf{y}'' - \mathbf{f}(x, \mathbf{y}),\ \mathbf{y}(\alpha) - \mathbf{a},\ \mathbf{y}(\beta) - \mathbf{b})$$

which maps functions from $C^2[\alpha, \beta]$ into $C[\alpha, \beta] \times \mathbf{R}^n \times \mathbf{R}^n$. The directional derivative of F at $\mathbf{y}$ in the direction $\boldsymbol{\xi}$ is the linear operator F' defined by

$$F'(\mathbf{y})\boldsymbol{\zeta} = (\boldsymbol{\zeta}'' - \mathbf{A}\boldsymbol{\zeta}, \boldsymbol{\zeta}(\alpha), \boldsymbol{\zeta}(\beta)) \tag{4.8}$$

where $\mathbf{A} = (\partial f_i(x, \mathbf{y})/\partial y_j)$. The Newton–Raphson method in this setting can be described as follows.

$$\text{Let } \mathbf{u}^{(0)} \in C^2[\alpha, \beta] \text{ be arbitrary.} \tag{4.9}$$

Define the sequence $\{\mathbf{u}^{(l)}(x)\}$ recursively by

$$\mathbf{u}^{(l+1)}(x) = \mathbf{u}^{(l)}(x) + \boldsymbol{\zeta}^{(l)}(x), \tag{4.10}$$

where $\boldsymbol{\zeta}^{(l)}(x)$ is the solution to the system of ordinary differential equations

$$F'(\mathbf{u}^{(l)})\boldsymbol{\zeta}^{(l)} + F(\mathbf{u}^{(l)}) = 0,$$

and F' is given by (4.8).

THEOREM 4.3. *Suppose* $\mathbf{f}$ *satisfies the assumptions made at the beginning of the section so that the Dirichlet problem* (4.1), (4.2) *has a unique solution,* $\mathbf{y}$. *Then if* $\mathbf{u}^{(0)}$ *is sufficiently close to* $\mathbf{y}(x)$ *in the sense of the* $C^2[\alpha, \beta]$ *norm, the sequence* $\{\mathbf{u}^{(l)}(x)\}$ *given by* (4.9) *and* (4.10) *converges uniformly to the solution.*

The proof is based on the formulation of the Newton–Raphson method in Banach spaces due to Kantorovich [5]. See also [1] and [3].

5. Concluding remarks. The above considerations carry over almost verbatim to other boundary value problems. The critical results underlying this analysis are existence and uniqueness theorems. In the case where $n = 1$, many more general existence and uniqueness theorems are available and consequently the corresponding convergence theorems are sharper and more general.

A similar philosophy can be developed for eigenvalue problems. We shall not go into this any further but simply remark that in this case, the eigenvalue problem is treated like a family of initial value problems. The resulting solution evaluated at, say, the right-hand endpoint is a function of the eigenvalue parameter λ and the eigenvalue problem amounts to finding values of λ where the function vanishes.

REFERENCES

1. L. Collatz, *Funktionalanalysis und numerische Mathematik*, Springer-Verlag, Berlin–Göttingen–Heidelberg (1964).

2. A. Granas and J. Dugundji, *Fixed point theory*, PWN-Polish Scientific Publishers, Polska Akademia Nauk, Instytut Matematyczny, Monografie Matematyczne Tom 61, Warszawa **1**, (1982).

3. A. Granas, R. B. Guenther, and J. W. Lee, *Nonlinear boundary value problems for ordinary differential equations*, Dissertationes Mathematicae **CCXLIV**, Warszawa, 1985.

4. R. B. Guenther and J. W. Lee, *Topological transversality and differential equations*, Fixed-point Theory and its Applications, Cont. Math. **72**, Amer. Math. Soc., Providence (1988), 121–130.

5. L. V. Kantorovich, *Functional analysis and applied mathematics* (in Russian), Vspeh: Mat. Nauk **3** (1948), 89–185, Trans. C. D. Benster, National Bureau of Standards.

OREGON STATE UNIVERSITY

Lectures in Applied Mathematics
Volume **26** (1990)

A Damped-Newton Method for the Linear Complementarity Problem

PATRICK T. HARKER[1] AND JONG-SHI PANG[2]

Abstract. This paper presents a damped-Newton method for solving the linear complementarity problem. The method is a specialization of a general Newton algorithm for solving B-differentiable equations, and a certain modification of the method resembles a direct block pivotal algorithm. An important feature of the method is that it contains a one-dimensional line-search step on which the global convergence crucially depends. The numerical results of an extensive set of computational experiments suggest that the method is potentially very efficient and promising for solving the large-scale problems which arise in applications.

1. Introduction. In a recent paper **[16]**, Pang has studied a global Newton method for solving a system of B-differentiable equations, and has discussed applications to several problems in mathematical programming. The objectives of this paper are twofold: (i) to analyze in detail the specialization of this method in the case of the linear complementarity problem, and (ii) to report on the results of extensive computational experiments using the method.

The literature on the linear complementarity problem is rich in solution algorithms (see **[14]**). Several of these solution algorithms are

Received by the editors February 1989.
1980 *Mathematics Subject Classification* (1985 *Revision*). Primary 90C33, 65H10; Secondary 65K05.
[1]This work was supported by the NSF Presidential Young Investigator Award ECE-8552773 and by the AT&T Program in Telecommunications Technology at the University of Pennsylvania.
[2]This work was based on research supported by the NSF under grant ECS-8717968.

related to the classical Newton method for solving systems of continuously differentiable (nonlinear) equations. In his doctoral thesis, Kostreva [**9**] described an *all-change* pivotal algorithm and showed that this algorithm can be interpreted as an application of Newton's method to a certain formulation of the linear complementarity problem as a system of piecewise linear equations. Since a piecewise linear function is not continuously differentiable in general, Kostreva was not able to establish the convergence of his method. Separately, Aganagić [**1**] described a Newton-type method for solving the linear complementarity problem with a hidden Z-matrix [**15**]. The convergence of this algorithm depends heavily on the hidden Z-property of the underlying matrix. Kojima and Shindo [**8**] discussed the local convergence properties of Newton's method for piecewise-continuous equations of which the linear complementarity problem is a special case, but provided no procedure which ensures global convergence. Recently, Subramanian [**24**] described a Gauss–Newton method for solving the nonlinear complementarity problem which is based on Mangasarian's [**11**] formulation of the complementarity problem as a continuously differentiable system of equations. Subramanian derived several convergence results for the method, but presented no theoretical or computational evidence as to the method's efficiency.

More recently, the polynomial-time, interior point algorithms for linear programming have been extended to the case of convex quadratic programs [**13**], [**26**], certain linear complementarity problems [**6**], [**12**], and the nonlinear complementarity problem [**7**]. The connection between these algorithms and the classical Newton method for nonlinear equations is well explained in [**6**].

The notion of a B-differentiable function on which our Newton algorithm is based is quite novel; the concept originated in two papers by S. M. Robinson [**17**], [**18**]. The properties of such a function are derived in these two papers as well as in [**19**], [**23**]. A related paper of Robinson [**20**] discusses Newton's method for a class of nonsmooth functions which include the B-differentiable ones. Suffice it to say that the (linear or nonlinear) complementarity problem can be conveniently formulated as a system of B-differentiable equations, and it is the primary purpose of this paper to investigate the benefits, if any, of this formulation and associated algorithms for the linear complementarity problem; subsequent research will be devoted to the nonlinear case.

It is well known that the linear complementarity problem can be solved by using piecewise-linear homotopies [**4**]. Our approach is also based on a piecewise-linear-equation formulation of the complementarity problem. However, an important feature of the algorithms presented in this paper is the use of a line-search step which provides for global convergence and significant computational efficiency. This line-search step is absent in all previous algorithms based on piecewise-linear equations.

The organization of this paper is as follows. The next section will review the basics of the damped-Newton method for B-differentiable systems of equations, and Section 3 discusses the specialization of the method to the linear complementarity problem and proposes a slight modification to the basic algorithm. Section 4 provides a further analysis of the algorithm for the special case of the linear complementarity problem with a Z-matrix, and the results of extensive computational experiments are reported in Section 5. The paper ends with a summary of the findings of this investigation as well as proposals for future research directions.

2. Review. We begin our discussion with the following definition:

DEFINITION 1. *A function $H : \Re^n \to \Re^n$ is said to be* B-differentiable *at the point $\mathbf{z}$ if* (i) *H is Lipschitz continuous in a neighborhood of $\mathbf{z}$, and* (ii) *there exists a positive homogeneous function $BH(\mathbf{z}) : \Re^n \to \Re^n$, called the* B-derivative *of H at $\mathbf{z}$, such that*

$$\lim_{\mathbf{v}\to\mathbf{0}} \frac{H(\mathbf{z}+\mathbf{v}) - H(\mathbf{z}) - BH(\mathbf{z})\mathbf{v}}{\|\mathbf{v}\|} = 0.$$

The function H is B-differentiable in set S if it is B-differentiable at every point in S. The B-derivative $BH(\mathbf{z})$ is said to be strong *if*

$$\lim_{(\mathbf{v},\mathbf{v}')\to(\mathbf{0},\mathbf{0})} \frac{H(\mathbf{z}+\mathbf{v}) - H(\mathbf{z}+\mathbf{v}') - BH(\mathbf{z})(\mathbf{v}-\mathbf{v}')}{\|\mathbf{v}-\mathbf{v}'\|} = 0.$$

Many facts are known about the B-derivative. Shapiro **[23]** has shown that if H is Lipschitzian in a neighborhood of $\mathbf{z}$ and if we confine our attention to finite-dimensional Euclidean space $\Re^n$, then H is B-differentiable at $\mathbf{z}$ if and only if it is directionally differentiable at that point. In such a case, the B-derivative coincides with the directional derivative. Pang **[16]** showed that if H is B-differentiable in a neighborhood of $\mathbf{z}$, then a strong B-derivative $BH(\mathbf{z})$ is equivalent to a strong F-derivative, which is further equivalent to the continuity of $BH(\cdot)$ at $\mathbf{z}$. In general, the above definition of a strong B-derivative is more stringent than the definition proposed by Robinson **[19]**.

Consider the solution of the system of nonlinear equations

$$H(\mathbf{z}) = 0, \tag{1}$$

where $H : \Re^n \to \Re^n$ is B-differentiable. Let $g : \Re^n \to \Re$ be the norm function of H; i.e.,

$$g(\mathbf{z}) = \frac{1}{2}H(\mathbf{z})^{\mathrm{T}}H(\mathbf{z}). \tag{2}$$

The following procedure describes a damped-Newton algorithm for solving the equation system (1).

DAMPED-NEWTON ALGORITHM. *Let $\mathbf{z}^0 \in \Re^n$ be an arbitrary initial vector. Let s, μ, and σ be given scalars with $s > 0, \mu \in (0,1)$, and $\sigma \in (0, 1/2)$. In general, given $\mathbf{z}^k$ with $H(\mathbf{z}^k) \neq 0$, we obtain the next iterate $\mathbf{z}^{k+1}$ by performing the following two steps:*

Step 1. *Solve the Newton equation*

$$H(\mathbf{z}^k) + BH(\mathbf{z}^k)\mathbf{d}^k = 0 \tag{3}$$

for the direction $\mathbf{d}^k$.

Step 2. *Let $\lambda_k = \mu^{m_k}s$, where m_k is the smallest nonnegative integer m for which the following condition holds:*

$$g(\mathbf{z}^k) - g(\mathbf{z}^k + \mu^m s\mathbf{d}^k) \geq 2\sigma\mu^m s g(\mathbf{z}^k). \tag{4}$$

Set $\mathbf{z}^{k+1} = \mathbf{z}^k + \lambda_k\mathbf{d}^k$ and check convergence.

The reader is referred to **[16]** for a detailed derivation and explanation of the above method.

Note that Step 2 of the damped-Newton algorithm is an Armijo line search applied to the norm function g since the directional derivative of g at the point $\mathbf{z}^k$ along the direction $\mathbf{d}^k$ is given by $g'(\mathbf{z}^k, \mathbf{d}^k) = -2g(\mathbf{z}^k)$. The line search begins at the iterate $\mathbf{z}^k + s\mathbf{d}^k$ and checks for the satisfaction of condition (4) at the vectors $\mathbf{z}^k + \mu^m s\mathbf{d}^k$ for $m = 0, 1, 2, \ldots$. A finite number of trials will identify the desired step length λ_k, and the new iterate $\mathbf{z}^{k+1}$ will be a point on the line segment joining $\mathbf{z}^k$ and $\mathbf{z}^k + s\mathbf{d}^k$ which yields sufficient decrease in the value of the norm function g. This damping of Newton's method is fairly common in the solution of systems of F-differentiable nonlinear equations, and it provides the basis for the global convergence of the method **[3]**. The main convergence result for the damped-Newton method when applied to a system of B-differentiable equations is given by the following theorem whose proof can be found in **[16]**:

THEOREM 1. *Let $H : \Re^n \to \Re^n$ be B-differentiable, and let $\mathbf{z}^0 \in \Re^n$ be an arbitrary initial point. Assume that the following two conditions hold:*

(a) *the level set*

$$\{\mathbf{z} \in \Re^n : \|H(\mathbf{z})\| \leq \|H(\mathbf{z}^0)\|\}$$

is bounded, and

(b) *each Newton equation* (3) *has a solution.*

Then the damped-Newton sequence $\{\mathbf{z}^k\}$ is well-defined, bounded, and satisfies

$$\|H(\mathbf{z}^k + 1)\| < \|H(\mathbf{z}^k)\|.$$

Moreover, if $\mathbf{z}^*$ is any accumulation point of $\{\mathbf{z}^k\}$ such that

(c) *the B-derivative $Bg(\mathbf{z}^*)$ is strong, and*

(d) *there exists a neighborhood N of $\mathbf{z}^*$ and a scalar $c > 0$ such that for all vectors $\mathbf{x} \in N$ and all $\mathbf{v} \in \Re^n$*

$$\|BH(\mathbf{x})\mathbf{v}\| \geq c\|\mathbf{v}\|,$$

then $H(\mathbf{z}^*) = 0$.

Note that this theorem does not assert that every accumulation point of the sequence $\{\mathbf{z}^k\}$ is a zero of the function H; this desired conclusion is guaranteed to hold if the accumulation point $\mathbf{z}^*$ satisfies the two properties (c) and (d). Such properties seem unavoidable at present and are very similar to the type of nonsingularity conditions which are required for the convergence of damped-Newton schemes for F-differentiable functions **[3]**. The meaning of these two conditions in the context of the linear complementarity problem will be explained in the sequel.

3. The linear complementarity problem. In this section, we shall specialize the discussion of the previous section to the case of the linear complementarity problem, which we shall denote by LCP($\mathbf{q}, \mathbf{M}$):

$$\mathbf{w} = \mathbf{q} + \mathbf{M}\mathbf{z} \geq \mathbf{0}, \qquad \mathbf{z} \geq \mathbf{0}, \qquad \mathbf{w}^{\mathrm{T}}\mathbf{z} = 0, \tag{5}$$

where $\mathbf{q} \in \Re^n$ is a given vector and $\mathbf{M}$ is a given $n \times n$ matrix. Recall that the matrix $\mathbf{M}$ is said to be *nondegenerate* if each of its principal submatrices are nonsingular, and is a *P-matrix* if each of its principal minors is positive.

It is well known (e.g., see **[16]**) that the LCP($\mathbf{q}, \mathbf{M}$) is equivalent to solving the following system of piecewise-linear equations:

$$H(\mathbf{z}) = \min(\mathbf{z}, \mathbf{q} + \mathbf{M}\mathbf{z}) = \mathbf{0}, \tag{6}$$

where the "min" operator is interpreted as the componentwise minimum. The following result summarizes the differentiability properties of the function H defined in equation (6); the proof of this result can be found in **[16]**:

THEOREM 2. *Let* $\mathbf{q}$ *and* $\mathbf{M}$ *be an n-vector and an* $n \times n$ *matrix, respectively. Let* H *be the function defined in equation* (6) *and* g *be the norm function of* H *defined in equation* (2). *Then,*

(1) *H is everywhere B-differentiable with the ith component of the B-derivative given by*

$$(BH(\mathbf{z})\mathbf{v})_i = \begin{cases} (\mathbf{M}_i)^{\mathrm{T}}\mathbf{v} & \text{if } i \in \alpha(\mathbf{z}) \\ \min((\mathbf{M}_i)^{\mathrm{T}}\mathbf{v}, v_i) & \text{if } i \in \beta(\mathbf{z}) \\ v_i & \text{if } i \in \gamma(\mathbf{z}) \end{cases}$$

where $\mathbf{M}_i$ *denotes the ith row of* $\mathbf{M}$ *and*

$$\alpha(\mathbf{z}) = \{i : (\mathbf{q} + \mathbf{Mz})_i < \mathbf{z}_i\},$$
$$\beta(\mathbf{z}) = \{i : (\mathbf{q} + \mathbf{Mz})_i = \mathbf{z}_i\},$$
$$\gamma(\mathbf{z}) = \{i : (\mathbf{q} + \mathbf{Mz})_i > \mathbf{z}_i\};$$

(2) *if for each* $i \in \beta(\mathbf{z})$, $\mathbf{z}_i = (\mathbf{q} + \mathbf{Mz})_i = \mathbf{0}$, *then the B-derivative* $Bg(\mathbf{z})$ *is strong*;

(3) *if* $\mathbf{M}$ *is nondegenerate, then* g *has bounded level sets.*

Following the terminology in **[16]**, we call the index set $\beta(\mathbf{z})$ the *degenerate set* and the indices in $\beta(\mathbf{z})$ the *degenerate indices.* If $\beta(\mathbf{z})$ is empty, then $\mathbf{z}$ is called a *nondegenerate vector.* According to conclusion (2) of Theorem 2, a sufficient condition for the B-derivative $Bg(\mathbf{z})$ to be strong is that for each degenerate index i, we have both $\mathbf{z}_i$ and $(\mathbf{q}+\mathbf{Mz})_i$ (which must themselves be equal) equal to zero. In particular, this condition will hold if $\mathbf{z}$ is a nondegenerate vector.

Consider now the application of the damped-Newton method to system (6). Let the iterate $\mathbf{z}^k$ be given. By the explicit formula for the B-derivative given in Theorem 2, it is easy to see that the vector $\mathbf{d}^k$ satisfying the Newton equation (3) can be obtained by (i) solving the following mixed linear complementarity problem for $(\mathbf{v}_\alpha, \mathbf{v}_\beta)$:

$$\mathbf{q}_\alpha + \mathbf{M}_{\alpha\alpha}\mathbf{v}_\alpha + \mathbf{M}_{\alpha\beta}\mathbf{v}_\beta = \mathbf{0}, \tag{7}$$

$$\mathbf{q}_\beta + \mathbf{M}_{\beta\alpha}\mathbf{v}_\alpha + \mathbf{M}_{\beta\beta}\mathbf{v}_\beta \geq \mathbf{0}, \tag{8}$$

$$\mathbf{v}_\beta \geq 0, \quad (\mathbf{v}_\beta)^{\mathrm{T}}(\mathbf{q}_\beta + \mathbf{M}_{\beta\alpha}\mathbf{v}_\alpha + \mathbf{M}_{\beta\beta}\mathbf{v}_\beta) = 0, \tag{9}$$

where α and β denote the index sets $\alpha(\mathbf{z}^k)$ and $\beta(\mathbf{z}^k)$, respectively, and (ii) setting (with $\gamma = \gamma(\mathbf{z}^k)$)

$$\mathbf{d}_\gamma^k = -\mathbf{z}_\gamma^k, \tag{10}$$

$$(\mathbf{d}_\alpha^k, \mathbf{d}_\beta^k) = (\mathbf{v}_\alpha, \mathbf{v}_\beta) - (\mathbf{z}_\alpha^k, \mathbf{z}_\beta^k). \tag{11}$$

A sufficient condition for system (7)–(9) to have a unique solution is that

(a) the matrix $\mathbf{M}_{\alpha\alpha}$ is nonsingular, and
(b) the Schur complement $\mathbf{M}_{\beta\beta} - \mathbf{M}_{\beta\alpha}(\mathbf{M}_{\alpha\alpha})^{-1}\mathbf{M}_{\alpha\beta}$ is a P-matrix.

In this case we shall call the vector $\mathbf{z}^k$ a *regular vector*. Note that if $\mathbf{M}$ itself is a P-matrix, then any vector $\mathbf{z}$ is regular.

If the matrix $\mathbf{M}_{\alpha\alpha}$ is nonsingular, then the system (7)–(9) can be converted into a standard LCP solely in terms of the variables $\mathbf{v}_\beta$. This conversion is achieved by solving for the $\mathbf{v}_\alpha$ variables in terms of $\mathbf{v}_\beta$ in equation (7) and then substituting for $\mathbf{v}_\alpha$ in condition (8)–(9). The cardinality of the degenerate set $\beta(\mathbf{z}^k)$ determines the size of the LCP resulting from this substitution, and can be interpreted as the count of degeneracy of the vector $\mathbf{z}^k$. In particular, if $\mathbf{z}^k$ is a nondegenerate vector, then this LCP becomes vacuous and thus, the system (7)–(9) reduces to the single system of linear equations

$$\mathbf{q}_\alpha + \mathbf{M}_{\alpha\alpha}\mathbf{v}_\alpha = \mathbf{0}. \tag{12}$$

A general convergence result for the damped-Newton method when applied to LCP $(\mathbf{q}, \mathbf{M})$ can be obtained from Theorem 1. According to (3) of Theorem 2, the norm function g of H defined in (2) has bounded level sets if the matrix $\mathbf{M}$ is nondegenerate. Thus, condition (a) of Theorem 1 is satisfied if $\mathbf{M}$ is nondegenerate. Condition (b) of Theorem 1 is satisfied if each iterate $\mathbf{z}^k$ is a regular vector. In particular, this condition will hold if $\mathbf{M}$ is a P-matrix. Condition (c) of the same theorem holds if for each degenerate index $i \in \beta(\mathbf{z}^*)$ we have $\mathbf{z}_i^* = (\mathbf{q}+\mathbf{M}\mathbf{z}^*)_i = \mathbf{0}$ (cf. (2) of Theorem 2). Finally, condition (d) of Theorem 1 is satisfied if $\mathbf{z}^*$ is a regular vector (see **[16]** for a proof). In what follows, we shall state a specialized version of Theorem 1 for the LCP$(\mathbf{q}, \mathbf{M})$ with $\mathbf{M}$ being a P-matrix. Among its conclusions, this result gives a necessary and sufficient condition for an accumulation point of the damped-Newton sequence to be a solution of the LCP.

THEOREM 3. *Let* $\mathbf{M}$ *be an* $n \times n$ *P-matrix, and let* $\mathbf{q}, \mathbf{z}^o \in \Re^n$ *be arbitrary vectors. Then, the damped-Newton sequence* $\{\mathbf{z}^k\}$ *is uniquely defined, bounded, and satisfies*

$$\| \min(\mathbf{z}^{k+1}, \mathbf{q} + \mathbf{M}\mathbf{z}^{k+1}) \| < \| \min(\mathbf{z}^k, \mathbf{q} + \mathbf{M}\mathbf{z}^k) \|. \tag{13}$$

Moreover, an accumulation point $\bar{\mathbf{z}}$ *of* $\{\mathbf{z}^k\}$ *solves the LCP* $(\mathbf{q}, \mathbf{M})$ *if and only if*

$$\bar{\mathbf{z}}_i = (\mathbf{q} + \mathbf{M}\bar{\mathbf{z}})_i = \mathbf{0} \quad \forall i \in \beta(\bar{\mathbf{z}}).$$

Notice that the above theorem does not assert that the damped-Newton sequence $\{\mathbf{z}^k\}$ will always converge to the unique solution $\mathbf{z}^*$ of the LCP$(\mathbf{q}, \mathbf{M})$. Whether or not an accumulation point exists which is not a solution to LCP$(\mathbf{q}, \mathbf{M})$ under the assumptions of Theorem 3 is an open question.

The descent property of the norm function in equation (13) is essential for the convergence of the overall method. This property is insured by the line-search step (4). An easy 2×2 example can be constructed to show that condition (4) is not always satisfied on the first trial with $m = 0$. In other words, the line search is indeed essential for the global convergence of the damped-Newton method in the sense of Theorem 3.

It is of interest to explore how the index sets $\alpha(\mathbf{z}^k), \beta(\mathbf{z}^k), \gamma(\mathbf{z}^k)$ are changing during the solution process. To this end, let us define

$$\mathbf{w}^k = \mathbf{q} + \mathbf{M}\mathbf{z}^k$$

for each iteration k. Using this definition, it is easy to show that the sequence $\{\mathbf{w}^k\}$ satisfies the following recursion:

$$\mathbf{w}^{k+1} = (1 - \lambda_k)\mathbf{w}^k + \lambda_k \mathbf{u}, \tag{14}$$

where

$$\mathbf{u} = \mathbf{q} + \mathbf{M}\mathbf{v}. \tag{15}$$

In the above equation, $\mathbf{v} = (\mathbf{v}_{\alpha(\mathbf{z}^k)}, \mathbf{v}_{\beta(\mathbf{z}^k)}, \mathbf{0})$ and $(\mathbf{v}_{\alpha(\mathbf{z}^k)}, \mathbf{v}_{\beta(\mathbf{z}^k)})$ satisfy conditions (7)–(9). By definition, the sequence $\{\mathbf{z}^k\}$ satisfies a similar recursion:

$$\mathbf{z}^{k+1} = (1 - \lambda_k)\mathbf{z}^k + \lambda_k \mathbf{v}. \tag{16}$$

It is easy to see that the two vectors $\mathbf{u}$ and $\mathbf{v}$ are complementary, i.e., $u_i v_i = 0$ for all indices i. Indeed, the pair $(\mathbf{u}, \mathbf{v})$ is a candidate solution of the LCP$(\mathbf{q}, \mathbf{M})$; this pair is a solution if and only if $\mathbf{u}_{\gamma(\mathbf{z}^k)}$ and $\mathbf{v}_{\alpha(\mathbf{z}^k)}$ are nonnegative. Note that besides being useful in the following analysis, the recursion (14) provides a computationally efficient method for updating the sequence $\{\mathbf{w}^k\}$, which in turn is needed to compute the sequence $\{H(\mathbf{z}^k)\}$.

Throughout the following discussion, we will fix the scalar s in the damped-Newton algorithm at 1. Then, each step length λ_k lies in the interval (0,1]. In this case, the new iterate $(\mathbf{w}^{k+1}, \mathbf{z}^{k+1})$ lies on the line segment joining the old iterate $(\mathbf{w}^k, \mathbf{z}^k)$ and the point $(\mathbf{u}, \mathbf{v})$. Note that if $(\mathbf{u}, \mathbf{v})$ is indeed a solution of the LCP($\mathbf{q}, \mathbf{M}$), the algorithm will detect this and terminate with $m = 0$ in condition (4).

Suppose that the index set $\beta(\mathbf{z}^k)$ is nonempty. In this case, there must be a change in the status of at least one index in the $(k+1)$st iteration as long as at least one of the u_i or v_i components for $i \in \beta(\mathbf{z}^k)$ is positive (we recall that for $i \in \beta(\mathbf{z}^k)$, u_i and v_i are nonnegative and complementary). Geometrically, the algorithm must move to a different $(\mathbf{u}, \mathbf{v})$ pair at this iteration.

Consider the case where the iterate $\mathbf{z}^k$ is nondegenerate (with a slight modification, the following discussion would also hold for the degenerate case where $\mathbf{u}_{\beta(\mathbf{z}^k)} = \mathbf{v}_{\beta(\mathbf{z}^k)} = \mathbf{0}$). Choose an index $i \in \alpha(\mathbf{z}^k)$. In this case, $u_i = 0$. Whether or not the index i will leave the set $\alpha(\mathbf{z}^k)$ depends critically on the sign of the quantity v_i. If $v_i \geq 0$, then i will remain in the set $\alpha(\mathbf{z}^{k+1})$. More generally, by defining the quantity

$$\rho = \min\{\rho_1, \rho_2\}$$

where

$$\rho_1 = \min\{\frac{z_i^k - w_i^k}{z_i^k - w_i^k - v_i} : i \in \alpha(\mathbf{z}^k), v_i < 0\}, \tag{17}$$

$$\rho_2 = \min\{\frac{w_i^k - z_i^k}{w_i^k - z_i^k - u_i} : i \in \gamma(\mathbf{z}^k), u_i < 0\}, \tag{18}$$

we see that as long as $\lambda_k < \rho$, there will be no change in the index sets α and γ, i.e., we have $\alpha(\mathbf{z}^{k+1}) = \alpha(\mathbf{z}^k)$ and $\gamma(\mathbf{z}^{k+1}) = \gamma(\mathbf{z}^k)$. In this case, the next iterate $\mathbf{z}^{k+1}$ will remain nondegenerate. Of course, the pair $(\mathbf{u}, \mathbf{v})$ will also remain unchanged if this situation arises. Geometrically, the next iteration continues to move on the line segment joining the iterate $(\mathbf{w}^k, \mathbf{z}^k)$ and the pair $(\mathbf{u}, \mathbf{v})$, except that the line search now occurs between $(\mathbf{w}^{k+1}, \mathbf{z}^{k+1})$ and $(\mathbf{u}, \mathbf{v})$.

In theory, it is possible for the algorithm to stay on this line segment from $(\mathbf{w}^k, \mathbf{z}^k)$ to $(\mathbf{u}, \mathbf{v})$ and for the step length never to exceed the quantity ρ. We term this situation a *jamming phenomenon.* When this phenomenon occurs, the algorithm *jams* and fails to solve the LCP($\mathbf{q}, \mathbf{M}$). Fortunately, there is an easy cure to overcome this failure. To explain this, observe that the scalar ρ is always less than one (unless the pair

$(\mathbf{u}, \mathbf{v})$ is already a solution of the LCP$(\mathbf{q}, \mathbf{M})$, which we assume is not the case). Moreover, for any $\lambda_k \in (0, \rho]$ and any $\sigma \in (0, 1/2)$ we must have

$$(19) \qquad g(\mathbf{z}^k) - g(\mathbf{z}^{k+1}) \geq 2\sigma\lambda_k g(\mathbf{z}^k),$$

i.e., the Armijo condition (4) must hold (with $s = 1$). Thus, we can set a lower bound for the step size λ_k to be equal to ρ. By doing so, we are guaranteed that there is at least one change in the index sets, and thus, a change in the $(\mathbf{u}, \mathbf{v})$ pair as well.

Note that even with the above anti-jamming scheme, we still can not establish that the sequence $\{\mathbf{z}^k\}$ converges to a solution of the LCP$(\mathbf{q}, \mathbf{M})$ under the conditions of Theorem 3. In the sequel, we discuss a variant of the damped-Newton method which is based on the above discussions. This variant aims at simplifying the task of solving the mixed complementarity subsystem (7)–(9) at each iteration.

As we have pointed out, if an iterate $\mathbf{z}^k$ is a nondegenerate vector, then the system (7)–(9) reduces to the system of linear equations (12). In general, the vector $\mathbf{z}^k$ computed from the damped-Newton algorithm described previously is not guaranteed to be nondegenerate. This nondegeneracy property, however, can be ensured by a simple perturbation step. The following lemma asserts that a nondegenerate iterate $\mathbf{z}^{k+1}$ can always be computed if $\mathbf{z}^k$ is nondegenerate. The proof of this lemma is straightforward and is therefore omitted.

LEMMA 1. *Let $\mathbf{z}^k$ be a nondegenerate vector which is not a solution of the LCP$(\mathbf{q}, \mathbf{M})$. Let $\sigma \in (0, 1/2)$, and let $\mathbf{d}^k$ be the direction vector*

$$\mathbf{d}^k = (\mathbf{d}^k_\alpha, \mathbf{d}^k_\gamma) = (\mathbf{v}_\alpha, \mathbf{0}) - (\mathbf{z}^k_\alpha, \mathbf{z}^k_\gamma),$$

where $\mathbf{v}_\alpha$ satisfies equation (12). Then there exists a $\delta > 0$ such that for every $t \in (\rho, \rho + \delta]$ the vector $\mathbf{z}_t = \mathbf{z}^k + t\mathbf{d}^k$ is nondegenerate and satisfies

$$g(\mathbf{z}^k) - g(\mathbf{z}_t) \geq 2\sigma t g(\mathbf{z}^k).$$

The above lemma ensures that if $\mathbf{z}^k$ is a nondegenerate iterate and not a solution of the LCP$(\mathbf{q}, \mathbf{M})$, then a nondegenerate iterate $\mathbf{z}^{k+1}$ with a sufficient decrease in the value of the norm function g can always be produced by slightly moving beyond the critical step length ρ. To summarize, let us state the following modified algorithm for solving the LCP$(\mathbf{q}, \mathbf{M})$. The algorithm is required to start at a nondegenerate vector $\mathbf{z}^0$. In general, such a vector is not freely available, although it is often not difficult to obtain. For example, if the vector $\mathbf{q}$ contains no zero components, then the zero vector is nondegenerate.

MODIFIED DAMPED-NEWTON ALGORITHM. *Let* $\mathbf{z}^0$ *be an arbitrary nondegenerate vector. Let* μ *and* σ *be given as before. In general, given a nondegenerate* $\mathbf{z}^k$ *with* $H(\mathbf{z}^k) \neq 0$, *we obtain the next iterate* $\mathbf{z}^{k+1}$ *by carrying out the following three steps*:

Step 1. *Solve the system of linear equations* (12) *for* $\mathbf{v}_\alpha$ *and let* $\mathbf{v}_\gamma = 0$. *As before,* $\alpha = \alpha(\mathbf{z}^k)$ *and* $\gamma = \gamma(\mathbf{z}^k)$. *Set*

$$\mathbf{d}^k = \mathbf{v} - \mathbf{z}^k,$$

where $\mathbf{v} = (\mathbf{v}_\alpha, \mathbf{0})$.

Step 2. *Test if the vector* $\mathbf{v}$ *solves the LCP*$(\mathbf{q}, \mathbf{M})$; *terminate if it does. If not, continue.*

Step 3. *Compute the scalar* ρ, *and set* $\lambda_k = \rho + \varepsilon$, *where* ε *is some positive quantity so that the iterate* $\mathbf{z}^{k+1} = \mathbf{z}^k + \lambda_k \mathbf{d}^k$ *is nondegenerate and condition* (19) *holds.*

Observe how the above algorithm resembles a direct block pivotal algorithm **[9]**, **[10]**, **[14]**. Each execution of Step 1 may be interpreted as a block pivot on the principal submatrix $\mathbf{M}_{\alpha\alpha}$. The index set $\alpha(\mathbf{z}^k)$ may be considered as consisting of the indices of the basic z-variables. The nondegeneracy requirement on each iterate $\mathbf{z}^k$ is reminiscent of the nondegeneracy assumption on the pivot steps in a pivotal algorithm, the latter being essential for the finite termination of most algorithms in this class. There exist two major differences between the damped-Newton algorithm and a typical pivotal algorithm. First, the former contains a line-search step which, although not at first glance a pivot rule, can in fact be interpreted as a different pivot rule from those which are typically employed. More precisely, the norm function of H in equation (2) is decreased throughout the solution process and serves to define the pivot rule in the damped-Newton scheme. Second, the damped-Newton algorithm is not proven to be finite at this time; rather, its convergence (in the sense of Theorem 3) is in the limit. It remains an open question as to whether or not the damped-Newton algorithm is finite in the general case.

We conclude this section by pointing out that although the modified damped-Newton algorithm generates a sequence of nondegenerate vectors, an accumulation point of the sequence produced is not necessarily nondegenerate. Indeed, any nondegenerate accumulation point, which is also regular, must solve the LCP$(\mathbf{q}, \mathbf{M})$ according to Theorem 1.

4. The case of a Z-matrix. In this section, we consider the convergence properties of the modified damped-Newton algorithm when applied to the LCP($\mathbf{q}, \mathbf{M}$) where $\mathbf{M}$ is a *Z-matrix*, i.e., where $\mathbf{M}$ has all its off-diagonal entries nonpositive. We shall also discuss the connection of this algorithm with Chandrasekaran's [2] greedy algorithm for solving such an LCP. In the following analysis, we shall make use of several known facts about an LCP($\mathbf{q}, \mathbf{M}$) with a Z-matrix $\mathbf{M}$; further details on these results can be found in [2].

Let $\mathbf{M}$ be a Z-matrix. Suppose that a vector $\mathbf{q}$ has no zero component (this is a nondegeneracy assumption), and let $\mathbf{z}^0$ be the zero vector. Consider the system (12) with $\alpha = \alpha(\mathbf{z}^0)$. From LCP theory, we know that either the problem LCP($\mathbf{q}, \mathbf{M}$) has no solution, or else the submatrix $\mathbf{M}_{\alpha\alpha}$ must be a Minkowski matrix, i.e., it must also be a P-matrix. Without loss of generality, we may therefore assume that $\mathbf{M}_{\alpha\alpha}$ is a P-matrix. In this case, the system (12) is uniquely solvable. Moreover, the solution vector $\mathbf{v}_\alpha = -(\mathbf{M}_{\alpha\alpha})^{-1}\mathbf{q}_\alpha$ is strictly positive. Inductively, suppose that $\mathbf{z}^k$ is a nondegenerate vector satisfying the following two properties for $\alpha = \alpha(\mathbf{z}^k)$: (i) the submatrix $\mathbf{M}_{\alpha\alpha}$ is a Minkowski matrix, and (ii) the vector $\mathbf{v}_\alpha = -(\mathbf{M}_{\alpha\alpha})^{-1}\mathbf{q}_\alpha$ is strictly positive. As mentioned above, the initial iterate $\mathbf{z}^0$ satisfies these properties. We shall now show that these properties hold throughout the course of the algorithm.

Consider the expressions (14)–(16), where the new vectors $\mathbf{z}^{k+1}$ and $\mathbf{w}^{k+1}$ are computed from the old vectors $\mathbf{z}^k$ and $\mathbf{w}^k$, respectively, and $\alpha = \alpha(\mathbf{z}^k), \gamma = \gamma(\mathbf{z}^k)$. Since the vector $\mathbf{v}_\alpha$ is strictly positive, it is easy to see that the new index set $\alpha(\mathbf{z}^{k+1})$ must contain (or be equal to) the old set $\alpha(\mathbf{z}^k)$. Moreover, an index i which was in the old set $\gamma(\mathbf{z}^k)$ moves into the set $\alpha(\mathbf{z}^{k+1})$ only if the corresponding u_i is negative. Thus, there are two possibilities. One is that the vector $\mathbf{u}_{\gamma(\mathbf{z}^k)}$ is nonnegative, in which case, the pair $(\mathbf{u}, \mathbf{v})$ solves the LCP($\mathbf{q}, \mathbf{M}$) and the algorithm terminates. The other possibility is that there is at least one negative u_i for some i in $\gamma(\mathbf{z}^k)$. In the latter case, the index set α expands by at least one index (we recall that the modified algorithm ensures that the next iterate $\mathbf{z}^{k+1}$ is nondegenerate). To verify that the two desired properties (i) and (ii) remain valid for the new index set $\alpha(\mathbf{z}^{k+1})$, we invoke the Z-property of $\mathbf{M}$ and the fact that if i is an index which has moved from $\gamma(\mathbf{z}^k)$ into $\alpha(\mathbf{z}^{k+1})$, then the corresponding u_i must be negative. Doing so easily allows us to conclude that either the LCP($\mathbf{q}, \mathbf{M}$) has no solution, or else the properties (i) and (ii) must hold with respect to the new index set $\alpha(\mathbf{z}^{k+1})$.

Consequently, since the set α can only grow larger (or else the algorithm would have terminated), the damped-Newton algorithm must terminate when or before the cardinality of the set α reaches n. Summarizing the above discussion, we have proven the following result:

THEOREM 4. *Let* $\mathbf{M}$ *be an* $n \times n$ *Z-matrix, and* $\mathbf{q} \in \Re^n$ *a vector with no zero components. With* $\mathbf{z}^0 \in \Re^n$ *taken to be the zero vector, the modified damped-Newton method will, in at most n iterations, terminate either with the conclusion that the LCP*$(\mathbf{q}, \mathbf{M})$ *is not solvable, or with a desired solution of the problem.*

The above discussion shares a great deal of resemblance with the argument used to establish the convergence of Chandrasekaran's **[2]** method for solving the same problem. In fact, both algorithms are very similar in nature except for the use of the line search in the damped-Newton scheme. There is one major difference, however; namely, our Newton scheme never takes a full step size of one (i.e., $\lambda_k < 1$ for all k unless termination occurs at a $(\mathbf{u}, \mathbf{v})$ solution), whereas each iteration in Chandrasekaran's algorithm always starts at the previous $(\mathbf{u}, \mathbf{v})$ vector. This difference is of course due to the fact that the Newton scheme maintains a steady decrease in the norm function g, to which Chandrasekaran's algorithm pays no attention. When $\mathbf{M}$ is a Z-matrix, both algorithms will converge in at most n steps. In the more general case, however, Chandrasekaran's algorithm does not extend to an arbitrary matrix $\mathbf{M}$. Computationally, we expect the two algorithms to be competitive. Chandrasekaran's algorithm may take fewer iterations, but the work in solving (12) is expected to be more due to the fact that this algorithm will make more changes in the α set at each iteration than will the damped-Newton scheme.

5. Computational results. In this section, the results of several numerical experiments with the damped-Newton scheme will be described. The damped-Newton algorithm was implemented in Fortran (no optimization) on an Apollo DN3000 workstation and uses the routines DGEFA-DGESL from LINPACK for the solution of the system of equations (12) when $\mathbf{M}_{\alpha\alpha}$ is dense, and LUSOL from the Systems Optimization Laboratory at Stanford University when this matrix is sparse. The method will be compared against J. Tomlin's implementation of Lemke's algorithm called LCPL, which is also available from the Systems Optimization Laboratory. In all the cases reported below, the following parameter values are used: $\mu = 0.5, \sigma = 0.1, s = 1.0$, and $\mathbf{z}^0$

is taken to be the zero vector. Note that in solving the equation system (12), no low-rank updates are used, and thus this implementation of the modified damped-Newton algorithm can be considered to be the "worst case" in terms of computational time since each linear system is solved with a full factorization. Finally, all CPU times reported are in seconds and exclude input/output times.

The first set of examples consists of LCP's with matrices **M** which are computed as follows. First, an $n \times n$ matrix **A** was randomly generated with uniformly distributed entries in the interval $(-5, +5)$, and a skew-symmetric matrix **B** was generated in the same interval. The matrix **M** was then computed as:

$$\mathbf{M} = \mathbf{A}^{\mathrm{T}}\mathbf{A} + \mathbf{B} + \operatorname{diag}(\eta_i),$$

where η_i is a uniformly distributed random variable in the interval $(0.0, 0.3)$. Thus, **M** is a random P-matrix. The vector **q** was generated from a uniform distribution in the interval $(-500, +500)$. Ten examples for each case were run; Table 1 contains the average results for this set.

As one can see from Table 1, the damped-Newton method takes very few iterations to converge to the solution of the LCP and in all cases outperforms Lemke's method in terms of total CPU time (with very substantial savings for the larger problems).

In order to test the "robustness" of these results, a set of examples designed to be "hard" for the damped-Newton algorithm to solve was generated. Specifically, the vector **q** is now generated to be uniformly distributed over the interval $(-500, 0)$. With **q** nonpositive, we expect the cardinality of the α set to be fairly large in each iteration, and thus more work will be required to solve (12). Table 2 contains the results of this set of examples. As one can see, the damped-Newton algorithm does indeed perform poorly relative to Lemke's method. The dramatic increase in the number of iterations (as compared with the problems in the previous set of experiments) is possibly due to the increase in the "degeneracy" of the problem; but we don't fully understand the true reason and how to deal with it in case the problem is indeed "highly degenerate."

While the damped-Newton algorithm takes a fairly large number of iterations on the "hard" examples relative to those reported in Table 1, the CPU times would not be as high as reported in Table 2 with a more sophisticated implementation. The current implementation does not include any low-rank, matrix-factorization updating techniques in

Table 1. Results for Random Examples with P-Matrices

n		Density of **M** %	Lemke's Algorithm Iterations	Lemke's Algorithm CPU sec.	Damped-Newton Algorithm Iterations	Damped-Newton Algorithm CPU sec.
50	Max.	100.0	32.0	4.636	5.0	3.431
	Avg.	100.0	25.4	2.134	3.8	2.102
	Min.	100.0	17.0	1.165	2.0	1.027
100	Max.	100.0	74.0	28.282	5.0	21.410
	Avg.	100.0	59.7	20.361	4.3	15.655
	Min.	100.0	51.0	14.239	4.0	11.213
150	Max.	100.0	102.0	88.674	7.0	56.063
	Avg.	100.0	81.1	53.854	4.9	43.304
	Min.	100.0	67.0	44.400	4.0	30.274
175	Max.	100.0	104.0	113.907	5.0	91.988
	Avg.	100.0	93.5	91.181	4.6	66.704
	Min.	100.0	83.0	63.585	4.0	42.988
175	Max.	3.4	100.0	21.924	4.0	5.021
	Avg.	3.3	95.8	20.100	4.0	4.603
	Min.	3.2	90.0	15.943	4.0	4.244
250	Max.	3.9	144.0	68.412	5.0	36.052
	Avg.	3.8	132.8	61.313	4.6	29.480
	Min.	3.7	122.0	53.521	4.0	27.373

Table 2. Results for "Hard" Examples

n		Density of **M** %	Lemke's Algorithm Iterations	Lemke's Algorithm CPU sec.	Damped-Newton Algorithm Iterations	Damped-Newton Algorithm CPU sec.
50	Max.	100.0	48.0	7.353	18.0	24.961
	Avg.	100.0	39.2	5.849	12.9	17.087
	Min.	100.0	33.0	5.042	11.0	11.897
100	Max.	100.0	99.0	58.856	36.0	252.913
	Avg.	100.0	85.2	41.546	24.2	202.741
	Min.	100.0	71.0	33.351	19.0	141.196

solving the system of linear equations (12). Table 3 illustrates a typical run of a "hard" example with $n = 100$. As one can see, except for a few, most iterations involve very small changes in the α set. Thus, a

Table 3. Changes in α for $n = 100$

Iteration	No. of Changes in α
1	100
2	42
3	2
4	4
5	2
6	2
7	3
8	3
9	1
10	1
11	2
12	14
13	1
14	1
15	1
16	1
17	1
18	1
19	1
20	2
21	1
22	1
23	1

more advanced equation-solver which incorporates good low-rank matrix updating schemes should provide a substantial reduction in total CPU times.

In order to test the performance of the algorithm on large-scale problems, the method for generating $\mathbf{M}$ was altered since the computation of $\mathbf{A}^T\mathbf{A}$ becomes prohibitive for large n. To generate large-scale problems, generate $\mathbf{A}$ and $\mathbf{B}$ as before, and let $\mathbf{M} = \mathbf{A} + \mathbf{B}$. In order to conform with the theory presented in Section 3, we adjust the diagonals of $\mathbf{M}$ so that it becomes diagonally dominant (and hence a P-matrix):

$$m_{ii} > \sum_{j \neq i} | m_{ij} |,$$

$$m_{ii} > \sum_{j \neq i} | m_{ji} | .$$

The vector $\mathbf{q}$ is generated with a uniform distribution over the interval $(-500, +500)$. As shown in Table 4, the damped-Newton algorithm performs quite well relative to Lemke's method on this series of examples.

To test the sensitivity of the results in Table 4 when the problem is made "harder" by making the vector $\mathbf{q}$ more negative, two examples of $n = 400$ with diagonally dominant $\mathbf{M}$ and $\mathbf{q}$ in (-500, 0) were run. In these cases, the damped-Newton algorithm required 5 and 7 iterations. Thus, the same pattern of an increased number of iterations is observed, although the increase is substantially less than the results shown in Table 2.

Table 4. Results for Random Examples with Diagonally Dominant Matrices

n		Density of $\mathbf{M}$ %	Lemke's Algorithm Iterations	Lemke's Algorithm CPU sec.	Damped-Newton Algorithm Iterations	Damped-Newton Algorithm CPU sec.
400	Max.	3.21	217.0	297.654	3.0	189.143
	Avg.	3.20	198.3	216.540	3.0	136.086
	Min.	3.19	175.0	148.741	3.0	84.735
475	Max.	3.17	251.0	525.603	3.0	466.401
	Avg.	3.16	240.3	450.714	2.8	303.958
	Min.	3.14	227.0	364.659	2.0	191.936

Finally, the algorithm was applied to two examples for which Lemke's algorithm is known to run in exponential time (see Chap. 6 in **[14]**). The first example consists of an n-vector $\mathbf{q}$ which has each component equal to -1 and the following matrix $\mathbf{M}$:

$$\mathbf{M} = \begin{pmatrix} 1 & 2 & 2 & \cdots & 2 \\ 0 & 1 & 2 & \cdots & 2 \\ 0 & 0 & 1 & \cdots & 2 \\ \vdots & \vdots & \vdots & \cdots & \vdots \\ 0 & 0 & 0 & \cdots & 1 \end{pmatrix}.$$

Table 5 reports the results of running the damped-Newton algorithm on this class of problems.

Table 5. Number of Iterations for Exponential-Time Examples

n	Example 1	Example 2
8	9	8
16	20	16
32	72	32
64	208	65
128	> 300	63

Note that the damped-Newton scheme performs poorly on the above example, and the rapid increase in the number of iterations seems to indicate that the complexity of the algorithm may not be polynomially bounded. However, the same table shows that the damped-Newton scheme performs very well on another exponential-time example for Lemke's algorithm. This example is defined by the same $\mathbf{q}$ and a matrix $\mathbf{M}$ given by:

$$\mathbf{M} = \begin{pmatrix} 1 & 2 & 2 & \cdots & 2 \\ 2 & 5 & 6 & \cdots & 6 \\ 2 & 6 & 9 & \cdots & 10 \\ \vdots & \vdots & \vdots & \cdots & \vdots \\ 2 & 6 & 10 & \cdots & 4(n-1)+1 \end{pmatrix}.$$

The numerical results presented in this section illustrate that the damped-Newton scheme is very effective, and with "fine tuning" (rank-one updates, more sophisticated line-search procedures, etc.), further improvements in efficiency can be achieved.

6. Summary and open research questions. This paper has shown how the application of Newton's method to solve systems of B-differentiable equations as described by Pang [**16**] can be successfully modified and applied to the case of the linear complementarity problem. Beyond the practical performance of the algorithm described in the previous section, this treatment of the LCP as a system of equations brings together this field with the classical discipline of nonlinear equations. While promising, the damped-Newton algorithm requires further study as there are still some unanswered questions pertaining to it.

First, the convergence theory for the damped-Newton algorithm embodied in Theorem 3 is not entirely satisfactory in that it leaves open the possibility that accumulation points exist which are not solutions of the

LCP($\mathbf{q}, \mathbf{M}$). While we cannot exclude this possibility at present, neither can we find a counterexample. If, in fact, a nonsolution accumulation point exists, then the next question is how to modify the algorithm in order for it to avoid such a point. One idea would be to try to generalize the results of solving nonlinear equations with singular Jacobians [**5**], [**22**] since degeneracy in our context is a form of singularity.

Second, the modified Newton method requires that each linear subproblem (12) has a solution; this condition is not guaranteed to be satisfied for a general positive semidefinite matrix $\mathbf{M}$. Thus, some procedure for dealing with this situation, such as the use of the *proximal point* concept [**21**], [**25**], must be studied.

Finally, the extension of the damped-Newton scheme presented in this paper to the case of the nonlinear complementarity problem as well as its numerical performance remains an area of on-going research by the authors.

References

1. M. Aganagić, *Newton's method for linear complementarity problems*, Math. Programming **28** (1984), 349–362.
2. R. Chandrasekaran, *A special case of the complementarity pivot problem*, Opsearch **7** (1970), 263–268.
3. J. E. Dennis, Jr. and R. B. Schnabel, *Numerical methods for unconstrained optimization and nonlinear equations*, Prentice-Hall, Englewood Cliffs, NJ, 1983.
4. B. C. Eaves, *A short course in solving equations with PL homotopies*, Nonlinear Programming, R.W. Cottle and C. E. Lemke (eds.), SIAM-AMS Proc. **9** (1976), 73–143.
5. A. Griewank, *On solving nonlinear equations with simple singularities or nearly singular solutions*, SIAM Review **27** (1985), 537–563.
6. M. Kojima, S. Mizuno, and A. Yoshise, *A polynomial-time algorithm for a class of linear complementarity problems*, Report No. B-193, Department of Information Sciences, Toyko Institute of Technology.
7. ——, *A new continuation method for complementarity problems with uniform P-functions*, Report No. B-194, Department of Information Sciences, Toyko Institute of Technology.
8. M. Kojima and S. Shindo, *Extension of Newton and quasi-Newton methods to systems of PC^1 equations*, J. Oper. Res. Soc. Japan **29** (1986), 352–374.
9. M. M. Kostreva, *Direct Algorithms for Complementarity Problems*, Ph.D. dissertation, Department of Mathematics, Rensselaer Polytechnic Institute, Troy, NY, 1976.
10. ——, *Block pivot methods for solving the complementarity problem*, Linear Algebra Appl. **21** (1978), 207–215.
11. O. L. Mangasarian, *Equivalence of the complementarity problem to a system of nonlinear equations*, SIAM J. Appl. Math. **31** (1976), 89–92.
12. S. Mizuno, A. Yoshise, and T. Kikuchi, *Practical polynomial time algorithms for linear complementarity problems*, Technical Report No. 13, Department of Industrial Engineering and Management, Tokyo Institute of Technology, Tokyo, 1988.

13. R. D. Monteiro and I. Adler, *Interior path following primal-dual algorithms, part II: quadratic programming*, Working paper, Department of Industrial Engineering and Operations Research, University of California-Berkeley, Berkeley, CA, 1987.

14. K. G. Murty, *Linear complementarity, linear and nonlinear programming*, Helderman-Verlag, Berlin, 1988.

15. J. S. Pang, *Hidden Z-matrices with positive principal minors*, Linear Algebra Appl. **23** (1979), 201–215.

16. ——, *Newton's method for B-differentiable equations*, Math. Oper. Res. (to appear).

17. S. M. Robinson, *Implicit B-differentiability in generalized equations*, Technical Summary Report No. 2854, Mathematics Research Center, University of Wisconsin, Madison, Wisconsin, 1985.

18. ——, *Local structure of feasible sets in nonlinear programming, part II: stability and sensitivity*, Math. Programming Stud. **30** (1987), 45–66.

19. ——, *An implicit function theorem for B-differentiable functions*, Working paper, Department of Industrial Engineering, University of Wisconsin, Madison, Wisconsin, 1988.

20. ——, *Newton's method for a class of nonsmooth functions*, Working paper, Department of Industrial Engineering, University of Wisconsin, Madison, Wisconsin, 1988.

21. R. T. Rockafellar, *Monotone operators and the proximal point algorithm*, SIAM J. Control Optim. **14** (1976), 877–898.

22. R. B. Schnabel and P. D. Frank, *Tensor methods for nonlinear equations*, SIAM J. Numer. Anal. **21** (1981), 815–843.

23. A. Shapiro, *On concepts of directional differentiability*, Research report 73/88(18), Department of Mathematics and Applied Mathematics, University of South Africa, Pretoria, South Africa, 1988.

24. P. K. Subramanian, *Gauss-Newton methods for the nonlinear complementarity problem*, Technical Summary Report No. 2845, Mathematics Research Center, University of Wisconsin, Madison, Wisconsin, 1985.

25. ——, *A note on the least two norm solutions of monotone complementarity problems*, Appl. Math. Letters, 1988 (to appear).

26. Y. Ye, *Interior Algorithms for Linear, Quadratic and Linearly Constrained Convex Programming*, Ph.D. dissertation, Department of Engineering-Economic Systems, Stanford University, Palo Alto, CA, 1987.

[1]University of Pennsylvania

[2]The Johns Hopkins University

Lectures in Applied Mathematics
Volume **26** (1990)

Some Superlinearly Convergent Methods for Solving Singular Nonlinear Equations

ANNEGRET HOY AND HUBERT SCHWETLICK

1. Introduction. Recently a lot of research has been done on investigating singular nonlinear systems

$$\mathbf{F}(\mathbf{x}) = \mathbf{0}, \qquad \mathbf{F}: \mathbf{R}^n \to \mathbf{R}^n, \tag{1.1}$$

and on developing rapidly convergent methods that are able to compute a singular root $\mathbf{x}^*$ of $\mathbf{F}(\mathbf{x})$.

We suppose that

$$\mathbf{F} \in \mathbf{C}^3, \tag{1.2}$$

$$\operatorname{rank}(\mathbf{F}'(\mathbf{x}^*)) = n - 1, \tag{1.3}$$

$$\mathbf{F}''(\mathbf{x}^*)\mathbf{vv} \notin R(\mathbf{F}'(\mathbf{x}^*)), \tag{1.4}$$

with a normalized vector $\mathbf{v}$ spanning the nullspace of $\mathbf{F}'(\mathbf{x}^*)$, i.e.,

$$N(\mathbf{F}'(\mathbf{x}^*)) = \operatorname{span}\{\mathbf{v}\}, \qquad \|\mathbf{v}\| = 1.$$

As usual, the nullspace and the range of a matrix $\mathbf{A}$ are denoted by $N(\mathbf{A})$ and $R(\mathbf{A})$, respectively.

Obviously, condition (1.4) is equivalent to

$$\mathbf{u}^{\mathrm{T}}\mathbf{F}''(\mathbf{x}^*)\mathbf{vv} = \mathbf{g}^{*\mathrm{T}}\mathbf{v} \neq 0 \tag{1.5}$$

with

$$\mathbf{g}^{*\mathrm{T}} := \mathbf{u}^{\mathrm{T}}\mathbf{F}''(\mathbf{x}^*)\mathbf{v}$$

1980 *Mathematics Subject Classification* (1985 *Revision*). Primary 65H10.

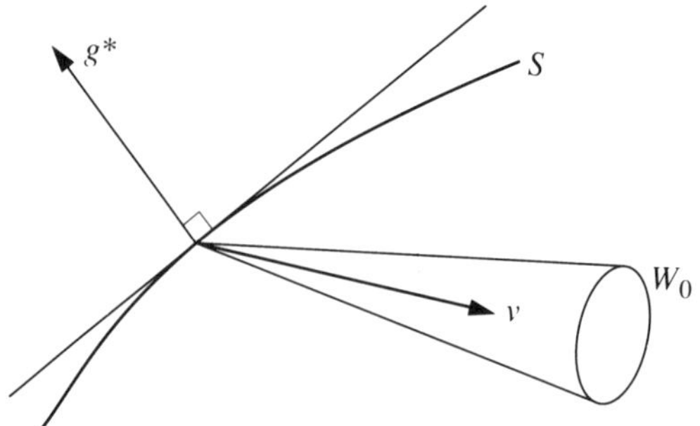

FIGURE 1. Singular manifold S and $N(\mathbf{F}'(\mathbf{x}^*)) = \text{span}\{\mathbf{v}\}$

and $\mathbf{u}$ being a normalized vector spanning the nullspace of $\mathbf{F}'(\mathbf{x}^*)^{\mathrm{T}}$, i.e.,

$$N(\mathbf{F}'(\mathbf{x}^*)^{\mathrm{T}}) = \text{span}\{\mathbf{u}\}, \qquad \|\mathbf{u}\| = 1.$$

It is useful to remember that $\mathbf{g}^*$ is a multiple of the gradient of the function

$$\gamma(\mathbf{x}) = \det(\mathbf{F}'(\mathbf{x})) \tag{1.6}$$

evaluated at $\mathbf{x}^*$. More precisely,

$$\nabla\gamma(\mathbf{x})_{\mathbf{x}=\mathbf{x}^*} = \rho(\mathbf{x}^*)\mathbf{g}^*, \tag{1.7}$$

where $\rho(\mathbf{x})$ is defined by

$$\rho(\mathbf{x}) = \|\text{adj}(\mathbf{F}'(\mathbf{x}))\|_{\mathbf{F}} \tag{1.8}$$

and the symbol "adj" denotes the adjoint of a matrix, i.e., the transpose of the matrix obtained by replacing each element by its cofactor; compare Griewank/Osborne **[GO2]**.

Assumption (1.3) ensures $\rho(\mathbf{x}^*) \neq 0$ and therefore (1.5) guarantees that $\mathbf{v}$ does not lie in the tangent space of the $(n-1)$-dimensional singular manifold $S = \{\mathbf{x}: \gamma(\mathbf{x}) = 0\}$ at $\mathbf{x}^*$; see Figure 1.

Consequently, $\mathbf{F}'(\mathbf{x})$ is invertible for all $\mathbf{x} \in W_0$ with W_0 specified as

$$W_0 = \{\mathbf{x}: \|\mathbf{P}_N(\mathbf{x}-\mathbf{x}^*)\| \leq \delta_0, \quad \|\mathbf{P}_X(\mathbf{x}-\mathbf{x}^*)\| \leq \varepsilon_0\|\mathbf{P}_N(\mathbf{x}-\mathbf{x}^*)\|\}, \tag{1.9}$$

with sufficiently small $\delta_0, \varepsilon_0 > 0$, where $\mathbf{P}_N$ denotes the (orthogonal) projection onto $N(\mathbf{F}'(\mathbf{x}^*))$, and $\mathbf{P}_X = \mathbf{I} - \mathbf{P}_N$. With the help of (1.5) and another important formula, namely

$$\mathbf{F}'(\mathbf{x})^{-1} = \tilde{J}^{\mathrm{T}}(\mathbf{x})/\gamma(\mathbf{x}) = [\rho(\mathbf{x}^*)\mathbf{v}\mathbf{u}^{\mathrm{T}} + \beta_1(\mathbf{x}-\mathbf{x}^*)]/\gamma(\mathbf{x}) \tag{1.10}$$

(compare **[G]**), where $\tilde{J}^{\mathrm{T}}(\mathbf{x}) = \text{adj}(\mathbf{F}'(\mathbf{x}))$ and the $\beta_i(\mathbf{y})$ denote quantities (real numbers as well as vectors or matrices) that are $O(\|y\|^i)$, we obtain some insight into the behavior of a Newton step calculated at $\mathbf{x}^0 \in W_0$. Premultiplication of the Taylor expansion of $\mathbf{F}(\mathbf{x})$ by $\mathbf{J}_0^{-1} = \mathbf{F}'(\mathbf{x}^0)^{-1}$ gives

$$\mathbf{J}_0^{-1}\mathbf{F}(\mathbf{x}^0) = \mathbf{x}^0 - \mathbf{x}^* - \tfrac{1}{2}\mathbf{J}_0^{-1}\mathbf{F}''(\mathbf{x}^*)(\mathbf{x}^0 - \mathbf{x}^*)(\mathbf{x}^0 - \mathbf{x}^*) + \mathbf{J}_0^{-1}\beta_3(\mathbf{x}^0 - \mathbf{x}^*).$$

Inserting (1.10) yields

$$\begin{aligned}\mathbf{x}^1 &= \mathbf{x}^0 - \mathbf{J}_0^{-1}\mathbf{F}(\mathbf{x}^0)\\ &= \mathbf{x}^* + \tfrac{1}{2}\rho(\mathbf{x}^*)\mathbf{v}\mathbf{u}^{\mathrm{T}}\mathbf{F}''(\mathbf{x}^*)(\mathbf{x}^0 - \mathbf{x}^*)(\mathbf{x}^0 - \mathbf{x}^*)/\gamma(\mathbf{x}^0) + \beta_2(\mathbf{x}^0 - \mathbf{x}^*).\end{aligned}$$

Representing $\mathbf{x}^0 - \mathbf{x}^*$ as

$$\mathbf{x}^0 - \mathbf{x}^* = \mathbf{P}_N(\mathbf{x}^0 - \mathbf{x}^*) + \mathbf{P}_X(\mathbf{x}^0 - \mathbf{x}^*) = \lambda_0\mathbf{v} + \mathbf{P}_X(\mathbf{x}^0 - \mathbf{x}^*), \tag{1.11}$$

we further obtain

$$\mathbf{x}^1 - \mathbf{x}^* = \tfrac{1}{2}\lambda_0\mathbf{v} + \beta_1(\mathbf{P}_X(\mathbf{x}^0 - \mathbf{x}^*)) + \beta_2(\mathbf{x}^0 - \mathbf{x}^*). \tag{1.12}$$

Now project (1.12) onto $N(\mathbf{F}'(\mathbf{x}^*))$ and $R(\mathbf{F}'(\mathbf{x}^*)^{\mathrm{T}})$, respectively. This leads to

$$\|\mathbf{P}_N(\mathbf{x}^1 - \mathbf{x}^*) - \tfrac{1}{2}\mathbf{P}_N(\mathbf{x}^0 - \mathbf{x}^*)\| \le c_1\|\mathbf{P}_X(\mathbf{x}^0 - \mathbf{x}^*)\| + c_2\|\mathbf{x}^0 - \mathbf{x}^*\|^2 \tag{1.13}$$

and

$$\|\mathbf{P}_X(\mathbf{x}^1 - \mathbf{x}^*)\| \le c_2\|\mathbf{x}^0 - \mathbf{x}^*\|^2. \tag{1.14}$$

These are just formulae (10) and (11) of **[GO1]**, which those authors found in a slightly different way. After it is shown that $\mathbf{x}^1 \in W_0$, the Q-linear convergence of the Newton iterates $\{\mathbf{x}^k\}$ with ratio $\frac{1}{2}$ follows immediately; see **[GO1]**.

The aim of this paper is to describe and motivate some recently developed alternatives to Newton's method that are Q-superlinearly convergent and possess further desirable properties. Out of the long list of contributions that study Newton's method in the singular case under different sets of assumptions, we want to mention **[GO1]** and **[GO2]** in the finite-dimensional case, and **[Ra]**, **[Re1]**, **[Re2]**, **[DK1]**, **[DK2]**, and **[DKK]** in a Banach space setting.

2. How to get superlinear convergence? In order to achieve a superlinear rate of convergence some modifications of Newton's method have been proposed; see **[G]** for a survey. The most prominent and pleasing technique seems to be what Griewank termed "bordering by a smooth singularity condition." The idea of the bordering approach is to apply Newton's method not directly to (1.1) but to an inflated system

$$\mathbf{T}(\mathbf{x}, \mathbf{c}) = \mathbf{T}(\mathbf{z}) = \mathbf{0}, \qquad \mathbf{T}\colon \mathbf{R}^{n+i} \to \mathbf{R}^{n+i}. \tag{2.1}$$

REMARK 2.2. The introduction of new equations and variables has some tradition in connection with the determination of critical points as turning or bifurcation points for parameter-dependent equations. We refer to Moore/Spence **[MS]**, Schwetlick **[Sc]**, Pönisch **[Pn]**, and the bibliographies therein. □

In the context of nonlinear equations the first proposal seems to be due to **[WW]**, who use

$$\mathbf{T}(\mathbf{x},\lambda,\mathbf{y}) := \begin{bmatrix} \mathbf{F}(\mathbf{x}) + \lambda\mathbf{y} \\ \mathbf{F}'(\mathbf{x})\mathbf{y} \\ \mathbf{y}^{\mathrm{T}}\mathbf{y} - 1 \end{bmatrix} = \mathbf{0}. \tag{2.3}$$

The new system (2.3) has the solution $\mathbf{z}^* = (\mathbf{x}^*, 0, \mathbf{v})^{\mathrm{T}}$, and $\mathbf{T}'(\mathbf{z}^*)$ is of full rank provided that in addition to (1.2), (1.3), and (1.4), the condition

$$\mathbf{v} \notin R(\mathbf{F}'(\mathbf{x}^*)) \tag{2.4}$$

is satisfied. Here and in the following we write $(\mathbf{a}, \mathbf{b})^{\mathrm{T}}$ instead of $(\mathbf{a}^{\mathrm{T}}, \mathbf{b}^{\mathrm{T}})^{\mathrm{T}}$ for brevity.

Because $\mathbf{T}(\mathbf{z}^*) = \mathbf{0}$ and $\mathbf{T}'(\mathbf{z}^*)$ is regular, Newton's method applied to (2.3) converges Q-quadratically in the triples $\{\mathbf{x}^k, \lambda^k, \mathbf{y}^k\}$. A disadvantage of (2.3) is that (2.4) has to be supposed in addition to (1.2), (1.3), and (1.4). In the following example due to **[WW]**, condition (1.4) is fulfilled, but (2.4) is not.

EXAMPLE 2.5.

$$x_1^2 - 2x_1 + \tfrac{1}{3}x_2^3 + \tfrac{2}{3} = 0,$$
$$x_1^3 - x_1x_2 - 2x_1 + \tfrac{1}{2}x_2^2 + \tfrac{3}{2} = 0$$

has the singular solution $\mathbf{x}^* = (1, 1)^{\mathrm{T}}$. □

Yamamoto **[Y]** overcomes this disadvantage by setting

$$\mathbf{T}(\mathbf{x},\lambda,\mathbf{y}) = \begin{bmatrix} \mathbf{F}(\mathbf{x}) + \lambda\mathbf{p} \\ \mathbf{F}'(\mathbf{x})\mathbf{y} \\ \mathbf{q}^{\mathrm{T}}\mathbf{y} - 1 \end{bmatrix} = \mathbf{0}. \tag{2.6}$$

In fact, he uses a coordinate-dependent formulation which, however, can readily be brought into the coordinate-free form (2.6). Now, $\mathbf{T}'(\mathbf{z}^*)$ is nonsingular if and only if (1.2), (1.3), and (1.4) hold, and $\mathbf{p}$ and $\mathbf{q}$ are chosen to satisfy

$$\mathbf{p} \notin R(\mathbf{F}'(\mathbf{x}^*)), \qquad \mathbf{q} \notin R(\mathbf{F}'(\mathbf{x}^*)^{\mathrm{T}}), \tag{2.7}$$

or, equivalently,

$$\mathbf{p}^{\mathrm{T}}\mathbf{u} \neq 0, \qquad \mathbf{q}^{\mathrm{T}}\mathbf{v} \neq 0.$$

The algorithmic choice of **p** and **q** proposed by Yamamoto, however, is somewhat complicated; Tsuchiya **[Ts]** describes a version of (2.6) that is easier to handle than Yamamoto's.

The fact that the system size is doubled appears at first glance to be a shortcoming of the extended mappings $\mathbf{T}(\mathbf{z})$ treated so far. However, the block structure of $\mathbf{T}'(\mathbf{z})$ can be exploited to organize the solution of the linear systems with coefficient matrix $\mathbf{T}'(\mathbf{z})$ in such a way that only one LU decomposition of an $n \times n$ matrix is necessary; see **[WW]** and **[MS]** for details.

[GR] and **[MP]** obviously independently noticed that

$$\mathbf{H}(\mathbf{x},\lambda) = \mathbf{F}(\mathbf{x}) + \lambda\mathbf{p} = \mathbf{0}, \qquad \mathbf{H}\colon \mathbf{R}^{n+1} \to \mathbf{R}^{n}, \tag{2.8}$$

possesses at $\mathbf{z}^* = (\mathbf{x}^*,\lambda^*)^{\mathrm{T}} = (\mathbf{x}^*,\mathbf{0})^{\mathrm{T}}$ a simple turning point if (1.2), (1.3), (1.4), and the first condition of (2.7) are valid. This means that the turning point methods available in the literature can be used to determine singular solutions of (1.1). Such turning point methods differ from each other in the characterizing equations that complete (2.8) to an $(n+1+j)\times(n+1+j)$ system; cf. **[Sc]**. Thus, each turning point method provides a special bordered system $\mathbf{T}(\mathbf{x},\mathbf{c}) = \mathbf{0}$ for our framework.

Menzel/Pönisch **[MP]** apply the turning point method of Pönisch/Schwetlick **[PS]** to (2.8): Choose $\mathbf{r} = (\mathbf{q},\rho)^{\mathrm{T}}$ with $\mathbf{r}^{\mathrm{T}}(\mathbf{v},\mathbf{0})^{\mathrm{T}} = \mathbf{q}^{\mathrm{T}}\mathbf{v} \neq 0$. Define $\mathbf{w} = (\mathbf{y},\varphi)^{\mathrm{T}}$ and require

$$\mathbf{H}(\mathbf{x},\lambda) = \mathbf{F}(\mathbf{x}) + \lambda\mathbf{p} = \mathbf{0}, \tag{2.9a}$$

$$\begin{gathered}\mathbf{H}'(\mathbf{x},\lambda)\mathbf{w} = \mathbf{F}'(\mathbf{x})\mathbf{y} + \varphi\mathbf{p} = \mathbf{0},\\ \mathbf{r}^{\mathrm{T}}\mathbf{w} = \mathbf{q}^{\mathrm{T}}\mathbf{y} + \rho\varphi = 1,\end{gathered} \tag{2.9b}$$

$$(\mathbf{e}^{n+1})^{\mathrm{T}}\mathbf{w} = \varphi = 0. \tag{2.9c}$$

The two equations (2.9b) form a linear system for $\mathbf{w} = (\mathbf{y},\varphi)^{\mathrm{T}}$ having the regular coefficient matrix

$$\mathbf{B}(\mathbf{x}) := \begin{bmatrix}\mathbf{H}'\\ \mathbf{r}^{\mathrm{T}}\end{bmatrix} = \begin{bmatrix}\mathbf{F}'(\mathbf{x}) & \mathbf{p}\\ \mathbf{q}^{\mathrm{T}} & \rho\end{bmatrix}.$$

Solving the system for

$$\mathbf{w}(\mathbf{x}) = \mathbf{B}(\mathbf{x})^{-1}\mathbf{e}^{n+1},$$

and substituting $\mathbf{w}(\mathbf{x})$ into (2.9c) yields the reduced system

$$\text{(2.10)} \qquad \mathbf{T}(\mathbf{x},\lambda) := \begin{bmatrix} \mathbf{H}(\mathbf{x},\lambda) \\ \varphi(\mathbf{x},\lambda) \end{bmatrix} = \begin{bmatrix} \mathbf{F}(\mathbf{x}) + \lambda\mathbf{p} \\ (\mathbf{e}^{n+1})^{\mathrm{T}}\mathbf{B}(\mathbf{x})^{-1}\mathbf{e}^{n+1} \end{bmatrix};$$

compare also **[G]**, who treats the special case $\rho = 0$.

The common characteristics of the methods described so far are:

C1: In each step linear systems of dimension n or $n+1$ with uniformly bounded inverses are to be solved. Remember that when Newton's method is applied to (1.1) without modifications, the inverses $\mathbf{F}'(\mathbf{x}^k)^{-1}$ grow as $\beta_{-1}(\mathbf{x}^k - \mathbf{x}^*)$.

C2: The algorithms can be implemented in such a way that second-order terms occur only in the form $\mathbf{F}''(\mathbf{x})\mathbf{aa}$ and $\mathbf{F}''(\mathbf{x})\mathbf{ab}$. Such terms can be approximated efficiently by divided differences, as proposed by **[PS]**.

$\widetilde{\mathrm{C}}3$: The extended sequences $\{\mathbf{x}^k, \mathbf{c}^k\}$ generated by applying Newton's method to (2.1) converge Q-quadratically, i.e.,

$$\left\| \begin{bmatrix} \mathbf{x}^{k+1} \\ \mathbf{c}^{k+1} \end{bmatrix} - \begin{bmatrix} \mathbf{x}^* \\ \mathbf{c}^* \end{bmatrix} \right\| \le c_3 \left\| \begin{bmatrix} \mathbf{x}^{k} \\ \mathbf{c}^{k} \end{bmatrix} - \begin{bmatrix} \mathbf{x}^* \\ \mathbf{c}^* \end{bmatrix} \right\|^2$$

with a constant $c_3 > 0$.

It is $\widetilde{\mathrm{C}}3$ that stimulated us to look for an alternative approach to the solution of (1.1), since superlinear convergence of the sequence $\{\mathbf{x}^k\}$ alone is desirable. The difference between the convergence behavior of a sequence $\{\mathbf{x}^k, \mathbf{c}^k\}$ and its x-part $\{\mathbf{x}^k\}$ is discussed in papers concerning so-called SQP methods for constrained optimization problems, where $\mathbf{c}^k$ plays the role of the Lagrange vector; see, e.g., Tapia **[Ta]**, Boggs/Tolle/Wang **[BTW]** and Nocedal/Overton **[NO]**.

In Section 3 we sketch a way to construct algorithms which, in addition to characteristics C1 and C2, possess the following property C3 as well:

C3: The sequence $\{\mathbf{x}^k\}$ produced by the algorithm converges Q-quadratically, i.e.,

$$\|\mathbf{x}^{k+1} - \mathbf{x}^*\| \le c_4 \|\mathbf{x}^k - \mathbf{x}^*\|^2$$

with a constant $c_4 > 0$.

In Section 4 this algorithm is modified in order to cover two further useful characteristics, namely:

C4: A possibly existing sparse structure of $\mathbf{F}'(\mathbf{x})$ can be exploited.

C5: Generalization to nonlinear least squares problems is possible.

The extension to nonlinear least squares problems is described in Section 5. In Section 6 we answer implementational questions and illustrate the behavior of the algorithm outlined in Section 4 by applying it to Example 2.5. Finally, in Section 7 we consider further research directions.

Let us remark that apart from the bordering techniques described above there are other approaches to singular equations, namely:

(i) Extrapolation of Newton's method; see, e.g., Kelley/Suresh **[KS]**.
(ii) Tensor models using quadratic approximations to $\mathbf{F}$; see, e.g., Frank/Schnabel **[FS]**.
(iii) Deflation techniques; see Ojika **[Oj]**.

However, these different approaches have a number of shortcomings. The methods of type (i) work with unbounded inverses and do not allow a high solution accuracy; see Griewank **[G]**. For the tensor methods (ii), a strong convergence analysis in the singular case has not yet been found, although the numerical results are encouraging; see Griewank [G] and Schnabel's paper in these proceedings. A strong theoretical foundation should be possible in the future, possibly after modification of the higher-order approximation scheme. The deflation techniques (iii) are suited for polynomial systems, but a generalization to general equations seems to be difficult.

3. A first new algorithm. The following algorithm was developed by investigating the overdetermined system of equations

$$\mathbf{T}(\mathbf{x}) := \begin{bmatrix} \mathbf{F}(\mathbf{x}) \\ \det(\mathbf{F}'(\mathbf{x})) \end{bmatrix} = \mathbf{0}, \qquad \mathbf{T}\colon \mathbf{R}^n \to \mathbf{R}^{n+1}; \tag{3.1}$$

cf. **[H1]**, i.e., the introduction of additional variables is avoided.

REMARK 3.2. Several authors, including Kubiček/Marek **[KM]**, Mittelmann/Weber **[MW]**, and Menzel **[M]**, have proposed overdetermined systems as defining equations for bifurcation points. None of them, however, arrived at an algorithm that satisfies the three conditions C1, C2, and C3. □

Equations (3.1) are consistent in $\mathbf{x}^*$. Assumptions (1.2), (1.3), and (1.4) ensure that the Jacobian $\mathbf{T}'(\mathbf{x}^*)$ has full rank. Consequently, the Gauss-Newton method

$$\mathbf{x}^{k+1} := \mathbf{x}^k - \mathbf{T}'(\mathbf{x}^k)^+\mathbf{T}(\mathbf{x}^k) \tag{3.3}$$

converges Q-quadratically, i.e., we have

$$\|\mathbf{x}^{k+1} - \mathbf{x}^*\| \leq c_5\|\mathbf{x}^k - \mathbf{x}^*\|^2 \tag{3.4}$$

with a fixed constant $c_5 > 0$.

The Gauss-Newton step (3.3) can be equivalently expressed as

$$\begin{aligned}\mathbf{p}_G^k := \mathbf{x}^k - \mathbf{x}^{k+1} &= (\mathbf{J}_k^{\mathrm{T}}\mathbf{J}_k + \nabla\nabla^{\mathrm{T}})^{-1}(\mathbf{J}_k^{\mathrm{T}}\mathbf{F}_k + \gamma_k \cdot \nabla) \\ &= \mathbf{J}_k^{-1}\mathbf{F}_k + \frac{\gamma_k - \nabla^{\mathrm{T}}\mathbf{J}_k^{-1}\mathbf{F}_k}{1 + \nabla^{\mathrm{T}}(\mathbf{J}_k^{\mathrm{T}}\mathbf{J}_k)^{-1}\nabla}(\mathbf{J}_k^{\mathrm{T}}\mathbf{J}_k)^{-1}\nabla\end{aligned} \tag{3.5}$$

where $\mathbf{J}_k := \mathbf{F}'(\mathbf{x}^k)$, $\mathbf{F}_k := \mathbf{F}(\mathbf{x}^k)$, $\gamma_k := \det(\mathbf{F}'(\mathbf{x}^k))$, $\nabla := \nabla\gamma(\mathbf{x})_{\mathbf{x}=\mathbf{x}^k}$; cf. **[H1]**. The Gauss-Newton direction (3.5) is not acceptable for practical computation because it requires the explicit evaluation of $\nabla\gamma(\mathbf{x})$, which, in general, depends on all entries of the second derivative $\mathbf{F}''(\mathbf{x})$.

A detailed analysis of the terms $(\mathbf{J}_k^{\mathrm{T}}\mathbf{J}_k)^{-1}\nabla$ and $\nabla^{\mathrm{T}}\mathbf{J}_k^{-1}\mathbf{F}_k$ leads to the estimate

$$\mathbf{p}_G^k = \mathbf{J}_k^{-1}\mathbf{F}_k + \frac{\gamma_k - \rho(\mathbf{x}^*)\mathbf{u}^{\mathrm{T}}\mathbf{F}''(\mathbf{x}^k)\mathbf{v}\mathbf{J}_k^{-1}\mathbf{F}_k}{\rho(\mathbf{x}^*)\mathbf{u}^{\mathrm{T}}\mathbf{F}''(\mathbf{x}^k)\mathbf{v}\mathbf{v}}\mathbf{v} + \beta_2(\mathbf{x}^k - \mathbf{x}^*); \tag{3.6}$$

see **[H1]**.

The new algorithm was worked out in **[H2]** by answering the following three questions that arise from considering formula (3.6):

Q1: How should $\mathbf{u}$ and $\mathbf{v}$ be approximated by certain vectors $\mathbf{p}^k$ and $\mathbf{q}^k$, respectively?

Q2: Is there a suitable approximation to $\gamma_k/\rho(\mathbf{x}^*)$ that could be substituted in (3.6)?

Q3: Is it possible to replace $\mathbf{J}_k^{-1}$ by an everywhere regular matrix to overcome the lack of invertibility of $\mathbf{J}(\mathbf{x})$ on the singular manifold?

We proceed with a description of the algorithm from **[H2]**:

ALGORITHM 3.7.

S0: *Choose* $\mathbf{x}^0 \approx \mathbf{x}^*$, $\mathbf{q}^0 \approx \mathbf{v}$, $\mathbf{p}^0 \approx \mathbf{u}$; *set* $k := 0$.

S1: *Set* $\mathbf{B}_k := \mathbf{F}'(\mathbf{x}^k) + \mathbf{p}^k(\mathbf{q}^k)^{\mathrm{T}}$.

S2: *Determine* $\mathbf{d}^k$ *from* $\mathbf{B}_k\mathbf{d}^k = \mathbf{F}(\mathbf{x}^k)$ *and* $\mathbf{v}^k$ *from* $\mathbf{B}_k\mathbf{v}^k = \mathbf{p}^k$.

S3: *Set*

$$\mathbf{x}^{k+1} := \mathbf{x}^k - \mathbf{d}^k - \frac{1 - (\mathbf{q}^k)^{\mathrm{T}}\mathbf{v}^k - (\mathbf{p}^k)^{\mathrm{T}}\mathbf{F}''(\mathbf{x}^k)\mathbf{q}^k\mathbf{d}^k}{(\mathbf{p}^k)^{\mathrm{T}}\mathbf{F}''(\mathbf{x}^k)\mathbf{q}^k\mathbf{q}^k}\mathbf{q}^k.$$

S4: *Set* $\mathbf{C}_{k+1} := \mathbf{F}'(\mathbf{x}^{k+1}) + \mathbf{p}^k(\mathbf{q}^k)^{\mathrm{T}}$.

S5: *Set* $\mathbf{q}^{k+1} := \mathbf{C}_{k+1}^{-1}\mathbf{p}^k/\|\mathbf{C}_{k+1}^{-1}\mathbf{p}^k\|_2$,

$$\mathbf{p}^{k+1} := [\mathbf{C}_{k+1}^{-1}]^{\mathrm{T}}\mathbf{q}^k/\|[\mathbf{C}_{k+1}^{-1}]^{\mathrm{T}}\mathbf{q}^k\|_2. \quad \square$$

THEOREM 3.8. *Algorithm* 3.7 *is well defined and* $\{\mathbf{x}^k\}$ *is Q-quadratically convergent provided that*

(i) *the assumptions* (1.2), (1.3), *and* (1.4) *are satisfied;*
(ii) *the starting vectors* $\mathbf{x}^0, \mathbf{p}^0$, *and* $\mathbf{q}^0$ *are chosen such that*

$$\|\mathbf{x}^0 - \mathbf{x}^*\| + \|\mathbf{p}^0 - \mathbf{u}\| + \|\mathbf{q}^0 - \mathbf{v}\| \le \delta_1$$

for a sufficiently small $\delta_1 > 0$.

For the proof of Theorem 3.8 see [H2].

Obviously, demands C1, C2, and C3 are fulfilled for Algorithm 3.7. However, the rank-one term $\mathbf{p}^k(\mathbf{q}^k)^{\mathrm{T}}$ in $\mathbf{B}_k = \mathbf{F}'(\mathbf{x}^k) + \mathbf{p}^k(\mathbf{q}^k)^{\mathrm{T}}$ destroys a possible special structure in $\mathbf{F}'(\mathbf{x}^k)$ since $\mathbf{p}^k$ and $\mathbf{q}^k$ and, therefore, the corrections $\mathbf{p}^k(\mathbf{q}^k)^{\mathrm{T}}$ are dense. This shortcoming is remedied in [H3] by working with fixed $\mathbf{p}$ and $\mathbf{q}$ so that $\mathbf{p}\mathbf{q}^{\mathrm{T}}$ can be made sparse.

4. A second new algorithm. We first describe the new algorithm step by step and then supply the major stages of the convergence analysis.

ALGORITHM 4.1.
S0: *Choose* $\mathbf{x}^0 \approx \mathbf{x}^*$, $\mathbf{q} \approx \mathbf{v}$, $\mathbf{p} \approx \mathbf{u}$; *set* $k := 0$.
S1: *Set* $\mathbf{B}_k := \mathbf{F}'(\mathbf{x}^k) + \mathbf{p}\mathbf{q}^{\mathrm{T}}$.
S2: *Determine* $\mathbf{d}^k$ *from* $\mathbf{B}_k\mathbf{d}^k = \mathbf{F}(\mathbf{x}^k)$, $\mathbf{v}^k$ *from* $\mathbf{B}_k\mathbf{v}^k = \mathbf{p}$, $\mathbf{u}^k$ *from* $\mathbf{B}_k^{\mathrm{T}}\mathbf{u}^k = \mathbf{q}$.
S3: *Set*

$$\mathbf{x}^{k+1} := \mathbf{x}^k - \mathbf{d}^k - \frac{1 - \mathbf{q}^{\mathrm{T}}\mathbf{v}^k - (\mathbf{u}^k)^{\mathrm{T}}\mathbf{F}''(\mathbf{x}^k)\mathbf{v}^k\mathbf{d}^k}{(\mathbf{u}^k)^{\mathrm{T}}\mathbf{F}''(\mathbf{x}^k)\mathbf{v}^k\mathbf{v}^k}\mathbf{v}^k. \quad \square$$

THEOREM 4.2. *Algorithm* 4.1 *is well defined and* $\{\mathbf{x}^k\}$ *converges Q-quadratically provided that*

(i) *the assumptions* (1.2), (1.3), *and* (1.4) *are satisfied;*
(ii) $\mathbf{p}$ *and* $\mathbf{q}$ *are chosen according to* $\mathbf{p}^{\mathrm{T}}\mathbf{u} \neq 0$ *and* $\mathbf{q}^{\mathrm{T}}\mathbf{v} \neq 0$, *respectively;*
(iii) *the starting vector* $\mathbf{x}^0$ *is sufficiently good, i.e.,* $\|\mathbf{x}^0 - \mathbf{x}^*\| \le \delta_2$ *for a sufficiently small* $\delta_2 > 0$. □

The proof of Theorem 4.2 is given in detail in [H3]. We will give here only the essential ideas in the form of certain lemmata.

LEMMA 4.3. *Let the assumptions of Theorem* 4.2 *be satisfied. Then the matrix* $\mathbf{B}(\mathbf{x}) = \mathbf{F}'(\mathbf{x}) + \mathbf{p}\mathbf{q}^{\mathrm{T}}$ *is regular for all* $\mathbf{x}$ *with* $\|\mathbf{x} - \mathbf{x}^*\| \le \delta_2$. □

LEMMA 4.4. *Let assumptions* (i) *and* (ii) *of Theorem* 4.2 *be satisfied. Then*

$$1 - \mathbf{q}^{\mathrm{T}}\mathbf{B}^{-1}(\mathbf{x}^*)\mathbf{p} = 0. \quad \square \tag{4.5}$$

With the help of Lemma 4.4, Algorithm 4.1 can be equivalently written as

$$\mathbf{x}^{k+1} = \mathbf{x}^k - \mathbf{d}^k - \frac{\mathbf{z}^{\mathrm{T}}(\mathbf{x}^k - \mathbf{x}^*) - \mathbf{z}^{\mathrm{T}}\mathbf{d}^k + \beta_2(\mathbf{x}^k - \mathbf{x}^*)}{\mathbf{z}^{\mathrm{T}}\mathbf{v}^k + \beta_1(\mathbf{x}^k - \mathbf{x}^*)}\mathbf{v}^k, \tag{4.6}$$

where

$$\mathbf{z} := \nabla(1 - \mathbf{q}^{\mathrm{T}}\mathbf{B}^{-1}(\mathbf{x})\mathbf{p})_{\mathbf{x}=\mathbf{x}^*} = [\mathbf{q}^{\mathrm{T}}\mathbf{B}^{-1}(\mathbf{x}^*)\mathbf{F}''(\mathbf{x}^*)\mathbf{B}^{-1}(\mathbf{x}^*)\mathbf{p}]^{\mathrm{T}}.$$

LEMMA 4.7. *Let the assumptions of Theorem* 4.2 *be satisfied. Then*

$$\mathbf{z}^{\mathrm{T}}\mathbf{B}^{-1}(\mathbf{x}^*)\mathbf{p} \neq 0. \quad \square \tag{4.8}$$

Of course, (4.8) implies that Algorithm 4.1 is well defined for sufficiently small $\|\mathbf{x}^k - \mathbf{x}^*\| \leq \delta_2$.

Expanding $\mathbf{F}(\mathbf{x})$ and dividing by $\mathbf{z}^{\mathrm{T}}\mathbf{v}^k$ yields

$$\mathbf{x}^{k+1} = \mathbf{x}^k - \mathbf{d}^k - \mathbf{q}^{\mathrm{T}}(\mathbf{x}^k - \mathbf{x}^*)\mathbf{v}^k + \beta_2(\mathbf{x}^k - \mathbf{x}^*). \tag{4.9}$$

Then premultiplication by $\mathbf{B}_k$ gives

$$\mathbf{B}_k(\mathbf{x}^k - \mathbf{x}^{k+1}) = \mathbf{F}(\mathbf{x}^k) + \mathbf{q}^{\mathrm{T}}(\mathbf{x}^k - \mathbf{x}^*)\mathbf{p} + \beta_2(\mathbf{x}^k - \mathbf{x}^*). \tag{4.10}$$

Now, define

$$\mathbf{G}(\mathbf{x}) := \mathbf{F}(\mathbf{x}) + \mathbf{q}^{\mathrm{T}}(\mathbf{x} - \mathbf{x}^*)\mathbf{p}, \tag{4.11}$$

which implies $\mathbf{G}(\mathbf{x}^*) = \mathbf{F}(\mathbf{x}^*)$, $\mathbf{G}'(\mathbf{x}) = \mathbf{F}'(\mathbf{x}) + \mathbf{p}\mathbf{q}^{\mathrm{T}}$, $\mathbf{G}'(\mathbf{x}^*) = \mathbf{B}(\mathbf{x}^*)$. In terms of $\mathbf{G}$, (4.10) reads as

$$\mathbf{G}'(\mathbf{x}^k)(\mathbf{x}^k - \mathbf{x}^{k+1}) = \mathbf{G}(\mathbf{x}^k) + \beta_2(\mathbf{x}^k - \mathbf{x}^*). \tag{4.12}$$

Lemma 4.3 guarantees $\mathbf{G}'(\mathbf{x}^*)$ to be regular so that Newton's method applied to $\mathbf{G}(\mathbf{x})$ converges Q-quadratically to $\mathbf{x}^*$. On the other hand, (4.12) shows that the steps $\mathbf{s}^k = \mathbf{x}^k - \mathbf{x}^{k+1}$ belonging to Algorithm 4.1 differ from the Newton steps for $\mathbf{G}(\mathbf{x}) = \mathbf{0}$ only in terms of order β_2. The quadratic convergence property is not affected by such perturbations; compare also the theory of the so-called "inexact Newton methods" given by Dembo/Eisenstat/Steihaug **[DES]**. In other words, Algorithm 4.1 can be considered as a slightly disturbed Newton method for the regular system $\mathbf{G}(\mathbf{x}) = \mathbf{0}$.

The method for finding appropriate vectors $\mathbf{p}$ and $\mathbf{q}$ will be presented in Section 6.

5. A generalization to least squares problems. Now, let $\mathbf{F}$ be a mapping with

$$\mathbf{F}: \mathbf{R}^n \to \mathbf{R}^m, \quad m \geq n, \qquad \mathbf{F}(\mathbf{x}^*) = \mathbf{0}. \tag{5.1}$$

As before, we suppose that the conditions (1.2), (1.3), and (1.4) are satisfied.

Whereas $N(\mathbf{F}'(\mathbf{x}^*)^{\mathrm{T}})$ is one-dimensional in the case $m = n$, it has dimension $m - n + 1 > 1$ in the strong least squares case $m > n$, and it is not obvious which vector from $N(\mathbf{F}'(\mathbf{x}^*)^{\mathrm{T}})$ should play the role of $\mathbf{u}$. However, there exists an orthonormal basis $\{\mathbf{u}^1, \dots, \mathbf{u}^{m-n+1}\}$ of $N(\mathbf{F}'(\mathbf{x}^*)^{\mathrm{T}})$, i.e.,

$$(5.2) \qquad N(\mathbf{F}'(\mathbf{x}^*)^{\mathrm{T}}) = \operatorname{span}\{\mathbf{u}^1, \dots, \mathbf{u}^{m-n+1}\}$$

with the following property:

$$(5.3) \qquad (\mathbf{u}^i)^{\mathrm{T}}\mathbf{F}''(\mathbf{x}^*)\mathbf{v}\mathbf{v} = \begin{cases} 0 & i = 1, \dots, m-n \\ \mathscr{H} \neq 0 & i = m-n+1 \end{cases},$$

and

$$(5.4) \qquad \mathbf{u} := \mathbf{u}^{m-n+1}$$

is, apart from the sign, uniquely determined by (5.2) and (5.3); see **[H3]** for a proof.

Then, as in the case $m = n$, the choice of $\mathbf{p}$ and $\mathbf{q}$ according to $\mathbf{p}^{\mathrm{T}}\mathbf{u} \neq 0$ and $\mathbf{q}^{\mathrm{T}}\mathbf{v} \neq 0$ ensures $\operatorname{rank}(\mathbf{B}_k) = n$. Consequently, $\mathbf{B}_k^+$ is given by

$$(5.5) \qquad \mathbf{B}_k^+ = [\mathbf{B}_k^{\mathrm{T}}\mathbf{B}_k]^{-1}\mathbf{B}_k^{\mathrm{T}}$$

and depends continuously on $\mathbf{x}^k$ in a neighborhood of $\mathbf{x}^*$. Therefore, an analogue of Algorithm 4.1 can be formulated as follows:

ALGORITHM 5.6. *Take* S0, *S*1, *and* S3 *as in Algorithm* 4.1. *Replace* S2 *of Algorithm* 4.1 *by the following step* S2* :

S2* : *Determine*

$$\begin{aligned} \mathbf{d}^k &:= \mathbf{B}_k^+\mathbf{F}(\mathbf{x}^k); \\ \mathbf{v}^k &:= \mathbf{B}_k^+\mathbf{p}; \\ \mathbf{u}^k &:= (\mathbf{B}_k^+)^{\mathrm{T}}\mathbf{q}. \quad \square \end{aligned}$$

This means that the first two linear systems of Step S2 of Algorithm 4.1, which define $\mathbf{d}^k$ and $\mathbf{v}^k$, are now solved in the least squares sense; the minimum-norm solution is taken as a solution of the consistent underdetermined third system that defines $\mathbf{u}^k$. Of course, for calculating $\mathbf{d}^k, \mathbf{v}^k$, and $\mathbf{u}^k$, there is no need for computing $\mathbf{B}_k^+$ explicitly. The notation in S2* is used only to show the analogy to S2. In fact, one can use the normal equation approach or some orthogonal decomposition of $\mathbf{B}_k$.

THEOREM 5.7. *Algorithm 5.6 is well defined and Q-quadratically convergent under the same hypothesis as formulated in Theorem 4.2.* □

Note that the proof given in **[H3]** is much more complicated than the proof for the case $m = n$.

6. Implementational questions. An important question is how to efficiently form the terms $\mathbf{F}''(\mathbf{x})\mathbf{aa}$ and $\mathbf{F}''(\mathbf{x})\mathbf{ab}$ occurring in the algorithms. **[PS]** proposed the use of difference approximations of the form

$$\mathbf{F}''(\mathbf{x})\hat{\mathbf{a}}\hat{\mathbf{a}} = \frac{1}{h^2}[\mathbf{F}(\mathbf{x}+h\hat{\mathbf{a}}) - 2\mathbf{F}(\mathbf{x}) + \mathbf{F}(\mathbf{x}-h\hat{\mathbf{a}})] + \beta_2(h), \tag{6.1}$$

$$\mathbf{F}''(\mathbf{x})\hat{\mathbf{a}}\hat{\mathbf{b}} = \frac{1}{h^2}[\mathbf{F}(\mathbf{x}+h(\hat{\mathbf{a}}+\hat{\mathbf{b}})) - \mathbf{F}(\mathbf{x}+h\hat{\mathbf{b}}) - \mathbf{F}(\mathbf{x}+h\hat{\mathbf{a}}) + \mathbf{F}(\mathbf{x})] + \beta_1(h), \tag{6.2}$$

where $\hat{\mathbf{a}} = \mathbf{a}/\|\mathbf{a}\|_2$, $\hat{\mathbf{b}} = \mathbf{b}/\|\mathbf{b}\|_2$, and $h \neq 0$ is an appropriate discretization stepsize.

In Algorithms 3.7, 4.1, and 5.6, we implemented (6.1). Instead of (6.2), however, we preferred the second-order approximation

$$\begin{aligned}\mathbf{F}''(\mathbf{x})\hat{\mathbf{a}}\hat{\mathbf{b}} = \frac{1}{h^2}\Big[&\mathbf{F}(\mathbf{x}-\frac{h}{2}(\hat{\mathbf{a}}+\hat{\mathbf{b}})) - \mathbf{F}\left(\mathbf{x}+\frac{h}{2}(\hat{\mathbf{a}}-\hat{\mathbf{b}})\right)\\ &+\mathbf{F}\left(\mathbf{x}+\frac{h}{2}(\hat{\mathbf{a}}+\hat{\mathbf{b}})\right) - \mathbf{F}\left(\mathbf{x}-\frac{h}{2}(\hat{\mathbf{a}}-\hat{\mathbf{b}})\right)\Big] + \beta_2(h).\end{aligned} \tag{6.3}$$

Figure 2 illustrates the three approximation formulae (6.1), (6.2), and (6.3). The points used in (6.1), (6.2), and (6.3) are marked by circles, squares, and triangles, respectively. As the discretization stepsize $h := h_k$ for (6.1) and (6.3) we used at the actual iterate $\mathbf{x}^k$

$$h = \min\{10^{-4}(\|\mathbf{x}^k\|_2 + 10^{-3}),\ \mathrm{HKRR}\},$$

where

$$\mathrm{HKRR} = \sqrt{\|\mathbf{x}^k - \mathbf{x}^{k-1}\|_2\|\mathbf{x}^{k-1} - \mathbf{x}^{k-2}\|_2} + 10^{-10}.$$

Another important question is how to choose $\mathbf{p}$ and $\mathbf{q}$ in Algorithms 4.1 and 5.6 appropriately. We set

$$\begin{aligned}\mathbf{q} &= \mathbf{q}(\mathbf{y}^{l-1}) = \mathbf{F}'(\mathbf{y}^{l-1})^+\mathbf{F}(\mathbf{y}^{l-1})/\|\mathbf{F}'(\mathbf{y}^{l-1})^+\mathbf{F}(\mathbf{y}^{l-1})\|_2,\\ \mathbf{p} &= \mathbf{p}(\mathbf{y}^{l-1}) = [\mathbf{F}'(\mathbf{y}^{l-1})^+]^{\mathrm{T}}\mathbf{q}/\|[\mathbf{F}'(\mathbf{y}^{l-1})^+]^{\mathrm{T}}\mathbf{q}\|_2,\end{aligned} \tag{6.4}$$

where $\mathbf{x}^0 = \mathbf{y}^l$ is reached by a sequence of l Newton steps (case $m = n$) or l Gauss-Newton steps (case $m > n$), respectively. When $\mathbf{F}'(\mathbf{x})$ is sparse, one should take as $\mathbf{q}$ and $\mathbf{p}$ certain coordinate vectors $\mathbf{e}^{i(\mathbf{q})}$ and

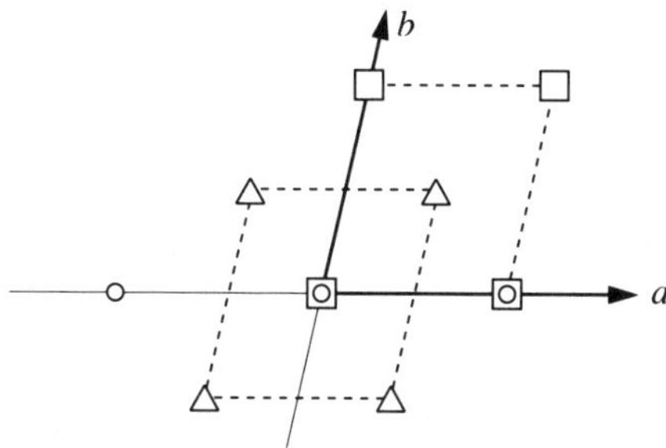

FIGURE 2. Approximation of directional derivatives $\mathbf{F}''\mathbf{aa}$ and $\mathbf{F}''\mathbf{ab}$.

$\mathbf{e}^{j(\mathbf{p})}$, respectively, where $i(\mathbf{q}) \in \{1, \ldots, n\}$, and $j(\mathbf{p}) \in \{1, \ldots, m\}$ are chosen such that $\mathbf{F}'$ has a nonzero entry at position $\{j(\mathbf{p}), i(\mathbf{q})\}$, and the elements $[\mathbf{e}^{i(\mathbf{q})}]^{\mathrm{T}}\mathbf{q}(\mathbf{y}^{l-1})$ and $[\mathbf{e}^{j(\mathbf{p})}]^{\mathrm{T}}\mathbf{p}(\mathbf{y}^{l-1})$ are sufficiently large in absolute value.

Illustrative numerical tests for Algorithm 3.7 can be found in **[H2]** and for Algorithm 5.6 in **[H3]**. We conclude this section by showing the performance of Algorithm 4.1 when it is applied to Example 2.5.

Start at $\mathbf{y}^0 = (2, -1)^{\mathrm{T}}$. Perform six Newton steps, leading to $\mathbf{y}^6 = (1.03488,\ 0.999943)^{\mathrm{T}}$. At $\mathbf{y}^6 = \mathbf{x}^0$ prepare for switching to Algorithm 4.1 by setting $\mathbf{q} = \mathbf{q}(\mathbf{y}^5)$ and $\mathbf{p} = \mathbf{p}(\mathbf{y}^5)$ according to (6.4). Then carry out Algorithm 4.1 until the desired solution accuracy is achieved.

The iterates, which were calculated in double-precision arithmetic with a 16-digit mantissa, are

$$
\begin{aligned}
\mathbf{x}^1 &= (0.9937, && 1.0013)^{\mathrm{T}} \\
\mathbf{x}^2 &= (0.99978, && 1.000030)^{\mathrm{T}} \\
\mathbf{x}^3 &= (0.99999975, && 1.000000037)^{\mathrm{T}} \\
\mathbf{x}^4 &= (0.99999999999967, && 1.00000000000005)^{\mathrm{T}} \\
\mathbf{x}^5 &= (1.0, && 1.0)^{\mathrm{T}} \qquad \text{to full precision.}
\end{aligned}
$$

These iterates clearly exhibit the quadratic convergence behavior predicted by Theorem 4.2.

7. Concluding remarks. The most commonly occurring singularities are apparently those satisfying the conditions (1.2), (1.3), and (1.4), i.e., the regular singularities of nullspace dimension one treated in our contribution. However, of interest in certain bifurcation problems (see, e.g., Reddien **[Re2]**, Decker/Kelley **[DK2]**, and Spence/Jepson **[SJ]**) and applications in optimization (see **[GO2]** and Jongen's paper in these proceedings) are irregular singularities, i.e., higher-order zeros.

A detailed analysis of Newton's method near irregular singularities is provided by **[GO2]**. A certain bordering method for irregular singularities is proposed by **[Ts]**. Along the lines of Section 4, the following algorithm for the singular case can be developed.

ALGORITHM 7.1. *Take* S0 *and* S1 *as in Algorithm* 4.1. *Append* S2 *of Algorithm* 4.1 *by the following calculations*:

ad S2: *Determine* $\mathbf{w}_1^k$ *from* $\mathbf{B}_k\mathbf{w}_1^k = \mathbf{F}''(\mathbf{x}^k)\mathbf{v}^k\mathbf{d}^k$;

$\mathbf{w}_2^k$ *from* $\mathbf{B}_k\mathbf{w}_2^k = \mathbf{F}''(\mathbf{x}^k)\mathbf{v}^k\mathbf{v}^k$.

Replace S3 *of Algorithm* 4.1 *by the following step* S3* :

S3* : *Determine* $\alpha_k = (\mathbf{u}^k)^{\mathrm{T}}\mathbf{F}'''(\mathbf{x}^k)\mathbf{v}^k\mathbf{v}^k\mathbf{v}^k - 3(\mathbf{u}^k)^{\mathrm{T}}\mathbf{F}''(\mathbf{x}^k)\mathbf{v}^k\mathbf{w}_2^k$.

Set

$$\mathbf{x}^{k+1} = \mathbf{x}^k - \mathbf{d}^k - \frac{1}{\alpha_k}[(\mathbf{u}^k)^{\mathrm{T}}\mathbf{F}''(\mathbf{x}^k)\mathbf{v}^k\mathbf{v}^k - (\mathbf{u}^k)^{\mathrm{T}}\mathbf{F}'''(\mathbf{x}^k)\mathbf{v}^k\mathbf{v}^k\mathbf{d}^k + 2(\mathbf{u}^k)^{\mathrm{T}}\mathbf{F}''(\mathbf{x}^k)\mathbf{v}^k\mathbf{w}_1^k + (\mathbf{u}^k)^{\mathrm{T}}\mathbf{F}''(\mathbf{x}^k)\mathbf{w}_2^k\mathbf{d}^k]\mathbf{v}^k. \quad \square$$

THEOREM 7.2. *Algorithm* 7.1 *is well defined and* $\{\mathbf{x}^k\}$ *converges Q-quadratically provided that assumptions* (ii) *and* (iii) *of Theorem* 4.2 *hold and assumption* (i) *holds with* (1.2) *and* (1.4) *replaced by*

$$\mathbf{F} \in \mathbf{C}^4 \ (\textit{compare } (1.2)); \tag{7.3}$$

$$\mathbf{F}'''(\mathbf{x}^*)\mathbf{vvv} - 3\mathbf{F}''(\mathbf{x}^*)\mathbf{v}[\mathbf{B}(\mathbf{x}^*)^{-1}\mathbf{F}''(\mathbf{x}^*)\mathbf{vv}] \notin R(\mathbf{F}'(\mathbf{x}^*)) \ (\textit{compare } (1.4)). \quad \square \tag{7.4}$$

Condition (7.4) can equivalently be written as

$$\mathbf{u}^{\mathrm{T}}\mathbf{F}'''(\mathbf{x}^*)\mathbf{vvv} - 3\mathbf{u}^{\mathrm{T}}\mathbf{F}''(\mathbf{x}^*)\mathbf{v}[\mathbf{F}'(\mathbf{x}^*)^{+}\mathbf{F}''(\mathbf{x}^*)\mathbf{vv}] \neq 0. \tag{7.5}$$

An analogue to (7.5) occurs as formula (2.6) in **[SJ]** in connection with the definition and determination of simple cubic turning points (also referred to as cusp points). A detailed derivation of Algorithm 7.1 and the proof of Theorem 7.2 will be submitted for publication in the near future.

Another interesting field for application and modification of the techniques described in the previous sections is inconsistent rank-deficient nonlinear least squares problems. Such problems are of importance in data-fitting (cf. Van Den Bos **[Vn]**) and are considered in bifurcation theory (cf. Menzel **[M]**, as well as in constrained optimization (cf. Gill/Murray **[GM]**). Some results concerning this topic are being prepared for publication.

REFERENCES

[BTW] P. T. Boggs, J. W. Tolle, and P. Wang, *On the local convergence of quasi-Newton methods for constrained optimization*, SIAM J. Control. Optim. **20** (1982), 161–171.

[DKK] D. W. Decker, H. B. Keller, and C. T. Kelley, *Convergence rates for Newton's method at singular points*, SIAM J. Numer. Anal. **20** (1983), 296–314.

[DK1] D. W. Decker and C. T. Kelley, *Newton's method at singular points* I, SIAM J. Numer. Anal. **17** (1980), 66–70.

[DK2] ——, *Newton's method at singular points* II, SIAM J. Numer. Anal. **17** (1980), 465–471.

[DES] R. S. Dembo, S. C. Eisenstat, and T. Steihaug, *Inexact Newton methods*, SIAM J. Numer. Anal **19** (1982), 400–408.

[FS] P. Frank and R. B. Schnabel, *Tensor methods for nonlinear equations*, SIAM J. Numer. Anal. **21** (1984), 815–843.

[GM] P. E. Gill and W. Murray, *Nonlinear least squares and nonlinearly constrained optimization*, Lecture Notes in Mathematics, no. 506, ed. by G. A. Watson, Springer-Verlag, Berlin–Heidelberg–New York, 1976, pp. 134–147.

[G] A. Griewank, *On solving nonlinear equations with simple singularities or nearly singular solutions*, SIAM Rev. **27** (1985), 537–563.

[GO1] A. Griewank and M. R. Osborne, *Newton's method for singular problems when the dimension of the nullspace is* > 1, SIAM J. Numer. Anal. **18** (1981), 145–149.

[GO2] ——, *Analysis of Newton's method at irregular singularities*, SIAM J. Numer. Anal. **20** (1983), 747–773.

[GR] A. Griewank and G. W. Reddien, *Characterization and computation of generalized turning points*, SIAM J. Numer. Anal. **21** (1984), 176–185.

[H1] A. Hoy, *A relation between Newton and Gauss–Newton steps for singular nonlinear equations*, Computing **40** (1988), 19–27.

[H2] ——, *An efficiently implementable Gauss–Newton-like method for solving singular nonlinear equations*, Computing **41** (1989), 107–122.

[H3] ——, *A superlinearly convergent Gauss–Newton-type method for rank-deficient consistent least squares problems*, Wiss. Z. d. Martin-Luther-Univ. Halle-Wittenberg (to appear 1990).

[KS] C. T. Kelley and R. Suresh, *A new acceleration method for Newton's method at singular points*, SIAM J. Numer. Anal. **20** (1983), 1001–1009.

[KM] M. Kubiček and M. Marek, *Evaluation of limit and bifurcation points for algebraic equations and nonlinear boundary-value problems*, Appl. Math. Comput. **5** (1979), 253–264.

[M] R. Menzel, *Numerical determination of multiple bifurcation points*, ISNM, **70**, ed. by T. Küpper, H. D. Mittelmann, and H. Weber, Birkhäuser, Basel, 1984, pp. 310–318.

[MP] R. Menzel and G. Pönisch, *A quadratically convergent method for computing simple singular roots and its application to determining simple bifurcation points*, Computing **32** (1984), 127–138.

[MW] H. D. Mittelmann and H. Weber, *Numerical methods for bifurcation problems—a survey and classification*, ISNM, vol. **54**, ed. by H. D. Mittelmann and H. Weber, Birkhäuser, Basel, 1980, pp. 1–45.

[MS] G. Moore and A. Spence, *The calculation of turning points of nonlinear equations*, SIAM J. Numer. Anal **17** (1980), 567–576.

[NO] J. Nocedal and M. L. Overton, *Projected Hessian updating algorithms for nonlinearly constrained optimization*, SIAM J. Numer. Anal. **22** (1985), 821–850.

[Oj] T. Ojika, *Modified deflation algorithm for the solution of singular problems* I. *A system of nonlinear algebraic equations*, J. Math. Anal. Appl. **123** (1987), 199–221.

[Pn] G. Pönisch, *Computing simple bifurcation points using a minimally extended system of nonlinear equations*, Computing **35** (1985), 277–294.

[PS] G. Pönisch and H. Schwetlick, *Computing turning points of curves implicitly defined by nonlinear equations depending on a parameter*, Computing **26** (1981), 107–121.

[Ra] L. B. Rall, *Convergence of the Newton process to multiple solutions*, Numer. Math. **9** (1966), 23–37.

[Re1] G. W. Reddien, *On Newton's method for singular problems*, SIAM J. Numer. Anal. **15** (1978), 993–996.

[Re2] ——, *Newton's method and high-order singularities*, Comput. Math. Appl. **5** (1979), 79–86.

[Sc] H. Schwetlick, *Algorithms for finite-dimensional turning point problems from viewpoint to relationships with unconstrained optimization*, ISNM, vol. **70**; Birkhäuser, Basel, ed. by T. Küpper, H. D. Mittelmann, and H. Weber, 1984, pp. 459–479.

[SJ] A. Spence and A. D. Jepson, *The numerical calculation of cusps, bifurcation points and isola formation points in two-parameter problems*, ISNM, vol. **70**; Birkhäuser, Basel, ed. by T. Küpper, H. D. Mittelmann, and H. Weber, 1984, pp. 502–513.

[Ta] R. A. Tapia, *Quasi-Newton methods for equality constrained optimization: Equivalence of existing methods and a new implementation*, Nonlinear Programming, vol. 3; ed. by O. Mangasarian, R. Meyer, and S. Robinson, Academic Press, New York, 1978, pp. 125–164.

[Ts] T. Tsuchiya, *Enlargement procedure for resolution of singularities at simple singular solutions of nonlinear equations*, Numer. Math. **52** (1988), 401–411.

[Vn] A. Van Den Bos, *Degeneracy in nonlinear least squares*, IEE-D Proc. **128** (1981), 109–116.

[WW] H. Weber and W. Werner, *On the accurate determination of nonisolated solutions of nonlinear equations*, Computing **26** (1981), 315–326.

[Y] N. Yamamoto, *Regularization of solutions of nonlinear equations with singular Jacobian matrices*, J. Inform. Proc. **7** (1984), 16–21.

Martin Luther University, German Democratic Republic

Lectures in Applied Mathematics
Volume **26** (1990)

Numerical Solutions of Some Nonlinear Dispersive Wave Equations

JOHN K. HUNTER

Abstract. We present numerical solutions and a related analysis of a number of nonlinear dispersive wave equations. The equations model long waves on a rotating fluid and sound waves.

1. Shallow water waves on a rotating fluid. Waves on the free surface of a fluid ("water waves") are a familiar and important example of nonlinear dispersive waves. Rotation of the fluid introduces additional dispersion. The exact water wave equations are a nonlinear free boundary value problem. If the slope of the free surface is small ("shallow water"), the exact equations can be approximated by the Boussinesq equations [13]. Including rotation, the Boussinesq equations are

$$
\begin{aligned}
&h_t + (hu)_x + (hv)_y = 0, \\
&u_t + uu_x + vu_y + gh_x + \tfrac{1}{3}c_0^2 h_0 \frac{\partial}{\partial x}(h_{xx} + h_{yy}) - fv = 0, \\
&v_t + uv_x + vv_y + gh_y + \tfrac{1}{3}c_0^2 h_0 \frac{\partial}{\partial y}(h_{xx} + h_{yy}) + fu = 0.
\end{aligned}
\tag{1.1}
$$

Here, $h(x,y,t)$ is the elevation of the free surface, and $u(x,y,t)$ and $v(x,y,t)$ are the x- and y-components of the fluid velocity. We use subscripts to denote partial derivatives. There are three independent parameters in (1.1)—the mean depth of the fluid, h_0; the gravitational acceleration, g; and the Coriolis frequency, f. The shallow water speed is $c_0 = (gh_0)^{1/2}$.

Work partially supported by the NSF under grant DMS-8810782.

We consider a wave propagating in the x-direction with wavelength λ. Let $0 < \varepsilon \ll 1$ be a small parameter, and suppose that

$$\begin{aligned} \frac{h-h_0}{h_0} &= O(\varepsilon^2) && \text{small amplitude,} \\ \frac{\lambda}{h_0} &= O(\varepsilon^{-1}) && \text{shallow water wave,} \\ \frac{h_0 f}{c_0} &= O(\varepsilon) && \text{slow rotation.} \end{aligned}$$

With these scaling assumptions, the effects of nonlinearity, shallow water dispersion, and rotational dispersion are all weak and of the same order of magnitude.

Let $\eta(x,t) = h(x,t) - h_0$. Singular perturbation methods show that, as $\varepsilon \to 0$, the disturbance η in the free surface satisfies

$$\frac{\partial}{\partial x}\left(\eta_t + c_0\eta_x + \frac{3c_0}{2h_0}\eta\eta_x + \frac{c_0 h_0^2}{6}\eta_{xxx}\right) - \frac{f^2}{2c_0}\eta = 0. \tag{1.2}$$

After making a Galilean transformation and rescaling variables, we can rewrite (1.2) as

$$\frac{\partial}{\partial x}(u_t + uu_x + u_{xxx}) + \sigma u = 0, \tag{1.3}$$

with $\sigma = -1$. The case $\sigma = +1$ arises for rotating fluids with large surface tension (specifically, for Bond numbers greater than one-third). Such values could be realized in low-gravity experiments, but do not occur in geophysical problems. Equation (1.2) was derived by Ostrovsky **[9]**, Redekopp **[10]**, and Grimshaw **[1]** for long internal waves.

There are several special cases of (1.3):

1. *No dispersion*: the inviscid Burgers equation,

$$u_t + uu_x = 0; \tag{1.4}$$

2. *No rotation*: the Korteweg–deVries (KdV) equation,

$$u_t + uu_x + u_{xxx} = 0; \tag{1.5}$$

3. *No long wave dispersion*: the short wave equation,

$$\frac{\partial}{\partial x}(u_t + uu_x) - u = 0. \tag{1.6}$$

Equations (1.4) and (1.5) are well known. We analyze (1.6) in the next section.

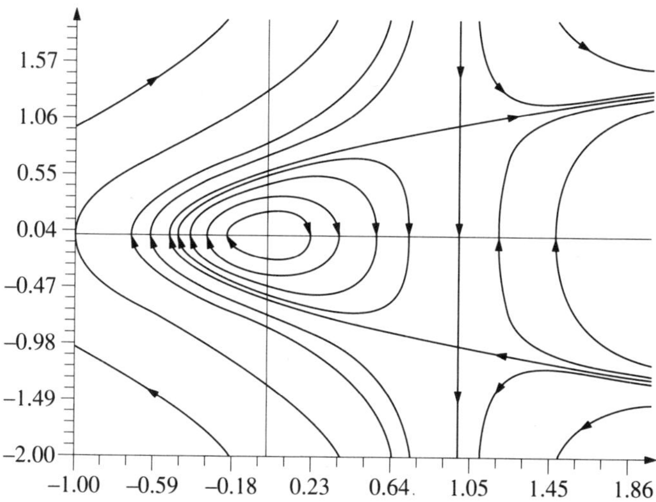

FIGURE 1. The phase plane of (2.1)

2. The short wave equation. Equation (1.6) models water waves on a very shallow rotating fluid when the wave frequency is much greater than the Coriolis frequency. It also models sound waves in a bubbly liquid when the wave frequency is much greater than the resonant frequency of bubble oscillations [**3**]. It is the canonical asymptotic equation for genuinely nonlinear waves that are nondispersive as their wavelength tends to zero [**4**].

The travelling wave solutions of (1.6) are $u = y(x - ct)$, where

$$(y - c)y'' + (y')^2 - y = 0. \tag{2.1}$$

The phase plane of (2.1), with $c = 1$, is shown in Figure 1. When $c > 0$, there is a family of periodic travelling waves. For each value of the wave speed c, there is a wave of maximum amplitude, which has a corner at its crest. A numerical solution of the periodic waves is shown in Figure 2. We discretized (2.1) with periodic boundary conditions and used Newton's method to solve the resulting algebraic equations. For the lowest amplitude wave, the initial guess was a sine wave. The solutions for larger amplitudes were obtained by continuation.

Next, we analyze spatially periodic solutions of (1.6). Such solutions have zero mean with respect to x. We write the initial value problem

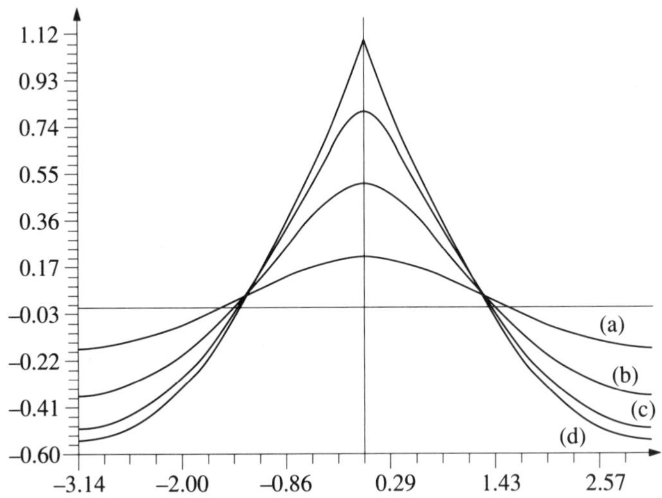

FIGURE 2. Travelling wave solutions of (1.6) with wave velocity one. Amplitude $y(0)$: (a) 0.2; (b) 0.5; (c) 0.8; (d) $\frac{1}{9}\pi^2$ (the limiting wave).

for (1.6) in the equivalent form

$$u_t + uu_x = L[u],\quad u(x,0) = u_0(x). \tag{2.2}$$

In (2.2), u_0 is a zero-mean, l-periodic function of x and

$$L[u](x,t) = \int_0^x u(y,t)\,dy - \frac{1}{l}\int_0^l \int_0^y u(z,t)\,dz\,dy. \tag{2.3}$$

Solutions of (2.2) need not remain smooth for all $t > 0$. The integral term $L[u]$ cannot prevent the nonlinear "breaking" of smooth solutions when their slope is sufficiently large. This fact is proved in the following proposition. If (2.2) models sound waves in a bubbly liquid, it is reasonable to continue solutions past the breakdown time by introducing shocks.

PROPOSITION. *Suppose that* $\sup|u_0(x)| = M$ *and* $\inf[u_0'(x)] = -m$. *Then if*

$$\frac{m^3}{m+4} > 4M, \tag{2.4}$$

smooth solutions of (2.2), *with* $l = 1$, *break down before* $t = 2/m$.

PROOF. Any smooth solution of (2.2) satisfies

$$\frac{\partial}{\partial t}(u^2) + \frac{\partial}{\partial x}\left[\frac{1}{3}u^3 - \frac{1}{2}u\int_0^x u(y,t)\,dy\right] = u\int_0^1\int_0^y u(z,t)\,dz\,dy. \tag{2.5}$$

Averaging (2.5) over a period implies that

$$\int_0^1 u^2(x,t)\,dx = \int_0^1 u_0^2(x)\,dx \le M^2.$$

Therefore,

$$|L[u](x,t)| \le 2\int_0^1 |u(x,t)|\,dx \le 2\left[\int_0^1 u^2(x,t)\,dx\right]^{1/2} \le 2M. \tag{2.6}$$

Next we write (2.2) in terms of characteristic coordinates (ξ, t). This gives

$$x = X(\xi,t), \quad u(x,t) = U(\xi,t), \quad L[u](x,t) = G(\xi,t),$$

where

$$\begin{aligned} X_t &= U, \qquad X(\xi,0) = \xi, \\ U_t &= G, \qquad U(\xi,0) = u_0(\xi). \end{aligned} \tag{2.7}$$

The solution of (2.7) is

$$\begin{aligned} X(\xi,t) &= \xi + \int_0^t U(\xi,s)\,ds, \\ U(\xi,t) &= u_0(\xi) + \int_0^t G(\xi,s)\,ds. \end{aligned} \tag{2.8}$$

If $X_\xi \ne 0$, we can solve for $\xi = F(x,t)$, and then $u(x;t) = U[F(x,t),t]$ is a smooth solution of (2.2). To prove that smooth solutions break down, we shall show X_ξ vanishes. Differentiating (2.8) with respect to ξ and eliminating U_ξ in the result shows that

$$X_\xi(\xi,t) = 1 + tu_0'(\xi) + \int_0^t\int_0^s G_\xi(\xi,s)\,ds. \tag{2.9}$$

From (2.3) and (2.6),

$$\begin{aligned} G_\xi &= X_\xi U, \\ |G(\xi,t)| &\le 2M. \end{aligned} \tag{2.10}$$

Combining (2.8b), (2.9), and (2.10a) gives an equation for G_ξ,

$$G_\xi = (1+tu_0')\left(u_0 + \int_0^t G\,ds\right) + \left(u_0 + \int_0^t G\,ds\right)\int_0^t\int_0^s G_\xi\,ds'\,ds. \tag{2.11}$$

We choose ξ_* so that

$$u_0'(\xi_*) = -m,$$

and define $\Gamma(t)$ by

$$\Gamma(t) = \sup_{0\le s\le t} |G_\xi(\xi_*, s)|. \tag{2.12}$$

We take the supremum of the absolute value of (2.11), and rearrange the result. Using (2.4), we find that

$$\Gamma(t) \le \frac{M(1+2t)}{1-\frac{1}{2}M(1+2t)t^2} \tag{2.13}$$

when $0 \le t \le 2/m$.

Equations (2.9) and (2.12) imply that

$$|X_\xi(\xi_*, t) - 1 + mt| \le \frac{1}{2}t^2\Gamma(t). \tag{2.14}$$

Use of (2.13) and (2.4) in (2.14) shows

$$\left| X_\xi\left(\xi_*, \frac{2}{m}\right) + 1 \right| < 1.$$

Therefore, $X_\xi(\xi_*, 2/m)$ is negative. Since $X_\xi = 1$ when $t = 0$, $X_\xi(\xi_*,\ t_*) = 0$ for some $0 < t_* < 2/m$. Moreover, from (2.10b),

$$U_\xi(\xi_*, t_*) = -m + \int_0^t G_\xi(\xi_*, s)\, ds.$$

Use of (2.12) and (2.13) implies that

$$|U_\xi(\xi_*, t_*) + m| < m.$$

Thus, $U_\xi(\xi_*, t_*) \neq 0$. Then, since $u_x = (X_\xi)^{-1}U_\xi$, $u_x[X(\xi_*, t_*), t]$ must be unbounded as $t \uparrow t_* < 2/m$. □

To solve (2.2) numerically, we used the Enquist–Osher scheme to difference the left-hand side **[8]**. We evaluated the right-hand side of (2.2) at the previous time step and treated it as an explicit source term.

In Figure 3, we show a numerical solution of (2.2) with

$$u(x, 0) = \sin x. \tag{2.15}$$

The wave profile steepens until a shock forms. The shock then weakens and the solution appears to approach a smooth travelling wave as $t \to +\infty$. It is not possible to tell if the shock disappears in finite time; the Enquist–Osher scheme is highly dissipative and smears out the shock. This long-time behavior contrasts with the behavior of solutions of the inviscid Burgers equation (1.4), which decay to zero as $t \to +\infty$.

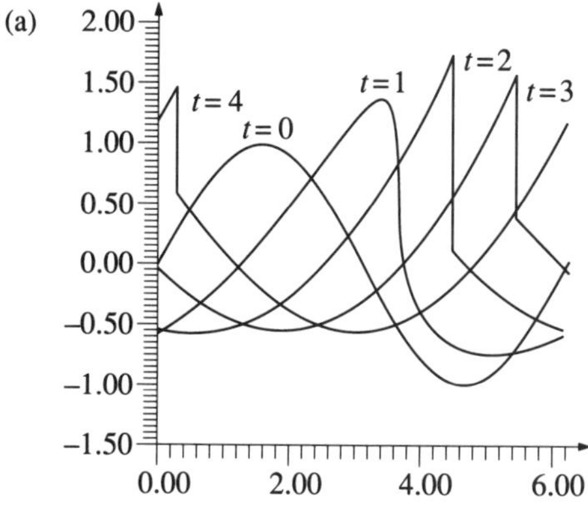

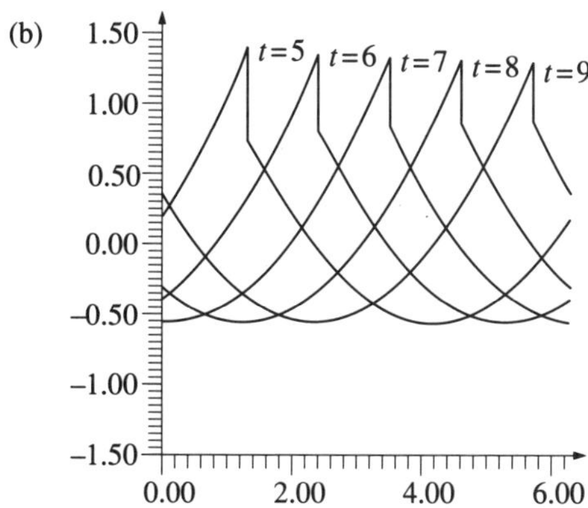

FIGURE 3. The numerical solution of (2.2) and (2.15)

If (1.6) models water waves, it is not appropriate to fit shocks into a solution, because the small slope approximation is invalid at the breakdown time. Instead, we can use (1.3) as a dispersive regularization of (1.6):

$$\frac{\partial}{\partial x}(u_t + uu_x + \varepsilon u_{xxx}) + \sigma u = 0, \qquad 0 < \varepsilon \ll 1. \tag{2.16}$$

To solve (2.16) numerically, we integrated it once with respect to x, as in (2.2), then used centered differences in space and a predictor–corrector step in time. In Figures 4 and 5, we show numerical solutions of (2.16) with initial data (2.15). Before the breakdown time, the solution is similar to the solution of (1.6). The third-derivative term

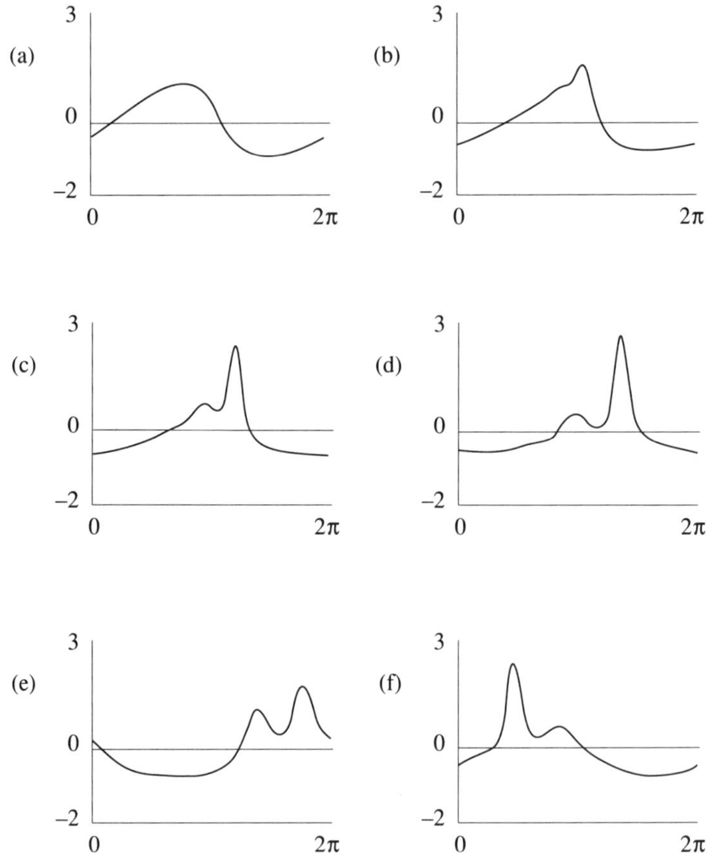

FIGURE 4. The solution of (2.16) and (2.13) with $\sigma = -1$ and $\varepsilon = 0.01$. Times: (a) 0.5; (b) 1; (c) 1.5; (d) 2; (e) 13; (f) 5.

becomes important when the derivative of the solution of (2.2) blows up, and oscillations spread out from the breakdown point. The solution for $\sigma = -1$ has a "soliton-like" appearance. The solution for $\sigma = +1$ is strongly oscillatory. The zero dispersion limit of integrable equations, like the KdV equation, has been characterized using the inverse scattering transform **[5]**, **[12]**. No similar results are known about the limiting behavior of solutions as $\varepsilon \to 0+$ for nonintegrable equations like (2.16). (Equation (2.16) is not integrable because the ordinary differential equation (ODE) for travelling wave solutions of (2.16) does not have the Painlevé property **[7]**.)

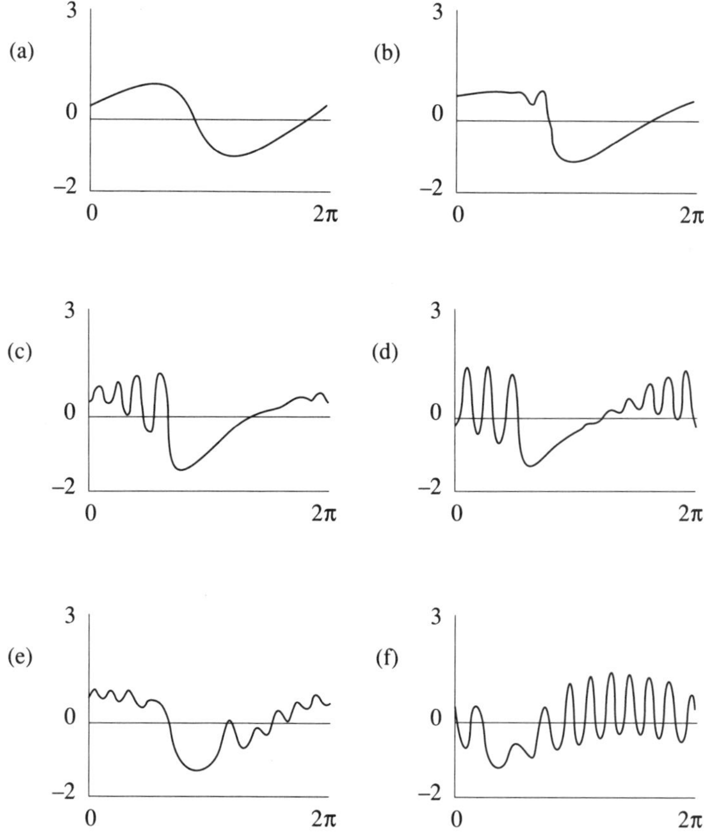

FIGURE 5. The solution of (2.16) and (2.13) with $\sigma = +1$ and $\varepsilon = 0.01$. Times: (a) 0.5; (b) 1; (c) 1.5; (d) 2; (e) 7; (f) 3.

3. Solitary waves. The KdV equation (1.5) has a solitary wave solution, i.e., a travelling wave that decays to zero with its derivatives as $|x| \to +\infty$. We shall show that (1.3) has solitary wave solutions which are perturbations of a KdV solitary wave when $\sigma = +1$, but not when $\sigma = -1$. This means that any amount of rotation destroys shallow water solitary waves.

It is convenient to make a change of variables in (1.3),

$$\begin{aligned} u &\to \varepsilon u, \\ x &\to \varepsilon^{-1/2} x, \\ t &\to \varepsilon^{-3/2} t. \end{aligned} \tag{3.1}$$

In the transformed variables, (1.3) is a perturbed KdV equation

$$\frac{\partial}{\partial x}(u_t + uu_x + u_{xxx}) + \sigma\varepsilon^2 u = 0. \tag{3.2}$$

The travelling wave solutions of (3.2), with wave velocity normalized to one, are $u = y(x - t; \varepsilon)$, where

$$(y'' + \frac{1}{2}y^2 - y)'' + \sigma\varepsilon^2 y = 0. \tag{3.3}$$

When $\varepsilon = 0$, (3.3) has the KdV solution $y = y_0$, where

$$y_0(x) = 3\,\mathrm{sech}^2\left(\frac{x}{2}\right). \tag{3.4}$$

We shall try to construct solitary wave solutions of (3.3), for $0 < \varepsilon \ll 1$, which are perturbations of (3.4). Although the small parameter ε multiplies a lower-order term in (3.3), this is a singular perturbation problem because it involves an infinite domain. Integrating (3.3) over $\mathbf{R}$ implies that a solitary wave solution satisfies

$$\int_{-\infty}^{+\infty} y(x)\,dx = 0 \tag{3.5}$$

if $\varepsilon \neq 0$. However, when $\varepsilon = 0$,

$$\int_{-\infty}^{+\infty} y_0(x)\,dx = 6.$$

Our asymptotic solution will consist of an inner solution, valid when $x = O(1)$ as $\varepsilon \to 0+$, and an outer solution, valid when $x = O(\varepsilon^{-1})$. For simplicity, we assume that y is an even function of x. To construct the inner solution, we expand y as

$$y(x;\ \varepsilon) = y_0(x) + \varepsilon y_1(x) + \varepsilon^2 y_2(x) + O(\varepsilon^3). \tag{3.6}$$

We require that the y_n's have at most algebraic growth as $|x| \to +\infty$. Using (3.6) in (3.3), equating coefficients of ε to zero, and integrating the resulting equations twice, implies that

$$\begin{aligned} y_1'' + (y_0 - 1)y_1 &= C_1, \\ y_2'' + (y_0 - 1)y_2 &= C_2 - \sigma \int_0^x \int_0^\xi y_0(\zeta)\,d\zeta\,d\xi, \end{aligned} \tag{3.7}$$

where C_1,C_2 are arbitrary constants of integration. Equation (3.7) implies that

$$\left.\begin{aligned} y_1 &\sim -C_1 \\ y_2 &\sim 6\sigma x \end{aligned}\right. \qquad \text{as } x \to +\infty. \tag{3.8}$$

Thus, (3.6) is nonuniform when $x = O(\varepsilon^{-1})$, because the second and third terms have the same order of magnitude.

To construct the outer expansion, we let

$$y(x;\ \varepsilon) = Y(\varepsilon x;\ \varepsilon), \qquad x > 0, \tag{3.9}$$

and expand Y as

$$Y(X;\ \varepsilon) = \varepsilon Y_1(X) + O(\varepsilon^2). \tag{3.10}$$

Using (3.9) and (3.10) in (3.3) and equating the coefficient of ε^2 to zero implies that

$$Y_1'' - \sigma Y_1 = 0. \tag{3.11}$$

The bounded solutions of (3.11) are

$$\begin{aligned} Y_1 &= A\sin X + B\cos X \qquad &&\text{if } \sigma = -1, \\ Y_1 &= Ae^{-X} &&\text{if } \sigma = +1, \end{aligned} \tag{3.12}$$

where A, B are constants of integration. Expanding (3.12) as $X \to 0+$ and writing the result in terms of the inner variable x shows that

$$\begin{aligned} &Y_1 \sim B + \varepsilon Ax && &&\text{if } \sigma = -1, \\ & &&\text{as } X \to +0, && \\ &Y_1 \sim A - \varepsilon Ax && &&\text{if } \sigma = +1. \end{aligned} \tag{3.13}$$

Matching (3.9), (3.10), and (3.13) with (3.6) and (3.8), we find that

$$\begin{aligned} A &= -6, \quad B = -C_1 \qquad &&\text{if } \sigma = -1, \\ A &= -6, \quad C_1 = 6 &&\text{if } \sigma = +1. \end{aligned} \tag{3.14}$$

Thus, we have the following asymptotic solution of (3.3) as $\varepsilon \to 0+$: for $\sigma = -1$,

$$y \sim \begin{cases} 3\operatorname{sech}^2\left(\frac{x}{2}\right) - \varepsilon C & \text{if } x = O(1), \\ -\varepsilon[6\sin(\varepsilon|x|) + C\cos(\varepsilon x)] & \text{if } x = O(\varepsilon^{-1}); \end{cases}$$

for $\sigma = +1$,

$$y \sim \begin{cases} 3\operatorname{sech}^2\left(\frac{x}{2}\right) - 6\varepsilon & \text{if } x = O(1), \\ -6\varepsilon\exp(-\varepsilon|x|) & \text{if } x = O(\varepsilon^{-1}). \end{cases} \tag{3.15}$$

When $\sigma = -1$, this solution is not a solitary wave. The KdV solitary wave sits on a train of long-wavelength periodic waves. (See **[2]** for a related problem.) When $\sigma = +1$, (3.2) has a solitary wave solution. There is a long shallow trough—of depth order ε and length order ε^{-1}—ahead of and behind the crest of the solitary wave. This allows the solution to satisfy (3.5), even though it is pointwise close to the KdV solitary wave solution.

In Figure 6(a), we show numerical solutions of solitary wave solutions of (1.3) with $\sigma = +1$. From (3.1), small values of ε correspond to large amplitudes. We discretized the ODE for travelling wave solutions of (1.3) on a truncated interval and used Newton's method to solve the resulting algebraic equations. The KdV solution was the initial guess for the largest-amplitude wave; solutions for smaller amplitudes were found by continuation. At amplitudes below a critical value between 5 and 10, the decay of the solitary wave is oscillatory instead of monotonic. In Figure 6(b), we compare the numerical solution with the asymptotic solution (3.15).

4. Nonlinear acoustics. In this section we describe a system of equations analyzed by Majda, Rosales, and Schonbeck **[6]**. The system is

$$
\begin{aligned}
u_t + uu_x &= v, \\
v_t + vv_x &= -u.
\end{aligned}
\tag{4.1}
$$

Equation (4.1) models the reflection of sound waves off periodic temperature variations in a gas. The linearization of (4.1) is the equation of a simple harmonic oscillator. Thus, (4.1) combines convective nonlinearity with lower-order "dispersive" terms. Solutions of (4.1) and (1.6) show some similar phenomena.

Equation (4.1) has a family of periodic travelling waves. For a given wave velocity, there is a wave of maximum amplitude. The profiles of u and v for the limiting wave have corners at their crests **[6]**.

Equation (4.1) also has separable sawtooth wave solutions,

$$
\begin{aligned}
u(x,t) &= a(t)S(x), \\
v(x,t) &= b(t)S(x),
\end{aligned}
\tag{4.2}
$$

where S is the sawtooth function,

$$S(x) = x, \quad |x| < \pi, \quad S(x + 2\pi) = S(x),$$

and a, b satisfy the ODEs

$$
\begin{aligned}
a' + a^2 - b &= 0, \\
b' + b^2 + a &= 0.
\end{aligned}
\tag{4.3}
$$

We show the phase plane of (4.3) in Figure 7.

The solution (4.2) contains shocks. The shocks are admissible **[11]** if and only if a and b are nonnegative. Figure 7 shows that if a and b are initially positive, then b changes sign at some later time. Majda, Rosales, and Schonbeck **[6]** conjecture that "cusped rarefaction waves," with square-root singularities, appear in v when b changes sign.

(a)

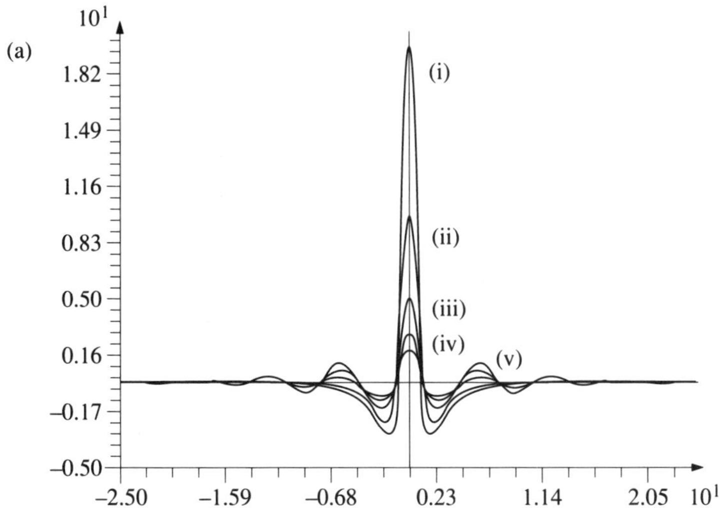

(b)

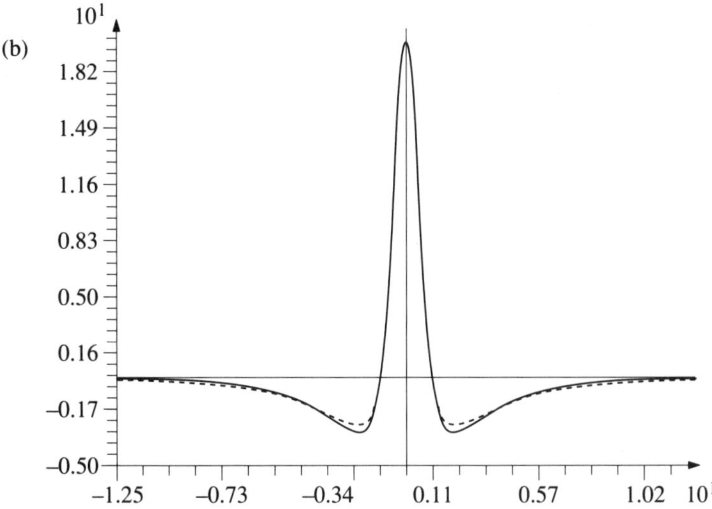

FIGURE 6. (a) Solitary wave solutions of (1.3) with $\sigma = +1$. Amplitude $y(0)$: (i) 20; (ii) 10; (iii) 5; (iv) 3; (v) 2.

(b) A comparison between a numerical solution and an asymptotic solution for the solitary wave solution of (1.3) with $y(0) = 20$.

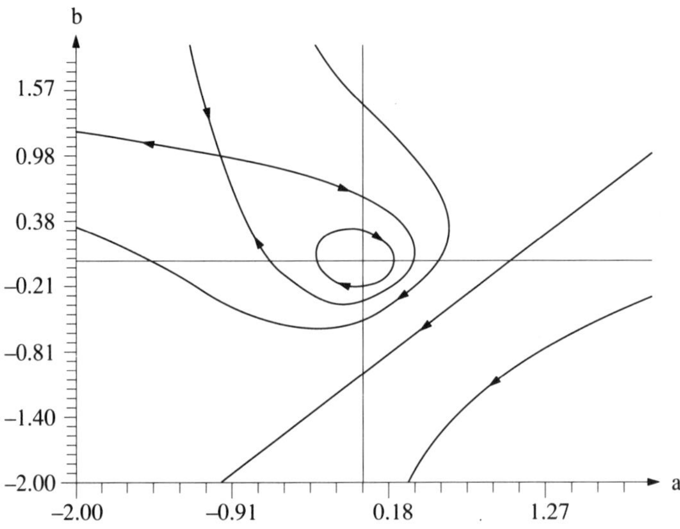

FIGURE 7. The phase plane of (4.3)

In Figure 8, we show a numerical solution of (4.1) with initial data

$$u(x,0) = S(x),\quad v(x,0) = 0. \tag{4.4}$$

These data arise from the sawtooth solution (4.2) if $a = 1$ when b passes through zero. The corresponding trajectory of (4.3) is $a - b = 1$. To compute the numerical solution, we used the Enquist–Osher scheme **[8]** to difference the left-hand side of (4.1) and treated the right-hand side as a source term.

Figure 8 shows that there is a complicated exchange of energy between u and v. Shocks form, disappear, and reform. The solutions for $t = 12$ and $t = 13$ are very similar to the solutions for $t = 6$ and $t = 7$. This suggests that the solution may approach a time-periodic standing wave as $t \to +\infty$.

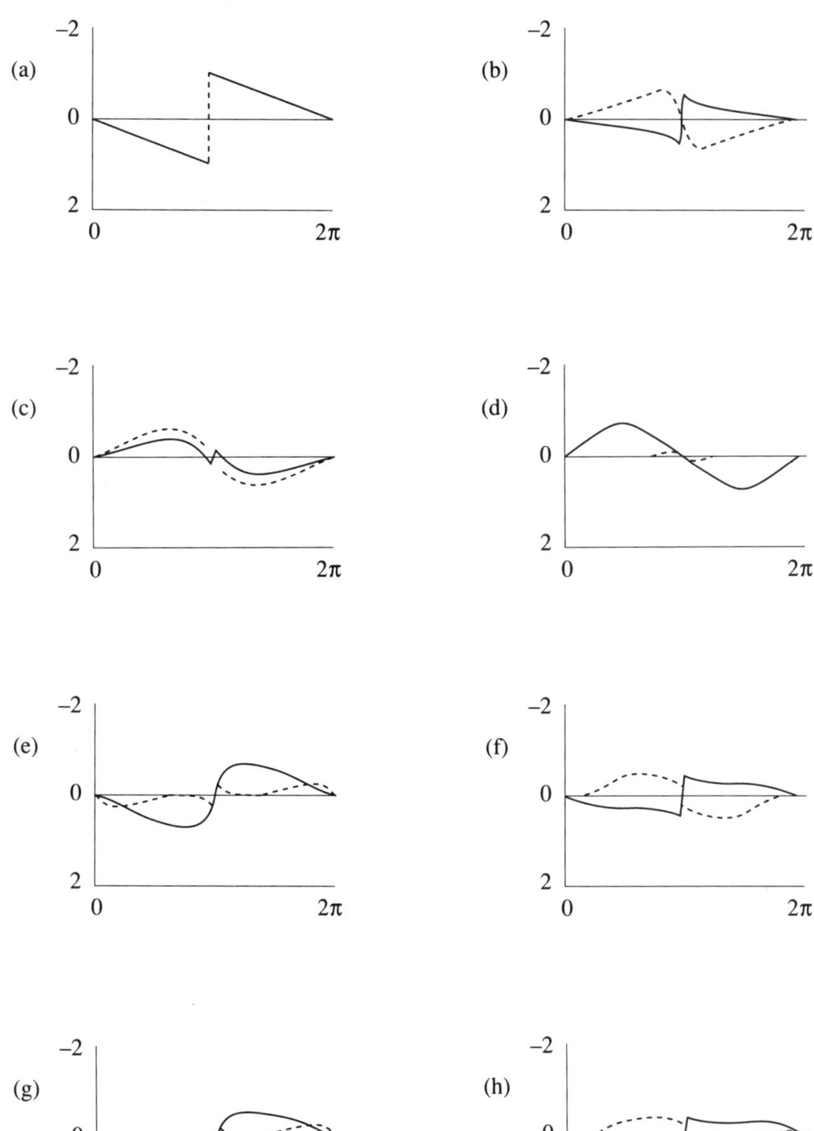

FIGURE 8. The solution of (4.1) and (4.4). The full line is u and the dotted line is v. Times: (a) 0; (b) 1; (c) 2; (d) 3; (e) 6; (f) 7; (g) 12; (h) 13. (The u–v-axis is reversed).

REFERENCES

1. R. Grimshaw, *Evolution equations for weakly nonlinear long internal waves in a rotating fluid*, Stud. Appl. Math. **73** (1985), 1–33.

2. J. K. Hunter and J. Scheurle, *Perturbed solitary wave solutions of a model equation for water waves*, Physica D **32** (1988), 253–268.

3. K. Tan and J. K. Hunter, *Nonlinear sound waves in a bubbly liquid*, work in progress.

4. ——, *Weakly dispersive short waves*, Proc. of the IVth International Conference on Waves and Stability in Continuous Media (Taormina, Sicily, 1987) (to appear).

5. P. D. Lax and C. D. Levermore, *The small dispersion limit for the KdV equation*, I, Comm. Pure Appl. Math. **36** (1983), 253–290; II, Comm. Pure Appl. Math. **36** (1983), 571–594; III, Comm. Pure Appl. Math. **36** (1983), 809–829.

6. A. Majda, R. Rosales, and M. Schonbeck, *A canonical system of integro-differential equations in nonlinear acoustics*, Stud. Appl. Math. **71** (1984), 149–179.

7. A. C. Newell, M. Tabor, and Y. B. Zeng, *A unified approach to Painlevé expansions*, Physica D **29** (1987), 1–68.

8. S. Osher, *Shock capturing algorithms for equations of mixed type*, Numerical Methods for Partial Differential Equations (1986), S. I. Hariharan and T. H. Moulton (eds.), Longman, Harlow, 1986.

9. L. A. Ostrovsky, *Nonlinear internal waves in a rotating ocean*, Oceanology **18** (1978), 181–191.

10. L. Redekopp, *Nonlinear waves in geophysics: Long internal waves*, Lectures in Appl. Math. **20** (1983), 59–78.

11. J. Smoller, *Shock waves and reaction–diffusion equations*, Springer-Verlag, New York, 1983.

12. S. Venakides, *The solution of completely integrable systems in the continuum limit of the spectral data*, Oscillation Theory, Computation, and Methods of Compensated Compactness, IMA Volumes in Mathematics **2** (1986), 337–355, C. Dafermos, J. L. Ericksen, D. Kinderlehrer, and M. Slemrod (eds.), Springer-Verlag, New York.

13. G. B. Whitham, *Linear and nonlinear waves*, Wiley, New York, 1974.

COLORADO STATE UNIVERSITY

Lectures in Applied Mathematics
Volume **26** (1990)

Parametric Optimization: Critical Points and Local Minima

H. TH. JONGEN

Abstract. We consider generic finite-dimensional optimization problems depending on one real parameter. For increasing values of the parameter we are interested in following a path of local minima. At certain parameter values a branch of local minima may have an endpoint and one is forced to jump to another branch. All such possible situations will be discussed and it is shown how a jump can be realized in the case that the underlying connected component of the feasible set remains nonempty.

0. Introduction. We consider an optimization problem $\mathscr{P}(t)$ depending on a real parameter t,

$$(1) \qquad \mathscr{P}(t): \quad \text{Minimize } f(\cdot, t) \text{ on } M(t), \qquad t \in \mathbf{R},$$

where the feasible set $M(t)$ is defined as follows:

$$(2) \quad \begin{gathered} M(t) = \{\mathbf{x} \in \mathbf{R}^n | h_i(\mathbf{x}, t) = 0, \quad i \in I, \quad g_j(\mathbf{x}, t) \geq 0, \quad j \in J\}, \\ I = \{1, \ldots, m\}, \quad m < n, \quad \text{and} \quad J = \{1, \ldots, s\}. \end{gathered}$$

We assume that the functions f, h_i, g_j belong to C^∞ $(\mathbf{R}^n \times \mathbf{R}, \mathbf{R})$, the space of smooth functions from $\mathbf{R}^n \times \mathbf{R}$ to $\mathbf{R}$. We are interested in following a path of local minima for $\mathscr{P}(t)$ as the parameter t increases. For almost all values of t one generally might expect that a local minimum $\mathbf{x}(t)$ for $\mathscr{P}(t)$ depends differentiably on t, mainly by virtue of the implicit function theorem. However, such a regular behavior cannot be expected for all values of t. The basic question now becomes: How can

1980 *Mathematics Subject Classification* (1985 *Revision*). Primary 90C31.

we handle such situations? In fact, the development of an answer to the latter question is the aim of our paper.

A first approach to it was presented in **[3]**. The main line is as follows. First, we introduce the notion of a *generalized critical point* (g. c. point). In particular, every local minimum is a g. c. point. For generic problems the structure of the set of g. c. points and its local interpretation are extensively studied and completely understood (cf. **[8]**, **[9]**). Next, we focus our attention on the subset of the g. c. point set consisting of *local minima.* By using *normal forms* as far as possible we can nicely deal with the basic question above. Normal forms reflect a certain situation in its easiest form and they facilitate the understanding of the underlying phenomenon. Here, normal forms describe the situations within appropriate new smooth local coordinates and, in a neighborhood of the point $(\bar{\mathbf{x}}, \bar{t}) \in \mathbf{R}^n \times \mathbf{R}$, the local coordinate transformation will be of the form (prime denotes differentiation)

$$\Psi\colon\ (\mathbf{x}, t) \mapsto (\psi_1(\mathbf{x}, t),\ \psi_2(t)), \quad \psi_2'(\bar{t}) > 0. \tag{3}$$

Formula (3) takes care of the special choice of t as a parameter (which increases).

We note that the result of our analysis below can also be obtained without introducing normal forms; in fact, we use partial derivatives only up to order 3, and so, for our aim C^3-differentiability of the underlying functions would be sufficient.

Let $\mathbf{z} = (\mathbf{x}, t)$ denote the general point in $\mathbf{R}^n \times \mathbf{R}$. A point $\mathbf{z}$ is called a *generalized critical point* (g. c. point) if $\mathbf{x} \in M(t)$ and if the set $\{\mathbf{D_x}f,\ \mathbf{D_x}h_i,\ i \in I,\ \mathbf{D_x}g_j,\ j \in J_0(\mathbf{z})\}|_{\mathbf{z}}$ is linearly *dependent.* Here, $\mathbf{D_x}f$ stands for the row vector of first partial derivatives with respect to $\mathbf{x}$, and $J_0(\mathbf{z})$ denotes the index set of active inequality constraints, i.e., $J_0(\mathbf{z}) = \{j \in J | g_j(\mathbf{z}) = 0\}$. If $\bar{\mathbf{x}} \in M(\bar{t})$ is a local minimum for problem $\mathscr{P}(\bar{t})$, then $\bar{\mathbf{z}} := (\bar{\mathbf{x}}, \bar{t})$ is a g. c. point. Let Σ denote the set of g. c. points. In **[9]** the local structure of Σ is completely described for $(f, \ldots, h_i, \ldots, g_j, \ldots)$ belonging to a C^3 open and dense subset $\mathscr{F}$ from the (product) space $C^\infty\ (\mathbf{R} \times \mathbf{R}, \mathbf{R})^{1+m+s}$. The set $\mathscr{F}$ is actually obtained by imposing certain natural transversality conditions on the function-tuples $(f, \ldots, h_i, \ldots, g_j, \ldots)$. The C^3 topology used refers to the strong (or Whitney-) C^3 topology (cf. **[4]**, **[7]**). For omitted details on the set $\mathscr{F}$ we refer to **[9]**. In the sequel we will make the following assumption.

MAIN ASSUMPTION. The function-tuple $(f, \ldots, h_i, \ldots, g_j, \ldots)$ belongs to the class $\mathscr{F}$.

The set Σ of g.c. points is pieced together from one-dimensional smooth manifolds and its points can be classified by means of five basic types. The points of Type 1 are the *nondegenerate* critical points. At these points the implicit function theorem is applicable and the set Σ can be parametrized locally by means of the parameter t. The set of points of Types 2–5, the *degenerate* ones, forms a discrete subset of Σ. In view of the above classification we can restrict ourselves to local considerations at the points of Types 1–5. In the rest of the paper we will separately discuss each type, and we end with some final remarks.

1. Points of Type 1. A generalized critical point $\bar{\mathbf{z}} = (\bar{\mathbf{x}}, \bar{t})$ is of Type 1 if the following conditions (4)–(7) hold:

$$\mathbf{D_x} f = \sum_{i\in I} \bar{\lambda}_i \mathbf{D_x} h_i + \sum_{j\in J_0(\bar{\mathbf{z}})} \mu_j \mathbf{D_x} g_j |_{\bar{\mathbf{z}}}. \tag{4}$$

(5) The set $\{\mathbf{D_x} h_i,\ i \in I,\ \mathbf{D_x} g_j,\ j \in J_0(\bar{\mathbf{z}})\}|_{\bar{\mathbf{z}}}$ is linearly independent (*linear independence constraint qualification*, LICQ).

(6) The numbers $\bar{\mu}_j$, $j \in J_0(\bar{\mathbf{z}})$, are unequal to zero (*strict complementarity*).

(7) $\mathbf{D}^2_{\mathbf{x}} L(\bar{\mathbf{z}})/T(\bar{\mathbf{z}})$ is nonsingular.

In the case where $\bar{\mathbf{z}}$ is of Type 1 we call the point $\bar{\mathbf{x}}$ a *nondegenerate critical point* for the problem $\mathscr{P}(\bar{t})$. The real numbers $\bar{\lambda}_i$, $\bar{\mu}_j$ are called *Lagrange multipliers.* Condition (7) needs some clarification: $\mathbf{D}^2_{\mathbf{x}} L$ is the matrix of second-order partial derivatives—with respect to $\mathbf{x}$—for the Lagrange function L, where

$$L = f - \sum_{i\in I} \bar{\lambda}_i h_i - \sum_{j\in J_0(\bar{\mathbf{z}})} \bar{\mu}_j g_j, \tag{8}$$

the numbers $\bar{\lambda}_i$, $\bar{\mu}_j$ being taken from (4). Futhermore, $T(\bar{\mathbf{z}})$ denotes the tangent space of $M(\bar{t})$ at $\bar{\mathbf{x}}$, i.e.,

$$T(\bar{\mathbf{z}}) = \bigcap_{i\in I} \operatorname{Ker} \mathbf{D_x} h_i(\bar{\mathbf{z}}) \cap \bigcap_{j\in J_0(\bar{\mathbf{z}})} \operatorname{Ker} \mathbf{D_x} g_j(\bar{\mathbf{z}}). \tag{9}$$

Now, $\mathbf{D}^2 L(\bar{\mathbf{z}})/T(\bar{\mathbf{z}})$ stands for $\mathbf{V}^{\mathrm{T}} \mathbf{D}^2_{\mathbf{x}} L(\bar{\mathbf{z}}) \mathbf{V}$, where $\mathbf{V}$ is a matrix whose columns (n-vectors) form a basis for $T(\bar{\mathbf{z}})$.

In **[11]** it is shown that a g.c. point is of Type 1 if and only if the Jacobian matrix of a certain associated system of equations is nonsingular. If $\bar{\mathbf{z}}$ is a point of Type 1, then the local behavior of $f(\cdot, t)|_{M(t)}$ around $\bar{\mathbf{z}} = (\bar{\mathbf{x}}, \bar{t})$ is completely determined by means of four characteristic numbers (or indices). In fact, taking the active constraint functions as new coordinate functions, a local diffeomorphism Ψ of the form (3) can be constructed such that in a neighborhood of $\bar{\mathbf{z}}$ the feasible set $M(t)$ transforms under Ψ to the *constant* set $\mathbf{H}^p \times \mathbf{R}^q$, $\mathbf{H}^p$ denoting the nonnegative orthant in $\mathbf{R}^p$, i.e.,

$$(10) \qquad \mathbf{H}^p = \{(\xi_1, \ldots, \xi_p) \in \mathbf{R}^p |\ \xi_i \geq 0, i = 1, \ldots, p\},$$

and where $p = |J_0(\bar{\mathbf{z}})|$, the number of active inequality constraints, and $q = n - |I| - |J_0(\bar{\mathbf{z}})|$. The latter coordinate transformation can be accomplished by virtue of LICQ (cf. (5)). Subsequently, by exploiting the extended Morse lemma (cf. **[6]**), one can transform the objective function into a standard (linear-) quadratic function. Altogether, there exists an open neighborhood U of $\bar{\mathbf{z}}$, an open neighborhood V of the origin in $\mathbf{R}^n \times \mathbf{R}$, and a C^∞-diffeomorphism Ψ: $U \to V$ of the form (3), sending $\bar{\mathbf{z}}$ onto the origin, such that

$$(11) \qquad \Psi[(M(t), t) \cap U] = (\mathbf{H}^p \times \mathbf{R}^q \times \{0_{n-p-q}\}, \psi_2(t)) \cap V,$$

$$(12) \qquad f_0 \Psi^{-1}(y_1, \ldots, y_{p+q}, 0, \ldots, 0, v) = c(v) - \sum_{i_1} y_{i_1} + \sum_{i_2} y_{i_2} - \sum_{i_3} y_{i_3}^2 + \sum_{i_4} y_{i_4}^2,$$

where $c(v) = f_0 \Psi^{-1}(0, \ldots, 0, v)$, $v \in (-\varepsilon, \varepsilon)$ for some positive ε, $\{1, \ldots, p+q\} = I_1 \dot\cup I_2 \dot\cup I_3 \dot\cup I_4$ and $i_j \in I_j$, $j = 1, \ldots, 4$, $p = |I_1| + |I_2|$; moreover, the number of negative/positive linear terms in (12) corresponds to the number of negative/positive numbers $\bar{\mu}_j$, $j \in J_0(\bar{\mathbf{z}})$, whereas the number of negative/positive quadratic terms corresponds to the number of negative/positive eigenvalues of $\mathbf{D}^2 L(\bar{\mathbf{z}})/T(\bar{\mathbf{z}})$. The latter four numbers are the aforementioned characteristic numbers.

Formulas (11) and (12) provide the normal form. We see that, in a neighborhood of $\bar{\mathbf{z}}$ we are dealing, up to smooth change of coordinates, with a standard type of a feasible set and a standard type of a quadratic objective function (*trivialization*). Moreover, in a neighborhood of $\bar{\mathbf{z}}$ the g. c. point set Σ can be parametrized by means of the parameter t; to see this, just note that $\Sigma \cap U$ equals $\{\Psi^{-1}(0, v) | v \in (-\varepsilon, \varepsilon)\}$. In particular, we are dealing with a local minimum branch if and only if both all $\bar{\mu}_j$, $j \in J_0(\bar{\mathbf{z}})$, are positive and all eigenvalues of $\mathbf{D}_\mathbf{x}^2 L(\bar{\mathbf{z}})/T(\bar{\mathbf{z}})$ are positive (i.e., $\mathbf{D}_\mathbf{x} L(\bar{\mathbf{z}})$ is positive definite on $T(\bar{\mathbf{z}})$). See Figure 1 for a picture of level sets in several situations.

So, if $\bar{\mathbf{x}} \in M(\bar{t})$ is a local minimum for $\mathscr{P}(\bar{t})$ and if $(\bar{\mathbf{x}}, \bar{t})$ is a point of Type 1, then, in a neighborhood of $(\bar{\mathbf{x}}, \bar{t})$ we can walk along Σ using some path-following technique, thereby staying on a branch of local minima.

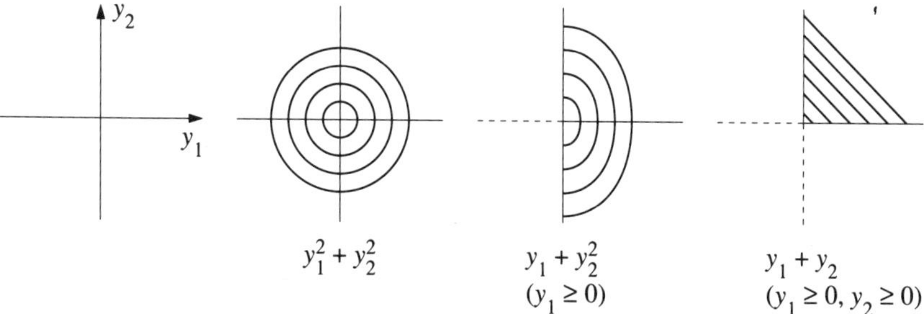

FIGURE 1

If $\bar{\mathbf{z}} \in \Sigma$ is not of Type 1, then three possible degeneracies (**D**1 – **D**3) can occur (cf. also [**11**]):

(**D**1) The strict complementarity is violated (Type 2).
(**D**2) $\mathbf{D}^2L(\bar{\mathbf{z}})/T(\bar{\mathbf{z}})$ is singular (Type 3).
(**D**3) The linear independence constraint qualification is violated (Type 4, Type 5).

We proceed with a discussion of the above degeneracies from the viewpoint of following a branch of local minima for increasing values of the parameter t.

2. Points of Type 2. A generalized critical point $\bar{\mathbf{z}} = (\bar{\mathbf{x}}, \bar{t})$ is of Type 2 if the *strict complementarity is violated* for exactly one active *inequality* constraint, say having index $\tilde{j} \in J_0(\bar{\mathbf{z}})$; however, LICQ is satisfied. We refer to [**9**] for a detailed description. The violation of the strict complementarity at $\tilde{j} \in J_0(\bar{\mathbf{z}})$ means that the corresponding Lagrange multiplier $\bar{\mu}_{\tilde{j}}$ in (4) vanishes. It follows that $(\bar{\mathbf{x}}, \bar{t})$ is a g. c. point for both $\mathscr{P}_1(\bar{t})$ and $\mathscr{P}_2(\bar{t})$. The difference between $\mathscr{P}(t)$ and $\mathscr{P}_1(t)$, $\mathscr{P}_2(t)$ consists in the fact that in $\mathscr{P}_1(t)$ the function $g_{\tilde{j}}$ is omitted as a constraint, whereas in $\mathscr{P}_2(t)$ the function $g_{\tilde{j}}$ is treated as an equality constraint. So, we obtain two curves of g. c. points, one from $\mathscr{P}_1(t)$ and one from $\mathscr{P}_2(t)$. Locally, the set Σ of g. c. points for $\mathscr{P}(t)$ is pieced together from the feasible part of the latter two curves; see Figure 2.

Next, we derive a normal form for the situation of Type 2.

LEMMA. *Let $\bar{\mathbf{z}} = (\bar{\mathbf{x}}, \bar{t})$ be a g.c. point of Type 2. Then there exists a diffeomorphism $\Psi\colon U \to V$ of the form* (3), *sending $\bar{\mathbf{z}}$ onto the origin*

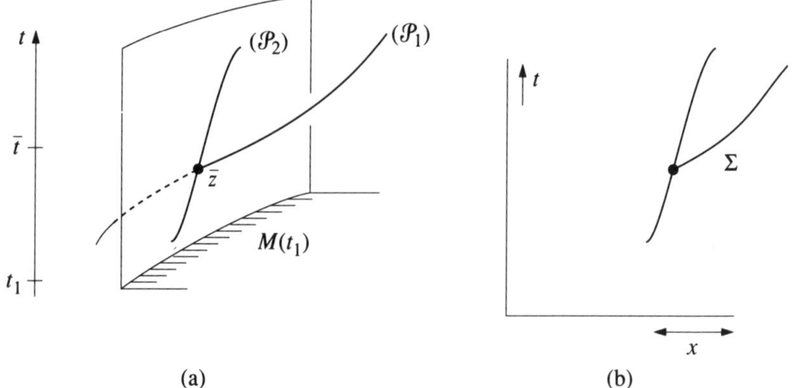

(a) (b)

FIGURE 2

such that (11) *holds, and, moreover,*

$$f_0\Psi^{-1}(y_1,\ldots,y_{p+q},0,\ldots,0,v), \tag{13}$$
$$= \alpha\gamma(v)(y_1-\beta v)^2+\delta(v)+\sum_{i=2}^{p}\pm y_i+\sum_{j=p+1}^{p+q}\pm y_j^2$$

with α, $\beta\in\{-1,1\}$ *and* $\gamma(\cdot)$, $\delta(\cdot)$ *smooth functions around the origin in* $\mathbf{R}$, *and* $\gamma(0)>0$.

PROOF. By means of the splitting argument performed in [7] (see pp. 525–529 of [7]), we may already assume that, in new smooth coordinates according to (3), we are in the following situation:

$$f(\mathbf{x},t)=F(x_1,t)+\sum_{i=2}^{p}\pm x_i+\sum_{j=p+1}^{p+q}\pm y_j^2, \tag{14}$$

$$x_i\geq 0,\quad i,\ldots,p, \tag{15}$$

where the origin is the point of Type 2. So, in our analysis we may focus on the situation of *one variable* x, i.e., we look at $F(x,t)$, $x\geq 0$, in a neighborhood of the origin in $\mathbf{R}\times\mathbf{R}$. In the next local change of coordinates the feasible set $\{x\in\mathbf{R}|x\geq 0\}$ is transformed to itself, and so, we take coordinate transformations of the following form (cf. (3)):

$$\Psi(x,t)=(x\psi_1(x,t),\psi_2(t)), \tag{16}$$

where

$$\psi_1(0)>0\quad\text{and}\quad\psi_2'(0)>0. \tag{17}$$

Note that, in view of the special structure of Ψ in (16), the origin $x = 0$ is mapped onto itself. Since we are dealing with a point of Type 2, it follows from its characterization (cf. **[9]**) that

$$F_x(0) = 0, \quad F_{xx}(0) \neq 0, \quad F_{xt}(0) \neq 0, \tag{18}$$

where F_x, F_{xx}, and F_{xt} denote the corresponding partial derivatives.

By virtue of (18), application of the implicit function theorem yields the existence of a local smooth function $t \mapsto \eta(t)$ such that

$$F_x(\eta(t), t) \equiv 0, \qquad \eta(0) = 0. \tag{19}$$

Put

$$\varphi(u) = F(\eta(t) + u(x - \eta(t)), t). \tag{20}$$

Then we have

$$\varphi(0) = F(\eta(t), t), \qquad \varphi(1) = F(x, t) \tag{21}$$

and it follows that

(22)
$$\begin{aligned}\varphi(1) - \varphi(0) &= \int_0^1 \frac{d\varphi(u)}{du}\, du = \int_0^1 F_x(\eta(t) + u(x - \eta(t)), t)(x - \eta(t))\, du \\ &= (x - \eta(t))G(x, t).\end{aligned}$$

From (19), (21), and (22) we see that $G(\eta(t), t) \equiv 0$, and hence, putting

$$\xi(u) = G(\eta(t) + u(x - \eta(t)), t), \tag{23}$$

we obtain, similarly as in (22):

$$G(x, t) = \xi(1) = \xi(1) - \xi(0) = \cdots = (x - \eta(t))H(x, t), \tag{24}$$

and, finally, combination of (21), (22), and (24) yields

$$F(x, t) = F(\eta(t), t) + (x - \eta(t))^2 H(x, t). \tag{25}$$

Based on the chain rule for differentiating $t \mapsto F_x(\eta(t), t)$, it follows from (18) that

$$\beta := \operatorname{sign}\left(\frac{d}{dt}\eta(0)\right) \neq 0.$$

Taking $t \mapsto \beta\eta(t)$ as a new t-coordinate, we may assume that F has the following form ($\delta(t) := F(\eta(t), t)$):

$$F(x, t) = \delta(t) + (x - \beta t)^2 w(x, t). \tag{26}$$

Finally, define

$$(27)\qquad \begin{cases} y = & (x - \beta t)\sqrt{|w(x,t)|}/\sqrt{|w(0,t)|} + \beta t, \\ v = & t. \end{cases}$$

Putting $\alpha = \operatorname{sign} w(0)$ and $\gamma(v) = |w(0, v)|$, it is easily checked that, at last, Formula (27) fits our aim. ■

Recall that we are interested in walking along a path of local minima for $\mathscr{P}(t)$, as the parameter t increases. From (11) and (13) it follows that the normal form (13) has only *positive* linear/quadratic terms for $i = 2, \ldots, p$ and $j = p+1, \ldots, p+q$, so we have

$$(28)\qquad \begin{aligned} &f_0\Psi^{-1}(y_1, \ldots, y_{p+q}, 0, \ldots, 0, v) \\ &\quad = \alpha\gamma(v)(y_1 - \beta v)^2 + \delta(v) + \sum_{i=2}^{p} y_i + \sum_{j=p+1}^{p+q} y_j^2. \end{aligned}$$

Now, it is easily seen that for α, β, there are precisely three possibilities:

$$(29)\qquad \left.\begin{array}{lll} \text{(I)} & \beta = -1, & \alpha = +1, \\ \text{(II)} & \beta = +1, & \alpha = +1, \\ \text{(III)} & \beta = -1, & \alpha = -1. \end{array}\right\}$$

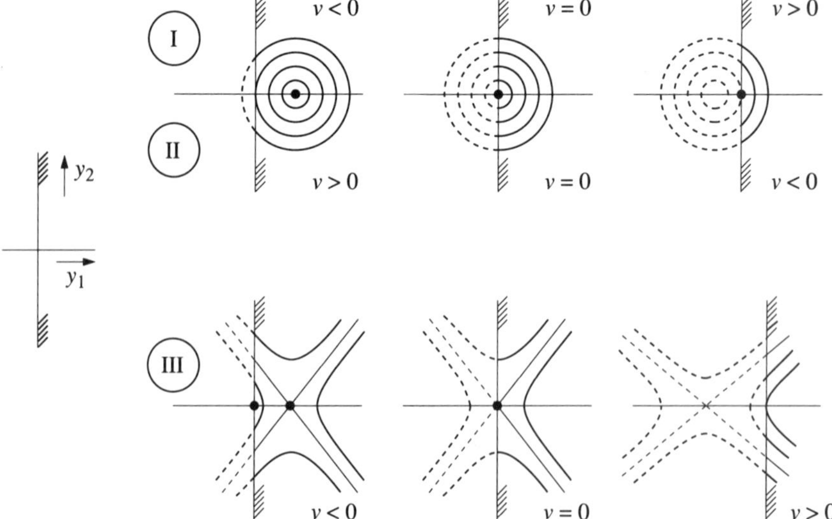

FIGURE 3

In Figure 3 the above possibilities are depicted in case $p = q = 1$, $\gamma(v) \equiv 1$; according to (28), the y-dimension equals 2, and the inequality constraint will be "$y_1 \geq 0$." Some level lines in y-space are depicted for negative, zero, and positive values of v.

In Figure 4, the above possibilities I, II, III are depicted from the viewpoint of Figure 2(b).

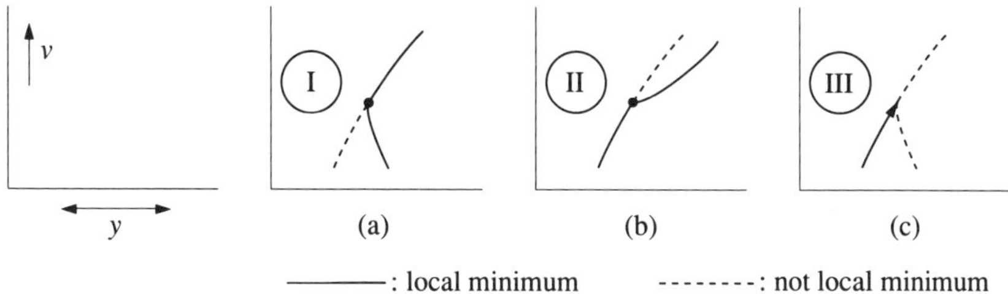

FIGURE 4

In terms of the original functions f, h_i, g_j, the difference between the Situations I, II (cf. (29)) on the one hand, and III on the other hand, can be explained as follows. Let $\tilde{j} \in J_0(\bar{\mathbf{z}})$ again denote the index of the active inequality constraint for which the Lagrange multiplier $\mu_{\tilde{j}}$ in (4) vanishes. Let the Lagrange function L and the tangent space $T(\bar{\mathbf{z}})$ be defined according to (8) and (9), respectively, and consider the *larger tangent space* $\widetilde{T}(\bar{\mathbf{z}})$:

$$(30) \qquad \widetilde{T}(\bar{\mathbf{z}}) := \bigcap_{i \in I} \operatorname{Ker} \mathbf{D}_{\mathbf{x}} h_i(\bar{\mathbf{z}}) \cap \bigcap_{j \in J_0(\bar{\mathbf{z}}) \setminus \{\tilde{j}\}} \operatorname{Ker} \mathbf{D}_{\mathbf{x}} g_j(\bar{\mathbf{z}}).$$

Then, in Situations I and II, both $\mathbf{D}^2 l(\bar{\mathbf{z}})/T(\bar{\mathbf{z}})$ and $\mathbf{D}^2 L(\bar{\mathbf{z}})/\widetilde{T}(\bar{\mathbf{z}})$ are positive definite; however, in III we have: $\mathbf{D}^2 L(\bar{\mathbf{z}})/T(\bar{\mathbf{z}})$ is positive definite but $\mathbf{D}^2 L(\bar{\mathbf{z}})/\widetilde{T}(\bar{\mathbf{z}})$ is not positive definite. In Situation III, we have

$$(31) \qquad \min\{\xi^{\mathrm{T}} \mathbf{D}^2 L(\bar{\mathbf{z}}) \xi \mid \|\xi\| = 1, \quad \xi \in \widetilde{T}(\bar{\mathbf{z}}), \quad \mathbf{D} g_{\tilde{j}}(\bar{\mathbf{z}}) \xi \geq 0\} < 0.$$

Thus, in Situations I and II, we can continue following a branch of local minima either by treating the constraint $g_{\tilde{j}}$ as an equality constraint (I) or by omitting $g_{\tilde{j}}$ as a constraint (II).

In Situation III the branch of local minima stops at $\bar{\mathbf{z}}$ as t increases. However, it is easily seen that, for $\mathbf{z} = (\mathbf{x}, t) \in \Sigma$ in a sufficiently small neighborhood of $\bar{\mathbf{z}}$, the point $\mathbf{x}$ cannot be a global minimum for $\mathscr{P}(t)$. In view of (31), at the point $\bar{\mathbf{z}}$ we can find a (quadratic) descent direction ξ for the problem $\mathscr{P}(\bar{t})$, and eventually we find a local minimum $\tilde{\mathbf{x}}$ for

$\mathscr{P}(\bar{t})$ if $M(\bar{t})$ is compact; of course, $\tilde{\mathbf{x}} \neq \bar{\mathbf{x}}$. Then, we can continue tracing the branch of local minima through the point $(\tilde{\mathbf{x}}, \bar{t})$, and so a jump to another branch has been realized.

3. Points of Type 3. A generalized critical point $\bar{\mathbf{z}} = (\bar{\mathbf{x}}, \bar{t})$ is of Type 3 if the linear independence constraint qualification and the strict complementarity hold, but one eigenvalue of $\mathbf{D}^2 L(\bar{\mathbf{z}})/T(\bar{\mathbf{z}})$ vanishes. We refer to **[9]** for a detailed description.

In $(\mathbf{x}, t)$-space the g. c. point set Σ exhibits a quadratic turning point (with respect to the t-direction). We are interested in following a branch of local minima for increasing values of t; so, if we approach the point $\bar{\mathbf{z}}$ along Σ, the path of local minima stops, and along Σ the local minimum switches into a saddlepoint.

As in the discussion of a g. c. point of Type 2, we can use a splitting argument in order to reduce our situation to a one-dimensional (unconstrained) problem. Then, application of the theory on unfoldings of singularities (cf. **[1]**) yields the following normal form in new local coordinates according to (3) (cf. also (28)):

$$\begin{aligned} &f_0\Psi^{-1}(y_1, \ldots, y_{p+q}, 0, \ldots, 0, v) \\ &\quad = \sum_{i=1}^{p} y_1 + y_{p+1}^3 + v y_{p+1} + \sum_{j=p+2}^{q} y_j^2 + \delta(v) \end{aligned} \tag{32}$$

where

$$y_1 \geq 0, \quad \ldots, \quad y_p \geq 0. \tag{33}$$

In Figure 5 the change of level lines of $f_0\Psi^{-1}$ is depicted with v as a variable, for the case $p = 0$, $q = 2$.

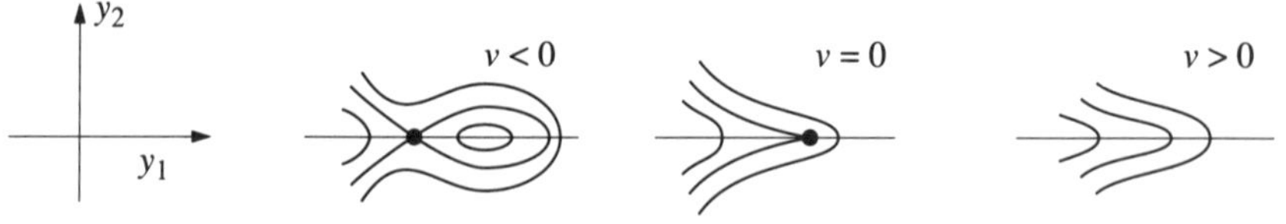

FIGURE 5

Now, suppose that the feasible set $M(\bar{t})$ for $\mathscr{P}(\bar{t})$ is compact. Then, we can jump from $\bar{\mathbf{z}}$ to another branch of local minima (note that for $(\mathbf{x}, t) \in \Sigma$ in a neighborhood of $\bar{\mathbf{z}}$ the point $\mathbf{x}$ cannot be a global minimum for $\mathscr{P}(t)$). In fact, at $\bar{\mathbf{z}}$ there exists a unique (tangential) direction of (cubic) descent; in (32) this will be the negative of the $(p+1)$th unit vector. Descending in that direction within $M(\bar{t})$ leads to a local minimum $\tilde{\mathbf{x}}$ ($\tilde{\mathbf{x}} \neq \bar{\mathbf{x}}$), and we can continue tracing the branch of local minima through the point $(\tilde{\mathbf{x}}, \bar{t}) \in \Sigma$.

In the original coordinates the latter direction of descent can be (approximately) obtained as follows; cf. Figure 6(a). For $t < \bar{t}$ the intersection of Σ with the hyperplane "$t =$ constant" consists, in a neighborhood of $\bar{\mathbf{z}}$, of exactly two points, one of them being a local minimum $(\mathbf{x}_{\mathrm{m}}(t), t)$ and the other one being a saddlepoint $(\mathbf{x}_{\mathrm{s}}(t), t)$. Then, the vector $(\mathbf{x}_{\mathrm{s}}(t) - \mathbf{x}_{\mathrm{m}}(t))/\|\mathbf{x}_{\mathrm{s}}(t) - \mathbf{x}_{\mathrm{m}}(t)\|$ converges to the desired direction of descent as t tends to $\bar{t}$. The latter vector also is the limit of the unit "forward" tangential vector along Σ taken at points $(\mathbf{x}_{\mathrm{m}}(t), t)$, as t tends to $\bar{t}$ (Figure 6(b)).

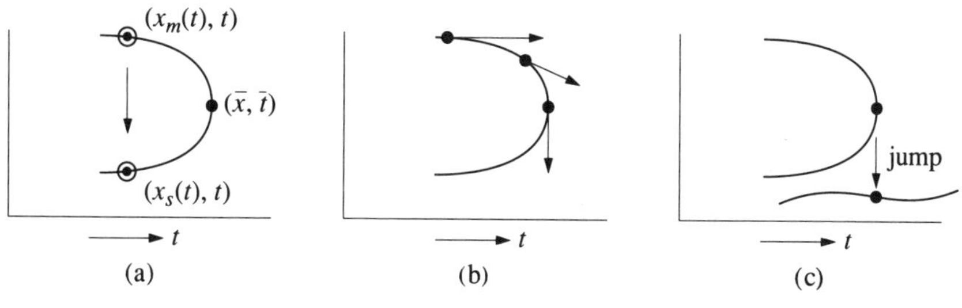

FIGURE 6

4. Points of Type 4 and Type 5. A generalized critical point $\bar{\mathbf{z}} = (\bar{\mathbf{x}}, \bar{t})$ is of Type 4 or Type 5 if the linear independence constraint qualification (LICQ) is violated. For a detailed description we refer again to [9]. Now, the feasible set $M(t)$ cannot be smoothly transformed to a standard set as in (11). Thus, we have to derive a new local normal form for $M(t)$ in which the parameter t plays an essential role. Such normal forms are studied in [5] and in Chapter 10 of [7]. It turns out that, for increasing t, $M(t)$ behaves locally as (lower-, upper-) level sets of an objective function in the neighborhood of a nondegenerate critical point (cf. Section 1). In fact, there exists a smooth coordinate transformation Ψ satisfying (3) such that $M(t)$ locally transforms as follows (we put $(y, v) = \Psi(\mathbf{x}, t)$ and we write down the (in)equalities that define $M(t)$ locally in the new coordinates):

SITUATION I. (Type 4, $J_0(\bar{\mathbf{z}}) = \emptyset$)

$$(34)\qquad \begin{cases} v = -\sum_{i_1=1}^{k} y_{i_1}^2 + \sum_{i_2=k+1}^{n-m+1} y_{i_2}^2 \qquad (m = |I|), \\ y_l = 0, \quad l = n-m+2, \ldots, n. \end{cases}$$

SITUATION II. (Type 4, $J_0(\bar{\mathbf{z}}) = \emptyset$)

$$(35)\qquad \begin{cases} \delta v \geq -\sum_{i_1=1}^{k} y_{i_1}^2 + \sum_{i_2=k+1}^{c} y_{i_2}^2 - \sum_{i_3=c+1}^{c+d} y_{i_3} + \sum_{i_4=c+d+1}^{c+d+e} y_{i_4}, \\ y_l \geq 0, \quad l = c+1, \ldots, c+d+e, \\ y_l = 0, \quad l = c+d+e+1, \ldots, n, \\ c = n - |I| - |J_0(\bar{\mathbf{z}})| + 1, \qquad \delta \in \{-1, 1\}, \\ d + e = |J_0(\bar{\mathbf{z}})| - 1, \qquad c \neq 0. \end{cases}$$

SITUATION III. (Type 5) As in situation II, but now $c = 0$; thus, the number of active constraints at $\bar{\mathbf{z}}$, being $|I| + |J_0(\bar{\mathbf{z}})|$, equals $n+1$.

After the description of the feasible set we now have to incorporate the objective function and the fact that we are interested in following a branch of local minima for increasing values of the parameter t. So, for $t < \bar{t}$ and $t \approx \bar{t}$ we assume that $\mathbf{x}(t)$ is a local minimum for $\mathscr{P}(t)$; in particular, we have $(\mathbf{x}(t), t) \in \Sigma$. Let us denote the objective function in the (y, v) coordinates again by f. We will systematically discuss Situations I–III.

SITUATION I. (Type 4, $J_0(\bar{\mathbf{z}}) = \emptyset$) We put $f_i = (\partial f / \partial y_i(0))$ for simplicity. From the characterization of points of Type 4 (cf. **[9]**) we have, according to (34):

$$(36)\qquad \Delta := -\sum_{i_1=1}^{k} f_{i_1}^2 + \sum_{i_2=k+1}^{n-m+1} f_{i_2}^2 \neq 0.$$

Since we focus on a branch of local minima for $v < 0$, $v \approx 0$, it is not difficult to see that exactly one of the following possibilities can occur:

$$(37)\qquad \begin{cases} \text{(Ia)} \qquad k = 1, \qquad \Delta < 0, \\ \text{(Ib)} \qquad k = n - m + 1. \end{cases}$$

It is shown in **[8]** that the set Σ of g. c. points exhibits a quadratic turning point (with respect to t). If we pass the point $\bar{\mathbf{z}}$ along Σ, then in case of Ia, Ib, our local minimum switches into a local maximum. Moreover, the value of f decreases (increases) in the case of Ia (Ib). This provides

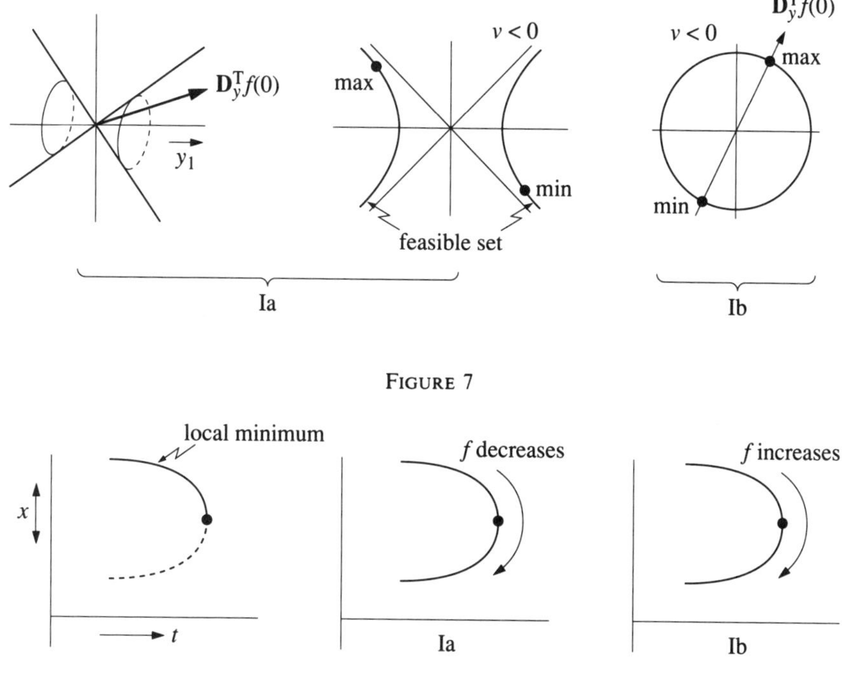

FIGURE 7

FIGURE 8

(also in the original coordinates) an easy way for checking whether we are in Situation Ia or Ib; see Figures 7 and 8.

In Situation Ia it is possible to jump to another branch of local minima. In fact, suppose that the feasible set $M(t)$ is compact for $t < \bar{t}$, $t \approx \bar{t}$. Then, compute a point on Σ beyond the turning point, say $(\mathbf{x}_{\max}(t), t)$ with $t < \bar{t}$; the point $\mathbf{x}_{\max}(t)$ is a local maximum for $\mathscr{P}(t)$ and we can start at $\mathbf{x}_{\max}(t)$ with a descent method in order to find a local minimum for $\mathscr{P}(t)$, say $\tilde{\mathbf{x}}_{\min}(t)$. Since f decreases along Σ, the point $\tilde{\mathbf{x}}_{\min}(t)$ differs from $\mathbf{x}_{\min}(t)$, the local minimum for $\mathscr{P}(t)$ corresponding to $\mathbf{x}_{\max}(t)$ on Σ (in the neighborhood of $\bar{\mathbf{z}}$); see Figure 9. However, in Situation Ib, the corresponding component of the feasible set becomes empty, and starting at $\mathbf{x}_{\max}(t)$ with a descent method as in Ia would drive us back to the same local minimum $\mathbf{x}_{\min}(t)$!

Up to now we have met two quadratic turning points for Σ, namely points of Type 3 and Type 4. We can easily detect whether we are approaching a point of Type 3 or Type 4. In fact, in the case of Type

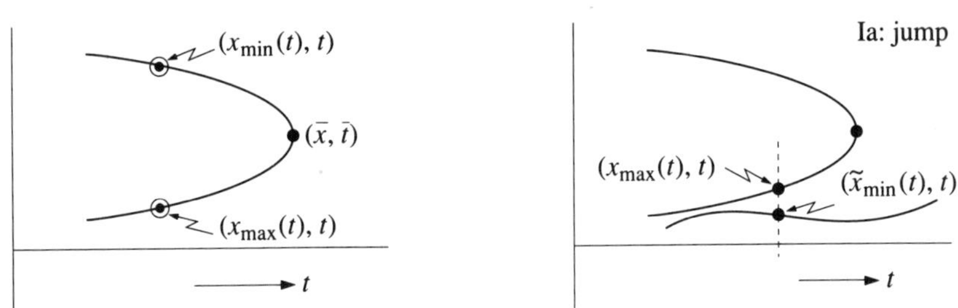

FIGURE 9

3 the vector of Lagrange multipliers corresponding to the traced local minima remains bounded, whereas in the case of Type 4 this vector becomes unbounded.

SITUATION II. (Type 4, $J_o(\bar{\mathbf{z}}) \neq 0$) Similarly as in Situation I, we have

$$\Delta := -\sum_{i_1=1}^{k} f_{i_1}^2 + \sum_{i_2=k+1}^{c} f_{i_2}^2 \neq 0. \tag{38}$$

Moreover, locally around the point $\bar{\mathbf{z}}$ all linear equalities in (35) are active. So, for our analysis we can proceed by considering only one inequality constraint:

$$\delta v \geq -\sum_{i_1=1}^{k} y_{i_1}^2 + \sum_{i_2=k+1}^{c} y_{i_2}^2, \qquad \delta \in \{-1, 1\}. \tag{39}$$

Since, in addition, we are walking on a branch of local minima for $v < 0$, $v \approx 0$, it is again not difficult to see that exactly one of the following possibilities can occur (it is not difficult to see that the number d in (35) necessarily vanishes):

(IIa) $\delta = 1, \quad k = 1, \quad \Delta < 0$,
(IIb) $\delta = -1, \quad k = 0$.

See Figure 10 for a picture analogous to Figure 7, where $|J_0(\bar{\mathbf{z}})| = 1$.

The question of whether a jump to another branch of local minima is possible (by means of a descent method) can be answered completely analogously to Situation I. In fact, assuming that $M(t)$ is compact for $t < \bar{t}$, $t \approx \bar{t}$, a jump is possible in Situation IIa, whereas it is not in Situation IIb. Also, in Situation IIb the current component of the feasible set becomes empty.

SITUATION III. (Type 5) Before we proceed we have to introduce another constraint qualification that is weaker than LICQ. Let $\bar{\mathbf{z}} = (\bar{\mathbf{x}}, \bar{t})$

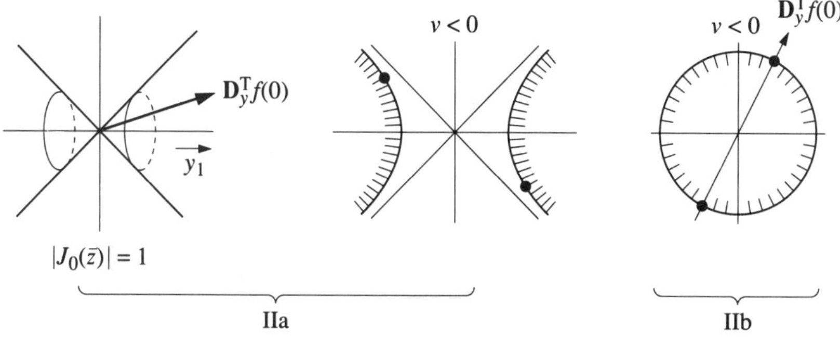

FIGURE 10

be feasible, i.e., $\bar{\mathbf{x}}$ belongs to $M(\bar{t})$. Then the Mangasarian–Fromovitz constraint qualification (MFCQ) is said to be fulfilled at $\bar{\mathbf{z}}$ if the next two conditions hold:

(MF1) The set $\{\mathbf{D_x}h_i(\bar{\mathbf{z}}),\ i \in I\}$ is linearly independent.

(MF2) There exists a vector $\xi \in \mathbf{R}^n$ satisfying

$$\begin{cases} \mathbf{D_x}h_i(\bar{\mathbf{z}})\xi = 0, & i \in I, \\ \mathbf{D_x}g_j(\bar{\mathbf{z}})\xi > 0, & j \in J_0(\bar{\mathbf{z}}). \end{cases}$$

A point $\bar{\mathbf{z}} \in \Sigma$ is called a Kuhn–Tucker point (KT-point) if there exist multipliers $\bar{\lambda}_i$, $i \in I$, and *nonnegative* $\bar{\mu}_j$, $j \in J_0(\bar{\mathbf{z}})$, such that (4) is satisfied. Note that the (Lagrange) multipliers at a KT-point need not be unique. However, it is well known (cf. **[2]**) that the set of Lagrange multipliers at a KT-point $\bar{\mathbf{z}}$ is bounded if and only if MFCQ is satisfied at $\bar{\mathbf{z}}$. Moreover, if $\bar{\mathbf{x}} \in M(\bar{t})$ is a local minimum for $\mathscr{P}(\bar{t})$ and if MFCQ is satisfied at $\bar{\mathbf{z}}$, then $\bar{\mathbf{z}}$ is a KT-point. In particular, if $\bar{\mathbf{z}} = (\bar{\mathbf{x}}, \bar{t})$ is a g. c. point of Type 1 and if the number of active constraints equals n, then $\bar{\mathbf{x}}$ is a local minimum for $\mathscr{P}(\bar{t})$ if and only if $\bar{\mathbf{z}}$ is a KT-point.

Let Σ_{KT} denote the closure of the subset of Σ consisting of all points of Type 1 that are Kuhn–Tucker points. Then, Theorem 4.1 in **[9]** tells us that Σ_{KT} is a one-dimensional (piecewise smooth) manifold with boundary; moreover, a point $\bar{\mathbf{z}} \in \Sigma_{\mathrm{KT}}$ is a boundary point if and only if at $\bar{\mathbf{z}}$ we have $J_0(\bar{\mathbf{z}}) \neq \emptyset$ and MFCQ is violated.

Now, if $\bar{\mathbf{z}} \in \Sigma$ is a point of Type 5, then exactly $n + 1$ constraints are active at $\bar{\mathbf{z}}$, and at all points $\mathbf{z} \in \Sigma$, $\mathbf{z} \neq \bar{\mathbf{z}}$, in some neighborhood $\mathscr{O}$ of $\bar{\mathbf{z}}$, the number of active constraints equals n. So, in particular, we can conclude from the foregoing that a branch of local minima stops at a point $\bar{\mathbf{z}}$ of Type 5 if and only if MFCQ is violated at $\bar{\mathbf{z}}$. But then, it is easily seen from (35) that $\delta = -1$ and $d = 0$. In addition, in the latter

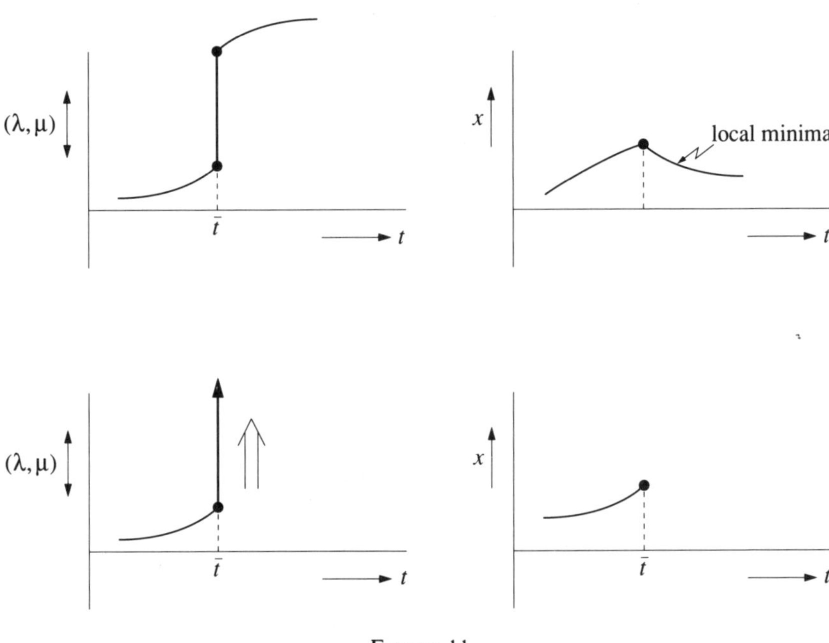

FIGURE 11

case, the feasible set becomes empty as the parameter t passes the value $\bar{t}$.

The Mangasarian–Fromovitz constraint qualification is also violated at a g.c. point of Type 4. In the latter case we are able to detect the approach of such a point along Σ since the vector of Lagrange multipliers tends to infinity. In case of a point of Type 5 this is different. Although at a point of Type 5 MFCQ might not be satisfied, the vector of Lagrange multipliers *does not tend to infinity as we approach the point*. Moreover, if we approach the point $\bar{\mathbf{z}}$ along a branch of local minima, then $\bar{\mathbf{z}}$ itself is also a KT-point, and *at the point* $\bar{\mathbf{z}}$ the set of Lagrange multipliers is either bounded (MFCQ is satisfied) or not (MFCQ is not satisfied). In order to continue our walk along a branch of local minima, we take the Lagrange multipliers (λ, μ) into account, and we follow a corresponding piecewise smooth curve in $(\mathbf{x}, \lambda, \mu, t)$-space, rather than in $(\mathbf{x}, t)$-space (cf. **[10]**). The projection of this path to (λ, μ, t)-space and $(\mathbf{x}, t)$-space, respectively, takes one of the forms as depicted in Figure 11.

5. Final remarks. In the foregoing we studied generic one-parameter families of optimization problems; it was our aim to discuss the possiblity of following a branch of local minima for *increasing values* of the parameter t. It turned out that at certain values of the parameter t the path of local minima stops, and, in order to follow local minima, we have to jump to another branch. In some cases such a jump is possible. Let us put things together, assuming that the feasible set $M(t)$ remains compact for all t (or, more weakly, assuming that the current component of $M(t)$ is compact at those points where we want to jump to another branch of local minima). Under generic conditions imposed on the underlying functions f, h_i, g_j, the set Σ of g. c. points, among them the local minima, consists of five basic types (Types 1–5).

TYPE 1. As long as the current point (local minimum) belongs to Type 1, we can follow the branch of local minima by means of any path-following method.

TYPE 2. When following a branch of local minima (in fact, Type 1), the arrival at a point $\bar{\mathbf{z}}$ of Type 2 will be announced in two different ways. First, a current Lagrange parameter $\mu_{\bar{j}}$ corresponding to an active inequality constraint $g_{\bar{j}}$ vanishes. In case of *positive definiteness* of the Hessian of the Lagrangian restricted to the (*larger*) tangent space $\widetilde{T}(\bar{\mathbf{z}})$ (cf. (30)), the constraint $g_{\bar{j}}$ should be omitted and we can continue tracing the branch of local minima passing through the point $\bar{\mathbf{z}}$; cf. Figure 4(b). In case of indefiniteness of the corresponding restricted Hessian, a jump can be performed to another branch of local minima, along the lines explained in Section 2; cf. Figure 4(c). Second, an inactive inequality constraint g_j becomes active. Then, we treat g_j as an equality constraint and we continue tracing the path of local minima through the point $\bar{\mathbf{z}}$; cf. Figure 4(a).

TYPE 3. The arrival at a point of Type 3 is announced by means of the fact that the least eigenvalue of the restricted Hessian of the Lagrangian vanishes. In fact, this can easily be recognized since the set Σ has a quadratic turning point and the vector in Lagrange multipliers remains bounded. A jump can be made along the lines explained in Section 3.

TYPE 4. If we approach a point of Type 4, then the vector of Lagrange multipliers tends to infinity. In the case where the objective function f decreases (as t increases), a jump is possible. However, if f increases, then the current component of the feasible set becomes empty and a jump (in a simple way) cannot be proposed.

TYPE 5. The arrival along Σ at a point of Type 5 is announced by the fact that the number of active constraints becomes $n+1$, where n is the dimension of $\mathbf{x}$-space. We can proceed following a branch of local minima passing through $\bar{\mathbf{z}}$ (in $(\mathbf{x}, \lambda, \mu, t)$-space) if the Mangasarian–Fromovitz constraint qualification is satisfied. However, if MFCQ is violated, then the current component of the feasible set becomes empty and a simple way for a jump cannot be proposed.

In summary, we see that we can continue following a branch of local minima (with or without jumps) as long as the current component of the feasible set remains nonempty. At first glance, one might think that some local perturbation of the functions f, h_i, g_j would possibly overcome the difficulties in case that the current component of the feasible set becomes empty. However, this is not true, since it can be seen that the *latter difficult phenomenon is stable* (i.e., persistent) under local C^2-perturbations of all functions involved. In case that the current component becomes empty, one might try to trace a branch of g. c. points (not local minima) backward and to go forward again at some point. In order that such a procedure be successful, it is obvious that a certain *global connectedness* of branches (and between branches via jumps) should be present. Here, we will not pursue this idea.

The remarks made in this section remain true if the functions f, h_i, g_j are only of differentiability class C^3, instead of C^∞. In fact, for the characterization of the points of Types 1–5 we need partial derivatives only up to order 3 (third-order is actually needed at points of Type 3).

We emphasize that the foregoing analysis is intimately related with the difficulties arising when one would try to find a feasible point of a set defined by (in)equality constraints by means of a path-following technique (and hence, with the analogous problem concerning global optimization). However, we will not further explore that point of view in this note.

ACKNOWLEDGMENT. I would like to thank Jan Rückmann from the Technical University at Leipzig for pointing out to me the eventual necessity of a jump at a point of Type 2. Moreover, I am indebted to an anonymous referee for the very precise report on the first version of the paper.

REFERENCES

1. Th. Bröcker, *Differentiable germs and catastrophes*, translated by L. Lander, London Math. Soc. Lecture Notes **17** (1975), Cambridge University Press.

2. J. Gauvin, *A necessary and sufficient regularity condition to have bounded multipliers in nonconvex programming*, Math. Programming **12** (1977), 136–138.

3. J. Guddat, H. Th. Jongen, and D. Nowack, *Parametric optimization: path-following with jumps*, Memorandum Nr. 607, University of Twente (1987), To appear in the proceedings of the Approximation and Optimization Conference at La Habana, Cuba, 1987.

4. M. W. Hirsch, *Differential topology*, Springer-Verlag, New York, 1976.

5. H. Th. Jongen, P. Jonker, and F. Twilt, *On one-parameter families of sets defined by (in)equality constraints*, Nieuw Archief v. Wiskunde (3), **XXX** (1982), pp. 307–322.

6. ——, *Nonlinear optimization in* $\mathbf{R}^n$, I. Morse Theory, Chebyshev Approximation, Peter Lang Verlag, Frankfurt a.M. Bern, New York, 1983.

7. ——, *Nonlinear optimization in* $\mathbf{R}^n$, II. Transversality, Flows, Parametric Aspects, Peter Lang Verlag, Frankfurt a.M. Bern, New York, 1986.

8. ——, *One-parameter families of optimization problems: Equality constraints*, J. Optim. Theory and Appl. **48** (1986), 141–161.

9. ——, *Critical sets in parametric optimization*, Math. Programming **34** (1986), 333–353.

10. M. Kojima and R. Hirabayashi, *Continuous deformations of nonlinear programs*, Math. Programming Stud. **21** (1984), 150–198.

11. A. B. Poore and C. A. Tiahrt, *Bifurcation problems in nonlinear parametric programming*, Math. Programming **39** (1987), 189–205.

AACHEN UNIVERSITY OF TECHNOLOGY AND UNIVERSITY OF HAMBURG, FEDERAL REPUBLIC OF GERMANY

Lectures in Applied Mathematics
Volume **26** (1990)

Interval Arithmetic Techniques in the Computational Solution of Nonlinear Systems of Equations: Introduction, Examples, and Comparisons

R. BAKER KEARFOTT

Abstract. Methods of interval arithmetic can be used to reliably find *all* solutions to nonlinear systems of equations and to reliably solve *global* optimization problems. Such methods can also be used in studies of the sensitivity of systems to certain parameters or in computing rigorous bounds on the range of behavior of a system as certain parameters vary. More generally, properly applied interval methods can give results that have *mathematical certainty*, since the effects of roundoff error are fully taken into account.

Here, for the nonexpert, we cite references and briefly review elementary aspects of interval arithmetic. We then describe a class of algorithms for finding all roots of nonlinear systems of equations within a box in n-space. Third, we discuss inherent differences between interval methods and alternate techniques for nonlinear systems. Finally, to illustrate the applicability of interval techniques, we show how interval arithmetic can be used to make the choice of predictor stepsize in continuation methods foolproof.

1. Introduction. Interval mathematics and computer algorithms involving interval arithmetic have been much studied. In [**12**] and [**13**], approximately 2,000 books, journal and conference proceedings articles, and technical reports on interval mathematics are listed. This interest is perhaps due to the fact that properly designed and implemented interval methods are *totally reliable* in the sense that they give results with

---

1980 *Mathematics Subject Classification* (1985 *Revision*). Primary 65H10; Secondary 90C30, 65G10.

0082-0717/90 $1.00 + $.25 per page

mathematical certainty. Here, we will use this property to solve the problem:

> Find, *with certainty, approximations to all solutions of the nonlinear system*
>
> $$f_i(x_1, x_2, \ldots, x_n) = 0, \qquad 1 \le i \le n, \tag{1.1}$$
>
> where bounds l_i and u_i are known such that
>
> $$l_i \le x_i \le u_i \qquad \text{for } 1 \le i \le n.$$

(We will denote the n-vector whose ith component is x_i by X and the n-vector whose ith component is f_i by $F(X)$.)

An interval algorithm will produce a list of solutions whose coordinates x_i are given as small intervals of uncertainty. If the proper algorithm (cf. Section 3) is correctly implemented with directed roundings (cf. Section 2), completion of this algorithm constitutes a computational but *mathematically rigorous proof* that all solutions of (1.1) are within the intervals given in the list.

Tasks other than (1.1) can be made totally reliable with interval arithmetic techniques. However, algorithms cannot necessarily be made both reliable and practical merely by replacing the usual arithmetic operations by intervals and interval operations. (That is, new algorithms need to be developed for interval arithmetic.)

In Section 2, we review some elementary definitions of interval arithmetic. In Section 3, we present a class of interval arithmetic algorithms for solving nonlinear algebraic systems and for nonlinear optimization. In Section 4, we briefly discuss some issues in the design of software for solving nonlinear algebraic systems with interval arithmetic, and we cite our portable package. In Section 5, we compare interval methods for the numerical analysis of nonlinear systems to alternate techniques. In Section 6, to illustrate interval mathematics as a general tool, we explain how it can be used in stepsize control for continuation methods.

2. Elementary facts about interval arithmetic. Thorough introductions to interval mathematics are given in the books **[1]** and **[41]**. (Also, Moore, who is attributed with inventing interval arithmetic, wrote **[38]** in 1966.) In particular, one can find details for this section in Chapters 1–3 of **[41]** or in Chapters 1–4 of **[1]**. Also see **[54]** and **[59]** if they are available. A conference on use of interval methods for scientific computing in general was recently held. This conference brought together experts in both interval techniques and non-interval techniques; in the proceedings **[42]**, the participants attempt to clarify the role of interval mathematics. Over the years, experts in the area have also presented technical details at conferences such as **[15]**, [37], **[48]**, and **[49]**.

Here, we will give an elementary explanation of some of the most important concepts. Throughout, interval quantities will be denoted by boldface.

Interval arithmetic is based on defining the four elementary arithmetic operations on intervals. Let $\mathbf{a} = [a_l, a_u]$ and $\mathbf{b} = [b_l, b_u]$ be intervals. Then, if $\text{op} \in \{+, -, *, /\}$, we define

$$\mathbf{a} \,\text{op}\, \mathbf{b} = \{x \,\text{op}\, y \mid x \in \mathbf{a} \text{ and } y \in \mathbf{b}\}. \tag{2.1}$$

For example,

$$\mathbf{a} + \mathbf{b} = [a_l + b_l, a_u + b_u].$$

In fact, all four operations can be defined in terms of addition, subtraction, multiplication, and division of the endpoints of the intervals, although multiplication and division may require comparison of several results. The result of these operations is an interval except when we compute $\mathbf{a} \,/\, \mathbf{b}$ and $0 \in \mathbf{b}$. In that case, we use *extended interval arithmetic* (cf., e.g., **[41]**, pp. 66–68) to get two semi-infinite intervals or else the whole real line. For example,

$$[10, 20] \,/\, [-2, 5] = (-\infty, -5] \cup [2, \infty).$$

A large part of the power of interval mathematics lies in the ability to compute *inclusion monotonic interval extensions* of functions. If f is a continuous function of a real variable, then an inclusion monotonic interval extension $\mathbf{f}$ is defined to be a function from the set of intervals to the set of intervals, such that, if $\mathbf{x}$ is an interval in the domain of $\mathbf{f}$,

$$\{f(x) \mid x \in \mathbf{x}\} \subset \mathbf{f}(\mathbf{x})$$

and such that

$$\mathbf{x} \subset \mathbf{y} \quad \text{implies} \quad \mathbf{f}(\mathbf{x}) \subset \mathbf{f}(\mathbf{y}).$$

Inclusion monotonic interval extensions of a polynomial may be obtained by simply replacing the dependent variable by an interval and by replacing the additions and multiplications by the corresponding interval operations. For example, if $p(x) = x^2 - 4$, then $\mathbf{p}([1, 2])$ may be defined by

$$\mathbf{p}([1, 2]) = ([1, 2])^2 - 4 = ([1, 4]) - [4, 4] = [-3, 0].$$

We emphasize here that the result of an elementary interval operation is precisely the range of values that the usual result attains as we let the operands range over the two intervals. However, the value of an interval extension of a function is not precisely the range of the function over its interval operand, but only contains this range; we hope to construct interval extensions whose values differ from the range as little as possible. For example, if we write p above as $p(x) = (x-2)(x+2)$, then the corresponding interval extension gives

$$\begin{aligned}\mathbf{p}([1,2]) &= ([1,2]-2)([1,2]+2) = [-1,0][3,4] \\ &= [-4,0],\end{aligned}$$

which is not as good as the previous extension. See **[58]** for a discussion of efficient ways of formulating interval extensions.

We may use the mean value theorem or Taylor's theorem with remainder formula to obtain interval extensions of transcendental functions. For example, suppose $\mathbf{x}$ is an interval and $a \in \mathbf{x}$. Then, for any $y \in \mathbf{x}$, we have

$$\sin(y) = \sin(a) + (y-a)\cos(a) - [(y-a)^2/2]\sin(c)$$

for some c between a and y. If a and y are both within a range where the sine function is nonnegative, then we obtain

$$\sin(y) \in \sin(a) + (\mathbf{x}-a)\cos(a) - (\mathbf{x}-a)^2/2.$$

The right side of this relationship gives the value of an interval extension of $\sin(x)$, albeit a somewhat crude one. We may also use rational approximations in cases where the function does not have a Taylor series; if the function is approximated in the uniform norm, then we may bound the error by a constant interval. See **[58]** for details.

Several computer packages (as in **[5]**, **[6]**, **[35]**, and **[72]**) are available for interval extensions of the elementary functions.

Mathematically rigorous interval extensions can be computed in finite-precision arithmetic via the use of *directed roundings*. Let x and y be machine-representable numbers, and assume op is one of the four elementary operations $+$, $-$, $*$, or $/$. Normally, $x \operatorname{op} y$ is not representable in the machine's memory, and there are various schemes of rounding. For example, we may always *round down* to the nearest machine number less than $x \operatorname{op} y$, or we may always *round up* to the nearest machine number greater than $x \operatorname{op} y$. In interval arithmetic with directed rounding, if

$$\mathbf{x} \operatorname{op} \mathbf{y} = [c, d],$$

then we always round the value for c down, and we always round the value for d up. In such computations, we first apply directed rounding to the initial data in order to store it.

Machine interval arithmetic with directed rounding does not involve deep concepts, but it can be quite powerful. For example, if interval arithmetic with directed rounding is used to compute an interval extension **f** of f,

$$[c,d] = \mathbf{f}([a,b]),$$

and $[c,d]$ does not contain zero, then this is a rigorous proof (regardless of the machine wordlength, etc.) that there is no root of f in $[a,b]$.

In [**34**], Kulisch and Miranker carefully present concepts of directed rounding and machine interval arithmetic. They include discussion related to producing interval extensions of functions which yield intervals that are as close as possible to the range of the function.

Various precompilers and compilers exist which support the interval datatype. See [**27**], etc., for details.

3. Interval methods in solving nonlinear systems of equations and in nonlinear optimization. Here, we discuss the problem (1.1) of finding *all* roots of a nonlinear system of equations subject to upper and lower bounds on the variables. We also discuss the related problem:

> Find, *with certainty, the* **global** *minimum* of the nonlinear objective function
>
> $$\Phi(x_1, x_2, \ldots, x_n) \tag{3.1}$$
>
> where bounds l_i and u_i are known such that
>
> $$l_i \le x_i \le u_i \qquad \text{for } 1 \le i \le n.$$

The problem (1.1) may be solved via generalized bisection in conjunction with interval Newton methods. This general technique is described in Chapters 19 and 20 of [**1**] and in Chapters 5 and 6 of [**41**]. An early paper on the technique is [**14**]. Other papers include [**16**], [**25**], [**26**], [**39**], [**40**], [**45**], [**47**], [**50**], and [**62**].

In what follows we denote the *box* in n-space described by

$$\{X = (x_1, x_2, \ldots, x_n) \mid l_i \le x_i \le u_i \text{ for } 1 \le i \le n\}$$

by **B**. Similarly, we denote vectors whose entries are intervals by capital boldface letters.

In interval Newton methods, we find a box $\overline{\mathbf{X}}_k$ that contains all solutions of the interval linear system

$$\mathbf{F}'(\mathbf{X}_k)(\overline{\mathbf{X}}_k - X_k) = -F(X_k), \tag{3.2}$$

where $\mathbf{F}'(\mathbf{X}_k)$ is a suitable interval extension of the Jacobian matrix of F over the box $\mathbf{X}_k$ (with $\mathbf{X}_0 = \mathbf{B}$), and where X_k is some point in $\mathbf{X}_k$.

(An elementwise interval extension of the Jacobian matrix is suitable.) We then define the next iterate $\mathbf{X}_{k+1}$ by

$$\mathbf{X}_{k+1} = \mathbf{X}_k \cap \overline{\mathbf{X}}_k. \tag{3.3}$$

The scheme based on solving (3.2) and performing (3.3) is termed an *interval Newton method*. The *convergence rate* of an interval Newton method is defined in terms of the ratios of the widths of the component intervals of $\mathbf{X}_{k+1}$ to the corresponding widths of $\mathbf{X}_k$.

If each row of $\mathbf{F}'$ contains all possible vector values that that row of the scalar Jacobian matrix takes on as X ranges over all vectors in $\mathbf{X}_k$, then it follows from the mean value theorem that all solutions of (1.1) in $\mathbf{X}_k$ must be in $\mathbf{X}_{k+1}$. If the coordinate intervals of $\mathbf{X}_{k+1}$ are smaller than those of $\mathbf{X}_k$, then we may iterate (3.2) and (3.3) until we obtain an interval vector the widths of whose components are smaller than a specified tolerance.

If the coordinate intervals of $\mathbf{X}_{k+1}$ are not smaller than those of $\mathbf{X}_k$, then we may *bisect* one of these intervals to form two new boxes; we then continue the iteration with one of these boxes, and push the other one onto a *stack* for later consideration. After completion of the current box, we pop a box from the stack, and apply (3.2) and (3.3) to it; we thus continue until the stack is exhausted. As is explained in **[40]**, **[25]**, and elsewhere, such a composite *generalized bisection algorithm* will reliably compute all solutions to (1.1) to within a specified tolerance.

The efficiency of the generalized bisection algorithm depends on

(1) the sharpness of the interval extension to the rows of the Jacobian matrix;
(2) the way we find the solution $\overline{\mathbf{X}}_k$ to (3.2); and
(3) how we select the coordinate directions in which to bisect.

In particular, iteration with formulas (3.2) and (3.3) should exhibit the quadratic local convergence properties of Newton's method, but repeated bisections are to be avoided if possible. We are thus interested in arranging the computations so that $\overline{\mathbf{X}}_k$ has coordinate intervals that are as narrow as possible.

An early method of solving (3.2) was the Krawczyk method, in which $\overline{\mathbf{X}}_k$ is given by

$$\overline{\mathbf{X}}_k = \mathbf{K}(\mathbf{X}_k) = X_k - Y_k F(X_k) + (I - Y_k \mathbf{F}'(\mathbf{X}_k))(\mathbf{X}_k - X_k), \tag{3.4}$$

where Y_k is a *preconditioner matrix* that is sometimes taken to be an approximation to $[F'(X_k)]^{-1}$. Moore observed in **[39]** that the Krawczyk method converged (without bisection)

$$\text{provided } \|I - Y_k \mathbf{F}'(\mathbf{X}_k)\| < 1, \tag{3.5}$$

where the norm is the usual one for interval matrices. (See **[1]** or **[41]**.) Condition (3.5) also serves as part of a computational existence and uniqueness test. Various researchers have subsequently weakened this condition. In fact, for many methods of solving (3.2), researchers have shown that

(3.6) if $\overline{\mathbf{X}}_k$ is *strictly* contained in $\mathbf{X}_k$, then the system of equations in (1.1) has a unique solution in $\mathbf{X}_k$, and Newton's method starting from any point in $\mathbf{X}_k$ will converge to that solution. Conversely, if $\mathbf{X}_k \cap \overline{\mathbf{X}}_k$ is empty, then there are no solutions of the system in (1.1) in $\mathbf{X}_k$.

(See **[20]**, **[45]**, and **[52]**.)

An interval version of the Gauss–Seidel method, with extended interval arithmetic, can also be used to bound the solution set to (3.2) (see, for example, **[19]**). In such a method, we often multiply both sides of (3.2) by the matrix Y_k before dividing by the diagonal element; this ensures local convergence, and results in a method that is often superior to the Krawczyk method. We will denote the ith component of X_k by x_i, the ith component of $Y_k F(X_k)$ by $\mathbf{k}_i$, and the entry in the ith row and jth column of $Y_k \mathbf{F}'(\mathbf{X}_k)$ by $\mathbf{G}_{i,j}$. The step for the ith row of the interval Gauss–Seidel method then becomes

$$\begin{aligned} \overline{\mathbf{x}}_i &= \mathbf{x}_i - \left[\mathbf{k}_i - \sum_{\substack{j=1\\ j\neq i}}^{n} \mathbf{G}_{i,j}(\mathbf{x}_j - x_j)\right] \Big/ \mathbf{G}_{i,i}; \\ \mathbf{x}_i^+ &\leftarrow \mathbf{x}_i \cap \overline{\mathbf{x}}_i. \end{aligned} \tag{3.7}$$

(It is understood that, if $i > 1$, we replace $\mathbf{x}_j$ by $\mathbf{x}_j^+$ for $j < i$ when we define $\overline{\mathbf{x}}_i$.)

A detailed example of an iteration of the interval Gauss–Seidel method is available from the author. In that example, because of the strict containments $\overline{\mathbf{x}}_1 \subset \mathbf{x}_1$ and $\overline{\mathbf{x}}_2 \subset \mathbf{x}_2$, we may use results in **[45]** to conclude that there is a unique root of F in the box $\mathbf{X}$, and that the usual Newton's method will converge to that root, starting from any point in $\mathbf{X}$.

For large, banded, or sparse systems, multiplication by an inverse is impractical. In **[61]** and **[63]**, Schwandt uses the interval Gauss–Seidel method without a preconditioner to solve systems like finite-difference discretizations of Poisson's equation with a nonlinear forcing term. In such cases, an interval generalization of diagonal dominance ensures convergence of repeated application of (3.7) when Y_k is the identity. An excellent exposition of this technique appears in [**67**].

Techniques for computing the rows of Y_k explicitly to minimize the widths of the intervals $\mathbf{x}_i^+$ (and thus maximize convergence rate) appear in [**31**]. These techniques involve solving linear programming problems for the elements of Y_k; these linear programming problems express optimality conditions for the width of $\overline{\mathbf{x}}_i$. Performance results for an interval Newton method using these techniques and for an interval Newton method with $Y_k = [F'(X_k)]^{-1}$ appear in [**31**]. Though the linear programming step is a major contributor to computation time, the technique is applicable to ill-conditioned and singular systems, and the linear programming problem can be altered to take account of structure or sparsity. Also, further study is yielding ways of making solution of the special linear programming problems arising here more efficient.

An alternate technique for applying interval Newton methods when the Jacobian is ill-conditioned or singular near the roots appears in [**30**].

The global nonlinear optimization problem (3.1) can be solved by solving (1.1), where the f_i are the components of $\nabla\Phi$. However, we may use the objective function directly to increase the algorithm's efficiency. If $\mathbf{p}$ and $\mathbf{q}$ are intervals, we say that $\mathbf{p} > \mathbf{q}$ if every element of $\mathbf{p}$ is greater than every element of $\mathbf{q}$. Suppose that $\mathbf{X}$ and $\mathbf{Y}$ are interval vectors in the stack described below (3.2), and let $\boldsymbol{\Phi}$ be an interval extension to Φ. Then, if $\boldsymbol{\Phi}(\mathbf{Y}) > \boldsymbol{\Phi}(\mathbf{X})$, we may discard $\mathbf{Y}$ from the stack.

Papers and reviews on solution of the global optimization problem include [**4**], [**17**], [**21**], [**23**], [**50**], [**56**], and [**60**]. Walster, Hansen, and Sengupta report performance results on their global optimization algorithm in [**69**]. Researchers not expert in interval mathematics also occasionally rediscover the exclusion principle just described without explicitly using the machinery of interval arithmetic.

4. Issues in implementing reliable interval Newton nonlinear equation solvers. As mentioned in Section 2, there are various language tools for working with interval arithmetic, and computer hardware supports interval arithmetic to varying degrees. Likewise, there are libraries of elementary and special functions for interval arithmetic. Also, it is possible through automatic differentiation (as in [**53**], [**55**], or [**66**]) or through symbolic manipulation (as in, for example, [**65**]) to shift the burden of programming the Jacobian matrix from the user to either the compiler or to the executable code. However, these arithmetic and differentiation tools are not yet universally available or standardized. (Future versions of the FORTRAN standard, if implemented, will help.)

Furthermore, there is also not yet a consensus about which interval Newton method is "best," nor is it clear how best to precondition the Jacobian matrix in all cases. Also, we are still doing practical investigations related to the choice of coordinate interval to bisect.

Another issue deals with how to handle the case where roots occur on or near boundaries of boxes. This problem could degrade the efficiency of the algorithm in higher dimensions when the root occurs near a low-dimensional boundary or vertex of the box. We are presently investigating refining the "expansion" technique described in **[25]** not only to eliminate redundancies, but also to significantly increase the algorithm's efficiency.

We are also pursuing additional work to handle systems where the Jacobian matrix is singular or ill-conditioned near the root. In such cases, (3.6) is usually not satisfied, and the predominant mode of the algorithm can be the costly coordinate bisection process. We are looking at the preconditioner technique mentioned above and in **[31]**, among other ideas.

Finally, as with many numerical tasks, there is an unclear interplay between the order of approximation used and the overall efficiency of the algorithm. For example, we could approximate elements of the Jacobian matrix to high order with centered forms (as in **[60]**). For many functions, this would allow completion of the algorithm with fewer function and Jacobian evaluations; however, we would encounter the additional complication of setting up second-derivative tensors and the additional cost of evaluating them.

Despite these problems and ambiguities, we have felt it important to make interval nonlinear equation software generally available. In **[29]**, we describe a portable, self-contained FORTRAN 77 package for interval Newton/bisection as in Section 3; we present performance results for an early version of this code in **[26]**, and the abstract structure of the algorithm in **[25]**. We can supply the FORTRAN source code, and also machine executable code for IBM PC compatibles, on MS-DOS $5\frac{1}{4}$ in. diskettes.

In this code, we have sacrificed some speed and generality for portability and ease of use. Our code is perhaps a factor of 20 slower than if interval arithmetic were available directly in the complier. However, it is still competitive with alternate techniques for many problems; see **[26]**.

In our code, we have used the interval Gauss–Seidel method, and we have preconditioned with the inverse of the matrix of midpoints of the elements of the interval Jacobian matrix. The interval Gauss–Seidel method is competitive in many instances (cf. **[18]** and **[19]**), while our choice of preconditioner seems popular and is also relatively good in practice. We have used a special scaling technique to choose coordinate intervals to be bisected; see **[29]**.

Our code simulates directed roundings, given a bound on the maximum number of units in the last place by which the result of an elementary machine arithmetic operation can be in error. It thus gives mathematically rigorous results.

5. Interval Newton/bisection methods and alternate nonlinear equation solvers. As the topics in this conference indicate, not only are there various techniques for solving nonlinear algebraic systems of equations, but the different techniques tackle different problems and have different goals. The class of interval methods outlined in Section 3 above is well suited to

(i) reliably finding *all* solutions within a given region of space (i.e., solving (1.1));
(ii) reliable *global* optimization;
(iii) analysis of the sensitivity of solutions to certain coefficients or parameters; and
(iv) the general study of parameter-dependent systems.

(The problem (1.1) is related to the global optimization problem since the latter can be solved by finding all critical points. However, global optimization is numerically easier since values of the objective function may also be used to eliminate some critical points early in the computation.)

Among alternate popular methods for finding all solutions are

(a) hybrid techniques (perhaps introduced with **[51]** and reviewed in **[10]**, **[11]**, or possibly notes to this conference), which include trust region algorithms with Newton's method or quasi-Newton methods;
(b) other bisection techniques, such as those involving computation of the topological degree; and
(c) continuation methods based on mathematical homotopies.

Techniques for solving the global optimization problem include, in addition to the above,

(d) schemes with a stochastic component, such as random function sampling, the tunneling method ([**36**]), and simulated annealing.

Below, we compare each of these four algorithmic classes with interval Newton/bisection.

5.1. *Comparison with hybrid methods.* Hybrid steepest descent/Newton-like algorithms are very efficient and reasonably (though not *rigorously*) reliable when we have sufficient knowledge of the problem. There is a large body of theory, and the methods have been used successfully in a wide range of applications; this includes very large problems and problems with a special structure. However, such methods are designed to find *just one* solution, or just a *local* optimum; the globalization techniques associated with these algorithms refer merely to an expansion of the domain of convergence of the underlying Newton-like method. Sensitivity information can be obtained indirectly from the algorithms' behavior, or local information can be obtained by decomposition of the Jacobian matrix at the solution.

5.2. *Comparison with topological degree-based bisection.* Topological degree-based bisection methods do not require interval arithmetic (and indeed do not need accurate function values), and do not require Jacobian evaluations. They are thus applicable to general nonsmooth functions (although interval methods are appropriate in some such cases). In [**68**] and elsewhere, Vrahatis has demonstrated that such methods can also be reasonably efficient.

For certain cases (such as for analytic functions of n complex variables), the underlying mathematics indicates that the algorithms will rigorously find all roots. However, computation of the topological degree is sometimes done heuristically, at the expense of reliability. The topological degree can be computed reliably when Lipschitz constants or moduli of continuity are used to bound the range of function components. (See other notes to this conference.) However, such range bounds can be viewed as values of an interval extension to the function, and thus are an example of application of interval techniques. Further investigation of the interplay between interval methods and the topological degree may yield discoveries of value.

In topological degree-based bisection methods, our goal should be perhaps merely to compute starting points for fast local methods. In interval Newton/bisection methods, once the box is small enough, the interval Newton method will exhibit quadratic convergence, and several mechanisms (including the intersection in (3.3)) can decrease the size of the box before then. In contrast, pure bisection without use of some kind of higher-order information is apt to be only linearly convergent. Also, interval Newton methods have been applied to relatively large systems; in contrast, most topological degree-based algorithms exhibit either computational or storage requirements that are exponential in the number of unknowns. (We do not claim this is an inherent property, though.)

5.3. *Comparison with continuation methods.* Continuation methods can do a good job of solving (1.1) in some cases. In particular, the theory and practice of finding all solutions to moderately sized polynomial systems is well developed. See **[43]** for an introduction, and see other lecture notes to this conference or **[44]**, etc., for examples of recent work.

One cannot say without additional information about the problems that continuation methods or interval Newton methods will be better suited to solve (1.1). There is abundant convergence theory for homotopy techniques only for polynomial systems; for other systems, homotopies may need to be constructed in an ad hoc fashion. (Interval Newton/bisection algorithms can be applied to any functions that have interval extensions.) In practice, the homotopy/continuation algorithms must find all complex roots to find all roots within a certain region; the number of such roots is equal to the total degree of the system, which is the product of the degrees of the individual components, but is somewhat less when the system has an m-homogeneous structure. (See other notes to this conference or **[44]**.) Thus, the algorithms may inefficiently compute large numbers of complex roots that are not of interest. Furthermore, unless most components are linear, or the system has a special structure, the total degree and amount of work will increase exponentially with the number of variables and equations.

On the other hand, the width of an interval image of a component $\mathbf{f}_i(\mathbf{X})$ will depend both on the number of arithmetic operations and the way that these operations are arranged. This could make interval Newton/bisection algorithms impractical for certain functions that have a large number of naively arranged operations. Higher-degree polynomials generally have wider interval values than lower-degree polynomials. However, this phenomenon is different from that of total degree, and the class of problems for which it occurs is neither contained in nor contains those for which the total degree is high.

Reliability is a second question. The underlying theory guarantees that the homotopies will lead to all roots to polynomial systems of equations, except if we are unlucky enough to choose certain parameters from "bad" sets of measure zero or of finite cardinality. The continuation methods themselves, however, can involve heuristics or can suffer from the effects of rounding errors. For example, if the predictor stepsize, which is usually heuristically chosen in predictor/corrector methods (see **[2]** for an introduction), is too large, then the steps may jump from one homotopy path to another. This results in roots listed more times than their multiplicity, and in some roots being totally missed. We have observed this behavior with some software (**[24]**). However, our experience with the code in **[43]** has convinced us that properly tuned and implemented algorithms can be very reliable in practice. Furthermore, we have a priori knowledge of the total number of solutions to a polynomial system, and we may check this against the number actually found and against rank deficiencies. Nonetheless, such reliability is qualitatively different from the *mathematical rigor* that can be achieved with interval Newton/bisection methods.

In Section 6, we describe a new use of interval arithmetic to control the stepsize in continuation methods. With that scheme, we believe we would have a rigorous guarantee that the predictor/corrector iteration would not jump from one homotopy path to another.

Both continuation methods and interval arithmetic techniques are useful in the study of sensitivities. In interval Newton methods, the studied coefficient or parameter would be separate from the unknown variables; it would appear in the formulas for evaluation of the $\mathbf{f}_i$ as a constant *interval*. Iteration of (3.3) would then result in convergence not to a point solution to the nonlinear system, but convergence to a box in n-space. That box would contain the set of all solutions to the system as the parameter ranged over its interval; the sharpness of the containment would depend on details of the interval extension.

With continuation methods, we can treat the studied parameter as a homotopy variable. We can then observe solutions to the system as we trace the homotopy paths. Using this technique, we successfully studied the amount of information that can be obtained with certain models of evoked cerebral potentials; see **[28]**.

Both continuation methods and interval mathematics can be used in the study of physical systems in which a parameter naturally varies. For example, Keller and others (e.g., **[32]**) study fluid flows as Reynolds number (or velocity or viscosity) varies. In this application of continuation methods, discretizations of partial differential equations result in very large systems. Researchers can take advantage of structure and sparsity since the predominant computation in such instances is the analysis of linear systems of equations in standard floating-point arithmetic; techniques for bordered systems, as in **[7]** or **[33]**, have also been inspired by this application.

Homotopy paths can also be resolved via interval Newton techniques; see **[46]**. However, large parameter-dependent systems have not yet been tackled.

5.4. *Comparison with methods with a stochastic component.* In contrast with hybrid methods, probabilistic methods make an effort to find a global optimum instead of just a single local one. However, none of these techniques can give mathematical assurance of their results, and random function evaluation can converge slowly even in a probabilistic sense. It may, however, be possible to use them on some very large or complicated optimization problems for which interval Newton methods embodying the present state-of-the-art would take too many operations to be practical.

The tunneling method is interesting since it combines a clever deterministic component with a stochastic component; see **[36]**. This method is able to "tunnel" under very large numbers of local minima on its way to the global minimum. However, interval Newton based methods also perform fairly well on some of these problems; see **[69]**.

6. An interval arithmetic stepsize control for continuation methods. We describe here a criterion for a foolproof stepsize control for predictor/corrector continuation methods. We first give an abstract algorithm (which would allow various predictor and corrector schemes) within which we can embed the stepsize control.

Let $H\colon \mathbf{R}^{n+1} \to \mathbf{R}^n$ denote the homotopy, and let $Y_k \in \mathbf{R}^{n+1}$ be such that $H(Y_k) = 0$. We then compute an approximation Y_{k+1}^p to another point on the zero manifold of H containing Y_k by

$$Y_{k+1}^p = Y_k + \delta_k B_k. \tag{6.1}$$

The direction vector B_k ($\|B_k\|_2 = 1$) may be obtained in a number of ways, while we will adaptively adjust and reset δ_k with our stepsize control. We then correct Y_{k+1}^p to obtain a point Y_{k+1} that is more nearly on the manifold. (See **[2]** for an introduction.) In other words, we have

ALGORITHM 6.1. (A SINGLE PREDICTOR/CORRECTOR STEP).

(1) *Compute a predictor step direction* B_k *and an initial predictor steplength* δ_k.
(2) *Compute* Y_{k+1}^p *by* (6.1).
(3) *Choose a corrector manifold* $r_k(Y)$, *where* $r_k : \mathbf{R}^{n+1} \to \mathbf{R}^n$ *and* $r_k(Y_{k+1}^p) = 0$.
(4) *See if corrector iteration* (*step 5*) *will converge starting with the present* Y_{k+1}^p.
    (a) *If it will not, then decrease* δ_k *and repeat steps 2, 3, and this step.*
    (b) *If it will, then perform step 5.*
(5) *Find* Y_{k+1} *by using a convergent method to solve the* $(n+1)$ *by* $(n+1)$ *system of equations*

$$\tilde{F}(Y) = \begin{bmatrix} r_k(Y) \\ H(Y) \end{bmatrix} = 0. \tag{6.2}$$

Generally, the corrector step is checked (step 4) using heuristic criteria; see **[3]**, **[22]**, **[64]**, etc. In such instances, if δ_k is chosen too large, the corrector iteration could converge to a point that is on a different manifold from that of Y_k; see Figure 1.

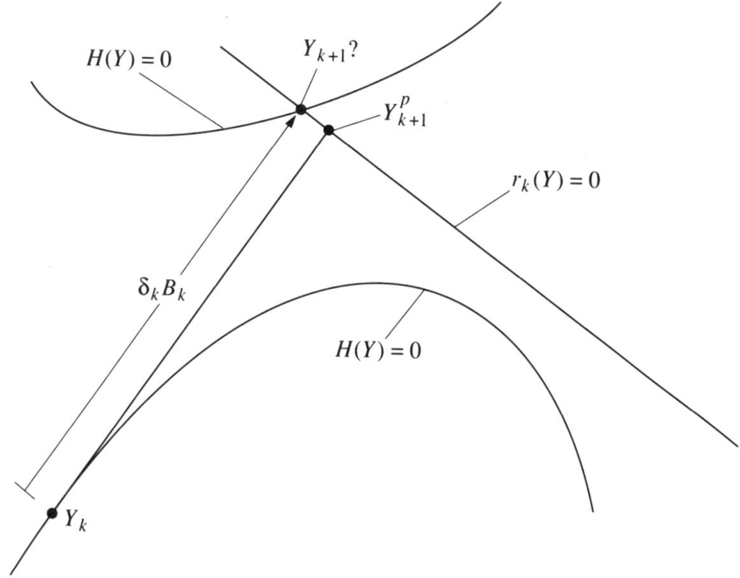

FIGURE 1. Even though the corrector iteration converges, iterates of the predictor/corrector algorithm may jump from one path to another.

In contrast, we may employ an interval Newton method in step (4), and use (3.5) or (3.6). If we include values of the Jacobian matrix over the entire line segment between Y_k and Y_{k+1}^p, then it is possible to construct an algorithm for which the phenomenon in Figure 1 cannot occur. In fact, with such interval arithmetic, the negative result in **[64]** is in many cases not relevant. (We do note, however, that we make some implicit assumptions when we use interval extensions.)

To construct the stepsize test, we first select a function $q(\delta)$ such that, for all sufficiently small δ with $\|Y - Y_k\|_2 = \delta$ and Y on the manifold of $H(Y) = 0$ that contains Y_k, the tangent of the angle that the line segment with Y_k and Y as endpoints makes with B_k is at most $q(\delta)$. We also require that $q(\delta) \to 0$ as $\delta \to 0$. For example, if there is a fixed $c > 0$ such that

$$\frac{B_k \circ (Y - Y_k)}{\|Y - Y_k\|_2} \geq c,$$

then $q(\delta) = \delta^{-\varepsilon}$ for any $\varepsilon > 0$ will do; see Figure 2.

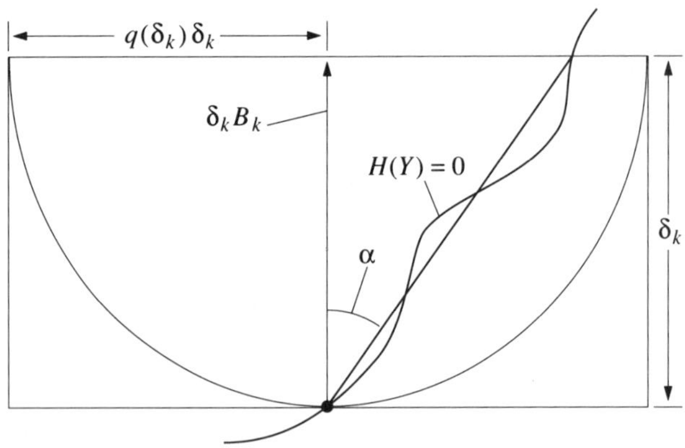

FIGURE 2. $q(\delta_k)$ is defined so that $\tan(\alpha) < q(\delta_k)$ for all sufficiently small δ_k.

For simplicity, we assume that $r_k(Y) = 0$ defines the linear manifold through Y_{k+1}^p perpendicular to B_k. Then, in the interval Newton method, we will extend the Jacobian matrix to contain values of the scalar Jacobian matrix over a box that contains the line segment between Y and Y_{k+1}^p and the portion of $r_k(Y) = 0$ within distance δ of Y_{k+1}^p; again refer to Figure 2. We may then take the point X_0 in (3.2) or in (3.7) to be Y_{k+1}^p. Our interval Newton method will operate on the manifold defined by r_k, so that it will solve a square n-by-n system. However, in order to ensure that path-jumping as in Figure 1 will not occur, the interval iterations will, in a sense, include all interval iterations corresponding to steplengths between 0 and δ_k.

To illustrate, let us examine

$$H(Y) = H(y_1, y_2) = y_2(y_1^2 - 4) + (1 - y_2)(y_1^2 - 1).$$

Let us take $Y_0 = (1,0)^{\mathrm{T}}$, and try $B_0 = (0,1)^{\mathrm{T}}$ and $\delta_0 = 0.1$. We will use $r_k(Y) = B_k \circ (Y - Y_{k+1}^p)$. We first characterize the manifold $r_k(Y) = 0$ for $\delta = \delta_0$ by

$$\{(x_1, \delta) | x_1 \in \mathbf{R}\}.$$

We may then solve $\widetilde{F}(Y) = 0$ by solving $F(x_1) = 0$, where

$$F(x_1) = H(x_1, \delta) = \delta(x_1^2 - 4) + (1 - \delta)(x_1^2 - 1).$$

Simplifying, we obtain

$$F(x_1) = x_1^2 - (1 + 3\delta), \quad \text{so} \quad F'(x_1) = 2x_1.$$

We make an interval extension to F by letting δ range in the interval $[0, \delta_0] = [0, .1]$; we use the function q then to define the interval for x_1 to be

$$\mathbf{X} = [.68, 1.32];$$

we may take $X = 1$ and $\delta_0 = 0.1$ to compute Y. (Throughout this example, we use directed roundings to represent quantities to two digits of accuracy.) We use (3.7) and (3.6). We then have

$$\mathbf{F}(X) = [0, .3],$$

where we get an interval value because we use an interval for δ_0. We also have

$$\mathbf{F}'(\mathbf{X}) = [1.36, 2.64], \quad Y = \tfrac{1}{2}, \quad \text{and} \quad \mathbf{G} = [.68, 1.32].$$

Finally,

$$\mathbf{k} = [0, .15] \quad \text{and} \quad \overline{\mathbf{X}} = [1, 1.23] \subset \mathbf{X}.$$

Since the inclusion is strict and since we have used intervals for δ_0 in the values of $\mathbf{F}$ and $\mathbf{F}'$, we believe we can conclude that there is a unique solution to (6.2) for each value of δ between 0 and δ_0, and that Newton's method starting at X will converge to that solution; the analysis would be similar to that in **[30]**.

Since we may control the stepsize, generalized bisection can be entirely avoided. Thus, the method should be applicable to problems in high-dimensional spaces.

A complete analysis, generalizations, and more involved numerical examples of this technique will appear elsewhere.

ACKNOWLEDGMENTS. The idea for the continuation method stepsize control came to Alexander Morgan and me over dinner. I am also indebted to Alec for putting the code in [**43**] into perspective.

REFERENCES

1. G. Alefeld and J. Herzberger, *Introduction to Interval Computations*, Academic Press, New York, 1983.

2. E. L. Allgower and K. Georg, *Simplicial and continuation methods for approximating fixed points and solutions to systems of equations*, SIAM. Rev. **22** (1) (1980), 28–55.

3. J. Avila, *The feasibility of continuation methods for nonlinear equations*, SIAM J. Numer. Anal. **11** (1974), 102–120.

4. E. Baumann, *Globale optimierung stetig differenzierbarer Functionen einer Variablen*, preprint, Freiburger Intervall-Berichte 86/6, Institut für Angewandte Mathematik der Universität Freiburg, 1986, pp. 1–89.

5. J. H. Bleher, S. M. Rump, U. Kulisch, M. Metzger, and W. Walter, *Fortran-SC—A study of a Fortran extension for engineering scientific computations with access to ACRITH*, Computing **39** (2) (1987), 93–110.

6. A. Bundy, *A generalized interval package and its use for semantic checking*, ACM Trans. Math. Software **10** (1984), 397–409.

7. T. F. Chan and Y. Saad, *Iterative methods for bordered systems with applications to continuation methods*, SIAM J. Sci. Statist. Comput. **6** (2) (1985), 438–451.

8. M. Clemmesen, *Interval arithmetic implementations using floating point arithmetic*, SIGNUM News, **19**(4), 1984, pp. 2–8.

9. F. Crary, *The AUGMENT precompiler*, Mathematics Research Center report no. 1470, the University of Wisconsin at Madison, 1976.

10. J. E. Dennis and J. J. Moré, *Quasi-Newton methods, motivation and theory*, SIAM Rev. **19** (1) (1977), 46–89.

11. J. E. Dennis and R. B. Schnabel, *Numerical Methods for Unconstrained Optimization and Nonlinear Least Squares*, Prentice-Hall, Englewood Cliffs, New Jersey, 1983.

12. J. Garloff, *Interval mathematics: a bibliography*, preprint, Freiburger Intervall-Berichte 85/6, Institut für Angewandte Mathematik der Universität Freiburg, 1985.

13. J. Garloff, *Bibliography on interval mathematics, continuation*, preprint, Freiburger intervall-Berichte 87/2, Institut für Angewandte Mathematik der Universität Freiburg, 1987.

14. E. R. Hansen, *On solving systems of equations using interval arithmetic*, Math. Comp. **22** (1968), 374–384.

15. E. R. Hansen (ed.), *Topics in interval analysis*, Oxford University Press, London, 1969.

16. E. R. Hansen, *Interval forms of Newton's method*, Computing **20** (1978), 158–163.

17. ——, *Global optimization using interval analysis–the multidimensional case*, Numer. Math. **34** (1980), 247–270.

18. E. R. Hansen and S. Sengupta, *Bounding solutions of systems of equations using interval arithmetic*, BIT **21** (1981), 203–211.

19. E. R. Hansen and R. I. Greenberg, *An interval Newton method*, Appl. Math. Comput. **12** (1983), 89–98.

20. E. R. Hansen and G. W. Walster, *Nonlinear equations and optimization*, accepted for publication in Math. Programming.

21. E. R. Hansen, *An Overview of Global Optimization Using Interval Analysis*, Reliability in Computing, R. E. Moore (ed.), Academic Press, New York, 1988.

22. C. den Heijer and W. C. Rheinboldt, *On steplength algorithms for a class of continuation methods*, SIAM J. Numer. Anal. **18** (5) (1981), 925–948.

23. K. Ichida and Y. Fujii, *An interval arithmetic method for global optimization*, Computing **23** (1) (1979), 85–97.

24. R. B. Kearfott, *On a general technique for finding directions proceeding from bifurcation points*, Numerical Methods for Bifurcation Problems, T. Küpper, H. D. Mittelmann, and H. Weber (eds.), Birkhäuser, Basel, 1984.

25. ——, *Abstract generalized bisection and a cost bound*, Math. Comput. **49** (179) (1987), 187–202.

26. ——, *Some tests of generalized bisection*, ACM Trans. Math. Software **13** (3) (1987).

27. ——, *Interval arithmetic methods for nonlinear systems and nonlinear optimization: an introductory review*, Impact of Recent Computer Advances on Operations Research (R. Sharda, B. L. Golden, E. Wasil, O. Balci, and W. Stewart, eds.), North-Holland, New York, 1989, pp. 533–542 (Proc. Conf., Williamsburg, Virginia, January–1989).

28. ——, *The role of homotopy techniques in biomedical modelling: a case study*, Twelfth IMACS World Congress on Scientific Computation (Proc. Conf., Paris, France, July 1988) (to appear).

29. R. B. Kearfott and M. Novoa, *A program for generalized bisection*, accepted for publication as an algorithm in ACM Trans. Math. Software.

30. R. B. Kearfott, *On handling singular systems with interval Newton methods*, Twelfth IMACS World Congress on Scientific Computation (Proc. Conf., Paris, France, July 1988).

31. ——, *Preconditioners for the interval Gauss–Seidel method*, accepted for publication in SIAM J. Numer. Anal. (June, 1990).

32. H. B. Keller, *Continuation methods in computational fluid dynamics*, Numerical and Physical Aspects of Aerodynamic Flows, T. Cebeci (ed.), Springer-Verlag, New York, 1982, pp. 3–13.

33. ——, *The bordering algorithm and path following near singular points of higher nullity*, SIAM J. Sci. Statist. Comput. **4** (4) (1983), 573–582.

34. U. W. Kulisch and W. L. Miranker, *The arithmetic of the digital computer*, SIAM Rev. **28** (1) (1986), 1–40.

35. T. D. Ladner and J. M. Yohe, *An interval arithmetic package for the Univac 1108*, Technical Summary Report no. 100, Mathematics Research Center, University of Wisconsin at Madison, 1970.

36. A. V. Levy and S. Gomez, *The tunneling method applied to global optimization*, Numerical Optimization 1984, P. T. Boggs, R. H. Byrd, and R. B. Schnabel (eds.), SIAM, Philadelphia, 1985, pp. 213–244.

37. W. L. Miranker (ed.), *Accurate scientific computations*, Lecture Notes in Computer Science no. 235, Springer-Verlag, New York, 1986.

38. R. E. Moore, *Interval Analysis*, Prentice-Hall, Englewood Cliffs, New Jersey, 1966.

39. ——, *A test for existence of solutions to nonlinear systems*, SIAM J. Numer. Anal. **14** (4) (1977), 611–615.

40. R. E. Moore and S. T. Jones, *Safe starting regions for iterative methods*, SIAM J. Numer. Anal. **14** (6) (1977), 1051–1065.

41. R. E. Moore, *Methods and applications of interval analysis*, SIAM, Philadelphia, 1979.

42. R. E. Moore (ed.), *Reliability in Computing*, Academic Press, New York, 1988.

43. A. P. Morgan, *Solving Polynomial Systems Using Continuation for Engineering and Scientific Problems*, Prentice-Hall, Englewood Cliffs, New Jersey, 1987.

44. A. P. Morgan and A. Sommese, *Computing all solutions to polynomial systems using continuation*, Appl Math. Comput. **24** (2) (1987), 115–138.

45. A. Neumaier, *Interval iteration for zeros of systems of equations*, BIT **25** (1) (1985), 256–273.

46. ——, *The enclosure of solutions of parameter-dependent systems of equations*, Reliability in Computing, R. E. Moore (ed.), Academic Press, New York, 1988.

47. K. Nickel, *On the Newton method in interval analysis*, Technical Summary Report no. 1136, Mathematics Research Center, University of Wisconsin at Madison, 1971.

48. K. Nickel (ed.), *Interval mathematics 1980*, International Symposium on Interval Mathematics (Proc. Conf.), Academic Press, New York, 1980.

49. ——, *Interval Mathematics 1985*, Lecture Notes in Computer Science vol. 212, Springer-Verlag, Berlin, 1986.

50. K. Nickel, *Optimization using interval mathematics*, preprint, Freiburger Intervall-Berichte 86/7, Institut für Angewandte Mathematik der Universität Freiburg, 1986, pp. 55–83.

51. M. J. D. Powell, *A hybrid method for nonlinear equations*, Numerical Methods for Nonlinear Algebraic Equations, P. Rabinowitz (ed.), Gordon and Breach, London, 1970.

52. L. Qi, *A note on the Moore test for nonlinear systems*, SIAM J. Numer. Anal. **19** (4) (1982), 851–857.

53. L. B. Rall, *Automatic differentiation: Techniques and applications*, Lecture Notes in Computer Science, Springer-Verlag, Berlin, Heidelberg, New York, 1981.

54. ——, *Interval analysis: A new tool for applied mathematics*, Mathematics Research Center report no. 2268, the University of Wisconsin at Madison, 1981.

55. ——, *Differentiation in Pascal-SC: type Gradient*, ACM Trans. Math. Software **10** (2) (1984), 161–184.

56. ——, *Global optimization using automatic differentiation and interval iteration*, Mathematics Research Center report no. 2832, the University of Wisconsin at Madison, 1985.

57. ——, *An introduction to the scientific computing language Pascal-SC*, Computers and Mathematics with Applications **14** (1) (1987), 53–69.

58. H. Ratschek and J. G. Rokne, *Computer Methods for the Range of Functions*, Horwood, Chichester, England, 1984.

59. H. Ratschek, *Interval mathematics*, preprint, Freiburger Intervall-Berichte 87/4, Institut für Angewandte Mathematik der Universität Freiburg (to appear as an entry in *Encyclopedia of computer science and technology*, A. G. Holzman, A. Kent, and J. G. Williams (eds.), Marcel Dekker, New York).

60. H. Ratschek and J. G. Rokne, *Efficiency of a global optimization algorithm*, SIAM J. Numer. Anal. **24** (5) (1987), 1191–1201.

61. H. Schwandt, *An interval arithmetic approach for the construction of an almost globally convergent method for the solution of the nonlinear Poisson equation*, SIAM J. Sci. Statist. Comput. **5** (2) (1984), 427–452.

62. ——, *Krawczyk-like algorithms for the solution of systems of nonlinear equations*, SIAM J. Numer. Anal. **22** (4) (1985), 792–810.

63. ——, *The solution of nonlinear elliptic Dirichlet problems on rectangles by almost globally convergent interval methods*, SIAM J. Sci. Statist. Comput. **6** (3) (1985), 617–638.

64. H. Schwetlick, *On the choice of steplength in path following methods*, Z. Angew. Math. Mech. **64** (9) (1984), 391–396.

65. J. M. Shearer and M. A. Wolfe, *ALGLIB, a simple symbol manipulation package*, Comm. ACM **28** (8) (1985), 820–825.

66. B. Speelpenning, *Compiling fast partial derivatives of functions given by algorithms*, Department of Computer Science technical report no. UIUCDCSR-80-1002, University of Illinois at Urbana-Champaign, 1980.

67. S. Thiel, *Intervalliterationsverfahren für diskretisierte elliptische Differentialgleichungen*, Diplomarbeit and preprint, Freiburger Intervall-Berichte 86/8, Institut für Angewandte Mathematik der Universität Freiburg, 1986, pp. 1–72.

68. M. N. Vrahatis and K. I. Iordanidis, *A rapid generalized method of bisection for solving systems of nonlinear equations*, Numer. Math. **49** (2) (1986), 123–138.

69. G. W. Walster, E. R. Hansen, and S. Sengupta, *Test results for a global optimization algorithm*, Numerical Optimization 1984, P. T. Boggs, R. H. Byrd, and R. B. Schnabel (eds.), SIAM, Philadelphia, 1985.

70. W. Walter and M. Metzger, *Fortran-SC, a Fortran extension for engineering/scientific computation with access to ACRITH*, Reliability in Computing, R. E. Moore (ed.), Academic Press, New York, 1988.

71. J. M. Yohe, *The interval arithmetic package*, Mathematics Research Center report no. 175, the University of Wisconsin at Madison, 1977.

72. ——, *Software for interval arithmetic: A reasonably portable package*, ACM Trans. Math. Software **5** (1) (1979), 50–63.

THE UNIVERSITY OF SOUTHWESTERN LOUISIANA

Lectures in Applied Mathematics
Volume **26** (1990)

Operator Prolongation Methods for Nonlinear Equations

C. T. KELLEY

Abstract. Mesh independence principles for the chord or modified Newton method lead to fast, linearly convergent algorithms for nonlinear equations if appropriate operator prolongations from coarse meshes to finer ones can be found. This paper discusses the properties that nonlinear maps on each mesh and prolongations should have. We show how quasi-Newton methods can improve the performance of these methods and even allow them to be effective for certain problems having singular Fréchet derivative at the solution.

1. Introduction. This paper considers variations of the modified Newton method, or chord method, for solution of nonlinear equations in Banach spaces. We shall write the nonlinear equation to be solved as

$$F(u) = 0, \tag{1.1}$$

where $F : \mathscr{X} \to \mathscr{Y}$ is a continuously Fréchet differentiable map between the Banach spaces $\mathscr{X}$ and $\mathscr{Y}$. We will let u^* denote the solution to (1.1). The chord method replaces the Fréchet derivative in the equation for the Newton step, $F'(u)s = -F(u)$, by another operator, A, that is computed only once. For example **[22]**, $A = F'(u_0)$ could be used. The transition from a current iterate, u_c, to a new iterate, u_+, is described by

(1.2)
(1) Evaluate $F(u_c)$.
(2) Compute the step by solving $As = -F(u_c)$.
(3) $u_+ = u_c + s$.

1980 *Mathematics Subject Classification* (1985 *Revision*). Primary 45D15, 65H10.
This research was supported by NSF grant #DMS-8601139 and by AFOSR grant #AFOSR-ISSA-860074.

The advantages of the chord method are that the approximate Fréchet derivative need be computed (and perhaps factored) only once. The price paid is that convergence is q-linear and not quadratic as is the case with Newton's method when the Fréchet derivative is Lipschitz continuous and nonsingular at the root.

When A is a good approximation to $F'(u^*)$ the convergence of the chord iterates is fast and the savings in derivative computation may outweigh the cost in the larger number of iterates required to converge. In addition, a good A is also a good starting approximation for quasi-Newton methods. If only a few iterates are required, as will be the case when the initial iterate is sufficiently near u^* and A is a good approximation to $F'(u^*)$, quasi-Newton updates that change A at each iterate by using the history of the iteration often allow computation of the step using A directly, a few stored vectors, and the Sherman–Morrison–Woodbury formula ([**14**], [**32**]).

The algorithms considered in this paper are motivated by work of Atkinson [**3**] on some two-grid methods for linear and nonlinear integral equations. In [**3**], A is an evaluation of $F'(u^*)$ on a coarse mesh that is extended to the finer mesh by a Nyström interpolation. The methods discussed in [**3**] perform most derivative evaluations and all matrix factorizations on the coarse mesh; the solution cost is dominated by fine mesh function evaluations and, for some of the methods, a few differentiations of the kernel of the nonlinear integral operator. The methods in [**3**] are comparable in efficiency to the multigrid methods of the second kind proposed in [**19**] and, as we shall see in this paper, can be extended to a wider class of problems.

The purpose of this paper is to consider extensions of the algorithms in [**3**]. The main directions of these generalizations are nested iteration and quasi-Newton methods for reduction in the number of iterates on the finer grids and quasi-Newton methods for preservation of good convergence properties for some singular problems. Rather than the two-grid method in [**3**], we advocate the use of more grids so that the number of function evaluations on the finer grids can be further reduced. In the case of problems for which the Fréchet derivative is nonsingular at the solution, the key feature of these algorithms, as in [**3**], is that on any of the finer meshes, the initial approximation to the Fréchet derivative on that mesh is a prolongation of an accurate derivative from a much coarser mesh.

The algorithms in this paper use more than two grids in order to reduce the number of function evaluations on the finer grids and augment the chord iterates with quasi-Newton updates at each grid level to accelerate convergence. The transfer of Fréchet derivative approximates between grids is done via prolongation maps as in [**29**]. For problems with nonsingular Fréchet derivative, unlike [**29**], we transfer only the coarsest grid approximate derivative and not the accumulation of quasi-Newton updates. The reasoning here, from the point of view of quasi-Newton methods, is that the coarse grid derivative is a good approximation and that iterative information from the current grid is of more value, and hence more worth storing, than the older information from the coarser meshes. The purpose of the quasi-Newton updates at each grid level is, aside from reduction of the number of iterates at each level, to allow the algorithms to be effective for some problems where $F'(u^*)$ is singular. For the singular problems, as we describe in more detail later, it may be necessary to carry quasi-Newton updates from one grid to the next. We did not observe this need in our experiments.

In Section 2 we establish the notation and give the convergence results for problems where the Fréchet derivative is nonsingular at the solution. In that section we also give some examples to illustrate the concepts. In Section 3 we consider problems where the derivative is singular at the solution. In Section 4 we report on numerical experiments.

2. Notation and results for nonsingular problems. The nonlinear equations considered here consist of an infinite-dimensional problem and a sequence of approximating problems. These approximating problems may be finite-dimensional or reducible to finite-dimensional problems. The underlying idea is that the approximate problems become more difficult as the approximation improves. We will refer to these approximations as being with a "mesh" and to better approximations as having "finer meshes." Let $F: \mathcal{X} \to \mathcal{Y}$ be a continuously Fréchet differentiable map between Banach spaces $\mathcal{X}$ and $\mathcal{Y}$. Consider a family of approximate maps $F_l: \mathcal{X}_l \to \mathcal{Y}_l$, $0 \le l \le \infty$, also continuously Fréchet differentiable, where $F_\infty = F$, $\mathcal{X}_\infty = \mathcal{X}$, and $\mathcal{Y}_\infty = \mathcal{Y}$. We assume that $F_l(u_l^*) = 0$ and, for the present, that $F_l'(u_l^*)$ is a nonsingular map from $\mathcal{X}_l$ to $\mathcal{Y}_l$. We let $\|\cdot\|_m$ denote the norm on $\mathcal{X}_m$, leaving the subscript off when $m = \infty$. Also we let $\mathcal{L}_l = \mathcal{L}(\mathcal{X}_l, \mathcal{Y}_l)$, the space of bounded linear operators from $\mathcal{X}_l$ to $\mathcal{Y}_l$.

It is important that the chord method behave well at all mesh sizes. Such a property was defined and studied in the case of quasi-Newton methods in the context of integral equations and optimal control problems in **[26]** and **[28]** and for Newton's method in a general setting in **[1]**. In **[1]** the name "mesh independence" was introduced to describe one such type of behavior. We intend here, as was done in **[2]** and **[30]**, to use this type of behavior to design grid refinement strategies that refine the grid as the iteration progresses. In order to proceed we must describe the nature of the mesh independence we need. The assumptions for linear convergence of chord-type methods are quite minimal, being uniform continuity of F_l' in a ball about u_l^* where the radius of the ball does not depend on l.

To make this precise we let $B_l(\rho)$ denote the ball in $\mathcal{X}_l$ of radius ρ about u_l^*. We assume that there are $M_\pm$ and $\overline{\rho} \in (0,1)$ and a decreasing continuous function $\varepsilon\colon [0,\overline{\rho}] \to [0,\infty)$ with $\varepsilon(0) = 0$ such that for all $l \geq 0$ and $u \in B_l(\overline{\rho})$

$$\|F_l'(u)\|_{\mathcal{L}_l} \leq M_+, \qquad \|F_l'(u)^{-1}\|_{\mathcal{L}(\mathcal{Y}_l,\mathcal{X}_l)} \leq M_-, \tag{2.1}$$

and for all $\rho \leq \overline{\rho}$ and $u,\ v \in B_l(\rho)$

$$\|F_l'(u) - F_l'(v)\|_{\mathcal{L}_l} \leq \varepsilon(\rho).$$

Note that (2.1) implies that $F_l'(u_l^*)$ is nonsingular.

We let $\{A_l\}$ be a sequence of approximations to $\{F_l'(u_l^*)\}$. The chord method on mesh l is given by

$$u_l^{(n+1)} = u_l^{(n)} - A_l^{-1} F_l(u_l^{(n)}). \tag{2.3}$$

The following theorem says that the accuracy requirements for A_l and the initial iterate on $\mathcal{X}_l$ needed for q-linear convergence at mesh level l are independent of l. In the statement of this theorem we let $\mathcal{O}_l(\rho)$ be the ball in $\mathcal{L}_l$ about $F_l'(u_l^*)$ of radius ρ.

THEOREM 2.1. *Assume that* (2.1) *and* (2.2) *hold. Then there is* ρ_c *and a decreasing continuous function* $\varepsilon_c\colon [0,\rho_c] \to [0,\infty)$ *with* $\varepsilon_c(0) = 0$, *such that for all* $l \geq 0$, $\rho \leq \rho_c$, *and* $(A_l, u_l^{(0)}) \in \mathcal{O}_l(\rho) \times B_l(\rho)$, *the chord iterates given for* $n > 0$ *by* (2.3) *converge* q*-linearly to* u_l^* *uniformly in* l. *That is,*

$$\|u_l^{(n+1)} - u_l^*\|_l \leq \varepsilon_c(\rho)\|u_l^{(n)} - u_l^*\|_l. \tag{2.4}$$

PROOF. Let l be given. We analyze the transition from a current iterate, $u_l^{(c)}$, to a new iterate, $u_l^{(+)}$. Letting $e = u_l - u_l^*$, we have, for

$(A_l, u_l^{(c)}) \in \mathscr{O}_l(\rho) \times B_l(\rho)$,

$$\begin{aligned} e_+ &= e_c - A_l^{-1} F_l(u_l^{(c)}) \\ &= e_c - A_l^{-1} \int_0^1 F_l'(u_l^* + te_c) e_c \, dt \\ &= A_l^{-1} \int_0^1 (A_l - F_l'(u_l^* + te_c)) e_c \, dt \\ &= A_l^{-1} \int_0^1 (A_l - F_l'(u_l^*)) e_c \, dt \\ &\quad + A_l^{-1} \int_0^1 (F_l'(u_l^*) - F_l'(u_l^* + te_c)) e_c \, dt. \end{aligned}$$

If $\rho \leq \overline{\rho}$ then $\|e_+\| \leq \|A_l^{-1}\|_{\mathscr{L}_l}(\rho + \varepsilon(\rho))\|e_c\|$. Since $A_l \in \mathscr{O}_l(\rho)$ and (2.1) holds, we have

$$\|A_l^{-1}\|_{\mathscr{L}_l} \leq \frac{M_-}{1-\rho} \leq \frac{M_-}{1-\overline{\rho}}.$$

Since $\overline{\rho} < 1$ we may complete the proof by using

$$\varepsilon_c(\rho) = \frac{M_-}{1-\overline{\rho}}(\rho + \varepsilon(\rho))$$

and any ρ_c such that $\varepsilon_c(\rho_c) < 1$. □

Note that nothing in the proof above requires that $A_l^{-1} F_l$ be computed exactly. As in **[13]**, sufficiently many steps of a linearly convergent iterative method for linear equations as an inner iteration would suffice to preserve linear convergence for the chord iterates.

The next step is to analyze the convergence of methods where the mesh is refined as the iteration progresses. Such methods are called "nested iteration" methods in **[19]**. Here we have a family of Fréchet derivative approximates, $\{A_l\}$, given at each mesh size, compute an approximation to u_l^* at level l, and extend it to level $l+1$ by some prolongation operator. As in **[30]** we expect r-linear convergence if the prolongations have reasonable continuity properties. As we are looking at the chord method only, however, one would expect weaker assumptions to suffice, and they do.

Let $\{\{P_{l\to m}\}_{l=0}^{\infty}\}_{m=l}^{\infty}$ be a family of continuous maps, not necessarily linear, such that $P_{l\to m}: \mathscr{X}_l \to \mathscr{X}_m$. These maps will be referred to as prolongations and carry the solution information from one mesh to the next. To measure the accuracy of the approximate solutions we define

$$\sigma_n = \sup_{l,m\geq n} \|P_{l\to m} u_l^* - u_m^*\|_m. \tag{2.5}$$

We will assume throughout that

$$\lim_{n\to\infty} \sigma_n = 0. \tag{2.6}$$

We shall assume that as l increases the approximate solutions, u_l^*, improve. We define

$$\overline{\sigma}_l = \|P_{l\to\infty}u_l^* - u^*\|, \tag{2.7}$$

and note that (2.6) implies that

$$\lim_{l\to\infty} \overline{\sigma}_l = 0. \tag{2.8}$$

We assume uniform continuity of the prolongations in the sense that there is $\rho_p \in (0, \rho_c)$ and a continuous decreasing function $\varepsilon_p : [0, \rho_p) \to (0, \infty)$ such that if $m > l \geq 0$, u, $v \in B_l(\rho)$, and $\|u - v\|_l \leq \delta$, then

$$\|P_{l\to m}u - P_{l\to m}v\|_l \leq \varepsilon_p(\delta). \tag{2.9}$$

Our nested iteration algorithm will require, in addition to the approximate Fréchet derivatives and prolongations, a decreasing sequence of tolerances for termination, $\{\omega_l\}$, where $\omega_l \to 0$. The steps in the algorithm are:

(2.10)

(1) Apply the chord method at level l until $\|F_l(u_l^{(n_l)})\|_{\mathscr{Y}_l} \leq \omega_l$.

(2) $u_{l+1}^{(0)} = P_{l\to l+1}u_l^{(n_l)}$.

We have

THEOREM 2.2. *Assume that* (2.6), (2.9), *and the assumptions for Theorem* 2.1 *hold. Then there is* ρ_ν *such that if* $u_0 \in B_0(\rho_\nu)$ *and* $A_l \in \mathscr{O}(\rho_c)$ *for all* $l \geq 0$ *and the sequences* $\{\omega_l\}$ *and* $\{\sigma_l\}$ *are sufficiently small, then the iterates given by* (2.10) *converge to* u^* *in the sense that* $P_{l\to\infty}u_l^{(n_l)} \to u^*$. *In addition, if* $\omega_l \to 0$ *r-linearly, then the sequence* $\{\|u_l^{(n_l)} - u_l^*\|\}$ *converges to zero r-linearly, and if both* $\overline{\omega}_l$ *and* σ_l *converge to zero r-linearly so does* $\|P_{l\to\infty}u_l^{(n_l)} - u_l^*\|$.

PROOF. We begin by requiring $\rho_\nu \leq \rho_p$ to be small enough so that

$$r = \frac{1 - M_{-}\varepsilon(\rho_\nu)}{M_{-}} > 0. \tag{2.11}$$

Next we require that $\{\omega_l\}$ and $\{\sigma_l\}$ be such that

$$\varepsilon_p(\omega_l/r) + \sigma_l < \rho_\nu \tag{2.12}$$

for all $l \geq 0$. Note that both of the estimates, (2.11) and (2.12), depend only on the continuity properties of the sequence $\{F_l\}$ and on the accuracy of the approximate solutions on each mesh, not the meshes themselves. (2.11) adds a bit more derivative information and has nothing to do with the meshes. The estimate (2.12) essentially demands that the coarsest mesh be sufficiently fine so that $P_{0\to\infty}u_0^*$ is a good approximation to u^* and that the accuracy demanded at each level not allow the resulting approximation to get too much worse. Note that as $\rho < \rho_c$, the iterates at level zero will converge to the desired tolerance. The question to be resolved is how the prolongations and the termination affect the errors on the next mesh.

For any $u \in B_l(u_l^*)$ we have, setting $e = u - u_l^*$,

$$F_l(u) = F_l'(u_l^*)e + \int_0^1 (F_l'(u_l^* + te) - F_l'(u_l^*))e\,dt$$

and hence $\|F_l(u)\|_{\mathscr{Y}_l} \geq r\|e\|_l$. This means that when the iteration at level l is terminated at iterate n_l we have, setting $e_{n_l} = u_l^{(n_l)} - u_l^*$, $\|e_{n_l}\|_l \leq \omega_l/r$, and, therefore,

$$\begin{aligned}\|P_{l\to\infty}u_l - u^*\|_l &\leq \|P_{l\to\infty}u_l - P_{l\to\infty}u_l^*\|_l + \|P_{l\to\infty}u_l^* - u^*\| \\ &\leq \varepsilon_p(\omega_l/r) + \overline{\sigma}_l \\ &\leq \varepsilon_p(\omega_l/r) + \sigma_l.\end{aligned}$$

To complete the proof we need only show that $u_{l+1}^{(0)} \in B_{l+1}(\rho_\nu)$. To see this we note that

(2.13)

$$\begin{aligned}\|u_{l+1}^{(0)} - u_{l+1}^*\|_{l+1} &\leq \|P_{l\to l+1}u_l^{(n_l)} - P_{l\to l+1}u_l^*\|_{l+1} + \|P_{l\to l+1}u_l^* - u_{l+1}^*\|_{l+1} \\ &\leq \varepsilon_p(\omega_l/r) + \sigma_l.\end{aligned}$$

This completes the proof in view of (2.12). □

Note that we may bound the number of iterates at each level from above in terms of $\varepsilon_c(\rho_c)$. We express this as a corollary.

Corollary 2.3. *Assume that the hypotheses of Theorem 2.2 hold. Then the number of iterates, m_l, required at each level satisfies*

$$m_l \leq \left|\left[\frac{\ln\left(\frac{\omega_{l+1}}{M_+(\varepsilon_p(\omega_l/r)+\sigma_l)}\right)}{\ln(\varepsilon_c(\rho_c))}\right]\right| + 1, \tag{2.14}$$

where $[\cdot]$ is the greatest integer function.

Proof. The assertion of the corollary is a consequence of the estimates (2.1), (2.4), and (2.13), which together imply that, for all $l \geq 0$,

$$\|F_{l+1}(u_{l+1}^{(n)})\|_{\mathscr{Y}_{l+1}} \leq M_+\varepsilon_c(\rho_c)^n(\varepsilon_p(\omega_l/r) + \sigma_l).$$

Requiring that $\|F_{l+1}(u_{l+1}^{(n)})\|_{\mathscr{Y}_{l+1}} < \omega_{l+1}$ and solving for n completes the proof. □

A simple implication of this corollary is that if the ω_l's are approximations to truncation error that converge linearly to zero, then a single iterate at each mesh size will suffice if the coarsest mesh is sufficiently fine. Hence, only one function evaluation will be required at each level. If more than one function evaluation is required at some levels, one may want to use A_l as the initial approximation to the derivative in a quasi-Newton method. Typical theorems for such methods in finite dimension ([**5**], [**6**], [**14**]) require that $A_l \in \mathscr{O}(\rho_Q)$ and $u_l^{(0)} \in B(\rho_Q)$, for some small ρ_Q, in order to obtain superlinear convergence. In infinite dimension ([**16**], [**17**], [**34**]), compactness conditions are needed as well. Convergence can fail to be superlinear if such conditions do not hold ([**35**]). When compactness conditions cannot be enforced, other operator norms, say, function space norms on the coefficients of operators, can be used to describe the sense in which A_l must be near $F_l'(u_l^*)$. Such compactness problems arise in boundary value problems ([**18**], [**20**], [**27**]), optimal control problems ([**28**], [**29**]), and integral equations ([**24**], [**25**], [**32**]). In all of these situations, the value of ρ_Q depends on $M_\pm$ and the Lipschitz constant of F'. If, in addition to the assumptions needed for the convergence of the chord method, F' is Lipschitz continuous, a superlinearly convergent quasi-Newton method of the type mentioned above will be effective for all mesh sizes.

This paper considers one particular choice of approximate derivatives, $\{A_l\}$. On level 0 we use Newton's method, rather than the chord method, or in some other way compute a good approximation to $F_0'(u_0^*)$ after the iteration on level 0 terminates. We express this by

$$A_0 = F'(u_0^*) + E_0. \tag{2.15}$$

Then

$$A_l = \mathscr{E}_l A_0, \tag{2.16}$$

where $\mathscr{E}_l$ is a continuous operator prolongation map from $\mathscr{Z}_0 \subset \mathscr{L}_0$ to $\mathscr{L}_l$. We assume that $F_0'(u_0^*) \in \mathscr{Z}_0$. $\mathscr{E}_l$ may also depend on current iterate information in a way we suppress. Such dependence may be through partial Fréchet derivative evaluation or, as in [**3**], use of the current derivative in formation of the prolongation. We will give an example of this latter type of dependence later in this section.

In general, $\mathscr{E}_l$ will not be a linear map. We ask only $\mathscr{E}_l$ satisfy continuity and approximation properties like the maps $P_{l\to m}$. In order to describe the accuracy of A_l at mesh level l we first define

$$\sigma_E = \sup_{l\geq 0} \|\mathscr{E}_l F_0'(u_0^*) - F_l'(u_l^*)\|_{\mathscr{L}_l}.$$

σ_E measures the effect of the operator prolongation and the accuracy of the approximation at coarsest mesh. If the mesh at level 0 is sufficiently fine and the operator prolongations reasonable, then σ_E will be small. In order to account for errors in the computation of $F_0'(u_0^*)$ we must require uniform continuity of $\mathscr{E}_l$. We do this by assuming that there is $\rho_E \in (0, \rho_\nu)$ and a continuous function $\varepsilon_E: [0, \rho_E] \to [0, \infty)$ with $\varepsilon_E(0) = 0$ such that if $A_0^{(1)}, A_0^{(2)} \in \mathscr{Z}_0 \cap B_0(\rho_E)$, then for all $l \geq 0$

$$\|\mathscr{E}_l A_0^{(1)} - \mathscr{E}_l A_0^{(2)}\|_{\mathscr{L}_l} \leq \varepsilon_E(\|A_0^{(1)} - A_0^{(2)}\|_{\mathscr{L}_0}). \tag{2.17}$$

We let the accuracy in the computation of $F_0'(u_0^*)$ be denoted by σ_A:

$$\sigma_A = \|A_0 - F_0'(u_0^*)\|_{\mathscr{L}_0} = \|E_0\|_{\mathscr{L}_0}.$$

We have the following trivial result, which we state without proof.

PROPOSITION 2.4. *If $\{\mathscr{E}_l\}$ and A_0 are such that $\varepsilon_E(\sigma_A) + \sigma_E < \rho_c$ and A_l is given by* (2.16), *then $A_l \in \mathscr{O}(\rho_c)$ for all $l \geq 0$ and hence the sequence $\{A_l\}$ may be used in the algorithm given by* (2.10) *to satisfy the hypotheses of Theorem* 2.2.

To close this section we consider some examples to illustrate the ideas. Our first example will use one of the methods suggested in [**3**]. Here $\mathscr{X}_l = C[0,1]$ for all l, and F is given by

$$F(u)(x) = u(x) - \int_0^1 k(x, y, u(y))\, dy. \tag{2.18}$$

In (2.18) $k \in C^\alpha([0,1] \times [0,1] \times \mathbf{R})$ with $\alpha > 0$ is given. We assume that a solution $u^* \in C[0,1]$ to $F(u) = 0$ exists and that $F'(u^*)$ is nonsingular. Our approximate equations, $F_l(u) = 0$, will be formed by replacing the integral with a quadrature rule. If $\{x_i^l\}_{i=1}^{N_l}$ and $\{w_i^l\}_{i=1}^{N_l}$ are the nodes and weights of a sequence of quadrature rules, we define

$$F_l(u)(x) = u(x) - \sum_{j=1}^{N_l} k(x, x_j^l, u(x_j^l)) w_j^l. \tag{2.19}$$

If the quadrature rules are sufficiently fine then $F_l(u) = 0$ will have a solution, u_l^*. This solution may be recovered from the solution to the finite-dimensional problem for the values of u_l^* at the nodes of the lth quadrature rule. This finite-dimensional problem, for a vector $\overline{u}_l \in \mathbf{R}^{N_l}$, is

$$(2.20) \qquad (\overline{F}_l(\overline{u}_l))_i = \overline{u}_{l,i} - \sum_{j=1}^{N_l} k(x_i^l, x_j^l, \overline{u}_{l,j}) w_j^l = 0.$$

Here $\overline{u}_{l,i}$ is the ith component of $\overline{u}_l$ and $(\overline{F}_l(\overline{u}_l))_i$ the ith component of the nonlinear map $\overline{F}_l$. If we let $E_l \colon C[0,1] \to \mathbf{R}^{N_l}$ be the evaluation map

$$(E_l u)_i = u(x_i^l),$$

we may regard a solution of (2.20) as a solution of (2.19) for $E_l u_l^*$ by observing that $E_l F_l = \overline{F}_l E_l$.

At level $l = 0$, the matrix $\overline{A}_0 \approx \overline{F}'_0(u_0^*)$ is factored and stored. This factorization may be used to solve equations of the form $A_0 s = g$ in $C[0,1]$ by solving $\overline{A}_0 E_l u = E_l g$ as follows. We create the matrix $\overline{A}_0$ from an approximation, $\tilde{u}$, to u_0^* by

$$(\overline{A}_0)_{i,j} = \delta_{i,j} - \frac{\partial k}{\partial u}(x_i^0, x_j^0, \tilde{u}_j) w_j^0.$$

We let $\tilde{v} = (\overline{A}_0)^{-1} E_0 g \in \mathbf{R}^{N_0}$ and then let

$$v(x) = (A_0^{-1} g)(x) = g(x) + \sum_{j=1}^{N_0} \frac{\partial k}{\partial u}(x, x_j^0, \tilde{u}_j) \tilde{v}_j w_j^0.$$

The approximate Fréchet derivative at level l is given by

$$(2.21) \qquad A_l = \mathscr{E}_l(A_0) = (I + A_0^{-1} K_l)^{-1}$$

where, for $v \in C[0,1]$, $K_l v$ is given by

$$\sum_{j=1}^{N_l} \frac{\partial k}{\partial u}(x, x_j^l, \hat{u}(x_j^l)) v(x_j^l) w_j^l.$$

Here, $\hat{u}$ is an approximation to u. We used $\hat{u} = u_l^{(0)}$ in the examples in Section 4. Note that $\hat{u}$, and hence K_l, is computed only once at each mesh level. Hence, our final formula for the step is

$$s(x) = -F_l(u)(x) - (A_0^{-1} K_l F_l(u))(x).$$

Note that in this formulation, A_l is never explicitly formed. The formula for the step uses only K_l and the factorization of $\overline{A}_0$. Quasi-Newton updates could use the Sherman–Morrison–Woodbury formula and therefore require storage of only a few vectors.

We let $\mathscr{Z}_0$ be all operators A such that Av is given by

$$(Av)(x) = v(x) - \sum_{i=0}^{N_0} a_i(x) v(x_i^0) w_i^0,$$

where $\{a_i\} \subset C[0,1]$. With this definition, (2.17) holds. We repeat that the gain in efficiency in this method is mostly in savings in linear algebra. Matrix factorizations need be done only at level 0. A gain in derivative evaluation is also realized as $\hat{u}$ need be computed only once and only the directional derivatives, $K_l F$, need be computed at each iterate. When A_l is given in this way, the hypotheses of the results of this section hold if the quadrature rule is reasonable and the grid at level 0 is sufficiently fine.

In the above we use $P_{l \to m} = I$, another choice that works equally well and is useful in moving between high-order quadrature rules at different levels in $u_{l+1}^{(0)} = \mathscr{K}_l(u_l^{(n_l)})$, where the nonlinear integral operator in (2.19) is given by

$$(\mathscr{K}_l u)(x) = \sum_{j=1}^{N_l} k(x, x_j^l, u(x_j^l)) w_j^l.$$

This is the form of $P_{l \to m}$ we use in the example in Section 4.

In **[3]** the use of $A_l = A_0$ is recommended as an alternative to (2.21). This eliminates the need to evaluate derivatives of the kernel on the fine grid completely. However, (2.17) does not hold for this alternative. In fact, the resulting method should not be viewed as the chord method; even for linear problems the resulting method is only two-step q-linear convergent on any given grid. We experimented with this approach and observed, as in **[3]**, greater efficiency but irregular convergence behavior. An alternative approach, not used in the examples in Section 4, is to use a piecewise linear interpolation of $\partial k/\partial u$ for coarse mesh points to compute K_l rather than explicit evaluation of partial derivatives of k; (2.17) would also hold in this case. This latter alternative looks especially attractive if kernel evaluations are expensive. This type of algorithm is easily parallelizable as the work is dominated by function evaluations at mesh points, dot products, and matrix-vector multiples.

As a simple example, suppose $k \in C^2$,

$$\|k\|_{C^2([0,1]\times[0,1]\times[0,1])} \leq \overline{M},$$
$$\max_{x,y \in [0,1], u \in \mathbf{R}} |k_u| \leq \mu < 1,$$

k_u is uniformly Lipschitz with Lipschitz constant γ, and the quadrature rule at level l is the trapezoid rule with mesh size h_l. If u^* is the solution, and $P_{l\to m}$ is piecewise linear interpolation, then $\sigma_n \le \widetilde{M} h_n^2$, where $\widetilde{M}$ is independent of n. (2.6) clearly holds. We may set $M_+ = \overline{M}$, $M_1 = 1/(1-\mu)$, $\varepsilon(\rho) = \overline{M}_\rho$, and $\varepsilon_p(\rho) = \rho$. It is clear that σ_A and σ_E approach 0 as $h_0 \to 0$; in fact, both are $O(h_0^2)$. In the last section of this paper we will consider a case where the quadrature rule is a composite Gauss rule and $k \in C^\alpha$ only for $\alpha < 1$. Here all we can say is that σ_A and σ_E approach 0 as the quadrature rule becomes finer.

This approach also extends to problems of the form

$$h(u(x)) - \int_0^1 k(x, y, u(y), u(x))\, dy = g(x),$$

where $h \in C^2$, as well as to systems of such equations. Here the Fréchet derivative is the sum of a multiplication operator and an integral operator. The prolongation maps are more complicated in this situation. We will explore these issues in a subsequent paper in the context of a specific application.

As our second example we consider the boundary value problem:

$$u''(x) + f(x, u(x), u'(x)) = 0, \qquad u(0) = u(1) = 0. \tag{2.22}$$

We assume that $u^* \in C^4[0,1]$ and that f is Lipschitz continuously differentiable in its arguments. Our approximations are the standard centered differences or, equivalently, piecewise linear finite elements on a sequence of equally spaced meshes. We will arrange this problem so that $\mathscr{X}$ and $\mathscr{Y}$ are the same. To do this we solve approximately on a coarse mesh to obtain a mesh function, $\overline{u}_0$, defined on points 0, h_0, $2h_0, \ldots, N_0 h_0 = 1$, with $u_0(0) = u_0(1) = 0$. We will assume that the meshes are nested. We let D_l be the finite centered difference approximation to the derivative on mesh l and define coefficient functions a_0 and b_0 on the set $\{ih_0\})_{i=1}^{N_0-1}$ by

$$(a_0)(ih_0) = \frac{\partial f}{\partial u}(ih_0, u_0(ih_0), D_0 u(ih_0))$$

and

$$(b_0)(ih_0) = \frac{\partial f}{\partial u'}(ih_0, u_0(ih_0), D_0 u(ih_0)).$$

Define $a_0(0) = a_0(h_0)$, $b_0(0) = b_0(h_0)$, $a_0(1) = a_0((N_0 - 1)h_0)$, and $b_0(1) = b_0((N_0 - 1)h_0)$. Finally, extend a_0 and b_0 to $[0,1]$ by piecewise linear interpolation.

Let E_l denote the evaluation function as in the previous example (with E_∞ understood to be the identity), and P_l be piecewise linear interpolation from the mesh functions at level l that vanish at 0 and 1 to $C[0,1]$. Our assumptions on u^* and f imply ([**21**]) that there is $\widetilde{M} > 0$ such that

$$\|P_0\overline{u}_0 - u^*\|_\infty \le \widetilde{M}h_0^2, \qquad \|P_0\overline{u}_0 - u^*\|_{H_0^1} \le \widetilde{M}h_0.$$

We define a discrete differential operator on each mesh by

$$Z_l = D_l^2 + (E_l b_0)D_l + (E_l a_0). \tag{2.23}$$

If h_0 is sufficiently small, D_l^2 will have an inverse G_l, so that $P_l G_l E_l$ and $P_l G_l D_l E_l$ are bounded operators on $H_0^1[0,1]$. We let $\mathscr{X}_l$ be the space of piecewise linear functions with nodes at $\{ih_l\}_{i=0}^{N_l}$ that vanish at 0 and 1. The norm on $\mathscr{X}_l$ is the usual H_0^1 norm. The idea is to premultiply each discrete boundary value problem by G_l and hence be able to set $\mathscr{Y}_l = \mathscr{X}_l$. We let $\mathscr{Z}_0$ be the set of operators on $\mathscr{X}_0$ of the form

$$A_0 = I + P_0 G_0((E_0 b_0)D_0 + (E_0 a_0)).$$

Then we set $A_l = \mathscr{E}_l A_0 = I + P_l G_l((E_l b_0)D_l + (E_l a_0))$ for all $l > 0$.

We write the discrete form of our boundary value problem as

$$\overline{F}_l(u_l) = D_l^2 u_l + E_l f(x, P_l u_l, P_l D_l u_l) = 0.$$

We define $F_l \colon \mathscr{X}_l \to \mathscr{X}_l$ by $F_l(u) = P_l G_l \overline{F}_l(E_l u)$. With these definitions, $F_l(u_l) = 0$ is equivalent to the original discrete boundary value problem in the sense that the chord method iterates for its solution will be the same.

With this notation we let $P_{l\to m}$ be the inclusion, which we may do as the grids are nested. Hence, $\sigma_E = O(h_0)$ and $\varepsilon_p = 0$. Hence, there is $\widetilde{M}_2 > 0$ such that

$$\begin{aligned} |M_+ - l| &\le \widetilde{M}_2 h_0, \\ |M_- - l| &\le \widetilde{M}_2 h_0, \\ \sigma_A &\le \widetilde{M}_2 h_0, \\ \varepsilon(\rho) &\le \widetilde{M}_2 h_0 \rho, \quad \text{and} \\ \sigma_n &\le \widetilde{M}_2 h_n, \quad \text{for all } n \ge 0. \end{aligned}$$

Hence, the hypotheses for the results of this section hold if h_0 is sufficiently small.

3. Convergence results for singular problems. If $F'(u^*)$ is singular or nearly so, the chord method should not be expected to perform well. In fact ([**10**]), the chord method is known to converge sublinearly in some such cases. In this section we consider a class of singular problems and show how one particular quasi-Newton method, Broyden's method ([**4**]), allows one to recover linear convergence. These results relate the choice of initial approximation to $F'(u^*)$ to the prolongation maps and allow for a simpler proof of linear convergence than that in [**12**]. The purpose of this section is to show that if Broyden's method is used at each grid level, with initial derivative approximation $\mathcal{E}_l F_0'(u_0^{(n_0)})$, and initial solution approximation coming from a prolongation of a solution at a lower level (perhaps $\neq 0$), then linear convergence can be obtained relative to a certain inner-product norm. On any given mesh level there are two levels of initial approximation, that of the derivative and that of the root. Our formalism will separate these two types of approximation. Hence, for the purposes of analysis, it suffices to consider only two levels, $l = 0$ and $l = \infty$. In practice, ∞ would be replaced by a large l. One should keep in mind, however, that the source for the initial approximation to the root could be a prolongation of a solution on a mesh level between 0 and ∞.

It is necessary to impose some additional structure on the nonlinear equations. We assume that $\mathcal{X} = \mathcal{Y}$ has a continuous inner product, which we will denote by $\langle \cdot, \cdot \rangle$. For example, if $\mathcal{X} = C[0, 1]$, this could be the L^2 inner product; if $\mathcal{X}$ is H_0^1, the H_0^1 inner product could be used. For Broyden's method, convergence in the norm induced by the inner product is the goal. As in [**26**], this norm may be weaker than the norm on $\mathcal{X}$, and recovering convergence in $\mathcal{X}$ may require additional interpolation. We will see such a problem in the examples. We assume that F can be extended to the closure of $\mathcal{X}$ in the norm given by the inner product and hence may, without loss of generality, assume that $\mathcal{X}$ is a Hilbert space and that $\langle \cdot, \cdot \rangle$ is the inner product.

We assume that $F \in C^3(B(\delta))$ for some δ sufficiently small and that $\widehat{F} = F'(u^*)$ is Fredholm of index zero and singular with one-dimensional null space, $N = \mathrm{span}(\varphi)$, and closed range X. We assume that $\|\varphi\| = 1$. As $\mathcal{X}$ is a Hilbert space we may define projection onto N parallel to X of the form

$$P_N u = \langle u, \varphi^* \rangle \varphi, \tag{3.1}$$

where φ^* is a unit vector in the (one-dimensional) null space of $\widehat{F}^*$, the adjoint of $\widehat{F}$. Without loss of generality we may also assume that N and X are orthogonal with respect to $\langle \cdot, \cdot \rangle$ and that $\widehat{F}$ is self-adjoint. If this is not the case we simply replace F by $(P_N + \widehat{F}^*)F$. We let $P_X = I - P_N$. By the orthogonality of N and X, $\widehat{F}$ may be viewed as an invertible map on X. By multiplying F by $\widehat{F}^{-1}+P_N$, if necessary, we may assume that $\widehat{F}=P_X$. Note that these premultiplications are merely for theoretical convenience. The actual Newton iterates produced will be the same whether they are done or not. Quasi-Newton iterates will be preserved as well if it is understood that A_0 is also premultiplied by the same maps that multiply F.

Our final assumption on the structure of the singularity is on its order. We assume that

$$P_N F''(u^*)(\varphi,\varphi)=\sigma\varphi \tag{3.2}$$

with $\sigma \neq 0$, and define $\hat{\xi} \in X$ by

$$\hat{\xi}=P_X F''(u^*)(\varphi,\varphi). \tag{3.3}$$

We let $\otimes$ denote the Hilbert space tensor product. That is, for $u, v, w \in \mathscr{X}$, $(u\otimes v)w=\langle v,w\rangle u$. For example, $P_N=\varphi\otimes\varphi$.

Domains of attraction near u^* for Newton-like methods are not balls centered at u^*, but rather cones of the form

$$W(\rho,\theta)=\{u \mid 0<\|P_N e\|<\rho, \|P_X e\|\leq\theta\|P_N e\|\}, \tag{3.4}$$

where, as before, $e=u-u^*$.

The next theorem is a synthesis of results from **[9]–[11]**, **[15]**, and **[33]**. It describes the convergence of Newton's method for singular problems of the type described above and the form of the derivative at such iterates. We would expect our prolongation operators to maintain this structure from level 0 to level ∞ and can use this information in the analysis of Broyden's method.

THEOREM 3.1. *Let $F(u^*)=0$, assume that $F\in C^3$ in a neighborhood of u^*, that $F'(u^*)$ is Fredholm of index zero, and that $F'(u^*)$ has one-dimensional null space $N=\operatorname{span}(\varphi)$ and range X. Assume that N is orthogonal to X and that $F'(u^*)=P_X=I-P_N$, where P_N is given by* (3.1). *Then if ρ and θ are sufficiently small and $u_0\in W(\rho,\theta)$ the Newton iterates remain in $W(\rho,\theta)$ and converge q-linearly to u^* with asymptotic linear ratio $\frac{1}{2}$. Moreover, there is $c_N>0$ such that if $n\geq 1$, we have, letting $P_N e_n=\rho_n\varphi$,*

$$\begin{aligned} e_n &= \frac{\rho_{n-1}\varphi}{2}+P_N a_1(e_{n_1})+\|e_{n_1}\|P_X a_2(e_{n-1}), \quad \textit{and} \\ F(u_n) &= \left(\frac{\sigma\varphi}{2}+\frac{\hat{\xi}}{2}\right)\rho_n^2+\|e_n\|a_3(e_n), \\ F'(u_n) &= P_X+\rho_n(\hat{\xi}\otimes\varphi+\sigma P_N)+E_1(e_n)P_X+\|e_{n+1}\|E_2(e_n)P_N, \end{aligned} \tag{3.5}$$

where

$$\|a_j(e_n)\| \leq c_N \|e_n\|^2 \qquad \text{for } j = 1, \ldots, 3$$

and

$$\|E_j(e_n)\|_{\mathscr{L}} \leq c_N \|e_n\| \qquad \text{for } j = 1, 2.$$

We formulate Broyden's method for singular problems as follows. From a current nonsingular approximation, A_c, to $F'(u_c)$ and an approximation, $u_c \in W(\rho, \theta)$, to u^*, we compute

(3.6)
(a) the step, $s = -A_c^{-1} F(u_c)$,
(b) the new iterate, $u_+ = u_c + s$, and
(c) an approximation to $F'(u_+)$, $A_+ = A_c + F(u_+) \otimes s / \|s\|^2$.

In **[12]** the convergence of Broyden's method was analyzed for finite-dimensional problems of the type considered in this section. The main result of that paper is that if the initial approximation to the derivative is sufficiently good, in the sense that the first iterate is like a Newton step, and the initial approximation to the solution is in $W(\rho, \theta)$ for ρ and θ sufficiently small, then the Broyden iterates converge q-linearly with asymptotic linear ratio $(\sqrt{5} - 1)/2$. The main result of that paper was

THEOREM 3.2. *Let the hypotheses of Theorem* 3.1 *hold and let* α *be given. Let* $u_0 \in W(\overline{\rho}, \theta)$ *and* A_0 *be such that*

$$\|(A_0 - F'(u_0))P_X\| \leq \alpha\rho, \qquad \|(A_0 - F'(u_0))P_N\| \leq \alpha\rho^2.$$

Then for ρ *and* θ *sufficiently small the Broyden iterates lie in* $W_2(\overline{\rho})$ *and converge to* u^* *q-linearly with asymptotic linear ratio* $(\sqrt{5} - 1)/2$.

The result in this section takes advantage of the fact that A_0 is an extension from a good approximation to F' on a coarse mesh and that u_0 also comes from a prolongation of an approximate solution to simplify the analysis in **[12]** and extend it to the infinite-dimensional case. This will imply that q-linear convergence will be observed at any mesh size if the solution and derivative at level 0 are sufficiently good. However, our assumptions are different from those in **[12]** and while our proof is simpler (but longer), our convergence rate estimates are not as precise. The reason for this is that we must allow more freedom in $A_0 P_X$ than is allowed in **[12]** to account for the fact that our approximate derivatives are extensions of derivatives from coarse mesh problems.

Another problem in the present algorithm that is not considered in **[12]** is that the first iterate may make very little progress toward the root. We must therefore show that the approximation to the derivative corrects itself as the iteration progresses. This self-correction is the point of Theorem 3.5.

From Theorem 3.1 we expect all Newton iterates after the first to satisfy $\|P_X e\| \leq \|P_N e\|^2$. As we expect that our initial Broyden iterate will be a Newton iterate from a problem on a coarse mesh, we will demand that our Broyden iterates lie in

$$W_2(\rho) = \{u \| P_N e\| < \rho, \|P_X e\| < c_W \|P_N e\|^2\}. \tag{3.7}$$

In (3.7) the constant c_W will depend on F and the prolongation method.

In **[10]** the sublinear convergence of the chord method was caused by the diminishing effect of the derivative as the iteration progresses. Our Broyden's method variation, using the same initial approximation at each level, will also exhibit poor performance in the first step. The next theorem shows that Broyden's update provides a self-correcting effect and that future approximate derivatives have the structure that linear convergence requires.

Since we expect the initial approximation to the derivative to be a prolongation of F_0', we demand more structure of the approximate derivative A. It requires proof that F_0' can be extended to preserve sufficient structure to verify our assumptions. We illustrate methods for such an extension after the proof of the theorem. Now, by our assumptions, any bounded operator, A, on $\mathscr{X}$ may be written as

$$A = (P_X + T) + \sigma\varepsilon P_N + \varepsilon\tilde{\xi} \otimes \varphi + \varphi \otimes \zeta, \tag{3.8}$$

where $T = P_X A P_X - P_X$, σ is given by (3.2), $\varepsilon = \langle\varphi, A\varphi\rangle/\sigma$, $\varepsilon\tilde{\xi} = P_X A\varphi$, and $\zeta = P_X A^*\varphi$. We assume that $A_c \in \mathscr{W}(\varepsilon_c, \delta_c, \eta_c)$, where

$$\mathscr{W}(\bar{\varepsilon}, \delta, \eta) = \{A | \varepsilon| < \bar{\varepsilon},\ \|T\|_{\mathscr{L}} < \delta,\ \|\zeta\| < \delta,\ \|\tilde{\xi} - \hat{\xi}\| < \eta\}. \tag{3.9}$$

In (3.9), T, ζ, and $\tilde{\xi}$ are related to A as described in (3.8); As in (3.3), $\hat{\xi} = P_X F''(u^*)(\varphi, \varphi)$. Our analysis of Broyden's method will compute the updated quantities, ε_+, δ_+, η_+, and T_+, corresponding to $A_+ \in \mathscr{W}(\varepsilon_+, \delta_+, \eta_+)$ given that $u_c \in W_2(\rho_c)$. We will show that ζ_+ and T_+ remain small and that η_+ can be bounded by a constant multiple of ρ_c. This will suffice to prove linear convergence.

The purpose of this section is to prove the following theorem.

THEOREM 3.3. *Let F satisfy the hypotheses of Theorem* 3.1. *Then there is $c_\nu > 0$ such that if $\rho_0 = P_N u_0$, $A_0 \in \mathscr{W}(|\varepsilon_0|, \delta_0, \eta_0)$, and $u_0 \in$*

$W_2(|\rho_0|)$, $|\varepsilon_0|$, η_0, δ_0, and $|\rho_0|$ are sufficiently small, and

$$\nu_0 = \rho_0/\varepsilon_0 \in \left(c_\nu\sqrt{\delta_0 + |\rho_0| + |\varepsilon_0|}, \tfrac{5}{4}\right), \tag{3.10}$$

then the Broyden iterates, $\{u_j\} \subset W_2(|\rho_0|)$, exist and converge q-linearly to u^*. Moreover, if

$$r = \overline{\lim_{n\to\infty}} \frac{\|e_{n+1}\|}{\|e_n\|},$$

then there is $c_r > 0$ such that

$$\left| r - \frac{\sqrt{5}-1}{2} \right| \le c_r(|\varepsilon_0| + \delta_0 + |\rho_0|). \tag{3.11}$$

There are some unusual requirements in the statement of Theorem 3.3. We require that ρ_0 and ε_0 have the same sign; this is certainly the case if $A_0 \approx F'(u_0)$. This condition holds at the level of the coarsest mesh in our case and will continue to hold as long as the sign of ρ does not change during the iteration. We shall see that this sign does not change in the course of the proof. Another requirement is that $\nu_0 = \rho_0/\varepsilon_0$ not be too small. Now ρ represents the accuracy in the initial approximation to u^* and ε_0 the accuracy in the derivative. This second requirement is that these initial accuracies not be too far out of balance. We indicate in remarks following the proof how this condition can be maintained.

The proof of Theorem 3.3 will be through a sequence of intermediate results. We begin with a lemma on the invertibility of $A \in \mathscr{W}(\varepsilon, \delta, \eta)$ for δ sufficiently small and η bounded.

LEMMA 3.1. *There are $c_I > 0$, $\overline{\delta}$ and $\overline{\eta}$ such that if $\delta \le \overline{\delta}$, $\eta < \overline{\eta}$, and $\varepsilon \ne 0$, and $A \in \mathscr{W}(\varepsilon, \delta, \eta)$ is given by (3.8), then A^{-1} exists and*

$$A^{-1} = P_X + \frac{1}{\sigma\varepsilon}P_N - \frac{\tilde{\xi} \otimes \varphi}{\sigma} + E, \tag{3.12}$$

where

$$\|P_N E\| \le c_I\delta/|\varepsilon| \quad \textit{and} \quad \|P_X E\| \le c_I\delta. \tag{3.13}$$

PROOF. Let $M = P_X + T$ and let $\overline{A} = M + \sigma\varepsilon P_N$. By hypothesis, if $\overline{\delta} < 1$, $\overline{A}^{-1}$ exists and

$$\overline{A}^{-1} = M^{-1} + \frac{1}{\sigma\varepsilon}P_N$$

where M^{-1} denotes the inverse of the restriction of M to X and

$$\|M^{-1}\|_{\mathscr{L}} \le (1 - \overline{\delta})^{-1}.$$

Let $Q = \overline{A} + \varepsilon\tilde{\xi} \otimes \varphi$. By the orthogonality of N and X and the Sherman–Morrison–Woodbury formula, we have

$$Q^{-1} = (I - \varepsilon(M^{-1}\tilde{\xi}) \otimes \varphi)\overline{A}^{-1}. \tag{3.14}$$

We use the orthogonality and the Sherman–Morrison–Woodbury formula again together with

$$Q^{-1}\varphi = \frac{\varphi - \varepsilon M^{-1}\tilde{\xi}}{\sigma\varepsilon}, \qquad \langle \zeta, Q^{-1}\varphi \rangle = -\langle \zeta, M^{-1}\tilde{\xi} \rangle / \sigma,$$

and

$$\|\tilde{\xi}\| \le \overline{\eta} + \|\hat{\xi}\|$$

to conclude that if

$$\overline{\delta}(\|\hat{\xi}\| + \overline{\eta})(1 - \overline{\delta})^{-1} < 1, \tag{3.15}$$

then $A = Q + \varphi \otimes \zeta$ has inverse

$$A^{-1} = \left(I - \frac{(Q^{-1}\varphi) \otimes \zeta}{1 - \langle \zeta, M^{-1}\tilde{\xi} \rangle / \sigma} \right) Q^{-1}. \tag{3.16}$$

For simplicity we will define c_a by

$$c_a = (1 - \overline{\delta}(\|\hat{\xi}\| + \overline{\eta})(1 - \overline{\delta})^{-1})^{-1}. \tag{3.17}$$

The remainder of the proof of the lemma is to verify (3.12) and (3.13). First we observe that $M^{-1} = P_X + E_a$, where $E_a \colon X \to X$, $E_a P_N = 0$, and $\|E_a\| \le \overline{\delta}(1 - \overline{\delta})^{-1}$. Hence,

$$\begin{aligned} Q^{-1} &= (I - \varepsilon\tilde{\xi} \otimes \varphi - \varepsilon E_a \tilde{\xi} \otimes \varphi)\overline{A}^{-1} \\ &= (I - \varepsilon\tilde{\xi} \otimes \varphi - \varepsilon E_a \tilde{\xi} \otimes \varphi)\left(P_X + \frac{1}{\sigma\varepsilon} P_N + E_a \right) \\ &= P_X + \frac{1}{\sigma\varepsilon} P_N - \frac{\tilde{\xi} \otimes \varphi}{\sigma} + E_b. \end{aligned} \tag{3.18}$$

In (3.18)

$$E_b = E_a - \frac{E_a \tilde{\xi} \otimes \varphi}{\sigma}.$$

We set

$$c_Q = 1 + \frac{\|\hat{\xi}\| + \overline{\eta}}{|\sigma|}. \tag{3.19}$$

We obtain the estimates

$$\|E_b\|_{\mathscr{L}} \le \overline{\delta} c_Q (1 - \overline{\delta})^{-1} \tag{3.20}$$

and

$$\|P_X Q^{-1}\|_{\mathscr{L}} \leq c_Q(1-\overline{\delta})^{-1}. \tag{3.21}$$

We let $\tilde{\zeta} = \zeta/(1 - \langle \zeta, M^{-1}\tilde{\xi}\rangle/\sigma)$ and note that $\|\tilde{\zeta}\| \leq \overline{\delta} c_a$, where c_a is given by (3.17). We may apply P_N to both sides of (3.16) and use (3.18) to obtain

$$P_N A^{-1} = \left(I - \frac{\varphi \otimes \tilde{\zeta}}{\sigma\varepsilon}\right) Q^{-1} = \frac{1}{\sigma\varepsilon} P_N + P_N E, \tag{3.22}$$

where

$$P_N E = -\frac{\varphi \otimes \tilde{\zeta}}{\sigma\varepsilon}(I + E_b) + \frac{\langle \tilde{\zeta}, \tilde{\xi}\rangle}{\sigma\varepsilon}.$$

Therefore, $\|P_N E\|_{\mathscr{L}} \leq c_I^{(1)}\overline{\delta}/|\varepsilon|$, where

$$c_I^{(1)} = c_a\left(1 + \frac{\|\hat{\xi}\| + \overline{\eta}}{|\sigma|} + \|E_b\|_{\mathscr{L}}\right) \Big/ |\sigma|,$$

and $\|E_b\|_{\mathscr{L}}$ can be estimated by (3.20).

We apply P_X to (3.16) and obtain

$$P_X A^{-1} = \left(P_X + \frac{M^{-1}\tilde{\xi} \otimes \tilde{\zeta}}{\sigma}\right) Q^{-1} = P_X - \frac{\tilde{\xi} \otimes \varphi}{\sigma} + P_X E, \tag{3.23}$$

where

$$P_X E = E_b + \frac{M^{-1}\tilde{\xi} \otimes \tilde{\zeta}}{\sigma} P_X Q^{-1}.$$

Hence, $\|P_X E\|_{\mathscr{L}} c_I^{(2)}\overline{\delta}$, where

$$c_I^{(2)} = c_Q(1-\overline{\delta})^{-1}(1 + c_a(c_Q - 1)(1-\overline{\delta})^{-1}).$$

Letting $c_I = \max(c_I^{(1)}, c_I^{(2)})$ completes the proof.

An immediate consequence of this lemma is that if $A \in \mathscr{W}(\varepsilon, \overline{\delta}, \overline{\eta})$, then

$$\|P_X A^{-1}\| \leq c_A = c_Q + c_I\overline{\delta}.$$

The next lemma is a synthesis of results from the literature on singular problems. We omit the proof.

Lemma 3.2. *There are $\overline{\rho}$ and c_S such that if $u \in W_w(\overline{\rho})$ and $P_N e = \rho\varphi$, then*

$$\|F(u)\| \leq c_S \rho^2,$$

and if β_3 on $\mathscr{X}$ is defined by

$$\beta_3(u) = F(u) - P_X e - \frac{\rho^2 \hat{\xi}}{2} - \frac{\rho^2 \sigma \varphi}{2}, \tag{3.24}$$

then $\|\beta_3(u)\| \leq c_2|\rho|^3$.

The next theorem in this section is that if $\overline{\rho}$, δ, and η are sufficiently small, $u \in W_2(\overline{\rho})$, and $A \in \mathscr{W}(\varepsilon, \delta, \eta)$, then the chord step remains in $W_2(\overline{\rho})$ and moves closer to the root. This is the result in **[10]** with enough additional information to allow us to analyze the effect of a Broyden update on A. An interesting aspect of this theorem is that if $P_N e = \rho\varphi$ then ρ and ε must have the same sign.

THEOREM 3.4. *Let $A \in \mathscr{W}(\varepsilon, \delta, \eta)$ with $\delta \leq \overline{\delta}$, $\eta \leq \overline{\eta}$, and let $u \in W_2(\overline{\rho})$ with $P_N u = \rho\varphi$ and*

$$\rho = \nu\varepsilon \tag{3.25}$$

for some $\nu \in (0, 2)$. Assume that $\overline{\rho}$, $\overline{\delta}$, and $\overline{\eta}$ are small enough so that

$$\frac{.5\overline{\eta} + (c_I\overline{\delta} + c_A\overline{\rho})\nu c_S}{I - .5\nu - (c_I\overline{\delta} + c_A\overline{\rho})\nu c_S} \leq 1. \tag{3.26}$$

Then if $u_+ = u - A^{-1}F(u)$, $u \in W_2(\overline{\rho})$ and

$$e_+ = u_1 - u^* = \rho\left(1 - \frac{\rho}{2\varepsilon}\right)\varphi + \frac{\rho^2}{2}(\tilde{\xi} - \hat{\xi}) + \mu_1, \tag{3.27}$$

where

(3.28)

$$\|P_X\mu_1\| \leq (c_I\delta + c_A|\rho|)c_S\rho^2 \quad \text{and} \quad \|P_N\mu_1\| \leq \frac{\rho}{\varepsilon}\left(\frac{|\rho|}{|\sigma|} + c_I\delta\right)c_S|\rho|.$$

PROOF. By (3.12) and (3.24),

$$A^{-1}F(u) = \frac{\rho^2}{2\varepsilon}\varphi + P_X e + \rho^2\frac{(\hat{\xi} - \tilde{\xi})}{2} - \mu_1, \tag{3.29}$$

where

$$\mu_1 = EF(u) = P_X A^{-1}\beta_3(u) + \frac{1}{\sigma\varepsilon}P_N\beta_3(u).$$

Hence, (3.27) holds. (3.28) is a consequence of (3.13) and Lemma 3.2. Therefore, if (3.26) holds, $u_+ \in W_2(\rho)$.

We can now describe how the approximate derivative is updated. Note the improvement in ε and η. This result will lead directly to the linear convergence result for Broyden's method.

THEOREM 3.5. *Let $A \in \mathscr{W}(\varepsilon, \delta, \eta)$ and $u \in W(\overline{\rho})$ with $|\varepsilon| < \overline{\varepsilon}$, $\delta < \overline{\delta}$, $\eta < \overline{\eta}$, and $P_N e = \rho\varphi$ such that $\nu = \rho/\varepsilon \in (0, 2)$ are such that the assumptions to Theorem 3.4 hold. Then if $\overline{\varepsilon}$ is sufficiently small there is $\overline{\kappa}$ such that if*

$$A_+ = A + \frac{F(u_+) \otimes s}{\|s\|^2},$$

then $A_+ \in \mathscr{W}(\varepsilon_+, \delta_+, \eta_+)$, where

$$\text{(3.30)} \quad \begin{aligned} &\text{(a)} \quad |\varepsilon_+ - \rho(1 - \nu/4)| \le \overline{\kappa}|\varepsilon|(|\varepsilon| + |\rho| + \delta) \\ &\text{(b)} \quad \eta_+ \le \overline{\kappa}|\varepsilon|(|\varepsilon| + |\rho| + \delta) \\ &\text{(c)} \quad \delta_+ \le \delta + \overline{\kappa}|\varepsilon|(|\varepsilon| + |\rho| + \delta). \end{aligned}$$

PROOF. By (3.24) and (3.27) we have

$$F(u_+) = \frac{\rho^2}{2}(\tilde{\xi} - \hat{\xi}) + \frac{\rho_+^2}{2}(\hat{\xi} + \sigma\varphi) + \mu_3, \tag{3.31}$$

where $\mu_3 = \beta_3(u_+) + P_X \mu_1$. Hence, by (3.28),

$$\|\mu_3\| \le \kappa_0 \rho^2(|\rho| + \delta), \tag{3.32}$$

where $\kappa_0 = c_s + c_A + c_I$.

By (3.29),

$$s = -\frac{\rho^2}{2\varepsilon}(\varphi + \mu_4), \tag{3.33}$$

where

$$\mu_4 = \frac{\varepsilon}{\rho^2}(P_X e - \rho^2(\hat{\xi} - \overline{\xi})/2 + \mu_1).$$

Hence,

$$\|\mu_4\| \le \kappa_1 |\varepsilon|, \tag{3.34}$$

where

$$\kappa_1 = 1 + \overline{\eta}/2 + c_S(2c_I\overline{\delta} + c_A\overline{\rho} + \overline{\rho}/|\sigma|).$$

Therefore, if we write $F(u_+) \otimes s/\|s\|^2 = \Delta_1 + E_1$, where

$$\Delta_1 = -\varepsilon\left((\tilde{\xi} - \hat{\xi}) + \frac{\varepsilon\rho_+^2}{\rho^2}(\hat{\xi} + \sigma\varphi)\right) \otimes \varphi,$$

and

$$E_1 = (1 - \|\varphi + \mu_4\|^{-2})\Delta_1 - \frac{2\varepsilon}{\rho^2\|\varphi + \mu_4\|^{-2}}(F(u_+) \otimes \mu_4 + \mu_3 \otimes (\varphi + \mu_4)),$$

we obtain the estimate

$$\|E_1\|_{\mathscr{L}} \le \kappa_3 |\varepsilon|(|\varepsilon| + |\rho| + \delta). \tag{3.35}$$

In (3.35) the constant κ_3 can be taken as $\kappa_3 = \tau\kappa_2 + (1+\tau\bar{\varepsilon})2c_S\kappa_1 + 2\kappa_0$, where $1 + \tau\varepsilon = (1 - \kappa_1\varepsilon)^{-2}$.

By (3.28) we may write $P_N\mu_1 = \rho m_1 \varphi$, where

$$|m_1| \le \kappa_4(|\rho| + \delta) \tag{3.36}$$

and κ_4 may be taken to be $2\max(|\sigma|^{-1}, c_I)c_S$. Hence,

$$\frac{\varepsilon\rho_+^2}{\rho^2} = \varepsilon\left(1 - \frac{\rho}{2\varepsilon}\right)^2 + m_2, \tag{3.37}$$

where $m_2 = 2\varepsilon(1-\rho/(2\varepsilon))m_1 + \varepsilon m_1^2$. Hence, $|m_2| \le \kappa_5|\varepsilon|(|\rho|+\delta)$, where $\kappa_5 = 2\kappa_4 + \kappa_4^2(\bar{\rho} + \bar{\delta})$. Hence, $\Delta_1 = \Delta_A + E_2$, where

$$\Delta_A = -\varepsilon(\tilde{\xi} - \hat{\xi} + (1 - \nu/2)^2(\hat{\xi} + \sigma\varphi)) \otimes \varphi \tag{3.38}$$

and $E_2 = -m_2(\sigma\varphi + \hat{\xi}) \otimes \varphi$. Therefore,

$$\frac{F(u_+) \otimes s}{\|s\|^2} = \Delta_A + E, \tag{3.39}$$

where $E = E_1 + E_2$ satisfies

$$\|E\|_{\mathscr{L}} \le \bar{\kappa}|\varepsilon|(|\varepsilon| + |\rho| + \delta). \tag{3.40}$$

In (3.40) we have $\bar{\kappa} = (1 + |\sigma|)\kappa_5 + \kappa_3$.

We can now complete the proof using (3.38), (3.39), and the estimate (3.40). Since $A_+ = A + F(u_+) \otimes s/\|s\|^2$ and $\varepsilon\nu = \rho$ we have that

$$A_+ = (P_X + T_+) + \varphi \otimes \zeta_+ + \varepsilon_+\tilde{\xi}_+ \otimes \varphi + \sigma\varepsilon_+ P_N,$$

where

$$T_+ = T + P_X E P_X, \qquad \zeta_+ = \zeta + P_X E^*\varphi,$$
$$\varepsilon_+ = \rho(1 - \nu/4) + \langle E\varphi, \varphi\rangle, \quad \text{and} \quad \tilde{\xi}_+ = \xi + P_X E\varphi.$$

The statement of the theorem follows from (3.40).

We are now able to complete the proof of Theorem 3.3. Let $\bar{\delta}$, $\bar{\eta}$, $\bar{\varepsilon}$, and $\bar{\rho}$ be such that the assumptions of Lemma 3.1, Lemma 3.2, Theorem 3.4, and Theorem 3.5 hold. We can continue the iteration and preserve the estimates in the previous results in this section provided that $A_j \in \mathscr{W}(\varepsilon_j, \delta_j, \eta_j)$ with $|\varepsilon_j| < \bar{\varepsilon}$, $\delta_j < \bar{\delta}$, and $\eta_j < \bar{\eta}$. We show how ε_0, δ_0, η_0, and ρ_0 must be chosen so that this happens. For j such that $A_j \in \mathscr{W}(\varepsilon_j, \delta_j, \eta_j)$ and $u_{j+1} \in W_2(\bar{\rho})$ we define, for $k \ge j+1$,

$$\tau_k = (|\rho_k| + |\varepsilon_k| + |\delta_k|).$$

We set $a = \overline{\kappa} + c_s(1/|\sigma| + c_I)$. We assume

$$(3.41)\qquad \begin{gathered} a\overline{\delta} \le \tfrac{1}{6}, \qquad (1 + 23\overline{\delta})\tau_0 \le \overline{\delta}, \\ \rho_0/\varepsilon_0 \in \left(\sqrt{12a\overline{\delta}},\ \tfrac{5}{4}\right), \quad \text{and} \quad |\rho_0| + a\overline{\delta}|\varepsilon_0| \le \overline{\varepsilon}. \end{gathered}$$

We show that (3.41) implies that the Broyden sequence exists and that the convergence is q-linear as asserted in Theorem 3.3. Then we will prove the more precise rate estimate in (3.11).

We first analyze the sequence $\{\nu_j\}$, where $\nu_j = \rho_j/\varepsilon_j$. To begin with we note that (3.27) implies that

$$\rho_1 = \rho_0\left(1 - \frac{\rho_0}{2\varepsilon_0}\right) + \nu_0\rho_0 c_0$$

and that (3.30) implies that $\varepsilon_1 = \rho_0(1 - \nu_0/4) + b_0\varepsilon_0$, where $|b_0|$ and $|c_0|$ are both $\le a\tau_0$. We divide and obtain

$$\nu_1 = \frac{\nu_0(1 - \nu_0/2 + c_0\nu_0)}{\nu_0(1 - \nu_0/4) + b_0}.$$

Hence, by (3.41) we have $\nu_1 \in (\frac{1}{6}, 1)$. Moreover,

$$\varepsilon_1 \le \rho_0 + a\overline{\delta}, \quad \delta_1 \le (1 + a\overline{\delta})\tau_0, \quad \rho_1 \le \rho_0, \quad \text{and} \quad \eta_1 \le a|\varepsilon_0|\overline{\delta}.$$

Hence, $A_1 \in \mathscr{W}(|\varepsilon_1|, \delta_1, \eta_1) \subset \mathscr{W}(\overline{\varepsilon}, \overline{\delta}, \overline{\eta})$ and we may continue the iteration.

We complete the proof by induction. Assume that $u_j \in W_2(|\rho_j|) \subset W_2(|\rho_0|)$ has been computed, that $\nu_j \in (\frac{1}{6}, 1)$, that $|\varepsilon_j| \le |\rho_{j-1}| + a\overline{\delta}|\varepsilon_{j-1}| \le \overline{\varepsilon}$, and that $\eta_j \le a|\varepsilon_{j-1}|\overline{\delta}$. We show that we can continue the iteration by showing that $\tau_{j+1} \le \overline{\delta}$. From the induction hypotheses, $u_{j+1} \in W_2(|\rho_j|)$ exists. As above we have $\nu_{j+1} \in (\frac{1}{6}, 1)$ and, in fact,

$$(3.42)\qquad |\rho_{j+1}| \le |\rho_j|(1 - (1 - a\overline{\delta})/12) \le 23|\rho_j|/24$$

and

$$(3.43)\qquad |\varepsilon_{j+1}| \le |\varepsilon_j|(\nu_j(1 - \nu_j/4) + a\overline{\delta}) \le 23|\varepsilon_j|/24.$$

Hence, $\tau_{j+1} \le \tau_j + a|\varepsilon_j|\overline{\delta} \le (1 + a\overline{\delta})\tau_0 + a|\varepsilon_1|\overline{\delta}\sum_{k=1}^{j}(\frac{23}{24})^k \le \overline{\delta}$. This completes the induction. Also note that q-linear convergence is implied by (3.42).

To complete the proof we note that, as $j \to \infty$,

$$(3.44)\qquad \nu_{j+1} = \frac{1 - \nu_j/2}{1 - \nu_j/4} + O(\overline{\delta}).$$

Let $N(\nu) = (1-\nu/2)/(1-\nu/4)$. $N(\nu)$ is a contraction on the interval $(\frac{1}{6}, 1)$ with fixed point $(3-\sqrt{5})$. Hence, as $j \to \infty$, $\nu_j - (3-\sqrt{5}) = O(\overline{\delta})$. This is the last assertion in Theorem 3.3.

We now indicate some ways in which the assumptions for the results above can be satisfied. We will assume, as in the examples, that $\mathscr{X}_l = \mathscr{Y}_l$ at all grid levels and that $\mathscr{X}_l$ is a subspace of $\mathscr{X}$. We first indicate that it is reasonable to assume that the initial approximate derivative at the fine mesh level, $\mathscr{E}A_0$, is in $\mathscr{W}(\varepsilon, \delta, \eta)$ for some sufficiently small ε, δ, and η. To do this we assume, as is the case in the examples at the end of the last section, that $A_0 = F_0'(\tilde{u})$, where $\tilde{u} \approx u_0^*$, the root of F_0, is nearly singular in the sense that A_0 can be written

$$A_0 = P_0 P_x P_0 + \sigma\tilde{\rho} P_N + \tilde{\rho}\tilde{\xi} \otimes \varphi + B_0. \tag{3.45}$$

In (3.45) $\tilde{\rho} \approx \|\tilde{u} - u_0^*\|$ and there is $c_B > 0$ such that

$$\|B_0 P_N\| \le c_B(\tilde{\rho}^2 + \tilde{\delta}), \quad \|B_0 P_X\| \le c_B(\tilde{\rho} + \tilde{\delta}), \quad \text{and} \quad s\|\tilde{\xi} - \hat{\xi}\| \le c_B\tilde{\delta}.$$

Here $\tilde{\delta} = \|u_0^* - u^*\|$ and P_0 is a projection onto $\mathscr{X}_0$. The idea is that $\tilde{\delta}$ measures truncation error at level 0 and $\tilde{\rho}$ measures the iteration error in $\tilde{u}$. We assume, again motivated by the examples, that the map $\mathscr{E}$ preserves this structure. That is,

$$A = \mathscr{E}A_0 = P_X + \sigma\tilde{\rho} P_N + \tilde{\rho}\tilde{\xi} \otimes \varphi + B. \tag{3.46}$$

In (3.46) B satisfies estimates like that for B_0. Hence, if $\sigma\tilde{\rho} \gg c_B(\tilde{\rho}^2 + \tilde{\delta})$, A will have an inverse and $A \in \mathscr{W}(\tilde{\rho}, c_A(\tilde{\rho} + \tilde{\delta}), c_A(\tilde{\rho} + \tilde{\delta}))$. The trick here is to find a root of F_0 to less than maximum accuracy (so that $\tilde{\rho} \gg c_B(\rho^2 + \tilde{\delta})$) in order to preserve the singular structure of the derivative.

We must also make sure that $\tilde{u} \in W_2(\overline{\rho})$ for some sufficiently small $\overline{\rho}$. If we wish to extend $\tilde{u}$ directly from mesh level 0 we may assume that if $\tilde{e} = \tilde{u} - \tilde{u}^*$, then there is c_N such that $\|P_X(\tilde{u} - u^*)\| \le c_N(\tilde{\rho}^2 + \tilde{\delta})$. This assumption reflects the near singularity of F_0. As we have no control over the direction of $u^* - \tilde{u}^*$, this estimate is the best that one can do in general. Hence, in order to force $\tilde{u} \in W_2(\overline{\rho})$ we would need $\tilde{\rho} = O(\overline{\delta}^2)$. The meaning of this is that an extremely low accuracy solution at a coarse grid level is needed to put $\tilde{u} \in W_2(\overline{\rho})$ if $\tilde{u}$ is simply taken to be the coarse mesh approximate solution. The prolongation in the next paragraph requires much less than $\tilde{\delta} = O(\tilde{\rho}^2)$.

A second approach, motivated by the examples in the previous section, is to write $F(u) = I - K(u)$, find $\tilde{u}$ and A_0 as before, and then take

$$u_0 = K(\tilde{u}) \tag{3.47}$$

at each mesh level. This approach allows us to solve the problems more accurately. At the lowest level, we can preserve the important structure of A_0 by requiring that $\tilde{\tilde{\delta}} \ll \tilde{\rho}$. This is weaker than $\tilde{\tilde{\delta}} \approx \tilde{\rho}^2$. For any $\tilde{u}$ near u^* if we let $\tilde{e} = \tilde{u} - u^*$ and $\tilde{r} = \|\tilde{e}\|$, we have $e = K(\tilde{u}) - u^* = K(\tilde{u}) - K(u^*)$ and so there is $c > 0$ such that $\|e - K'(u^*)\tilde{e}\| \le c\tilde{r}^2$. As $K'(u^*) = I - F'(u^*) = P_N$ and $\tilde{r} \approx \|e\|$, we have $u \in W_2(\tilde{r})$ with $c_W = c/(1-c\tilde{r})$. Note that if K were approximated up to accuracy $O(r)$ or better the estimate would not change. An approximate solution as good as the truncation error can be used as we need make no effort to force $y \in W_2(\rho)$ as the prolongation will take care of that.

Finally, we consider the requirement that ν not be too small. As we saw above, at the beginning $\rho \approx \varepsilon$ and so $\nu \approx 1$. However, if we move u through several meshes and leave A_0, and hence ε, unchanged, ν could become too small. However, on any given grid level $\varepsilon_j \to 0$. Hence, if one were to carry A_0 and the accumulated information of the rank-one updates to the finer meshes, ν would remain safely bounded away from zero. We found no need to do this for the singular problem in the examples.

4. Numerical examples. As an example we consider Chandrasekhar's H-equation ([7]),

$$\mathcal{F}(H)(x) = H(x) - (1 - c\mathcal{T}(H)(x))^{-1} = 0, \tag{4.1}$$

where $H \in C[0, 1]$ is the unknown function, $c \in (0, 1]$ is a parameter, and the integral operator $\mathcal{T}$ is given by $\mathcal{T}(u)(x) = \int_0^1 xu(y)/(x+y)\,dy$. This equation arises in the theory of neutron and radiative transport. For each $c \in (0, 1]$ there is a solution, H, to (4.1) that is unique in the class of analytic $C[0, 1]$ valued functions of c. The Fréchet derivative of $\mathcal{F}$ is singular if $c = 1$ and nonsingular if $c < 1$ (**[23]**, **[31]**). Note that this is a slightly different problem than that considered in Section 2 in that instead of $\int_0^1 k(x, y, u(y))\,dy$ we have a nonlinear term of the form $\mathcal{G}(\int_0^1 k(x, y)u(y)\,dy)$ where $\mathcal{G}$ is a nonlinear function. The discussion in Section 2 carries over to this case in an obvious way.

We used composite 20-point Gaussian quadratures for integration. The mesh levels $l = 0, 1, 2, 3$ corresponded to $N_l = 20, 40, 80$, and 160 unknowns. The subintervals were of equal length. (2.21) defines the operator prolongation. We define the prolongation of functions by (3.47) for both nonsingular and singular problems. We solved the coarse mesh problem by Newton's method, terminating the iteration when the step length was $< 10^{-1}$ in the L^∞ norm. For the nonsingular problems we used $\omega_l = 10^{-3(1+l)}$ for $l = 1, 2, 3$. As a nonsingular problem we used $c = .9$. In Table 1, we tabulate the norms of stepsizes and function values and the ratios of new and old norms of steps and functions for each mesh level. Note that two iterates at some mesh levels were needed. The reason for this was that $F_0'(u_0^{n_0})$ was a poor approximation to $F'(u^*)$. In Table 2 we tabulate similar results, the only difference being that the iteration at the coarse mesh was terminated when the step size was $< 10^{-3}$. In that case only one step was computed on each mesh. In Table 1 and the tables that follow, $r_n = \|s_n\|/\|s_{n-1}\|$ and $R_n = \|F(u_n)\|/\|F(u_{n-1})\|$ for $n \geq 1$. The norm used in Tables 1 and 2 is the max norm. In Table 3, where convergence properties of Broyden's method are observed, the L^2 norm is used.

For the singular problem, $c = 1$, we tabulate the same information in Table 3. Here we use Broyden's method to recover linear convergence. Note that after the first few iterations the rate is always near $(\sqrt{5}-1)/2$. We terminated the iteration at level 0 when the norm of the step was $< 10^{-2}$ and then at level l when the norm of the function was $< 10^{3l+3}$. Note that the ratios of successive step lengths may increase in the early stages of the iteration. The reason for this is that the first chord steps are too small to provide sufficient decrease in the error. Broyden's method compensates for this, as the theory predicts.

Table 1. Nested iteration, poor approximation

n	N_l	$\|s_{n-1}\|_\infty$	r_{n-1}	$\|F_l(u_n)\|_\infty$	R_n
0	40			0.42d-03	
1	40	0.80d-03		0.33d-05	0.79d-02
2	40	0.63d-05	0.79d-02	0.26d-07	0.79d-02
0	80			0.51d-04	
1	80	0.51d-04		0.27d-08	0.53d-04
2	80	0.51d-08	0.10d-03	0.21d-10	0.77d-02
0	160			0.26d-04	
1	160	0.26d-04		0.66d-09	0.26d-04
2	160	0.12d-08	0.49d-04	0.51d-11	0.77d-02
3	160	0.98d-11	0.77d-02	0.40d-13	0.78d-02

Table 2. Nested iteration, improved approximation

n	N_l	$\|s_{n-1}\|_\infty$	r_{n-1}	$\|F_l(u_n)\|_\infty$	R_n
0	40			0.10d-03	
1	40	0.10d-03		0.12d-09	0.12d-05
0	80			0.51d-04	
1	80	0.51d-04		0.51d-10	0.30d-06
0	160			0.26d-04	
1	160	0.26d-04		0.31d-11	0.12d-06
2	160	0.26d-11	0.10d-06	0.22d-15	0.72d-04

Table 3. Singular problem

n	N_l	$\|s_{n-1}\|_2$	r_{n-1}	$\|F_l(u_n)\|_2$	R_n
0	40			0.18d-04	
1	40	0.16d-02		0.10d-04	0.56d+00
2	40	0.20d-02	0.13d+01	0.39d-05	0.39d+00
3	40	0.96d-03	0.48d+00	0.16d-05	0.40d+00
4	40	0.65d-03	0.68d+00	0.59d-06	0.37d+00
0	80			0.27d-05	
1	80	0.43d-04		0.49d-06	0.18d+00
2	80	0.52d-03	0.12d+02	0.12d-06	0.25d+00
3	80	0.15d-03	0.28d+00	0.55d-07	0.44d+00
4	80	0.13d-03	0.86d+00	0.20d-07	0.35d+00
5	80	0.69d-04	0.55d+00	0.77d-08	0.39d+00
6	80	0.45d-04	0.65d+00	0.29d-08	0.38d+00
7	80	0.27d-04	0.61d+00	0.11d-08	0.38d+00
8	80	0.17d-04	0.62d+00	0.43d-09	0.38d+00
0	160			0.96d-06	
1	160	0.96d-06		0.39d-09	0.41d-03
2	160	0.31d-07	0.32d-01	0.39d-09	0.10d+01
3	160	0.14d-04	0.44d+03	0.11d-09	0.28d+00
4	160	0.46d-05	0.34d+00	0.48d-10	0.44d+00
5	160	0.37d-05	0.79d+00	0.17d-10	0.36d+00
6	160	0.21d-05	0.56d+00	0.67d-11	0.39d+00
7	160	0.13d-05	0.64d+00	0.25d-11	0.38d+00
8	160	0.80d-06	0.61d+00	0.97d-12	0.38d+00

REFERENCES

1. E. L. Allgower, K. Böhmer, F. A. Potra, and W. C. Rheinboldt, *A mesh-independence principle for operator equations and their discretizations*, SIAM J. Numer Anal. **23** (1986), 160–169.

2. E. L. Allgower and K. Böhmer, *Application of the mesh independence principle to mesh refinement strategies*, SIAM J. Numer. Anal. **24** (1987), 1335–1351.

3. K. Atkinson, *Iterative variants of the Nyström method for the numerical solution of integral equations*, Numer. Math. **22** (1973), 17–31.

4. C. G. Broyden, *A class of methods for solving simultaneous equations*, Math. Comp. **19** (1965), 577–593.

5. ——, *The convergence of single-rank quasi-Newton methods*, Math. Comp. **24** (1970), 365–382.

6. C. G. Broyden, J. E. Dennis, and J. J. Moré, *On the local and superlinear convergence of quasi-Newton methods*, J.I.M.A. **12** (1973), 223–246.

7. S. Chandrasekhar, *Radiative Transfer*, Dover, New York, 1960.

8. D. W. Decker, H. B. Keller, and C. T. Kelley, *Convergence rates for Newton's method at singular points*, SIAM J. Numer. Anal. **20** (1983), 296–314.

9. D. W. Decker and C. T. Kelly, *Newton's method at singular points* I, SIAM J. Numer. Anal. **17** (1980), 66–70.

10. ——, *Sublinear convergence of the chord method at singular points*, Numer. Math. **42** (1983), 147–154.

11. ——, *Expanded convergence domains for Newton's method at nearly singular roots*, SIAM J. Sci. Stat. Comp. **6** (1985), 951–966.

12. ——, *Broyden's method for a class of problems having singular Jacobian at the root*, SIAM J. Numer. Anal. **22** (1985), 566–574.

13. R. S. Dembo, S. C. Eisenstat, and T. Steihaug, *Inexact Newton methods*, SIAM J. Numer. Anal. **19** (1982), 400–408.

14. J. E. Dennis and R. B. Schnabel, *Numerical Methods for Nonlinear Equations and Unconstrained Optimization*, Prentice-Hall, Englewood Cliffs, New Jersey, 1983.

15. A. Griewank, *Analysis and modification of Newton's method at singularities*, thesis, Australian National University, 1980.

16. ——, *The local convergence of Broyden-like methods on Lipschitzian problems in Hilbert space*, SIAM J. Numer. Anal. **24** (1987), 684–705.

17. ——, *Rates of convergence for secant methods on nonlinear problems in Hilbert space*, Springer Lecture Notes no. 1230, J. P. Hennart (ed.), Springer-Verlag, New York, 1987, pp. 138–157.

18. ——, *The solution of boundary value problems by Broyden-based secant methods*, in Computational Techniques and Applications (J. Noye and R. May, eds.), CTAC (Proc. Conf., Melbourne, August 1985), North-Holland, 1986, pp. 309–321.

19. W. Hackbusch, *Multigrid methods of the second kind*, Multigrid Methods for Integral and Differential Equations, D. J. Paddon and H. Holstein (eds.), Oxford University Press, Oxford, 1985.

20. W. E. Hart and S. O. W. Soul, *Quasi-Newton methods for discretized nonlinear boundary problems*, J. Inst. Appl. Math. **11** (1973), 351–359.

21. E. Isaacson and H. B. Keller, *Analysis of Numerical Methods*, Wiley, New York, 1966.

22. L. V. Kantorovich and G. P. Akilov, *Functional Analysis in Normed Spaces*, Pergamon, New York, 1964.

23. C. T. Kelley, *Solution of the Chandrasekhar H-equation by Newton's method*, J. Math. Phys. **21** (1980), 1625–1628.

24. C. T. Kelley and J. I. Northrup, *Pointwise quasi-Newton methods and some applications*, Distributed Parameter Systems, F. Kappel, K. Kunisch, and W. Schappacher (eds.), Springer-Verlag, New York, 1987, pp. 167–180.

25. ——, *A pointwise quasi-Newton method for integral equations*, SIAM J. Numer. Anal. **25** (1988), 1138–1155.

26. C. T. Kelley and E. W. Sachs, *Broyden's method for approximate solution of nonlinear integral equations*, J. Int. Eqs. **9** (1985), 25–44.

27. ——, *A quasi-Newton method for elliptic boundary value problems*, SIAM J. Numer. Anal. **24** (1987), 516–531.

28. ——, *Quasi-Newton methods and optimal control problems*, SIAM J. on Control and Optimization **25** (1987), 1503–1517.

29. ——, *A pointwise quasi-Newton method for unconstrained optimal control problems*, Numer. Math. **55** (1989), 159–176.

30. ——, *Approximate quasi-Newton methods*, to appear in Math. Prog., Ser. B.

31. T. W. Mullikin, *Neutron branching processes*, J. Math. Anal. Appl. **8** (1961), 520–525.

32. J. I. Northrup, *Pointwise quasi-Newton methods and integral equations*, thesis, North Carolina State University, 1988.

33. G. W. Reddien, *On Newton's method for singular problems*, SIAM J. Numer. Anal. **15** (1978), 993–996.

34. E. Sachs, *Broyden's method in Hilbert space*, Math. Prog. **35** (1986), 71–82.

35. J. Stoer, *Two examples on the convergence of certain rank-2 minimization methods for quadratic functionals in Hilbert space*, Lin. Alg. Appl. **28** (1984), 37–52.

North Carolina State University

Lectures in Applied Mathematics
Volume **26** (1990)

Smooth Penalty Functions and Continuation Methods for Constrained Optimization

BRUCE N. LUNDBERG, AUBREY B. POORE, AND BING YANG

Abstract. Several path-following algorithms for constrained optimization along with numerical test results are presented. These algorithms combine three smooth penalty functions (the quadratic penalty for equality constraints and the quadratic loss and log barrier functions for inequality constraints) and their modern counterparts (augmented Lagrangian or multiplier methods) with predictor-corrector continuation methods and sequential quadratic programming. In the first phase an unconstrained or linearly constrained penalty function, or augmented Lagrangian, is minimized. A homotopy path generated from these functions is then followed to optimality using efficient predictor-corrector continuation methods. Simple continuation steps are asymptotic to those taken by sequential quadratic programming, which then can be used in the final steps. Numerical test results show the methodology to be efficient, robust, and a competitive alternative to sequential quadratic programming. Many well-known optimization and continuation techniques are used as part of these algorithms and are reviewed to explain their connection with the current procedures.

1. Introduction. Path-following algorithms for constrained optimization have been revitalized in recent years, due in no small part to the claims and success of the Karmarkar algorithm in linear programming.

1980 *Mathematics Subject Classification* (1985 *Revision*). Primary 49D37, 65K05, 90C30.

The work of the first author was partially supported by the National Aeronautics and Space Administration through NASA Grant #NGT-06-002-802. The second and third authors were partially supported by the Air Force Office of Scientific Research through Grant #AFOSR-88-0059 and by the National Science Foundation through Grant #DMS-87-04679.

The ones presented here can be viewed as combinations of various elements and techniques in nonlinear programming: three smooth penalty functions (quadratic penalty for equality constraints, quadratic loss and log barrier for inequality constraints), their more modern counterparts (augmented Lagrangian or multiplier methods), sequential quadratic programming (Newton's method), and predictor-corrector continuation methods for efficient path-following. The objective in this work then is to describe this class of algorithms and to present numerical evidence of their efficiency, robustness, and potential.

One view [**29**] of these algorithms starts with the three aforementioned smooth penalty functions. One first performs an unconstrained or linearly constrained optimization of the penalty function. The minimizer then satisfies a set of first-order necessary conditions (the gradient of the penalty function is zero when all constraints are incorporated into the penalty function) from which one can define an equivalent system of parametrized nonlinear equations, which does not have the inevitable ill-conditioning present in these smooth penalty funtions. This system represents a perturbation of the Karush–Kuhn–Tucker first-order necessary conditions, and the solution is followed to optimality using efficient predictor-corrector continuation methods. The simplest predictor-corrector steps are asymptotic to those taken in sequential quadratic programming, which may be used in the final steps of the algorithm. When shifts and weights are added to these three penalty functions and are adaptively chosen or updated during the optimization phase, one has the class of multiplier or augmented Lagrangian methods. Theoretically, one can expect a shorter path through the use of augmented Lagrangians [**15**, Theorem 12.2.1], which suggests that these updates in the shifts and weights may be used to generate good paths to optimality.

A different perspective of these algorithms evolves from sequential quadratic programming (SQP) itself. The SQP methods perform exceptionally well in minimizing function evaluations, but may be slow since the combinatorial complexity of the inequality constraints is reflected in the subproblems at each step. Furthermore, they are currently restricted to small to medium-size problems with promise for large-scale applications [**15**], [**33**]. Augmented Lagrangian methods, on the other hand, are currently used effectively for large-scale problems with structured sparsity arising, for example, from discretized differential equations in control and nonlinear variational inequalities [**33**, p. 190]; [**3**, pp. 223–224]; [**17**], [**18**]. Although augmented Lagrangians are often used as merit functions for globalizing sequential quadratic programming, the minimizer at any given stage is not a solution of the original problem, and thus the homotopy between the minimizer of the augmented Lagrangian and sequential quadratic programming may be viewed as an intermediate globalization technique.

Having given this view, several qualifications are needed. For many large-scale optimization problems there is generally a lack of sparsity in the Hessian of the augmented Lagrangian due to the term $\mathbf{AA}^T$, where $\mathbf{A}$ is the active set constraint Jacobian matrix, even though $\mathbf{A}$ may be sparse. Thus, one must store and factor full $n \times n$ matrices, thus rendering these methods ineffective. However, for the aforementioned discretized differential equations, the matrix $\mathbf{A}$ is frequently banded and the term $\mathbf{AA}^T$ is also banded but with a larger bandwidth. Thus, the augmented Lagrangians and penalty functions for these applications do not suffer the same degradation in sparsity. This does not mean that sequential quadratic programming could not be used just as effectively, particularly if one could use updating procedures appropriately. Finally, if a good initial approximation is known or if a good guess for the active set is available, the sequential quadratic programming may well be the method of choice.

In Sections 2–4, we briefly outline the methodology for the mathematical programming problem

$$\min\{f(\mathbf{x})|\mathbf{h}(\mathbf{x}) = \mathbf{0},\ \mathbf{g}(\mathbf{x}) \geq \mathbf{0}\} \tag{1.1}$$

where $f\colon \mathbf{R}^n \to \mathbf{R}^1$, $\mathbf{h}\colon \mathbf{R}^n \to \mathbf{R}^q$, and $\mathbf{g}\colon \mathbf{R}^n \to \mathbf{R}^p$ are assumed to be twice continuously differentiable in an open set Ω containing the feasible region $\mathscr{R} = \{\mathbf{x}|\mathbf{h}(\mathbf{x}) = \mathbf{0},\ \mathbf{g}(\mathbf{x}) \geq \mathbf{0}\}$. The relation to other methods based on ordinary differential equations and to sequential quadratic programming is discussed in Sections 5 and 6. Section 7 contains the results of our current numerical testing, which demonstrates the robustness, efficiency, and potential of the general methodology.

2. Homotopy methods for constrained optimization. The idea of a homotopy method is to continuously embed a "difficult" problem into a parametrized set of problems such that at one parameter value the problem is "easy" to solve and at another value one recovers the "difficult" problem. One then continues the solution of this parametrized system from the "easy" problem to the "difficult" one. For nonlinear equations $\mathbf{F}(\mathbf{x}) = \mathbf{0}$, two commonly used homotopies are $\mathbf{G}(\mathbf{x}, t) = t\mathbf{F}(\mathbf{x}) + (1-t)(\mathbf{F}(\mathbf{x}) - \mathbf{F}(\mathbf{a}))$ (the global Newton homotopy) and $\mathbf{G}(\mathbf{x}, t) = t\mathbf{F}(\mathbf{x}) + (1-t)(\mathbf{x} - \mathbf{a})$ (the regularizing homotopy). Indeed, at $t = 0$ these are easy to solve, while at $t = 1$ one recovers the original problem. These homotopy methods tend to be quite robust, but currently are not as efficient as the use of a modified or quasi-Newton method with a merit function. The homotopy methods discussed here are generally very efficient, but the "easy" problem is not as easy as in the above homotopies in that the "easy" problem requires the solution of an unconstrained or linearly constrained optimization problem.

The first homotopy to be considered is based on the mixed quadratic penalty–log barrier function

$$P(\mathbf{x}, r) = f + \left(\frac{1}{2r}\right)\sum h_i^2(\mathbf{x}) - r\sum \ln(g_j(\mathbf{x})) \tag{2.1}$$

or in the more general form

$$\begin{aligned} P(\mathbf{x}, \nu, \hat{\lambda}, \sigma, \omega, \gamma, \delta, r) = \nu f + \left(\frac{1}{2r}\right)\sum \sigma_i(\gamma_i h_i(\mathbf{x}) - r\sigma_i^{-1}\hat{\lambda}_i)^2 \\ - r\sum \omega_j \ln(g_j(\mathbf{x}) + r\delta_j), \end{aligned} \tag{2.2}$$

where weights σ_i and ω_j, scales ν and γ_i, and shifts $\hat{\lambda}_i$ and δ_i have been introduced.

To explain how one can derive a homotopy, we consider the simpler form (2.1). At $r = r_0$, a minimizer, $\mathbf{x}_0$, of this penalty function satisfies

$$\nabla P = \nabla f + \sum \nabla h_i(h_i/r) - \sum (r/g_j)\nabla g_j = 0,$$

which, along with the definitions $\lambda_i = -h_i/r$ and $\mu_j = r/g_j$, yields an equivalent system of parametrized nonlinear equations

$$\mathbf{F}(\mathbf{x}, \boldsymbol{\lambda}, \boldsymbol{\mu}; r) = \begin{pmatrix} \nabla\mathscr{L}(\mathbf{x}, \boldsymbol{\lambda}, \boldsymbol{\mu}) \\ \mathbf{h}(\mathbf{x}) + r\boldsymbol{\lambda} \\ M\mathbf{g}(\mathbf{x}) - r\mathbf{e} \end{pmatrix} = \mathbf{0} \tag{2.3}$$

where $\boldsymbol{\mu} > \mathbf{0}$ and $\mathbf{g}(\mathbf{x}) > \mathbf{0}$ for $r > 0$, $\mathscr{L} = f - \mathbf{h}^{\mathrm{T}}\boldsymbol{\lambda} - \mathbf{g}^{\mathrm{T}}\boldsymbol{\mu}$, $M = \mathrm{diag}(\boldsymbol{\mu})$, $\boldsymbol{\mu} = (\mu_1, \ldots, \mu_p)^{\mathrm{T}}$, $\boldsymbol{\lambda} = (\lambda_1, \ldots, \lambda_q)^{\mathrm{T}}$, and $\mathbf{e} = (1, \ldots, 1)^{\mathrm{T}}$. This expansion (2.3) of $\nabla P = 0$ removes the ill-conditioning caused by the penalty function P [**29**]. A solution to this parametrized system at $r = r_0$ is given by $\boldsymbol{\lambda} = -\mathbf{h}(\mathbf{x}_0)/r$ and $\boldsymbol{\mu} = r/\mathbf{g}(\mathbf{x}_0)$, componentwise. Furthermore, these equations represent a perturbation of the Karush–Kuhn–Tucker first-order necessary conditions and $\boldsymbol{\mu}(r) > \mathbf{0}$ for $r > 0$. Once the optimization phase is complete, continuation techniques can be used to track the solution to optimality at $r = 0$. A similar homotopy can be derived for (2.2).

Another class of homotopies can be based on the quadratic penalty-loss function

$$P(\mathbf{x}, r) = f + \left(\frac{1}{2r}\right)\sum h_i^2(\mathbf{x}) + \left(\frac{1}{2r}\right)\sum [g_j^-(\mathbf{x})]^2 \tag{2.4}$$

or more generally, the augmented Lagrangian function or shifted quadratic penalty-loss function,

(2.5)

$$\begin{aligned} P(\mathbf{x}, \nu, \hat{\lambda}, \hat{\mu}, \sigma, \phi, \gamma, \beta, r) = \nu f &+ \left(\frac{1}{2r}\right)\sum \sigma_i(\gamma_i h_i(\mathbf{x}) - r\sigma_i^{-1}\hat{\lambda}_i)^2 \\ &+ \left(\frac{1}{2r}\right)\sum \phi_j[(\beta_j g_j(\mathbf{x}) - r\phi_j^{-1}\hat{\mu}_j)^-]^2, \end{aligned}$$

where $g_j^-(\mathbf{x}) = \min\{g_j(\mathbf{x}), 0\}$, and weights ν, σ_i, and ϕ_j, shifts $\hat{\lambda}_i$ and $\hat{\mu}_j$, and scales γ_i and β_j have been introduced. At a minimizer of the augmented Lagrangian

$$\begin{aligned} \nabla P = \nu\nabla f &+ \left(\frac{1}{r}\right)\sum \nabla(\gamma_i h_i)(\sigma_i\gamma_i h_i(\mathbf{x}) - r\hat{\lambda}_i) \\ &+ \left(\frac{1}{r}\right)\sum \nabla(\beta_j g_j)[\phi_j\beta_j g_j(\mathbf{x}) - r\hat{\mu}_j)^-] = 0. \end{aligned} \tag{2.6}$$

The definitions $\boldsymbol{\lambda} = -(\mathbf{S\Gamma h} - r\hat{\boldsymbol{\lambda}})/r$ and $\boldsymbol{\mu} = -(\mathbf{\Phi B g} - r\hat{\boldsymbol{\mu}})^-/r$ along with this equation yield the equivalent system

$$\begin{aligned} &\nabla\mathscr{L} = 0, \qquad \mathscr{L} = \nu f - (\mathbf{\Gamma h})^{\mathrm{T}}\boldsymbol{\lambda} - (\mathbf{Bg})^{\mathrm{T}}\boldsymbol{\mu}, \\ &\mathbf{\Gamma h} + \mathbf{S}^{-1}r(\boldsymbol{\lambda} - \hat{\boldsymbol{\lambda}}) = \mathbf{0}, \\ &\beta_j g_j(\mathbf{x}) + \phi_j^{-1}r(\mu_j - \hat{\mu}_j) = 0 \quad \text{if } \beta_j g_j(\mathbf{x}) < \phi_j^{-1}r\hat{\boldsymbol{\mu}}, \end{aligned} \tag{2.7}$$

where $\mathbf{S} = \mathrm{diag}(\boldsymbol{\sigma})$ and $\mathbf{\Phi} = \mathrm{diag}(\phi)$ represent the weights and $\mathbf{\Gamma} = \mathrm{diag}(\gamma)$ and $\mathbf{B} = \mathrm{diag}(\boldsymbol{\beta})$ the scales. The usual updates for the multipliers $\hat{\boldsymbol{\lambda}}$ and $\hat{\boldsymbol{\mu}}$ **[15]** can be used in the optimization phase.

Note that the use of different scales, weights, and updated multiplier approximations all change the homotopy path, and should be used to balance the objectives of generating a "good" homotopy path and moderating the difficulty of solving the unconstrained optimization problem. Furthermore, the ill-conditioning present in the penalty method is no longer present in the homotopies generated from these penalty functions. There are two important modifications to these homotopies. First, the minimization of the penalty function is solved only to low to moderate accuracy and the equation $\nabla\mathscr{L} = 0$ is replaced by $\nabla\mathscr{L} = (r/r_0)\nabla P(\mathbf{x}_0)$. Second, the Fritz John normalization of the multipliers **[29]** is used to prevent multipliers from tending to infinity, which happens generically when the linear independence constraint qualification is violated. When this normalization is used, the homotopy becomes a perburbation of the Fritz John first-order necessary conditions.

Finally, one should observe that at the end of the optimization phase the active set at the minimizer of the constrained problem may or may not have been correctly identified. If it has, then the path, in the absence of any singularities, is C^k provided the functions f, $\mathbf{h}$, and $\mathbf{g}$ are C^{k+1}. On the other hand, if the correct active set has not been identified, the path is continuous with a piecewise continuous derivative. This generally implies that predictor-corrector continuation procedures should be restarted when a constraint is "activated" or "deactivated" according to the relation of $\beta_j g_j(\mathbf{x})$ to $\phi_j^{-1} r\hat{\boldsymbol{\mu}}$ in (2.7), but one does not need to reoptimize the penalty function or the augmented Lagrangian.

3. The penalty or homotopy paths. The primary objective of this section is to state local conditions that ensure the existence of the penalty path and to discuss briefly the associated bifurcation problems. Central to these conditions and discussion are the first-order necessary and second-order sufficient conditions, which are intimately related to the robustness of these homotopies. As an example, we consider the homotopy path generated by the quadratic penalty-log barrier function. The quadratic penalty-loss function can be treated similarly **[29]**.

THEOREM 3.1 **[19, 29, 30]**. *Let $(\mathbf{x}_0, \boldsymbol{\lambda}_0, \boldsymbol{\mu}_0; 0)$ be a solution of $F(\mathbf{x}, \boldsymbol{\lambda}, \boldsymbol{\mu}; r) = \mathbf{0}$. Assume f, $\mathbf{g}$, and $\mathbf{h}$ are twice continuously differentiable in a neighborhood of $\mathbf{x}_0$ and define two index sets $\overline{\mathscr{A}}$ and $\mathscr{A}$ and a corresponding tangent space $\overline{T}$ by*

$$\overline{\mathscr{A}} = \{i \colon 1 \le i \le p,\ g_i(\mathbf{x}_0) = 0\}, \quad \mathscr{A} = \{i \in \overline{\mathscr{A}} \colon \mu_i^0 \neq 0\},$$
$$\overline{T} = \{\mathbf{y} \in \mathbf{R}^n \colon D_x\mathbf{h}(\mathbf{x}_0)\mathbf{y} = 0, \quad D_x g_i(\mathbf{x}_0)\mathbf{y} = 0\ (i \in \overline{\mathscr{A}})\}.$$

Then a necessary and sufficient condition that $D_{(x,\lambda,\mu)}\mathbf{F}(\mathbf{x}_0, \boldsymbol{\lambda}_0, \boldsymbol{\mu}_0; 0)$ be nonsingular is that each of the following three conditions hold:

(a) $\overline{\mathscr{A}} = \mathscr{A}$;

(b) $\{\{\nabla_x g_i(\mathbf{x}_0)\}_{i \in \overline{\mathscr{A}}} \cup \{\nabla_x h_j(\mathbf{x}_0)\}_{j=1}^q\}$ *is a linearly independent collection of $q + |\overline{\mathscr{A}}|$ vectors, where $|\overline{\mathscr{A}}|$ denotes the cardinality of $\overline{\mathscr{A}}$;*

(c) *the Hessian of the Lagrangian $\nabla_x^2 \mathscr{L}$ is nonsingular on the tangent space $\overline{T}$ at $(\mathbf{x}_0, \boldsymbol{\lambda}_0, \boldsymbol{\mu}_0)$.*

Suppose $D_{(\mathbf{x},\lambda,\mu)}\mathbf{F}(\mathbf{x}_0,\boldsymbol{\lambda}_0,\boldsymbol{\mu}_0;0)$ *is nonsingular. Then there exist (open) neighborhoods* $\mathscr{B}$ *of* $(\mathbf{x},\boldsymbol{\lambda},\boldsymbol{\mu};r) = (\mathbf{x}_0,\boldsymbol{\lambda}_0,\boldsymbol{\mu}_0;0)$ *and* $\mathscr{I}$ *of* $r = 0$ *and a solution* $(\mathbf{x},\boldsymbol{\lambda},\boldsymbol{\mu}) = (\mathbf{x}(r),\boldsymbol{\lambda}(r),\boldsymbol{\mu}(r)) \in C^1(I;\mathbf{R}^{n+p+q})$ *of* $\mathbf{F} = \mathbf{0}$ *such that any solution of* $\mathbf{F} = \mathbf{0}$ *that lies in* $\mathscr{B}$ *is on this curve of solutions.*

Furthermore, if (b) *remains valid,* $\boldsymbol{\mu}_0 \geq \mathbf{0}$, $\mathbf{g}(\mathbf{x}_0) \geq \mathbf{0}$, *and* (a) *and* (c) *are strengthened to*

(a′) $\mu_i^0 > 0$ *for all* $i \in \overline{\mathscr{A}}$,

(c′) *the Hessian of the Lagrangian* $\nabla_x^2\mathscr{L}$ *is positive definite on the tangent space* $\overline{T}$ *at* $(\mathbf{x}_0,\boldsymbol{\lambda}_0,\boldsymbol{\mu}_0)$,

then $D_{(x,\lambda,\mu)}F(\mathbf{x}_0,\boldsymbol{\lambda}_0,\boldsymbol{\mu}_0)$ *is nonsingular,* $\mathbf{x}_0$ *is a local minimum of the nonlinear programming problem* (1.1), *and* $\nabla^2P(\mathbf{x}(r),r)$ *is positive definite for* r *sufficiently small and positive.*

The condition (a′) is called strict complementarity, (b) is the linear independence constraint qualification, and (c′) is a second-order sufficient condition. This theorem gives conditions under which the penalty path exists, at least locally. When the conditions (a)–(c) are relaxed, $D_{(\mathbf{x},\lambda,\mu)}\mathbf{F}(\mathbf{x}_0,\boldsymbol{\lambda}_0,\boldsymbol{\mu}_0;0)$ is singular and the question of the existence of the penalty path comes into question. Jittorntrum and Osborne **[21]** have treated the case of loss of strict complementarity and have shown the persistence of the penalty path with the behavior that the solution $\mathbf{x}(r) = O(\sqrt{r})$, which is essentially the behavior of a fold or limit point within the context of singularity theory. Using the Fritz John normalization and singularity theory, one can establish a similar result for the loss of the linear independence constraint qualification. For both cases the generic behavior is that of a fold point singularity and the continuation procedures introduced by Keller **[22]** are particularly well suited to deal with this singularity. Furthermore, for such a singularity there are generically, at least in the zero-codimension case, two penalty paths which exist and the Hessian of the penalty function is positive definite for sufficiently small positive r.

The situation for a singularity arising from the singularity of the Hessian of the Lagrangian on the tangent space to the active constraints is far more complex. For example, the following situations are possible in the presence of this singularity: the constrained problem may have a local minimum but the penalty path may exist locally only for negative r; the critical point may not be a local minimizer, but the penalty path may exist for small positive r with the Hessian of the penalty function being positive definite. In the former case one cannot get to the minimizer by following the penalty path for r positive in the vicinity of the minimizer and in the latter one can follow a local minimum of the penalty function only to arrive at a saddlepoint. The existence and behavior of the penalty path in the case of low-codimension singularities are treated through the use of bifurcation theory in the work of Hasan and Poore **[19]**.

Although the probability of encountering such singularities in applied problems at $r = 0$ is small, perturbations of these singularities may be observed. For example, given $\varepsilon > 0$, a penalty path may exist only for the penalty parameter $r \geq \varepsilon$ or only for $r \leq \varepsilon$, may not exist at all, or may diverge. Path-following offers a way of recovering from such difficulties. If a fold or bifurcation point is encountered in the penalty path, numerical bifurcation techniques **[22]** can be used to switch branches at a bifurcation point or continue around a fold point. We have not encountered this difficulty or opportunity in the test problems in Section 7.

4. Continuation methods. The system of parametrized equations posed in the previous section can be written as $G(\mathbf{z}, r) = 0$, where the homotopy parameter r is arranged so that it goes from $r_0 > 0$ to 0. The primary objective of this section is to describe briefly the methodology of traversing the path from $r = r_0$ to optimality at $r = 0$. The idea is to generate a sequence of points $\{(\mathbf{z}_i, r_i)\}_{i=0}^{n}$ with $r_n = 0$. To get from $(\mathbf{z}_i, r_i)$ to $(\mathbf{z}_{i+1}, r_{i+1})$, one first predicts a new point $(\mathbf{z}_{i+1}^p, r_{i+1}^p)$ near the curve and then corrects back to the curve to obtain the desired $(\mathbf{z}_{i+1}, r_{i+1})$. Prediction is based on extrapolation of current and previous information about the solution. Extrapolation via polynomial interpolation of the solution values has been used for some time, but extrapolation of the tangents to the curve as is used here appears to be numerically more robust and efficient. A brief explanation of this methodology is given in the remainder of this section, but a more comprehensive explanation can be found in the works of Keller **[22]**, Shampine and Gordon **[32]**, and Lundberg and Poore **[24]**.

A formal differentiation of $G(\mathbf{z}, r) = 0$ with respect to a third variable s yields the Davidenko differential equation

$$D_z\mathbf{G}\frac{d\mathbf{z}}{ds} + D_r\mathbf{G}\frac{dr}{ds} = 0,$$

where s can be chosen to be arclength by adding the normalization

$$\left\|\frac{d\mathbf{z}}{ds}\right\|^2 + \left|\frac{dr}{ds}\right|^2 - 1 = 0.$$

If $D_z\mathbf{G}$ is nonsingular, and once the orientation (sign of dr/ds) is known, one can write this system as the differential equation

$$\frac{d\mathbf{w}}{ds} = f(\mathbf{w}).$$

Thus, given a point $(\mathbf{w}_k, s_k)$ on the curve, the differential equation can be integrated to obtain

$$\mathbf{w}(s_k + \Delta s) = \mathbf{w}_k + \int_{s_k}^{s_k+\Delta s} f(\mathbf{w})\, ds$$

where $f(\mathbf{w})$ denotes the tangent to the curve at the point $\mathbf{w}(s)$. If a polynomial $P_{k,m}$ of degree at most m is used to interpolate f at $(\mathbf{w}_{k-j}, s_{k-j})$ for $j = 0, \ldots, m$, the predicted solution is taken to be

$$\mathbf{w}_{k+1}^p = \mathbf{w}_k + \Delta s \mathbf{D}(\Delta s), \quad \mathbf{D}(\Delta s) = \frac{1}{\Delta s} \int_{s_k}^{s_k+\Delta s} P_{k,m}(s)\, ds.$$

For a given error tolerance, one can vary the order of this formula to achieve the largest stepsize Δs possible. This method varies from the standard Adams–Bashforth technique in that the stepsize and order are varied at each step.

Once a predicted point is obtained the correction back to the curve can be obtained in several ways. Two popular ones are the vertical correction, wherein the system of equations $\mathbf{G}(\mathbf{z}, r) = \mathbf{0}$ is solved with r fixed at the predicted value, and the correction in a hyperplane orthogonal to the predictor direction. In this latter method one solves the augmented system of equations

$$\mathbf{G}(\mathbf{w}) = \mathbf{0},$$
$$\mathbf{N}(\mathbf{w}) = \mathbf{0}$$

where $\mathbf{N}(\mathbf{w}) = \mathbf{D}^{\mathrm{T}}(\mathbf{w} - \mathbf{w}_{k+1}^p)$ represents the plane orthogonal to the predictor direction and passing through the predicted point $\mathbf{w}_{k+1}^p$ and $\mathbf{w} = (\mathbf{z}, r)$.

5. Relation to other ODE-based methods. In recent times the analogy between discrete (iterative) processes and their continuous realizations has become a popular subject [**12**]. In optimization various examples ranging from reduced and projected gradients [**7**], [**8**], [**34**] to methods based on smooth penalty functions [**9**], [**11**], [**25**] have been proposed, and some of these are briefly discussed in this section to explain the relation to our homotopy-based methods.

Continuous versions of steepest descent and Newton's method for the unconstrained optimization problem $\min f(\mathbf{x})$ where $f \in C^2(\mathbf{R}^n, \mathbf{R})$ are represented, respectively, by the dynamical systems

$$(5.1) \qquad \text{(a)} \quad \frac{d\mathbf{x}}{dt} = -\nabla f(\mathbf{x}), \qquad \text{(b)} \quad \nabla^2 f(\mathbf{x})\frac{d\mathbf{x}}{dt} = -\nabla f(\mathbf{x}),$$

where the prescription of the initial condition $\mathbf{x}(t_0) = \mathbf{x}_0$ completes the formulation. One may recover the discrete version by using an Euler step with increment $\Delta t = 1$. Also, the stepsize $\Delta t < 1$ corresponds to a restricted step similar to that obtained with a line search. A local minimizer $\hat{\mathbf{x}}$ of the optimization problem $\min f(\mathbf{x})$ that satisfies the second-order sufficient condition that $\nabla^2 f(\hat{\mathbf{x}})$ be positive definite is an asymptotically stable critical point of the above autonomous systems. The hope then is that the region of this attractor is larger than that of Newton's method.

The philosophy behind this approach is that one should follow more or less accurately the trajectory of the dynamical system, as opposed to using a line search or trust region globalization technique for Newton's method. An argument against this approach is that the use of ordinary differential equation (ODE) solvers for optimization is much too expensive in comparison to conventional methods. The experience of Brown and Bartholomew-Biggs **[9]**, **[10]**, however, suggests that ODE techniques can be implemented in such a way as to be competitive with currently available unconstrained techniques. It seems to us that the real value of the dynamical systems approach may well lie in the generation of different iterative techniques suggested from discretizations of the initial value problems (5.1). Certainly, to be competitive in function evaluation counts one would need a continuous version of a quasi-Newton method such as the BFGS update **[13]**.

For smooth penalty functions $P(\mathbf{x}, r)$ such as the quadratic penalty-log barrier function $P = f + \mathbf{h}^T\mathbf{h}/(2r) - r\sum \ln(g_i(\mathbf{x}))$ for the constrained problem $\min\{f(\mathbf{x}) : \mathbf{h}(\mathbf{x}) = \mathbf{0},\ \mathbf{g}(\mathbf{x}) \geq \mathbf{0}\}$ the continuous versions of the steepest descent and Newton's method for $\min P(\mathbf{x}, r)$ are

$$(5.2) \quad \text{(a)} \quad \frac{d\mathbf{x}}{dt} = -\nabla P(\mathbf{x}, r), \qquad \text{(b)} \quad \nabla^2 P(\mathbf{x}, r)\frac{d\mathbf{x}}{dt} = -\nabla P(\mathbf{x}, r)$$

and the discussion of the previous two paragraphs applies. An interesting variation on this scheme is to let r now depend parametrically on the variable t through some functional relation $r = f(t)$ with $r = 1/t$ being a popular choice. The computation of $d\mathbf{x}/dt$ from (5.2b) now runs into the inevitable ill-conditioning since $\kappa_2(\nabla^2 P(\mathbf{x}, r)) = O(1/r)$ as

r tends to 0^+; however, the accurate computation can be obtained from the equivalent expanded system

$$\text{(5.3)}\qquad \begin{bmatrix} \mathbf{B} & -D_x\mathbf{h}^T & -D_x\mathbf{g}^T \\ D_x\mathbf{h} & r\mathbf{I} & \mathbf{0} \\ r\mathbf{G}^{-1}D_x\mathbf{g} & \mathbf{0} & \mathbf{G} \end{bmatrix} \begin{bmatrix} \frac{d\mathbf{x}}{dt} \\ \boldsymbol{\lambda} + \frac{d\boldsymbol{\lambda}}{dt} \\ \boldsymbol{\mu} + \frac{d\boldsymbol{\mu}}{dt} \end{bmatrix} = - \begin{bmatrix} \nabla f \\ \mathbf{h} \\ -r\mathbf{e} \end{bmatrix}$$

where $\mathbf{B} = \nabla^2 f + \nabla^2\mathbf{h}^T(\mathbf{h}/r) - \nabla^2\mathbf{g}^T(r\mathbf{G}^{-1}\mathbf{e})$, $\mathbf{e} = (1, \ldots, 1)^T$, and $\mathbf{G} = \operatorname{diag}(\mathbf{g})$. The definitions $\boldsymbol{\lambda}+d\boldsymbol{\lambda}/dt = -r^{-1}(D_x\mathbf{h}d\mathbf{x}/dt+\mathbf{h})$ and $\boldsymbol{\mu}+d\boldsymbol{\mu}/dt = -\mathbf{G}^{-1}(r\mathbf{G}^{-1}D_x\mathbf{g}d\mathbf{x}/dt - r\mathbf{e})$ give the second and third equations. An Euler step of this system is within $O(r)$ of a step of sequential quadratic programming, which suggests possibly different BFGS updates for the minimization of the penalty function P. When $r = f(t)$ is introduced into (5.2) or (5.3), the convergence of the resulting differential equation is less certain than when following a pure penalty path wherein $\nabla P(\mathbf{x}, r) = 0$. (The time scale $r = f(t)$ may be too fast or too slow and inefficient.) In any event, the differential equations formulation along with $r = f(t)$ suggests an adaptive technique for updating the penalty parameter r.

The approach taken in this work is to first minimize $P(\mathbf{x}, r)$ at some value $r = r_0$, which results in $\nabla P = 0$. Then the penalty path as defined by $\nabla P(\mathbf{x}, r) = 0$ is followed to optimality at $r = 0$. Along this path

$$\text{(5.4)}\qquad \nabla^2 P(\mathbf{x}, r)\frac{d\mathbf{x}}{dr} + D_r\nabla P(\mathbf{x}, r) = 0,$$

which when expanded yields a system similar to that obtained in (5.3). As discussed in the previous section, a predictor-corrector method is used to follow the penalty path to $r = 0$. A continuous realization of this predictor-corrector procedure is to augment (5.4) with the addition of a portion of $-\nabla P$, i.e.,

$$\text{(5.5)}\qquad \nabla^2 P(\mathbf{x}, r)\frac{d\mathbf{x}}{dr} + D_r\nabla P(\mathbf{x}, r) = -\alpha\nabla P(\mathbf{x}, r),$$

where the constant of proportionality α needs to be chosen in some way, e.g., $\alpha = 1$. Again, one can expand the system to remove the ill-conditioning. Thus, equation (5.4) represents a combination of the pure dynamical systems approach (5.2) and the homotopy approach (5.3), but requires that an approximate minimization of P be performed initially, possibly with an adaptive strategy for choosing r during this phase.

In recent work on projective SUMT, McCormick [25] starts with equation (5.4), uses $\nabla P = 0$ to simplify $D_r \nabla P$, and then uses the resulting differential equation away from the penalty path to generate discrete iterative schemes for solving convex programming problems. The projective part arises from the treatment of linear (affine) constraints directly. An interesting aspect of this work is that the directions of search are similar to those seen in the Karmarkar algorithm for linear programming.

6. Relation to sequential quadratic programming. The predictor-corrector method of Section 5 gives steps toward the optimal solution that are, in fact, asymptotic to those obtained by sequential quadratic programming, and thus the continuation phase may be viewed as an intermediate globalizer for Newton's method. To explain the connection between the continuation phase and sequential quadratic programming, we confine our attention in this section to the homotopy generated by the quadatic penalty-log barrier function in equation (2.3).

That the unrestricted steps of sequential quadratic programming are asymptotic (as r tends to 0^+) to an Euler predictor and a vertical correction can most easily be explained by considering the following expansion of $\mathbf{F}$:

$$\begin{aligned} \mathbf{0} &= \mathbf{F}(\mathbf{x} + \Delta\mathbf{x}, \boldsymbol{\lambda} + \Delta\boldsymbol{\lambda}, \boldsymbol{\mu} + \Delta\boldsymbol{\mu}; r + \Delta r) \\ &= \begin{pmatrix} \nabla^2 \mathscr{L} \Delta\mathbf{x} - D_x\mathbf{h}^{\mathrm{T}} \Delta\boldsymbol{\lambda} - D_x\mathbf{g}^{\mathrm{T}} \Delta\boldsymbol{\mu} + \nabla \mathscr{L} \\ D_x\mathbf{h}\Delta\mathbf{x} + \mathbf{h} + (r + \Delta r)(\boldsymbol{\lambda} + \Delta\boldsymbol{\lambda}) \\ (M + \Delta M)(D_x\mathbf{g}\Delta\mathbf{x} + \mathbf{g}) - (r + \Delta r)\mathbf{e} \end{pmatrix} + O(\Delta^2) \end{aligned} \tag{6.1}$$

where $O(\Delta^2)$ refers to terms that are second-order and higher in the increments $(\Delta\mathbf{x}, \Delta\boldsymbol{\lambda}, \Delta\boldsymbol{\mu})$. Note that the quadratic increments $\Delta M D_x \mathbf{g} \Delta\mathbf{x}$ and $\Delta r \Delta\boldsymbol{\lambda}$ have been retained. Setting $\Delta r = -r$ or removing r and Δr from the equations and setting the bracketed quantity to zero yields

$$\begin{aligned} &\nabla^2 \mathscr{L} \Delta\mathbf{x} - D_x\mathbf{h}^{\mathrm{T}} \Delta\mathbf{x} - D_x\mathbf{g}^{\mathrm{T}} \Delta\mathbf{x} = -\nabla \mathscr{L}, \\ &D_x\mathbf{h}\Delta\mathbf{x} + \mathbf{h} = \mathbf{0} \\ &(M + \Delta M)(D_x\mathbf{g}\Delta\mathbf{x} + \mathbf{g}) = \mathbf{0} \end{aligned}$$

or

$$\begin{aligned} \text{MIN } & \mathbf{f} + \nabla\mathbf{f}^{\mathrm{T}} \Delta\mathbf{x} + \tfrac{1}{2} \Delta\mathbf{x}^{\mathrm{T}} \nabla^2 \mathscr{L} \Delta\mathbf{x} \\ \text{ST } & D_x\mathbf{h}\Delta\mathbf{x} + \mathbf{h} = \mathbf{0} \\ & D_x\mathbf{g}\Delta\mathbf{x} + \mathbf{g} \geq \mathbf{0}, \end{aligned}$$

where the latter quadratic programming problem is derived from the system of equations above it by imposing dual feasibility, i.e., $\boldsymbol{\mu} + \Delta\boldsymbol{\mu} \geq \mathbf{0}$.

This quadratic subproblem is the one solved sequentially in sequential quadratic programming.

The Euler tangent with a stepsize of $\Delta r = -r$ is obtained from (6.1) by setting the bracketed expression in (6.1) to zero, removing those terms which are not homogeneous of degree one in $\Delta\mathbf{x}$, $\Delta\boldsymbol{\lambda}$, $\Delta\boldsymbol{\mu}$, and Δr, solving the resulting equations for $(\Delta\mathbf{x}, \Delta\boldsymbol{\lambda}, \Delta\boldsymbol{\mu})$ in terms of Δr, and setting $\Delta r = -r$. Let this solution be denoted by $\boldsymbol{\Delta}_1 = (\Delta\mathbf{x}_1, \Delta\boldsymbol{\lambda}_1, \Delta\boldsymbol{\mu}_1)$. The vertical correction is similarly obtained by setting $\Delta r = 0$, removing the quadratic term, and solving the resulting equation. Let this solution be denoted by $\boldsymbol{\Delta}_2 = (\Delta\mathbf{x}_2, \Delta\boldsymbol{\lambda}_2, \Delta\boldsymbol{\mu}_2)$. If $\boldsymbol{\Delta} = (\Delta\mathbf{x}, \Delta\boldsymbol{\lambda}, \Delta\boldsymbol{\mu})$ denotes the solution obtained from the quadratic subproblem, then under the conditions of Theorem 3.1

$$\boldsymbol{\Delta} = \boldsymbol{\Delta}_1 + \boldsymbol{\Delta}_2 + O(r\|\Delta\boldsymbol{\lambda}_2\|, \|\boldsymbol{\Delta}\|^2) \quad \text{as } r \to 0 \text{ and } \boldsymbol{\Delta} \to \mathbf{0}.$$

A similar result applies for the quadratic penalty–loss function with the slight modification

$$\boldsymbol{\Delta} = \boldsymbol{\Delta}_1 + \boldsymbol{\Delta}_2 + O(r\|\Delta\boldsymbol{\lambda}_2, \Delta\boldsymbol{\mu}_2\|, \|\boldsymbol{\Delta}\|^2) \quad \text{as } r \to 0 \text{ and } \boldsymbol{\Delta} \to \mathbf{0}.$$

This suggests that as soon as the predicted value reaches $r = 0$, one could just as effectively switch to sequential quadratic programming without a globalizer.

The predictor-corrector steps taken in the continuation phase are also closely related to the quadratic subproblems developed in the work of Murray [**26**], Biggs [**2**], [**4**], [**5**], and Murray and Wright [**27**]. In fact, the methods developed by these authors may be viewed as loosely following the penalty path or penalty homotopy. The difference is that in the continuation phase the line-search globalization technique is replaced by the path-following procedure with an efficient extrapolation technique for the choice of the penalty parameter.

7. Numerical examples. In this section we present some of the test results for three different implementations of the algorithms discussed in Sections 2–4. The first few paragraphs of this section contain a discussion of some of the details of the three implementations which are based on the quadratic penalty-log barrier function, the quadratic penalty-loss function, and the quadratic penalty-loss function with multiplier updates and called PENCON, LOSSCON, and MLTCON, respectively. For a numerical comparison we have solved many problems from the book of Hock and Schittkowski [**20**] and the paper by Brown and Bartholomew-Biggs [**11**], and present the various methods and codes used in Table 1 and a comparative summary of the function evaluation count in Tables 2–4. These results indicate the robustness, efficiency, and potential of the methodology.

The implementation PENCON uses the quadratic penalty-log barrier function. Since an initial point satisfying strict inequality in the inequality constraint $\mathbf{g} \geq \mathbf{0}$ is required for the log barrier function, we first use the loss function $(\mathbf{g}^-)^T\mathbf{g}^-$ to generate a point $\hat{\mathbf{x}}$ at which $\mathbf{g}(\hat{\mathbf{x}}) > \mathbf{0}$ or is at least close to the feasible region $\{\mathbf{x}: \mathbf{g}(\mathbf{x}) \geq \mathbf{0}\}$ and then introduce a shift $\boldsymbol{\delta}$ so that $\mathbf{g}(\hat{\mathbf{x}}) + \boldsymbol{\delta} > \mathbf{0}$. Then a quasi-Newton method with a BFGS update is used to minimize the penalty function $P(\mathbf{x}, r) = f(\mathbf{x}) + \mathbf{h}^T(\mathbf{x})\mathbf{h}(\mathbf{x})/(2r) - r\sum \ln(g_i(\mathbf{x}) + r\delta_i/r^0)$ at some value of the penalty parameter, say r_0, at which the problem is reasonably well conditioned. A quadratic-cubic line search and an Armijo stopping criterion **[13]**, modified to maintain feasibility ($\mathbf{g}(\mathbf{x}) + \boldsymbol{\delta} > \mathbf{0}$), have been used to globalize the quasi-Newton method. Once the minimization problem is solved, continuation techniques are used to track the solution to optimality at $r = 0$. The initial value of $r_0 = 0.1$ has been used in the numerical experiments reported in Table 2 under the heading PENCON, and scaling has not been used. Additional information can be found in **[29]**.

The implementation that uses the quadratic penalty-loss function is called LOSSCON. Since the loss function is an exterior penalty function, an initial feasible point for the inequality constraints is not required. The minimization of this penalty function (2.5) with $\hat{\boldsymbol{\lambda}}$ and $\hat{\boldsymbol{\mu}}$ set to $\mathbf{0}$ uses a quasi-Newton method with a BFGS update and a bracketing-sectioning line search of Fletcher **[15]** with the stopping criteria $\|\nabla P(\mathbf{x}+t\,\mathbf{d})^T\mathbf{d}\| \leq \sigma\|\nabla P(\mathbf{x})^T\,\mathbf{d}\|$, where $\mathbf{d}$ is the direction of search and σ controls the exactness of the line search. The value $\sigma = 0.5$ has been used in our current implementation. Also, following Fletcher **[15]**, a model for the penalty function that employs a quadratic interpolator to the objective function and each of the constraints has been used along with the usual cubic fit to the penalty function. Our experience has been that the former can in some, but not all, examples significantly reduce the function evaluation count.

During the minimization phase, an adaptive choice of the scales ν, γ, and β, which occur in this quadratic penalty-loss function (2.5), has been employed to improve efficiency as well as robustness on badly scaled and large multiplier problems. The use of these scales can significantly reduce the length of the penalty path as well as the number of steps to optimality. Although many different strategies have been implemented and tested, a simple strategy which currently seems to work well is to choose the scales ν, γ, and β such that the norms $\|\nu\nabla f\|$, $\|\gamma_i\nabla h_i\|$, and $\|\beta_j\nabla g_j\|$ are not too different, which we quantify as follows: If $\|\nu\nabla f\| \notin [L_0, U_0]$, then choose ν so that $\|\nu\nabla f\| = 1$; if $\|\nu\nabla f\|/\|\gamma_i\nabla h_i\| \notin [L, U]$, then choose γ_i so that $\|\nu\nabla f\|/\|\gamma_i\nabla h_i\| = 1$; and if $\|\nu\nabla f\|/\|\beta_j\nabla g_j\| \notin [L, U]$, choose β_j so that $\|\nu\nabla f\|/\|\beta_j\nabla g_j\| = 1$. The values of L_0, U_0, L, and U used in this work are 0.01, 100, 0.05, and 15, respectively, and these rules are appropriately safeguarded to avoid extreme cases such as when a gradient is near zero.

The third implementation, MLTCON, utilizes multiplier updates, hopefully to shorten the penalty path. The penalty function (2.5) with $\hat{\boldsymbol{\lambda}} = \mathbf{0}$, $\sigma_i = 1$, and $r = 0.1$ is minimized to moderate accuracy ($\|\nabla P\| < 0.001$). If, after the initial optimization, we are left with $|\gamma_i h_i(\mathbf{x})| > 1.5\|\mathbf{\Gamma h}\|_1/n$, we set $\sigma_i =: 10\sigma_i$ and reoptimize. This step is often skipped, and in most cases takes only a few additional function evaluations. Then one step of an augmented Lagrangian method is performed using the first-order update

$$\hat{\boldsymbol{\lambda}} =: \hat{\boldsymbol{\lambda}} - r^{-1}\mathbf{S\Gamma h}.$$

This involves one further unconstrained optimization, but usually few additional function evaluations. In both of the additional unconstrained optimizations, information from the previous unconstrained optimization is carried forward, including the Hessian approximation which has been built up.

Both line-search algorithms mentioned above have been implemented with MLTCON, a backtracking scheme using an Armijo rule **[13]** and the more sophisticated bracketing-sectioning line search due to Fletcher **[15]**, both using quadratic-cubic interpolation to estimate the minimizer of P along the search direction. The latter is more reliable, but sometimes uses significantly more gradient evaluations. The former is used to obtain the results for MLTCON in Tables 3 and 4.

For equality constrained problems, the BFGS update formula for the Hessian approximation (beginning with $\mathbf{B}^0 = \mathbf{Id}$) is the usual formula $\mathbf{B}^{k+1} = \mathbf{B} + (\mathbf{y}^T\mathbf{y})/(\mathbf{y}^T\Delta\mathbf{x}) - (\mathbf{d}^T\mathbf{d})/(\mathbf{d}^T\Delta\mathbf{x})$, where $\Delta\mathbf{x} = \mathbf{x}^{k+1} - \mathbf{x}^k$, $\mathbf{d} = \mathbf{B}\Delta\mathbf{x}$, but where the usual definition $\mathbf{y} = \nabla P(\mathbf{x}^{k+1}) - \nabla P(\mathbf{x}^k)$ is replaced by $\mathbf{y} = \nu(\nabla f^{k+1} - \nabla f^k) + r^{-1}(\mathbf{\Gamma} D_x\mathbf{h}^{k+1})^T\mathbf{S}(\mathbf{\Gamma} D_x\mathbf{h}^{k+1})\Delta\mathbf{x} + r^{-1}(\mathbf{\Gamma} D_x\mathbf{h}^{k+1} - \mathbf{\Gamma} D_x\mathbf{h}^k)^T\mathbf{S}(\mathbf{\Gamma h}^{k+1} - r\mathbf{S}^{-1}\hat{\boldsymbol{\lambda}})$, which is suggested by the nonlinear least squares update of Fletcher **[15]**. This more complex choice of $\mathbf{y}$ typically results in 10–20% savings in function evaluations in the unconstrained phase over the usual choice of $\mathbf{y}$.

At the end of the unconstrained optimization phase, we convert to the expanded system (2.7) as discussed in Section 2 and treat the active inequality constraints as equality constraints thereafter. This situation is monitored and corrected as discussed at the end of Section 2. We have found, however, that with the aforementioned scaling, the active set was correctly identified for the current test problems with the choice of $r = 0.1$. We now turn to the numerical test problems.

Several standard test problems from the book by Hock and Schittkowski **[20]** and the paper of Brown and Bartholomew-Biggs **[11]** have been solved using a variety of codes in addition to PENCON, LOSSCON, and MLTCON. These codes, along with the corresponding authors and methods, are given in Table 1.

Table 1. Summary of codes compared.

CODE	AUTHOR	METHOD
VF02AD, VMCON, VMCWD	Powell	Quadratic approximation
OPRQP, OPXRQP	Bartholomew-Biggs	Quadratic approximation
E04VDF	Gill et al.	Quadratic approximation
GRGA	Abadie	Generalized reduced gradient
VF01A	Fletcher	Multiplier
FUNMIN	Kraft	Multiplier
FMIN, IMPBOT	Kraft, Lootsma	Penalty
CONIMP, CONRK1	Brown, B.-B. **[11]**	Continuous Newton–ODE
RKLAG	Brown, B.-B. **[11]**	Penalty trajectory–ODE
MODGMP	Brown, B.-B. **[11]**	Projected gradient–ODE
PENCON	Al-Hassan, Poore	Penalty–continuation
LOSSCON	Lundberg, Poore, Yang	Loss function–continuation
MLTCON	Lundberg, Poore, Yang	Multiplier–continuation

A comparative summary of the number of function evaluations for various codes and problem sets is presented in Tables 2, 3, and 4. The function evaluation counts for the codes other than PENCON, LOSSCON, and MLTCON are taken from **[11]**, **[20]**. Consistent with those function evaluation counts, we count the evaluation of a p-dimensional vector as p function evaluations; however, we do not count upper and lower bounds on variables since they are handled directly in the code and the gradient evaluations of linear constraints are counted only once. In particular, the n-dimensional gradient of a scalar function is counted as n function evaluations. The approximation of the Hessian of the Lagrangian in the continuation phase is based on finite differences **[13]**.

Table 2. Function evaluation count for inequality constrained problems from Hock and Schittkowski (** indicates a failure).

Prob. no.	LOSSCON	PENCON	VF02AD	OPRQP	GRGA	VF01A	FUNMIN	FMIN
H5	23	48	24	23	136	48	46	234
H10	97	124	50	128	644	282	556	689
H12	99	210	72	184	320	442	578	362
H13	348	448	270	450	217	844	1,098	4,610
H14	140	156	44	149	118	226	774	887
H19	438	615	117	2,361	370	1,217	3,936	750
H20	140	407	240	296	130	510	852	4,501
H29	310	375	104	362	926	823	662	570
H34	321	427	75	405	377	906	**	2,801
H38	444	233	545	462	359	530	140	964
H43	590	1,199	240	508	2,342	1,000	1,596	3,053
H63	252	289	84	231	895	417	672	1,239
H65	209	310	88	4,746	281	600	3,044	397
H71	742	866	75	471	531	1,391	2,403	8,043
H118	4,296	6,752	**	1,860	4,237	**	**	**
Average rank								
	2.6	4.2	2.0	3.7	4.1	5.7	6.2	7.0

Along with each table is a ranking obtained by assigning one to the lowest function evaluation count, two to the next, etc., for each problem, and then averaging over all problems. In case of a failure, the failed codes are all given the rank plus one of the successful code having the largest function evaluation count.

Table 2 contains a comparative study of the function evaluation count for various codes for typical inequality constrained problems from Hock and Schittkowski [**20**]. These comparisons show that on these test problems LOSSCON as currently implemented has an average ranking of 2.6, which is the second highest of all codes compared, surpassing one of the sequential quadratic programming codes and surpassed only by the SQP code VF02AD of Powell. In addition, the methodology is quite robust, primarily because of the robustness of penalty paths leading to optimality and the robustness of the continuation methodology. The answers given by PENCON and LOSSCON have approximately 12 digits of accuracy on a 14+ digit CDC Cyber 180/840, whereas the remaining answers listed above are computed on a 10-digit machine [**20**] and those for VF02AD are generally to low accuracy. Furthermore, the quadratic

Table 3. Function evaluation counts: Equality constrained problems from Brown and Bartholomew-Biggs (** indicates a failure).

Prob. no.	MLTCON°	MLTCON	LOSSCON	CONIMP	CONRK1	RKLAG
B1	1,782	1,683	9,775	6,328	1,075	**
B3	201	212	303	4,808	810	206
B4	89	88	58	714	878	166
B5	56	51	809	10,481	2,114	281
B6	131	91	161	615	703	4,773
B7	968	870	623	1,966	**	4E4L
B8	353	383	207	482	**	598L
B9	1,030	1,036	271	12,341	**	315
B11	677	689	157	23,720	835	448
B13	567	585	330	30,927	1,502L	4,723
Average rank	4.1	4.2	4.0	8.5	8.4	6.9

Prob. no.	IMPBOT	MODGPM	OPXRQP	VMCON	VMXWD	E04DF
B1	2,541	13E4	**	**	**	**
B3	2,082	736	161	3L	3L	258
B4	224	520	112	6	6	42
B5	23,307	979	82	81	81	123
B6	777	235	402	207	27	21
B7	1,135	3,769	**	**	**	**
B8	424L	1,306	186	217	168	**
B9	**	589	320	**	**	5
B11	**	610	**	**	**	**
B13	362L	934	**	**	**	57
Average rank	7.8	7.2	6.3	5.8	5.1	5.4

penalty-loss function tends to perform better than the quadratic penalty-log barrier function with the current implementations. Scaling yields a significant improvement in some problems by reducing the number of function evaluations required in the unconstrained optimization phase and can reduce the number of steps taken in the continuation phase.

In Tables 3 and 4 we have solved only equality constrained problems, but now include in the comparison the results of MLTCON, which employs multiplier updates. This same code is used without multiplier updates and the results are reported under the heading MLTCON^0.

Table 4. Function evaluation counts: Equality constrained problems from Hock and Schittkowski.

Prob. no.	MLTCON°	MLTCON	LOSSCON	VF02AD	OPRQP	GRGA
H6	325	331	429	60	70	42
H7	167	179	231	72	184	270
H26	301	317	541	152	210	444
H27	246	247	292	200	346	1,714
H39	1,269	1,253	805	195	411	2,557
H40	492	431	526	120	300	2,099
H46	942	571	1,028	252	624	4,314
H47	839	943	1,159	480	928	959
H56	1,649	1,974	1,448	440	9,800	5,943
H77	1,172	1,226	798	288	691	3,621
H78	627	747	873	216	588	**
H79	947	1,071	957	240	374	1,479
			Average Rank			
	3.5	3.9	4.7	1.1	3.3	7.2

Prob. no.	VF01A	FUNMIN	FMIN	CONIMP	CONRK1	RKLAG
H6	492	430	620	—	—	—
H7	258	972	302	—	—	—
H26	1,560	2,276	622	—	—	—
H27	435	960	556	—	—	—
H39	1,110	1,638	2,295	682	758	1,914
H40	800	1,724	2,912	—	—	—
H46	2,664	996	1,212	—	—	—
H47	1,728	1,860	760	—	—	—
H56	**	3,285	5,030	2,107	1,914	1,976
H77	1,422	1,560	1,833	823	723	807
H78	1,248	1,648	2,072	1,080	1,052	583
H79	1,272	1,912	1,756	506	506	678
			Average rank			
	6.8	7.4	7.3	—	—	—

Notes:

1. L indicates that the terminal point is a KT point different from that obtained by most of the other codes.
2. ** indicates a failure, i.e., that $\|\nabla\mathscr{L}\|$ at the terminal point was unacceptably large.
3. — indicates that results were not reported for this code.
4. The problems labelled Bn, and the associated results for the ODE and SQP codes, were obtained from [**11**]. The label Hn signifies a problem from [**20**].
5. The results reported in [**29**] for PENCON counted a gradient evaluation as one function evaluation; equivalent function evaluation counts consistent with the counting scheme of Brown and Bartholomew-Biggs [**11**] were used in Tables 2–4.
6. If n codes are successful on a problem, all codes that failed are given the rank $n+1$.

The ten test problems of Table 3 are particularly difficult, as is reflected in the large number of failures of the SQP codes (last four columns in the table). The penalty-continuation algorithm solved each problem successfully with an average rank significantly superior to the ODE and SQP codes, regardless of the implementation (MLTCON^0 or LOSSCON). It should be remarked that the function evaluation counts for some of the problems were dramatically lower for MLTCON if the constraints and objective were dynamically scaled by $\gamma_i = 1/\max(\|\nabla h_i\|, K)$, $\nu = 1/\max(\|\nabla f\|, K)$, where K is an appropriate bound such as the square root of machine precision, or if the scaling scheme mentioned above is used. However, in a few of the problems such scaling schemes led to divergence in the unconstrained optimization phase in the code MLTCON. Thus, in order to obtain a fair and consistent comparison, and for robustness, no scaling was used to obtain the results for MLTCON in these tables. The initial value of r_0 can also have a dramatic impact on the function evaluation count. For example, if $r_0 = 0.001$ is used for problem B1 in Table 3, the function evaluation count for LOSSCON drops from 9,775 to 663; and, if $r_0 = 100$ is used for problem H6 in Table 4, the function evaluation count for MLTCON^0 drops from 325 to 108.

Given the superior performance of these penalty-continuation algorithms on the previous problem sets, we decided to go back to the equality constrained problems in Hock and Schittkowski and rerun the various codes. It seems to us that these problems are relatively simple and many are quite close to quadratic programming problems in that the nonlinearities are relatively mild in comparison to those of Brown and Bartholomew-Biggs. Table 4 indicates the results of this comparison. Again the penalty-continuation algorithms performed very well, but the code VF02AD of Powell clearly came in first.

8. Summary and concluding remarks. The robustness and efficiency of a class of penalty-continuation algorithms based on three smooth penalty functions, augmented Lagrangians, the derived homotopy, and predictor-corrector continuation techniques, have been illustrated in Section 7 for the test problems from the book by Hock and Schittkowski [**20**] and the paper by Brown and Bartholomew-Biggs [**11**]. The numerics of Section 7 show the ranking of these algorithms relative to other ODE-based methods, straight penalty methods, generalized reduced gradients, multiplier or augmented Lagrangian methods, and sequential quadratic programming (SQP). These penalty-continuation algorithms are among the best in terms of robustness and are competitive with SQP algorithms in function evaluation counts. This conclusion agrees with the recent work of Brown and Bartholomew-Biggs, who conclude that ODE-based methods "can perform very much better than some well-known and successful sequential quadratic programming techniques," and deserve further study [**11**, p. 20]. In Table 2, LOSSCON ranks second and competes favorably with Powell's VF02AD. On the very difficult equality constrained problems in Table 3 both MLTCON and LOSSCON are clearly superior in function evaluation counts with respect to both the ODE and SQP codes. The problems in Table 4 are generally close to quadratic programming problems in that the nonlinearities are not nearly as severe as those in Table 3, but MLTCON and LOSSCON still rank third through fifth out of nine codes. This is particularly encouraging when one considers the possible areas of improvement.

Scaling can have a considerable impact on the function evaluation count in the minimization phase of these algorithms, but other than the simple strategy discussed in Section 7, we have not succeeded in developing a strategy that is uniformly the "best." The BFGS update performs quite well, as expected; however, updating techniques that take advantage of the special structure of the penalty functions have not been fully explored. We have implemented in MLTCON a variation as discussed in Section 7 and noted a 10–20% decrease in the function evaluation count during the minimization phase. Another possibility is to update only part of the Hessian of the penalty function and retain those terms depending on first-order derivatives. This possibility has been explored in the context of nonlinear least squares [**13**] and may be helpful here.

The continuation phase could be made more efficient by using updates on the Hessian of the Lagrangian as is done in SQP starting from the update at the end of the optimization phase. Since we currently use finite differences to approximate the Hessian of the Lagrangian, the function evaluation count should be reduced. Furthermore, in light of the discussion in Section 6, the use of an SQP-type algorithm in the final steps may add efficiency.

One of the disappointments in our numerical testing was the lack of improvement from linear multiplier updates. As expected, the length of the penalty path was shortened with the use of a single multiplier update, but the function evaluation count generally increased or remained the same, principally because reoptimization is required after an update and because the predictor-corrector continuation phase is so efficient. However, since our test set is composed entirely of problems of small dimension, these results may not reflect the relative efficiency of penalty-continuation versus multiplier-continuation for larger-scale problems, where the use of $O((1+q)n^2)$ function evaluations for the approximation of the Hessian of the Lagrangian in each continuation step may catch up with the expense of the unconstrained optimization, which uses no second derivatives. Further testing on larger problems is needed to evaluate the methodologies fully, and a secant approximation to the Hessian of the Lagrangian in the continuation phase should be explored. Finally, other than the adaptive scaling which indirectly affects the penalty parameter, we have not used an adaptive choice of this parameter during the minimization, but such a technique could potentially help. Indeed, our numerical experience indicates that the function evaluation counts on the test problems could usually be reduced, sometimes drastically, by finding an optimal value of r_0 for each problem. Penalty function methods are indeed sensitive to the initial choice of r.

As with any method a word of caution is appropriate. One can construct simple examples illustrating the following situations for penalty paths: given $\varepsilon > 0$, a penalty path may exist only for the penalty parameter $r \geq \varepsilon$ or only for $r \leq \varepsilon$, may not exist at all, may diverge, or may exist for $r > 0$ but the limit point at $r = 0$ is not a local minimum of the original problem. (When this last situation occurs, the Karush–Kuhn–Tucker equations have a singularity.) In spite of these examples, penalty path following is a mathematically robust method of solving constrained optimization problems, as is illustrated by the examples in Hock and Schittkowski **[20]** and Brown and Bartholomew-Biggs **[11]**, and these homotopy methods appear to make them efficient and a competitive alternative to sequential quadratic programming.

ACKNOWLEDGMENT. We wish to thank a referee for many insightful and constructive comments.

REFERENCES

1. J. Abadie and J. Carpentier, *Generalization of the Wolfe reduced gradient method to the case of non-linear constraints*, Optimisation (R. Fletcher, ed.), Academic Press, London, 1969, pp. 37–48.

2. M. C. Bartholomew-Biggs, *Recursive quadratic programming methods for nonlinear constraints*, Nonlinear Optimization 1981 (M. J. D. Powell, ed.), Academic Press, London, 1982.

3. D. P. Bertsekas, *Lagrangian and exact penalty funds*, Nonlinear Optimization 1981 (M. J. D. Powell, ed.), Academic Press, London, 1982.

4. M. C. Biggs, *Constrained minimization using recursive equality quadratic programming*, Numerical methods for non-linear optimization (F. A. Lootsma, ed.), Academic Press, New York, 1972.

5. ——, *Constrained minimization using recursive quadratic programming: some alternative subproblem formulations*, Towards global optimization (L. C. W. Dixon and G. P. Szego, ed.), North-Holland, Amsterdam, 1975.

6. P. T. Boggs, *The solution of nonlinear systems of equations by A-stable integration techniques*, SIAM J. Numer. Anal. **8** (1971), 767–785.

7. C. A. Botsaris, *A Newton-type curvilinear search method for constrained optimisation*, J. Math. Anal. Appl. **69** (1979), 372–397.

8. ——, *Constrained optimisation along geodesics*, J. Math. Anal. Appl. **79** (1981), 296–306.

9. A. A. Brown, *Optimisation methods involving the solution of ordinary differential equations*, Ph.D. thesis, The Hatfield Polytechnic, 1986.

10. A. A. Brown and M. C. Bartholomew-Biggs, *Some effective methods for unconstrained optimisation based on the solution of systems of ordinary differential equations*, Tech. report 178, Numerical Optimisation Centre, The Hatfield Polytechnic, 1986.

11. ——, *ODE vs SQP methods for constrained optimisation*, Tech report 179, Numerical Optimisation Centre, The Hatfield Polytechnic, 1987.

12. M. T. Chu, *On the continuous realization of iterative processes*, SIAM Rev. **30** (1988), 375–387.

13. J. E. Dennis and R. B. Schnabel, *Numerical Methods for Unconstrained Optimization and Nonlinear Equations*, Prentice-Hall, Englewood Cliffs, New Jersey, 1983.

14. L. C. W. Dixon and G. P. Szego (eds.), *Towards Global Optimization*, North-Holland, Amsterdam, 1975.

15. R. Fletcher, *Practical Methods of Optimization*, second edition, Wiley, New York, 1987.

16. R. Fletcher (ed.), *Optimization*, Academic Press, New York, 1969.

17. R. Glowinski, *Numerical Methods for Nonlinear Variational Problems*, Springer-Verlag, New York, 1984.

18. R. Glowinski and P. Le Tallec, *Augmented Lagrangian methods for the solution of variational problems*, MRC technical summary report #2965, Mathematics Research Center, University of Wisconsin-Madison, 1989.

19. M. Hasan and A. B. Poore, *A bifurcation analysis of the quadratic penalty-log barrier function*, in preparation.

20. W. Hock and K. Schittkowski, *Test Examples for Nonlinear Programming Code*, Springer-Verlag, New York, 1981.

21. K. Jittorntrum and M. R. Osborne, *Trajectory analysis and extrapolation in barrier function methods*, J. Austral. Math. Soc. (Series B) **20** (1978), 352–369.

22. H. B. Keller, *Numerical solution of bifurcation and nonlinear eigenvalue problems*, Applications of Bifurcation Theory (P. Rabinowitz, ed.), Academic Press, New York, 1977, pp. 359–384.

23. F. A. Lootsma (ed.), *Numerical Methods for Non-linear Optimization*, Academic Press, New York, 1972.

24. B. N. Lundberg and A. B. Poore, *Variable order Adams–Bashforth predictors with error-stepsize control for continuation methods*, in preparation.

25. G. P. McCormick, *The projective SUMT method for convex programming*, Tech. report GWU/IMSE/Serial T-518/87, George Washington University, 1987.

26. W. Murray, *An algorithm for constrained optimization*, Optimization (R. Fletcher, ed.), Academic Press, New York, 1969.

27. W. Murray and M. H. Wright, *Projected Lagrangian methods based on the trajectories of penalty and barrier methods*, Report SOL 78-23, Department of Operations Research, Stanford University, 1978.

28. M. R. Osborne and D. M. Ryan, *A hybrid algorithm for non-linear programming*, Numerical methods for non-linear optimization (F. A. Lootsma, ed.), Academic Press, New York, 1972.

29. A. B. Poore and Q. Al-Hassan, *The expanded Lagrangian system for constrained optimization problems*, SIAM J. Control and Optimization **26** (1988), 417–427.

30. A. B. Poore and C. A. Tiahrt, *Bifurcation problems in nonlinear parametric programming*, Mathematical Programming **39** (1987), 189–205.

31. M. J. D. Powell (ed.), *Nonlinear Optimization*, 1981, Academic Press, London, 1982.

32. L. F. Shampine and M. K. Gordon, *Computer Solution of Ordinary Differential Equations : The Initial Value Problem*, W. H. Freeman and Co., San Francisco, 1975.

33. J. Stoer, *Principles of sequential quadratic programming for solving nonlinear programs*, Computational mathematical programming (K. Schittkowski, ed.), NATO ASI Series, Vol. F15, Springer-Verlag, Heidelberg, 1985.

34. K. Tanabe, *Differential geometric methods in nonlinear programming*, Math. Software **10** (1979), 299–316.

COLORADO STATE UNIVERSITY

Lectures in Applied Mathematics
Volume **26** (1990)

Application of the Fast Adaptive Composite Grid Method to Nonlinear Partial Differential Equations

S. McKAY AND J. W. THOMAS

1. Introduction. The need for local resolution in the numerical solution of partial differential equations occurs often in practice. Special local features of the forcing function, operator coefficients, boundary and boundary conditions can demand resolution in restricted regions of the domain that is much finer than is required in the rest of the domain. It is important that the discretization and solution processes account for this locally, that is, that the local phenomena do not precipitate a dramatic increase in overall computation. Unfortunately, the objective of efficiently adapting to local features is often in conflict with the solution process: equation solvers can degrade or even fail in the presence of varying discretization scales. In addition, data structures that account for irregular grids can be cumbersome, and this may negate the advantages of most of the available vector/parallel computer architectures.

This problem is even more severe when solving nonlinear partial differential equations. The solving techniques are generally less robust, which increases the likelihood of degradation or failure of the solver. The increased difficulty of the solution schemes in conjunction with the cumbersome data structures make it that much more difficult to use the vector/parallel capabilities of the available supercomputers. And finally, the complexity of the solutions makes it much more difficult to

1980 *Mathematics Subject Classification* (1985 *Revision*). Primary 65N20.
This work was partially supported by the Air Force Office of Scientific Research under grant AFOSR-86-0126 and by the NSF under grant DMS-8813406.

predict, based on the coefficients, boundaries, boundary data and forcing function, where additional resolution may be needed.

The fast adaptive composite grid method (FAC) **[1]**–**[4]** is a discretization and solution method designed to achieve local resolution by systematically constructing the discretization based on various regular grids and using them as a basis for fast solution. It shares all of the attributes of the composite grid and multigrid methods. The flexibility of the method allows it to be fully automated for problem solution. The basic structure of FAC, which is based on rectangular grid patches within the domain of the problem, allows FAC to use any of the available nonlinear solution techniques and efficiently use the capabilities of the available large, multi-processor computers. This same patch structure of FAC makes it relatively easy to use the current approximation to the solution to decide whether more resolution is needed to solve the problem to a predetermined order of accuracy and where the additional resolution is needed.

The goal of this paper is to illustrate the effectiveness of FAC and the associated self-adaptive FAC algorithm on nonlinear problems. The FAC algorithm for nonlinear equations (based on the full approximation scheme of multigrid) is presented in Section 2. In Section 3 we present the self-adaptive scheme used with the FAC algorithm. The self-adaptive scheme is based on approximating the error by using a Richardson extrapolation of solutions on different grids. And finally, the numerical solution of a model nonlinear partial differential equation is discussed in Section 4.

2. FAC algorithm. The FAC algorithm is designed to solve partial differential equations where there are one or more local regions that require a high degree of resolution and accuracy. Several different schemes and general-purpose software for these schemes have been or are currently being developed. The first of these schemes is the linear FAC algorithm, which is based on residual correction and is described in **[1]**. In **[1]** it is shown that, under reasonably mild hypotheses (i.e., that the composite grid operator is essentially symmetric and positive definite), the FAC algorithm is convergent with rates that depend on certain regularity and approximation properties. These results include the case when the grid equations are only solved approximately. The numerical results reported in **[1]**–**[3]** show that the rate of convergence of FAC is very good and that it is applicable to a wide variety of problems not covered by the theory in **[1]**.

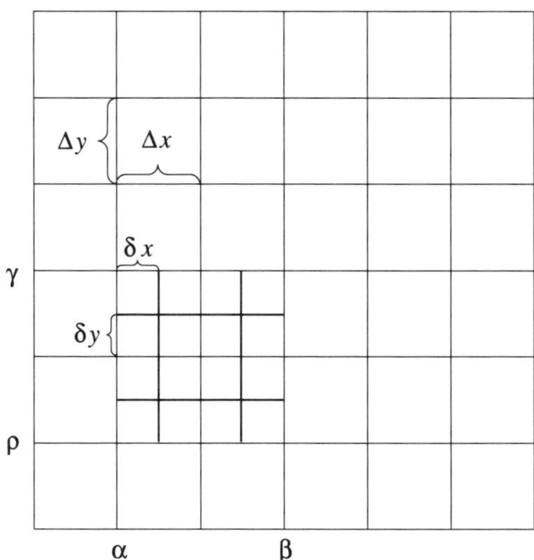

FIGURE 1. Example of a composite grid

In this paper we shall describe a nonlinear FAC algorithm that is based on the full approximation scheme of multigrid. Here, for simplicity, we formulate FAC using one refinement region and one local grid. Though theory is not yet available for this FAC scheme, observed convergence rates are very similar to those for the corresponding linear problems.

Suppose the partial differential equation we wish to solve is given by $H(v) = g$, including boundary conditions on the domain Ω. Suppose for convenience that Ω is a rectangle in $\mathbf{R}^2$ containing a proper rectangular subregion, Ω_F, and that Ω_F requires a finer resolution (grid spacing) than the rest of Ω. Then, given a coarse grid $\mathbf{G}$ with grid spacing $\Delta x = \Delta y$, on Ω_F (which is assumed to be aligned with the coarse grid) we place a finer grid $\mathcal{G}_F$ with grid spacing $\delta x = \delta y$. We assume that $\Delta x = \Delta y = m\delta x = m\delta y$, where the mesh ratio m is a positive integer. The composite grid $\mathcal{G}$ is defined to be the union of $\mathbf{G}$ and $\mathcal{G}_F$ as illustrated by Figure 1.

Consider an approximation to the given partial differential equation and boundary conditions on the composite grid $\mathcal{G}$. Assume that this approximation can be written as

$$\mathcal{L}(u) = f, \tag{1}$$

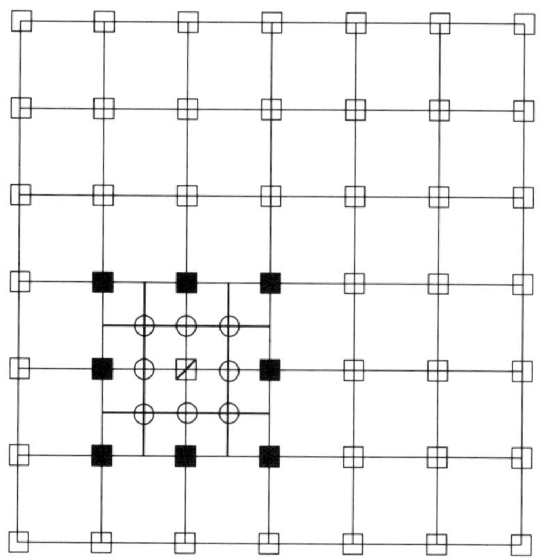

FIGURE 2. $\mathscr{G} = \square \cup \blacksquare \cup \boxslash \cup \bigcirc$, $\mathscr{G}_C = \square$, $\mathscr{G}_I = \blacksquare$, $\mathscr{G}_F = \boxslash \cup \bigcirc$, $\mathbf{G} = \square \cup \blacksquare \cup \boxslash$, $\mathbf{G}_C = \square$, $\mathbf{G}_I = \blacksquare$, and $\mathbf{G}_F = \boxslash$.

where $\mathscr{L}$ is the nonlinear operator resulting from a discrete approximation of the partial differential operator H, and f represents the inhomogeneous terms in the equation and boundary conditions.

The composite grid can be partitioned so that $\mathscr{G} = \mathscr{G}_C \cup \mathscr{G}_I \cup \mathscr{G}_F$, where $\mathscr{G}_I$ consists of the coarse grid points along the boundary of Ω_F, $\mathscr{G}_F$ the fine grid points inside Ω_F, and $\mathscr{G}_C$ the coarse grid points outside of Ω_F. The partitioning is illustrated in Figure 2.

The coarse grid, $\mathbf{G}$, can be partitioned similarly as $\mathbf{G} = \mathbf{G}_C \cup \mathbf{G}_I \cup \mathbf{G}_F$, where $\mathbf{G}_C$ consists of the coarse grid points outside of Ω_F ($\mathbf{G}_C = \mathscr{G}_C$), $\mathbf{G}_I$ the coarse grid points on the boundary of $\Omega_F(\mathbf{G}_I = \mathscr{G}_I)$, and $\mathbf{G}_F$ the coarse grid points inside $\Omega_F(\mathbf{G}_F \subset \mathscr{G}_F)$.

The principal grids used in the FAC algorithm are $\mathscr{G}$, $\mathscr{G}_F$, and $\mathbf{G}$. To be able to pass information between them, assume that a prolongation or interpolation operator, I, and a restriction operator, I^{T}, have been defined so that

(2) $$I : \mathbf{G} \longrightarrow \mathscr{G}$$

and

(3) $$I^{\mathrm{T}} : \mathscr{G} \longrightarrow \mathbf{G}.$$

Using I and I^{T}, an operator, $\mathbf{L}$, on the coarse grid can be defined according to the Galerkin condition

$$\mathbf{L} = I^{\mathrm{T}} \circ \mathscr{L} \circ I. \tag{4}$$

Based on the above partitioning of the grids $\mathscr{G}$ and $\mathbf{G}$, we can partition u, $\mathbf{u}$, $\mathscr{L}$, and $\mathbf{L}$ as

$$u = (u_C, u_I, u_F)^{\mathrm{T}}, \tag{5}$$

$$\mathbf{u} = (\mathbf{u}_C, \mathbf{u}_I, \mathbf{u}_F)^{\mathrm{T}}, \tag{6}$$

$$\mathscr{L}(u) = \begin{pmatrix} \mathscr{L}_{CC} & \mathscr{L}_{CI} & \Theta \\ \mathscr{L}_{IC} & \mathscr{L}_{II} & \mathscr{L}_{IF} \\ \Theta & \mathscr{L}_{FI} & \mathscr{L}_{FF} \end{pmatrix} (u), \tag{7}$$

and

$$\mathbf{L}(\mathbf{u}) = \begin{pmatrix} \mathbf{L}_{CC} & \mathbf{L}_{CI} & \Theta \\ \mathbf{L}_{IC} & \mathbf{L}_{II} & \mathbf{L}_{IF} \\ \Theta & \mathbf{L}_{FI} & \mathbf{L}_{FF} \end{pmatrix} (\mathbf{u}). \tag{8}$$

We should note that for nonlinear problems the operators used in the above description of $\mathbf{L}$, $\mathscr{L}$, $\mathbf{L}_{CC}$, $\mathscr{L}_{CC}$, $\ldots$, will generally depend on u. This notation is related to and is useful for the special forms of the operators encountered when numerically solving nonlinear partial differential equations.

We also note here that $\mathscr{L}_{CC} = \mathbf{L}_{CC}, \mathscr{L}_{IC} = \mathbf{L}_{IC}$, and $\mathscr{L}_{CI} = \mathbf{L}_{CI}$. $\mathscr{L}_{FF}$ is similar to $\mathbf{L}_{FF}$ except that there are more grid points, and thus more entries, in the $\mathscr{L}_{FF}$ block. The significant difference between $\mathbf{L}$ and $\mathscr{L}$ is in blocks IF and FI, where $\mathbf{L}$ and $\mathscr{L}$ "reach" for grid points in $\mathscr{G}_I$ and in $\mathscr{G}_F$, respectively. How this is done is what ultimately defines the composite grid operator $\mathscr{L}$ and determines how well the coarse grid problem approximates the composite grid problem.

With this machinery in place, an FAC cycle that allows for nonlinear operators can be written as follows.

$$\left.\begin{array}{ll} \text{Step 1.} & \text{Solve } \mathbf{L}(\mathbf{u}) = I^{\mathrm{T}}(f - \mathscr{L}(u)) + \mathbf{L}(I^{\mathrm{T}}u) \\ \text{Step 2.} & u \leftarrow u + I\{\mathbf{u} - I^{\mathrm{T}}u\} \\ \text{Step 3.} & \text{Solve } \mathscr{L}_{FF}(u_F) = f_F - \mathscr{L}_{FI}(u_I) \end{array}\right\} \tag{9}$$

Beginning with a zero initial guess for u (or any other guess if one is available), in Step 1 the coarse grid is solved as if there were no fine patch. After the first cycle, the right-hand side of Step 1 is augmented with the residual to make a correction to the most recent approximation to the composite grid solution.

If the composite grid problem (1) is expanded using the form of $\mathscr{L}$, u, and f given in (5)–(8), it is easy to see that Step 1 is an approximate solver for the first two equations and that Step 3 solves the third equation using the previous approximations for u_C and u_I.

2. Self-adaptivity. The first step in generating an FAC grid is to use a traditional grid generation scheme to generate a grid that will resolve any irregular boundaries in the problem. This grid can be generated by a variety of schemes (for example, **[5]** or **[6]**), a simple scheme can be coded as a part of the system, or, in the case of a rectangular domain, a uniform, coarse grid is placed on the region. Besides resolving any irregular boundaries, the grid should be generated so that it is reasonably uniform and sufficiently fine that it can be used to resolve the problem in regions where the solutions are smooth and will not require a high degree of resolution. If the latter criterion is not met, the self-adaptive scheme will inform us that a finer *coarse* grid is needed by flagging most of the region for refinement. For the remainder of this discussion we will assume such a coarse grid has been generated and that the problem to be solved has been mapped onto a rectangular region.

After there is a coarse grid on the region, the next step is to decide where more refinement is needed. There are two basic approaches that we have investigated: using Richardson extrapolation on the two solutions found on the coarse grid and on a coarser grid to approximate the error; and using Richardson extrapolation on the two solutions found using different-order numerical schemes to approximate the error. Since most of our numerical work has been done using the first of these schemes, we will describe this scheme in detail.

Let (i, j), (I, J), and (x, y) all denote the same physical point in a given grid, a coarser grid, and the domain, respectively. Let u_{ij}, U_{IJ}, and $u(x, y)$ denote the solutions to their respective problems. Also let the grid spacing on the two grids be denoted by Δx, Δy and δx, δy, where we assume that $\Delta x = m_1 \delta x$ and $\Delta y = m_2 \delta y$, where m_1 and m_2 are integers. Then, assuming that we have a kth-order numerical scheme, we can represent the error at the given point for each of the two grids by

$$\begin{aligned} e_{ij} &= u(x, y) - u_{ij} \\ &= \delta x^k F_{ij}^{(1)} + \delta x^{k+1} F_{ij}^{(2)} + \cdots \end{aligned} \tag{10}$$

and

$$e_{IJ} = u(x,y) - U_{IJ} \tag{11}$$
$$= \Delta x^k F^{(1)}_{IJ} + \Delta x^{k+1} F^{(2)}_{IJ} + \cdots ,$$

where $F^{(l)}_{ij}$ and $F^{(L)}_{IJ}$ are generally undetermined coefficients involving the appropriate partial derivatives and expressions involving $\delta y/\delta x$, $\Delta y/\Delta x$, etc., which were used to reduce the dependence of the above expressions to only one of the grid spacings. Subtracting equation (10) from equation (11) gives

$$e_{IJ} - e_{ij} = u_{ij} - U_{IJ} \tag{12}$$
$$= (m_1^k F^{(1)}_{IJ} - F^{(1)}_{ij})\delta x^k + (m_1^{k+1} F^{(2)}_{IJ} - F^{(2)}_{ij})\delta x^{k+1} + \cdots .$$

Assuming that $F^{(1)}_{IJ} = F^{(1)}_{ij}$ (they are Taylor coefficients evaluated at the same points), then division by $(m_1^k - 1)$ yields

$$\frac{u_{ij} - U_{IJ}}{m_1^k - 1} = \delta x^k F^{(1)}_{ij} + \frac{\delta x^{k+1}(m_1^{k+1} F^{(2)}_{IJ} - F^{(2)}_{ij})}{m_1^k - 1} + \cdots . \tag{13}$$

Comparing equations (10) and (13), we note that

$$e_{ij} = \frac{u_{ij} - U_{IJ}}{m_1^k - 1} + \mathscr{O}(\delta x^{k+1}). \tag{14}$$

Thus, we use

$$\mathscr{E}_{ij} = \frac{u_{ij} - U_{IJ}}{m_1^k - 1} \tag{15}$$

as our $(k+1)$st-order approximation to the error. Of course, depending on the size of some of the $F^{(l)}_{ij}$'s and $F^{(L)}_{IJ}$'s, the order of approximation of the error may be greater than the above analysis indicates.

The difficulty of using equation (15) as our approximation in most settings is that solutions on two grids are needed. In the FAC algorithm, to solve on one additional grid it is necessary to apply the above approximation to the original coarse grid. But since this grid is chosen to be coarser than the given coarse grid, the expense of this extra solution is not great. When the above approach is then used to determine where there should be higher levels of refinement (i.e., where further refinement is needed in the previously refined region), the present grid already has an underlying coarser grid that can be used for the Richardson approximation of the error.

The major shortcoming of the above-described scheme is that points are flagged only on the coarser of the two grids. Hence, to be conservative in our refinement scheme, we have assumed that if a particular point is flagged for refinement, the neighboring eight points should also be flagged as needing refinement. This conservative approach of flagging points needing refinement may add up to m_i, $i = 1, 2$, additional fine grid points in each direction.

A method to overcome this difficulty is to use the higher-order scheme for flagging that was mentioned earlier. To use the higher-order scheme we proceed much as we did before. We let (x, y) and (i, j) denote the same point in physical space, where the (i, j)-coordinates are with respect to a given coarse grid. Then let $u(x, y)$, u_{ij}, and U_{ij} denote the solution, a kth-order numerical solution, and a $(k+1)$st (or higher) order numerical solution, respectively. Then the error at the given point for each of the schemes is given by

$$\begin{aligned} e_{ij} &= u(x,y) - u_{ij} \\ &= \delta x^k f_{ij}^{(1)} + \delta x^{k+1} f_{ij}^{(2)} + \cdots \end{aligned} \tag{16}$$

and

$$\begin{aligned} E_{ij} &= u(x,y) - U_{ij} \\ &= \delta x^{k+1} F_{ij}^{(2)} + \cdots , \end{aligned} \tag{17}$$

where as before the f's and F's are undetermined coefficients depending on the partial derivatives of the appropriate orders and the m_i's. Then subtracting (17) from (16) we get

$$\begin{aligned} e_{ij} - E_{ij} &= U_{ij} - u_{ij} \\ &= \delta x^k f_{ij}^{(1)} + \mathcal{O}(\delta x^{k+1}), \end{aligned}$$

so we can approximate e_{ij} by

$$\mathcal{E}_{ij} = U_{ij} - u_{ij}, \tag{18}$$

which as in the previous scheme will be a $(k+1)$st-order approximation of the error.

As was stated earlier, the advantage of the higher-order flagging scheme is that the error is approximated on the entire coarse grid. The disadvantage of the higher-order flagging scheme is that it is necessary to solve the problem twice on each level, once to the kth order of accuracy and once at some higher order. However, this does not need to be excessively expensive. First consider the linear case and suppose that we wish to compare second-order and fourth-order schemes for solving the problem. Let the second- and fourth-order operators associated with the problem be given by $L^{(2)}$ and $L^{(4)}$, respectively. If we already have a vector $\mathbf{u}$ that satisfies

$$L^{(2)}\mathbf{u} = \mathbf{f},$$

we can calculate the difference between the fourth-order solution and the second-order solution by solving

$$L^{(4)}\Delta\mathbf{u} = \mathbf{f} - L^{(4)}\mathbf{u},$$

where $\mathbf{u}+\Delta\mathbf{u}$ is then the fourth-order solution. If the solution technique being used is one of the traditional iterative solution schemes, solving for this higher-order correction can be expensive. Multigrid techniques for higher-order solution schemes, [7], would be very efficient for this purpose. In the case with nonlinear problems the situation looks different but is essentially the same. For most solution techniques the existence of a good initial guess will speed up the solution process. The second-order solution will always provide an excellent starting guess for the fourth-order problem.

Now that we have decided at which points of the domain more resolution is needed, the next step is to form these points into blocks of refined grids. This problem appears to be easy, but it is very important that it be done correctly. If one strives for a scheme that produces very efficient blocks (efficient in terms of the percentage of flagged points in any block), it is not difficult for the resulting algorithm to be $\mathcal{O}(n^2)$ where n is the number of composite grid points. Part of our design plans for developing the self-adaptive scheme is that the work done by the scheme be $\mathcal{O}(n)$, or the same order of a multigrid solver [**4**].

The method by which we limit the complexity of our scheme is to search the data once, doing it efficiently in small blocks.

(i) The scheme begins by sweeping through the region in some logical order, say $i-j$, looking for a flagged point.

(ii) When a flagged point is encountered, say (i_0, j_0), a small region is swept out (i_0+k, j_0+l) for $0 < k, l \leq d$, and we determine the size of the minimum rectangular that will contain the flagged points in this region. The definition of small, d, is left up to the user as an input parameter. However, the potential efficiency of the block is a function of the size of d.

(iii) The size of the block is then increased by one in each direction. This buffer zone is added because by our flagging procedure we have no information about the error between the last flagged point and the first unflagged point. If we knew that no refinement was needed in this buffer region, it would be sufficient to leave the blocks as they were defined in Step (ii). Since we have no information in this buffer region, the conservative approach is to assume that refinement is needed up to the next point, where we know no further refinement is necessary.

(iv) The scheme next determines whether this new block should be combined with any of the previous blocks that have been constructed. (Of course, if the block just created was the first block, this step is skipped.) If the block is combined with some other block, the data describing the other block is altered. If the block is not combined with any other block, the appropriate data concerning the block is stored. The criteria chosen for whether or not a block should be combined with another block is the percentage of overlap of the sides of the two blocks and is also a user-supplied parameter. Requiring a large percentage would give efficient blocks and a complex block structure, while requiring only a small percentage would give fewer, large, inefficient blocks.

(v) Finally, we move on to search for the next flagged point that is not an interior point of a previous block and return to Step (ii) when such a point is encountered.

It should be mentioned again that the efficiency of the scheme is related to keeping d small and making all of the decisions concerning the blocks during the one sweep through the points.

At this point the regions where more resolution is needed have been determined. The one last step in generating the FAC grid is to determine how much refinement is needed in the various blocks. This can be accomplished using the information that we used to flag the points needing refinement. Let ε_{IK} denote the error in the calculation in the refined region that has been flagged by the Richardson extrapolation technique. Then, analogous to the expressions (10) and (11), ε_{IK} can be written as

$$\varepsilon_{IK} = dx^k F_{IK}^{(1)} + \mathcal{O}(dx^{k+1}),$$

where dx represents the grid spacing (which is yet to be determined) in the refined grid. As was done in the previous analysis, we assume that $F_{IK}^{(1)} = F_{ij}^{(1)}$. But then we note by observing equations (10), (14), and (15) that our approximation of e_{ij} will also provide us with an approximation of $F_{ij}^{(1)}$. Hence, we see that

$$\varepsilon_{IK} = \frac{\mathcal{E}_{ij}}{\delta x^k} dx^k + \mathcal{O}(dx^{k+1}).$$

If we then require that ε_{IK} be less than some prescribed error tolerance η, we can solve for dx as

$$dx = \sqrt[k]{\frac{\eta}{\mathscr{E}_{ij}}}\,\delta x$$

to the $\frac{k+1}{k}$th order. This expression has been used very successfully for our adaptive calculations.

The grid refinement work developed above has all been done in the logically rectangular coordinate system made possible because we began our generation scheme by mapping our domain in (x, y)-space into a (ξ, η)-space using a traditional grid generation scheme as was described in the beginning of this section. The patched grids that the above algorithm has defined are in the (ξ, η)-space. Thus we must define the mapping between the two spaces for these additional fine grid points. But this can be done easily by a variety of methods, say, using either linear or quadratic interpolation in the (ξ, η)-space. It seems that because of the fineness of the grids in this region, more smoothness should not be needed. If it is decided that more smoothness is necessary, the interpolated grids can be smoothed using the same differential equation techniques used in the usual grid generation schemes.

3. Numerical results. In this section some sample computations are presented that illustrate the use of the self-adaptive FAC algorithm on nonlinear problems. While there are many problems that could have been chosen, the example presented illustrates the use of FAC for solving nonlinear problems, the use of the self-adaptive algorithm to determine where more refinement is needed, and the use of both FAC and the self-adaptive algorithm in a setting where two levels of refinement are used. Other FAC calculations generally involving linear problems can be found in **[1]**–**[3]**.

The specific problem we solve is

$$[(K - M\phi_x)\phi_x]_x + \phi_{yy} = 0 \quad \text{in} \quad \mathbf{R} = (-1, 2) \times (0, 1)$$

$$\frac{\partial \phi}{\partial n} = g \quad \text{on} \quad \partial R.$$

K and M, positive constants, are chosen so that $K - M\phi_x > 0$, but very small. The function g was set equal to zero except for $y = 0$ and $0 \le x \le 1$, where we set $g = \alpha(x - .5)$.

The solution to the above problem is strongly affected by this boundary condition. The results given below (Figure 3) show that the region where additional refinement is needed is near the interval where the boundary condition is nonzero. It should be noted that the algorithm does not refine along the entire nonzero boundary condition. The algorithm determines that more resolution is needed near the beginning and the end of the region with a nonzero boundary condition. The solution is also affected strongly by the choice of K and M. The problem is elliptic but is very close to having a hyperbolic region in the domain since K and M were chosen so that $K - M\phi_x$ is small. It is the choice of the boundary condition and K and M that drive the problem.

In the tests run for this problem we began by choosing a coarse grid (41 horizontal and 15 vertical points) and a desired error tolerance. The code was then allowed to decide where more refinement was necessary and how many levels of refinement were needed (when to quit). The solutions were then compared to solutions on an extensive fine grid. The mesh on the extensive fine grid was chosen to be as fine as the finest mesh used in any of the patches. This extensive fine grid was chosen for comparison because using this extensive fine grid is one alternative (an expensive one) for solving the problem to the desired accuracy. For FAC to be successful, it must be able to solve the problem more cheaply and as accurately as can be done on the extensive fine grid.

In Figure 3(a) we show the domain with the fine grid patch (where $\delta x = \delta y = \Delta x/2$) that has been chosen by the self-adaptive algorithm for a particular prescribed error tolerance. For this tolerance value, no third level was necessary. As was stated earlier, it should be noted that the algorithm chooses to refine in regions that cover the beginning and the end of the nonzero boundary condition, not along the entire boundary region having a nonzero boundary condition. Figures 4 and 5 show the difference between the extensive fine grid solution and the coarse grid solution and the difference between the extensive fine grid solution and the composite grid solution, respectively. A comparison of the two figures shows how much is gained by using FAC and a composite grid to solve the problem as compared to just the coarse grid. Also, Figure 5 shows that the difference between the extensive fine grid solution and the composite grid solution is very small.

In a second test solution a smaller error tolerance is used. In this case the algorithm chooses to include a second level of refinement. Again the mesh ratio of each of the patches is two. The domain along with the patches is shown in Figure 3(b). Again the difference between the extensive fine grid solution and the three-level composite grid solution is minimal (Figure 6). Finally, the composite grid solution based on three levels is presented in Figure 7.

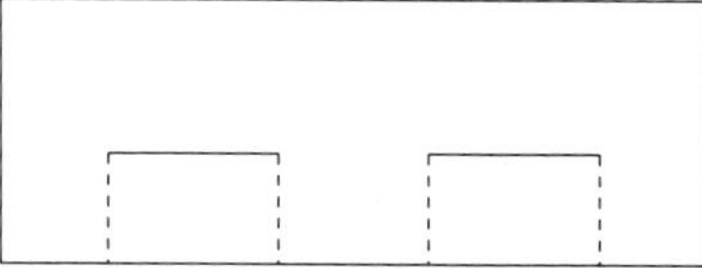

FIGURE 3(a). Domain and second-level patch

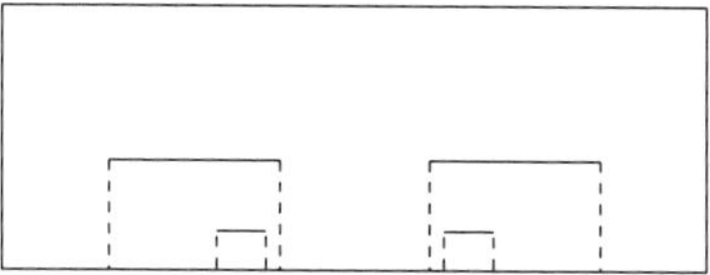

FIGURE 3(b). Domain with second- and third-level patches

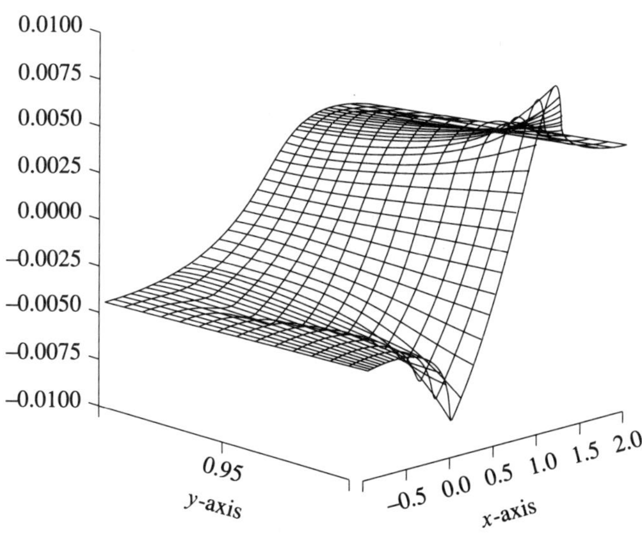

FIGURE 4. Difference between the extensive fine grid and the coarse grid solutions

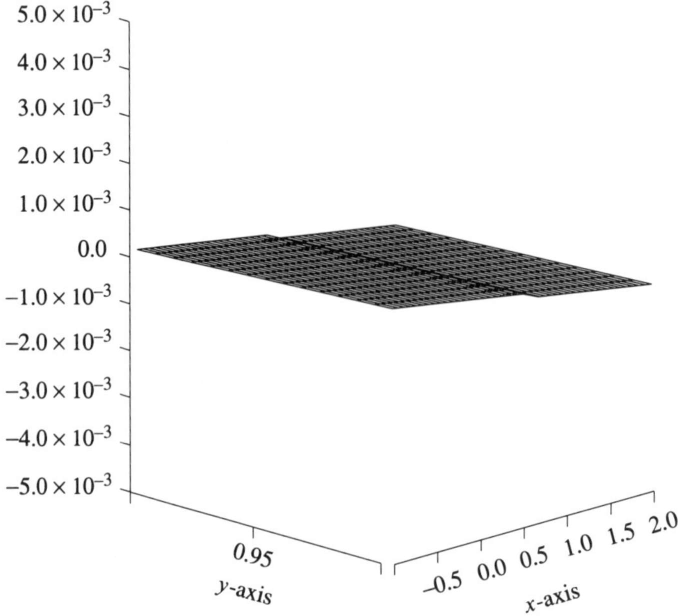

FIGURE 5. Difference between the extensive fine grid and composite grid solutions

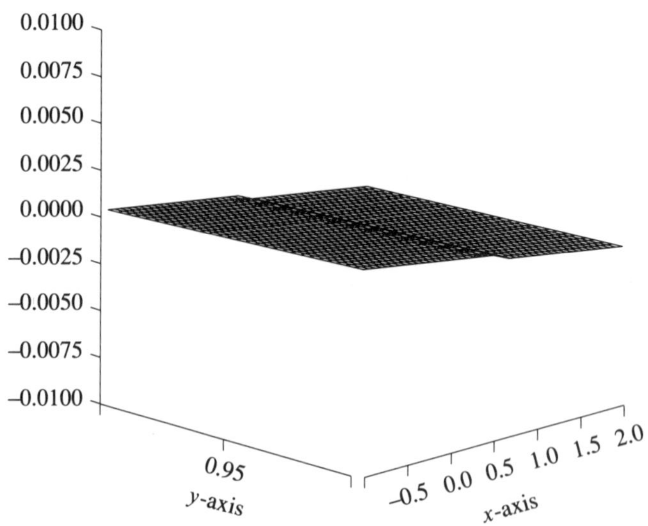

FIGURE 6. Difference between the extensive fine grid and the three-level composite grid solution

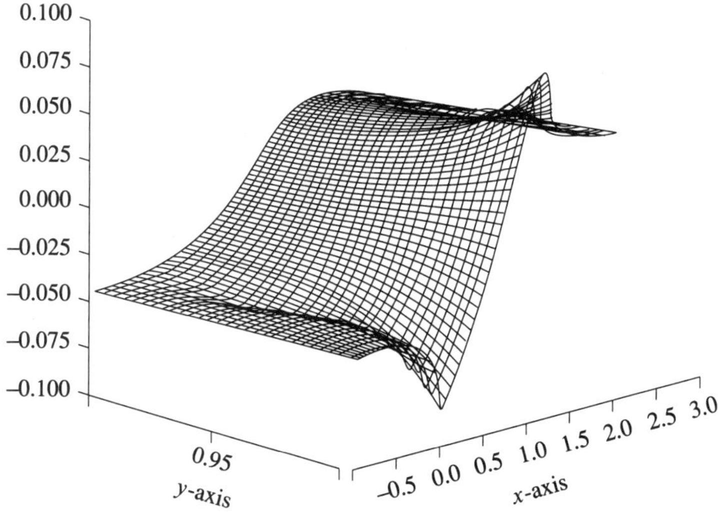

FIGURE 7. Three-level composite grid solution

4. Conclusions. In previous papers FAC along with the self-adaptive algorithm has been shown to be an effective tool for numerically solving partial differential equations. In this paper we have shown that FAC along with the self-adaptive algorithm can also be used effectively for solving nonlinear partial differential equations. Since for nonlinear problems it is more difficult to predict where additional refinement might be necessary than is the case for linear problems, it is more important to be able to use the self-adaptive feature for nonlinear problems than for linear problems.

REFERENCES

1. S. McCormick and J. Thomas, *The fast adaptive composite grid* (*FAC*) *method for elliptic equations*, Math. Comp. **46** (1986), 439–456.

2. J. W. Thomas, M. Heroux, S. McKay, S. McCormick, and A. M. Thomas, *Application of the fast adaptive composite grid method to computation fluid dynamics*, Numerical Methods in Laminar and Turbulent Flow (C. Taylor, W. G. Habashi, and M. M. Hafez, eds.), Pineridge Press, Swansea, U.K., 1987, pp. 1071–1082.

3. M. Heroux, S. McCormick, S. McKay, and J. W. Thomas, *Applications of the fast adaptive composite grid method*, Lecture Notes in Pure and Applied Mathematics, Marcel Dekker **110** (May, 1988).

4. Steve F. McCormick, Steven M. McKay, and J. W. Thomas, *Complexity of the fast adaptive composite grid (FAC) method*, (to appear) in *Applied Numerical Mathematics.*

5. J. F. Thompson, *A composite grid generation code for general 3-D regions*, AIAA-87-0275, AIAA 25th Aerospace Sciences Meeting, Reno, Nevada, 1987.

6. P. R. Eiseman, *A control point form of algebraic grid generation*, Numerical Methods in Laminar and Turbulent Flow (C. Taylor, W. G. Habashi, and M. M. Hafez, eds.), Pineridge Press, Swansea, U.K., 1987, 1083–1091.

7. S. Schaeffer, *High-order multigrid methods*, Ph. D. Thesis, Colorado State University, 1982.

COLORADO STATE UNIVERSITY

Lectures in Applied Mathematics
Volume **26** (1990)

Interactive Program for Continuation of Solutions of Large Systems of Nonlinear Equations

RAYMOND MEJIA

1. Introduction. We describe an interactive program for continuation of solutions of large systems of nonlinear equations that arise in many applications in science and engineering. We consider, in particular, systems that, when linearized, result in a structured, sparse coefficient matrix. Examples include discretized convection-diffusion and reaction-diffusion equations that describe chemical reactors and countercurrent exchange models of the mammalian kidney [**16**].

In Section 2 we describe the continuation algorithm used by CONKUB [**12**], the path-following program. The algorithm is based on work by Keller [**9**] and Kubiček [**10**]. We describe the procedure for proceeding along a solution branch and identifying turning points and bifurcations, which is based on work by Crandall and Rabinowitz [**4**], [**5**] and by Bunow and Kernevez [**2**].

Much progress has been made recently on the numerical continuation of solutions of nonlinear equations. In addition to papers and references appearing in these proceedings, see surveys by Allgower and Georg [**1**], Mittelmann and Weber [**18**], and Watson [**29**]. Algorithms for root finding and continuation of solutions include those by Chow, Mallet-Paret and Yorke [**3**], Jürgens, Peitgen and Saupe [**8**], Moré and Cosnard [**19**], [**20**], Watson and Fenner [**30**], Morgan [**21**], Rheinboldt and Burkhardt [**25**], Doedel [**7**], Deuflhard, Fiedler and Kunkel [**6**] and Watson and Scott [**31**].

In Section 3 we describe a convection-diffusion model of the mammalian kidney that, when discretized, results in a large, structured system of nonlinear algebraic equations [**27**]. In Section 4 we show how

0082-0717/90 $1.00 + $.25 per page

we discretize this problem and decompose the domain in order to trace steady-state solutions effectively. In Section 5 we pose a physiological question [17] that we will attempt to answer using the model described in Section 3. We demonstrate how to use CONKUB to step along the solution surface as a function of parameters of the model. We then use the continued solution of the model to answer the question posed. We conclude in Section 6 with observations about the strengths and weaknesses of interactive continuation in the solution of large systems of nonlinear equations. Of special interest are problems where the solution is sought as a function of one or more parameters in a model, and problems where turning points and/or bifurcations are to be identified.

2. Continuation. Given a nonlinear system of algebraic equations:

$$F(x,\alpha) = 0, \qquad F : \mathbf{R}^m \times \mathbf{R}^q \to \mathbf{R}^m, \tag{2.1}$$

with unknowns x and parameters α and F sufficiently smooth, we use the assumed continuity and differentiability of F with respect to x and α to write the differential equations

$$\frac{dx}{d\alpha_\ell} + F_x^{-1}\frac{\partial F}{\partial \alpha_\ell} = 0, \qquad x(\alpha^0) = x^0, \tag{2.2}$$

with

$$F_x \equiv \frac{\partial F_i}{\partial x_j}, \qquad 1 \le i,\ j \le m,$$

$$\frac{\partial F}{\partial \alpha_\ell} \equiv \left(\frac{\partial F_1}{\partial \alpha_\ell}, \dots, \frac{\partial F_m}{\partial \alpha_\ell}\right)^{\mathrm{T}}, \qquad 1 \le \ell \le q,$$

$$F(x^0, \alpha^0) = 0,$$

and α_k fixed for $k \neq \ell$.

If F_x is nonsingular in a neighborhood of $(x(\widehat{\alpha}), \widehat{\alpha})$, then the solution $x(\widehat{\alpha})$ of equations (2.2) is a solution of equations (2.1). However, when F_x is singular (or numerically ill-conditioned), such as at a turning point or bifurcation, this approach fails.

CONKUB avoids this difficulty by first parameterizing with respect to the arclength of the solution locus in $\mathbf{R}^{m+1}$, as follows:

$$\sum_{j=1}^{m}\left(\frac{dx_j}{ds}\right)^2 + \left(\frac{d\alpha_\ell}{ds}\right)^2 = 1. \tag{2.3}$$

Differentiation of F with respect to s yields the equations

$$\frac{dF_i}{ds} = \sum_{j=1}^{m} \frac{\partial F_i}{\partial x_j}\frac{dx_j}{ds} + \frac{\partial F_i}{\partial \alpha_\ell}\frac{d\alpha_\ell}{ds} = 0, \tag{2.4}$$

with $i = 1, 2, \ldots, m$ and $1 \leq \ell \leq q$.

Let x_k be an independent variable and $x_{m+1} \equiv \alpha_\ell$ be an unknown. Solution of (2.4) yields

$$\frac{dx_i}{ds} = \beta_i \frac{dx_k}{ds}, \qquad i = 1, 2, \ldots, k-1, k+1, \ldots, m+1, \tag{2.5}$$

and substitution into (2.3) yields

$$\frac{dx_k}{ds} = \pm \left[1 + \sum_{\substack{i=1 \\ i \neq k}}^{m+1} \beta_i{}^2 \right]^{-1/2} \tag{2.6}$$

CONKUB integrates (2.5) and (2.6) to obtain the next value of the solution using a variable order Adams–Bashforth multistep method (up to order 4), and chooses the sign of (2.6) that preserves the orientation of the curve **[10]**. A Newton method is used to correct the truncation error to remain within a specified distance of the solution. A change of sign (or a change of direction near zero) of the determinant of F_x indicates a possible singularity. Bisection is used to obtain $(\tilde{x}, \tilde{\alpha})$ such that $\det F_x(\tilde{x}, \tilde{\alpha}) \approx 0$. Thus, turning points are readily identified.

CONKUB identifies and traces arcs at branch points in a manner suggested by Crandall and Rabinowitz **[4]**, **[5]** and Keller **[9]**, and treats all bifurcations as if simple. Let the dimension of the null space of F_x at $(\tilde{x}, \tilde{\alpha})$, $\dim N(F_x^{\sim}) = 1$. Then $N(F_x^{\sim})^{\mathrm{T}} = \operatorname{span} \Psi$, and $|\Psi^{\mathrm{T}} F_{\alpha_\ell}^{\sim}| \leq \delta$ with δ small indicates that $(\tilde{x}, \tilde{\alpha})$ is a bifurcation point. Since the tangent of the original branch is known, CONKUB uses the bifurcation condition $F_{\alpha_\ell}^{\sim} \in R(F_x^{\sim})$ to obtain the tangent of the bifurcating branch. Small steps in each direction yield solutions that are then corrected onto the bifurcating branch with Newton's method **[2]**. These may in turn be used to initiate tracing a branch of solutions.

3. Mathematical model. The principal function of the kidney is to maintain the chemical composition of body fluids within the narrow limits compatible with life **[22]**. The mammalian kidney is composed of nephrovascular units that may be described as a countercurrent exchanger **[11]** that serves to conserve water and solutes as needed, and where a small single effect can be multiplied to produce a concentrated urine. A mathematical model that merges the vasculature and the interstitium into a central core was first considered by Stephenson [26]. Such a model consists of convection-diffusion equations for solutes and fluid, and equations of motion [17] that are as follows:

$$\frac{\partial}{\partial t}(A_i C_{ik}) + \frac{\partial}{\partial \xi}\left[F_{iv} C_{ik} - A_i D_{ik} \frac{\partial C_{ik}}{\partial \xi}\right] = -J_{ik}, \tag{3.1}$$

$$\frac{\partial A_i}{\partial t} + \frac{\partial F_{iv}}{\partial \xi} = -J_{iv}, \tag{3.2}$$

$$\frac{\partial P_i}{\partial \xi} + R_{iv} F_{iv} = 0, \tag{3.3}$$

for $0 \le i \le I$ tube segments and $1 \le k \le K$ solute species. t is time; A_i is the cross-sectional area of the ith tube; C_{ik} is the concentration of the kth solute in the ith tube; ξ is axial distance along each tube; F_{iv} is the axial volume flow; D_{ik} is the diffusion coefficient of the kth solute in the ith tube; J_{ik} is the transmembrane solute flux; J_{iv} is the transmembrane volume flux; P_i is the hydrostatic pressure; and R_{iv} is the resistance to flow.

Transmural volume and solute fluxes are given as follows:

$$\begin{aligned} J_{iv} &= 2\pi \rho_i P_{f,i} V_w \left[\Delta P_i - RT \sum_{k=1}^{K} \sigma_{ik} \Delta C_{ik}\right], \\ J_{ik} &= 2\pi \rho_i P_{ik} \Delta C_{ik} + (1 - \sigma_{ik}) \overline{C}_{ik} J_{iv} + \frac{V_{m,i}^k}{1 + K_{m,i}^k / C_{ik}}, \end{aligned} \tag{3.4}$$

where ρ_i is the radius of the ith tube; $P_{f,i}$ is the osmotic water permeability; V_w is the partial molar volume of water; $\Delta P_i = P_i - P_q$, where P_q is the hydrostatic pressure outside the ith tube; R is the gas constant; T is absolute temperature; σ_{ik} is the Staverman reflection coefficient of the kth solute in the ith tube, $0 \le \sigma_{ik} \le 1$; $\Delta C_{ik} = C_{ik} - C_{qk}$; P_{ik} is the solute permeability; $\overline{C}_{ik}$ is the concentration at the tube wall — if $C_{ik} \gg C_{qk}$, $\overline{C}_{ik} = (C_{ik} - C_{qk}) / (\ln C_{ik} - \ln C_{qk})$; otherwise, $\overline{C}_{ik} = (C_{ik} + C_{qk})/2$. $V_{m,i}^k$ is the maximum rate of active transport of the kth solute out of the ith tube using Michaelis–Menten kinetics; and $K_{m,i}^k$ is the Michaelis constant, the concentration at which active transport is one-half the maximum.

Consider a model with five nephron populations as shown schematically in Fig. 3.1. Filtration of water and small solutes from the vasculature into each nephron occurs at the glomerulus, which is depicted as

• in the figure. Each nephron segment empties into the next, and the final output is urine. Each nephron segment has been shown experimentally in vitro to have distinct transluminal transport properties, with the exchange of solutes and water between tubes occurring through a well-mixed vascular core in the cortical labyrinth and through a medullary central core that is homogeneous at each depth of the medulla. Initial conditions are:

$$(3.5) \qquad C_{ik}(\xi,0) = C_{ik}^0, \qquad F_{iv}(\xi,0) = F_{iv}^0, \qquad P_i(\xi,0) = P_i^0,$$

$\forall i,\ k,$ and ξ.

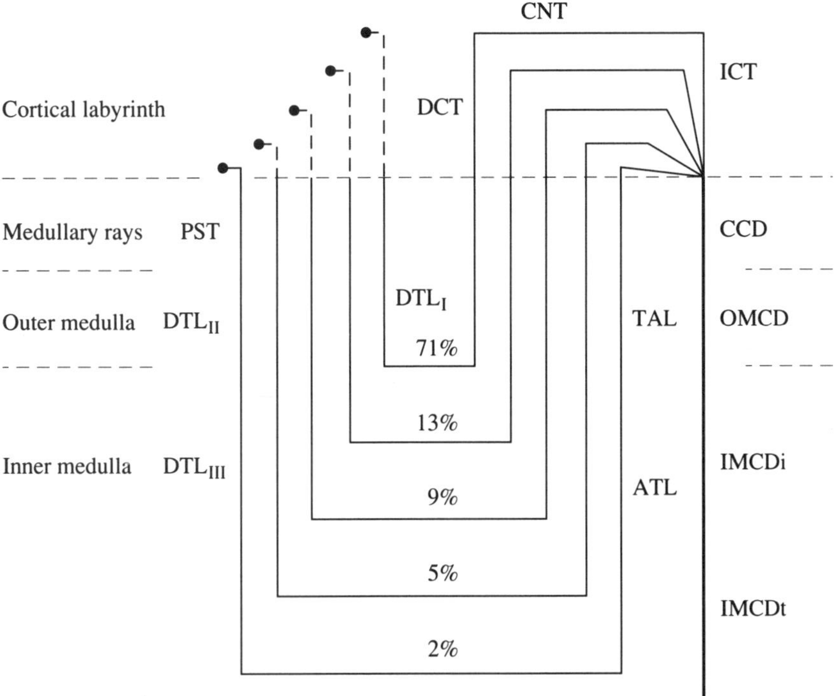

FIGURE 3.1. Schematic diagram of a mathematical model of the rat kidney showing the percent of total nephrons assigned to each of five populations [17]. • shows the origin in the cortex of each population. Tubular segments with distinct transtubular transport properties are: proximal straight tubules, PST; descending limbs of Henle, DTL$_{\text{I}}$, DTL$_{\text{II}}$ and DTL$_{\text{III}}$; ascending thin limbs, ATL, and thick ascending limbs of Henle, TAL; distal convoluted tubules, DCT; connecting tubules, CNT; initial collecting tubules, ICT; cortical collecting ducts, CCD; outer medullary collecting ducts, OMCD; initial and terminal inner medullary collecting ducts, IMCDi and IMCDt.

Boundary conditions for each nephron population at $t \geq 0$ are as follows:

$$\begin{aligned} C_{1k}(0,t) &= C_{1k}^0, \qquad 1 \leq k \leq K, \\ F_{1v}(0,t) = F_{1v}^0, \qquad P_1(0,t) &= P_1^0, \qquad P_I(\xi_{Id},t) = P^b, \end{aligned} \tag{3.6}$$

where P^b is the exit (bladder) pressure, and subscripts id denote the distal end of tube i. Given I_h tube segments contributing to the hth nephron population, we have for all h that

$$\begin{aligned} C_{ik}(\xi_{ip},t) &= C_{(i-1)k}(\xi_{(i-1)d},t), \qquad \forall k, \\ F_{iv}(\xi_{ip},t) &= \pm F_{(i-1)v}(\xi_{(i-1)d},t), \\ P_i(\xi_{ip},t) &= P_{i-1}(\xi_{(i-1)d},t), \end{aligned} \tag{3.7}$$

for $1 < i < I_h$. Assuming $D_{ik} \equiv 0, \forall k$, and $i > 0$,

$$\begin{aligned} F_{Iv}(\xi_{Ip},t) &= \sum_h F_{(I_h-1)v}(\xi_{(I_h-1)d},t), \\ F_{Iv}C_{Ik}(\xi_{Ip},t) &= \sum_h F_{(I_h-1)v}C_{(I_h-1)k}(\xi_{(I_h-1)d},t), \qquad \forall k, \end{aligned} \tag{3.8}$$

where subscripts ip denote the proximal end of tube i, and flow away from the cortex and toward the papilla is defined to be positive.

In the cortex, the boundary conditions for the core are:

$$C_{qk} = C_k^p, \qquad P_q = P^p, \tag{3.9}$$

for $q = c$, and all k. Superscript p denotes the value of the associated variable in the blood plasma, which is specified. The medullary central core is considered to be a tube that is open at the border of the cortical labyrinth and the medullary rays and is closed at the papilla. Hence, equations (3.1)–(3.3) apply, and boundary conditions are:

$$\begin{aligned} A_0(L)\frac{\partial C_{0k}}{\partial t}(L,t) &= C_{0k}(L,t)J_{0v}(L,t) - J_{0k}(L,t), \\ F_{0v}(L,t) &= F_{0k}(L,t) = 0, \\ P_0(0,t) &= P_c, \end{aligned} \tag{3.10}$$

for $1 \leq k \leq K$, and

$$F_{0k} = F_{0v} - A_0 D_{0k}\frac{\partial C_{0k}}{\partial \xi}. \tag{3.11}$$

Finally, volume and mass conservation require that

$$
\begin{aligned}
J_{0v}(\xi, t) &= -\sum_{i\in\Omega_\xi} J_{iv}(\xi, t),\\
J_{0k}(\xi, t) &= -\sum_{i\in\Omega_\xi} J_{ik}(\xi, t),
\end{aligned}
\tag{3.12}
$$

for $0 \le \xi \le L$ and all k, and where $\Omega_\xi = \{i : \xi_{ip} \le \xi \le \xi_{id} \text{ or } \xi_{id} \le \xi \le \xi_{ip}\}$.

4. Method of solution. Equations (3.1)–(3.3) are discretized using a box scheme in space and a backward Euler approximation in time. A typical differential equation of the form

$$\frac{\partial(A_{ik}C_{ik})}{\partial t} + \frac{\partial F_{ik}}{\partial \xi} = -J_{ik} \tag{4.1}$$

is discretized as

$$
\begin{aligned}
\frac{F_{ik}^n(j) - F_{ik}^n(j-1)}{\Delta\xi} = -\frac{1}{2}\Bigg[& J_{ik}^n(j) + J_{ik}^n(j-1)\\
&+ \frac{A_i(j)\big(C_{ik}^n(j) - C_{ik}^{n-1}(j)\big)}{\Delta t}\\
&+ \frac{A_i(j-1)\big(C_{ik}^n(j-1) - C_{ik}^{n-1}(j-1)\big)}{\Delta t}\Bigg]
\end{aligned}
\tag{4.2}
$$

for $1 \le j \le J$, $J = L/\Delta\xi$, $1 \le k \le K$, and $t_n = n\Delta t$, $n \ge 1$. This finite difference scheme has been shown to be $O(\Delta\xi^2)$ accurate and stable for renal models [13].

The system of nonlinear algebraic equations that results from the discretization of equations (3.1)–(3.3), with initial conditions (3.5) and boundary conditions (3.6)–(3.11), may be partitioned and solved as a smaller boundary value problem and a sequence of small initial value problems [14], [28], [15] as follows:

$$
\begin{aligned}
F_i(x_i^n, x_i^{n-1}, x_G^n; \alpha) &= 0, \qquad i = 1, \dots, I,\\
F_G(x_1^n, x_2^n, \dots, x_I^n, x_G^n, x_G^{n-1}; \alpha) &= 0.
\end{aligned}
\tag{4.3}
$$

x_i^n is the vector of unknowns in the ith tube at the nth time step; x_G^n is the vector of unknowns in the central core and variables associated with continuity conditions where two or more tubes merge (for example, where the five ICT empty into the CCD in Fig. 3.1). α is the vector of parameters available for continuation.

For continuation, we write the steady-state equations as in (2.1), and for $x_i \in \mathbf{R}^{r_i}$, $x_G \in \mathbf{R}^g$, and $\alpha \in \mathbf{R}^q$ domain decomposition as in (4.3) yields **[16]**:

$$(4.4) \qquad F_i(x_i, x_G; \alpha) = 0, \qquad F_i : \mathbf{R}^{r_i+g} \times \mathbf{R}^q \to \mathbf{R}^{r_i},$$

with $i = 1, 2, \ldots, I$, $3 \le r_i \le K + 2$, and

$$(4.5) \qquad F_G(x_1, x_2, \ldots, x_I, x_G; \alpha) = 0, \qquad F_G : \mathbf{R}^{r+g} \times \mathbf{R}^q \to \mathbf{R}^g,$$

with $g \ll r = \sum_{i=1}^{I} r_i$.

5. A physiological question. Consider a hormone, atrial natriuretic factor (ANF), that is secreted by the cardiac atria and induces a significant increase in water and NaCl excretion by the kidney when its circulating level rises. Several possible sites of renal action have been identified experimentally. We shall use the model described in Section 3 to address a specific issue: namely, how might one distinguish between an increase in volume flow into the PST segment of the nephrons (see Fig. 3.1) and a reduction in water and NaCl reabsorption from the CD? In other words, how does one distinguish between increased $F_{1v}^0 = F_{\mathrm{PST}v}^0$ in boundary conditions (3.6) and reduced $P_{f,\mathrm{CD}}$ and $V_{m,\mathrm{CD}}^{\mathrm{NaCl}}$ in equations (3.4)? For clarity, we shall use the tube acronym PST for the proximal straight tubule, CD for the collecting duct, and solute NaCl for sodium chloride, rather than their numerical indices in the model. We proceed by studying the model as a function of the boundary condition and each of these parameters **[17]**.

Assume that we have used CONKUB to save an approximate solution for a set of parameters, initial and boundary conditions in file ANF869. When prompted by CONKUB for the next command with a '/', use the restart (RE), set bounds (SB), data (DT), and go (GO) commands (see the Appendix for a list and description of the interactive commands available in CONKUB) to verify that we have a solution as follows:

```
CONKUB ALGORITHM FOR PATH FOLLOWING 100587
TYPE HE FOR HELP.
/RE
ANF869
/SB
HH =
1.0000000E-01 ... 1.0000000E-01
HMAX =
1.0000000E-01 ... 1.0000000E-01
```

/DT
IDATA=
-1 0 0 0 0
RDATA=
0.0000000E+00 ... 0.0000000E+00
/GO 1

We have set both the starting and maximum step to be taken by CONKUB to 0.1; set IDATA(1)= −1 to allow the subroutine that evaluates the model to interrupt the calculation and instructed CONKUB to take one step. The model prompts for the 'next' command (see the Appendix for a description of the interactive commands available in the model). We will now identifiy $V_{m,\mathrm{CD}}^{\mathrm{NaCl}}$ (kparm = 4071) as the parameter to be followed, give it an initial multiplier par = 0.9, and proceed:
next = 2
kparm = 4071
next = 3
par = 9.000000000E-01
next = 0
RESIDUAL IMPROVEMENT FACTOR = 0.0E+00 kparm = 4071 par = 9.00000E-01
RESIDUAL IMPROVEMENT FACTOR = 1.0E+00 kparm = 4071 par = 1.00000E+00
THE VECTOR NORM IS: 2.2647E+00; PAR(1)= 1.0000E+00
CONKUB : NEWTON ITER 1 ERR = 7.068E-16, ||Xnew-Xold|| = 2.571E-12
RESIDUAL IMPROVEMENT FACTOR = 1.8E+01 kparm = 4071 par = 1.00000E+00
THE VECTOR NORM IS: 2.2647E+00; PAR(1)= 1.0000E+00
CONKUB : NEWTON ITER 2 ERR = 3.849E-17, ||Xnew-Xold|| = 2.713E-13
CONKUB: NEWTONS CONVERGED; K=171
RESIDUAL IMPROVEMENT FACTOR = 9.8E-01 kparm = 4071 par = 1.00000E+00
DET= 0.3351*2** -1072 K=171

After the command to the model to proceed, CONKUB took a step to PAR(1)=1.0E+0, and two quasi-Newton iterations to a solution; det F_x at the solution is positive for a value of $1 \times V_{m,\mathrm{CD}}^{\mathrm{NaCl}}$, and the index of the continuation parameter is K=171, the multiplier of $V_{m,\mathrm{CD}}^{\mathrm{NaCl}}$. In these calculations, if the RESIDUAL IMPROVEMENT FACTOR is 50 or larger, the previous Jacobian matrix has been reused. If it is less than 50 but 10 or larger, a rank one update has been performed. If it is less than 10, the Jacobian matrix has been calculated using difference quotients. At this point, we usually might wish to solve the time-dependent equations (4.3) to verify that we start at a time-stable solution. Since we have restarted the calculation, we have previously ascertained this.

We now proceed to set the maximum step permitted to 0.2 using command SB and then target to a parameter value of 0.5 — that is, to $0.5 \times V_{m,\mathrm{CD}}^{\mathrm{NaCl}}$ — as follows:

```
/SB
HMAX =
2.0000000E-01 ... 2.0000000E-01
/TP
TRGPT= 5.0000000E-01
RESIDUAL IMPROVEMENT FACTOR = 7.8E-01 kparm = 4071 par
= 1.00000E+00
RESIDUAL IMPROVEMENT FACTOR = 2.9E-12 kparm = 4071 par
= 9.27545E-01
RESIDUAL IMPROVEMENT FACTOR = 7.5E+01 kparm = 4071 par
= 9.27545E-01
RESIDUAL IMPROVEMENT FACTOR = 6.0E+00 kparm = 4071 par
= 9.27545E-01
RESIDUAL IMPROVEMENT FACTOR = 1.1E+04 kparm = 4071 par
= 9.27545E-01
RESIDUAL IMPROVEMENT FACTOR = 2.5E-01 kparm = 4071 par
= 9.27545E-01
RESIDUAL IMPROVEMENT FACTOR = 1.6E+05 kparm = 4071 par
= 9.27545E-01
RESIDUAL IMPROVEMENT FACTOR = 1.4E+00 kparm = 4071 par
= 9.27545E-01
RESIDUAL IMPROVEMENT FACTOR = 3.8E-12 kparm = 4071 par
= 8.08704E-01
.
.
.
RESIDUAL IMPROVEMENT FACTOR = 2.5E+00 kparm = 4071 par
= 8.08704E-01
RESIDUAL IMPROVEMENT FACTOR = 4.8E-12 kparm = 4071 par
= 6.24557E-01
.
```

```
.
.
RESIDUAL IMPROVEMENT FACTOR = 9.7E+00 kparm = 4071 par
= 6.24557E-01
RESIDUAL IMPROVEMENT FACTOR = 2.5E-11 kparm = 4071 par
= 5.13885E-01
.
.
.
RESIDUAL IMPROVEMENT FACTOR = 6.6E-01 kparm = 4071 par
= 5.13885E-01
RESIDUAL IMPROVEMENT FACTOR = 1.5E-10 kparm = 4071 par
= 5.00000E-01
.
.
.
RESIDUAL IMPROVEMENT FACTOR = 2.8E+03 kparm = 4071 par
= 5.00000E-01
THE VECTOR NORM IS: 2.2019E+00; PAR( 1)= 5.0000E-01
CONKUB : NEWTON ITER 6 ERR = 3.179E-16, ||Xnew-Xold|| =
3.650E-13
CONKUB: NEWTONS CONVERGED; K=171
RESIDUAL IMPROVEMENT FACTOR = 8.8E-01 kparm = 4071 par
= 5.00000E-01
DET= 0.4114*2** -1096 K=171
/PV
X =
0.0000000E+00 ... 0.0000000E+00
1.9512055E-01 ... 5.4208890E-01
1.6076196E-03 ... 1.5149268E-01
0.0000000E+00 ... 0.0000000E+00
-4.4286914E-03 ... -7.8544556E-06
0.0000000E+00 ... 0.0000000E+00
PAR =
5.0000000E-01
/DT
IDATA=
-1 0 0 0 0
RDATA=
0.0000000E+00 ... 0.0000000E+00
```

/TP
TRGPT= 1.0000000E-01
next = 6
next = 0

Note that CONKUB took five steps in the parameter: 0.927545, 0.808704, 0.624557, 0.513885, and 0.5. For brevity, we have omitted intermediate print-outs at all but the first step. Six quasi-Newton iterations were required after each step to satisfy the error tolerance, 1.E-12. The last commands shown are DT to permit the model to interrupt; TP to target to a parameter value of 0.1; and the model writing data included in the graphs shown in Figures 5.1 and 5.2 before proceeding.

Figure 5.1 shows the ratio of urine to plasma NaCl concentration versus $F^0_{\mathrm{PST}v}$ in panel (a) and versus $V^{\mathrm{NaCl}}_{m,\mathrm{CD}}$ and $P_{f,\mathrm{CD}}$ in panel (b). An increase of 5–20% in $F^0_{\mathrm{PST}v}$ is shown in panel (a), while panel (b) shows reductions of 50% and 80% in $P_{f,\mathrm{CD}}$ (long dashes), 50% and 90% in $V^{\mathrm{NaCl}}_{m,\mathrm{CD}}$ (solid line), and decreased $V^{\mathrm{NaCl}}_{m,\mathrm{CD}}$ in addition to an 80% reduction in $P_{f,\mathrm{CD}}$ (short dashes). The NaCl ratio for the combined reduction in the transport parameters is indistinguishable from that observed for increased $F^0_{\mathrm{PST}v}$. Consequently, this is not sufficient to distinguish between these possible effects of ANF. However, Figure 5.2 shows that the NaCl mass flow leaving the ICT of superficial nephrons expressed as percent of filtered load is significantly larger for increased $F^0_{\mathrm{PST}v}$ than for any change in transport parameters considered. Hence, the model suggests that measurement of NaCl mass flow from superficial nephrons into the collecting duct, which is possible experimentally, will distinguish between increased $F^0_{\mathrm{PST}v}$ and reduced $P_{f,\mathrm{CD}}$ and $V^{\mathrm{NaCl}}_{m,\mathrm{CD}}$.

6. Conclusions. We have described an algorithm and a program for interactive continuation of solutions of large systems of nonlinear equations. We have also described a physiological model that uses the program to address a specific question. In this way, we have shown how it is possible to study a complex phenomenon as a function of key parameters. Where solution of a model for many combinations of parameter values may be difficult due to a multipliticy of solutions, tracing the solution as a function of the parameters will yield the shape of a solution surface for the desired parameter domain.

The ability to trace and retrace solution branches when necessary, to determine their time-stability, and to obtain a portion of a solution manifold are of particular scientific importance. Graphic visualization of the solution as it is obtained is a present need, and the advent of

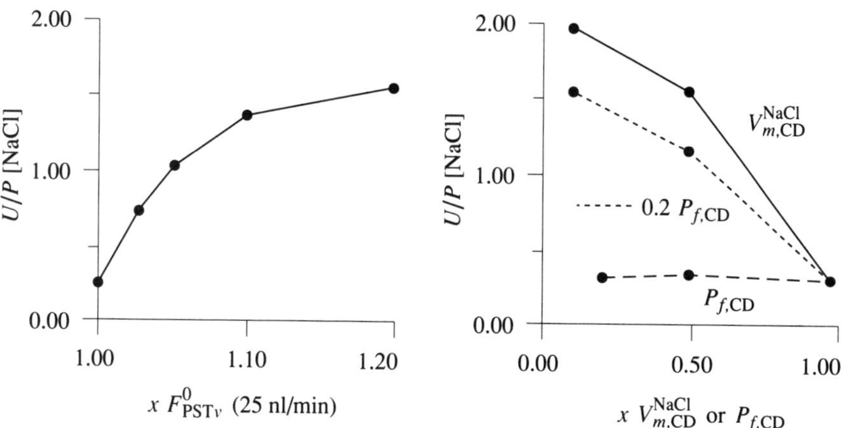

FIGURE 5.1. Urine to plasma NaCl concentration versus (a) volume flow entering proximal straight tubule, $F^0_{\mathrm{PST}v}$, (b) change in maximum rate of NaCl transport from the collecting duct, $V^{\mathrm{NaCl}}_{m,\mathrm{CD}}$, and osmotic water permeability of the collecting duct, $P_{f,\mathrm{CD}}$.

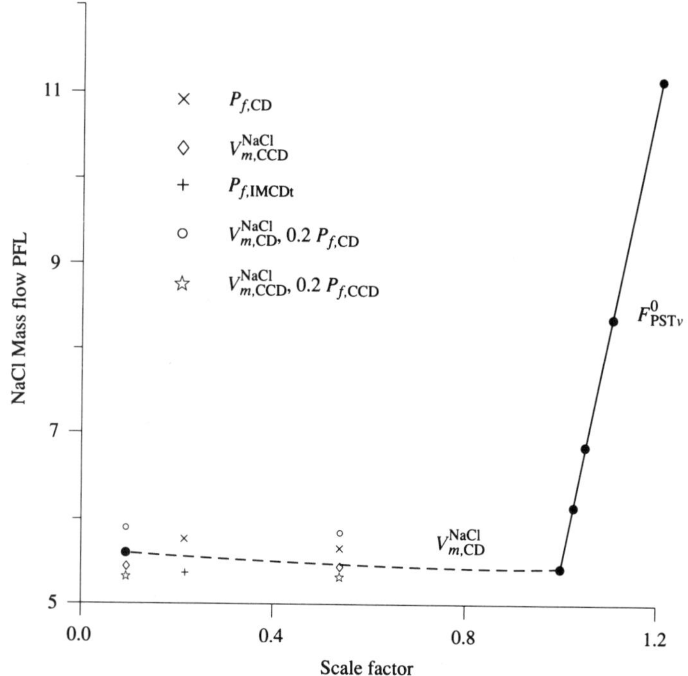

FIGURE 5.2. NaCl mass flow leaving the initial collecting tubule, ICT, of superficial nephrons versus change in thermodynamic parameters of the collecting duct and volume flow entering the proximal straight tubule.

work stations with the capability to handle multiple windows should make this soon possible. Recently, Rheinboldt **[23]**, **[24]** has described an algorithm, based on the computation of orthonormal moving frames on the manifold, that he uses to compute efficiently the vertices of a simplicial triangulation of portions of a solution manifold.

The treatment of turning points and bifurcations has been discussed briefly. Folds traced with CONKUB are illustrated in **[16]** and **[12]**, and an example with multiple bifurcations is treated in **[12]**.

Acknowledgments. Computations were done using the CRAY X-MP computer at the Advanced Scientific Computing Laboratory of the National Cancer Institute. I wish to thank J. E. Fletcher, J. M. González-Fernández, and M. V. A. Mascagni for their critical reading of this manuscript and for their suggestions for improvements. Many thanks also to H. M. Madine for her patience and editorial assistance.

Appendix. The following is a copy of the file available to provide help during a CONKUB session **[12]**:

```
*********************************************************
CONKUB ALGORITHM FOR PATH FOLLOWING 100587
*********************************************************
TYPE HE FOR HELP.
ILLEGAL ENTRY—TYPE HE FOR HELP.
ERROR OCCURRED WHILE READING FROM A FILE — CON-
TROL RETURNED TO YOU.

YOU ARE IN INITIALIZATION MODE.
VALID COMMANDS ARE:
SB PB ME GO
SV PV RE SK
RB PN HE
RV CP ST
SF DT CM
WHEN THEY ARE NEEDED, YOU WILL BE ASKED FOR VALUES
FOR:
NVBL—NUMBER OF VARIABLES
NPAR—NUMBER OF PARAMETERS
TYPE XX? FOR INFO ON COMMAND XX.
```

YOU ARE IN COMMAND CONTROL MODE. VALID COMMANDS ARE:
LS—LEAVE THE VARIABLE AS IT IS
AI—ASSIGN EACH VARIABLE(I) INDIVIDUALLY
SB—INFORMATION ON VARIABLES AND DEFAULT VALUES
ST—RETURN TO INITIALIZATION MODE OR RUN MODE.
ONCE IN THE AI MODE:
LS—LEAVE THE VARIABLE THE SAME
ST—GO ON TO THE NEXT VARIABLE.
ENTER THE VALUE ITSELF IF ALL VARIABLES(I) ARE TO HAVE THE SAME VALUE.

YOU ARE IN RUN MODE.
VALID COMMANDS ARE:
SB PB ME GO
RB PV RE TU
SF PN HE LE
SK CP ST TP
DT CM OV
TYPE XX? FOR INFO ON COMMAND XX.

SET VECTOR—SV
YOU WILL BE ASKED TO INPUT INITIAL VALUES FOR THE VARIABLE VECTOR X(I) AND THE PARAMETER VECTOR PAR(I) (FORMAT D10.4).

READ VECTOR—RV
READ IN THE START VECTORS FROM A FILE. YOU WILL BE ASKED FOR THE FILE NAME. THE FIRST TWO VALUES MUST BE NVBL, THEN NPAR, ONE PER LINE, FORMAT I6. VECTOR VALUES WILL BE READ 5 PER LINE, FORMAT D12.4,2X. CONTROL COMMANDS ARE NOT VALID—INCLUDE ALL VALUES.

SET BOUNDS—SB
YOU WILL BE ASKED TO INPUT VALUES(FORMAT D10.4,I6 OR L5) FOR:
HH(I)—INTEGRATION STEP ALONG THE ARC LENGTH OF THE SOLUTION LOCUS. —DEFAULT .1
HMAX(I)—APPROXIMATE UPPER BOUNDS FOR INCREMENT OF X(I) IN ONE INTEGRATION STEP. —DEFAULT 1.

EPS—ACCURACY DESIRED IN NEWTON ITERATIONS. CONVERGENCE CRITERION IS: SUM OF THE ABSOLUTE VALUE OF THE CHANGE IN EACH COMPONENT OF X IS ≤EPS. —DEFAULT .0001
NCORR—THE NUMBER OF NEWTON CORRECTIONS TO GET BACK NEAR THE CURVE. IF THE CRITERION IS NOT MET, THE STEP SIZE WILL BE HALVED AND WILL TRY AGAIN. THIS WILL BE REPEATED NCORR TIMES. —DEFAULT 4
NDIR(I)—DIRECTION OF CHANGE IN X(I)
1—POSITIVE -1—NEGATIVE. —DEFAULT 1
XUPP(I)—UPPER BOUNDS ON X(I). —DEFAULT 40.0
XLOW(I)—LOWER BOUNDS ON X(I). —DEFAULT -40.0
FACT—BIFURCATION CRITERION. CRITERION IS THAT THE INNER PRODUCT OF PSI1 AND THE PARTIAL DERIVATIVE OF FUNCTION F, WITH RESPECT TO A PARAMETER IS ≤FACT, WHERE THE SPAN OF PSI1 IS THE NULL SPACE OF DERIVATIVE OF F WITH RESPECT TO X. ALSO, SIZE OF STEP TAKEN ON EITHER SIDE OF A BIFURCATION IN SEARCH OF BIFURCATING BRANCH. —DEFAULT .001
BSW—SET TRUE FOR BIFURCATION TEST ON CHANGE OF DIRECTION; SET FALSE FOR BIFURCATION TEST ON CHANGE OF SIGN ONLY. —DEFAULT FALSE
LPRNT—SET TRUE TO PRINT RESULT OF EACH NEWTON ITERATION. —DEFAULT FALSE
THE ITH PARAMETER BOUND CORRESPONDS TO X(NVBL+I).

READ BOUNDS—RB
READ IN THE BOUNDS FROM A FILE. YOU WILL BE ASKED FOR THE FILE NAME. THE FIRST TWO VALUES MUST BE NVBL AND NPAR, ONE PER LINE, FORMAT I6. ALL INTEGER VALUES WILL BE IN FORMAT I6; REAL VALUES WILL BE IN FORMAT D12.4; LOGICAL VALUES IN L5. ENTER REAL VECTOR VALUES 5 PER LINE, INTEGER VECTOR VALUES 10 PER LINE, 2X IN BETWEEN VALUES. ENTER SCALAR VALUES ONE PER LINE. CONTROL COMMANDS ARE NOT VALID— INCLUDE ALL VALUES. VALUES WILL BE READ IN THE SAME ORDER AS SB.

HELP—HE

STOP—ST

IF IN INITIALIZATION MODE, COMPLETES THE RUN AND RETURNS TO THE MONITOR. IF IN RUN MODE, RETURNS TO INITIALIZATION MODE.

PRINT BOUNDS—PB
PRINT OUT THE CURRENT VALUES OF THE BOUNDS.

PRINT VECTOR—PV
PRINT OUT THE VECTOR OF VARIABLES AND PARAMETERS.

PRINT NORM—PN
PRINT THE NORM OF THE VARIABLE VECTOR AND PRINT THE PARAMETERS THAT ARE BEING FOLLOWED.

GO
IN INITIALIZATION MODE:
DOES INITIAL NEWTON ITERATIONS TO GET NEAR A ROOT.
IN THE RUN MODE:
GO NP—CALCULATE THE NEXT NP POINT(S).
INITIAL DEFAULT FOR NP IS 1. THEN DEFAULT IS PREVIOUS NP.

RESTART—RE
READ IN A PREVIOUSLY STORED POINT CREATED BY THE MEMORY COMMAND. YOU WILL BE ASKED FOR THE FILE NAME. IF YOU WISH TO GET OUT OF THIS COMMAND, INPUT ST.

MEMORY—ME
STORES THE CURRENT POINT AND STATUS IN A FILE WITH THE NAME OF YOUR CHOICE. IF YOU WISH TO GET OUT OF THIS COMMAND, INPUT ST. A MEMORY FILE WILL AUTOMATICALLY BE PRODUCED FOR POINTS ON BIFURCATED BRANCHES.

CHANGE PRINT—CP
YOU WILL BE GIVEN THE OPTION TO PRINT OUT THE NORM, THE VECTOR, OR BOTH.

SET FOLLOWING PARAMETERS—SF NPF

SPECIFY WHICH PARAMETER(S) TO FOLLOW, WHERE NPF ($\leq$ 5)IS THE NUMBER OF PARAMETERS. INDICES OF THE FOLLOWING PARAMETERS WILL BE STORED IN IFOLO(I). IF IFOLO IS NOT INITIALIZED ALL PARAMETERS MAY BE FOLLOWED. IF NPF=NPAR IFOLO WILL BE SET AUTOMATICALLY. —DEFAULT VALUE FOR NPF IS 1.
LS (LEAVE SAME) AND ST(STOP) ARE VALID, HOWEVER THEY WILL BE IGNORED IF AN ERROR OCCURS.

COMMENT—CM NL
INSERT A COMMENT IN THE SESSION LOG, WHERE NL IS THE NUMBER OF LINES. EACH LINE MAY BE UP TO 75 CHARACTERS LONG. —DEFAULT NL IS 1.

DATA—DT
INPUT REAL VALUES FOR RDATA(1)-RDATA(5), AND INTEGER VALUES FOR IDATA(1)-IDATA(5), FOR OPTIONAL USE IN THE FUNCTION SUBROUTINE. TO USE THESE ARRAYS, THE FUNCTION SUBROUTINE MUST HAVE THE STATEMENT:
COMMON/IDATA/RDATA(5),IDATA(5)

TURN—TU NP
FOLLOW THE PATH IN THE OPPOSITE DIRECTION. NP IS THE NUMBER OF POINTS. —INITIAL DEFAULT IS 1. THEN DEFAULT BECOMES THE PREVIOUS VALUE FOR NP.

LEVEL STEPS—LE NP
FOLLOW X(K) IN LEVEL STEPS ALONG THE CURVE. NP IS THE NUMBER OF POINTS. —INITIAL DEFAULT IS 1; SUBSEQUENTLY THE DEFAULT IS THE PREVIOUSLY USED VALUE; INITIAL DEFAULT ON INCREMENT = HMAX; INITIAL DEFAULT ON ITERATIONS = 10. IF K CHANGES OR ANY ERROR OCCURS, LEVEL MODE WILL STOP.

TARGET POINT—TP
FIND THE SOLUTION FOR A PARTICULAR VALUE OF X(K). YOU WILL BE ASKED FOR THE VALUE, TRGPT. IF K CHANGES OR ANY ERROR OCCURS, TARGET POINT MODE WILL STOP.

OUTPUT CURRENT VECTOR—OV

STORES THE CURRENT POINT IN A BINARY FILE WITH A NAME OF YOUR CHOICE. EACH RECORD MAY BE READ AS FOLLOWS:
READ(FILENAME)(X(I),I=1,NXTOT)
WHERE NXTOT=NVBL+NPAR. THIS IS USEFUL FOR SAVING DATA FOR PLOTTING.

SET K—SK
RESTRICT FOLLOWING TO VARIABLE K, FOR $0 < K \leq$ NXTOT. YOU WILL BE ASKED FOR K. K = 0, FREES CONKUB TO SELECT K (DEFAULT).
TURNING POINT OR BIFURCATION CALCULATIONS FREES SELECTION OF K.

The following interactive commands available within the model **[17]** may be passed through IDATA(1) or entered upon the prompt 'next' when IDATA(1)= -1:
-1 — accept input again asap, proceed
0 — nothing to be input, proceed
1 — to set iread
2 — to set kparm
3 — to set param
4 — kpa set to pa (for this run only)
5 — (kpa,pa) to be shown
6 — write output (by calling output subrtn)
7 — to set vrbse to true/false
8 — each element of vector kpa set (for this run only)
9 — same as 6 and also prints on tty
10 — set tube (global) eps
11 — show eps
12 — set transitent parameter delti
13 — show delti
14 — set rdata(1)=damping factor (0,1], < 0 for s-s tubes
15 — show rdata(1)
50 — set idata(3) = maxts
97 — same as 9 but brief
98 — set error=.true. to get back to CONKUB
99 — same as 6 with vrbse=.true. and stop

REFERENCES

1. E. Allgower and K. Georg, *Simplicial and continuation methods for approximating fixed points and solutions to systems of equations*, SIAM Rev. **22** (1980), 28–85.

2. B. Bunow and J. Kernevez, *Numerical exploration of bifurcating branches of solutions to reaction-diffusion equations from immobilized enzyme kinetics*, Instabilities, Bifurcations, and Fluctuations in Chemical Systems, L. E. Reichl and W. C. Schieve, eds., University of Texas Press, Austin, 1982, pp. 61–80.

3. S.-N. Chow, J. Mallet-Paret, and J. A. Yorke, *A homotopy method for locating all zeros of a system of polynomials*, Functional Differential Equations and Approximation of Fixed Points, H.-O. Peitgen and H.-O. Walther, eds., **730**, Springer-Verlag Lecture Notes in Math., New York, 1979, pp. 77–80.

4. M. G. Crandall and P. H. Rabinowitz, *Bifurcation from simple eigenvalues*, J. Funct. Anal. **8** (1971), 321–340.

5. ——, *Bifurcation, perturbation of simple eigenvalues, and linearized stability*, Arch. Rat. Mech. Anal. **52** (1973), 161–180.

6. P. Deuflhard, B. Fiedler, and P. Kunkel, *Efficient numerical pathfollowing beyond critical points*, SIAM J. Numer. Anal. **24** (1987), 912–927.

7. E. Doedel, *AUTO: A program for the automatic bifurcation analysis of autonomous systems*, Congr. Numer. **30** (1981), 265–284.

8. H. Jürgens, H.-O. Peitgen, and D. Saupe, *Topological perturbations in the numerical study of nonlinear eigenvalue and bifurcation problems*, Analysis and Computation of Fixed Points, Academic Press, New York, 1980, pp. 139–181.

9. H. B. Keller, *Numerical solution of bifurcation and nonlinear eigenvalue problems*, Applications of Bifurcation Theory, P. H. Rabinowitz, ed., Academic Press, New York, 1977, pp. 359–384.

10. M. Kubiček, *Algorithm 502: Dependence of solution of nonlinear systems on a parameter*, ACM Trans. Math. Software **2** (1976), 98–107.

11. W. Kuhn and K. Ryffel, *Herstellung Konzentrierter Lösungen aus verdunnten durch bloße Membranwirkung. Ein Odellversuch zur Funktion der Niere*, Hoppe-Seylers Physiol. Chem. **276** (1942), 145–178.

12. R. Mejia, *CONKUB: A conversational path-follower for systems of nonlinear equations*, J. Comput. Phys. **63** (1986), 67–84.

13. R. Mejia, R. B. Kellogg, and J. L. Stephenson, *Comparison of numerical methods for renal network flows*, J. Comput. Phys. **23** (1977), 53–62.

14. R. Mejia and J. L. Stephenson, *Numerical solution of multinephron kidney equations*, J. Comput. Phys. **32** (1979), 235–246.

15. ——, *Symbolics and numerics of a multinephron kidney model*, Proceedings of 1979 MACSYMA User's Conference, V. E. Lewis, ed., Washington, DC, MIT Laboratory of Computer Science, 545 Technology Square, Cambridge, MA, 1979, pp. 596–603.

16. ——, *Solution of a multinephron, multisolute model of the mammalian kidney by Newton and continuation methods*, Math. Biosci. **68** (1984), 279–298.

17. R. Mejia, J. Sands, J. L. Stephenson, and M. A. Knepper, *Renal action of atrial natriuretic factor: A mathematical modelling study*, Amer. J. Physiol. **257** (Renal Fluid Electrolyte Physiol. **26**) (1989), F1140–F1151.

18. H. D. Mittelmann and H. Weber, *Numerical methods for bifurcation problems - A survey and classification*, Bifurcation Problems and Their Numerical Solution, H. D. Mittelmann and H. Weber, eds., **54**, Birkhäuser Verlag, Basel, 1980, pp. 1–45.

19. J. J. Moré and M. Y. Cosnard, *Numerical solution of nonlinear equations*, ACM Trans. Math. Software **5** (1979), 64–85.

20. ———, *Algorithm 554: BRENTM, a Fortran subroutine for the numerical solution of systems of nonlinear equations*, ACM Trans. Math. Software **6** (1980), 240–251.

21. A. P. Morgan, *A method for computing all solutions to systems of polynomial equations*, ACM Trans. Math. Software **9** (1983), 1–17.

22. R. F. Pitts, *Physiology of the Kidney and Body Fluids, 3rd. ed.*, Year Book Med. Pub., Chicago, 1974.

23. W. C. Rheinboldt, *On a moving frame algorithm and the triangulation of equilibrium manifolds*, Bifurcation: Analysis, Algorithms, Applications, T. Küpper, R. Seydel, and H. Troger, eds., **79**, Birkhäuser Verlag, Basel, 1987, pp. 256–267.

24. ———, *On the computation of multi-dimensional solution manifolds of parametrized equations*, Numer. Math. **53** (1988), 165–181.

25. W. C. Rheinboldt and J. V. Burkhardt, *Algorithm 596: A program for a locally parameterized continuation process*, ACM Trans. Math. Software **9** (1983), 236–241.

26. J. L. Stephenson, *Concentrating engines and the kidney: I. Central core model of the renal medulla*, Biophys. J. **13** (1973), 512–545; II. *Multisolute central core system*, Biophys. J. **13** (1973), 546–567.

27. J. L. Stephenson, R. Mejia, and R. P. Tewarson, *Model of solute and water movement in the kidney*, Proc. Nat. Acad. Sci. **73** (1976), 252–256.

28. R. P. Tewarson, J. L. Stephenson, and L. L. Juang, *A note on solution of large sparse systems of nonlinear equations*, J. Math. Anal. Appl. **63** (1978), 439–445.

29. L. T. Watson, *Numerical linear algebra aspects of globally convergent homotopy methods*, SIAM Rev. **28** (1986), 529–545.

30. L. T. Watson and D. Fenner, *Algorithm 555: Chow-Yorke algorithm for fixed points or zeros of C^2 maps*, ACM Trans. Math. Software **6** (1980), 252–259.

31. L. T. Watson and L. R. Scott, *Solving Galerkin approximations to nonlinear two-point boundary value problems by a globally convergent homotopy method*, SIAM J. Sci. Stat. Comput. **8** (1987), 768–789.

NATIONAL INSTITUTES OF HEALTH, BETHESDA, MD 20892

Lectures in Applied Mathematics
Volume **26** (1990)

Nonlinear Parametrized Equations: New Results for Variational Problems and Inequalities

HANS D. MITTELMANN

Abstract. Nonlinear equations which depend on one or several parameters are a generic problem in applications. Frequently the equations have the form of one or several coupled differential equations which depend on physical parameters of the system they describe. These parameters may enter in a variety of ways. The differential operators, the right-hand sides, or the boundary conditions may depend on them but also, for example, on the underlying domain. An additional difficulty arises if the problems cannot be described through equations unless a free boundary is introduced, but side conditions on the solution yield variational inequalities. Several recent results of both theoretical as well as algorithmic nature are surveyed for these classes of problems.

1. Introduction. Basically two different problems will be considered in the following. The first may in abstract form be written as

$$G(u,\lambda) = 0 \quad , \quad G : \mathbf{X}^m \times \Re^p \to \mathbf{X}^m. \tag{1.1}$$

Here $\mathbf{X}$ denotes a function space while λ is the vector of parameters on which the problem depends. For an introduction into these problems and a survey of earlier results, see **[1]**, **[18]**, **[27]**, **[28]**.

Frequently, the only way to explore the complex solution manifold of (1.1) is to continue with respect to one or very few of the parameters for fixed values of the others. In the case $p = 1$ methods for this purpose

1980 *Mathematics Subject Classification* (1985 *Revision*). Primary 65N25, 65N30, 35J85, 49A29.

Supported by the Air Force Office of Scientific Research under Grant AFOSR 84-0315.

0082-0717/90 $1.00 + $.25 per page

are well-known, while for $p > 1$ only recently some approaches have been proposed. Let us concentrate on the case $p = 1$, $m = 1$, although $m > 1$ has also been considered previously **[3]**. In the following section we outline a new method for continuation along solution paths of (1.1). Here, for simplicity, we even assume $\mathbf{X} = \Re^n$ so that (1.1) could, for example, be the discretization of a differential equation problem, or a genuinely finite-dimensional problem.

In case, however, (1.1) is a boundary value problem associated with a partial differential equation on a two-dimensional domain, then its solution $u(x, y)$ may only describe manifolds in (x, y, u)-space that are graphs. In Section 3 we give an example of a nonlinear second-order elliptic boundary value problem the solution of which cannot be represented in the above nonparametric form but, denoting u by z, the solution manifold in (x, y, z)-space has to be parametrized on a two-dimensional parameter domain.

In Sections 4 and 5 we address parameter-dependent variational inequalities. Here, we restrict consideration to the case:

Find a $u \in V$ such that

$$(f'(u), v - u) \geq \lambda(g'(u), v - u) \tag{1.2}$$

holds for all $v \in V$, where V is a closed convex subset of a Hilbert space H, $V \neq \{0\}$, $0 \in V$, and f, g are given functionals on H. This class includes, for example, obstacle problems from mechanics.

The fundamental problems associated with continuation along solution curves, while well understood for (1.1), have only recently been addressed for (1.2). First results were obtained in **[19]**. A specific class of second-order elliptic problems was studied in **[9]**. In Section 4 we quote new results for the general problem (1.2) and a continuation parameter different from the one used in **[8]**, **[9]**, **[22]**, **[26]**. Numerical results obtained with a corresponding algorithm will also be presented.

In the context of continuation a sequence of nonlinear problems has to be solved, in general, a nontrivial computational task. The most efficient numerical methods for partial differential equations have thus been utilized as, for example, preconditioned conjugate gradient and multi-grid methods. For variational inequalities of the form (1.2) this had not been done. In Section 5 a multi-grid method is outlined that has proved efficient for variational inequalities.

2. A new continuation algorithm. The finite-dimensional version of (1.1) may be written as

$$G(u) = 0, \quad G : \Re^{n+1} \to \Re^n. \tag{2.1}$$

Here, all components of $\lambda \in \Re^p$ but one are held fixed and the latter is incorporated into the vector u. From now on we denote by u_0 a regular solution of (2.1). Standard continuation methods make use of the tangent vector in u_0 given by (up to the sign)

$$G_u(u_0)\dot{u}_0 = 0, \quad \dot{u}_0^{\mathrm{T}}\dot{u}_0 = 1 \tag{2.2}$$

to predict a new solution, and then they augment (2.1) by a normalization

$$N(u) - \sigma = 0 \tag{2.3}$$

in order to define such solutions and compute them in an iterative way from (2.1), (2.3) (corrector). While other predictor-corrector continuation methods have been proposed, in the following we will stay exactly in the above framework. The augmented system will be denoted by

$$H(u) = \begin{pmatrix} G(u) \\ N(u) - \sigma \end{pmatrix} = 0. \tag{2.4}$$

We assume that $N(u)$ is differentiable and that $N(u_0) = 0$. The Euler predictor

$$u_p = u_0 + \sigma\dot{u}_0 \tag{2.5}$$

is frequently combined with the pseudo-arclength normalization

$$N(u) = \dot{u}_0^{\mathrm{T}}(u - u_0) \tag{2.6}$$

so that (2.3) is satisfied for u_p. Here and in the following "$\cdot$" denotes differentiation with respect to arclength. If subsequently a Newton iteration starting with u_p is applied to (2.4), it is in general not clear if this converges to a solution. For small enough σ, however, this will be the case at least away from singular points **[18]**. Consequently, step-picking strategies have been developed and implemented to determine a suitable sequence of σ-values; see e.g. **[10]**. In the special case of homotopy continuation for nonlinear algebraic systems, damped Newton techniques as utilized below have apparently been used for the first time in **[19]**.

The principal idea of the new method **[4]** is to use known techniques for this problem **[5]**, namely, the generation of converging Newton sequences, in order to avoid having to find steplengths for which the corrector iteration may be expected, though in general not be guaranteed to converge. Theoretically, the proposed corrector iteration will converge under mere regularity assumptions, but since this convergence may initially be slow it is proposed to already modify the predictor step.

$N(u_0) = 0$ implies that the residual

$$||H||^2 = ||G||^2 + |N - \sigma|^2 \tag{2.7}$$

satisfies $||H(u_0)|| = |\sigma|$. We consider the linearization of $H(u)$ at u_0,

$$\begin{bmatrix} G_u \\ N_u \end{bmatrix} \Delta = \begin{bmatrix} 0 \\ \sigma \end{bmatrix}, \tag{2.8}$$

and solve for Δ:

$$\Delta = \sigma \dot{u}_0 / \dot{N}_0 \tag{2.9}$$

where $\dot{N}_0 = 1$ for (2.6). Instead of satisfying (2.3) through (2.5) in the case of (2.6) we propose to apply a damping factor $t \in (0, 1]$,

$$u_p = u_0 + t\Delta. \tag{2.10}$$

Then, a sufficient decrease criterion may be fulfilled,

$$||H(u_p)|| \leq (1 - t\delta)||H(u_0)|| = (1 - t\delta)|\sigma| \tag{2.11}$$

for any fixed δ, $0 < \delta < 1$, as was shown in **[4]**. If, analogously, damping is applied in the corrector iteration

$$u_{j+1} = u_j + t_j \Delta_j \quad , \quad j = 1, 2, \dots ,$$

where $t_j \in (0, 1]$, $u_1 = u_p$, and

$$H_u(u_j)\Delta_j = -H(u_j), \tag{2.12}$$

then convergence may be shown for suitable sequences of the t_j. In Theorem 2 of **[4]** an explicit formula is given for the quantities K_j in the special form

$$t_j = (1 + K_j ||H(u_j)||)^{-1} \tag{2.13}$$

such that convergence takes place. We note, however, that in practice it is not necessary to evaluate this or a similar formula, but that merely a definition as (2.13) may be combined with some heuristic method to find the K_j as long as in each step a sufficient decrease criterion analogous to (2.11) is satisfied. For a more detailed analysis and numerical demonstration we again refer to **[4]**.

3. Computation of a parametrized solution. In this section we use the word *parametrized* in a different sense. Here it does not describe the fact that the problem and its solution depend on one or several parameters, although this will also be the case below. The solution of a boundary value problem for a single partial differential equation on a two-dimensional domain may be written as $u(x, y)$ and may be graphed as a surface in space. This is called a nonparametric representation of the surface, and usually u is assumed to be a univalued function of (x, y). Surfaces that do not fit into this framework have to be parametrized, i.e., their three components in space, say, (x_1, x_2, x_3) are each given as functions of two variables, say, (r, s).

In the following we present the approach recently used in **[16]**, **[17]**, avoiding the somewhat lengthy nomenclature needed there for the specific (capillarity) problem. Let us assume that an (energy) functional $E(X)$ has to be minimized, where $X = (x_1, x_2, x_3)$ denotes the solution surface. The sought solution has to satisfy some boundary conditions. If these are of Dirichlet type as, for example, for minimal surfaces, then local variations in the interior may be used to minimize E; see **[13]**. Here, we allow nonhomogeneous natural (Neumann) boundary conditions and additionally impose an equality constraint, say, $H(X) = 0$.

The basic idea for the numerical approximation is to proceed analogously to the classical calculus of variations technique. The feasible surfaces X are represented through a reference surface $\Gamma_0 \subset \Re^3$,

$$\Gamma(u) = \{X = Y + u(Y)Z(Y), \quad Y \in \Gamma_0\}, \tag{3.1}$$

where Z denotes a fixed vector field whose directions are such that $\Gamma(u)$ in fact describes all the permissible surfaces as functions of $u(Y) : \Gamma_0 \to \Re$. While for Dirichlet boundary conditions Z can always be chosen to be normal to Γ_0, this is not possible here.

The next step is then to consider the Lagrange functional

$$L(u, \lambda) = E(u) - \lambda H(u), \tag{3.2}$$

where we have switched to the argument u. Subsequently, this problem is discretized. Without change of notation u then denotes a vector of finitely many parameters describing the variation of the discrete surface. The necessary conditions satisfied by the approximate solution are

$$\frac{\partial}{\partial u_i} L(u, \lambda) = 0, \quad i = 1, \dots, n. \tag{3.3}$$

This is a nonlinear system that may be solved, for example, by Newton's method.

Several questions remain to be addressed:

(i) construction of the discrete reference surface or "skeleton" Γ_0;

(ii) definition of the vector field Z in the discrete case in such a way that for finer and finer discretizations it yields smoothly varying feasible surfaces $X(u)$;

(iii) generation of starting values for the Newton iteration.

Since the vector field Z is assumed to be fixed, it is clear that in order to be able to describe and thus also to approximate a specific minimum X^* of E, the skeleton has to be close enough to this surface, and then Z has to be chosen in such a way that X^* is in $\Gamma(u)$ (see (3.1)). In order to explain how the questions (i)–(iii) above were solved in **[16]**, **[17]** let us consider the case of a drop in the corner of the nonnegative octant in space. For the specific form of the E in this capillarity problem, see **[16]**. For the constraint H we have here $H = V - V_0$, where V is the volume occupied by the drop and V_0 a prescribed volume. Since the drop is confined to the corner, Z will have to be tangent to the (x_1, x_2), (x_2, x_3), and (x_1, x_3) planes for points Y in the intersection of Γ_0 and these "walls," while for $Y \in \text{int}\,(\Gamma_0)$ the elements of Z will have to be nontangential with respect to the surface X. A smooth transition between these cases will have to be assured for increasing fineness of the discretization.

Questions (i) and (ii) are solved simultaneously by a recursive construction of the discretization, a triangulation. Assume that a coarsest discretization consists of a singular triangular surface whose vertices are a fixed distance from the origin on the positive x_1, x_2, and x_3 axes. Those vertices uniquely describe the surface Γ_0. In each of them a direction Z has to be defined and it has to point in the direction of the axes.

This coarse approximation is now recursively refined by connecting the midpoints of the sides of each triangle and by defining as direction Z in those midpoints the linear interpolants of the directions used in the two endpoints of the corresponding side. In this way a triangulated surface is generated that, when used to approximate the drop, may not yield an approximately uniform subdivision of its surface. Thus, at some or all the stages of the above refinement process the new midpoints may be projected on a suitable curved surface. At least at the end of the refinement one such projection has to be done.

This last issue is related to question (iii). In the absence of gravitational forces the solution to the above capillarity problem is a spherical surface which intersects the walls with the contact angle that holds for the given solid-liquid-gas combination. This corresponds to the natural boundary condition that the surface has to satisfy.

Since the differential equation holding in this case is that the mean curvature is equal to $-\lambda$, λ as in (3.2), and the mean curvature of a spherical surface is known, the triangular surface constructed above will be arbitrarily close to a solution of the associated zero-gravity problem if only the discretization is fine enough. Convergence of Newton's method can thus be guaranteed.

In general, however, the Lagrange multiplier λ will not be known, and the volume V_0 will instead be explicitly prescribed. In this case the augmented system

$$\begin{aligned} \partial_\lambda L(u,\lambda) &= 0, \\ \partial_u L(u,\lambda) &= 0, \end{aligned} \tag{3.4}$$

can be solved and continuation procedures may be applied with respect to a varying volume V_0. For other geometric configurations, slightly more complicated procedures have to be used in order to obtain starting values. For plots for a large variety of cases, see **[17]**.

4. Continuation for variational inequalities. In this and the following section we turn to variational inequalities of the form

$$(f'(u), v-u) \geq \lambda(g'(u), v-u), \quad \forall v \in V. \tag{4.1}$$

Here V is a closed convex subset of a Hilbert space H, $V \neq \{0\}$, $0 \in V$, and f, g are functionals on H. While the theoretical results quoted below apply to this general form, we give a specific class of (4.1) for illustration purposes. For this case numerical results will also be presented below.

$$\begin{gathered} V = \{v \in H_0^1(\Omega);\ v \leq \psi \text{ on } \Omega\}, \quad \psi \geq 0 \text{ on } \Omega, \\ f(u) = \frac{1}{2}\int_\Omega |\nabla u|^2\, dx, \qquad g(u) = \int_\Omega h(u)\, dx, \end{gathered} \tag{4.2}$$

where $\Omega \subset \Re^2$ and $H_0^1(\Omega)$ denotes the usual Sobolev space. Without the upper bound ψ on the solution, this leads to the semilinear elliptic boundary value problem

$$\begin{aligned} -\Delta u &= \lambda h(u) \quad && \text{in } \Omega, \\ u &= 0 && \text{on } \partial\Omega. \end{aligned} \tag{4.3}$$

This is an important instance of a reaction-diffusion equation and includes many interesting applications. The question that arises for problems of type (4.1)–(4.2) is how the solution curves in a suitable bifurcation diagram look in the variational inequality case, i.e., $u \not< \psi$. For practical purposes one would like to know the analog of the tangent $\dot{u}_0$ in (2.2) in order to be able to define a predictor for a continuation method. Finally, for the corrector a numerical method has to be developed for the solution of (4.1).

If along the variational inequality branch there were no singular points, then λ could be used as a continuation parameter. Corresponding theoretical results were given in **[9]**, **[21]**, **[22]**. If, however, fold points or bifurcation points have to be expected, then a different parameter must be used. In **[26]** the value of one of the functionals was used for continuation; a numerical algorithm was given and tested. More detailed results were presented in **[15]**, **[23]**, and a theoretical analysis was given in **[24]**. While this theory does not apply to the singular case, for which some results were given in **[9]** for problems of type (4.1), (4.2), it holds for the general case (4.1). The associated numerical methods of **[15]**, **[23]**, **[26]** permitted continuation along branches with many fold points. For numerical results associated with **[9]**, see **[8]**.

We quote the basic existence results for continuation of solutions to (4.1). For the proof, as well as additional results on uniqueness and monotonicity, we refer to **[24]**. Since the assumptions are natural but partly somewhat technical we do not list them here but only cite an important hypothesis and show how it can be verified in typical applications. For given $u \in V$ and $t > 0$ we set $V_t(u) = \{w \in H, u+tw \in V\}$ and assume that $V_t(u) \neq 0$ for all t with $0 < t \leq t_0$. For a given constant A not depending on t,

$$V_{t,A}(u) = \{w \in V_t(u);\ f''(u)(w,w) \leq 1, |f'(u)(w)| \leq At \text{ and } |g'(u)(w)| \leq At\},$$

and let $(u, \lambda) \in V \times \Re_+$ be a solution to (4.1). We make the following hypothesis:

(**A**) Let $t_n \downarrow 0$, $t_n > 0$, and $w_n \in V_{t_n,A}(u)$ be a weakly convergent sequence $w_n \rightharpoonup w$. Then it follows that $w \in K_u$ for a given closed convex cone with the vertex at zero.

For a definition of K_u in the general case, see **[24]**; we will give examples below. We are now in a position to state a theorem on continuation in the neighborhood of a solution (u_0, λ_0).

THEOREM. *Under several technical assumptions* ([**23**]) *and if hypothesis* (**A**) *holds, there exists a constant* $\eta > 0$ *such that for every* r, $|r - r_0| < \eta$, $r_0^2 = f(u_0)$, *there exists a solution* $(u(r), \lambda(r))$ *of the variational inequality* (4.1) *with* $f(u(r)) = r^2$. *Moreover, there exists a constant* c *which does not depend on* r *such that*

$$\|u(r) - u_0\| \le c|r - r_0|^{1/2} \quad \textit{and} \quad |\lambda(r) - \lambda_0| \le c|r - r_0|^{1/2}.$$

In case of the example (4.2) we define the coincidence set

$$I_u = \{x \in \bar{\Omega};\ u(x) = \psi(x)\}$$

and

$$K_u = \left\{ w \in H_0^1(\Omega \backslash I_u);\ \int_{\Omega \backslash I_u} \nabla u \,\nabla w \, dx = 0 \text{ and } \int_{\Omega \backslash I_u} e^u w \, dx = 0 \right\}.$$

Then it is easy to check that the above hypothesis holds for this K_u.

The case of fourth-order problems is more involved. Beams or plates, for example, may be considered with constraints on their deflection. A way to verify (**A**) is then the following. First it is shown that the w from (**A**) satisfies $w = 0$ and $\nabla w = 0$ on the boundary of the coincidence set and $w \le 0$ in the coincidence set [**20**, Lemma 2]; [**21**, Lemma 3.3]. One then defines a sequence $\{w_\epsilon\}$ of functions in K_u with $w_\epsilon \to w$, and since K_u is closed this finally yields $w \in K_u$. See [**22**] for the beam.

We finish this section by presenting graphically results obtained by applying the numerical continuation method from [**26**] to two typical variational inequalities. This numerical method is of predictor-corrector type. It makes heavy use of the theory developed in [**23**]. Continuation is done with respect to r. An Euler predictor is followed by a corrector iteration which is a feasible direction method with an active set strategy and Newton as basic algorithm. Results for the continuation along solution branches of problems that exhibit no bifurcation were given in [**23**] and [**26**]. Here we consider problems with bifurcations from the trivial solution. In particular, we follow the branch bifurcating from the principal eigenvalue. For a definition of the method, further explanations, and discussions we refer to [**23**], [**26**].

For $g(u)$ in (4.2) we choose either of the following:

$$g(u) = \int_\Omega \sinh(u)\, dx, \tag{4.4a}$$

$$g(u) = \int_{\Omega} \sin(u)\, dx. \tag{4.4b}$$

Here Ω is the unit square in $\mathbf{R}^2$.

It is well known that for $\psi > 0$, ψ the obstacle function, a branch bifurcates from the principal eigenvalue $\lambda = 2\pi^2$, and that this branch turns to the left (right) for (4.4a) ((4.4b)), asymptotically approximating the r-axis in the first case, where $r = ||u||$, $||\cdot||$ the norm in $H_0^1(\Omega)$, and the solution is graphed in the (λ, r)-plane.

If now an obstacle, for simplicity let $\psi \equiv$ const., is in the way, the curve will show a different behavior. The point where the solution contacts the obstacle for the first time was called a transition point in **[9]**. In addition to this point, the curve for a discretization of (4.1)–(4.2) will exhibit transition points whenever the "discrete" contact set changes as was first observed in **[26]**. Again, we refer to this paper as well as **[23]** for more details.

A standard finite difference discretization was applied on a uniform grid of width h. Figures 4.1–4.3 show the results for problem (4.4a),

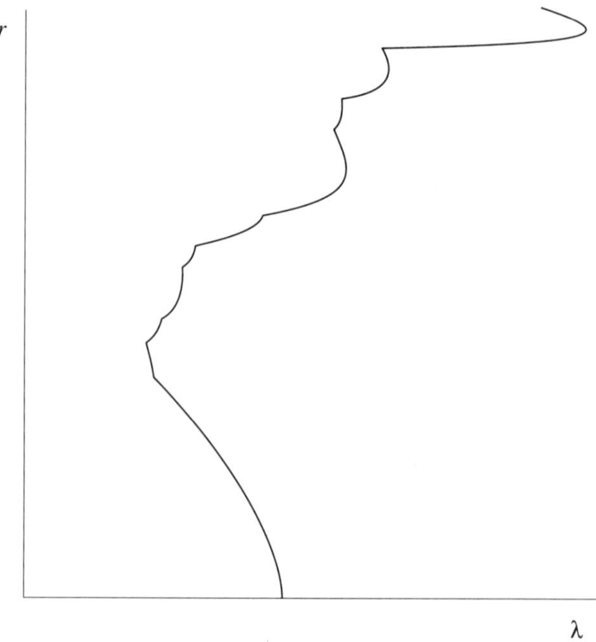

FIGURE 4.1. $h(u) = \sinh(u)$; $\psi \equiv 3$; $h = \frac{1}{8}$; $\lambda_{\max} = 43$; $r_{\max} = 127$

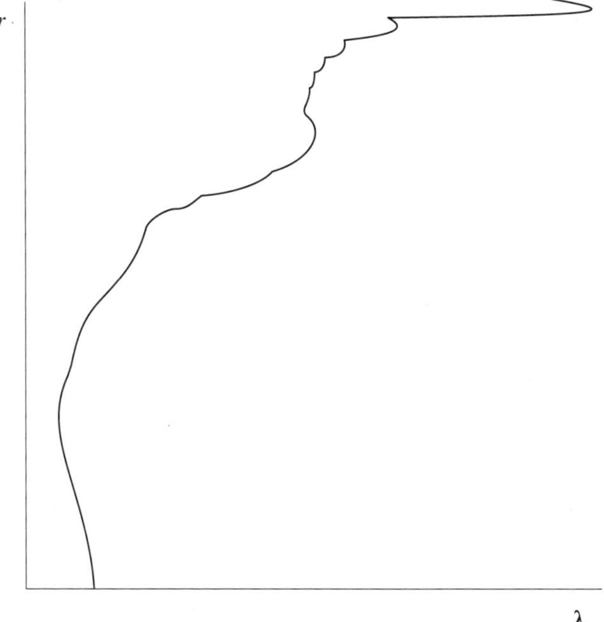

FIGURE 4.2. $h(u) = \sinh(u)$; $\psi \equiv 3$; $h = \frac{1}{16}$; $\lambda_{\text{max}} = 170$; $r_{\text{max}} = 186$

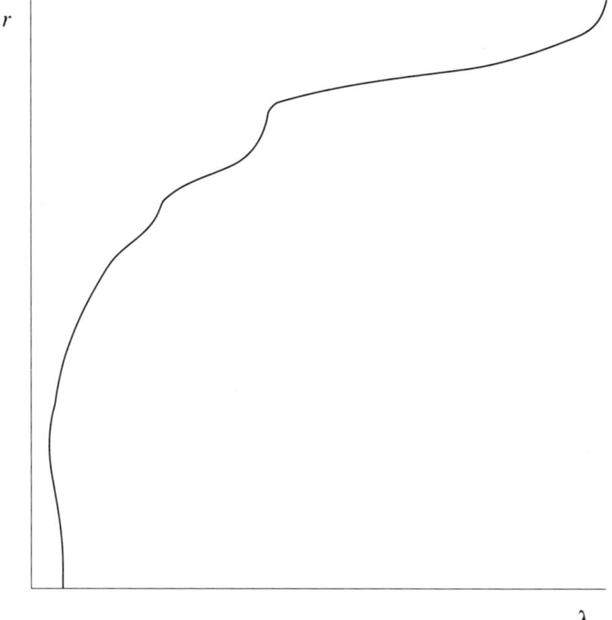

FIGURE 4.3. $h(u) = \sinh(u)$; $\psi \equiv 3$; $h = \frac{1}{32}$; $\lambda_{\text{max}} = 339$; $r_{\text{max}} = 215$

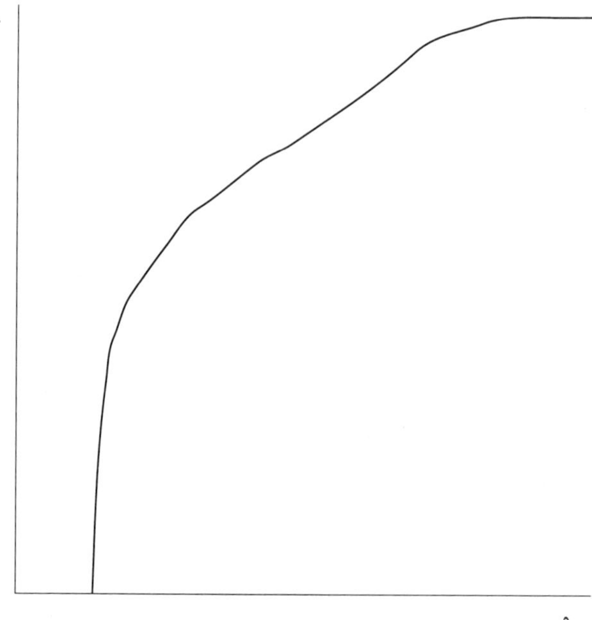

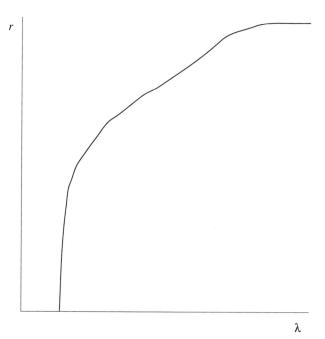

FIGURE 4.4. $h(u) = \sin(u)$; $\psi \equiv 1.5$; $h = \frac{1}{8}$; $\lambda_{\max} = 143$; $r_{\max} = 62$

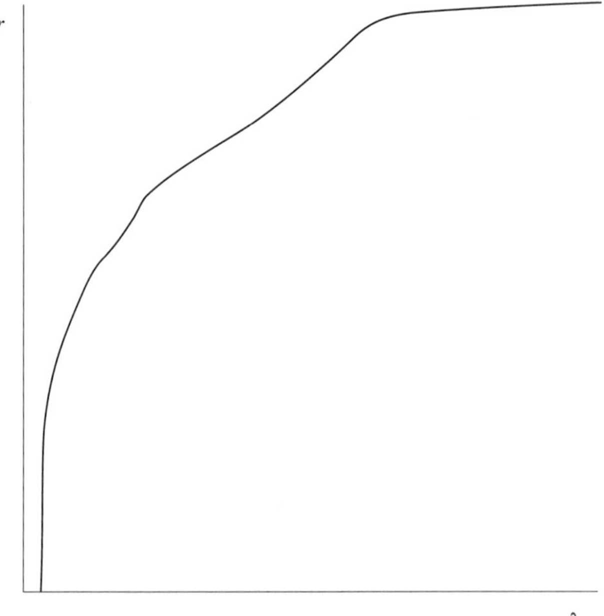

FIGURE 4.5. $h(u) = \sin(u)$; $\psi \equiv 1.5$; $h = \frac{1}{16}$; $\lambda_{\max} = 654$; $r_{\max} = 92$

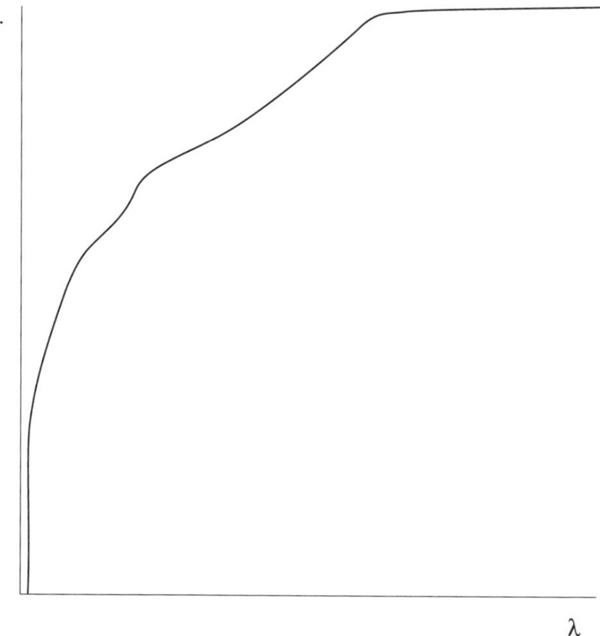

FIGURE 4.6. $h(u) = \sin(u)$; $\psi \equiv 1.5$; $h = \frac{1}{32}$; $\lambda_{\max} = 2567$; $r_{\max} = 133$

while Figures 4.4–4.6 correspond to problem (4.4b). For the first problem we see a strong effect caused by the discretization and the resulting transition points. Along the r–axis more and more values (at gridpoints) of the solution contact the obstacle. Each nonsmoothness of the curve corresponds to another set of values reaching ψ. The curve ends when all values are in contact. For smaller h-values the curves get increasingly smoother. We refer to **[23]** for a discussion of the similar problem $h(u) = \exp(u)$ and to an argument as to why, even for small h, the curve exhibits an arc with a left-turning fold point.

For the problem (4.4b) again some of the discrete transition points can be observed along the curve, but here the effect is less pronounced because the asymptotic behavior for this nonlinearity is different from the one for (4.4a). It should be noted that the (λ, r)-ranges depicted always start at $(0, 0)$ but extend to different maximal values for different h. The main reason for this is that finer uniform grids have points closer to the boundary of the region. It appears from Figures 4.4–4.6 that for this problem the overall shape is similar for different h, although the scale is very different.

5. Multi-grid for Parametrized Variational Inequalities. There are two fundamental papers on the application of multi-grid techniques to variational inequalities. Linear inequalities are considered in **[6]**, while **[12]** covers the general nonlinear case. On the other hand there are several papers on multi-grid methods for continuation problems [7], **[11]**, **[25]**. In **[15]** the combination of these two situations is dealt with successfully.

The complete definition of the multi-grid method used in **[15]** is beyond the scope of this paper. We instead concentrate on several important features, in particular those that are novel. For simplicity, we consider variational inequalities of the form (4.1), (4.2). We assume that the problem is discretized on a sequence of nested grids resulting in the nonlinear complementarity problems

$$\max(G_k(u_k;\lambda_k), u_k - \psi_k) = 0, \qquad k = 0, \ldots, \ell, \tag{5.1}$$

where $G_k(u_k;\lambda_k) = A_k u_k - \lambda_k h_k(u_k)$ is the discretization of the corresponding equation (4.3). The system (5.1) is augmented by the scalar equation $u_k^{\mathrm{T}} A_k u_k = r_k^2$ where $r = r_\ell$ plays the role of the continuation parameter r as used in Section 4. We recall that in the case of equations one can define the (generalized) Rayleigh-quotient

$$\lambda_k = \frac{u_k^{\mathrm{T}} A_k u_k}{u_k^{\mathrm{T}} h_k(u_k)}. \tag{5.2}$$

In generalization of (5.2) we define for (5.1)

$$\begin{aligned} \Lambda_k(u_k;r_k) &= \frac{\sum_{i\notin I_k} u_{k,i}(A_k u_k)_i}{\sum_{i\notin I_k} u_{k,i}(h_k(u_k))_i} \\ &= \frac{r_k^2 - \sum_{i\in I_k} u_{k,i}(A_k u_k)_i}{\sum_{i\notin I_k} u_{k,i}(h_k(u_k))_i}, \end{aligned} \tag{5.3}$$

where I_k denotes the set of indices i for which $u_{k,i} = \psi_{k,i}$.

The smoothing of the proposed multi-grid method is accomplished through the application of projected Gauss–Seidel–Newton to (5.1). After each smoothing step λ is updated through $\Lambda_k(\tilde{u}_k;r_k)$, where $\tilde{u}_k$ is the smoothed u_k. The coarse grid correction process has the form of a full approximation scheme. A special restriction operator is applied with pointwise restriction near the free boundary and fully weighted restriction away from it. Bilinear interpolation is used. A nested iteration or full multi-grid scheme serves as predictor with several smaller continuation steps on coarser grids. For a full definition utilizing the standard recursive quasi-ALGOL form, see **[15]**.

The method was applied to examples similar to those in **[23]** and Section 4. The results obtained are, of course, analogous. It is interesting to evaluate the performance of the multi-grid method. Denoting by $e_\ell^\nu(r_\ell) = u_\ell^\nu(r_\ell) - u_\ell^{\nu-1}(r_\ell)$ the difference of two subsequent iterates one can define a convergence rate through

$$\gamma_\ell^{\mathrm{MG}}(r_\ell) = (\|e_\ell^{\nu^*}(r_\ell)\|_{\ell,2}/\|e_\ell^1(r_\ell)\|_{\ell,2}) * *(1/[(\nu^* - 1) * N_{\mathrm{WU}}]),$$

where $\nu^* > 1$ is the iterate for which machine accuracy has been reached and N_{WU} is the number of work units required for the execution of one multi-grid cycle. $\|\cdot\|_{\ell,2}$ denotes the discrete L^2-norm, and a work unit was taken to be one projected Gauss–Seidel–Newton iteration on the level ℓ. Over the entire range of r-values used, the convergence rate was very similar with a mean value of .6 for $\ell = 3$ and .62 for $\ell = 4$. These rates are almost in the same range as those obtained for multi-grid algorithms applied to other free boundary problems (**[6]**, **[14]**).

REFERENCES

1. E. L. Allgower and K. Georg, *Predictor-corrector and simplicial methods for approximating fixed points and zero points of nonlinear mappings*, Mathematical Programming: The State of the Art, Springer, New York, 1983.

2. R. E. Bank, *PLTMG User's Guide, Edition 5.0*, Tech. Rep., Department of Mathematics, University of California, San Diego, 1988.

3. R. E. Bank and H. D. Mittelmann, *Continuation and multi-grid for nonlinear elliptic systems*, Multigrid Methods II, W. Hackbusch and U. Trottenberg (eds.), Lect. Notes in Math. 1228, Springer, Berlin, 1986.

4. ——, *Stepsize selection in continuation procedures and damped Newton's method*, J. Comput. Appl. Math. **26** (1989), 67–77.

5. R. E. Bank and D. J. Rose, *Global approximate Newton methods*, Numer. Math. **37** (1981), 279–295.

6. A. Brandt and C. W. Cryer, *Multigrid algorithms for the solution of linear complementarity problems arising from free boundary problems*, SIAM J. Sci. Stat. Comp. **4** (1983), 655–684.

7. T. F. Chan and H. B. Keller, *Arc-length continuation and multi-grid techniques for nonlinear elliptic eigenvalue problems*, SIAM J. Sci. Stat. Comp. **3** (1982), 173–194.

8. F. Conrad, R. Herbin, and H. D. Mittelmann, *Approximation of obstacle problems by continuaton methods*, SIAM J. Numer. Anal. **25** (1988), 1409–1431.

9. F. Conrad, F. Issard-Roch, Cl.-M. Brauner, and B. Nicolaenko, *Nonlinear eigenvalue problems in elliptic variational inequalities: a local study*, Comm. Partial Differential Equations **10** (1985), 151–190.

10. C. Den Heijer and W. C. Rheinboldt, *On steplength algorithms for a class of continuation methods*, SIAM J. Numer. Anal. **18** (1981), 925–947.

11. W. Hackbusch, *Multi-grid solution of continuation problems*, Iterative Solution of Nonlinear Systems, R. Ansorge, T. Meis, and W. Törnig (eds.), Lect. Notes in Math. 953, Springer, Berlin, 1982.

12. W. Hackbusch and H. D. Mittelmann, *On multi-grid methods for variational inequalities*, Numer. Math. **42** (1983), 239–254.

13. M. Hinata, M. Shimasaki, and T. Kiyono, *Numerical solution of Plateau's problem by a finite element method*, Math. Comp. **28** (1974), 45–60.

14. R. H. W. Hoppe and R. Kornhuber, *Multi-grid solution of the two-phase Stefan problem*, Multigrid Methods, S. F. McCormick (ed.), Lecture Notes in Pure and Appl. Math., vol. 110, Marcel Dekker, New York and Basel, 1988.

15. R. H. W. Hoppe and H. D. Mittelmann, *A multi-grid continuation strategy for parameter-dependent variational inequalities*, J. Comput. Appl. Math. **26** (1989), 35–46.

16. U. Hornung and H. D. Mittelmann, *A finite element method for capillary surfaces with volume constraints*, J. Comput. Phys. (to appear).

17. ——, *The augmented skeleton method for parametrized surfaces of liquid drops*, J. Colloid Interface Sci. (to appear).

18. H. B. Keller, *Numerical solution of bifurcation and nonlinear eigenvalue problems*, Applications of Bifurcation Theory, P. Rabinowitz (ed.), Academic Press, New York, 1977.

19. D. Leder, *Automatische Schrittweitensteuerung bei global konvergenten Einbettungsmethoden*, ZAMM **54** (1974), 319–324.

20. E. Miersemann, *Zur Lösungsverzweigung bei Variationsungleichungen mit einer Anwendung auf den geraden Knickstab mit begrenzter Durchbiegung*, Math. Nachr. **102** (1981), 7–15.

21. ——, *Stabilitätsprobleme für Eigenwertaufgaben bei Beschränkungen für die Variationen mit einer Anwendung auf die Platte*, Math. Nachr. **106** (1982), 211–221.

22. E. Miersemann and H. D. Mittelmann, *On the continuation for variational inequalities depending on an eigenvalue parameter*, Math. Methods Appl. Sci. **11** (1989), 95–104.

23. ——, *Continuation for parametrized nonlinear variational inequalities*, J. Comput. Appl. Math. **26** (1989), 23–34.

24. ——, *Extension of Beckert's continuation method to variational inequalities*, Math. Nachr. (to appear).

25. H. D. Mittelmann, *Multi-level continuation techniques for nonlinear boundary value problems with parameter-dependence*, Appl. Math. Comp. **19** (1986), 265–282.

26. ——, *On continuation for variational inequalities*, SIAM J. Numer. Anal. **24** (1987), 1374–1381.

27. ——, *Continuation methods for parameter-dependent boundary value problems*, AMS Lectures in Appl. Math. (to appear).

28. W. C. Rheinboldt, *Numerical analysis of parametrized nonlinear equations*, John Wiley and Sons, New York, 1986.

ARIZONA STATE UNIVERSITY

Lectures in Applied Mathematics
Volume **26** (1990)

Generically Nonsingular Polynomial Continuation

ALEXANDER P. MORGAN AND ANDREW J. SOMMESE

Abstract. We present a new theory of homotopy continuation for solving polynomial systems. It features a focus on "generically nonsingular" solutions in a very general polynomial context. The new homotopy formulation is established using results from modern algebraic geometry. The practical usefulness of the approach is established by the solution of some problems from engineering.

1. Introduction. In 1977, Garcia and Zangwill in [**10**] and Drexler in [**5**] independently presented theorems suggesting that homotopy continuation could be used to find numerically the full set of geometrically isolated solutions to a system of n polynomial equations in n unknowns. Garcia and Zangwill used theorems from differential topology to establish their results, while Drexler used algebraic geometry. A number of papers have followed.

Although Drexler's paper [**6**] appeared in 1978, most published work on the structure of homotopies for polynomial continuation has developed along the line established by Garcia and Zangwill, that is, using topological arguments. See Brunovský and Meravý [**3**], Chow, Mallet-Paret, and Yorke [**4**], Garcia and Li [**9**], Garcia and Zangwill [**11–13**], Li [**20**], Morgan [**24–26**], and Wright [**47**]. There have also been some significant physical applications. See Richter and De Carlo [**39**], Meintjes and Morgan [**23**], Morgan and Wampler [**35**], Morgan and Sarraga [**29**], Safonov [**40**], Tsai and Morgan [**41**], and Wampler et al. [**42**]. Recently the algebraic geometry approach of Drexler has been revived in papers by Li, Sauer, and Yorke [**21**], [**22**], Morgan and Sommese [**30–32**], and

1980 *Mathematics Subject Classification* (1985 *Revision*). Primary 65H10; Secondary 14-02, 65-02.

Zulehner [**48**], [**49**]. The survey Morgan [**28**] summarizes some of this recent work. There has been little published on the computational aspects of polynomial continuation. A good deal of Morgan [27] touches on this topic, and it is the main focus of Morgan et al. [**34**]. In contrast, path tracking for continuation to solve general nonlinear systems has received a good deal of attention. See, for example, Allgower [**1**], Allgower and Georg [**2**], Garcia and Zangwill [**14**], Rheinboldt and Burkardt [**38**], Watson [**43**], [**44**], Watson et al. [**45**], and Watson and Fenner [**46**].

Most recent research in polynomial homotopy construction focuses on exploiting special structures in the system; thus, the emphasis is not on "general systems" but rather "special systems." This is justified by progress in solving classes of engineering problems. (See [**31**], [**32**], [**33**], [**35**], and [**42**].) In this paper we continue this line of development, as well as summarize some of the previous work. We define "generically nonsingular solutions" to polynomial systems and present a theorem that suggests an efficient approach to finding these solutions using homotopy continuation. Naturally, the question arises: "To what extent do the generically nonsingular solutions include the physically meaningful solutions?" The answer, we suggest, is that this is commonly the case if the physical modeling of the problem has been carried out in a reasonable manner. See Section 5.

In Section 2, we discuss polynomial systems and the general approach of *polynomial continuation.* Then in Section 3 we present *coefficient parameters* and *coefficient parameter continuation*, ideas that we adapt from [32]. In Section 4 we discuss compactification of C^n, Bezout's theorem, and the projective transformation. Then, in Section 5, we define *generically nonsingular solutions* and state a theorem on finding such solutions using polynomial continuation. In Section 6 we discuss implementation considerations, and Section 7 contains a summary with conclusions. The Appendix includes a brief summary of some facts from algebraic geometry and states the powerful main theorem from [32], which is the basis of our approach.

2. Polynomial continuation. Let

$$f_j(C, z) = \sum_{k=1}^{r_j} C_{j,k} \prod_{l=1}^{n} z_l^{\delta_{j,l,k}} \tag{1}$$

denote a polynomial system in the n variables $z_1, \ldots, z_n$ with the jth equation having r_j terms, where the kth term has (complex number) coefficient $C_{j,k}$ and in which z_l is raised to the $\delta_{j,l,k}$th power. Here

$C = (C_{j,k})$. We will generally consider systems of $n + N$ equations, where $N \geq 0$. That is, we let $j = 1, \ldots, n + N$. Sometimes we call the equations indexed by $j = n + 1, \ldots, N$ *side conditions*.

Traditionally, polynomial continuation has not considered side conditions, focusing on systems of n equations in n unknowns. Further, the traditional goal of polynomial continuation has been to compute all geometrically isolated solutions to $f(z) = 0$. However, we wish to refine and make more flexible the goal of polynomial continuation: allow more equations than unknowns and compute a distinguished subset of the solutions to $f(z) = 0$. We shall focus in this paper on the *generically nonsingular* solutions. However, this is not the only case of interest. For example, frequently classes of solutions at infinity are uninteresting, and it is desirable to omit them from the solution subset being computed. (This is the motivation for the *m-homogeneous* approach to polynomial continuation, introduced in **[30]**, **[31]**.)

In the literature for polynomial continuation, homotopies are typically constructed by letting the $C_{j,k}$ in (1) vary via a homotopy parameter t. Thus $C_{j,k}(t) : [0, 1] \to \mathbf{C}^r$ where $r = \sum_{j=1}^{n} r_j$, and (1) with coefficients $C_{j,k}(0)$ is "generic" and can be "easily solved," and (1) with coefficients $C_{j,k}(1)$ is the system we want to solve. Thus the homotopy proceeds from $t = 0$ to $t = 1$. We call the system with $t = 0$ the *start system* and with $t = 1$ the *target system*. When such a homotopy is constructed "generic enough" so that $h^{-1}(0) \subseteq \mathbf{C}^n \times [0, 1]$ is a collection of well-defined paths, then we proceed with the numerical business of tracking the paths, beginning at the solutions to the start system and ending with the desired subset of the solutions to the target system. (See Fig. 1.)

For example, consider the homotopy

$$z_1^2 + 3tz_2 + 1 = 0,$$
$$z_2^2 + 2tz_1 - 5 = 0.$$

Here, we can make the coefficient structure explicit by writing:

$$f(C_{1,1}, C_{1,2}, C_{1,3}, C_{2,1}, C_{2,2}, C_{2,3}; z_1, z_2) = \begin{bmatrix} C_{1,1}z_1^2 + C_{1,2}z_2 + C_{1,3} \\ C_{2,1}z_2^2 + C_{2,2}z_1 + C_{2,3} \end{bmatrix},$$

and the coefficient homotopy is defined by $C(t) = (1, 3t, 1, 1, 2t, -5)$.

Basically, we can distinguish two main steps in polynomial continuation:

(1) Define a homotopy, $h(z, t)$.
(2) Choose a numerical method for tracking the paths defined by $h(z, t) = 0$.

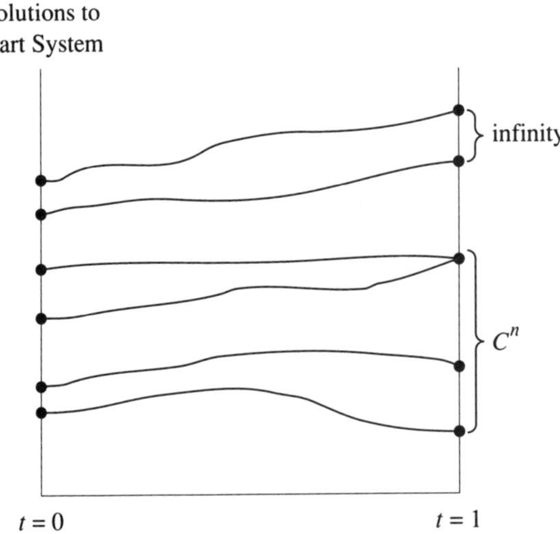

FIGURE 1a. Schematic showing possible polynomial-continuation path behavior. Note that paths can merge when $t = 1$ and can go to infinity.

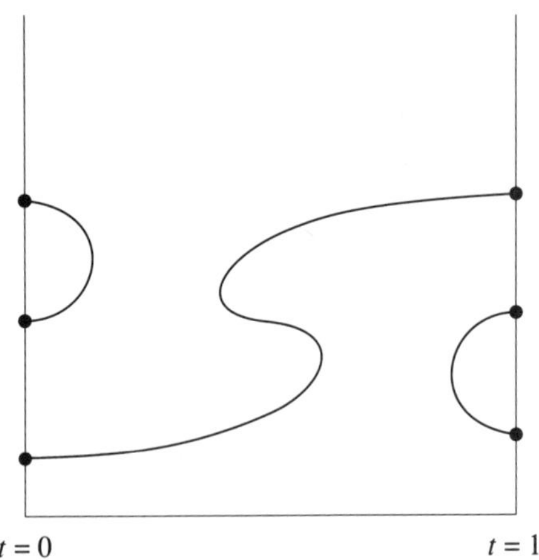

FIGURE 1b. Schematic showing impossible polynomial-continuation path behavior. Thus paths cannot "turn back in t."

Step (1) is guided by results from differential topology and algebraic geometry, while Step (2) is based on methods for the numerical solution of ordinary differential equations and local methods for the solution of nonlinear systems. Both parts are nontrivial and important. In this paper, however, we focus on Step (1).

3. Coefficient parameter polynomial continuation. When polynomial systems arise in engineering, the *coefficients* are usually not the *physical parameters*. To mirror this reality in the mathematics, we assume the coefficients are holomorphic functions on an irreducible complex analytic variety Q (see the Appendix), and the continuation will be generated in "parameter space," Q, rather than "coefficient space," $\mathbf{C}^r$. Thus

$$C_{j,k} = c_{j,k}[q], \tag{2}$$

where we call (2) the *coefficient parameter formulas*. The homotopy will be defined by the composition

$$C_{j,k}(t) = c_{j,k}[q(t)], \tag{3}$$

where $q(t) : [0, 1] \to Q$ and $c_{j,k} : Q \to C^1$. Generally, the dimension of the parameter space will be much less than that of the coefficient space. The special structure of the solutions of (1) induced by (2) is naturally acknowledged with a homotopy like (3). The result (we will see) is that fewer paths have to be tracked to solve the system, significantly reducing in some cases the total numerical cost. We further allow $c_{j,k}[q]$ to be a general holomorphic function of q, rather than merely a polynomial in q. The need for this generality arises, for example, in mechanical engineering, where it is common for trigonometric functions to arise.

Consider the following simple example:

$$\begin{aligned} z_1 z_2 + q_2 z_1 + q_1 z_2 + q_1 q_2 &= 0, \\ z_1^2 + (q_1 + q_3) z_1 + q_1 q_3 &= 0. \end{aligned} \tag{4}$$

Here z_1 and z_2 are the variables, and q_1, q_2, q_3 are the coefficient parameters. Thus (4) can be written

$$\begin{aligned} c_{1,1} z_1 z_2 + c_{1,2} z_1 + c_{1,3} z_2 + c_{1,4} &= 0, \\ c_{2,1} z_1^2 + c_{2,2} z_1 + c_{2,3} &= 0, \end{aligned} \tag{5}$$

where

$$\begin{aligned} &c_{1,1} = 1, \quad c_{1,2} = q_2, \quad c_{1,3} = q_1, \quad c_{1,4} = q_1 q_2, \\ &c_{2,1} = 1, \quad c_{2,2} = q_1 + q_3, \quad c_{2,3} = q_1 q_3, \end{aligned} \tag{6}$$

Now, construct a parameter homotopy so that when $t = 0$, $q_1 = 0$, $q_2 = 1$, and $q_3 = 1$, and when $t = 1$, $q_1 = 1$, $q_2 = 2$, and $q_3 = 3$; for example, $q(t) = [(1 - t), (1 + t), (1 + 2t)]$. Then, the coefficients will continue via the formulas:

$$c[q(t)] = [1, 1+t, 1-t, (1-t)(1+t), 1, (1-t)+(1+2t), (1-t)(1-t)(1+2t)].$$

Thus the coefficient parameter continuation is not an affine transformation with respect to the homotopy parameter t.

4. Compactifications of $\mathbf{C}^n$. The polynomial equations f defined by equation (1) naturally can be viewed as a map

$$f : \mathbf{Q} \times \mathbf{C}^n \to \mathbf{C}^{n+N} \tag{7}$$

with the solution set

$$\mathbf{X}_+ \equiv f^{-1}(0) \subseteq \mathbf{Q} \times \mathbf{C}^n. \tag{8}$$

However, $\mathbf{X}_+$ accounts only for the *finite* solutions of $f = 0$, whereas it is convenient to expand the context of the system in such a way that additional *solutions at infinity* are defined. It is further convenient to be able to do this in a variety of ways. What we have in mind is to find algebraic compactifications $\mathbf{P}$ of $\mathbf{C}^n$ and extensions ϕ of f with $\phi : \mathbf{Q} \times \mathbf{P} \to \mathbf{E}$, for appropriately defined $\mathbf{E}$. It will follow that $\mathbf{X}_+ \subseteq \phi^{-1}(0) \equiv \mathbf{X}$, where $\mathbf{X}$ therefore depends on the choice of compactification and the extension of f, but $\mathbf{X}_+$ does not. (See the Appendix for details.)

The two most common ways of defining $\mathbf{P}$ and ϕ are

(1) $\mathbf{P} = \mathbf{P}^n$, n-dimensional complex projective space; ϕ the "homogenization of f," and

(2) $\mathbf{P} = \mathbf{P}^{k_1} \times \cdots \times \mathbf{P}^{k_m}$; ϕ the "m-homogenization of f."

These two are sufficiently important that we will spend some time discussing them.

Complex projective space, $\mathbf{P}^k$, consists of the lines through the origin in $\mathbf{C}^{k+1}$, denoted $[(z_0, \ldots, z_k)]$, where $(z_0, \ldots, z_k) \in \mathbf{C}^{k+1} - \{0\}$; that is, $[(z_0, \ldots, z_k)]$ is the line through the origin that contains $(z_0, \ldots, z_k)$. It is natural to view $\mathbf{P}^k$ as a disjoint union of points $[(z_0, \ldots, z_k)]$ with $z_0 \neq 0$ (identified with Euclidean space via $[(z_0, \ldots, z_k)] \to (z_1/z_0, \ldots, z_k/z_0))$ and the "points at infinity," the $[(z_0, \ldots, z_n)]$ with $z_0 = 0$.

We partition the variables $\{z_1, \ldots, z_n\}$ into m nonempty collections. It will be notationally simpler here if we rename the variables with double subscripts. Thus, let $\{z_1, \ldots, z_n\} = \bigcup_{j=1}^m \{z_{1,j}, \ldots, z_{k_j,j}\}$, where $\sum_{j=1}^m k_j = n$. Now choose homogeneous variables $z_{0,j}$ for $j = 1$ to m and define $Z_j = \{z_{0,j}, z_{1,j}, \ldots, z_{k_j,j}\}$ for $j = 1$ to m. Then evoke the substitution $z_{i,j} \leftarrow z_{i,j}/z_{0,j}$ for $i = 1$ to k_j and $j = 1$ to m, generating a system $f' = 0$ of n equations in $n + m$ unknowns (after we clear the denominators of powers of the $z_{0,j}$). Now $f' = 0$ naturally has solutions in $\mathbf{P} \equiv \mathbf{P}^{k_1} \times \mathbf{P}^{k_2} \times \cdots \times \mathbf{P}^{k_m}$. (See **[30]** and **[31]**.) We say f' is *m-homogeneous* because the variables are partitioned into m collections, $Z_1, \ldots, Z_m$, so that f' is homogeneous as a system in the variables of any one of the collections. We let $d_{j,l}$ denote the jth degree of the lth polynomial; that is, with all variables held fixed except those in Z_j, f'_l

has homogeneous degree $d_{j,l}$. Note that "1-homogeneous" is the same as "homogeneous," so theorems about m-homogeneous polynomial systems apply essentially to all polynomial systems. Generally, we abuse the notation by not distinguishing f from its m-homogenization f'.

The *Bezout number*, d, of an m-homogeneous polynomial system is defined to be the coefficient of $\prod_{j=1}^{m} \alpha_j^{k_j}$ in the product

$$D = \prod_{l=1}^{n} \sum_{j=1}^{m} d_{j,l} \alpha_j. \tag{9}$$

Then we have the classical

THEOREM OF BEZOUT. *Let d denote the Bezout number of the system $f = 0$, and assume $f = 0$ does not have an infinite number of solutions in* $\mathbf{P}$. *Then $f = 0$ has exactly d solutions in* $\mathbf{P}$, *counting multiplicities.*[1]

The numerical significance of the Bezout number is that it is an upper bound on the number of homotopy continuation paths we will track in the space $\mathbf{P} \times [0, 1]$ (Theorem 1, below). The smaller d is, the better. Frequently, the m-homogenization of f for $m > 1$ has a (much) smaller Bezout number than the 1-homogenization. If $m = 1$, then $d = d_1 \cdots d_n$, the total degree of f, where $d_j = \deg(f_j)$. This is the "traditional" number of paths to track in polynomial continuation.

We acknowledge $\mathbf{P} = \mathbf{P}^n$ and $\mathbf{P} = \mathbf{P}^{k_1} \times \cdots \times \mathbf{P}^{k_m}$ computationally via *the projective transformation*, first proposed in **[26]** and extended to the m-homogeneous case in **[31]**. We sketch what is involved here. Let us first consider the 1-homogeneous case. Thus h is a homogeneous system of n equations in the $n+1$ variables $z_0, \ldots, z_n$, and $h(z) = 0$ has solutions in $\mathbf{P}^n$. Let constants $b_0, \ldots, b_n$ be given with $b_{k_0} \neq 0$. Define a (nonhomogeneous) system $h'(z)$ of $n + 1$ equations in $n + 1$ unknowns by

$$h'_l(z_0, \ldots, z_n) = h_l(z_0, \ldots, z_n)$$

for $l = 1$ to n, and

$$h'_0(z_0, \ldots, z_n) = L(z_0, \ldots, z_n) - 1,$$

[1] Fully clarifying the concept of *multiplicity* leads deep into algebraic geometry. However, it may be helpful to observe that if we perturb the system by arbitrarily small random perturbations of the coefficients, all the solutions to the perturbed system will be nonsingular. Further, near any solution z^0 of the original system will be a number of solutions into which z^0 has been *resolved*. This number is the multiplicity of z^0, which will be 1 for each nonsingular solution and 2 or more for each geometrically isolated singular solution. By "counting multiplicities" we mean that we count the multiplicities of the solutions rather than the solutions themselves.

where

$$L(z) \equiv \sum_{j=0}^{n} b_j z_j.$$

Then the solutions to $h(z) = 0$ in $\mathbf{U}_L = \{[z] \in \mathbf{P}^n \mid L(z) \neq 0\}$ are in one-to-one correspondence with solutions to $h'(z) = 0$ in $\mathbf{C}^{n+1}$. The multiplicity of solutions (and all other local properties) are preserved under this correspondence. Further, we may view $h'_1(z) = 0, \ldots, h'_n(z) = 0$ as a system in the n variables $z_0, \ldots, z_{k_0-1}, z_{k_0+1}, \ldots, z_n$ with

$$z_{k_0} = \frac{1 - (b_0 z_0 + \cdots + b_{k_0-1} z_{k_0-1} + b_{k_0+1} z_{k_0+1} + \cdots + b_k z_k)}{b_{k_0}}.$$

It is this system of n equations in n unknowns that we call "the projective transformation of h." We want to use this system for computations. Its solutions in $\mathbf{C}^n$ are in one-to-one correspondence with the solutions to $h'(z) = 0$ in $\mathbf{C}^{n+1}$. In Theorem 3 of **[31]** it is proven that the continuation paths are contained in $\mathbf{U}_L$ with probability one if the parameters are chosen at random. This theorem establishes the validity of the projective transformation.

In creating a computer code to implement the projective transformation, our usual procedure is to let $k_0 = 0$ and write a subroutine for h as a system of n equations in the $n+1$ variables $z_0, \ldots, z_n$, but include the formula

$$z_0 = \sum_{j=1}^{n} \beta_j z_j \ + \ \beta_0,$$

which makes z_0 an implicitly defined function of the other variables. The partial derivatives of the projective transformation with respect to $z_1, \ldots, z_n$ are then generated from those of h with respect to $z_0, \ldots, z_n$ using the chain rule. To make use of Theorem 3 from **[31]**, the $\beta_0, \ldots, \beta_n$ are chosen at random.

If h is m-homogeneous, we may evoke the projective transformation on each component of $\mathbf{P}$. Thus, with m-homogeneous h in the variables $z_{i,j}$ for $i = 0$ to n and $j = 1$ to m, we define

$$z_{0,j} = \sum_{i=1}^{k_j} \beta_{i,j} z_{i,j} \ + \ \beta_{0,j}$$

for $j = 1$ to m. (Theorem 3 of **[31]** is proven in this generality.) The finite solutions of $f(z) = 0$ are recovered via $z_{i,j} \leftarrow z_{i,j}/z_{0,j}$ for $i = 1$ to k_j and $j = 1$ to m. (If any $z_{0,j} = 0$, then the solution is at infinity.)

EXAMPLE 1. Consider the system:

$$(10)\qquad \begin{aligned} z_1^2 + z_2^2 - 25 &= 0, \\ z_1^2 + z_2^2 - 16z_2 + 39 &= 0. \end{aligned}$$

This is the intersection of two circles of radius 5 with centers at the origin and at $(0, 8)$. We homogenize (that is, 1-homogenize) via the substitutions $z_1 \leftarrow z_1/z_0$ and $z_2 \leftarrow z_2/z_0$, yielding

$$(11)\qquad \begin{aligned} z_1^2 + z_2^2 - 25z_0^2 &= 0, \\ z_1^2 + z_2^2 - 16z_2z_0 + 39z_0^2 &= 0. \end{aligned}$$

We obtain the solutions at infinity by solving the system with $z_0 = 0$ and $z_1 = 1$. This reduces (11) to $1 + z_2^2 = 0$. Thus, $z_2 = \pm i$, with $i = \sqrt{-1}$. (In finding solutions in projective space, we always set one of the variables equal to 1.) The projective transformation of (10) is (11) with $z_0 \equiv \beta_1 z_1 + \beta_2 z_2 + \beta_0$, where the β_j are chosen at random.

EXAMPLE 2. Consider the following system:

$$(12)\qquad \begin{aligned} z_1z_2z_3z_4 + 1 &= 0, \\ z_1z_3 + z_2z_4 + z_1z_4 &= 0, \\ 4z_1z_3z_4 - 2z_2z_3z_4 + 1 &= 0, \\ z_1 + z_2 &= 0. \end{aligned}$$

Now by grouping the variables of (12) into different sets, we create different m-homogeneous structures and Bezout numbers. Normally, we would want to solve such a system with the m-homogeneous structure that gives the smallest Bezout number. For each grouping of variables, we will form the combinatorial product, D, defined in (9) above, and then pick out the distinguished coefficient that gives the Bezout number, d. Thus:

EXAMPLE 2.1. Group variables as: $\{z_1, z_2\} \cup \{z_3, z_4\}$. Then, $D = (2\alpha_1 + 2\alpha_2)(\alpha_1 + \alpha_2)(\alpha_1 + 2\alpha_2)(\alpha_1 + 0\alpha_2)$, and $d = \text{Coef}[D, \alpha_1^2\alpha_2^2]$ (i.e., the coefficient of the $\alpha_1^2\alpha_2^2$ term of D). Thus, $d = 10$.

EXAMPLE 2.2. Group variables as: $\{z_1, z_2\} \cup \{z_3\} \cup \{z_4\}$. Then, $D = (2\alpha_1 + \alpha_2 + \alpha_3)(\alpha_1 + \alpha_2 + \alpha_3)^2(\alpha_1 + 0\alpha_2 + 0\alpha_3)$, and $d = \text{Coef}[D, \alpha_1^2\alpha_2\alpha_3] = 8$.

EXAMPLE 2.3. Group variables as: $\{z_1\} \cup \{z_2\} \cup \{z_3, z_4\}$. Then, $D = (\alpha_1 + \alpha_2 + 2\alpha_3)^2(\alpha_1 + \alpha_2 + \alpha_3)(\alpha_1 + \alpha_2 + 0\alpha_3)$, and $d = \text{Coef}[D, \alpha_1\alpha_2\alpha_3^2] = 16$.

We see that Example 2.2 gives the smallest Bezout number. Thus, while the 1-homogeneous (traditional) polynomial continuation yields a 24-path homotopy (i.e., the number of the total degree), we can (easily) find a 3-homogeneous 8-path homotopy. Such a savings in computer work (i.e., by a factor of 1/3) can be significant in some applications. The projective transformation for Example 2.2 is easy to construct: First, we 3-homogenize (12) via $z_1 \leftarrow z_1/z_5$, $z_2 \leftarrow z_2/z_5$, $z_3 \leftarrow z_3/z_6$, and $z_4 \leftarrow z_4/z_7$, where we take z_5, z_6, and z_7 to be the three homogeneous variables. This yields

$$\begin{aligned} z_1z_2z_3z_4 + z_5^2z_6z_7 &= 0, \\ z_1z_3 + z_2z_4 + z_1z_4 &= 0, \\ 4z_1z_3z_4 - 2z_2z_3z_4 + z_5z_6z_7 &= 0, \\ z_1 + z_2 &= 0, \end{aligned} \tag{13}$$

with $z_5 \equiv \beta_1 z_1 + \beta_2 z_2 + \beta_5$, $z_6 \equiv \beta_3 z_3 + \beta_6$, and $z_7 \equiv \beta_4 z_4 + \beta_7$, where the β_j are chosen at random.

5. A theorem about generically nonsingular solutions. In this section we state Theorem 1, a corollary to the very general theorem cited in the Appendix (from [**32**]). We focus on this special case because

(1) it highlights an important practical result;
(2) it illustrates the more general theory without requiring a background in algebraic geometry.

Recall that we are considering the system $f(c[q], z) = 0$ of $n + N$ equations in n unknowns with $q \in \mathbf{Q}$ and $z \in \mathbf{P}$. We make the following assumptions:

(1) $\mathbf{Q} = \mathbf{C}^s$, for some s;
(2) $c : \mathbf{Q} \to \mathbf{C}^r$ is polynomial;
(3) $\mathbf{P} = \mathbf{P}^{k_1} \times \cdots \times \mathbf{P}^{k_m}$.

We emphasize that these assumptions are not necessary, but are merely taken to allow a more elementary presentation. We need the following definition.

DEFINITION. The solution z to $f(c[q], z) = 0$ is *generically nonsingular* if there is a dense open $\mathbf{Q}_0 \subseteq \mathbf{Q}$ such that for every neighborhood $\mathbf{B}$ about z in $\mathbf{P}$ there is a neighborhood $\mathbf{B}'$ about q in $\mathbf{Q}$ so that if $q' \in \mathbf{B}' \cap \mathbf{Q}_0$, then $f(c[q'], z) = 0$ has a nonsingular solution z' with $z' \in \mathbf{B}$.

Thus a solution z is generically nonsingular if almost all nearby systems have a nonsingular solution near z. Now we have

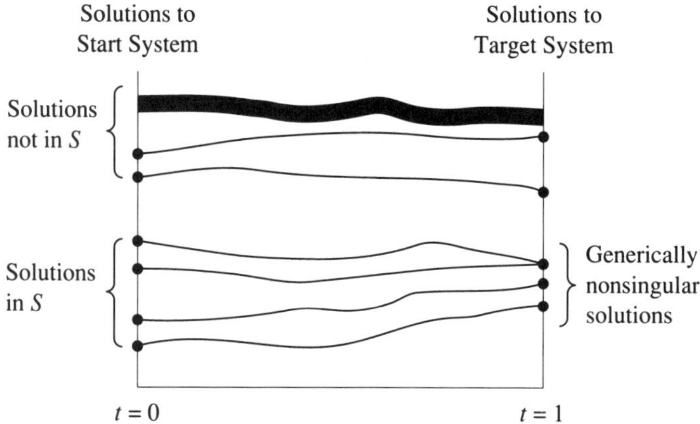

FIGURE 2. Schematic showing path behavior for "generically nonsingular polynomial continuation," as determined by Theorem 1. Note that the paths with start points in S find all generically nonsingular solutions. Also note that outside of S the start system (and the target system) may have positive-dimensional solution components.

THEOREM 1. *Given $q^1 \in Q$, there is a dense open full measure $\mathbf{Q}_0 \subseteq \mathbf{Q}$ such that if $q^0 \in \mathbf{Q}_0$ and S is the (finite) set of nonsingular solutions to $f(c[q^0], z) = 0$ in $\mathbf{P}$, then the homotopy*

$$h(z, t) = f(c[(1-t)q^0 + tq^1], z)$$

with start points in S and $(z, t) \in \mathbf{P} \times [0, 1]$ will have well-defined homotopy paths in $\mathbf{P}$ that are smooth and strictly increasing in t as a function of arclength, and the endpoints will include all the geometrically isolated generically nonsingular solutions of $f(c[q^1], z) = 0$.

What this theorem says is that if we choose almost any system (given by $q^0 \in \mathbf{Q}_0$), solve it somehow (say, by a traditional m-homogeneous continuation), and pick out the nonsingular solutions (the set S), then we may find the generically nonsingular solutions that are geometrically isolated of any other system (given by q^1) via a path-tracking approach with fewer paths (because we can restrict the choice of start points to S). Figure 2 illustrates this result schematically. There are three key observations about the "real" world that make this theorem important:

1. In engineering and scientific practice, small polynomial systems tend to arise in families (as indexed by $\mathbf{Q}$). It makes sense to solve one (given by q^0) at some expense, if the rest can then be solved cheaply.

2. The number of points in the set S, which is the full Bezout number for general systems, is often much less than the Bezout number for actual physical systems.

3. The physically meaningful solutions tend to be included among the generically nonsingular solutions.

In our experience, polynomial models with physically meaningful solutions that are not generically nonsingular are degenerate, often because a special singular case has arisen in a general model. In these cases, it is usually best to solve the more appropriate special model; no numerical method behaves well on the inappropriate general model. We note also that when physically meaningful solutions are not geometrically isolated, a similar type of model degeneracy has usually arisen. If such degeneracies are unavoidable (rarely the case), then using "random real" parameters as discussed in Section 5.1 can help, because then polynomial continuation will tend to find some real solutions, which is usually what is being sought. We cannot "prove" these observations about reality. We cite the examples given in [32], [33], [35], and [42] for supporting evidence.

The proof of Theorem 1 is given in the Appendix. However, we sketch the idea of the proof here. Define $\mathbf{X} = \{(q, z) \in \mathbf{Q} \times \mathbf{P} \mid f(c[q], z) = 0\}$. Consider the natural map $\pi\colon \mathbf{X} \to \mathbf{Q}$ defined by $\pi(q, z) = q$. We then can prove that there is an open dense $\mathbf{Q}_0 \subseteq \mathbf{Q}$ and a smooth manifold $\mathbf{X}_0 \subseteq \mathbf{X}$ such that $\pi_0 \equiv \pi \mid \mathbf{X}_0\colon \mathbf{X}_0 \to \mathbf{Q}_0$ is a finite-to-one *covering space*, and $\mathbf{X}_0$ can be identified with the geometrically isolated solutions of $f = 0$ when $q \in \mathbf{Q}_0$. Further, $\mathbf{X}_0 = \mathbf{X}_{ns} \cup \mathbf{X}_s$, where $\mathbf{X}_{ns}$ and $\mathbf{X}_s$ denote the nonsingular and singular solutions, respectively, and $\pi \mid \mathbf{X}_{ns}\colon \mathbf{X}_{ns} \to \mathbf{Q}_0$ and $\pi \mid \mathbf{X}_s\colon \mathbf{X}_s \to \mathbf{Q}_0$ are both covering spaces. In addition we show that if $\mathbf{K} \subseteq \mathbf{Q}$ is a polynomial (complex) curve and $\mathbf{K}_0 \equiv \mathbf{K} \cap \mathbf{Q}_0$, then $\mathbf{K} - \mathbf{K}_0$ is finite. This means that all but a finite number of points in $\mathbf{K}$ are in $\mathbf{Q}_0$. For example, $\mathbf{K}$ might be taken to be $\{(1-t)q^0 + tq^1 \mid t \in \mathbf{C}^1\}$. To define a standard homotopy, we choose $\mathbf{K}$ a complex curve, and then choose $\alpha\colon [0, 1] \to \mathbf{K}$ such that $\alpha\colon [0, 1) \to \mathbf{K}_0$, which is easy to do because all but a finite number of points in $\mathbf{K}$ are in $\mathbf{K}_0$. (Note that the image of α is one-real-dimensional in the two-real-dimensional $\mathbf{K}$.) Now consider the diagram

$$\begin{array}{ccccccc} \pi^{-1}(\alpha[0,1]) & \subseteq & \pi^{-1}(\mathbf{K}) & \subseteq & \mathbf{X} & \subseteq & \mathbf{Q}\times\mathbf{P} \\ \downarrow & & \downarrow & & \downarrow{\scriptstyle \pi} & & \\ \alpha[0,1] & \subseteq & \mathbf{K} & \subseteq & \mathbf{Q} & & \end{array}$$

The image of α, which is a path in the parameter space $\mathbf{Q}$, is lifted via π to $\mathbf{Q} \times \mathbf{P}$. This lifted path is a collection of paths in the "parameter-solution" space $\mathbf{Q} \times \mathbf{P}$. These paths are the continuation paths that we will track numerically.

Let us reconsider the example from Section 2, equations (4), (5), and (6). The total degree is 4. The 2-homogeneous Bezout number is 2. But the number of paths to be tracked for the parameter homotopy is 1. Thus one path will find all the geometrically isolated generically nonsingular solutions, by Theorem 1.

6. Implementation considerations.

6.1. *Coefficient parameter continuation.* Let us continue with the special assumptions from the previous section. In particular, we take $\mathbf{Q} = \mathbf{C}^s$, but remind the reader that the theory developed in **[32]** allows a more general context. (See the Appendix.)

We take the implementation problem to be: Given the coefficient parameter polynomial system $f(c[q], z)$, we want to solve a sequence of systems $f(c[q^k], z) = 0$ for $k = 1, 2, 3, \ldots$. We recognize two steps:

1. Choose $q^0 \in \mathbf{Q}$, solve $f(c[q^0], z) = 0$, and pick out the set of nonsingular solutions, S.

2. Track the solution paths of $h(z, t) = f(c[(1-t)q^0 + tq^k], z) = 0$, beginning at the points in S.

There are two ways to choose q^0:

(1) Choose q^0 "at random" from $\mathbf{Q}$.

(2) Choose q^{00} "at random" from $\mathbf{Q} \cap \mathbf{R}^s$, choose γ "at random" from $\mathbf{C}^1$, and take $q^0 = \gamma q^{00}$. (This "random-real" alternative has advantages for certain physical problems. The proof of the validity of choosing q^0 in this way is given at the end of the Appendix.)

6.2. *Variants on coefficient parameter continuation.* The coefficient parameter formulas for a particular problem may be complicated and involve many transcendental functions. This makes the function and Jacobian matrix evaluations expensive, and typically polynomial continuation involves thousands, if not millions, of these evaluations. Sometimes variants of the coefficient parameter polynomial homotopies have a similar small number of paths and are much more efficient in implementation. We sketch some of these variants here. We have particularly found the *secant homotopy* useful. What these variants amount to is that, from the given coefficient parameter formulas, we can define corollary coefficient parameter formulas that have a simpler form but admit a larger class of systems. Generally, this means that there will be more generically nonsingular solutions; therefore, "weakening" the coefficient parameter formulas in this way would generally seem to be a bad idea. But often the trade-off between efficiencies makes a variant preferable.

Let us consider a *hierarchy of homotopies*:

(1) *Traditional*: $h(z,t) = (1-t)g(z) + tf(c[q^k], z)$, where $g(z)$ has d solutions.
(2) *Coefficient*: $h(z,t) = f((1-t)C_0 + tc[q^k], z)$, where $C_0 \in \mathbf{C}^r$ is a random choice of coefficients.
(3) *Secant*: $h(z,t) = f((1-t)c[q^0] + tc[q^k], z)$, where q^0 is a random choice of parameters. Note that here the start system is defined by coefficient parameter formulas, but the intermediate systems (for $0 < t < 1$) are not given by coefficient parameters.
(4) *Parameter*: $h(z,t) = f(c[(1-t)q^0 + tq^k], z)$, where q^0 is a random choice of parameters. Note that taking the convex combination of q^0 and q^k is natural when $\mathbf{Q} = \mathbf{C}^s$, but almost any path in $\mathbf{Q}$ from q^0 to q^k will do.

We can easily see that the traditional, coefficient, and secant homotopies are "coefficient parameter" homotopies for appropriately defined coefficient parameter formulas. These variant coefficient parameter formulas are derived from the "original" coefficient parameter formulas, which we conceptualize as having been given by physically meaningful relations.

6.3. *Side conditions.* In this subsection we explore the surprising and useful fact that *algebraic relations persist along homotopy paths (generically).* This "falls out" of the fact that the theory is developed for $n + N$ equations in n unknowns, where $N \geq 0$. The theory of polynomial continuation up until now has taken $N = 0$, but we can recognize considerable implementation advantages from the more general context.

We have called the equations indexed by j for $j = n+1$ to $n+N$ "side conditions," although in the structure of the theory they are not distinguished from the other equations. Now, however, it will be better for the exposition if we call the side conditions and the rest of the system by different names. Thus let $f'(c[q], z)$ denote a system of n polynomials in n unknowns, with coefficient parameters $q \in \mathbf{Q}$, and let $s(c[q], z)$ denote a system of N polynomials in the same n unknowns with the same coefficient parameter set $\mathbf{Q}$. The result we want to point out is as follows: Suppose we choose q^0 at random from $\mathbf{Q}$ and suppose z^0 is a geometrically isolated solution to $f'(c[q^0], z) = 0$ that also satisfies $s(c[q^0], z) = 0$. Let $(q(t), z(t))$ denote the homotopy path of $h(z,t) = f'(c[(1-t)q^0 + q^1], z)$ with start point (q^0, z^0). Then $s(q(t), z(t)) = 0$ for $0 \le t \le 1$.

In other words, if the side conditions hold for a solution to the start system $f'(c[q^0], z) = 0$, then the side conditions will hold along the associated path and, in particular, at the corresponding solution to the target system $f'(c[q^1], z) = 0$. Therefore, if the side conditions represent conditions that we *do not want* to hold, we can omit the start points that *satisfy* the side conditions, and only track paths from the resulting smaller set of start points. Thus, we can track fewer paths than we would otherwise need to track. (Without the side condition result, we might track all the paths whose start points are given by solutions to $f'(c[q^0], z) = 0$ and then omit the resulting endpoints that satisfy $s(c[q^1], z) = 0$.) For example, solutions at infinity of $f'(c[q], z) = 0$ are distinguished by obeying the additional relation $s(z) = 0$, where $s(z)$ denotes the product of the homogeneous coordinates of z. If we are not interested in solutions that are generically at infinity, we can omit any start points that are at infinity, since we know that they will end up at infinity.

The converse idea is generally true, but there are some subtleties. That is, if we *want* the side conditions to hold, it is a reasonable approach to omit the paths whose start points are solutions to $f'(c[q^0], z) = 0$ that do not also obey $s(c[q^0], z) = 0$. This will certainly yield only solutions that obey the side conditions. However, certain types of solutions may be missed. In many cases, this is not a difficulty. However, for completeness, let us consider exactly what is involved.

1. It might happen, for a choice of $q^1 \notin \mathbf{Q}_0$, that $s(c[q^1], x) = 0$ might be satisfied in the limit of a continuation path as $t \to 1$ but not along the path. Generally, we can argue that these *unstable solutions* are nonphysical.

2. $f'(c[q^0], z) = 0$ could have some positive-dimensional solution sets that yield isolated solutions when the side conditions are included. We will miss such solutions if we find merely the isolated solutions of $f'(c[q^0], z) = 0$ and then eliminate those not satisfying $s(c[q^0], z) = 0$. But usually we are not able to find the positive-dimensional solution sets of $f'(c[q^0], z) = 0$. So to be absolutely rigorous we can find instead the isolated solutions of

$$f'_j(c[q^0], z) + \sum_{k=1}^{N} c_{j,k} s_k(c[q^0], z) = 0 \qquad \text{for } j = 1 \text{ to } n$$

that also satisfy

$$s_j(c[q^0], z) = 0 \qquad \text{for } j = 1 \text{ to } N,$$

where the $c_{j,k}$ are random numbers.

See Section 3.2 of [**32**] for a further discussion of side conditions.

6.4. *Path tracking.* Tracking a homotopy path amounts to solving an initial value problem, but because we have a formula for the integral (namely, $h = 0$) there are options for path tracking not generally available in solving ordinary differential equations. Path tracking has been much investigated; see the references cited in the Introduction. Here we note some considerations special to polynomial continuation.

To carry out Step 1 from Section 6.1 we must be able to solve $f(c[q^0], z) = 0$ and distinguish the singular solutions. Often, we would like to determine whether the singular solutions satisfy some side conditions of physical significance (usually to establish that these solutions have no physical significance). However, computing singular endpoints accurately is difficult to do [**15**]. The Jacobian matrix at singular endpoints is always real-rank deficient by at least two[2], ruling out the most investigated and easiest-to-control case. It would be very helpful to polynomial continuation to develop more accurate and efficient ways to compute singular endpoints.

The homotopy paths for polynomial continuation are strictly increasing in t as a function of arclength. (This is proven, for example, in [**30**].) Most standard path trackers don't make use of this fact. (See for example [**45**]. The path tracker in [**27**] is an exception.) Presumably, greater stability and computational efficiency could be realized if the fact that "paths don't turn back in t" were explicitly acknowledged by the path tracker.

In polynomial continuation the main numerical work of the method is to solve a large number of small linear systems. Further, there are many paths (rather than just one), and the possibility of path crossing must be considered. In these considerations polynomial continuation generally differs from homotopy methods for solving large nonlinear systems. Thus common numerical practices and rules of thumb may need to be reconsidered for polynomial continuation.

[2]In terms of path tracking, it is the rank of the realification of the (complex) Jacobian matrix that determines the rank of a solution. (See Appendix 3 of [**27**].) Essentially, if a solution is singular, the realified Jacobian matrix is deficient for the real and imaginary part of some deficient row of the Jacobian matrix.

7. Summary and conclusions. Coefficient parameter polynomial continuation includes and extends all current approaches to polynomial continuation. Variants include traditional (1-homogeneous and m-homogeneous), coefficient, secant, and parameter homotopies. Alternative compactifications of $\mathbf{C}^n$ and side conditions have proven to be especially useful. Letting the generic parameters be chosen "random real" rather than complex is often convenient.

Coefficient parameters arise naturally in engineering problems. Acknowledging them directly increases efficiency and reliability of polynomial continuation. Focusing on generically nonsingular solutions is

(1) computationally advantageous and
(2) justifiable as a modeling philosophy.

The examples in **[32]** from the kinematics of mechanisms and biomedical engineering illustrate a number of the ideas we have presented. See also **[33]**.

Matching the structure of the homotopy to the physical problem will continue to be a principal inspiration for the theory of homotopy construction. The main numerical issue raised by polynomial continuation is how to compute singular end points efficiently and accurately.

ACKNOWLEDGMENTS. The second author would like to thank the General Motors Research Laboratories, the National Science Foundation (DMS 87 22330), and the University of Notre Dame for their support.

Appendix. The goal of this appendix is to explain the statement of Theorem 3 of **[32]**, and show how Theorem 1 in Section 5 follows as a corollary. We would like to stress here that our efforts with respect to this theorem in both this paper and **[32]** are mainly interpretive. The hard mathematics underlying the theorem is contained in the theorems on generic flatness of algebraic and proper holomorphic maps due to Grothendieck and Frisch, respectively. This last point will be clearer by the end of this appendix.

To achieve our goal we must summarize some of the basics of analytic and algebraic varieties. The appendix of **[32]** contains a more complete summary. For full details we refer to **[7]**, **[16]**, **[17]**, **[36]**, **[37]**. The basic objects of study in algebraic and analytic geometry are algebraic and analytic varieties, respectively. Locally, an algebraic variety is defined as the zero set of finitely many polynomials on $\mathbf{C}^n$; locally an analytic variety is defined as the zero set of finitely many holomorphic functions on a neighborhood of the origin in $\mathbf{C}^n$. We will usually suppress the qualifying adjectives, algebraic and analytic, when the statement applies to both types of varieties. Algebraic varieties are analytic varieties, but not conversely.

Varieties are in general not smooth manifolds. Letting $\mathbf{V}$ be a variety, $\mathbf{V}_{\text{reg}}$ denotes the points where $\mathbf{V}$ is biholomorphic to a smooth manifold, the *smooth* or *regular* points of $\mathbf{V}$, and $\text{Sing}(\mathbf{V})$ denotes $\mathbf{V} - \mathbf{V}_{\text{reg}}$, the *singular* points of $\mathbf{V}$. $\mathbf{V}_{\text{reg}}$ forms an open dense subset of $\mathbf{V}$, and $\text{Sing}(\mathbf{V})$ is a subvariety. A variety $\mathbf{V}$ is said to be irreducible if $\mathbf{V}_{\text{reg}}$ is connected.

Let $\mathbf{V}$ be a variety. The usual topology induced by the metric topology in $\mathbf{C}^n$ or $\mathbf{P}^n$ is called the *complex topology*. Defining the set of closed sets of $\mathbf{V}$ to be the set of subvarieties of $\mathbf{V}$ defines the *Zariski topology* on $\mathbf{V}$. This topology is coarser than the usual topology. Except in trivial cases, it is not even Hausdorff. We define *Zariski open subsets* and *Zariski dense subsets* of $\mathbf{V}$ to be subsets of $\mathbf{V}$ that are open or dense, respectively, in the Zariski topology. A set of the form $\mathbf{V} - \mathbf{W}$, where $\mathbf{W}$ is a subvariety of $\mathbf{V}$, is Zariski open in $\mathbf{V}$, and every Zariski open subset can be expressed in this form. Let $\mathbf{Y}$ be a Zariski dense subset of $\mathbf{V}$. It follows that if $\mathbf{W}$ is a subvariety of $\mathbf{V}$ containing $\mathbf{Y}$, then $\mathbf{W} = \mathbf{V}$. Conversely, any subset $\mathbf{Y}$ with this property is Zariski dense in $\mathbf{V}$. We noted above that $\mathbf{V}_{\text{reg}}$ is open in $\mathbf{V}$. It is, in fact, Zariski open.

Zariski open sets are open in the complex topology. We call sets that are both Zariski open and Zariski dense *Zariski open dense*. Zariski open dense sets are open and dense in the complex topology. If $\mathbf{V}$ is an irreducible variety, and $\mathbf{Y} \subseteq \mathbf{V}$ is Zariski open and nonempty, then $\mathbf{Y}$ is Zariski dense.

However, Zariski dense subsets of $\mathbf{V}$ are not necessarily dense in the complex topology. For example, the open unit ball, or even merely its boundary, is Zariski dense in $\mathbf{C}^n$. The set of n-tuples with rational (or integer) entries is Zariski dense in $\mathbf{C}^n$. Any set dense (in the complex topology) in an open subset of $\mathbf{R}^n$ is Zariski dense in $\mathbf{C}^n$. Thus, the fractions between 0 and 1 are Zariski dense in the complex plane. Let $\mathbf{Q}$ be an irreducible variety, and $\mathbf{Q}_R$ be the fixed points for some antiholomorphic involution. (For example, $\mathbf{Q} = \mathbf{C}^n$ with complex conjugation, giving $\mathbf{Q}_R = \mathbf{R}^n$.) Let $\mathbf{U}$ be a nonempty Zariski open subset of $\mathbf{Q}$. Then $\mathbf{U} \cap \mathbf{Q}_R \cap \mathbf{Q}_{\text{reg}}$ is Zariski dense in $\mathbf{Q}$, unless $\mathbf{Q}_R \cap \mathbf{Q}_{\text{reg}}$ is empty. Also $\mathbf{U} \cap \mathbf{Q}_R \cap \mathbf{Q}_{\text{reg}}$ is open and dense (in the complex topology) in $\mathbf{Q}_R \cap \mathbf{Q}_{\text{reg}}$. (See Section 2c of **[37]**.) These results often allow us in homotopy continuation to choose our generic start systems with real coefficients. For more on this see the end of this appendix and Section 3.4 of **[32]**.

The dimension of an irreducible variety $\mathbf{V}$ is defined as the complex dimension of the smooth points. If $\mathbf{A}$ is a variety, $\mathbf{A}_{\text{reg}}$ is the union of connected components, $\{\mathbf{U}_i | i \in I\}$, each of which is Zariski open in its closure, $\mathbf{V}_i$, which is an irreducible subvariety of $\mathbf{A}$. The $\mathbf{V}_i$ are called the irreducible components of $\mathbf{A}$. This decomposition of $\mathbf{A}$ into irreducible components $\{\mathbf{V}_i | i \in I\}$ is fundamental. The dimension of $\mathbf{A}$ is defined as the maximum of the dimensions of the irreducible components. $\mathbf{A}$ is said to be pure dimensional if all its irreducible components have the same dimension. We deal exclusively with finite-dimensional varieties. If $\mathbf{A}$ is algebraic, then the number of irreducible components is finite; if $\mathbf{A}$ is analytic, then there are finitely many components meeting any compact set.

Let $\mathbf{X}$ be a compact complex variety that contains $\mathbf{C}^n$ as a Zariski open set. We say that $\mathbf{X}$ induces the usual algebraic structure on $\mathbf{C}^n$ if, given any polynomial on $\mathbf{C}^n$ and any point x of $\mathbf{X} - \mathbf{C}^n$, there is a neighborhood $\mathbf{U}$ of x in $\mathbf{X}$ such that there are two holomorphic functions f and g on $\mathbf{U}$ with the polynomial restricted to $\mathbf{U} \cap \mathbf{C}^n$ equal to f/g. In other words, the polynomial extends to $\mathbf{X}$ as a meromorphic function. Or we say that the polynomial extends to $\mathbf{X}$ with poles at infinity. For example, $\mathbf{P}^1 \times \mathbf{P}^1$ and the Hirzebruch surfaces $\mathbf{F}_r$ (see **[18]**) induce the usual algebraic structure on $\mathbf{C}^2$, as does $\mathbf{P}^2$.

Let $\pi: \mathbf{X} \to \mathbf{Y}$ be a holomorphic map between analytic varieties. If $\mathbf{X}$ and $\mathbf{Y}$ are algebraic, then π is said to be algebraic if there are compact algebraic varieties $\overline{\mathbf{X}}$, $\overline{\mathbf{Y}}$ with $\mathbf{X}$ Zariski open in $\overline{\mathbf{X}}$, $\mathbf{Y}$ Zariski open in $\overline{\mathbf{Y}}$, and such that π extends to a holomorphic map $\bar{\pi}: \overline{\mathbf{X}} \to \overline{\mathbf{Y}}$. There are a number of alternate (but equivalent) definitions of what an algebraic map is. For example, if $\mathbf{X}$ and $\mathbf{Y}$ are both Euclidean space, then π is algebraic with respect to this definition if and only if π is given by polynomial functions in the coordinates.

Assume that π is proper, i.e., that the inverse image under π of any compact set is compact. A basic fact about proper maps, $\pi: \mathbf{X} \to \mathbf{Y}$, referred to as Remmert's proper mapping theorem, is that the image of any analytic variety under a proper map is an analytic variety. If π is proper and algebraic, then the image of any algebraic subvariety is algebraic. Given a proper holomorphic map $\pi: \mathbf{X} \to \mathbf{Y}$ from a variety $\mathbf{X}$ onto a variety $\mathbf{Y}$, it follows that there is a Zariski open set $\mathbf{U}$ of $\mathbf{Y}$ such that for any $y \in \mathbf{U}$, $\pi^{-1}(y)$ is pure $\dim \mathbf{X} - \dim \mathbf{Y}$ dimensional. This is closely related to the following fundamental fact about proper maps, referred to as the upper semicontinuity of dimension. Letting $\pi: \mathbf{X} \to \mathbf{Y}$ be a proper holomorphic map between analytic (algebraic) varieties, it follows that for any given nonnegative integer k, the locus of points $y \in \mathbf{Y}$ such that $\dim[\pi^{-1}(x)] \geq k$ is an analytic (algebraic) subvariety of $\mathbf{Y}$. A variant of this, also referred to as the upper semicontinuity of dimension, is that, given a holomorphic map π (not necessarily proper) from one analytic variety $\mathbf{X}$ to another $\mathbf{Y}$, and a point x in $\mathbf{X}$, there is a neighborhood $\mathbf{U}$ of x in the complex topology such that $\dim[\pi^{-1}(\pi(z)) \cap \mathbf{U}] \leq \dim[\pi^{-1}(\pi(x)) \cap \mathbf{U}]$ for all z in $\mathbf{U}$. Another useful fact about proper maps is the Stein factorization theorem. Let $\pi: \mathbf{X} \to \mathbf{Y}$ be a proper holomorphic map. Then π factors $\pi = s \circ r$, where r is a holomorphic map with connected fibres from $\mathbf{X}$ to some analytic variety $\mathbf{W}$, and s is a finite-to-one holomorphic proper map from $\mathbf{W}$ to $\mathbf{Y}$. In general $\mathbf{W}$ is not smooth, even if $\mathbf{X}$ and $\mathbf{Y}$ are (though $\mathbf{W}$ does inherit some regularity properties from $\mathbf{X}$ and $\mathbf{Y}$, e.g., normality). If $\mathbf{X}$ and $\mathbf{Y}$ are projective algebraic, then so is $\mathbf{W}$. There are many versions of the Stein factorization theorem, e.g., see Corollary 11.5 of **[18]** for a very general algebraic version.

There is a useful and deep collection of results in complex algebraic (analytic) geometry known as desingularization theorems (mainly due to Zariski and Hironaka) that often let us replace an arbitrary algebraic (analytic) variety, $\mathbf{X}$, by a projective (complex) manifold. Given such an $\mathbf{X}$, there exists a complex manifold $\overline{\mathbf{X}}$ and a proper surjective holomorphic map $\pi: \overline{\mathbf{X}} \to \mathbf{X}$ such that π gives a biholomorphism from $\pi^{-1}(\mathbf{X}_{\text{reg}})$ to $\mathbf{X}_{\text{reg}}$. If $\mathbf{X}$ is compact and algebraic, then $\overline{\mathbf{X}}$ can be chosen to be a projective manifold.

The geometric objects defined above are centered around the notion of a point set defined by a collection of functions. From this point of view there is no difference between the variety defined in the complex plane by $z^n = 0$ for $n = 1$ or for some integer $n \geq 2$. For the problem of finding the solutions of systems of polynomials, it is essential to keep track of multiplicities. Though a simple numerical notion of multiplicity carries us quite far, it is insufficient to deal with the systems that come up in practice. We need the general notion of an *analytic space*; these are often called *complex spaces* also. A good general reference is Fischer [7]. Given a set of holomorphic functions $f_1, \dots, f_k$ on $\mathbf{C}$, we let $\mathbf{V}$ denote the variety defined by these functions. At each point $v \in \mathbf{V}$ we associate the local ring of germs at $v \in \mathbf{V}$ of holomorphic functions on $\mathbf{V}$: $\mathbf{O}_{\mathbf{V},v} = \mathbf{O}_{\mathbf{C}^n,v}/(f_1, \dots, f_k)$, where $\mathbf{O}_{\mathbf{C}^n,v}$ denotes the local ring of convergent power series at $v \in \mathbf{C}^n$, and $(f_1, \dots, f_k)$ denotes the ideal of this ring generated by the functions $\{f_1, \dots, f_k\}$. These rings, $\mathbf{O}_{\mathbf{V},v}$, "patch together" to give the structure sheaf, $\mathbf{O}_{\mathbf{V}}$, of the analytic space,

$(\mathbf{V}, \mathbf{O_V})$, associated to $\{f_1, \ldots, f_k\}$. A general analytic space is "patched together" out of such local models, in the same way that a manifold is covered by Euclidean patches. Let $\mathbf{N_V}$ denote the subsheaf of $\mathbf{O_V}$ consisting of all elements, g, of the local rings $\mathbf{O}_{\mathbf{V},v}$ such that $g^m = 0$ for some positive integer m. This sheaf captures the "multiplicity structure" of the solution set of the polynomial system that we are studying. When we are interested only in the set-theoretic properties of the solution set and wish to ignore the multiplicities, it is often useful to work with the reduction of the nonreduced space. We define the reduction of $(\mathbf{V}, \mathbf{O_V})$ [or the reduced analytic space associated to $(\mathbf{V}, \mathbf{O_V})$] to be $(\mathbf{V}, \mathbf{O_V}/\mathbf{N_V})$. This space is denoted as $(\mathbf{V}, \mathbf{O}_{\mathbf{V}_{\text{red}}})$, or $\mathbf{V}_{\text{red}}$ for short. An analytic space $(\mathbf{V}, \mathbf{O_V})$ is said to be reduced if it is equal to its own reduction. Every variety, $\mathbf{V}$, has a structure sheaf $\mathbf{O_V}$ so that $(\mathbf{V}, \mathbf{O_V})$ is a reduced analytic space. This reduced structure is defined as follows. $\mathbf{V}$ is covered by open sets $\mathbf{U}_i$ such that $\mathbf{U}_i$ is the point set in some open set $\mathbf{W} \subset \mathbf{C}^n$ where some finite set of holomorphic functions on $\mathbf{W}$ vanish. Consider the ideal in the ring of holomorphic functions consisting of all holomorphic functions on $\mathbf{W}$ that vanish on $\mathbf{U}_i$. The open sets $\mathbf{U}_i$ can be chosen so that this ideal is generated by a finite set $f_1, \ldots, f_k$. The structure sheaf associated to this basis by the construction earlier in the paragraph is independent of the basis and the associated analytic space $(\mathbf{V}, \mathbf{O_V})$ is reduced. Following standard conventions we usually suppress the structure sheaf when dealing with reduced analytic spaces. For emphasis we often refer to possibly nonreduced analytic spaces to remind the reader that there may be nilpotents in the structure sheaf.

One place analytic spaces naturally come up is when we must deal with fibres of maps, even between manifolds. For example let $w = z^2$ be a map from the complex plane to itself. Then the fibre over 0 is defined as $(0, \mathbf{O}_0)$, where the structure sheaf $\mathbf{O}_0$ is the quotient of the ring of convergent power series in z by the ideal generated by z^2, the pullback under the map of the maximal ideal in the ring of convergent power series in w. By the same procedure in general, given a holomorphic map $f : \mathbf{X} \to \mathbf{Y}$ between possibly nonreduced analytic spaces, the inverse image of any point in $\mathbf{Y}$ possesses a structure sheaf making it into a possibly nonreduced analytic space. An important case is when we have a system f of holomorphic functions. The solution set $\mathbf{V}$ has a possibly nonreduced structure sheaf $\mathbf{O}_f$. To appreciate its importance, assume that this system is defined on $\mathbf{C}^n$ and has an isolated point x as a solution. Then the stalk of the structure sheaf at x, $\mathbf{O}_{f,x}$, is a finite-dimensional ring over the complex numbers with $\dim \mathbf{O}_{f,x}$ equal to the multiplicity of the solution x of f.

Flatness is an extremely useful geometric condition which captures very precisely the notion of fibres of a map changing continuously without any "jumps." A surjective holomorphic map between complex manifolds is flat if and only if the map is open, which in turn is equivalent to the map having equal-dimensional fibres. The general definition is technical, and we refer the reader to (Lecture 6, Section 3 of **[36]**) for a beautiful discussion of the concept in a slightly more algebraic setting. Flatness intervenes in our work by means of Frisch's theorem (**[7]**, pp. 155–156, and **[8]**; see also **[19]** and Lecture 8 of **[36]** for a very general algebraic analogue of this theorem due to Grothendieck) which says that given a proper holomorphic map $\pi: \mathbf{X} \to \mathbf{Q}$ between possibly nonreduced analytic spaces, there is a Zariski open dense set $\mathbf{U} \subset \mathbf{Q}$ such that $\pi_{\pi^{-1}(U)}$ is flat. The standard consequence of flatness that we want to draw (see the discussion in **[36]** on pp. 42–44) is that if $\pi_{\mathbf{H}_{\mathrm{red}}}$ is a covering in the sense of differential topology, where $\mathbf{H}_{\mathrm{red}}$ is a manifold and a connected component of $\pi^{-1}(\mathbf{U})_{\mathrm{red}}$, then there is an integer $n_{\mathbf{H}}$ such that given any $x \in \mathbf{H}$, the multiplicity of x in the fibre of $\pi_{\mathbf{H}}$ over $\pi_{\mathbf{H}}(x)$ is $n_{\mathbf{H}}$.

Now let us reconsider the discussion at the beginning of Section 4. We have equation (7), where we take f to be a holomorphic map with $\mathbf{Q}$ an irreducible and reduced complex analytic variety. Assume that for any given $q \in \mathbf{Q}$, f_q, the restriction of f to $\{q\} \times \mathbf{C}^n$ is algebraic. We think of f as a complex analytic family of systems of $n+N$ polynomials on $\mathbf{C}^n$ (with coefficients $c[q]$) parameterized by $\mathbf{Q}$. The set $\mathbf{X}_+$ defined by (8) gives the finite solutions of $f = 0$. If $\mathbf{P}$ is some algebraic compactification[3] of $\mathbf{C}^n$, then unless the functions $f_1(c[q], z), \ldots, f_{n+N}(c[q], z)$ that are obtained as the composition of f with the coordinates of $\mathbf{C}^{n+N}$, are constant as functions of z, f does not extend to a holomorphic map $f: \mathbf{Q} \times \mathbf{P} \to \mathbf{C}^{n+N}$.

We can however find an extension ϕ of f to a holomorphic section, $\phi: \mathbf{Q} \times \mathbf{P} \to \mathbf{E}$, of a rank $n+N$ holomorphic vector bundle $\Psi: \mathbf{E} \to \mathbf{Q} \times \mathbf{P}$. This is proven in **[32]** (see Theorem 2). It comes down to noting that there is a positive integer i_0 such that the $n+N$ equations $f_1(c[q], z), \ldots, f_{n+N}(c[q], z)$ can each be written as a sum of terms of the form $c_I[q]z^I$ where I is a multi-index $(i_1, \ldots, i_n)$ with $i_1 + \cdots + i_n \leq i_0$,

[3]For example, $\mathbf{P}$ could be $\mathbf{P}^n$ or a product of projective spaces, but $\mathbf{P}$ could also be some other algebraic compactification of $\mathbf{C}^n$.

$z^I = z_1^{i_1} \cdots z_n^{i_n}$, and $c_I[q]$ is a holomorphic function on $\mathbf{Q}$. From this it follows that each f_j gives rise to a homogeneous polynomial f_j' of some degree $\leq i_0$ for each $q \in \mathbf{Q}$.

It follows that $\phi^{-1}(0) \equiv \mathbf{X} \supseteq \mathbf{X}_+ = \mathbf{X} \cap (\mathbf{Q} \times \mathbf{C}^n)$. This vector bundle and section are not unique, so $\mathbf{X} - \mathbf{X}_+$ is not unique, i.e., the solutions at infinity depend on the choice of infinity. In specific cases, $\mathbf{P}$, Ψ, and ϕ are chosen to suit the application context. Then we define $\pi : \mathbf{X} \to \mathbf{Q}$ as the restriction of the natural map $\mathbf{Q} \times \mathbf{P} \to \mathbf{Q}$.

It is essential to consider $\mathbf{X}_+$ and $\mathbf{X}$ equipped with their natural nonreduced structures. The above theorem of Frisch–Grothendieck applied to this map π gives Theorem 3 of **[32]**, which contains most of the numerically important basic facts about systems, in very great generality.

THEOREM 3 **[32]**. *Assume* $\mathbf{X}$ *is a possibly nonreduced analytic space and* $\mathbf{Q}$ *is an irreducible and reduced analytic space. Let* $\pi : \mathbf{X} \to \mathbf{Q}$ *be a proper holomorphic map (not necessarily onto). Then we have the following:*

1. *There is a Zariski open dense subset* $\mathbf{Q}_0 \subseteq \mathbf{Q}$*, an integer* $n_0 > 0$*, and a smooth manifold* $\mathbf{X}_0 \subseteq [\pi^{-1}(\mathbf{Q}_0)]_{\text{red}}$ *such that*
    a. $\pi_0 : \mathbf{X}_0 \to \mathbf{Q}_0$ *is a finite-to-one covering space with* n_0 *sheets, where* π_0 *denotes the restriction of* π *to* $\mathbf{X}_0$*. Further, given any irreducible*[4] *component* $\mathbf{J} \subseteq \mathbf{X}_0$*, the multiplicity of a point* $x \in \mathbf{J}$ *as a point of* $\pi^{-1}(\pi(x))$ *is a constant, depending only on* $\mathbf{J}$*. Thus* $\mathbf{X}_0 = \bigcup_{j=1}^{\infty} \mathbf{X}_j$*, where* $\mathbf{X}_j$ *consists of points of multiplicity* j*,* $\mathbf{X}_j \cap \mathbf{X}_{j'} = \emptyset$ *for* $j \neq j'$*, and* π_0 *restricted to any (nonempty)* $\mathbf{X}_j$ *is a covering space over* $\mathbf{Q}_0$*.*
    b. *For any* $q^0 \in \mathbf{Q}_0$*, if* $x \in \pi^{-1}(q^0)$ *is a geometrically isolated point of* $\pi^{-1}(q^0)$*, then* $x \in \mathbf{X}_0$*.*
    c. *The topological closure of* $\mathbf{X}_0$*,* $\overline{\mathbf{X}}_0$*, is an analytic variety in which* $\mathbf{X}_0$ *is Zariski open and dense. Further,* $\overline{\mathbf{X}}_0$ *is a finite union of irreducible components of* $\mathbf{X}$*.*
2. *Given* $q^0 \in \mathbf{Q}_0$*, and* $q^1 \in \mathbf{Q}$*, and a one-dimensional irreducible complex analytic subvariety* $\mathbf{K}$ *of* $\mathbf{Q}$ *containing* q^0 *and* q^1*, all but a countable number of points* $\mathbf{K}_*$ *in* $\mathbf{K}$ *are in* $\mathbf{Q}_0$*. The points in* $\mathbf{K}_*$ *are geometrically isolated. Thus* $\mathbf{K}_*$ *has no cluster points, and, in fact, if* $\mathbf{X}$*,* $\mathbf{Q}$*,* π*,* $\mathbf{K}$*, and* c *are algebraic, then* $\mathbf{K}_*$ *is finite. Let* $\mathbf{K}_0 \equiv \mathbf{K} - \mathbf{K}_*$*. Then the restriction of* π_0 *to* $\pi_0^{-1}(\mathbf{K}_0)$ *is a covering space with* n_0 *sheets.*

---
[4]Since X_0 is smooth, "irreducible" is equivalent to "connected."

3. *If $q \in \mathbf{K}_0$, then $\pi^{-1}(q) \cap \mathbf{X}_0$ is the set of geometrically isolated points in $\pi^{-1}(q)$. If $q \in \mathbf{K}_*$, then $\pi^{-1}(q) \cap \overline{\mathbf{X}_0}$ contains all the stable geometrically isolated points in $\pi^{-1}(q)$.*
4. *Let α: $[0, 1] \to \mathbf{K}$ be continuous, and assume that α is smooth on $[0, 1)$ with $\alpha([0, 1)) \subseteq \mathbf{K}_0$. Let $\alpha_0 \equiv \alpha_{[0,1)}$. Then $\alpha_0 : [0, 1) \to \mathbf{K}_0$, and α_0 lifts to $\mathbf{X}_0$ as n_0 smooth paths, $\tilde{\alpha}_0 : [0, 1) \to \mathbf{X}_0$, each having a unique continuous extension $\tilde{\alpha} : [0, 1] \to \overline{\mathbf{X}_0}$. These paths, $\tilde{\alpha}$, are distinct (nonoverlapping, nonbifurcating), except some may share common endpoints in $\pi^{-1}(\alpha(1))$, if $\alpha(1) \notin \mathbf{K}_0$. They are strictly increasing in t as a function of arclength. If x^1 is a stable geometrically isolated point in $\pi^{-1}(\alpha(1))$, then there is a geometrically isolated point x^0 in $\pi^{-1}(\alpha(0))$ with a lifted path $\tilde{\alpha}$ so that $\tilde{\alpha}(0) = x^0$ and $\tilde{\alpha}(1) = x^1$.*

This completes the statement of Theorem 3.

To see what the theorem says, note that, given $q \in \mathbf{Q}_0$, the Zariski open set in $\mathbf{Q}$, the set $\pi^{-1}(q) \cap \mathbf{X}_0$ consists of the isolated solutions of $f_q(z) = 0$. Further $\pi^{-1}(q) \cap \mathbf{X}_j$ consists of the multiplicity j solutions of $f_q(z) = 0$. In particular $\pi^{-1}(q) \cap \mathbf{X}_1$ consists of the nonsingular solutions of $f_q(z) = 0$. For an arbitrary $q \in \mathbf{Q}$, $\pi^{-1}(q) \cap \mathbf{X}_1$ consists of the generically nonsingular solutions of $f_q(z) = 0$. We have an analogous interpretation of $\pi^{-1}(q) \cap \mathbf{X}_0$ (respectively $\pi^{-1}(q) \cap \mathbf{X}_j$) as those solutions of $f_q(z) = 0$ which are the limits of the isolated (respectively multiplicity j) solutions of $f_{q'}(z) = 0$ for $q' \in \mathbf{Q}_0$.

Now we will show how Theorem 1 follows from Theorem 3 of [32]. Assume that $f(c[q], z) = 0$ satisfies the hypotheses of Theorem 1. Thus as in Theorem 1 assume that $\mathbf{Q} = \mathbf{C}^s$. Given the system $f(c[q^1], z) = 0$ that we wish to solve, choose a general point $q^0 \in \mathbf{Q}$. This point will lie in $\mathbf{Q}_0$, since $\mathbf{Q} - \mathbf{Q}_0$ is small in the very strong sense that it is a countable union of $2s - 2$ real-dimensional manifolds. We can take $\mathbf{K}$ to be the complex analytic curve $\{(1 - \lambda)q^0 + \lambda q^1 \mid \lambda \in \mathbf{C}\}$. Take α in part 4 of Theorem 3 to be given by $c[(1 - t)q^0 + tq^1]$ with $t \in [0, 1]$. Reading over part 4 of Theorem 3, we see that indeed we have the conclusions of Theorem 1.

To appreciate the power of the above results, continue with the hypotheses of Theorem 1, e.g., assume that $\mathbf{Q} = \mathbf{C}^s$. Assume that the coefficient function $c[q]$ takes real vectors to real coefficients; this happens more often than not in applications. By the discussion on real points earlier in this appendix, the point q^0 can be assumed to be random real, and therefore the conclusions of Theorem 1 follow from using the homotopy $h(z,t) = f(c[(1-t)\gamma q^0 + tq^1], z)$, where γ is a random complex number. For more on this see Section 3.4 of [**32**].

Theorem 5 of [**32**] shows that the limit points of the homotopies depend only on the complex curve **K** and not on the path $\alpha(t)$. Theorem 6 of [**32**], which depends on Theorem 3, gives information on the dependence of the limit points on the curve **K**, and in particular on when a singular solution of a system of polynomials lies on a positive-dimensional solution component. For more details and applications of Theorem 3 see [**32**].

REFERENCES

1. E. L. Allgower, *A survey of homotopy methods for smooth mappings*, Numerical Solution of Nonlinear Equations (E. L. Allgower, K. Glashoff, and H.-O. Peitgen, eds.), Lecture Notes in Mathematics no. 878, Springer-Verlag, New York, 1981.

2. E. L. Allgower and K. Georg, *Simplicial and continuation methods for approximating fixed points and solutions to systems of equations*, SIAM Rev. **22** (1980), 28–85.

3. P. Brunovský and P. Meravý, *Solving systems of polynomial equations by bounded and real homotopy*, Numer. Math. **43** (1984), 397–418.

4. S. N. Chow, J. Mallet-Paret, and J. A. Yorke, *A homotopy method for locating all zeros of a system of polynomials*, Functional Differential Equations and Approximation of Fixed Points (H.-O. Peitgen and H. O. Walther, eds.), Lecture Notes in Math. no. 730, Springer-Verlag, New York, 1979, pp. 228–237.

5. F. J. Drexler, *Eine Methode zur Berechnung sämtlicher Lösunger von Polynomgleichungessystemen*, Numer. Math. **29** (1977), 45–58.

6. ——, *A homotopy method for the calculation of zeros of zero-dimensional polynomial ideals*, Continuation Methods (H. G. Wacker, ed.), Academic Press, New York, 1978, pp. 69–93.

7. G. Fischer, *Complex analytic geometry*, Lecture Notes in Mathematics no. 538, Springer-Verlag, New York, 1977.

8. J. Frisch, *Points de platitude d'un morphisme d'espace analytiques complexes*, Invent. Math. **4** (1967), 118–138.

9. C. B. Garcia and T. Y. Li, *On the number of solutions to polynomial systems of equations*, SIAM J. Numer. Anal. **17** (1979), 159–176.

10. C. B. Garcia and W. I. Zangwill, *Global continuation methods for finding all solutions to polynomial systems of equations in N variables*, Center for Math. Studies in Business and Economics Report no. 7755, University of Chicago, 1977.

11. ——, *Finding all solutions to polynomial systems and other systems of equations*, Math. Programming **16** (1979), 159–176.

12. ——, *An approach to homotopy and degree theory*, Math. Oper. Res. **4** (1979), 390–405.

13. ——, *Determining all solutions to certain systems of nonlinear equations*, Math. Oper. Res. **4** (1979), 1–14.

14. ——, *Pathways to Solutions, Fixed Points, and Equilibria*, Prentice-Hall, Englewood Cliffs, N.J., 1981.

15. A. Griewank, *On solving nonlinear equations with simple singularities or nearly singular solutions*, SIAM Rev. **27** (1985), 537–563.

16. P. Griffiths and J. Harris, *Principles of Algebraic Geometry*, John Wiley and Sons, New York, 1978.

17. R. Gunning and H. Rossi, *Analytic functions of several complex variables*, Prentice Hall, Englewood Cliffs, N.J., 1965.

18. R. Hartshorne, *Algebraic Geometry*, Graduate Texts in Mathematics, Springer-Verlag, New York, 1977.

19. R. Kiehl, *Note zu der Arbeit von J. Frisch*, Invent. Math. **4** (1967), 139–141.

20. T. Y. Li, *On Chow, Mallet-Paret and Yorke homotopy for solving system of polynomials*, Bull. Inst. Math. Academica Sinica **11** (1983), 433–437.

21. T. Y. Li, T. Sauer, and J. A. Yorke, *Numerical solution of a class of deficient polynomial systems*, SIAM J. Numer. Anal. **24** (1987), 435–451.

22. ———, *Numerically determining solutions of systems of polynomial equations*, Bull. Amer. Math. Soc. **18** (1988), 173–177.

23. K. Meintjes and A. P. Morgan, *A methodology for solving chemical equilibrium systems*, Appl. Math. Comput. **22** (1987), 333–361.

24. A. P. Morgan, *A method for computing all solutions to systems of polynomial equations*, ACM Trans. Math Software **9** (1983), 1–17.

25. ———, *A homotopy for solving polynomial systems*, Appl. Math. Comput. **18** (1986), 87–92.

26. ———, *A transformation to avoid solutions at infinity for polynomial systems*, Appl. Math. Comput. **18** (1986), 77–86.

27. ———, *Solving Polynomial Systems using Continuation for Scientific and Engineering Problems*, Prentice-Hall, Englewood Cliffs, N.J., 1987.

28. ———, *Polynomial continuation*, Impacts of Recent Computer Advances on Operations Research (R. Sharda, ed.), Publications in Operations Research Series, North Holland, 1989.

29. A. P. Morgan and R. F. Sarraga, *A method for computing three surface intersection points in GMSOLID*, ASME paper 82-DET-41 (1982).

30. A. P. Morgan and A. J. Sommese, *A homotopy for solving general polynomial systems that respects m-homogeneous structures*, Appl. Math. Comput. **24** (1987), 101–113.

31. ———, *Computing all solutions to polynomial systems using homotopy continuation*, Appl. Math. Comput. **24** (1987), 115–138.

32. ———, *Coefficient-parameter polynomial continuation*, Appl. Math. Comput. **29** (1989), 123–160.

33. A. P. Morgan, A. J. Sommese, and C. W. Wampler, *Polynomial continuation for mechanism design problems*, Proc. AMS-SIAM Summer Seminars on Computational Solution of Nonlinear Systems of Equations (Colorado State Univ., July 18–29, 1988), AMS Lectures in Applied Mathematics, this volume.

34. A. P. Morgan, A. J. Sommese, and L. T. Watson, *Finding all solutions to polynomial systems using HOMPACK*, ACM Trans. Math. Software **15** (1989), 93–122.

35. A. P. Morgan and C. W. Wampler, *Solving a planar 4-Bar design problem using continuation*, Research Pub. GMR-6509, Math. Dept., General Motors Research Laboratories, Warren, Mich. 48090, 1988.

36. D. Mumford, *Lectures on curves on an algebraic surface*, Ann. of Math. Stud., no. 59, Princeton University Press, Princeton, N.J., 1966.

37. ———, *Algebraic geometry I; complex projective varieties*, Grundlehren der mathematischen Wissenschaften 221, New York, 1976.

38. W. C. Rheinboldt and J. V. Burkardt, *Algorithm 596: A program for a locally parameterized continuation process*, ACM Trans. Math. Software **9** (1983), 236–241.

39. S. Richter and R. De Carlo, *A homotopy method for eigenvalue assignment using decentralized state feedback*, IEEE Trans. Auto. Control **AC-29** (1984), 148–158.

40. M. G. Safonov, *Exact calculation of the multivariable structured-singular-value stability margin*, IEEE Control and Decision Conference, Las Vegas, NV, Dec. 12-14, 1984.

41. L.-W. Tsai and A. P. Morgan, *Solving the kinematics of the most general six- and five-degree-of-freedom manipulators by continuation methods*, ASME J. of Mechanisms, Transmissions and Automation in Design **107** (1985), 189–200.

42. C. W. Wampler, A. P. Morgan, and A. J. Sommese, *Numerical continuation methods for solving polynomial systems arising in kinematics*, Research Pub. GMR-6372, General Motors Research Laboratories, Warren, Mich. 48090, 1988. Also, to appear in ASME J. Mechanisms, Transmissions, and Automation in Design.

43. L. T. Watson, *A globally convergent algorithm for computing fixed points of* $\mathbf{C}^2$ *maps*, Appl. Math. Comput. **5** (1979), 297–311.

44. ——, *Numerical linear algebra aspects of globally convergent homotopy methods*, SIAM Review **28** (1986), 529–545.

45. L. T. Watson, S. C. Billups, and A. P. Morgan, *HOMPACK: A suite of codes for globally convergent homotopy algorithms*, ACM Trans. Math. Software **13** (1987), 281–310.

46. L. T. Watson and D. Fenner, *Chow-Yorke algorithm for fixed points or zeros of* $\mathbf{C}^2$ *maps*, ACM Trans. Math. Software **6** (1980), 252–260.

47. A. H. Wright, *Finding all solutions to a system of polynomial equations*, Math. Comp. **44** (1985), 125–133.

48. W. Zulehner, *A simple homotopy method for determining all isolated solutions to polynomial systems*, Math. Comp. **50** (1988), 167–177.

49. W. Zulehner, *On the solutions to polynomial systems obtained by homotopy methods*, preprint.

MATHEMATICS DEPARTMENT, GENERAL MOTORS RESEARCH LABORATORIES, WARREN, MICHIGAN 48090-9057

MATHEMATICS DEPARTMENT, UNIVERSITY OF NOTRE DAME, NOTRE DAME, INDIANA 46556

Lectures in Applied Mathematics
Volume **26** (1990)

Polynomial Continuation for Mechanism Design Problems

ALEXANDER P. MORGAN, ANDREW J. SOMMESE, AND CHARLES W. WAMPLER

Abstract. The area of mechanical engineering devoted to the design of mechanisms generates polynomial systems for which the full set of (real) solutions is sought. This occurs both in the analysis and synthesis phases of mechanism design. We give some sample problem types and discuss their solution via the numerical method of polynomial continuation. This problem area provides examples of nonlinear algebraic systems with more than one physically meaningful solution.

1. Introduction. The purpose of this paper is to introduce to a mathematical audience an engineering area which is a rich source of polynomial systems that require numerical solution. These systems are "small," usually involving many less than 100 variables. On the other hand, there are frequently many physically meaningful solutions, and computing the full set of (real) solutions is desirable. In this paper we shall show that the numerical method of polynomial continuation suits this problem area well (see [**14**] in this volume for an overview and many references; see also [**3**] and [**9**]). However, other solution techniques that compute the full solution set may also prove useful, e.g., generalized bisection ([**4**], [**6**], [**7**], [**8**]) or a methodology based on Gröbner bases ([**2**]).

The type of design problems we will focus on are from the kinematics of mechanisms. Such problems express length and angle relationships between linkages. These relationships naturally take mathematical form

---

1980 *Mathematics Subject Classification* (1985 *Revision*). Primary 65H10; Secondary 14-02, 65-02.

as polynomial and trigonometric equations. Frequently, the trigonometry can be aliased with pure polynomials. The real solutions to the resulting polynomial systems, including the solutions at infinity, have physical meaning. *Analysis problems* are formulated so that the solutions tell how a mechanism behaves. Different solutions represent different possible behaviors. For *synthesis problems* the solutions yield design parameters, such as link lengths, for mechanisms meeting specified motion criteria. Different solutions represent different designs that meet the given criteria. The references **[1]**, **[5]**, and **[17]** give introductory material on the kinematics of mechanisms, and **[15]**, **[18]**, and **[19]** specifically address solving kinematics problems using polynomial continuation.

In Section 2 we discuss *inverse kinematics problems*, including planar and spatial manipulators. These illustrate analysis problems. Section 3 is devoted to *precision point problems*, focusing on planar 4-bars and the planar and spatial Burmester problems. These are examples of synthesis problems. Section 4 contains a summary and conclusions.

The style of presentation of this paper is informal, and the content is elementary. The references direct the reader to more technical material.

2. Inverse kinematics problems. Manipulators are chains of links. We restrict ourselves to manipulators with revolute joints (that is, joints that turn on a pivot) as opposed, say, to prismatic joints (joints that slide). The material in this section is adapted from Chapter 10 of **[9]**.

2.1. *Revolute joint planar manipulators.* Planar manipulators are chains of links in the plane. Let us consider the simple "manipulator" shown in Fig. 1. It consists of two links and two joints. The links are long, narrow pieces of cardboard, and the two joint axes are thumbtacks, one stuck into the table and the other sticking up through the two links. The end of the second link, the one without a thumbtack, is called "the hand." Its position has coordinates (x, y). We measure the first angle, θ_1, counterclockwise from the horizontal, and the second angle, θ_2, counterclockwise from the line segment extending the second link. The basic question about this mechanism is: Given (x, y), what values of θ_1 and θ_2 put the hand at (x, y)? This is the "inverse position problem" (IPP). Generally, there is more than one solution (Fig. 2).

To turn this particular problem into a polynomial system and at the same time develop a general formalism for this type of problem, we proceed as follows. In Fig. 3 we have reduced the width of the links

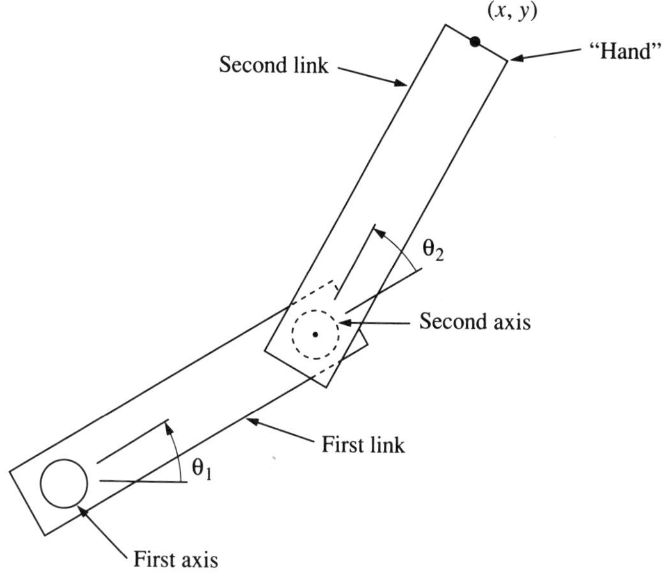

FIGURE 1. Thumbtack and cardboard manipulator. Note θ_2 is measured from the extension of the first link.

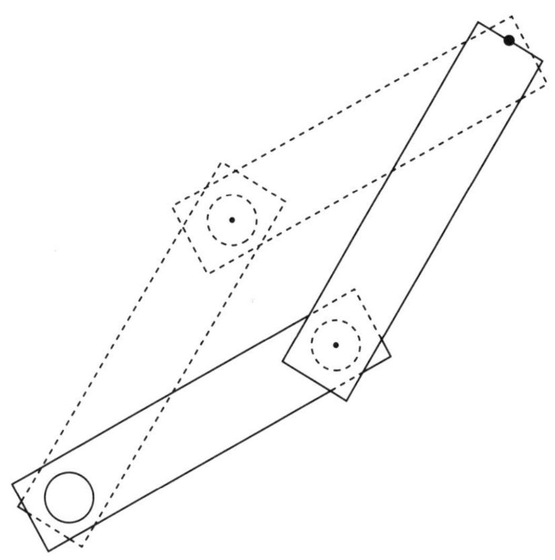

FIGURE 2. The inverse position problem can have two solutions.

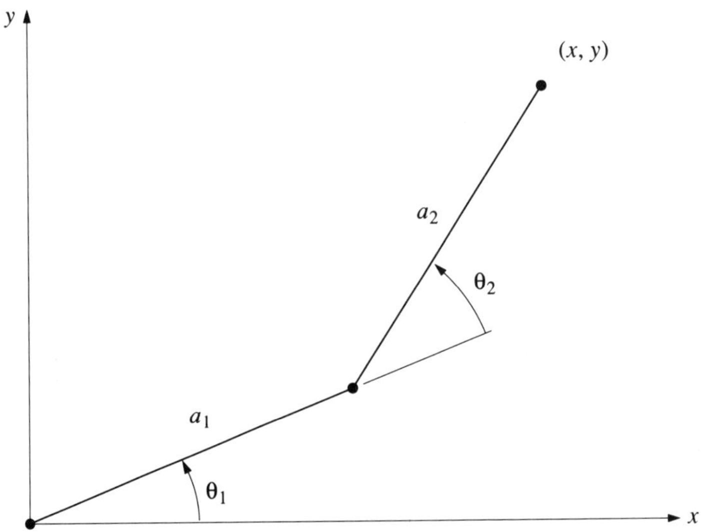

FIGURE 3. Two-link planar manipulator

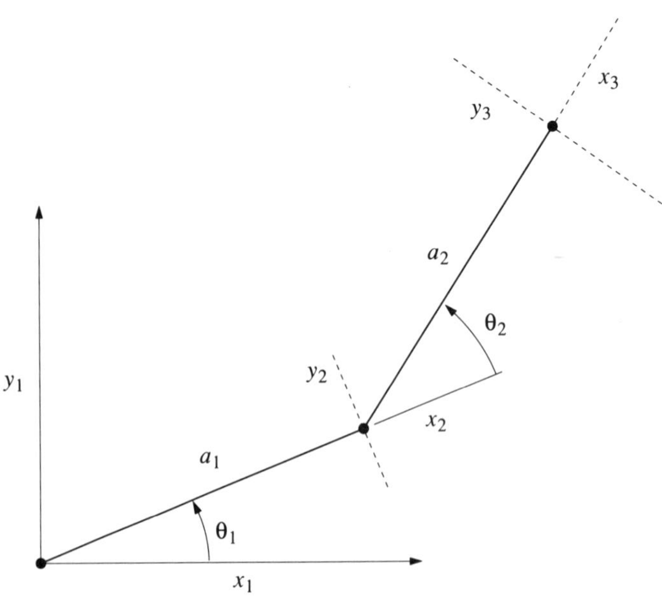

FIGURE 4. Two-link planar manipulator showing the three coordinate systems

to zero, placed the first axis at the origin of a global coordinate system, and labeled the link lengths a_1 and a_2. Further, we introduce coordinate systems at the second joint and at the hand, with the x-axis of the new coordinate systems extended along the previous link (Fig. 4). Now, if $\mathbf{p}$ is a point in the plane, we can express $\mathbf{p}$ in any of the given three coordinate systems. Denote these coordinates for $\mathbf{p}$ in the first, second, and third coordinate systems by $\mathbf{z}_j = (x_j, y_j)$ for $j = 1$ to 3, respectively. Now we observe that if a rigid motion transformation (rotation with translation) is defined by

$$\mathbf{A}_j = \begin{bmatrix} c_j & -s_j & c_j a_j \\ s_j & c_j & s_j a_j \\ 0 & 0 & 1 \end{bmatrix}, \tag{1}$$

then we have

$$\mathbf{Z}_j = \mathbf{A}_j \mathbf{Z}_{j+1}, \tag{2}$$

where $c_j = \cos(\theta_j)$, $s_j = \sin(\theta_j)$, and $\mathbf{Z}_j$ is defined from (and identified with) $\mathbf{z}_j$ via

$$\mathbf{Z}_j = \begin{bmatrix} x_j \\ y_j \\ 1 \end{bmatrix}.$$

Now we can express the fact that $(0, 0)$, the origin of the third coordinate system, the hand position, equals $\mathbf{z} = (x, y)$ in the first coordinate system (the global coordinate system) via

$$\mathbf{Z} = \mathbf{A}_1 \mathbf{A}_2 \mathbf{Z}_0, \tag{3}$$

where

$$\mathbf{Z} = \begin{bmatrix} x \\ y \\ 1 \end{bmatrix}$$

and

$$\mathbf{Z}_0 = \begin{bmatrix} 0 \\ 0 \\ 1 \end{bmatrix}.$$

Equation (3) expresses the IPP in standard matrix form (that is, in terms of rigid motions), but here it is written out as a polynomial system:

$$\begin{aligned} a_2 c_1 c_2 - a_2 s_1 s_2 + a_1 c_1 &= x, \\ a_2 s_1 c_2 + a_2 c_1 s_2 + a_1 s_1 &= y, \\ c_1^2 + s_1^2 &= 1, \\ c_2^2 + s_2^2 &= 1, \end{aligned} \tag{4}$$

where we have converted c_j and s_j into algebraic variables by appending the equations expressing $\sin^2 + \cos^2 = 1$.

In (4) the unknowns are c_1, s_1, c_2, and s_2, while a_1, a_2, x, and y are the problem parameters. When we solve for the unknowns, we will have solved the IPP, recovering the angles from their sines and cosines.

We refer the reader to **[14]** in this volume for the basic definitions (and concepts) associated with the numerical method of polynomial continuation. In particular, we need the definitions of *total degree*, *m-homogeneous*, *Bezout number*, *coefficient parameters*, *coefficient parameter continuation*, and *side conditions*. (See also **[10]**, **[13]**, and **[19]**.) The system (4) has a total degree of 16, this being the number of paths we would have to track to solve the system using a traditional 1-homogeneous continuation. However, if we group the variables into two sets (namely, $\{c_1, s_1\}$ and $\{c_2, s_2\}$), then we see that the associated 2-homogeneous Bezout number is 8, the number of paths we track using a 2-homogeneous continuation. However, the number of paths for the coefficient parameter continuation is 2, using the four parameters noted above. We cannot reduce the number of paths further, since some IPP problems have two solutions (Fig. 2). We see, then, that the coefficient parameter approach captures the structure of this problem very well.

If we add a link to the two-link manipulator, the IPP becomes: Solve

$$\mathbf{Z} = \mathbf{A}_1\mathbf{A}_2\mathbf{A}_3\mathbf{Z}_0, \tag{5}$$

where A_3 is defined by (1). However, (5) generally has an infinite number of solutions, because (5) has more unknowns than equations. Thus, the IPP is not well posed in this case. However, we can generalize the IPP to recover a system with the same number of equations as unknowns. At the same time, we will ask a more sophisticated question. We specify not only the position of the hand but also the alignment of the last joint. That is, we specify a point (x, y) and an arrow $\mathbf{v}$. A different $\mathbf{v}$ (say, $\mathbf{v}'$) with the same (x, y) generally yields a different set of solutions, as indicated in Fig. 5. Since $\mathbf{v}$ gives the x-axis of the last coordinate system, we can express this new condition as: The x-axis in coordinate system 4 should be $\mathbf{v}$ in coordinate system 1:

$$\begin{bmatrix} c_0 \\ s_0 \\ 0 \end{bmatrix} = \mathbf{A}_1\mathbf{A}_2\mathbf{A}_3 \begin{bmatrix} 1 \\ 0 \\ 0 \end{bmatrix}, \tag{6}$$

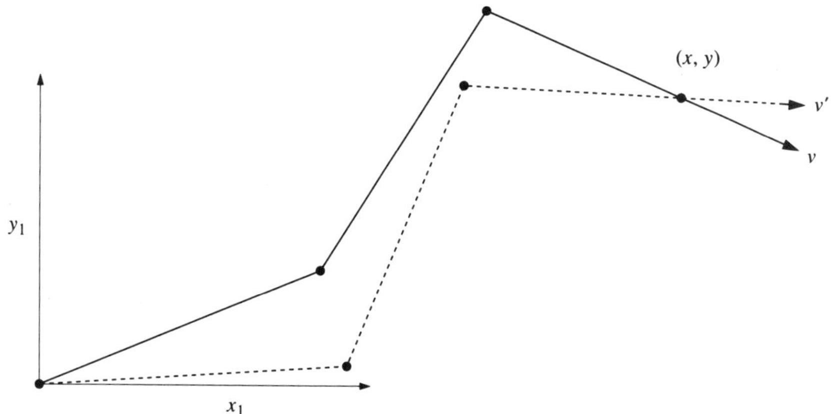

FIGURE 5. Three-link planar manipulator with hand position (x, y) and orientation $\mathbf{v}$

where $\mathbf{v} = (c_0, s_0)$, and we take $c_0^2 + s_0^2 = 1$. Alternatively, we can combine (5) and (6) into a single matrix equation:

$$\mathbf{A}_1\mathbf{A}_2\mathbf{A}_3 = \mathbf{A}_0, \tag{7}$$

where

$$\mathbf{A}_0 = \begin{bmatrix} c_0 & -s_0 & x \\ s_0 & c_0 & y \\ 0 & 0 & 1 \end{bmatrix}.$$

The resulting system (7) of nine equations in six unknowns[1] is overspecified, but any three of the four equations given by the first and second columns of $\mathbf{A}_0$ can be omitted. This yields a system of six equations in six unknowns, the solutions to which give the solutions to the generalized IPP.

This system consists of three cubics and three quadrics. Therefore, its total degree is 216. However, we can, by a simple maneuver, reduce this system to a form with a smaller total degree. In fact, we simply write (7) as

$$\mathbf{A}_1\mathbf{A}_2 = \mathbf{A}_0\mathbf{A}_3^{-1}. \tag{8}$$

This system consists only of quadrics. Let us consider it in more detail. Define $\varepsilon_1, \ldots, \varepsilon_6$ by

$$\mathbf{A}_1\mathbf{A}_2 - \mathbf{A}_0\mathbf{A}_3^{-1} \equiv \begin{bmatrix} \varepsilon_1 & \varepsilon_2 & \varepsilon_5 \\ \varepsilon_3 & \varepsilon_4 & \varepsilon_6 \\ 0 & 0 & 1 \end{bmatrix}.$$

[1]We always include $c_j^2 + s_j^2 = 1$ for each j.

Thus we have the (main) polynomial system:

$$\varepsilon_2 = 0, \qquad \varepsilon_5 = 0, \qquad \varepsilon_6 = 0,$$

with

$$c_j^2 + s_j^2 - 1 = 0 \quad \text{for } j = 1 \text{ to } 3,$$

along with side conditions

$$\varepsilon_1 = 0, \qquad \varepsilon_3 = 0, \qquad \varepsilon_4 = 0.$$

Thus,

$$\begin{aligned} \varepsilon_2 &= -c_1 s_2 - s_1 c_2 - c_0 s_3 + s_0 c_3, \\ \varepsilon_5 &= a_2 c_1 c_2 - a_2 s_1 s_2 + a_1 c_1 + [c_0 a_3 - x], \\ \varepsilon_6 &= a_2 c_1 s_2 + a_2 s_1 c_2 - a_1 s_1 + [s_0 a_3 - y], \end{aligned}$$

and (for example)

$$\varepsilon_4 = c_1 c_2 - s_1 s_2 - s_0 s_3 - c_0 c_3.$$

Here the c_j and s_j for $j = 1$ to 3 are the variables, and a_j, c_0, s_0, x, and y are the parameters.

Thus, we see that the total degree of the system is $2^6 = 64$, but the 3-homogeneous Bezout number associated with the natural grouping of variables:

$$\{c_1, s_1\}, \qquad \{c_2, s_2\}, \qquad \{c_3, s_3\}$$

is 16, which we verify via

$$\text{Coef}[(\alpha_1 + \alpha_2 + \alpha_3)(\alpha_1 + \alpha_2)^2 (2\alpha_1)(2\alpha_2)(2\alpha_2), \alpha_1^2 \alpha_2^2 \alpha_3^2] = 2^4,$$

where Coef$[A, \omega]$ denotes the coefficient of A for the term whose variables are ω. (See **[14]**, Eq. (9).) Further, the parameter homotopy (associated with the natural parameters noted above) generates 4 paths for the system without side conditions, but only 2 paths if the single side condition $\varepsilon_4 = 0$ is included. We know there are sample systems with 2 solutions, so reducing to 2 paths is the best that can be done.

2.2. *Revolute-joint spatial manipulators.* Spatial manipulators are chains of links in space. As in the planar case, we restrict ourselves to revolute joints. The basic ideas for defining spatial manipulators, for generating polynomial systems to solve the inverse position problem and for reducing and solving these systems, are analogous to those for

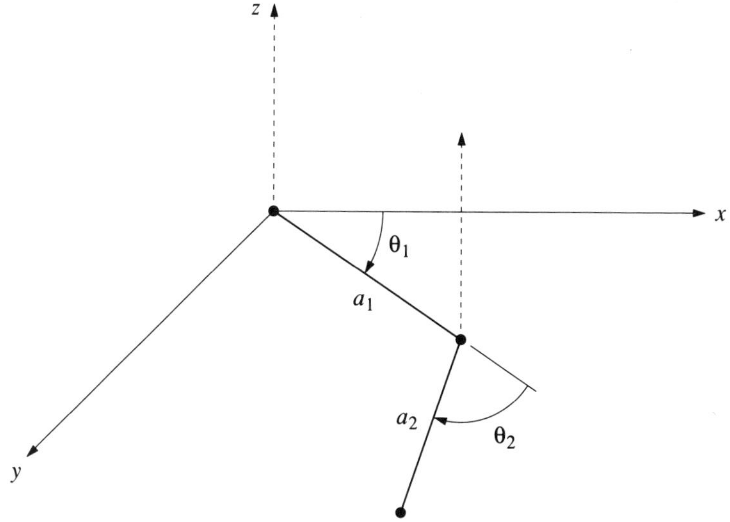

FIGURE 6. Two-link planar manipulator with z-axes indicated

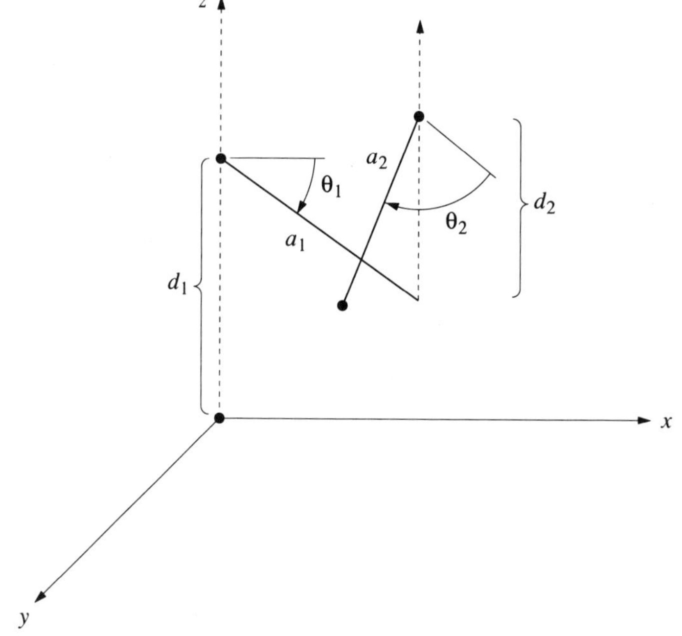

FIGURE 7. Two-link spatial manipulator derived from two-link planar manipulator by offsetting the links along the z-axes

planar manipulators. The resulting models, however, are more complex, harder to visualize, and yield less trivial polynomial systems.

We extend the planar manipulator model to three dimensions in two steps. Looking at our two-link planar model in Fig. 3, we may visualize the joints pivoting on axes that stick straight out of the page. Imagine long nails instead of thumbtacks (Fig. 6). These axes are parallel, we can say, to the z-axis. For the first step, extend the links along these axes by distances d_1 and d_2, as indicated in Fig. 7. Looking straight down, our model seems the same as before, but from an oblique view, we see that the links are now in different but parallel planes. For the second step, we "twist" the second axis by an angle τ_1 in the plane normal to the first link (Fig. 8). We establish coordinate systems at each joint, as before, with the joint being the z-axis, the link being the x-axis, and the y-axis chosen perpendicular to these. The basic schematic is provided in Fig. 9. Then the coordinate transformation matrices from one joint coordinate system to the next provide us with our fundamental algebraic relation, just as in the planar case. Thus we have

$$\mathbf{A}_j = \begin{bmatrix} c_j & -s_j\lambda_j & s_j\mu_j & a_jc_j \\ s_j & c_j\lambda_j & -c_j\mu_j & a_js_j \\ 0 & \mu_j & \lambda_j & d_j \\ 0 & 0 & 0 & 1 \end{bmatrix},$$

where $c_j = \cos\theta_j$, $s_j = \sin\theta_j$, $\lambda_j = \cos\tau_j$, $\mu_j = \sin\tau_j$, and

$$\mathbf{p}_j = \mathbf{A}_j\mathbf{p}_{j+1},$$

where "$\mathbf{p}_k$" denotes "point $\mathbf{p}$'s coordinates in the kth coordinate system." (Note that the first coordinate system has no link to define its x-axis. We chose this axis arbitrarily. Also, the last coordinate system (that of the hand) has no actual joint associated with it. Therefore, for the last coordinate system, we may choose $\mathbf{a} = 0$ and $\mathbf{d} = 0$. The z-axis is arbitrary also.)

Physically, links may not look much like line segments in the normal planes of the joint axes (as we have here). However, we lose no generality with this model (see Hartenberg and Denavit [5]). Thus we are not distinguishing between links and the common normals between joint axes.

Naturally, we may have as many (or few) joints as we like. However we will study six-joint manipulators, because a manipulator generally needs six joints to reach any point in its workspace with any orientation.

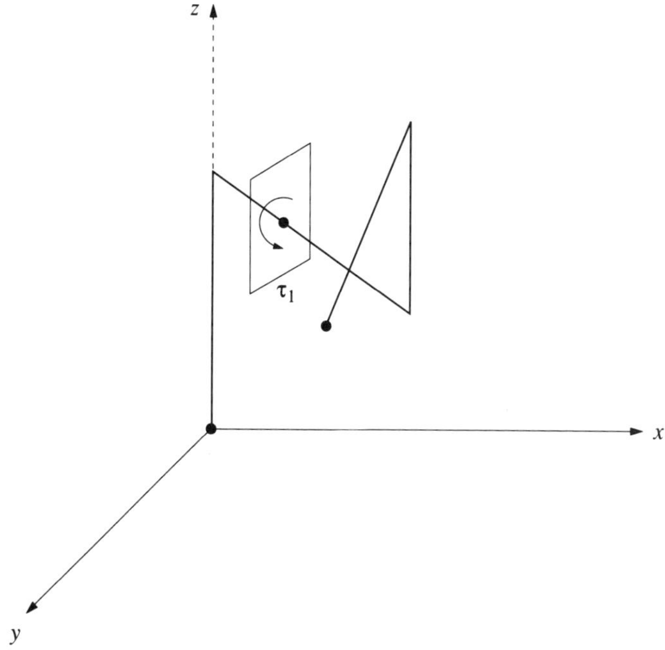

FIGURE 8. Two-link spatial manipulator showing twist angle τ_1

Thus six-joint manipulators are standard "general purpose" industrial robots.

Here is a formal statement of the spatial IPP for six revolute joint manipulators: *Given*:

$d_1, \ldots, d_6$ (offsets),

$a_1, \ldots, a_6$ (link lengths),

$\tau_1, \ldots, \tau_6$ (twist angles),

(p_x, p_y, p_z) (desired hand position),

$$\mathbf{O}_j = \begin{bmatrix} l_x & m_x & n_x \\ l_y & m_y & n_y \\ l_z & m_z & n_z \end{bmatrix} \quad \text{(desired hand orientation)},$$

find $\theta_1, \ldots, \theta_6$ such that

$$\mathbf{A}_1\mathbf{A}_2 \cdots \mathbf{A}_6 = \mathbf{A}_0, \tag{9}$$

where

$$\mathbf{A}_0 = \begin{bmatrix} l_x & m_x & n_x & p_x \\ l_y & m_y & n_y & p_y \\ l_z & m_z & n_z & p_z \\ 0 & 0 & 0 & 1 \end{bmatrix}.$$

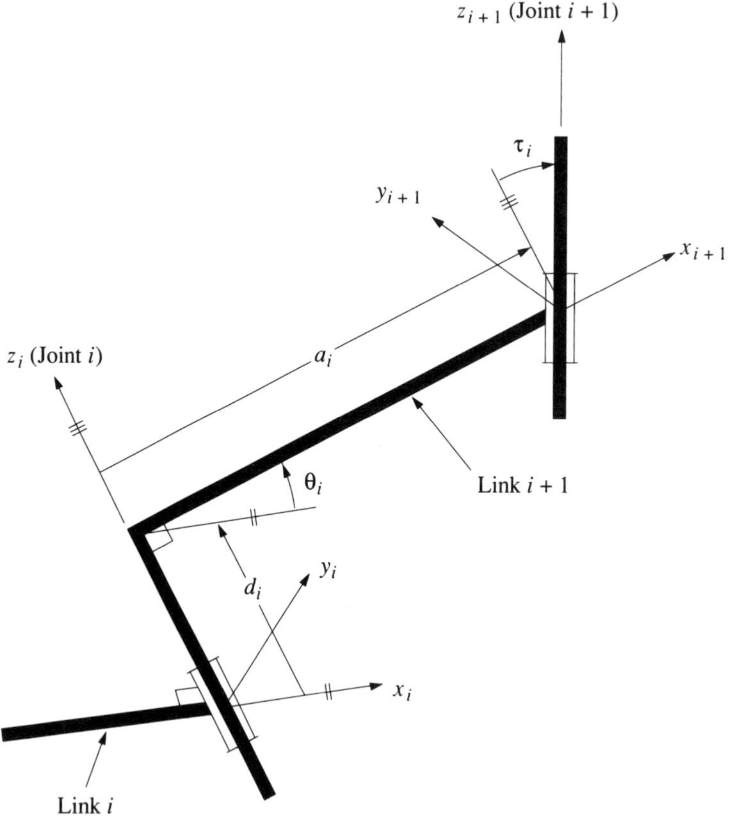

FIGURE 9. Generic manipulator notation. Parallel lines are indicated by tick marks.

Here $\mathbf{O}_j$ is an orthogonal matrix.

As in the planar case, we append the algebraic relations

$$c_j^2 + s_j^2 - 1 = 0 \qquad \text{for } j = 1 \text{ to } 6, \tag{10}$$

and, since (9) consists of 12 polynomial equations in 12 unknowns, we see that we must set aside 6 of these equations as side conditions. We may do this by choosing 6 of the 9 equations in the upper left 3×3 block of (9) to be the side conditions. Then the resulting "main system" consisting of the rest of (9) with (10) will (generally) have only a finite number of solutions. Note that this system consists of 4 sixth-degree, 2 fifth-degree, and 6 second-degree equations, so its total degree is 2,073,600. Obviously, we don't want to deal with the system in this form! In Section 2.1 we replaced (7) by (8) to achieve a reduction in system complexity, and the same idea works here also. Thus, define

$$\mathbf{P} = \mathbf{A}_3\mathbf{A}_4\mathbf{A}_5, \qquad \mathbf{Q} = \mathbf{A}_2^{-1}\mathbf{A}_1^{-1}\mathbf{A}_0\mathbf{A}_6^{-1},$$

and then (9) is equivalent to $\mathbf{P} - \mathbf{Q} = \mathbf{0}$. This equation with (10) yields a system of cubics and quadrics, which has a much lower total degree than (9) with (10). However, we can do better by considering some further corollary relations. (This reduction strategy is used in **[18]** and is discussed in detail in Chapter 10 of **[9]**.) The corollary relation that has proven to be particularly useful is $\mathbf{P}^{\mathrm{T}}\mathbf{P} - \mathbf{Q}^{\mathrm{T}}\mathbf{Q} = \mathbf{0}$, where "T" denotes matrix transpose. In fact, consider

$$(11) \qquad \mathbf{P} - \mathbf{Q} \equiv \begin{bmatrix} \varepsilon_1 & \varepsilon_2 & \varepsilon_3 & \varepsilon_{10} \\ \varepsilon_4 & \varepsilon_5 & \varepsilon_6 & \varepsilon_{11} \\ \varepsilon_7 & \varepsilon_8 & \varepsilon_9 & \varepsilon_{12} \\ 0 & 0 & 0 & 1 \end{bmatrix} = \mathbf{0},$$

and

$$(12) \qquad \mathbf{P}^{\mathrm{T}}\mathbf{P} - \mathbf{Q}^{\mathrm{T}}\mathbf{Q} \equiv \begin{bmatrix} 0 & 0 & 0 & \delta_1 \\ 0 & 0 & 0 & \delta_2 \\ 0 & 0 & 0 & \delta_3 \\ \delta_1 & \delta_2 & \delta_3 & \delta_4 \end{bmatrix} = \mathbf{0}.$$

We observe that from the many polynomial equations in (10), (11), and (12), we may extract the following system of 8 second-degree equations in 8 unknowns (that generally has a finite number of finite solutions):

$$(13) \qquad \begin{aligned} \varepsilon_{11} &= 0, & \varepsilon_{12} &= 0, \\ \delta_3 &= 0, & \delta_4 &= 0, \\ c_j^2 + s_j^2 - 1 &= 0 & &\text{for } j = 1, 2, 4, 5. \end{aligned}$$

We take the rest of the $\varepsilon_j = 0$ and $\delta_j = 0$ as side conditions. (Note that (13) omits the variables c_3, s_3, c_6, and s_6.) It is instructive to look at one of these equations to get a sense of the coefficient and parameter structure of the system. The coefficient structure of δ_3 is given by

$$\begin{aligned} \delta_3 = {} & r_1c_1c_2 + r_2c_1s_2 + r_3s_1c_2 + r_4s_1s_2 \\ & + r_5c_4c_5 + r_6c_4s_5 + r_7s_4c_5 + r_8s_4s_5 \\ & + r_9c_1 + r_{10}s_1 + r_{11}c_2 + r_{12}s_2 \\ & + r_{13}c_4 + r_{14}s_4 + r_{15}c_5 + r_{16}s_5 \\ & + r_{17}, \end{aligned}$$

with the coefficient parameter formulas:

$$
\begin{aligned}
r_1 &= m_x\mu_6a_2 + n_x\lambda_6a_2\\
r_2 &= m_y\mu_6a_2\lambda_1 + n_y\lambda_6a_2\lambda_1\\
r_3 &= m_y\mu_6a_2 + n_y\lambda_6a_2\\
r_4 &= -m_x\mu_6a_2\lambda_1 - n_x\lambda_6a_2\lambda_1\\
r_5 &= -d_3\mu_5\lambda_4\mu_3\\
r_6 &= \mu_5a_3\\
r_7 &= \mu_5\lambda_4a_3\\
r_8 &= d_3\mu_5\mu_3\\
r_9 &= -d_2m_y\mu_6\mu_1 - d_2n_y\lambda_6\mu_1 + m_x\mu_6a_1 + n_x\lambda_6a_1\\
r_{10} &= d_2m_x\mu_6\mu_1 + d_2n_x\lambda_6\mu_1 + m_y\mu_6a_1 + n_y\lambda_6a_1\\
r_{11} &= 0\\
r_{12} &= m_z\mu_6a_2\mu_1 + n_z\lambda_6a_2\mu_1\\
r_{13} &= -d_3\lambda_5\mu_4\mu_3\\
r_{14} &= \lambda_5\mu_4a_3\\
r_{15} &= -d_3\mu_5\mu_4\lambda_3 - d_4\mu_5\mu_4\\
r_{16} &= \mu_5a_4\\
r_{17} &= d_1m_z\mu_6 + d_1n_z\lambda_6 + d_2m_z\mu_6\lambda_1\\
&\quad + d_2n_z\lambda_6\lambda_1 + d_3\lambda_5\lambda_4\lambda_3 + d_4\lambda_5\lambda_4\\
&\quad + d_5\lambda_5 + d_6 - m_xp_x\mu_6 - n_xp_x\lambda_6\\
&\quad - m_yp_y\mu_6 - n_yp_y\lambda_6 - m_zp_z\mu_6 - n_zp_z\lambda_6.
\end{aligned}
$$

It turns out that (13) includes 25 parameters and 59 nonzero coefficients. Its total degree is 256. However, the 2-homogeneous Bezout number associated with the grouping of variables

$$\{c_1, s_1, c_4, s_4\}, \qquad \{c_2, s_2, c_5, s_5\}$$

is 96. Further, the natural coefficient homotopy yields 64 paths (as proven in [**12**]), the parameter homotopy without side conditions yields 32 paths and with side conditions 16 paths. This latter result is best possible, since the problem can have 16 physical solutions.

3. Precision point problems. Precision point problems arise in the design of "motion generators." For example, one might formulate a precision point problem to design a mechanism that converts rotational motion into straight line motion, or more generally, into motion along

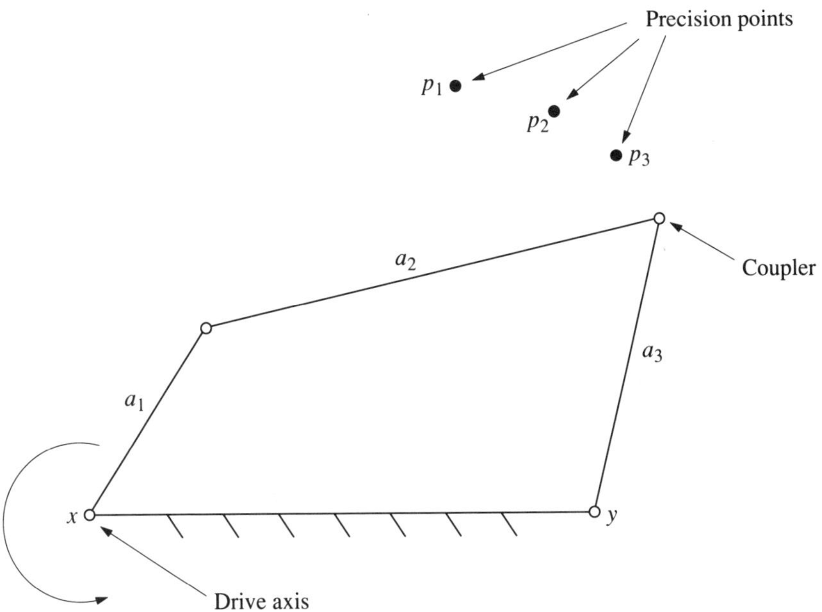

FIGURE 10. Four-bar showing drive axis, coupler, link lengths, and precision points

a specified curve. Precision point problems have the following general structure:

Given:

- the basic structure of a mechanism, with a designated distinguished point (e.g., a "coupler point");
- a set of points in space.

Find:

- each mechanism of the given structure whose distinguished point can pass through the given precision points.

We shall consider several types of precision point problems in this section. In 3.1 we discuss a very simple example, which illustrates the natural way in which solutions at infinity with physical meaning can arise. Section 3.2 (based on material from **[19]**) addresses a problem of more substance. The reference **[15]** also addresses a precision point problem.

3.1. *4-Bars*. A *4-bar* is a mechanism in the plane consisting of four links, with (generally) rotational joints in which the last axis coincides with the first (Fig. 10). As indicated by the hatching in Fig. 10, one

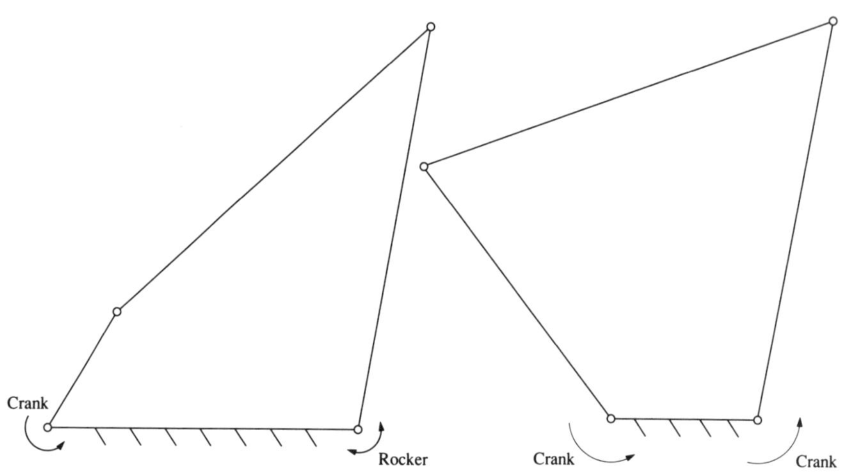

FIGURE 11a. Crank-rocker 4-bar

FIGURE 11b. Crank-crank 4-bar

link is viewed as attached to the *ground*. There is a *drive axis*, which we may imagine, for example, as being driven in a constant circular motion. The *coupler* is to trace out the prescribed motion. A finite number of precision points are specified for the coupler to pass through, the idea being that if the coupler passes through these points, it will approximate the desired motion. The drive axis is a *crank* if it can turn 360°; otherwise, it is a *rocker*. The coupler axis is also designated a crank or a rocker. Grashof's law ([**17**]) formulates simple inequality relations on the link lengths that determine whether the drive axis and coupler are cranks or rockers. See, for example, Fig. 11. A simple precision point problem involving a 4-bar is as follows (refer to Fig. 10):

> *Given*: points $\mathbf{p}_j$ for $j = 1$ to n.
> *Find*: points $\mathbf{x}$ and $\mathbf{y}$, and link lengths a_1, a_2, and a_3 such that the coupler point equals $\mathbf{p}_j$ for some position of the 4-bar for $j = 1$ to n.

For this simple problem we see that:

1. The $\mathbf{p}_j$ fall on a circle of radius a_3 and center $\mathbf{y}$.

2. We can find the other parameters by elementary means once a_3 and $\mathbf{y}$ are determined.

In fact, we can find $\mathbf{y}$ as the intersection of perpendicular bisectors between any two sets of adjacent $\mathbf{p}_j$ (Fig. 12). This shows how solutions at infinity can have physical meaning in precision point problems: If three adjacent precision points are colinear, then $\mathbf{y}$ is at infinity. But this corresponds to a rotational joint of infinite radius, i.e., a sliding joint (Fig. 13).

3.2. *Burmester problems.* Let us consider a less trivial precison point example. Suppose we wish to design a mechanism which will carry a rigid body through a series of given positions. We might do so by finding points of the body that lie on a special curve or surface that can be easily implemented in linkwork, such as a sphere, circle, or line. The study of such special points is called Burmester theory, after the 19th century engineer who first solved the following problem. Suppose we are given a series of positions of a rigid body in the plane (Fig. 14a). If we can find a point of the body that stays on a circle through each of the positions, then we could pin a rigid link to that point and the center of the circle and still pass through the given positions (Fig. 14b). If we do the same for a second point, we will have found a 4-bar linkage that carries the body through the specified positions (Fig. 14c). Burmester's classical result (circa 1876) shows that for five given positions there are at most four such points, any two of which yield a possible 4-bar mechanism. The classical method of determining the number of solutions involves an intricate study of intersections between curves of points which fall on circles for a subset of the given positions, and so on. (See, for example, [**1**], p. 249.) We will show how to obtain these results in a more direct manner. The strength of our approach is further illustrated when the same analysis is applied to a spatial analogue of the problem.

Consider first the planar problem. Let $\mathbf{x}^j$ be the position of a point when the body is in position j. Taking the initial position of the body as our reference, subsequent positions are given by a rotation matrix $\mathbf{R}_j$ and a translation $\mathbf{d}^j$. Then the position of a point at the jth location of the body is given in terms of its initial position as

$$\mathbf{x}^j = \mathbf{R}_j\mathbf{x}^0 + \mathbf{d}^j. \tag{14}$$

Suppose that for each of the given positions the point lies on a circle whose center is $\mathbf{p}$ and whose radius is r. Then we must have

$$(\mathbf{x}^j - \mathbf{p})^2 = r^2. \tag{15}$$

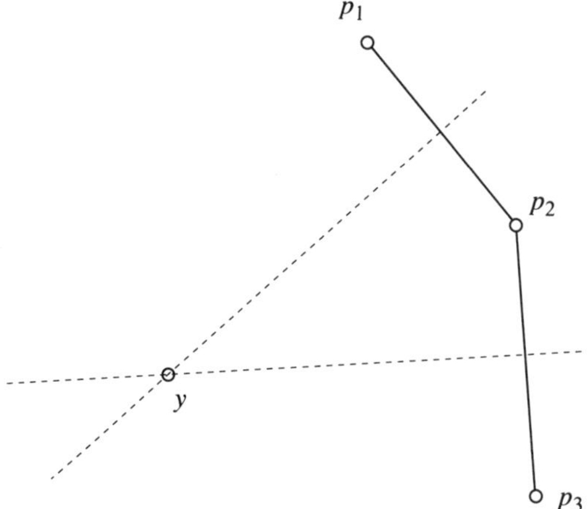

FIGURE 12. Intersection of perpendicular bisectors between two sets of $\mathbf{p}_j$ gives point $\mathbf{y}$.

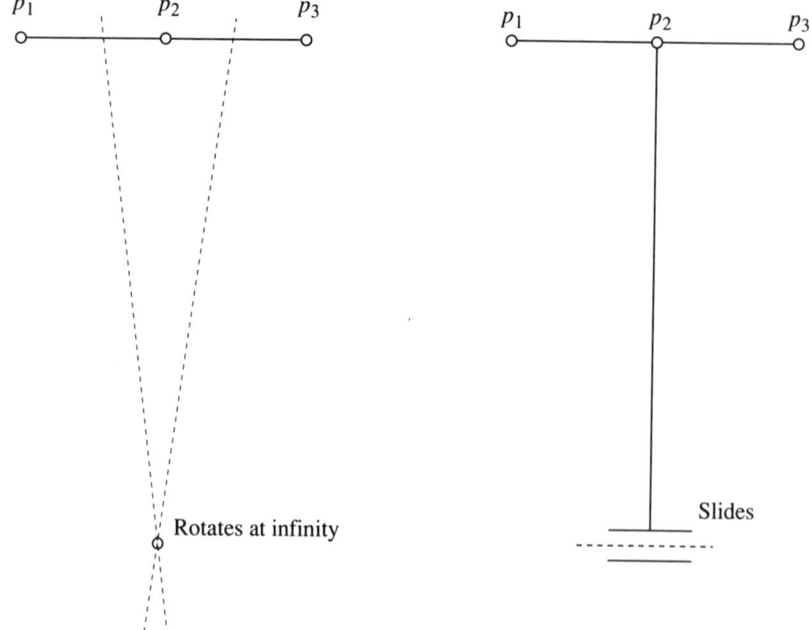

FIGURE 13. Rotational joint of infinite radius is equivalent to a sliding joint.

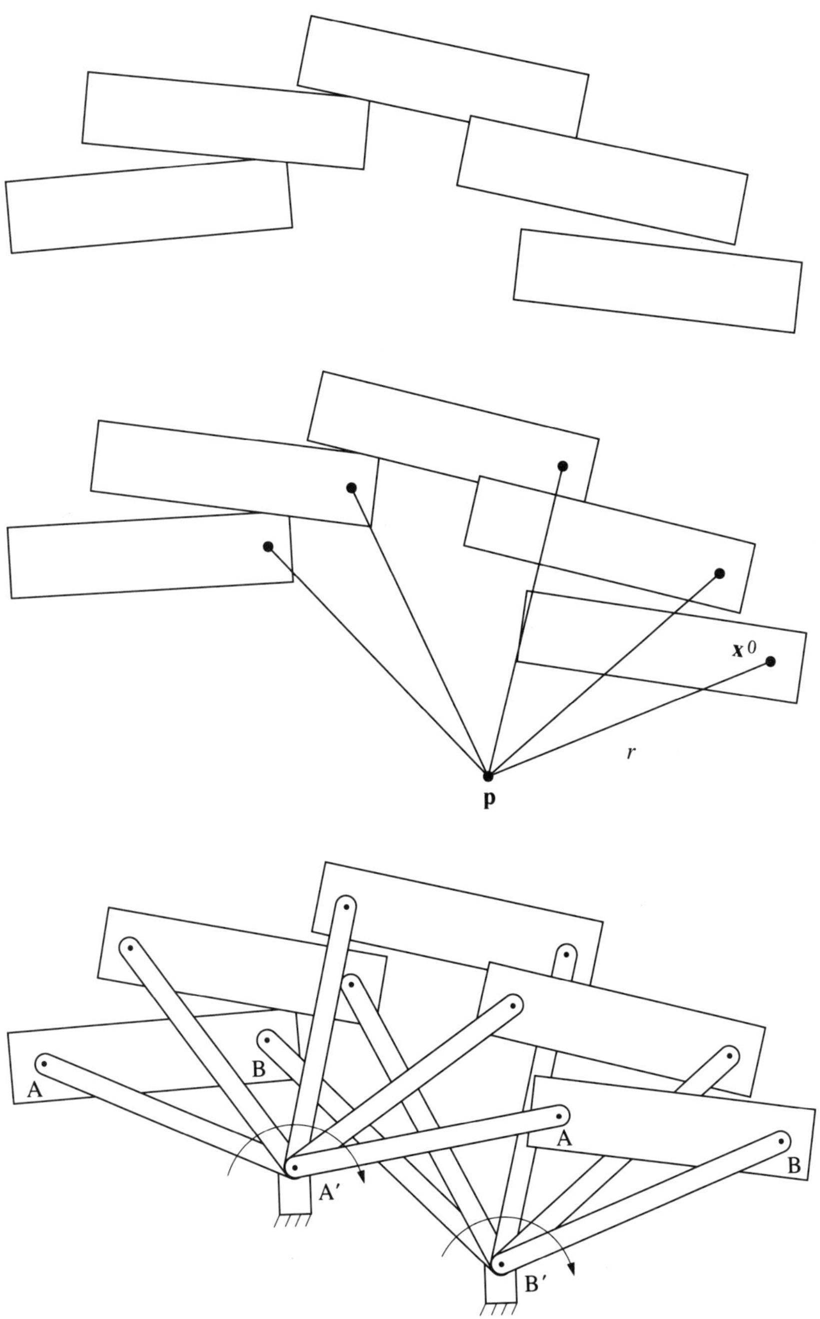

FIGURE 14. Planar Burmester problem (see Section 3.1).

Substituting from (14) into (15), we have a system of equations in 5 variables: r and two components each in $\mathbf{x}^0$ and $\mathbf{p}$. Thus, in general five positions will determine a finite number of solutions, unless the associated polynomial system happens to have positive-dimensional solution sets. In fact, since the equations are quadric, there are at most $2^5 = 32$ isolated solutions. However, we can eliminate r by subtracting (15) for $j = 0$ from the remaining four ($j = 1, \dots, 4$) to obtain

$$(16) \qquad (\mathbf{x}^j - \mathbf{p})^2 - (\mathbf{x}^0 - \mathbf{p})^2 = 0 \qquad (j = 1, \dots, 4).$$

Substituting for $\mathbf{x}^j$ from (14) and noting that $\mathbf{R}_j^{\mathrm{T}}\mathbf{R}_j = \mathbf{I}$, where $\mathbf{I}$ is the identity matrix, we find that all powers of 2 cancel to give

$$(17) \quad (\mathbf{x}^0)^{\mathrm{T}}(\mathbf{I}-\mathbf{R}_j^{\mathrm{T}})\mathbf{p}+(\mathbf{d}^j)^{\mathrm{T}}\mathbf{R}_j\mathbf{x}^0-(\mathbf{d}^j)^{\mathrm{T}}\mathbf{p}+(\mathbf{d}^j)^{\mathrm{T}}\mathbf{d}^j/2 = 0 \quad (j = 1, \dots, 4).$$

Now these are four quadrics, so the total degree is reduced to 16. However, these equations are linear in both $\mathbf{x}^0$ and $\mathbf{p}$, so it is beneficial to treat the system as 2-homogeneous. Introducing homogeneous variables $\mathbf{x}_0^0$ and $\mathbf{p}_0$ associated with $\mathbf{x}^0$ and $\mathbf{p}$, respectively, we get a grouping of the variables into two sets:

$$\{\mathbf{x}_0^0, \mathbf{x}_1^0, \mathbf{x}_2^0\}, \qquad \{\mathbf{p}_0, \mathbf{p}_1, \mathbf{p}_2\},$$

and

$$(18) \quad (\mathbf{x}^0)^{\mathrm{T}}(\mathbf{I} - \mathbf{R}_j^{\mathrm{T}})\mathbf{p} + (\mathbf{d}^j)^{\mathrm{T}}\mathbf{R}_j\mathbf{x}^0\mathbf{p}_0 - (\mathbf{d}^j)^{\mathrm{T}}\mathbf{p}\mathbf{x}_0^0 + \mathbf{x}_0^0\mathbf{p}_0(\mathbf{d}^j)^{\mathrm{T}}\mathbf{d}^j/2 = 0 \qquad (i = 1, \dots, 4).$$

Each of these four equations has degree 1 in each set, so the 2-homogeneous Bezout number is 6. There is still a discrepancy of 2 between this and the classical result of 4, but the difference can be resolved by looking for solutions at infinity. It turns out that since a 2×2 rotation matrix always has the form

$$(19) \qquad \mathbf{R}_j = \begin{pmatrix} c_j & -s_j \\ s_j & c_j \end{pmatrix},$$

the system always has the pair of complex conjugate solutions $(\mathbf{x}_0^0, \mathbf{x}_1^0, \mathbf{x}_2^0; \mathbf{p}_0, \mathbf{p}_1, \mathbf{p}_2) = (0, 1, i; 0, 1, i)$ and $(0, 1, -i; 0, 1, -i)$, independent of the actual positions $\mathbf{R}_j, \mathbf{d}^j$.[2] Therefore, there can be at most 4 isolated real solutions, the so-called Burmester points.

A spatial analog of the preceding planar problem is to find points of a body that fall on spheres for a series of spatial positions of the

[2]Although this case is very simple, we formally eliminate two solutions via the side conditions: $(\mathbf{x}_0^0, \mathbf{x}_1^0, \mathbf{x}_2^0; \mathbf{p}_0, \mathbf{p}_1, \mathbf{p}_2) = (0, 1, \pm i; 0, 1, \pm i)$, demonstrating that the concept of "side conditions" occurs in this example also.

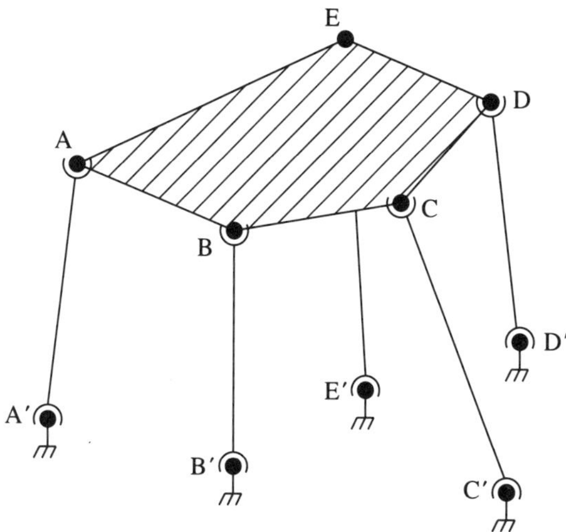

FIGURE 15. Spatial Burmester problem (see Section 3.2).

body. If five such points can be found, a one-degree-of-freedom 7-bar mechanism can be built that carries the body through the specified positions. The spherical motions are obtained using rigid links with ball-and-socket joints to connect each of the five points to their corresponding centers (Fig. 15). The spatial problem can be analyzed exactly as its planar equivalent, except now $\mathbf{x}^0$ and $\mathbf{p}$ are points in three-space, with concomitant modifications in the dimensions of $\mathbf{R}_j$ and $\mathbf{d}^j$. We have initially seven variables r, $\mathbf{x}^0$, and $\mathbf{p}$, so seven positions will determine isolated points. In 1886, Shoenflies first showed that there are at most 20 points with seven positions on a sphere. Both that original proof and the revised one reported by Roth **[16]** depend on a detailed examination of curves and surfaces of points that lie on spheres for fewer than seven positions. We can however show the same result by computing the 2-homogeneous Bezout number for (18), noting that now there are 3 variables in each variable set and the equations run $j = 1, \ldots, 6$. Thus, the Bezout number is 20.

4. Summary and conclusions. Mechanism problems generate (small) polynomial systems. Computing all (real) solutions, including solutions at infinity, is important to the engineering applications. Polynomial continuation is well suited to this problem area, and in particular the technique of coefficient parameter continuation (**[14]**) yields the most efficient variant of this numerical method.

ACKNOWLEDGMENTS. This work represents a cooperative interdisciplinary effort. The order of the names given on this paper has no significance. There is no primary author. The second author would like to thank the General Motors Research Laboratories, the National Science Foundation (DMS 87 22330), and the University of Notre Dame for their support.

REFERENCES

1. O. Bottema and B. Roth, *Theoretical kinematics*, North-Holland, New York, 1979.

2. B. Buchberger, *Gröbner bases: an algorithmic method in polynomial ideal theory*, in *Recent Trends in Multidimensional Systems Theory*, N. K. Bose, Ed., Reidel, 1983.

3. C. B. Garcia and W. I. Zangwill, *Pathways to solutions, fixed points, and equilibria*, Prentice-Hall, Englewood Cliffs, N.J., 1981.

4. E. R. Hansen and S. Sengupta, *Bounding solutions of systems of equations using interval arithmetic*, BIT **21** (1981), 203–211.

5. R. S. Hartenberg and J. Denavit, *Kinematic synthesis of linkages*, McGraw-Hill, 1964.

6. R. B. Kearfott, *Some tests of generalized bisection*, ACM Trans. Math. Software **13** (1987), 197–220.

7. ——, *Interval arithmetic techniques in the computational solution of nonlinear systems of equations: introduction, examples, and comparisons*, Proc. AMS-SIAM Summer Seminars on Computational Solution of Nonlinear Systems of Equations (Colorado State Univ., July 18–29, 1988), AMS Lectures in Applied Mathematics, to appear.

8. R. E. Moore and L. Qi, *A successive interval test for nonlinear systems*, SIAM J. Numer. Anal. **14** (1977), 1051–1065.

9. A. P. Morgan, *Solving polynomial systems using continuation for scientific and engineering problems*, Prentice-Hall, Englewood Cliffs, N.J., 1987.

10. ——, *Polynomial continuation*, in *Impact of Recent Computer Advances on Operations Research*, R. Sharda, Ed., Publications in Operations Research Series, North Holland, to appear.

11. A. P. Morgan and A. J. Sommese, *A homotopy for solving general polynomial systems that respects m-homogeneous structures*, Appl. Math. Comput. **24** (1987), 101–113.

12. ——, *Computing all solutions to polynomial systems using homotopy continuation*, Appl. Math. Comput. **24** (1987), 115–138.

13. ——, *Coefficient parameter polynomial continuation*, Appl. Math. Comput. **29** (1989), 123–160.

14. ——, *Generically nonsingular polynomial continuation*, Proc. AMS-SIAM Summer Seminars on Computational Solution of Nonlinear Systems of Equations (Colorado State Univ., July 18–29, 1988), AMS Lectures in Applied Mathematics, to appear.

15. A. P. Morgan and C. W. Wampler, *Solving a planar 4-bar design problem using continuation*, in Advances in Design Automation—1989: Volume Three, Mechanical Systems Analysis, Design and Simulation (B. Ravan, ed.), ASME, DE-Vol. 19-3, 1989, pp. 409–416.

16. B. Roth, *The kinematics of motion through finitely separated positions*, ASME J. of Applied Mechanics **34E-3** (1967), 591–598.

17. J. Shigley and J. Uicker, *Theory of machines and mechanisms*, McGraw-Hill, 1980.

18. L.-W. Tsai and A. P. Morgan, *Solving the kinematics of the most general six- and five-degree-of-freedom manipulators by continuation methods*, ASME J. of Mechanisms, Transmissions and Automation in Design **107** (1985), 189–200.

19. C. W. Wampler, A. P. Morgan, and A. J. Sommese, *Numerical continuation methods for solving polynomial systems arising in kinematics*, Research Pub. GMR-6372, General Motors Research Laboratories, Warren, Mich. 48090, 1988. Also, to appear in ASME J. Mechanisms, Transmissions, and Automation in Design.

A. P. Morgan and C. W. Wampler: Mathematics Department, General Motors Research Laboratories, Warren, Michigan 48090

A. J. Sommese: Mathematics Department, University of Notre Dame, Notre Dame, Indiana 46556

Lectures in Applied Mathematics
Volume **26** (1990)

A Lagrangian Method for Collisional Kinetic Equations

G. RUSSO

Abstract. A particle method for collisional kinetic equations is presented. It is an extension of the particle schemes commonly used for the Vlasov equation. The general features of the method are described in one and several dimensions, and an application to the diffusion operator is shown.

1. Introduction. The purpose of this paper is to give a new method for treating kinetic equations with a collisional operator on the right-hand side.

The kinetic description of matter has been formulated by physicists to describe a system with a large number of degrees of freedom—for example, a gas or a plasma. It is not possible to study the dynamics of such a system by solving the equations of motion for all the particles, which involves a fairly large number of equations (typical numbers are in the wide range 10^{12}–10^{30}). Under a certain physical hypothesis (assumption of "total disorder" [7]) the system is described in a probabilistic way by means of a density function $f(\mathbf{x}, \mathbf{v}, t)$ in the phase space, which gives the fraction of particles which are in the unit volume around $\mathbf{x}$, $\mathbf{v}$.

The prototype equation for f is of the form

$$\frac{\partial f}{\partial t} + \mathbf{v} \cdot \nabla_{\mathbf{x}} f + \mathbf{F} \cdot \nabla_{\mathbf{v}} f = Q[f]. \tag{1}$$

1980 *Mathematics Subject Classification* (1985 *Revision*). Primary 65M99, 82A40, 60J60.

Partially supported by U.S. Department of Energy.

The second and third terms on the left-hand side are, respectively, the convective and drift terms, and Q is the collisional operator acting on f.

In a particle scheme the function f is approximated by a discrete set of particles, whose density in the phase space is proportional to f. This statement can be formulated more precisely by saying that the absolutely continuous measure μ with density f is approximated by a discrete measure μ_N, according to a certain metric in the space of measures. The partial differential equation for f translates into a system of ordinary differential equations for the particles. The system is then solved and the function f is reconstructed, at desired times, from the particle distribution.

After a description of a typical particle scheme in a collisionless case, we describe the general features of the method for the collisional equation. Then we shall describe the one-dimensional case and, as a specific example, we shall treat the diffusion equation. The relation between linear and quasilinear heat equations is briefly described. The problems arising in more dimensions are discussed and a technique is proposed based on the use of Voronoi diagrams. Applications of the method to other problems are mentioned.

2. General description. Let us rewrite the equation for f in the collisionless case and consider the initial value problem:

$$\frac{\partial f}{\partial t} + \mathbf{v} \cdot \nabla_{\mathbf{x}} f + \mathbf{F} \cdot \nabla_{\mathbf{v}} f = 0, \tag{2}$$

$$f(\mathbf{x}, \mathbf{v}, 0) = f_0(\mathbf{x}, \mathbf{v}). \tag{3}$$

Here $(\mathbf{x}, \mathbf{v}) \in \mathbf{R}^m$, $m = m_x + m_v$, and m_x, $m_v \leq 3$ specify the dimension of the phase space. The equation can be rewritten in a characteristic form:

$$\frac{df}{dt} = 0 \tag{4}$$

along the curves in the phase space:

$$\frac{d\mathbf{x}}{dt} = \mathbf{v}, \qquad \frac{d\mathbf{v}}{dt} = \mathbf{F}. \tag{5}$$

Equations (5) are solved with the initial conditions

$$\mathbf{x}(t = 0) = \mathbf{x}^0, \qquad \mathbf{v}(t = 0) = \mathbf{v}^0,$$

and the solution of (2) is obtained in a parametric form:

$$f(\mathbf{x},\mathbf{v},t) = f_0(\mathbf{x}^0,\mathbf{v}^0),$$
$$\mathbf{x} = \mathbf{x}(t;\mathbf{x}^0,\mathbf{v}^0), \qquad \mathbf{v} = \mathbf{v}(t;\mathbf{x}^0,\mathbf{v}^0).$$

In many cases the field $\mathbf{F}$ is a functional of f or is related to f through an equation and does not depend explicitly on $\mathbf{v}$. A typical case is the Vlasov–Poisson system where $\mathbf{F}$ is the electric field generated by the charge distribution and satisfies the equations

$$\begin{aligned} &\nabla_{\mathbf{x}} \cdot \mathbf{F} = C \int (1 - f(\mathbf{x},\mathbf{v},t))\, d\mathbf{v}, \\ &\nabla_{\mathbf{x}} \times \mathbf{F} = 0. \end{aligned} \tag{6}$$

A "particle solution" of equation (2) is a discrete measure with density of the form

$$f_{(N)}(\mathbf{x},\mathbf{v},t) = \frac{1}{N}\sum_{i=1}^{N} \delta(\mathbf{x} - \mathbf{x}_i(t))\delta(\mathbf{v} - \mathbf{v}_i(t)). \tag{7}$$

This is a solution of (2) provided $\mathbf{x}_i(t)$ and $\mathbf{v}_i(t)$ satisfy equations (5). The approximation of the initial density $f_0(\mathbf{x},\mathbf{v})$ with a distribution of particles $\mathbf{x}_i^0$, $\mathbf{v}_i^0$ ($i = 1, \ldots, N$) could be obtained by extracting N random points in the phase space with a probability proportional to $f_0(\mathbf{x},\mathbf{v})$. However, this method is affected by statistical fluctuations and a large number of points is required in order to obtain a satisfactory approximation. A better accuracy is obtained with the following three steps:

PROCEDURE 1.

(a) *Fill the unit cube* $[0,1]^m$ *with a uniform distribution of particles* (ξ_i, ζ_i). *There are techniques based on number theory, which can be used to obtain an "optimal" distribution* **[15]**.

(b) *Consider a mapping* $T\colon \mathbf{R}^m \to [0,1]^m$ *for which the Jacobian,* $\partial(\xi,\zeta)/\partial(\mathbf{x},\mathbf{v})$, *is equal to the function* $f_0(\mathbf{x},\mathbf{v})$. *It is possible to consider a mapping which is globally invertible in the region where* $f_0(\mathbf{x},\mathbf{v}) > 0$.

(c) *Invert the mapping in the points* (ξ_i, ζ_i) *and determine* $(\mathbf{x}_i^0, \mathbf{v}_i^0)$.

This technique can be used to generate a random variable with an assigned distribution as a function of a random variable with a uniform distribution.

Once the initial locations of the particles are determined, their evolution is described by equations (5):

$$\dot{\mathbf{x}}_i = \mathbf{v}_i, \qquad \dot{\mathbf{v}}_i = \mathbf{F}_i, \tag{8}$$

where $\mathbf{F}_i = \mathbf{F}(\mathbf{x}_i, t)$.

System (8) is then solved with an ordinary differential equation solver. A scheme which is often used is the *leap-frog scheme*:

$$\begin{aligned} \mathbf{v}_i^{n+\frac{1}{2}} &= \mathbf{v}_i^{n-\frac{1}{2}} + \mathbf{F}_i^n \Delta t, \\ \mathbf{x}_i^{n+1} &= \mathbf{x}_i^n + \mathbf{v}_i^{n+\frac{1}{2}} \Delta t, \end{aligned} \tag{9}$$

which is second-order in time and is very fast **[2]**. Typical numbers of particles in actual numerical computations are of the order of 10^4–10^5. One of the main problems in collisionless codes is the computation of $\mathbf{F}_i$. We shall not consider this problem here because our main concern is to treat the collision term. This problem is extensively treated in **[11]**, **[2]**, **[10]**.

Once the evolution of the particle has been computed, the function $f(\mathbf{x}, \mathbf{v}, t)$ (or other macroscopic quantity such as spatial density, momentum, and so on) can be obtained by a suitable "counting" of the particles in a fixed grid. In Figure 1 a counting procedure, the so-called "weighted area rule," is shown in the case of a two-dimensional phase space. The particle is smeared into a rectangle of the same size of the cell and is counted in four neighbor cells, with a fraction proportional to the overlapping area. This is equivalent to taking the convolution of $f_{(N)}$ with some regularizing function $g(\mathbf{x}, \mathbf{v})$. The function corresponding to the "weighted area rule" is, for example:

$$g(x,v) = \begin{cases} \left(1 - \frac{|x|}{\Delta x}\right)\left(1 - \frac{|v|}{\Delta v}\right) & \text{if } |x| < \Delta x, |v| < \Delta v \\ 0 & \text{otherwise.} \end{cases}$$

There are rigorous estimates on the error between the numerical discrete solution and the exact solution, which are based on a proper metric in the space of measures and on the theory of ordinary differential equations **[13]**, **[12]**.

Let us consider now the complete collisional equation. The idea of the method is to modify the equation of motion (5) by adding a term that takes into account collisions:

$$\frac{d\mathbf{x}}{dt} = \mathbf{v} + \mathbf{p}, \qquad \frac{d\mathbf{v}}{dt} = \mathbf{F} + \mathbf{q}. \tag{10}$$

An expression of $\mathbf{q}$ and $\mathbf{p}$ is obtained in the following way. We look for a transformation

$$\mathbf{x} = \varphi(t, \mathbf{x}^0, \mathbf{v}^0), \qquad \mathbf{v} = \psi(t, \mathbf{x}^0, \mathbf{v}^0), \tag{11}$$

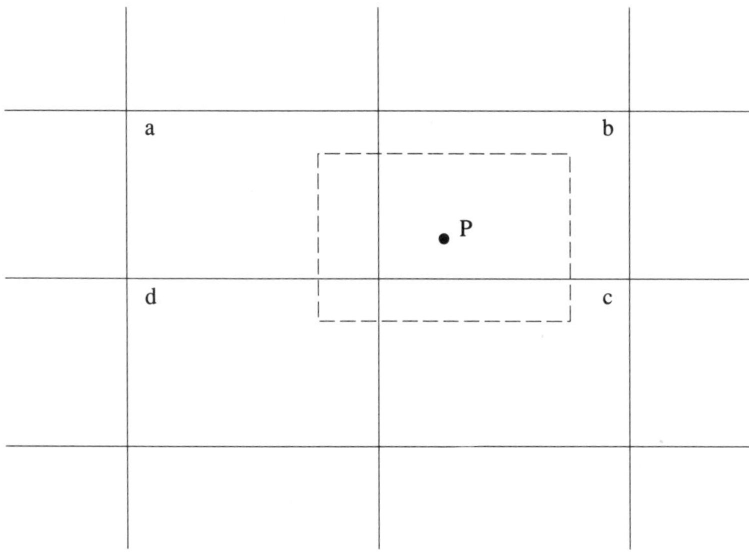

FIGURE 1. Weighted area rule: The particle P is counted in the four cells a, b, c, d with a fraction proportional to the overlapping between the dashed rectangle and the cells.

such that, for each volume V moving according to this transformation with $\mathbf{x}^0$, $\mathbf{v}^0$ constant, the integral of f is constant, i.e.,

$$\frac{d}{dt}\int_{V(t)} f(\mathbf{x},\mathbf{v},t)\,d\mathbf{x}\,d\mathbf{v} = 0. \tag{12}$$

Performing the derivative in (12) yields

$$\int_V \left[\frac{\partial f}{\partial t} + \nabla_{\mathbf{x}}\cdot\left(f\frac{\partial\varphi}{\partial t}\right) + \nabla_{\mathbf{v}}\cdot\left(f\frac{\partial\psi}{\partial t}\right)\right] d\mathbf{x}\,d\mathbf{v} = 0. \tag{13}$$

Combining this with equation (1), we obtain

$$\nabla_{\mathbf{x}}\cdot\left[\left(\mathbf{v}-\frac{\partial\varphi}{\partial t}\right)f\right] + \nabla_{\mathbf{v}}\cdot\left[\left(\mathbf{F}-\frac{\partial\psi}{\partial t}\right)f\right] = Q[f].$$

The general solution of this equation is given by

$$\frac{\partial\varphi}{\partial t} = \mathbf{v} + \frac{1}{f}(\nabla_{\mathbf{x}}\times\mathbf{A}+\mathbf{h}), \tag{14}$$

$$\frac{\partial\psi}{\partial t} = \mathbf{F} + \frac{1}{f}(\nabla_{\mathbf{v}}\times\mathbf{B}+\mathbf{k}), \tag{15}$$

where $\mathbf{A}$ and $\mathbf{B}$ are arbitrary vectors and ($\mathbf{h}$, $\mathbf{k}$) satisfy the equations

$$\nabla_{\mathbf{x}}\cdot\mathbf{h} + \nabla_{\mathbf{v}}\cdot\mathbf{k} = -Q[f], \tag{16}$$

(17) $$\nabla_{\mathbf{x}} \times \mathbf{h} = 0, \qquad \nabla_{\mathbf{v}} \times \mathbf{k} = 0.$$

Equations (16–17) do not uniquely define the two fields **h** and **k**. A particular choice is, for example,

(18) $$\mathbf{h} = 0, \qquad \nabla_{\mathbf{v}} \cdot \mathbf{k} = -Q[f].$$

Different choices of the fields **A**, **B**, **h**, **k** give different parametrizations of the solution. Equations (14–17) define a transformation of the form (11) that has the property (12). The function $f(\mathbf{x}, \mathbf{v}, t)$ is related to the transformation (11) by

$$f(\mathbf{x}, \mathbf{v}, t) = f(\mathbf{x}^0, \mathbf{v}^0, 0)/J(\mathbf{x}^0, \mathbf{v}^0, t),$$

where

$$J = \left| \frac{\partial(\varphi, \psi)}{\partial(\mathbf{x}^0, \mathbf{v}^0)} \right|$$

is the Jacobian of the transformation. In the case of no boundary in the velocity space, the solution to equations (18) can be expressed in terms of Green's functions of the divergence operator:

$$k = -\int_{-\infty}^{v} Q(\mathbf{x}, v', t)\, dv',$$

$$\mathbf{k} = \frac{1}{2\pi} \int_{\mathbf{R}^2} \frac{Q(\mathbf{x}, \mathbf{v}', t)(\mathbf{v} - \mathbf{v}')}{|\mathbf{v} - \mathbf{v}'|^2}\, d\mathbf{v}',$$

$$\mathbf{k} = \frac{1}{4\pi} \int_{\mathbf{R}^3} \frac{Q(\mathbf{x}, \mathbf{v}, t)(\mathbf{v} - \mathbf{v}')}{|\mathbf{v} - \mathbf{v}'|^3}\, d\mathbf{v}',$$

in the case, respectively, of one, two, and three dimensions in velocity space. Here $Q[f]$ is considered a function of $(\mathbf{x}, \mathbf{v}, t)$.

3. The one-dimensional, space-homogeneous case. In one dimension the procedure is easily described in terms of a Lagrangian formulation of the problem. Let us concentrate on the collisional term and consider the initial value problem for a function $f(x, t)\colon \mathbf{R} \times [0, T] \to \mathbf{R}^+$:

(19) $$\frac{\partial f}{\partial t} = Q[f],$$

(20) $$f(x, 0) = f_0(x),$$

and suppose that the initial condition is integrable:

$$\int_{-\infty}^{+\infty} f_0(x)\, dx = C.$$

We define a Lagrangian coordinate X as

$$X \equiv \int_{-\infty}^{x} f(y,t)\,dy. \tag{21}$$

Under the assumption that $f(x, t) > 0$, relation (21) defines x as a function of (X, t). Let us consider now X and t as independent coordinates and perform the derivative of (21) with respect to t and X. This gives

$$0 = \int_{-\infty}^{x} \frac{\partial f}{\partial t}(y,t)\,dy + f(x,t)\frac{\partial x}{\partial t}, \tag{22}$$

$$1 = f(x,t)\frac{\partial x}{\partial X}. \tag{23}$$

Making use of equation (19) for f gives the equation of motion for the Eulerian coordinate:

$$\frac{\partial x}{\partial t} = -\frac{\partial x}{\partial X}\int_{-\infty}^{x} Q[f](y,t)\,dy. \tag{24}$$

This equation can be discretized to obtain the equation of motion of "particles." We divide the initial range of the Lagrangian coordinate by N^0 equally spaced points:

$$X_i = C\frac{2i-1}{2N^0}, \qquad i = 1, \ldots, N^0.$$

Then the initial position of the particles is defined by

$$X_i = \int_{-\infty}^{x_i^0} f_0(x)\,dx.$$

The Jacobian of the transformation is approximated as

$$\frac{\partial x}{\partial X} \approx (x_{i+1} - x_{i-1})\frac{N^0}{2C}. \tag{25}$$

The discretization of the collisional term depends on the particular expression of the collisional operator. If Q is a linear integral operator of the form

$$Q[f] = \int_{-\infty}^{+\infty} K(x,y)f(y,t)\,dy,$$

the integral term on the right-hand side of (24) is computed with the substitution

$$f(x,t) \rightarrow \frac{C}{N}\sum_{i=1}^{N}\delta(x - x_i(t)).$$

If the collisional operator has the property

$$\int_{-\infty}^{+\infty} Q[h]\,dx = 0$$

for every function $h(x)$ of the proper class, the integral of f is conserved and the total number of particles is constant. If this is not the case then the range of the Lagrangian coordinate X depends on time and, consequently, the number of particles is not constant, but particles are created or annihilated at the boundary during the computation. The creation-annihilation procedure allows the treatment of initial-boundary value problems associated with equation (19), as described in **[17]**.

As a specific example we shall treat the diffusion equation. This is an important application because it constitutes a deterministic method to diffuse particles and, in one dimension, it is a good test problem for the method. This technique could be applied to treat the diffusion term in the Fokker–Planck equation in plasma physics and to diffuse the "vorticity particles" in the vortex method applied to the Prandtl equations **[8]**. Consider the equation

$$\frac{\partial f}{\partial t} = \frac{\partial}{\partial x}\left(\nu \frac{\partial f}{\partial x}\right), \tag{26}$$

where $\nu(x)$ is the diffusion coefficient. Equation (24) becomes

$$\frac{\partial x}{\partial t} = -\frac{\nu}{f}\frac{\partial f}{\partial x}, \tag{27}$$

and is supplemented by equation (23):

$$\frac{\partial x}{\partial X} = \frac{1}{f}.$$

The transformation between Lagrangian and Eulerian coordinates gives, for constant ν, the relation between the heat equation and the quasilinear heat equation:

$$\frac{\partial \tilde{f}}{\partial t} = \nu \tilde{f}^2 \frac{\partial^2 \tilde{f}}{\partial X^2}, \tag{28}$$

where $\tilde{f}(X,t) \equiv f(x(X,t),t)$. This is obtained by differentiating (27) with respect to X. Note that the domain of $\tilde{f}(\cdot,t)$ is the interval $(0,C)$. This transformation is a particular case of the Bäcklund transformation and provides the exact parametric solution for equation (28).

Equation (27) can be discretized by a finite difference scheme in X. A second-order discretization is given by

$$\dot{x}_i = \nu_i \left(\frac{1}{x_i - x_{i-1}} - \frac{1}{x_{i+1} - x_i}\right). \tag{29}$$

This system describes the one-dimensional motion of particles as a first-order dynamic in which the neighboring particles repel each other with a force inversely proportional to the distance. The system may be closed by setting $x_0 = -\infty$, $x_{N+1} = +\infty$.

It is remarkable that the solution of this system maintains some properties of the solution of the heat equation even with only two particles. Consider the following initial condition:

$$N = 2, \quad x_1 = -1, \quad x_2 = 1.$$

Solving the equation of motion, we obtain the expression for the distance of the particles as a function of time:

$$x_2 - x_1 = 2\sqrt{\nu t},$$

which is exactly the distance between two points under which the area is constant in the exact solution of the heat equation with an initial Gaussian profile.

We mention here some properties of system (29) with the periodic conditions

$$x_0 = x_N - 1, \qquad x_{N+1} = x_0 + 1.$$

(1) The minimum length of the intervals $x_i - x_{i-1}$ is a nondecreasing function of time.

(2) The maximum length of the intervals $x_i - x_{i-1}$ is a nonincreasing function of time.

(3) The uniform distribution, $x_i = iC/N$, $i = 1, \dots, N$, is a stable solution of (29).

(4) System (29) is consistent with (27) and the accuracy is second-order in $1/N$.

The stability of point (3) is asymptotic stability and is proved by constructing a Lyapunov function for the system, which is the discrete version of

$$\int_{-\infty}^{+\infty} f^2(x,t)\, dx.$$

The result on consistency can be stated more precisely as follows. Let $\hat{x}_i \equiv x(X_i, t)$ denote the solution of (27) at (X_i, t) and

$$R_i \equiv \frac{1}{\hat{x}_i - \hat{x}_{i-1}} - \frac{1}{\hat{x}_{i+1} - \hat{x}_i} + \frac{1}{f}\frac{\partial f}{\partial x}(\hat{x}_i, t).$$

Under the assumptions

$$\frac{1}{\widetilde{C}} < f < \widetilde{C}, \quad |f_x| < \widetilde{C}, \quad |f_{xx}| < \widetilde{C}, \quad |f_{xxx}| < \widetilde{C},$$

with $\widetilde{C}$ a positive number, the following estimate holds:

$$\max_{1\le i\le N}|R_i| < \frac{15}{4}\widetilde{C}^{16}\frac{C^2}{N^2} + \frac{13}{6}\widetilde{C}^{17}\frac{C^3}{N^3},$$

where $C \equiv \int_0^1 f(x,t)\,dx$. The proofs of these properties are in **[17]**.

We shall consider now some simple numerical scheme for the system (29). As we shall see, explicit schemes suffer a limitation on the time step due to stability conditions. This problem can be easily overcome with implicit or linearly implicit schemes.

Let us consider first the forward Euler scheme:

$$(30)\qquad x_i^{n+1} = x_i^n + \Delta t\left(\frac{1}{x_i^n - x_{i-1}^n} - \frac{1}{x_{i+1}^n - x^n + i}\right).$$

We state the results of a linear stability analysis. Let $\{\mathring{x}_i^n\}$ and $\{x_i^n\}$ be two solutions of system (30). The linearized equation for the "error" $e_i^n \equiv x_i^n - \mathring{x}_i^n$ is

$$(31)\qquad e_i^{n+1} = e_i^n + \Delta t\left[\frac{e_{i+1}^n - e_i^n}{(\mathring{x}_{i+1}^n - \mathring{x}_i^n)^2} - \frac{e_i^n - e_{i-1}^n}{(\mathring{x}_i^n - \mathring{x}_{i-1}^n)^2}\right].$$

This may be written in a vector form:

$$\mathbf{e}^{n+1} = \mathbf{A}\mathbf{e}^n,$$

where $\mathbf{A}$ is an $N\times N$ matrix. By applying the Gershgorin theorem (which gives bounds to the eigenvalues of a matrix) to $\mathbf{A}$, one obtains a sufficient condition for linear stability:

$$\Delta t \le \frac{1}{2}\min_j(\mathring{x}_{j-1}^n - \mathring{x}_j^n)^2.$$

This condition is analogous to the von Neumann condition for the finite difference scheme **[16]**.

Consider now the implicit scheme for system (29):

$$(32)\qquad x_i^{n+1} = x_i^n + \Delta t\left(\frac{1}{x_i^{n+1} - x_{i-1}^{n+1}} - \frac{1}{x_{i+1}^{n+1} - x_i^{n+1}}\right)\nu_i^{n+1}.$$

This scheme requires the solution of a nonlinear system of equations. A linearized version of scheme (32) is given by

$$(33)\qquad x_i^{n+1} = x_i^n + \nu_i^n\frac{x_{i+1}^{n+1} - 2x_i^{n+1} + x_{i-1}^{n+1}}{(x_i^n - x_{i-1}^n)(x_{i+1}^n - x_i^n)}\Delta t.$$

Both these schemes are unconditionally stable, as is easy to prove by the use of the "maximum principle" **[16]**. They are first-order accurate in time.

4. The one-dimensional, space-nonhomogeneous case. In this case equation (1) is

$$\frac{\partial f}{\partial t} + v\frac{\partial f}{\partial x} + F\frac{\partial f}{\partial \varphi} = Q. \tag{34}$$

The equations of motion of the points are

$$\dot{x} = v,$$
$$\dot{v} = F(x, v, t) - \frac{1}{f}\int_{-\infty}^{v} Q[f](x, w, t)\, dw.$$

The discretization of the equations is based on the consideration that $Q[f]$ acts only in velocity space; therefore, particles at different positions do not interact. We use the following scheme:

Step 1. Initialize particle positions and velocities $\{x_i, v_i\}$.

Step 2. Divide x-space into K strips $S_1, \ldots, S_K$.

Step 3. Compute the collisional contribution to the acceleration, solving the space-homogeneous problem in each strip.

Step 4. Update the velocity of the "split particles" in the strips.

Step 5. Compute the quantities that are necessary to evaluate F_i (for example, the spatial density in each strip related to the electric field via the Poisson equation in the Vlasov–Poisson system).

Step 6. Compute the acceleration of each particle F_i.

Step 7. Update the particle velocities and positions.

As a simple application we consider the equation

$$\frac{\partial f}{\partial t} + v\frac{\partial f}{\partial x} = \frac{\partial^2 f}{\partial v^2}, \tag{35}$$

which is a model of a kinetic equation with diffusion in velocity. We look for $f(x, v, t)$: $[0, 1] \times \mathbf{R} \times \mathbf{R}^+ \to \mathbf{R}$ periodic in x. An exact solution of (35) is given by

$$\begin{aligned} f(x, v, t) &= \frac{1}{2\sqrt{\pi(t+t_0)}} \exp\left[-\frac{v^2}{4(t+t_0)}\right] \\ &\times \left(1 - B\exp\left[-\frac{(2\pi l)^2(t+t_0)^3}{12}\right]\cos\left[2\pi l\left(x - v\frac{t+t_0}{2}\right)\right]\right), \end{aligned} \tag{36}$$

where l is an integer. The numerical solution is obtained by dividing the x-space into K strips and solving the equations

$$\dot{x}_{k,i} = v_{k,i},$$
$$\dot{v}_{k,i} = \frac{1}{v_{k,i} - v_{k,i-1}} - \frac{1}{v_{k,i+1} - v_{k,i}},$$

where now i is the index of the ith particle (ordered in velocity) in the strip k. Note that the indexing changes at each time step. The equation for $v_{k,i}$ is solved using the linearly implicit scheme (33):

$$v_{k,i}^{n+1} = v_{k,i}^{n} + v_{k,i}^{n} \frac{v_{k,i+1}^{n+1} - 2v_{k,i}^{n+1} + v_{k,i-1}^{n+1}}{(v_{k,i+1}^{n+1} - v_{k,i}^{n+1})(v_{k,i}^{n+1} - v_{k,i-1}^{n+1})} \Delta t. \tag{37}$$

The comparison with the exact solution is shown in Figures 2 and 3, where the exact solution is obtained by applying Procedure 1 to (36). The parameters used in the computation are $t_0 = 0.1$, $l = 2$, $N = 987$, $B = 0.9$, and the number of strips is $K = 30$.

5. The multi-dimensional case. In order to discretize the equations (14–15) we need to compute a discrete approximation of $f(\mathbf{x}_i, \mathbf{v}_i, t)$ from the particle distribution. This is one of the basic questions in particle methods and still there is no satisfactory answer. The problem can be stated as follows: Given a set of points that are generated from a continuous distribution $f_0(\mathbf{x}, \mathbf{v})$ (for example, following Procedure 1), associate with each point a "local density of points" which is as close as possible to $f_0(\mathbf{x}, \mathbf{v})$.

The "weighted area rule" or other smoothing procedures obtained by performing the convolution of $f_{(N)}(\mathbf{x}, \mathbf{v}, t)$ with some function $g(\mathbf{x}, \mathbf{v})$ have the disadvantage that they smooth too much in regions with a large density of points (with the consequence of losing a lot of information) and they give an extremely poor accuracy in regions where the points are sparse.

Here we propose an alternative method for the computation of f_i, based on a particular geometrical structure associated with the set of points, namely the *Voronoi diagrams* **[3]**, **[5]**.

Given a set $S = \{\mathbf{x}_i \in \mathbf{R}^2,\ i = 1, \ldots, N\}$ of points in the plane, to each point is associated a convex polygon defined by

$$P_i \equiv \{\mathbf{x} \in \mathbf{R}^2 \colon |\mathbf{x} - \mathbf{x}_i| \leq |\mathbf{x} - \mathbf{x}_j|, \forall j = 1, \ldots, N\}.$$

It follows from the definition that each edge of a polygon stays on the axial line of the segment joining a pair of points of S. The diagrams can be used to compute the "local density of points." From property (12) it follows that the product of the function and the element of volume is constant along the particle path; then we define f_i in such a way that

$$f_i V_i = \text{const}. \tag{38}$$

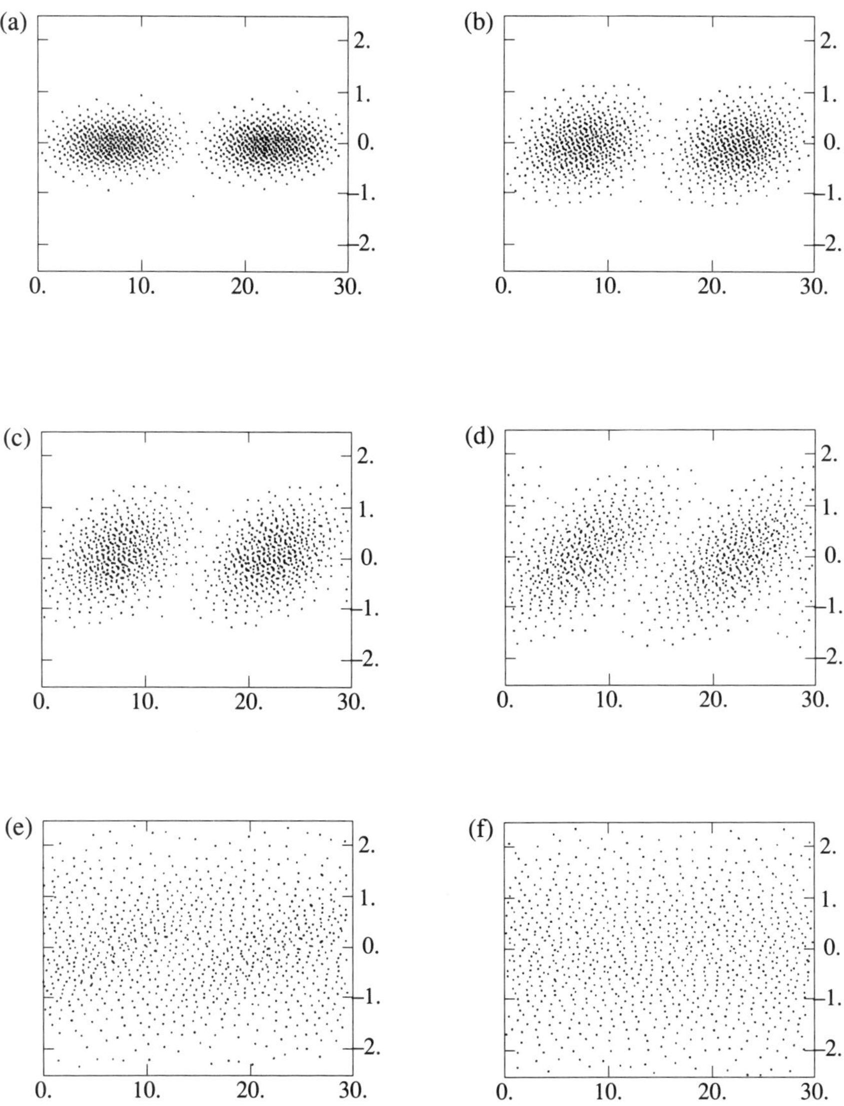

FIGURE 2. Numerical results for the model Fokker–Planck equation at different times: (a) $t = 0$; (b) $t = 0.06$; (c) $t = 0.1$; (d) $t = 0.2$; (e) $t = 0.44$; (f) $t = 0.6$.

The Voronoi mesh allows the computation of a discrete approximation of differential operators on an irregular grid and has been used in Lagrangian fluid dynamics codes by several authors **[5]**, **[4]**, **[1]**.

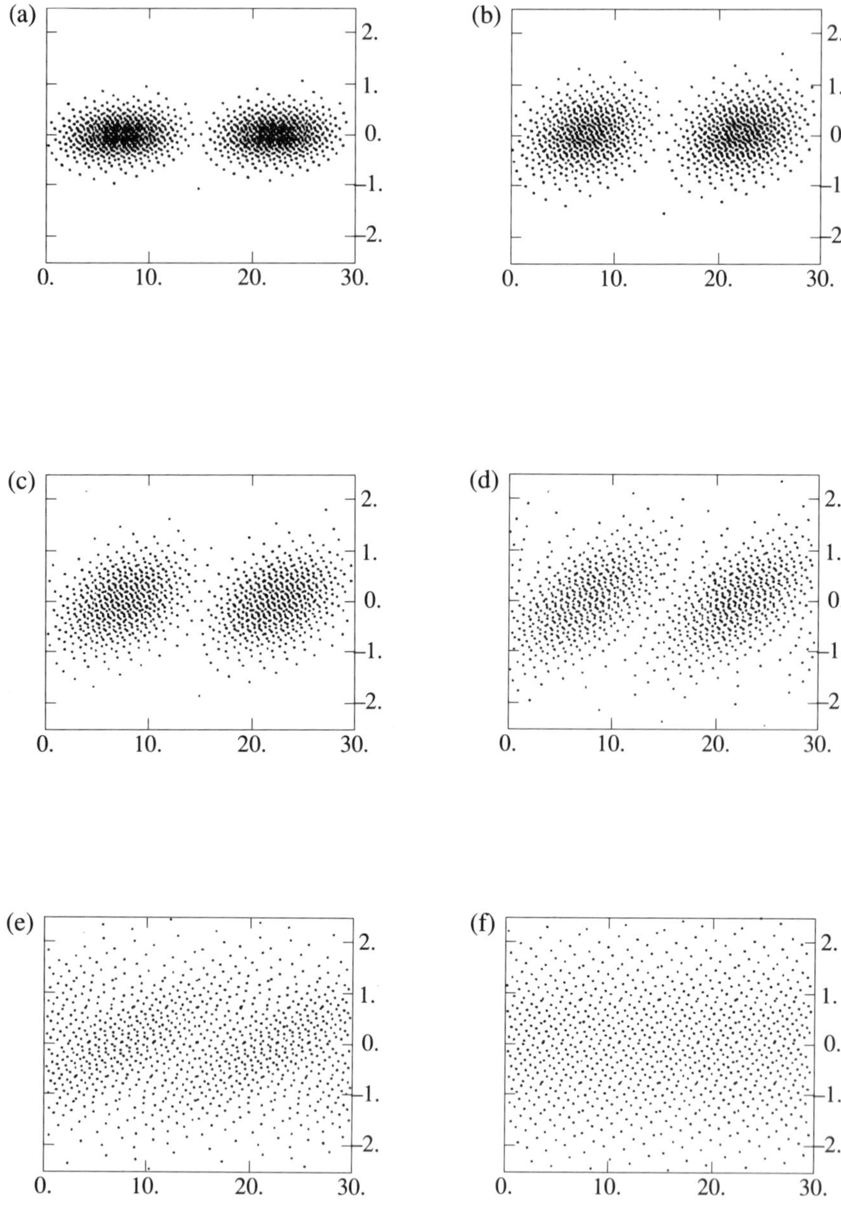

FIGURE 3. Particle locations corresponding to the exact solution (36) of the model Fokker–Planck equation. The points are obtained using Procedure 1. The times are the same as in Figure 2.

There are fast parallel algorithms for the computation of these diagrams, as well as algorithms which update a Voronoi grid for a set of moving points in $O(N)$ operations (see **[5]** and references).

Voronoi diagrams are also used as a tool to construct the Delaunay triangulation associated with the set S, which is a useful triangulation for finite element methods.

We describe how the differential operators on irregular grids can be approximated by the use of a Voronoi mesh.

For $\mathbf{x}_i \in S$ let A_{ij} denote the length of the common edge of P_i and P_j, and $\mathbf{c}_{ij}$ the middle point of the same edge. $A_{ij} = 0$ if P_i and P_j are not adjacent to each other.

The discrete Laplacian L is defined by

$$V_j(L\varphi)(\mathbf{x}_j) = \sum_{i \neq j} A_{ij} \frac{\varphi(\mathbf{x}_i) - \varphi(\mathbf{x}_j)}{|\mathbf{x}_i - \mathbf{x}_j|}.$$

This definition is motivated by the relation

$$\int_P \Delta\varphi(\mathbf{x})\, d\mathbf{x} = \int_{\partial P} \frac{\partial \varphi}{\partial \mathbf{n}}\, ds.$$

The discrete divergence D is defined by

$$V_j(D\mathbf{u})(\mathbf{x}_j) = \sum_{i=1}^{N} \frac{\partial V_j}{\partial \mathbf{x}_i} \cdot \mathbf{u}_i. \tag{39}$$

The motivation of this definition is the following. The continuity equation for the element of volume of a field flow is

$$\frac{dJ}{dt} = J\nabla \cdot \mathbf{u},$$

which, in discrete terms, leads to the definition

$$\frac{dV_j}{dt} = V_j(D\mathbf{u})(\mathbf{x}_j),$$

which is equivalent to (39).

The discrete gradient G is defined as the negative adjoint of D with respect to the discrete L^2 inner product:

$$(\varphi, \psi) \equiv \sum_{j=1}^{N} \varphi_j \psi_j V_j,$$

where $\varphi_j \equiv \varphi(\mathbf{x}_j)$. This leads to the formula

$$V_j(G\varphi)(\mathbf{x}_j) = -\sum_{i=1}^{N} \frac{\partial V_i}{\partial \mathbf{x}_j} \varphi_i. \tag{40}$$

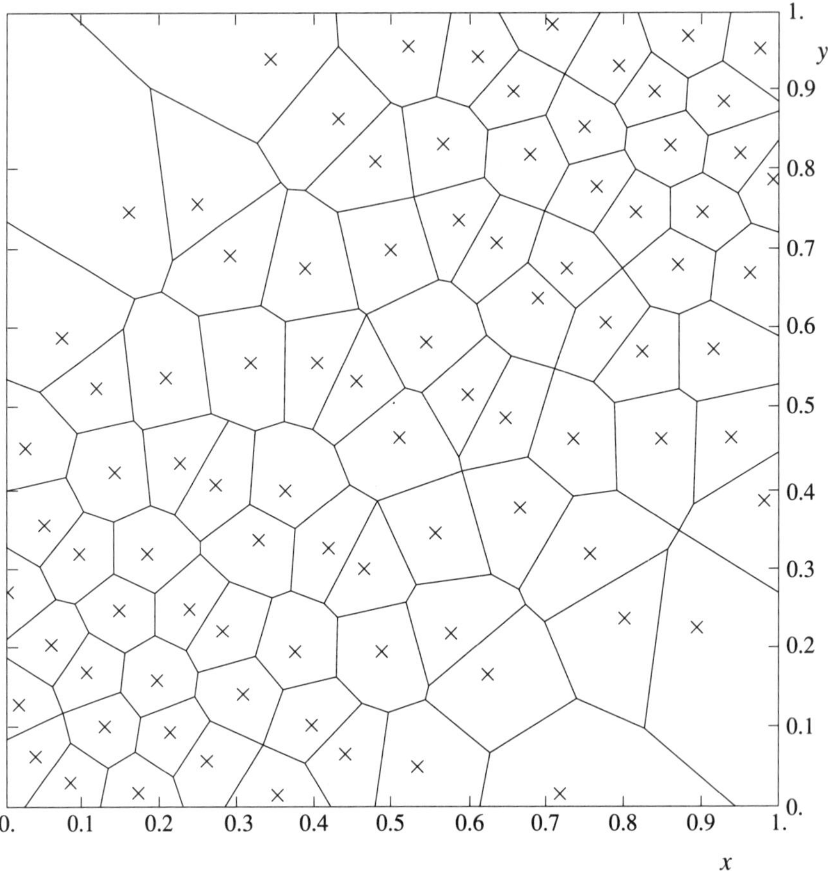

FIGURE 4a. Initial location of the particles (×) and corresponding Voronoi diagram

It can be proved that all these three operators are weakly consistent to first order to the corresponding differential operators, but they are not pointwise consistent [**5**].

We consider now the specific case of the heat equation

$$\frac{\partial f}{\partial t} = \nabla^2 f.$$

The equation of motion (14) (with $\mathbf{A} = 0$) becomes

$$\dot{\mathbf{x}} = -\frac{\nabla f}{f}. \tag{41}$$

The discretization of this equation using the Voronoi mesh gives

$$\dot{\mathbf{x}}_i = \sum_{j=1}^{N} \frac{1}{V_j} \frac{\partial V_j}{\partial \mathbf{x}_i}.$$

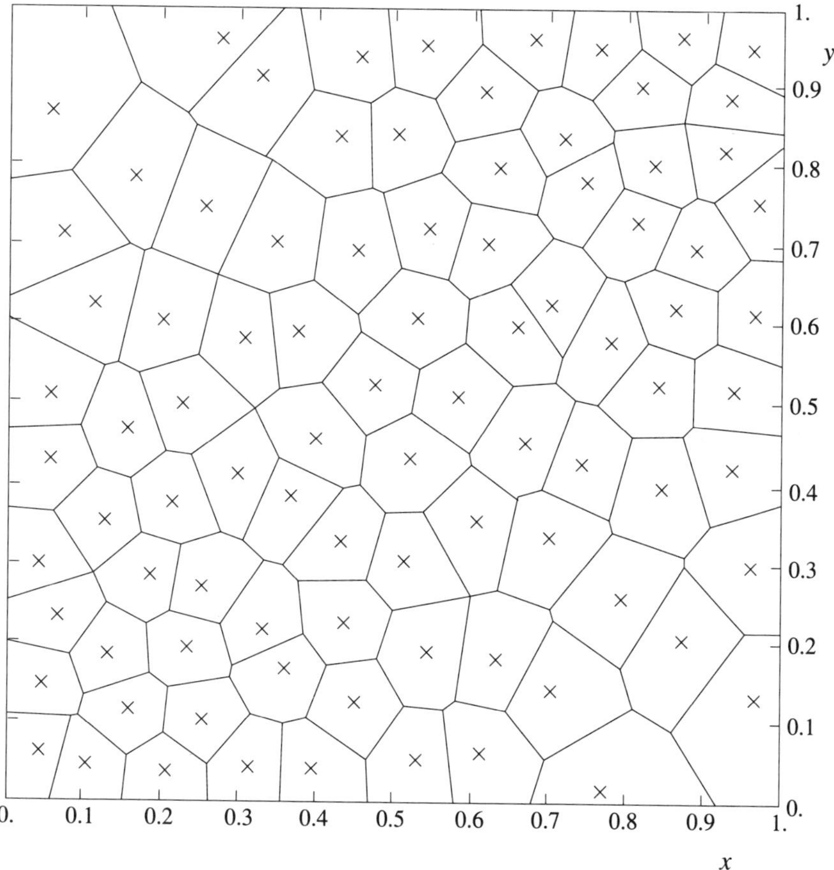

FIGURE 4b. Particle distribution at time $t = 0.028$

There are formulas that give an explicit expression of $(\partial V_j/\partial \mathbf{x}_i)$ as a function of the position of the points $\mathbf{x}_i$ and the corners of the diagram [5]:

$$\frac{\partial V_j}{\partial \mathbf{x}_i} = \frac{\mathbf{x}_i - \mathbf{c}_{ij}}{|\mathbf{x}_i - \mathbf{x}_j|} A_{ij}$$

if $i \neq j$ and

$$\frac{\partial V_i}{\partial \mathbf{x}_i} = -\sum_{j \neq i} \frac{\partial V_j}{\partial \mathbf{x}_i} = -\sum_{j \neq i} \frac{\partial V_i}{\partial \mathbf{x}_j}.$$

The final expression for the velocity of the points is

$$\dot{\mathbf{x}}_i = \sum_{j \neq i} \left(\frac{1}{V_j} - \frac{1}{V_i} \right) \frac{\mathbf{x}_i - \mathbf{c}_{ij}}{|\mathbf{x}_i - \mathbf{x}_j|} A_{ij} \equiv \mathcal{F}_i(\{\mathbf{x}_j\}). \tag{42}$$

In Figure 4 we show a sequence of Voronoi diagrams corresponding to the numerical solution of (42) at different times. The initial condition

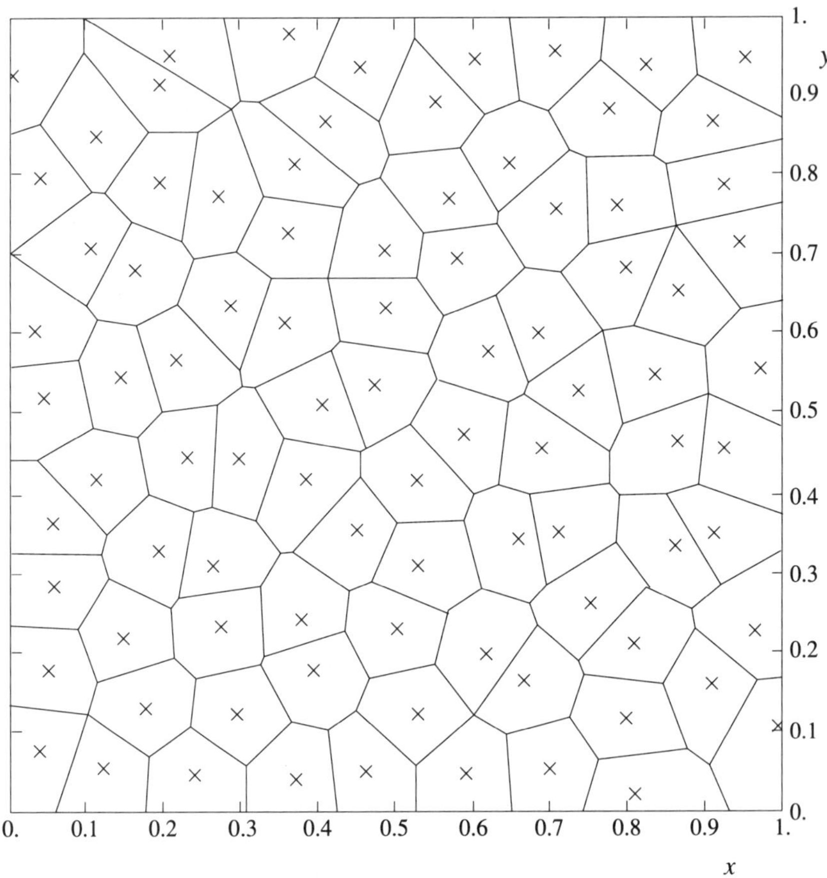

FIGURE 4c. Particle distribution at time $t = 0.176$

is

$$f_0(x, y) = 1 + A\cos(\pi x)\cos(\pi y).$$

Here $N = 89$, $A = 0.96$. Equations (42) are solved by the Euler scheme. After some time the distribution of points relaxes to the uniform distribution and all the polygons become roughly of the same size.

6. Alternative schemes and other applications. A problem that arises with the use of Voronoi diagrams in the solution of the heat equation is the stability restriction on time step for explicit schemes. From numerical experiments it appears that stability is satisfied if

$$\Delta t \leq C_s \min_i V_i,$$

where C_s is a constant of order 1. An implicit scheme for (42) seems hopeless, because the expression for $\mathcal{F}_i(\{\mathbf{x}_j\})$ is not known explicitly, but is computed numerically after the construction of the Voronoi diagram. The dependence on the position of the points is nonlinear and, though the Jacobian matrix $(\partial \mathcal{F}_i/\partial \mathbf{x}_j)$ is sparse, the structure of the matrix is not regular and changes at each time step.

In view of these considerations we are presently working on other approaches to the problem, based on a different type of structure to index the points of an irregular grid.

Given $N = N_1 \cdot N_2$ arbitrary points in the plane, there exists an indexing $\mathbf{x}(i_1, i_2)$, $1 \le i_1 \le N_1$, $1 \le i_2 \le N_2$, such that

$$x_1(i_1, i_2) \le x_1(i_1+1, i_2), \quad x_2(i_1, i_2) \le x_2(i_1, i_2+1), \qquad \forall i_1, i_2.$$

This indexing is called a monotonic logical grid **[6]** and could be used to diffuse the particles alternately in the x and y direction (a sort of alternate direction implicit method for an irregular grid).

The advantage of this method is that the scheme (37) can be applied in each direction, thereby overcoming stability problems. The result of an application of this method is shown in Figure 5, where the points "diffuse" until they distribute in a regular mesh. Reflecting boundary conditions have been used.

Numerical experiments show that the method introduces an anisotropy in the motion of the points. Further study is required to overcome this problem.

We conclude by mentioning a possible application of a Lagrangian treatment of the heat equation.

The particle-diffusion method can be used to transform an initial distribution of particles into another. Consider the initial-value problem for the heat equation:

$$\frac{\partial f}{\partial t} = \Delta(f - g(\mathbf{x})), \qquad f(\mathbf{x}, 0) = f_0(\mathbf{x}), \tag{43}$$

with $\mathbf{x} \in \mathbf{R}^m$, $g(\mathbf{x})$ integrable, and

$$\int_{\mathbf{R}^m} f_0(\mathbf{x})\, d\mathbf{x} = \int_{\mathbf{R}^m} g(\mathbf{x})\, d\mathbf{x}.$$

Then

$$\lim_{t \to \infty} f(\mathbf{x}, t) = g(\mathbf{x})$$

uniformly in $\mathbf{R}^m$. The equation of motion of a *particle* corresponding to (43) is

$$\dot{\mathbf{x}} = -\frac{1}{f}\nabla f + \frac{1}{f}\nabla g.$$

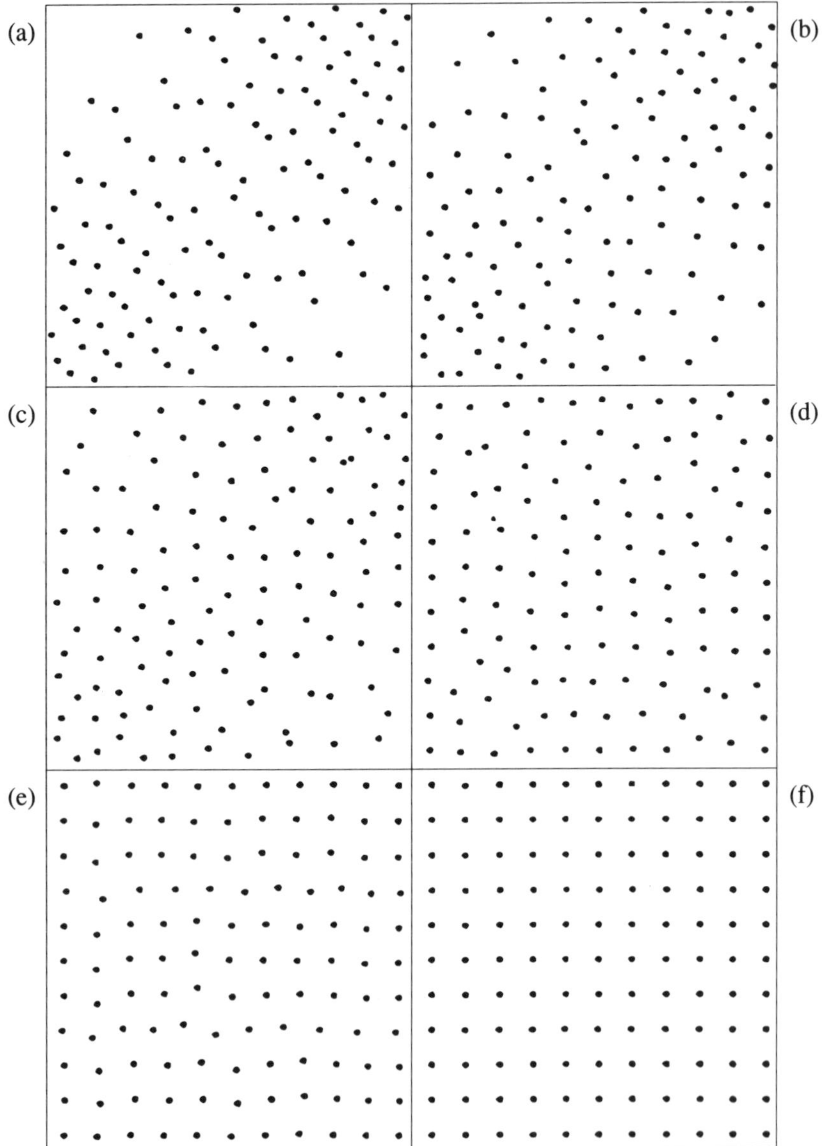

FIGURE 5. Particle solution of the heat equation with the use of monotonic logical grids. The initial distribution of particles (a) relaxes to the uniform grid.

If $f_0(\mathbf{x})$ is approximated by a set of points $\{\mathbf{x}_i\}$ and these points are moved according to the equations

$$\dot{\mathbf{x}}_i = -\frac{1}{f(\mathbf{x}_i, t)} \nabla f(\mathbf{x}_i, t) + \frac{1}{f(\mathbf{x}_i, t)} \nabla g(\mathbf{x}_i),$$

then their distribution will approximate the function $g(\mathbf{x})$. An analogous result is obtained by solving the equations

$$\dot{\mathbf{x}}_i = -\frac{\nabla f}{f} + \frac{\nabla g}{g}.$$

This procedure can be used as a tool to generate a transformation $(\xi, \zeta) \to (\mathbf{x}, \mathbf{v})$ with a given Jacobian $(\partial(\xi, \zeta)/\partial(\mathbf{x}, \mathbf{v}))$ and is an alternative technique to distribute particles according to a prescribed density $g(\mathbf{x})$.

References

1. J. M. Augenbaum, *A Lagrangian method for the shallow water equations based on Voronoi mesh-flows on a rotating sphere, The Free Lagrange Method*, Lecture Notes in Physics, vol. 238 (M. J. Fritts, W. P. Crowley, and H. Trease, eds.), Springer-Verlag, Berlin, 1985, p. 54.
2. C. K. Birdsall and A. B. Langdon, *Plasma physics via computer simulation*, McGraw-Hill, New York, 1985.
3. C. Börgers, *A Lagrangian fractional step method for incompressible Navier–Stokes equations*, Ph.D. thesis, New York University, 1985.
4. C. Börgers and C. S. Peskin, *A Lagrangian method based on the Voronoi diagram for the incompressible Navier–Stokes equations on a periodic domain, The Free Lagrange Method*, Lecture Notes in Physics, vol. 238 (M. J. Fritts, W. P. Crowley, and H. Trease, eds.), Springer-Verlag, Berlin, 1985, p. 87.
5. C. Börgers and C. S. Peskin, *A Lagrangian fractional step method for the incompressible Navier–Stokes equations on a periodic domain*, J. Comput. Phys. **70**, No. 2 (1987), p. 397.
6. J. Boris, *A vectorized "near neighbors" algorithm of order N using a monotonic logical grid*, J. Comput. Phys. **66** (1986), p. 1.
7. C. Cercignani, *Theory and application of the Boltzmann equation*, Elsevier, New York, 1975.
8. A. J. Chorin, *Vortex sheet approximation of boundary layers*, J. Comput. Phys. **27** (1978), p. 428.
9. R. J. Gelinas, S. K. Doss, and K. Miller, *The moving finite element method: Applications to general partial differential equations with multiple large gradients*, J. Comput. Phys. **40** (1981), 202–249.
10. L. Greengard and V. Rokhlin, *A fast algorithm for particle simulations*, J. Comput. Phys. **73**, No. 2 (1987), 325–348.
11. R. W. Hockney and J. W. Eastwood, *Computer simulations using particles*, McGraw-Hill, New York, 1981.
12. R. Kress and J. Wick, *Mathematical methods of plasma physics* (Proc. conf. Oberwolfach, 1979), Verlag, Peter D. Lang, Frankfurt, 1980.

13. H. Neunzert, *An introduction to the nonlinear Boltzmann–Vlasov equation, Kinetic theories and the Boltzmann equation*, Lecture Notes in Mathematics **1048**, Springer-Verlag, Berlin, 1984, pp. 60–110.

14. H. Neunzert and J. Wick, *Die Theorie der asymptotischen Verteilung und die numerische Lösung von Integrodifferentialgleichungen*, Numer. Math. **21** (1973), 234–243.

15. H. Niederreiter, *Quasi-Monte Carlo methods and pseudo-random numbers*, Bull. Amer. Math. Soc. **84** (1978), 957.

16. R. D. Richtmeyer and K. W. Morton, *Difference Methods for Initial Value Problems*, 2nd edition, Wiley-Interscience, New York, 1967.

17. G. Russo, *A particle method for collisional kinetic equations.* I. *Basic theory and one-dimensional results*, accepted for publication in J. Comput. Phys.

COURANT INSTITUTE OF MATHEMATICAL SCIENCES, NEW YORK UNIVERSITY

Lectures in Applied Mathematics
Volume **26** (1990)

On the Number of Solutions of Semilinear Elliptic Problems at Resonance: Some Numerical Experiments

RENATE SCHAAF AND KLAUS SCHMITT

Abstract. We consider Dirichlet problems for semilinear elliptic equations whose nonlinear term is oscillatory and whose linear part is at resonance. We show that such problems may have infinitely many positive and infinitely many negative solutions on certain types of domains. We conjecture that such a result fails in case the domain is a ball in dimension greater than three and provide a sequence of numerical experiments to support this conjecture.

1. Introduction. Let Ω be a bounded domain in $\mathbb{R}^n$ with smooth boundary, and let λ_1 be the principle eigenvalue of the Dirichlet problem

$$(1) \qquad \begin{cases} \Delta u + \lambda u = 0, & x \in \Omega, \\ u = 0, & x \in \partial\Omega. \end{cases}$$

Then λ_1 is simple and has an associated eigenfunction ϕ with the properties

$$\phi(x) > 0, \quad x \in \Omega, \ \frac{\partial \phi(x)}{\partial \nu} < 0, \ x \in \partial\Omega,$$

where $\frac{\partial}{\partial \nu}$ is the exterior normal derivative to $\partial\Omega$. (We normalize ϕ so that $\phi_{\max} = 1$.)

1980 *Mathematics Subject Classification* (1985 *Revision*). Primary 35B15, 47H15, 58E07.

Consider the *resonant* nonlinear problem

$$\text{(2)} \qquad \begin{cases} \Delta u + \lambda_1 u + g(u) = h(x), & x \in \Omega, \\ u = 0, & x \in \partial\Omega, \end{cases}$$

where $h : \overline{\Omega} \to \mathbb{R}$ and $g : \mathbb{R} \to \mathbb{R}$ are Hölder continuous functions and satisfy

$$\text{(3)} \qquad \int_\Omega h\phi\, dx = 0,$$

$$\text{(4)} \qquad g(s+T) = g(s), \quad -\infty < s < \infty, \qquad \int_0^T g(s)ds = 0, \quad g \not\equiv 0,$$

where T is the period of g.

Problems of this type have been studied in our earlier work **[CJSS]**, **[SS1]** (more general types of oscillatory functions g are permissible in these papers), and **[SS2]**, where we have shown that such problems have infinitely many positive and infinitely many negative solutions in case x is a one-dimensional variable and also in higher dimensions, whenever Ω is a shell-type domain whose inner and outer radii satisfy certain restrictions. In the case where x is a two-dimensional variable, we were also able to show in **[CJSS]** that the result holds whenever Ω is a disc, and using a somewhat more refined analysis (see **[SS2]**), in case Ω is a convex bounded domain in the plane. Numerical experiments (**[CJSS]**) indicate that the latter result does not hold for Ω a ball in dimensions greater than 3.

In this paper we shall present the main results of the above papers and provide additional numerical experiments, based on different numerical methods, which support the above conjecture.

The method of attack is to embed problem (2) into the one-parameter family of problems

$$\text{(5)} \qquad \begin{cases} \Delta u + \lambda u + g(u) = h(x), & x \in \Omega, \\ u = 0, & x \in \partial\Omega, \end{cases}$$

and employ bifurcation and continuation techniques to study the solution set of (5), and then consider λ_1-sections of this solution set to obtain the desired result. For our numerical calculations we will not use (5) as an embedding for (2) but rather

$$\text{(5}'\text{)} \qquad \begin{cases} \Delta u + \lambda^2 \left(u + \dfrac{1}{\lambda_1} g(u) \right) = h, \\ u|_{\partial\Omega} = 0, \end{cases}$$

so any solution of (5′) with $\lambda = \sqrt{\lambda_1}$ will give a solution of (2).

We first shall present the results from bifurcation theory which are needed in deriving the main existence results. These will be followed by existence theorems guaranteeing infinitely many solutions of (2) for various types of domains, and finally we present some numerical experiments which support the conjecture that if Ω is a ball in dimension greater than three such a result will no longer be true.

2. On bifurcation from infinity. In this section we shall state an abstract result about bifurcation from infinity which we shall need in our discussion. We refer to **[SS1]** for proofs (see also **[PS]** and **[R]**).

Let $\mathbf{X}$ be a real Banach space with norm $\|\cdot\|$. We consider the equation

$$u = K(\lambda)u + k(\lambda, u), \quad u \in \mathbf{X}, \tag{6}$$

where $K : [a,b] \subset \mathbf{R} \to \mathscr{B}(\mathbf{X})$ is a differentiable family of compact linear operators on $\mathbf{X}$ and $k : [a,b] \times \mathbf{X} \to \mathbf{X}$ is a completely continuous mapping satisfying

$$\frac{k(\lambda, u)}{\|u\|} \to 0 \quad \text{as } \|u\| \to \infty, \tag{7}$$

uniformly on $[a,b]$.

In this setting we have:

LEMMA 1. *Let $\lambda_1 \in (a,b)$ be such that*

$$\ker(\mathrm{id} - K(\lambda_1)) = \mathrm{span}\phi, \qquad \|\phi\| = 1, \tag{8}$$

$$K'(\lambda_1)\phi \notin \mathrm{range}(\mathrm{id} - K(\lambda_1)), \tag{9}$$

and $\mathbf{P} \subset \mathbf{X}$ is an open cone containing ϕ. Then there exist ε_0 and a continuum (i.e. a closed, connected set) $\mathscr{C} \subset [a,b] \times \mathbf{P}$ of solutions of (6) *with the property that for any $0 < \varepsilon \leq \varepsilon_0$ we can find a subcontinuum $\mathscr{C}_\varepsilon \subset \mathscr{C}$ such that*

$$\mathscr{C}_\varepsilon \subset \mathscr{U}_\varepsilon := \{(\lambda, u) : |\lambda - \lambda_1| < \varepsilon,\ \|u\| > 1/\varepsilon\},$$

and $\mathscr{C}_\varepsilon$ connects (λ_1, ∞) to $\partial\mathscr{U}_\varepsilon$. Moreover, if $\{(\lambda_n, u_n)\} \subset \mathscr{C} \cap \mathscr{U}_\varepsilon$ is such that $\|u_n\| \to \infty$, then

$$\lambda_n \to \lambda_1 \quad \text{and} \quad \frac{u_n}{\|u_n\|} \to \phi. \tag{10}$$

COROLLARY 2. *Let the assumptions of Lemma* 1 *hold and assume that $K(\lambda)$, $k(\lambda, \cdot)$ map $\mathbf{X}$ continuously into a Banach space $\mathbf{Y} \subset \mathbf{X}$ which is* *compactly embedded in* $\mathbf{X}$ *and that* $K : [a,b] \to \mathscr{B}(\mathbf{X},\mathbf{Y})$, $k : [a,b]\times\mathbf{X} \to \mathbf{Y}$ *are continuous with*

$$\frac{k(\lambda,u)}{\|u\|} \to 0, \quad \text{in } Y, \quad \text{as } \|u\| \to \infty, \tag{11}$$

uniformly on $[a,b]$. *Then if* $\{(\lambda_n,u_n)\} \subset \mathscr{C} \cap \mathscr{U}_{\varepsilon_0}$, *is such that* $\|u_n\| \to \infty$, *we get*

$$\lambda_n \to \lambda_1, \quad \text{and} \quad \left\| \frac{u_n}{\|u_n\|} - \phi \right\|_{\mathbf{Y}} \to 0. \tag{12}$$

In particular, if $\widetilde{\mathbf{P}} \subset \mathbf{Y}$ *is any open cone containing* ϕ, *then, by decreasing* $\varepsilon_0 > 0$, *if necessary, we obtain that*

$$\mathscr{C} \subset [a,b] \times \widetilde{\mathbf{P}}. \tag{13}$$

3. The semilinear problem. We shall now consider the resonant Dirichlet problem (2) given in the introduction and assume that all terms satisfy the hypotheses stated there. We embed (2) into the one-parameter problem (5) which we restate as

$$\begin{cases} \Delta u + \lambda u + g(u) = h(x), & x \in \Omega, \\ u = 0, & x \in \partial\Omega. \end{cases} \tag{14}$$

Let $K : \mathbf{C}(\overline{\Omega}) \to \mathbf{C}(\overline{\Omega})$ denote the operator defined by $Kf = u$ if and only if u solves

$$\begin{cases} \Delta u = f, & x \in \Omega, \\ u = 0, & x \in \partial\Omega. \end{cases}$$

As is well known, K is a bounded linear operator from $\mathbf{C}(\overline{\Omega})$ to $\mathbf{C}_0^1(\overline{\Omega})$ and hence

$$K : \mathbf{C}(\overline{\Omega}) \to \mathbf{C}(\overline{\Omega})$$

is compact. Further, by regularity theory, we have

$$K : \mathbf{C}^{\mu}(\overline{\Omega}) \to \mathbf{C}_0^{2+\mu}(\overline{\Omega}),$$

continuously. Our problem (14) is hence equivalent to the operator equation

$$u = \lambda K u + K(g(u) + h), \tag{15}$$

in the space $\mathbf{X} = \mathbf{C}_0(\overline{\Omega})$. We may hence apply Lemma 1 and Corollary 2 with $K(\lambda) = \lambda K$ and $k(\lambda,u) = K(g(u)+h)$, and letting $\mathbf{Y} = \mathbf{C}_0^1(\overline{\Omega})$, $\mathbf{P} = \{u \in \mathbf{X} : \int_\Omega u\phi\,dx > 0\}$, $\widetilde{\mathbf{P}} = \{u \in \mathbf{Y} : u > 0, \text{ in } \Omega, \frac{\partial u}{\partial \nu} < 0 \text{ on } \partial\Omega\}$. We hence obtain $\varepsilon_0 > 0$ and a continuum $\mathscr{C} \subset \mathbf{R}\times\widetilde{\mathbf{P}}$ of solutions of (15) such that $\mathscr{C} \cap \mathscr{U}_\varepsilon \neq \varnothing$ for any $0 < \varepsilon \leq \varepsilon_0$ and such that if $(\lambda_n, u_n) \in \mathscr{C}$ with $|\lambda_n - \lambda_1| < \varepsilon_0$ and $\|u_n\| = \max|u_n| \to \infty$, then

$$\lambda_n \to \lambda_1 \quad \text{and} \quad \frac{u_n}{\max u_n} \to \phi, \quad \text{in } \mathbf{C}_0^1(\overline{\Omega}).$$

In fact, regularity theory and arguments as used in Corollary 2 imply

$$\frac{u_n}{\max u_n} \to \phi \quad \text{in } \mathbf{C}_0^{2+\mu}(\overline{\Omega}). \tag{16}$$

4. Shells and balls. We shall now study the asymptotic behavior of the continuum, whose existence was derived in the previous section, in the case of special types of domains Ω. If $(\lambda, u) \in \mathscr{C}$ we multiply (14) by ϕ and integrate by parts to obtain

$$(\lambda_1 - \lambda)\int_\Omega u\phi\, dx = \int_\Omega g(u)\phi\, dx. \tag{17}$$

Since $u \in \mathbf{P}$ it follows that the right-hand side of (17) determines the sign of $\lambda_1 - \lambda$.

We now proceed formally, but note that the computations to follow make sense, provided the eigenfunction ϕ is suitably well-behaved, as will be the case for Ω a ball or a spherical shell.

Let u be a solution with $\max_\Omega u = \alpha$. Then

$$\begin{aligned}\int_\Omega g(u)\phi\, dx &= \int_\Omega \nabla(G(u) - G(\alpha)) \cdot \frac{\phi\nabla u}{|\nabla u|^2} dx \\ &= \int_\Omega \operatorname{div}\left((G(u) - G(\alpha))\frac{\phi\nabla u}{|\nabla u|^2}\right) dx \\ &\quad + \int_\Omega (G(\alpha) - G(u))\operatorname{div}\frac{\phi\nabla u}{|\nabla u|^2} dx,\end{aligned} \tag{18}$$

where $G(s) = \int_0^s g(t)\, dt$. By the divergence theorem, we obtain that the term in the second line of (18) vanishes. We hence must determine the sign of

$$\int_\Omega (G(u) - G(\alpha))\operatorname{div}\frac{\phi\nabla u}{|\nabla u|^2} dx. \tag{19}$$

We compute and find

$$\operatorname{div}\frac{\phi\nabla u}{|\nabla u|^2} = \frac{1}{|\nabla u|^2}(\nabla\phi\cdot\nabla u + \phi\Delta u - 2\phi H_u(\xi, \xi)), \tag{20}$$

where $\xi = \frac{\nabla u}{|\nabla u|}$ and

$$H_u(\xi, \xi) = \sum_{i,j=1}^n u_{x_i x_j}\xi_i\xi_j$$

is the quadratic form associated with the Hessian matrix $(u_{x_i x_j})$ of u.

In view of the identity

$$\Delta\phi - H_\phi(\theta,\theta) = |\nabla\phi| \operatorname{div}\frac{\nabla\phi}{|\nabla\phi|}, \qquad \theta = \frac{\nabla\phi}{|\nabla\phi|}, \tag{21}$$

and the fact that

$$\operatorname{div}\left(\frac{\nabla\phi}{|\nabla\phi|}\right)(x) = -(n-1)H(x) \tag{22}$$

(see **[GT**, p. 383**]**), where $H(x)$ is the mean curvature with respect to the normal direction $-\nabla\phi$ at $y = x$ of the level surface $\phi(y) = \phi(x)$, it follows that the sign of

$$\operatorname{div}\frac{\phi\nabla\phi}{|\nabla\phi|^2}$$

is given by the sign

$$Q(\phi) = |\nabla\phi|^2 - \phi\Delta\phi - 2(n-1)\phi|\nabla\phi|H, \tag{23}$$

which by the convergence considerations of the previous section (i.e. (16)) will give the sign of (20) and hence of (17). Hence, if we assume that $Q(\phi) > 0$, we have that (20) is positive as long as $\max u$ is large enough. We now choose a sequence $\{\alpha_n\}_{n=1}^{\infty}$ with

$$1 \ll \alpha_1 < \alpha_2 < \cdots \to \infty$$

and solutions $\{(\lambda_n, u_n)\} \subset \mathscr{C}$ with

$$\max_\Omega u_n = \alpha_n$$

and

$$G(\alpha_{2n}) \geq G(t), \quad \alpha_{2n} \geq t,$$
$$G(\alpha_{2n+1}) \leq G(t), \quad \alpha_{2n+1} \geq t$$

(this may be done, since G is periodic). It follows that the integrals

$$\int_\Omega g(u_n)\phi\, dx$$

alternate in sign infinitely often. Since $\mathscr{C}$ is a continuum it follows that there exists $\tilde{u}_n$, $\max u_n \leq \max \tilde{u}_n \leq \max u_{n+1}$ such that $(\lambda_1, \tilde{u}_n) \in \mathscr{C}$.

Let us now assume that $h = 0$ and that

$$\Omega = \mathbf{B}_1(0) = \{x \in \mathbb{R}^n : |x| < 1\},$$

where we have chosen the radius equal to 1 for simplicity's sake. As argued above, we obtain infinitely many solutions, once we can show that either $Q(\phi) > 0$, or else $\inf \frac{1}{|\nabla\phi|^2} Q(\phi) > 0$, in Ω. Since ϕ is radially symmetric, it satisfies

$$\Delta\phi + \nu_1^2\phi = \phi_{rr} + \frac{n-1}{r}\phi_r + \nu_1^2\phi = 0,$$

where ν_1 is the first zero of the Bessel function $J_{\frac{n-2}{2}}$ and ϕ is given by

$$\phi(r) = c_n r^{-(\frac{n-2}{2})} J_{\frac{n-2}{2}}(\nu_1 r),$$

where c_n is a suitable constant so $\phi(0) = 1$. Let us now distinguish the cases $n = 2$ and $n \geq 3$.

THE CASE $n = 2$. One easily computes

$$\nabla\phi = \phi_r \frac{x}{r}, \qquad |\nabla\phi|^2 = \phi_r^2,$$

$$\theta = \frac{\nabla\phi}{|\nabla\phi|} = -\frac{x}{r}, \qquad H_\phi(\theta, \theta) = \phi_{rr},$$

and hence

$$A(r) = \frac{1}{\phi_r^2} Q(\phi)(r) = \frac{1}{\phi_r^2}(\phi_r^2 - \nu_1^2\phi^2 - 2\phi\phi_{rr}).$$

Now $Q(\phi)(0) = 0$ since $\phi(0) = 1$, $\phi_r(0) = 0$, and $\phi_{rr}(0) = -\frac{\nu_1^2}{2}$. Further we have

$$\frac{d}{dr} Q(\phi)(r) = 2\phi \left(\frac{\phi_r}{r}\right)_r \tag{24}$$

and hence (24) becomes

$$Q'(\phi)(r) = 2\nu_1^2 r^{-1} J_0(\nu_1 r) J_2(\nu_1 r) > 0, \quad 0 < r \leq 1,$$

and hence $Q(\phi)(r) > 0$, if $0 < r \leq 1$. Also we may compute

$$A(0) = \lim_{r\to 0} \frac{Q'(\phi)(r)}{2\phi_r\phi_{rr}} = \lim_{r\to 0} \frac{Q''(\phi)(r)}{2\phi_{rr}^2 + 2\phi_r\phi_{rr}} = \frac{Q''(\phi)(0)}{2\phi_{rr}^2(0)},$$

where

$$Q''(\phi)(0) = \frac{\nu_1^4}{2}\left(\nu_1 - \frac{1}{2}\right) > 0;$$

hence

$$A(0) = \nu_1 - \frac{1}{2} > 0, \quad \text{since } \phi_{rr}^2(0) = \frac{\nu_1^4}{4}.$$

We may therefore conclude that in case the domain Ω is a disk in the plane or a small perturbation of such, there will be infinitely many positive solutions of (2).

THE CASE $n \geq 3$. As before we have that

$$Q(\phi)(r) = \phi_r^2 - \nu_1^2\phi^2 - 2\phi\phi_{rr}.$$

We compute that

$$\phi(0) = \frac{c_n}{\Gamma(\frac{n}{2})}\left(\frac{\nu_1}{2}\right)^{\frac{n-2}{2}}, \qquad \phi_{rr}(0) = -\frac{2c_n}{\Gamma(\frac{n+2}{2})}\left(\frac{\nu_1}{2}\right)^{\frac{n+2}{2}}.$$

Hence, using the above , we obtain

$$Q(\phi)(0) = \frac{\nu_1^n c_n^2}{2^{n-2}\Gamma(\frac{n}{2})^2}(2-n).$$

Thus $Q(0) < 0$, since $n \geq 3$, and our method of proof may not be applied to balls in dimension ≥ 3. In fact, our numerical computations seem to indicate that a multiplicity result of the above type fails in dimensions greater than 3 (see our experiments below).

We next consider the case that Ω is a spherical shell. In this case Q takes the form

$$Q(\phi) = |\nabla\phi|^2 + \lambda_1\phi^2 - 2(n-1)\phi|\nabla\phi|H, \tag{25}$$

and hence for any $\delta > 0$, we obtain that

$$Q(\phi) \geq |\nabla\phi|^2 + \lambda_1\phi^2 - \frac{(n-1)^2H^2\phi^2}{\delta} - \delta|\nabla\phi|^2,$$

and if $\delta < 1$ we obtain that $Q(\phi) > 0$, whenever

$$\lambda_1 > \frac{(n-1)^2H^2}{\delta}. \tag{26}$$

It is clear that (26) may be satisfied on domains Ω of the form

$$\Omega = \{x : a < |x| < b\},$$

provided $a \gg 1$ and $b - a$ is small, and thus we obtain infinitely many positive solutions for such spherical shell-type domains and small perturbations of such.

5. Convex domains. We next consider the case that the domain Ω is convex.

Let $\|u\| = \max u$ denote the norm in the space $\mathbf{X}$, and instead of (17) we shall consider

$$\|u\|^2(\lambda_1 - \lambda)\int_\Omega \frac{u}{\|u\|}\phi\, dx = \|u\|\int_\Omega g(u)\phi\, dx \tag{27}$$

and determine the sign of that quantity for large $\|u\|$.

Let us now consider a sequence of solutions $\{(u_k, \lambda_k)\} \subset \mathscr{C}$ with

$$\|u_k\| = a_k + kT, \quad 0 \leq a_k \leq T,\ k \geq 2,$$

where a_k will be chosen appropriately, and T is the period of g.

We let

$$v_k = \frac{u_k}{\|u_k\|},$$

and recall that $v_k \to \phi$ in $\mathbf{C}_0^{2+\mu}(\overline{\Omega})$.

Since Ω is assumed convex it follows from a result in **[BL]** that $\nabla\phi$ only vanishes at a single point, where ϕ assumes its maximum and $D^2\phi$ is negative definite there. Therefore the same will be true for v_k for all sufficiently large k. (This is the only consequence of convexity which is needed in our discussion, and we hence could replace the convexity assumption by this implication as an assumption, certainly a somewhat less restrictive requirement. Certain types of symmetry conditions on the domain as used in **[CS]**, for example, will also be sufficient.) We now use the co-area formula (see **[B]** or **[C]**) and find that

$$\|u_k\| \int_\Omega g(u_k)\phi\, dx = \|u_k\| \int_0^{\|u_k\|} g(t) \int_{u_k=t} \frac{\phi}{|\nabla u_k|}\, dS_t\, dt, \tag{28}$$

where dS_t denotes the Riemannian $(n-1)$-density on the level sets $\{u_k = t\}$. The latter may be rewritten as

$$\int_0^{\|u_k\|} g(t) \int_{v_k=\frac{t}{\|u_k\|}} \frac{\phi}{|\nabla v_k|}\, dS_t\, dt. \tag{29}$$

If we define

$$f_k(s) = \int_{v_k=s} \frac{\phi}{|\nabla v_k|}\, dS_s,$$

then (29) becomes

$$\int_0^{\|u_k\|} g(t) f_k\left(\frac{t}{\|u_k\|}\right) dt. \tag{30}$$

The latter integral we now write as the sum of the integrals

$$\int_0^{kT} g(t) f_k\left(\frac{t}{\|u_k\|}\right) dt = I_1 \tag{31}$$

and

$$\int_{kT}^{a_k+kT} g(t) f_k\left(\frac{t}{\|u_k\|}\right) dt = I_2. \tag{32}$$

We first consider the integral I_2. Using the periodicity of g we find that (32) may be rewritten as

$$I_2 = \int_0^{a_k} g(t) f_k\left(\frac{t+kT}{a_k+kT}\right) dt. \tag{33}$$

We next observe that each f_k for k sufficiently large will be of class $\mathbf{C}^{1+\mu}$ on any given compact subinterval of $[0, 1)$ (recall $\phi_{\max} = 1$), and we may conclude that

$$f_k \to f$$

in $\mathbf{C}^1$ on any compact subinterval of $[0, 1)$, where f is given by

$$f(s) = \int_{\phi=s} \frac{\phi}{|\nabla\phi|}\, dS_s.$$

It follows also from the nondegeneracy of v_k and ϕ at their maxima that f is continuous and that $f_k \to f$ in $\mathbf{C}^0[0, 1]$. From these observations it follows that we may pass to the limit in (24) and conclude that

$$I_2 \to \int_0^a g(z)\, dz\, f(1), \tag{34}$$

where a has been preassigned in $[0, T]$, and the sequence $\{a_k\}$ was chosen so that $a_k \to a$.

We next consider the integrals I_1. We first integrate by parts and obtain

$$\begin{aligned} I_1 &= -\int_0^{kT} G(t) \frac{1}{a_k + kT} f_k'\left(\frac{t}{a_k + kT}\right) dt \\ &= -\int_0^{kT/(a_k+kT)} G(s(a_k + kT)) f_k'(s)\, ds, \end{aligned} \tag{35}$$

where

$$G(s) = \int_0^s g(t)\, dt,$$

is periodic.

As $k \to \infty$, I_1 has the same limit as

$$\begin{aligned} J_1 &= -\int_0^{kT} G(t) \frac{1}{a_k + kT} f'\left(\frac{t}{a_k + kT}\right) dt \\ &= -\int_0^{kT/(a_k+kT)} G(s(a_k + kT)) f'(s)\, ds, \end{aligned} \tag{36}$$

provided we can show that

$$\lim_{k\to\infty} \int_0^{kT/(a_k+kT)} |f_k'(s) - f'(s)|\, ds = 0. \tag{37}$$

In case $n = 2$, the convergence of the sequence $\{f_k\}$ in the norm of $\mathbf{H}^{1,1}$ is by no means obvious; we refer to **[SS2]** for the proof.

We next use the periodicity of G to rewrite (36) as

$$-\frac{1}{T}\int_0^T G(t)\sum_{j=1}^{k}\frac{T}{a_k+kT}f'\left(\frac{t+(j-1)T}{a_k+kT}\right)dt. \tag{38}$$

Letting $k\to\infty$ in (29), we obtain

$$-\frac{1}{T}\int_0^T G(t)\int_0^1 f'(\tau)d\tau dt,$$

which equals

$$-\frac{1}{T}\int_0^T G(t)(f(1)-f(0))dt.$$

We hence have that

$$\begin{aligned} I_1+I_2 &= f(1)\left\{\int_0^a g(s)ds-\frac{1}{T}\int_0^T G(s)ds\right\} \\ &= f(1)[G(a)-\overline{G}], \end{aligned} \tag{39}$$

since $f(0)=0$ and where $\overline{G}$ is the mean value of G. We therefore may, once we know that $f(1)\neq 0$, determine the sign of $\lambda-\lambda_1$ by examining the sign of $G(a)-\overline{G}$. On the other hand, as a varies from 0 to T, the function $G(a)-\overline{G}$ will change sign, and we will consequently find infinitely many positive solutions of (2) on the continuum $\mathscr{C}$.

To determine whether $f(1)\neq 0$ we proceed as follows. Since $\phi(x)>0$, $x\in\Omega$, it suffices to consider the function

$$h(s)=\int_{\phi=s}\frac{1}{|\nabla\phi|}dS_s,$$

since $f(s)=sh(s)$. That $h(1)>0$ for $n=2$ and $h(1)=0$ for $n\geq 3$ follows immediately from the fact that $D^2\phi$ is negative definite, where $\phi(x)=1$. But the result, for $n=2$, also holds without this assumption.

We introduce the following notation:

$$A(s)=\int_{\phi=s}dS_s$$

and

$$V(s)=\int_{\phi\geq s}dx.$$

The *isoperimetric* inequality (see **[B]**) states that

$$A^2(s)\geq n^2\omega_n^{\frac{2}{n}}V^{2-\frac{2}{n}}(s),$$

where ω_n is the volume of the unit ball in $\mathbb{R}^n$. Hence we obtain that

$$(40)\qquad \begin{cases} n^2\omega_n^{\frac{2}{n}} V^{2-\frac{2}{n}}(s) \le A^2(s) \\ \qquad = \left(\int_{\phi=s} dS_s\right)^2 \\ \qquad \le \int_{\phi=s} \frac{1}{|\nabla\phi|} dS_s \int_{\phi=s} |\nabla\phi|\, dS_s \\ \qquad = h(s)\int_{\phi>s} -\nabla\cdot\nabla\phi\, dx \\ \qquad = \lambda_1 h(s)\int_{\phi>s} \phi\, dx \\ \qquad \le \lambda_1 h(s)\int_{\phi>s} \phi_{\max}\, dx \\ \qquad \le \lambda_1 h(s) V(s). \end{cases}$$

From (40) it follows that

$$(41)\qquad n^2\omega_n^{\frac{2}{n}} V^{1-\frac{2}{n}}(s) \le \lambda_1 h(s),$$

which, in case $n = 2$, implies that

$$(42)\qquad \frac{4\pi}{\lambda_1} \le h(s).$$

Inequality (42) implies that $h(1) > 0$, which implies the desired result.

Since $f(1) = 0$, for $n \ge 3$ and convex domains, we can only conclude in this case that

$$\|u\|^2(\lambda_1 - \lambda) \to 0,$$

as $\|u\| \to \infty$, as an estimate for the order of convergence as $\lambda \to \lambda_1$.

6. Some experiments. If we only consider radially symmetric $u = u(r)$ on the unit ball of dimension n, omit the factor $1/\lambda_1$, take $h = 0$, and scale $t = \lambda r$, then $(5')$ becomes

$$(43)\qquad \begin{cases} u'' + \dfrac{n-1}{t}u' + u + g(u) = 0, \\ u'(0) = 0, \qquad u(\lambda) = 0. \end{cases}$$

This way λ will become a function of $\|u\| = \max u$:

Let u be the solution of equation (43) with initial values $u(0) = \|u\|$, $u'(0) = 0$. Then $\lambda = \Lambda(\|u\|)$ has to be the first time $t > 0$ such that $u(t) = 0$. The function $\|u\| \mapsto \Lambda(\|u\|)$ is also called the time map of (43).

In order to get a numerical picture of $\Lambda(\|u\|)$ we solve

$$
\begin{aligned}
u' &= v, \\
v' &= \begin{cases} -\dfrac{n-1}{t}v - u - g(u) & \text{for } t > 0, \\[2ex] -\dfrac{1}{n}(u + g(u)) & \text{for } t = 0, \end{cases} \\
u(0) &= \|u\|, \qquad v(0) = 0,
\end{aligned}
\tag{44}
$$

until $u(t)$ becomes nonpositive for the first meshpoint t, or until $v(t)$ becomes nonnegative for the first time. In the latter case $\Lambda(\|u\|)$ is not defined.

In case of $u(t) \leq 0$ we have to find an approximation of the value $t - \Delta t$ where $u(t - \Delta t) = 0$. By expanding

$$0 = u(t - \Delta t) = u(t) - \Delta t v(t) + \frac{(\Delta t)^2}{2} v'(t)$$

up to order 2 we can solve for Δt using the differential equations. If $v'(t)$ is numerically very small we use the first-order expansion.

As an integrator for (44) we use the Runge–Kutta method. For reasons of better accuracy a step size control should be employed which takes care of the varying curvature of orbits in the phase plane. However, we did choose a fixed stepsize for each orbit since this saves computing time. Since $|\Lambda(\|u\|) - \sqrt{\lambda_1}|$ becomes very small (up to order 10^{-8} in our examples) programs have to run very long anyway, and a fixed stepsize gives nicer pictures in less time because the error is more systematic. Comparisons between fixed and variable stepsize pictures have made us believe that we can trust the result if the graph of Λ shows some sort of self similarity as $\|u\|$ increases. (There is of course no theoretical justification for this statement.)

It is important to have very accurate values for $\sqrt{\lambda_1}$ for the various space dimensions (the broken lines in the pictures) in order to get a hint whether the branch really does or does not cross this level. Since we could not find any tables which are accurate enough, we have calculated these values by the same method, using $g(u) = 0$, with a stepsize that gives at least 10^{-9} precision. The values computed are:

(1) $n = 1$: 1.5707963267948966
(2) $n = 2$: 2.4048255575712759
(3) $n = 3$: 3.1415926535725905
(4) $n = 4$: 3.8317059701305754
(5) $n = 5$: 4.4934094578017930.

In our numerical examples we have scaled $g(u)$ to $\frac{1}{a}g(au)$. This corresponds to a scaling $u \mapsto \frac{u}{a}$. The calculations give rise to some speculations:

For $n = 3$ we could not find a function for which crossing of the line $\lambda = \sqrt{\lambda_1}$ stops (see Figure 1).

A completely different behavior seems to be true for $n = 4$: Looking at Figures 2–6 one might dare to guess that repeated crossing or final noncrossing takes place according to the sign of $g(0)$.

For $n = 5$ (Figures 7–9) oscillation of the branch seems to stop for $\cos u$ and $-\cos u$; branches are likely to stay above (below) the level $\lambda = \sqrt{\lambda_1}$. In the picture for $\sin u$ its branches look like they should stay below $\lambda = \sqrt{\lambda_1}$, in contrast to the behavior for $n = 4$. We suspect that there is a phase shift for the cosine such that repeated crossing takes place, since the branch of $\cos u$ can be homotoped into the one of $-\cos u$ via shifting of the phase.

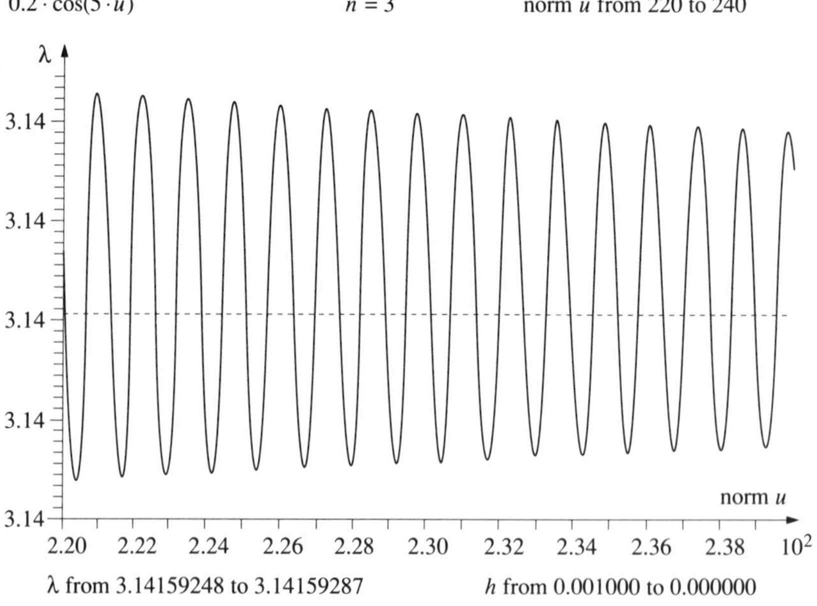

FIGURE 1

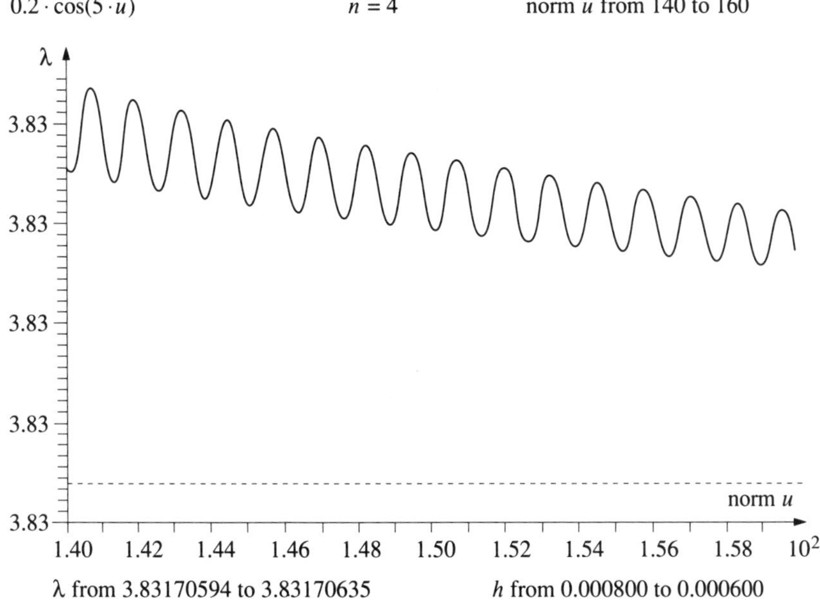
0.2 · cos(5 · u)
n = 4
norm u from 140 to 160
λ
3.83
3.83
3.83
3.83
3.83
norm u
1.40 1.42 1.44 1.46 1.48 1.50 1.52 1.54 1.56 1.58 10^2
λ from 3.83170594 to 3.83170635
h from 0.000800 to 0.000600

FIGURE 2

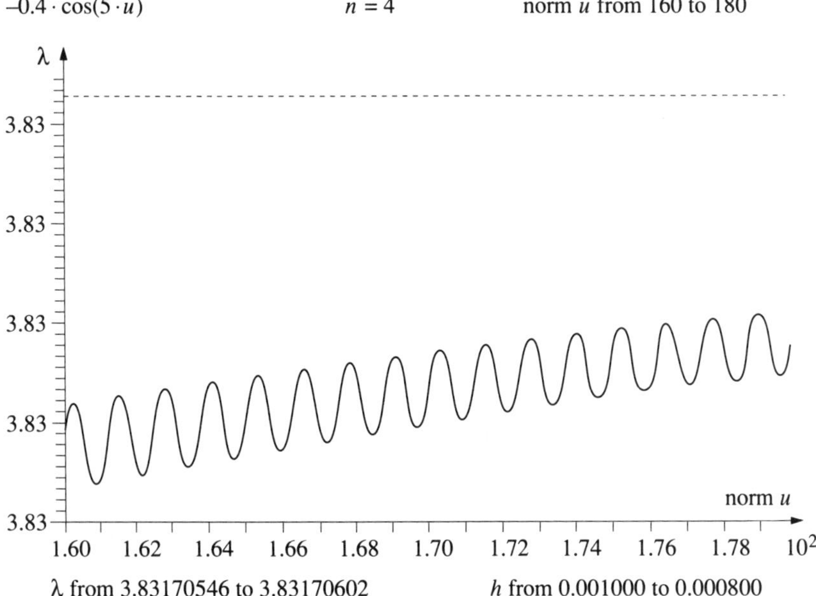
−0.4 · cos(5 · u)
n = 4
norm u from 160 to 180
λ
3.83
3.83
3.83
3.83
3.83
norm u
1.60 1.62 1.64 1.66 1.68 1.70 1.72 1.74 1.76 1.78 10^2
λ from 3.83170546 to 3.83170602
h from 0.001000 to 0.000800

FIGURE 3

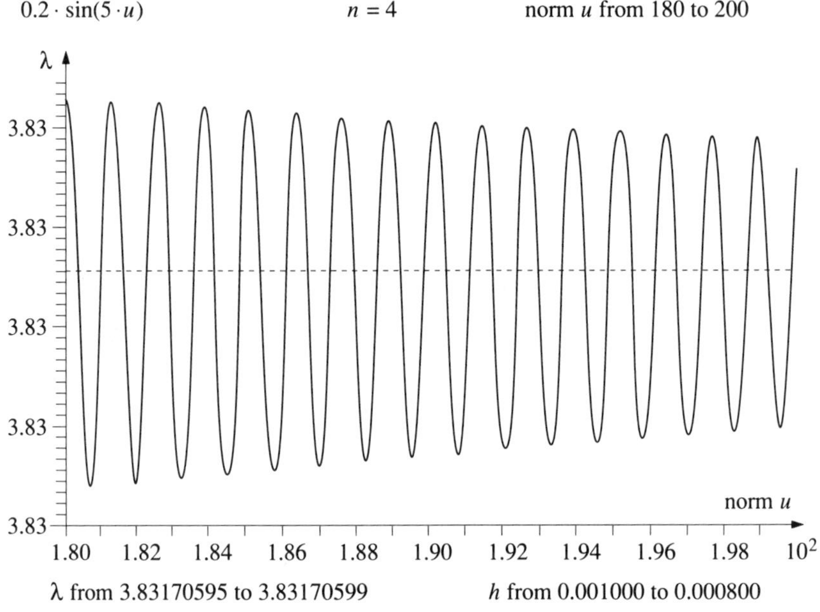
$0.2 \cdot \sin(5 \cdot u)$
$n = 4$
norm u from 180 to 200
λ
3.83
3.83
3.83
3.83
3.83
norm u
1.80
1.82
1.84
1.86
1.88
1.90
1.92
1.94
1.96
1.98
10^2
λ from 3.83170595 to 3.83170599
h from 0.001000 to 0.000800

FIGURE 4

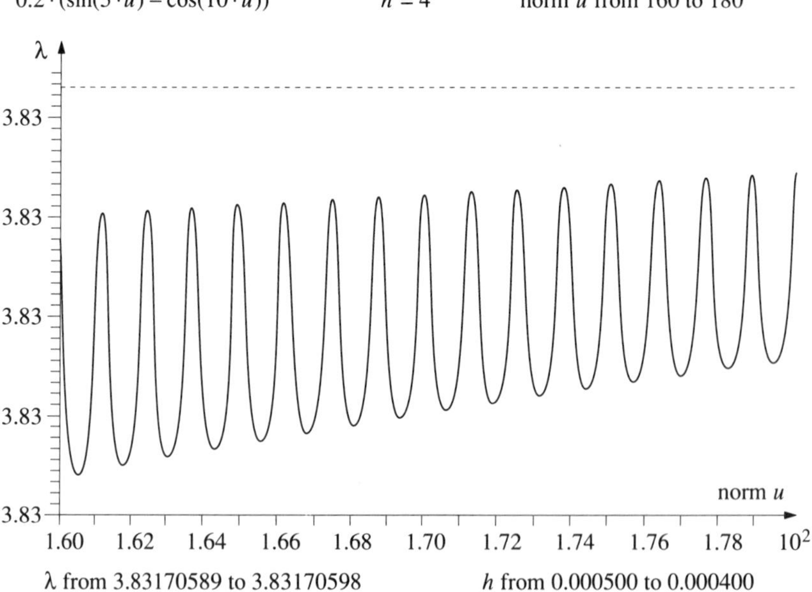
$0.2 \cdot (\sin(5 \cdot u) - \cos(10 \cdot u))$
$n = 4$
norm u from 160 to 180
λ
3.83
3.83
3.83
3.83
3.83
norm u
1.60
1.62
1.64
1.66
1.68
1.70
1.72
1.74
1.76
1.78
10^2
λ from 3.83170589 to 3.83170598
h from 0.000500 to 0.000400

FIGURE 5

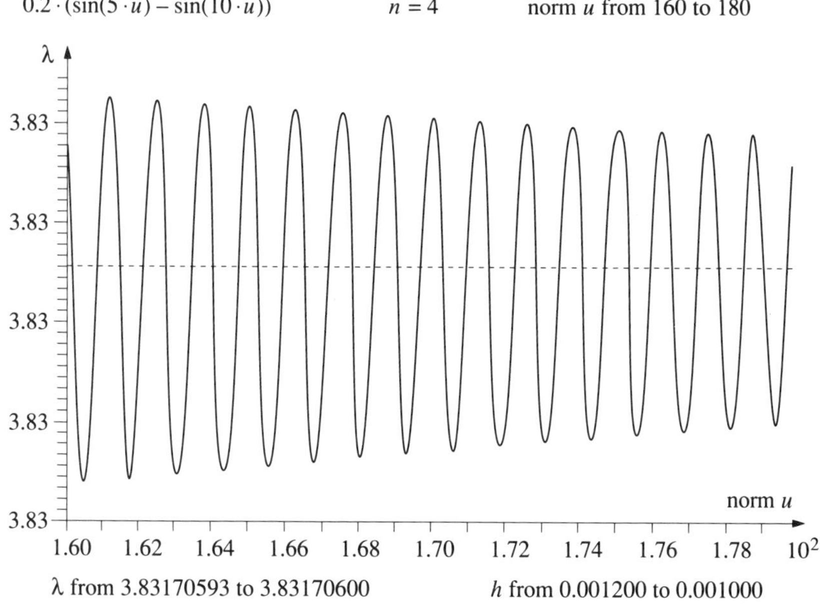

FIGURE 6

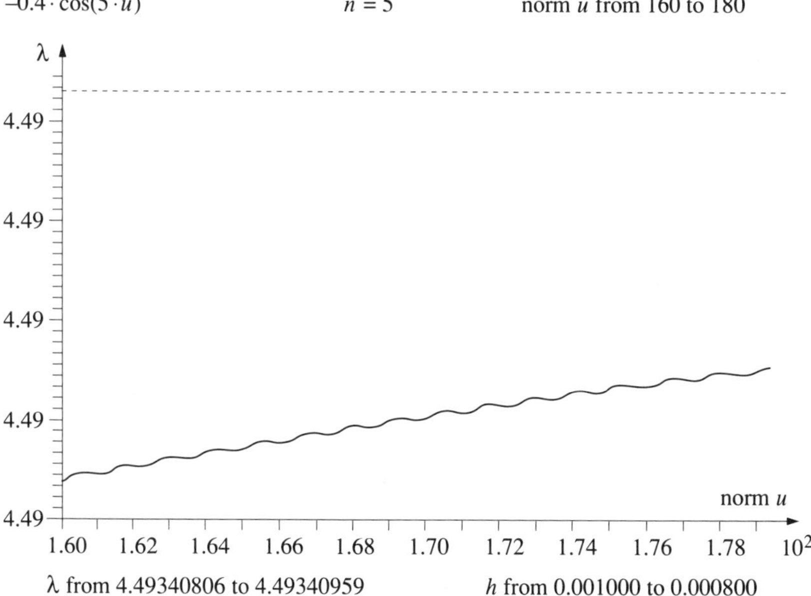

FIGURE 7

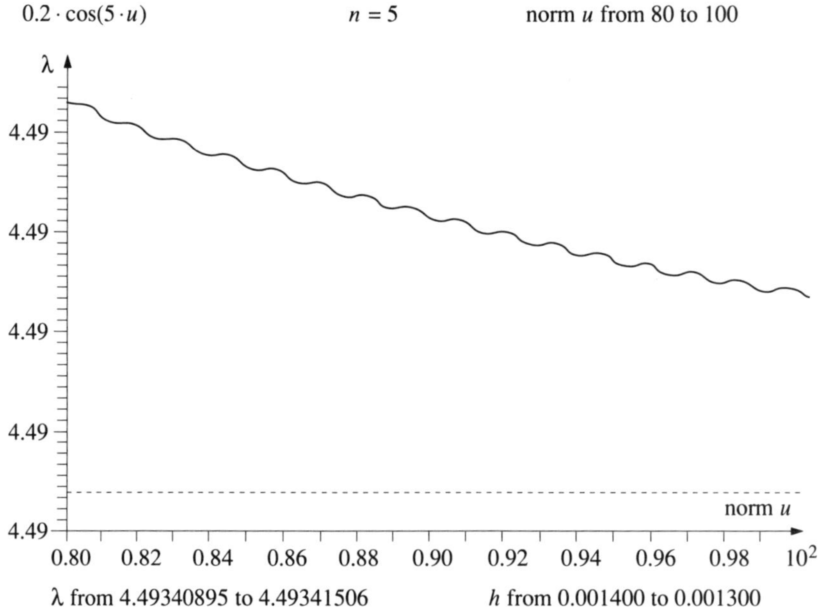

FIGURE 8

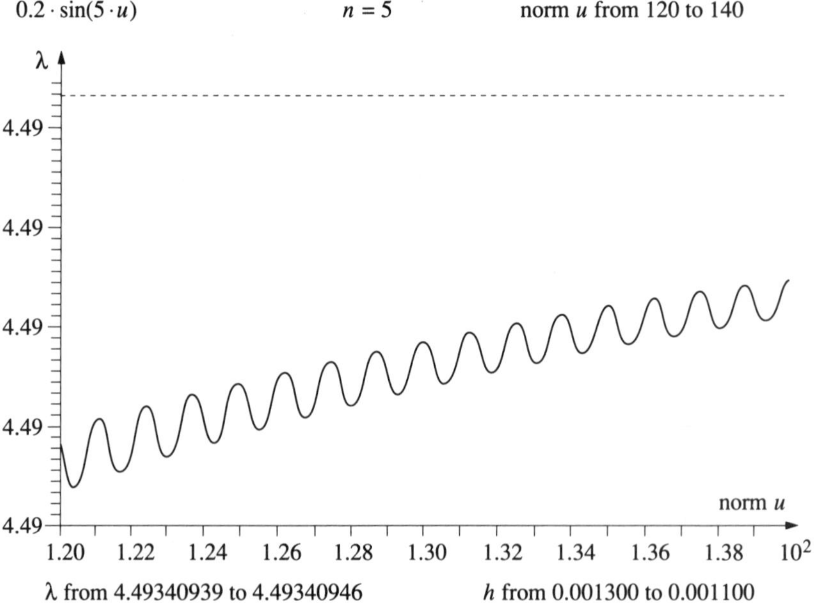

FIGURE 9

REFERENCES

[B] C. Bandle, *Isoperimetric Inequalities and Applications*, Pitman, Boston, London, 1980.

[BL] H. Brascamp and E. Lieb, *On extensions of the Brunn–Minkowski and Prekopa–Leindler theorems, including inequalities for log concave functions, and with an application to a diffusion equation*, J. Functional Analysis **22** (1976), 366–389.

[C] I. Chavel, *Eigenvalues in Riemannian Geometry*, Academic Press, New York, 1984.

[CS] C. Cosner and K. Schmitt, *On the geometry of level sets of positive solutions of semilinear elliptic equations*, Rocky Mtn. J. Math. **18** (1988), 277–286.

[CJSS] D. Costa, H. Jeggle, R. Schaaf, and K. Schmitt, *Oscillatory perturbations of linear problems at resonance*, Results in Math. **14** (1988), 275–287.

[GNN] B. Gidas, W. Ni, and L. Nirenberg, *Symmetry and related properties via the maximum principle*, Comm. Math. Phys. **68** (1979), 209–243.

[GT] D. Gilbarg and S. Trudinger, *Elliptic Partial Differential Equations of Second Order*, Springer-Verlag, Berlin, New York, 1977.

[PS] H. Peitgen and K. Schmitt, *Global analysis of two-parameter elliptic eigenvalue problems*, Trans. Amer. Math. Soc. **283** (1984), 57–95.

[R] P. Rabinowitz, *On bifurcation from infinity*, J. Diff. Eqs. **14** (1983), 462–475.

[SS1] R. Schaaf and K. Schmitt, *A class of nonlinear Sturm–Liouville problems with infinitely many solutions*, Trans. Amer. Math. Soc. **306** (1988), 853–859.

[SS2] R. Schaaf and K. Schmitt, *Periodic perturbations of linear problems at resonance on convex domains*, Rocky Mtn. J. Math. (to appear).

SFB 123, UNIVERSITÄT HEIDELBERG, IM NEUENHEIMER FELD 294, D-6900 HEIDELBERG, WEST GERMANY

DEPARTMENT OF MATHEMATICS, UNIVERSITY OF UTAH, SALT LAKE CITY, UT 84112, USA

Lectures in Applied Mathematics
Volume **26** (1990)

Splitting of Separatrices and Chaos

JÜRGEN SCHEURLE

Abstract. We consider plane dynamical systems with a saddle point and a corresponding homoclinic orbit. Under external forcing such a separatrix generally splits, i.e., the stable and unstable manifolds of the perturbed saddle point separate. Transverse intersections of these invariant manifolds imply chaotic behaviour of the system. The purpose of this article is to discuss this phenomenon on the basis of recent work by the author and joint work with others.

1. Introduction. We say that the behaviour of a dynamical system is chaotic if it is complicated and depends sensitively on the initial data. Thus, a long-term prediction of such behaviour is practically impossible although the system is completely deterministic. In science and engineering there are lots of systems, respectively mathematical models, that exhibit this phenomenon. Therefore, one would like to understand it, i.e., one is looking for underlying mechanisms and certain structures of order.

In this paper we describe one such mechanism, namely, splitting of separatrices under external forcing (Section 2), and the resulting form of chaos for certain forcing terms (Section 3). Also we describe a criterion that enables one to detect this phenomenon in specific examples a priori (Section 4). Finally, in Section 5 we briefly discuss the so-called phenomenon of exponentially small splitting of separatrices, which arises in a lot of interesting applications. An example is coupled nonlinear oscillators at resonance.

To make things as easy as possible, we explain everything in terms of very simple examples here. More general results and more sophisticated

1980 *Mathematics Subject Classification* (1985 *Revision*). Primary 58, 34.

applications are mentioned only occasionally. Proofs are completely omitted. For them we refer to the literature.

2. Splitting of separatrices. In this section we explain what we mean by splitting of separatrices. To begin, let us consider the simple pendulum equation in dimensionless form:

$$\ddot{\varphi} + \sin\varphi = 0. \tag{2.1}$$

As is well known, solutions $\varphi = \varphi(t)$ of (2.1) describe the motion of a rigid, straight, weightless rod of fixed length l with point mass m at one end, under the action of gravitation. The rod is free to rotate about a horizontal axis at the point of suspension. Here φ is the angle between the pendulum and the vertically downward direction, say measured counterclockwise and t denotes time scaled by the factor $\sqrt{g/l}$, where g is the gravitational acceleration. Dots denote derivatives with respect to t.

The motion of the pendulum is uniquely determined by (2.1) for all times, provided that the position angle φ and the angular velocity $\dot{\varphi}$ are known at a time t_0. In this sense, (2.1) generates a dynamical system in the $(\varphi, \dot{\varphi})$-plane, the so-called phase plane. Geometrically, motions of the pendulum are described by orbits in this plane, given by $(\varphi(t), \dot{\varphi}(t))$, $t \in \mathbf{R}$, where $\varphi(t)$ is a solution of (2.1). In particular, there is a *stable equilibrium point* at $\varphi = \dot{\varphi} = 0$ representing the vertically downward position of rest of the pendulum. Closed orbits around this point describe periodic swings of the pendulum. Moreover, there are two more equilibrium points at $\varphi = \pm\pi$, $\dot{\varphi} = 0$, which can actually be identified because of the 2π-periodicity of the angular variable φ. These points represent the vertically upward position of rest of the pendulum, which is not stable. They are so-called *hyperbolic saddle points*. For each of them there exist orbits that start arbitrarily near and leave a certain neighbourhood as time increases. Also there are orbits that approach these points from far away. The leaving (approaching) orbits together with the points themselves form the so-called *unstable* (*stable*) *manifolds*. In the present case we have a specific situation. Namely, the unstable (stable) manifold of the saddle point $(-\pi, 0)$ coincides with the stable (unstable) manifold of the saddle point $(\pi, 0)$ along an orbit Γ^+ (Γ^-). Such orbits are the kind of *separatrices* we consider here. One can also think of $\Gamma^\pm$ as being so-called *homoclinic orbits*. In fact, if we identify the two points $(\pm\pi, 0)$, then both Γ^+ and Γ^- connect this one saddle point with itself. $\Gamma^\pm$ correspond to motions of the pendulum where it starts from the vertically upward position and rotates by an angle of 360° either clockwise or counterclockwise.

For later use we remark that the separatrices $\Gamma^{\pm}$ admit the closed-form analytic representation

$$\begin{aligned} \varphi &= \overline{\varphi}(t) = \pm 2 \arctan(\sinh(t)) \\ & \qquad\qquad\qquad\qquad\qquad (-\infty < t < \infty). \\ \dot{\varphi} &= \dot{\overline{\varphi}}(t) = \pm 2 \operatorname{sech}(t) \end{aligned}$$

Now we modify equation (2.1) by adding a T-periodic forcing term with amplitude $\delta > 0$:

$$\ddot{\varphi} + \sin\varphi = \delta f(t), \qquad f(t+T) = f(t). \tag{2.2}$$

We assume that the function $f\colon \mathbf{R} \to \mathbf{R}$ is continuous. Then the mapping $P_\delta(t_0)\colon \mathbf{R}^2 \to \mathbf{R}^2$, which takes the initial data for (2.2) at time t_0 to the value of the solution at time $t_0 + T$, is a good tool to describe the resulting dynamics. Iteration by this mapping in the $(\varphi, \dot{\varphi})$-plane yields the values of the solution at times $t = t_0 + jT$, $j \in \mathbf{Z}$. So $P_\delta(t_0)$ represents a discretized version of (2.2). An *orbit of* $P_\delta(t_0)$ in the $(\varphi, \dot{\varphi})$-plane is a smooth curve that is invariant under iteration by $P_\delta(t_0)$. In particular, any orbit of (2.1) is an orbit of $\mathbf{P}_0(t_0)$. So the question arises, how do these orbits change when δ becomes different from 0?

Next we discuss this question for the separatrix Γ^+. For small $\delta > 0$, the saddle points $(\pm\pi, 0)$ of (2.1) are perturbed to nearby fixed points of $P_\delta(t_0)$, which correspond to T-periodic solutions of (2.2). Furthermore, close to Γ^+ there are orbits W^- and W^+ of $P_\delta(t_0)$, along which the perturbed fixed points are approached under backward, respectively forward, iteration (see, e.g., **[5]**). W^- is a piece of the unstable manifold of $P_\delta(t_0)$ corresponding to the fixed point close to $(-\pi, 0)$, and W^+ is a piece of the stable manifold corresponding to the fixed point close to $(\pi, 0)$. However, in general W^- and W^+ do not coincide for $\delta > 0$, i.e., the separatrix Γ^+ splits under the influence of the forcing term in (2.2). Of course, W^- and W^+ can still intersect or touch each other at certain isolated points. Let us define the *distance function* $d_\delta(t_0)$ which measures the Euclidean distance between W^- and W^+ along the $\dot{\varphi}$-axis in the phase plane:

$$d_\delta(t_0) = \operatorname{dist}(W^- \cap \dot{\varphi}\text{–axis}, W^+ \cap \dot{\varphi}\text{–axis}).$$

We point out that $d_\delta(t_0)$ is a T-periodic, continuously differentiable function of t_0 for each fixed, sufficiently small δ. Obviously, W^- and W^+ intersect on the $\dot{\varphi}$-axis if and only if $d_\delta(t_0) = 0$. Moreover, one can show that $d_\delta(t_0) = 0$ and $d'_\delta(t_0) \neq 0$ imply that W^- and W^+ intersect transversally, i.e., Γ^+ splits. Here $d'_\delta(t_0)$ is the derivative with respect to t_0. Note that in the definition of $d_\delta(t_0)$ the $\dot{\varphi}$-axis could be replaced by any straight line that intersects Γ^+ transversally.

3. Chaos. It is well known that for a dynamical system generated by a diffeomorphism in the plane with a saddle point, transverse intersection of the corresponding stable and unstable manifolds implies chaos. Already Poincaré knew that this phenomenon implies the existence of complicated orbits [**15**]. However, as far as the author knows, Smale [**21**] was the first one to suggest looking for a certain order in this chaos. Using his horseshoe construction, he showed that the Bernoulli shift on a bi-infinite sequence of two symbols can be embedded as a subsystem into the dynamical system generated by a certain power of the diffeomorphism. Later Moser [**12**] generalized this result to sequences of any finite number of symbols. We do not explain here what this means for the motion of the pendulum. Rather, we next give a more detailed description of the chaotic motion resulting from the transverse intersection of W^- and W^+ as discussed in the previous section. This description is based on results by Palmer [**14**], which he proved directly for differential equations (even in $\mathbf{R}^n$) rather than for corresponding diffeomorphisms. This approach can even be generalized to almost periodic forcing terms, where no mapping like $P_\delta(t_0)$ is available to describe the dynamics (see Scheurle [**19**]). An example would be, if in (2.2) the amplitude δ were a periodic function of t with period rationally independent of T. For the general definition of an almost periodic function see, e.g., Bohr [**2**].

But let us now come back to the periodic case and assume that $d_\delta(t_0) = 0$ and $d'_\delta(t_0) \neq 0$ hold for some t_0 and small $\delta > 0$. Then any motion of the following type is possible for the pendulum described by equation (2.2). Starting from the vertically upward position, it does a complete clockwise revolution around the point of suspension, then it swings around the vertically upward position for a certain period of time, then it does another complete clockwise revolution, and so on. So these motions look very erratic. The times when the revolutions start and the length of time intervals between two consecutive revolutions are not completely arbitrary, though. In fact, the time at which the pendulum is in the vertically downward position during the kth revolution is approximately given by $t = t_0 + j_k T$, where the j_k are integers such that $j_{k+1} - j_k$ is sufficiently large, but otherwise arbitrary. One can think of the motion during each revolution to follow approximately the separatrix Γ^+ of the unperturbed problem. It is also possible that there is no complete revolution at all before or after a certain time, i.e., the pendulum swings around the vertically upward position asymptotically as $t \to -\infty$ or as $t \to +\infty$.

However, the complicated structure of these solutions alone would not justify the term chaotic motion, since each solution is still uniquely determined by initial data. What makes the motion really chaotic is the sensitive dependence on these data. This means that for each solution in the described class there is another one in this class arbitrarily close initially, but far away eventually as time increases, i.e., the time intervals of complete revolutions of the pendulum approximately agree initially, but differ eventually. Hence, although the motion is deterministic in principle, a long-term prediction is practically impossible, since measurements of data are not exact in general. This is exactly what deterministic chaos is all about.

So far we have discussed only the splitting of the separatrix Γ^+ and its consequences for the motion of the pendulum. But everything remains true if we replace Γ^+ by Γ^-, except that the complete revolutions of the pendulum around the point of suspension are then counterclockwise. On the other hand, in general both Γ^+ and Γ^- split simultaneously under external forcing. Then even solutions of the described type with any combination of clockwise and counterclockwise complete revolutions are possible.

Finally, we mention that in case of an almost periodic forcing term, similar solutions are possible. However, the times of revolutions of the pendulum are not determined as exactly as in the periodic case. One can prove only the existence of a certain time length such that within any time interval of this length the pendulum can start a complete revolution. So, there is a little less order in the structure of the solutions than in the periodic case.

4. Melnikov theory. At this point the question remains: Is there a simple way to verify the above conditions for the transverse intersection of W^- and W^+? In principle it is clear how to proceed. One has to compute the distance function $d_\delta(t_0)$ or an appropriate approximation of it and to look for simple zeros with respect to t_0. For instance, this can be achieved using the so-called *Melnikov function* $M(t_0)$. This is nothing other than the first-order coefficient in the Taylor expansion of $d_\delta(t_0)$ with respect to δ about $\delta = 0$:

$$d_\delta(t_0) = \delta M(t_0) + O(\delta^2) \quad \text{as } \delta \to 0. \tag{4.1}$$

Obviously, if δ is sufficiently small, then simple zeros of $M(t_0)$ imply simple zeros of $d_\delta(t_0)$. So we can state the following.

CRITERION. *Let δ be sufficiently small and assume that $M(t_0) = 0$ and $M'(t_0) \neq 0$ hold for some $t_0 \in \mathbf{R}$. Then splitting of the separatrix Γ^+ and chaos in the sense described in the previous sections follow.*

Thus, the search for chaos in a problem like (2.2) reduces to the search for simple zeros of the scalar function $M(t_0)$. This approach goes back to Melnikov [**11**]; see also the book of Guckenheimer and Holmes [**4**] and the references therein. Of course, how useful this criterion is depends on how complicated it is to compute $M(t_0)$. But this is no problem in principle. In fact, there is a simple integral formula for $M(t_0)$ involving only the representation of the unperturbed separatrix and terms from the equation itself. Actually we can define a Melnikov function for general equations of the form

$$\dot{\mathbf{u}} = g(\mathbf{u}) + \delta h(\mathbf{u}, t, \delta), \qquad \mathbf{u} \in \mathbf{R}^2, \tag{4.2}$$

under the following assumptions (see Palmer [**14**]).

DEFINITION. Let g and h in (4.2) be two-times continuously differentiable functions in suitable domains of their arguments. Assume that h together with its derivatives up through second order is uniformly bounded for $t \in \mathbf{R}$. Moreover, let the unperturbed equation

$$\dot{\mathbf{u}} = g(\mathbf{u}) \tag{4.3}$$

have a hyperbolic saddle point $\mathbf{u} = \mathbf{u}_0$, i.e., $g(\mathbf{u}_0) = 0$ and the Jacobian matrix $Dg(\mathbf{u}_0)$ has a positive and a negative real eigenvalue. Also assume that (4.3) has a homoclinic orbit given by $\mathbf{u} = \bar{\mathbf{u}}(t)$, $t \in \mathbf{R}$, with $\lim_{t \to \pm\infty} \bar{\mathbf{u}}(t) = \mathbf{u}_0$. Then the Melnikov function is defined to be

$$M(t_0) = \int_{-\infty}^{\infty} g(\bar{\mathbf{u}}(t)) \wedge h(\bar{\mathbf{u}}(t), t + t_0, 0) \exp\left[-\int_0^{t+t_0} \operatorname{trace} Dg(\bar{\mathbf{u}}(\tau))\, d\tau\right] dt$$

where $\mathbf{u} \wedge \mathbf{v} = u_1 v_2 - u_2 v_1$ is the usual wedge product for vectors $\mathbf{u}$, $\mathbf{v}$ in $\mathbf{R}^2$ with components u_i and v_i $(i = 1, 2)$.

REMARK. If, in addition, the forcing function h in (4.2) is almost periodic in t, then the criterion is still valid, although in the nonperiodic case the geometric interpretation of the Melnikov function is completely different from the one in (4.1) (see Scheurle [**19**]). This is due to the fact that a map like $P_\delta(t_0)$ is available only in the case of periodic forcing terms.

Of course, the evaluation of the improper integral in the definition of $M(t_0)$ can cause trouble. Sometimes the method of residues applies to do it in closed form. But in general one might have to use numerical methods. Also, one might have to use numerical methods to determine simple zeros of $M(t_0)$. In any case, Melnikov's method is a good means to detect chaos in specific problems (see also **[6]**, **[7]**, **[16]**, and **[18]**).

We finish this section with a simple example.

EXAMPLE. Consider equation (2.2) with $f(t) = \sin \omega t$, $\omega \in \mathbf{R}$. Rewriting it as a first-order system yields an equation of type (4.2) with

$$\mathbf{u} = \begin{pmatrix} \varphi \\ v \end{pmatrix}, \quad g(\mathbf{u}) = \begin{pmatrix} v \\ -\sin \varphi \end{pmatrix}, \quad h(\mathbf{u}, t, \delta) = \begin{pmatrix} 0 \\ \sin \omega t \end{pmatrix},$$

where $\dot{\varphi} = v$. According to the discussion in Section 2, the unperturbed equation (4.3) has a saddle point and a corresponding homoclinic orbit given by

$$\mathbf{u}_0 = \begin{pmatrix} \pm\pi \\ 0 \end{pmatrix} \quad \text{and} \quad \Gamma = \Gamma^+ : \begin{cases} \overline{\varphi}(t) = 2\arctan(\sinh(t)) \\ \overline{v}(t) = 2\,\mathrm{sech}(t) \end{cases}.$$

Hence, using the definition and the theorem of residues, we obtain for the Melnikov function

$$M(t_0) = 2\int_{-\infty}^{\infty} \mathrm{sech}(t) \sin \omega(t + t_0)\, dt = 2\pi\, \mathrm{sech}\left(\frac{\pi\omega}{2}\right) \sin \omega t_0.$$

Obviously, for $\omega \neq 0$ it has a simple zero at $t_0 = 0$. Thus, the criterion applies.

5. Exponentially small splitting of separatrices. To explain this phenomenon let us now consider the rapidly forced pendulum equation

$$\ddot{\varphi} + \sin \varphi = \delta \sin(t/\varepsilon), \tag{5.1}$$

where ε is supposed to be a small, positive parameter. In the limit as ε tends to 0, the period of the forcing term in (5.1) tends to 0 and its frequency to ∞. In this sense, we now have a rapidly forced system. Of course, for fixed ε this is the same problem as in the previous example. Just set $\omega = 1/\varepsilon$. So the Melnikov function is given by

$$M_\varepsilon(t_0) = 2\pi\, \mathrm{sech}\left(\frac{\pi}{2\varepsilon}\right) \sin\left(\frac{t_0}{\varepsilon}\right).$$

Since it has a simple zero at $t_0 = 0$, the criterion applies again. It implies splitting of the separatrix Γ^+ and chaos in the dynamics described by (5.1), if δ is sufficiently small. However, as $\varepsilon \to 0$, the upper bound for δ might shrink to 0, since $M_\varepsilon(t_0)$ depends on ε. In fact, $M_\varepsilon(t_0)$ is of order $O(e^{-\pi/2\varepsilon})$ as $\varepsilon \to 0$. Therefore, by (4.1) Melnikov's theory appears to be valid only if δ is exponentially small with respect to ε. As far as the author knows, for problems like (5.1) this was first observed by Sanders **[17]**. The following theorem tells us that for (5.1) it is really difficult to detect a splitting of the separatrices $\Gamma^\pm$ for values of δ that are not exponentially small with respect to ε.

THEOREM (HOLMES, MARSDEN, AND SCHEURLE **[8]**). *Consider equation* (5.1) *and, for each fixed* $\varepsilon > 0$, *let the distance function* $d_{\delta,\varepsilon}(t_0)$ *be defined as in Section* 2. *Then, given any* $\eta > 0$, *there exist constants* $\delta_0 > 0$ *and* $C = C(\eta, \delta_0) > 0$, *such that*

$$d_{\delta,\varepsilon}(t_0) \leq \delta C e^{-(\pi/2-\eta)/\varepsilon}$$

holds for all $t_0 \in \mathbf{R}$, $\varepsilon \in (0,1]$, *and* $\delta \in [0, \delta_0]$.

So, for parameter values in a full neighbourhood of the origin in the (δ, ε)-plane, the distance between W^- and W^+ along the $\dot{\varphi}$-axis is exponentially small with respect to ε, i.e., splitting of Γ^+, if it occurs at all, is an exponentially small phenomenon, beyond all algebraic orders in ε. On the one hand, in this case splitting of the separatrix Γ^+ really affects the dynamics only with respect to very large time scales. Thus, if one is interested only in the evolution of the system with respect to moderate time scales, then one can just ignore it. On the other hand, as far as the asymptotic behaviour as $t \to \pm\infty$ is concerned, splitting of Γ^+ is important, but very hard to predict in case of exponential smallness. In particular, a power expansion of the distance function with respect to ε does not help either. For a first positive result see **[8]** and **[20]**.

The theorem is a special case of a corresponding result for the general equation

$$\dot{\mathbf{u}} = g(\mathbf{u}) + \delta h(\mathbf{u}, t/\varepsilon, \delta), \qquad \mathbf{u} \in \mathbf{R}^2, \tag{5.2}$$

with the following properties. The functions g and h are analytic; h is 2π-periodic in $\theta = t/\varepsilon$. For $\delta = 0$, (5.2) has a hyperbolic saddle point $\mathbf{u} = \mathbf{u}_0$ and a corresponding homoclinic orbit Γ: $\mathbf{u} = \overline{\mathbf{u}}(t)$ which is analytic in the strip $|\operatorname{Im} t| < r$ of the complex t-plane. Equation (5.2) is either a Hamilton system or is time-reversible, i.e., invariant under a transformation $t \mapsto -t$, $\mathbf{u} \mapsto R\mathbf{u}$, where R is a linear reflection operator in $\mathbf{R}^2$ with eigenvalues ± 1. An additional technical assumption on the first variation equation $\dot{\mathbf{v}} = Dg(\overline{\mathbf{u}}(t))\mathbf{v}$ is satisfied. We note that all these assumptions are satisfied for equation (5.1) considered as a first-order system. In particular, this is invariant under the transformation

$t \mapsto -t$, $\varphi \mapsto -\varphi$, and $\dot{\varphi} \mapsto \dot{\varphi}$. So it is time-reversible. Also, the closed-form representation for Γ^+ given in Section 2 is analytic in the strip $|\operatorname{Im} t| < \pi/2$. This is where the term $\pi/2$ in the exponent of the upper bound for $d_{\delta,\varepsilon}(t_0)$ comes from in the theorem. For similar theorems see Neisthadt **[13]** and Fontich and Simó **[3]**. Also see the discussion in Arnol'd (**[1]**, pp. 395, 407).

Finally, we show that equations like (5.2) really come up in applications. To this end we consider a system of two weakly coupled nonlinear oscillators with Hamilton function

$$\widetilde{H}(\theta, I, \varphi, J, \varepsilon) = I - I^2 + J + \varepsilon I \sin^2\theta \cos\varphi$$

written in action angle variables. Here ε is a small positive parameter. Restriction to a three-dimensional energy surface given by $\widetilde{H} = \text{const.}$, elimination of the J variable, and replacing the time t by φ as the independent variable lead to the following one-degree-of-freedom reduced system

$$\begin{aligned} \frac{dI}{d\varphi} &= -\varepsilon 2I \sin\theta \cos\theta \cos\varphi, \\ \frac{d\theta}{d\varphi} &= 1 - 2I + \varepsilon \sin^2\theta \cos\varphi \end{aligned} \tag{5.3}$$

with Hamilton function

$$H(\theta, I, \varphi, \varepsilon) = I - I^2 + \varepsilon I \sin^2\theta \cos\varphi.$$

Obviously, for $\varepsilon = 0$ the energy surface $\widetilde{H} = \text{const.}$ is filled with nested 2-tori given by $I = \text{const.}$ The flow on these tori is either periodic or quasiperiodic, depending on whether the number $1 - 2I$ is rational or irrational. According to the theory of Kolmogorov, Arnol'd, and Moser (**[1]**, **[12]**), most of the quasiperiodic tori persist for $\varepsilon \neq 0$, while the periodic tori typically break down. In the fixed energy surface, the persisting quasiperiodic 2-tori are separated by so-called resonance zones, which replace the destroyed periodic tori.

It turns out that the motion in these resonance zones is governed by equations of type (5.2). For example, let us consider the torus given by $I = 1/4$ for $\varepsilon = 0$. This is periodic. The periodic orbits on it are given by $\theta = \varphi/2 + \text{const.}$, i.e., there is a $2:1$ resonance of the angular frequencies. To study the motion in the corresponding resonance zone for $\varepsilon \neq 0$ (the so-called $2:1$ resonance zone), we introduce local coordinates p, Ψ via

$$I = 1/4 - \sqrt{\varepsilon} p, \qquad \theta = \varphi/2 + \Psi.$$

Introducing these in (5.3) yields

$$\tag{5.4} \begin{aligned} \frac{dp}{d\varphi} &= \sqrt{\varepsilon}\,[(-1/8)\sin 2\Psi + \sin 2(\Psi + \varphi)] \\ &\quad + \varepsilon\,[(p/2)\sin 2\Psi + \sin 2(\Psi + \varphi)], \\ \frac{d\Psi}{d\varphi} &= 2\sqrt{\varepsilon} p + \varepsilon\,[(1/4)\cos 2\Psi + \cos 2(\Psi + \varphi) - (1/2)\cos\varphi]. \end{aligned}$$

Finally, by the averaging transformation

$$\begin{aligned} \Psi &= \tilde{\Psi}, \\ p &= \tilde{p} - (\sqrt{\varepsilon}/16)\cos(\tilde{\Psi} + \tilde{\varphi}/\sqrt{\varepsilon}), \\ \varphi &= \tilde{\varphi}/\sqrt{\varepsilon} \end{aligned}$$

we obtain the equivalent system of equations

$$\tag{5.5} \begin{aligned} \frac{d\tilde{p}}{d\tilde{\varphi}} &= -\,(1/18)\sin 2\Psi + \sqrt{\varepsilon} P(\tilde{\Psi}, \tilde{p}, \tilde{\varphi}/\sqrt{\varepsilon}, \sqrt{\varepsilon}), \\ \frac{d\tilde{\Psi}}{d\tilde{\varphi}} &= 2\tilde{p} \qquad\qquad\qquad + \sqrt{\varepsilon} Q(\tilde{\Psi}, \tilde{p}, \tilde{\varphi}/\sqrt{\varepsilon}, \sqrt{\varepsilon}). \end{aligned}$$

Here both the functions P and Q are 2π-periodic in $\varphi = \tilde{\varphi}/\sqrt{\varepsilon}$. Hence, this system is of type (5.2) if we replace both δ and ε by $\sqrt{\varepsilon}$ there. Note that δ is not exponentially small with respect to ε in this case.

Setting $\sqrt{\varepsilon} = 0$ in (5.5), we obtain a first-order approximation for the dynamics in the 2 : 1 resonance zone. But this leads to a problem that is equivalent to the simple pendulum equation (2.1) modulo some scaling. In particular, there are separatrices like $\Gamma^{\pm}$ in the phase plane. So it is likely that these separatrices split under the influence of the terms of order $O(\sqrt{\varepsilon})$ in (5.5) and that we actually have chaos in the 2 : 1 resonance zone (cf. Zehnder **[22]**). However, according to our theory the splitting is exponentially small with respect to ε, if it occurs at all. Actually this is also the reason for the difficulties that arise in proving nonintegrability of systems like coupled oscillators. Already Poincaré **[15]** faced this difficulty when he tried to prove nonintegrability of the three-body problem.

For more applications involving equations like (5.2), see Holmes, Marsden, and Scheurle **[9]** and Hunter and Scheurle **[10]**.

References

1. V. I. Arnol′d, *Mathematical Methods of Classical Mechanics*, Springer-Verlag, New York, 1978.

2. H. Bohr, *Almost Periodic Functions*, Chelsea, New York, 1947.

3. E. Fontich and C. Simó, *The splitting of separatrices for analytic diffeomorphisms*, preprint, 1988.

4. J. Guckenheimer and P. Holmes, *Nonlinear Oscillations Dynamical Systems, and Bifurcations of Vector Fields*, Springer-Verlag, New York, 1983.

5. M. W. Hirsch, C. C. Pugh, and M. Shub, *Invariant manifolds*, Lecture Notes in Mathematics, **583**, Springer-Verlag, New York, 1977.

6. P. Holmes and J. Marsden, *Bifurcations to divergence and flutter in flow-induced oscillations: An infinite-dimensional analysis*, Automatica **14** (4) (1978), 367–384.

7. ——, *A partial differential equation with infinitely many periodic orbits: Chaotic oscillations of a forced beam*, Arch. Rat. Mech. Anal. **76** (1981), 135–166.

8. P. Holmes, J. Marsden, and J. Scheurle, *Exponentially small splitting of separatrices*, Proc. Nat. Acad., to appear.

9. ——, *Exponentially small splittings of separatrices in KAM theory and degenerate bifurcations*, Cont. Math. **81** (1988), 213–244.

10. J. Hunter and J. Scheurle, *Existence of perturbed solitary wave solutions to a model equation for water waves*, Physica D. **32** (1988), 253–268.

11. V. K. Melnikov, *On the stability of the center for time periodic perturbations*, Trans. Moscow Math. Soc. **12** (1963), 1–57.

12. J. Moser, *Stable and random motions in dynamical systems*, Princeton University Press, Princeton, New Jersey, 1973.

13. A. Neisthadt, *The separation of motions in systems with rapidly rotating phase*, P.M.M. USSR **48** (1984), 133–139.

14. K. Palmer, *Exponential dichotomies and transversal homoclinic points*, J. Differential Equations **55** (2), (1984), 225–256.

15. H. Poincaré, *Sur la problème des trois corps et les équations de la dynamique*, Acta Math. **13** (1890), 1–271.

16. F. M. A. Salam and S. S. Sastry, *Dynamics of the forced Josephson junction circuit: The regions of chaos*, IEEE Trans. Circuits and Systems, **CAS-32** (1985), 784–796.

17. J. Sanders, *Melnikov's method and averaging*, Celestial Mech. **28** (1982), 171–181.

18. S. Schecter, *Melnikov's method at a saddle-node and the dynamics of the forced Josephson junction*, SIAM J. Math. Anal. **18** (6) (1987), 1699–1715.

19. J. Scheurle, *Chaotic solutions of systems with almost periodic forcing*, ZAMP **37** (1986), 12–26.

20. ——, *Chaos in the rapidly forced pendulum equation*, Cont. Math., to appear.

21. S. Smale, *Diffeomorphisms with many periodic points*, Differential and combinatorial topology (S. S. Cairns, ed.), Princeton University Press, Princeton, New Jersey, 1963, pp. 63–80.

22. E. Zehnder, *Homoclinic points near elliptic fixed points*, Comm. Pure Appl. Math. **26** (1973), 131–182.

INSTITUT FÜR ANGEWANDTE MATHEMATIK, UNIVERSITÄT HAMBURG

DEPARTMENT OF MATHEMATICS, COLORADO STATE UNIVERSITY

Lectures in Applied Mathematics
Volume **26** (1990)

PL Methods for Constructing a Numerical Implicit Function

PHILLIP H. SCHMIDT

Given a mapping $\mathbf{f}: \mathbf{R}^{N+K} \longrightarrow \mathbf{R}^N$ the problem of interest is to determine and numerically represent the zero set

$$\mathscr{M} = \{\mathbf{x} \in \mathbf{R}^{N+K} : \mathbf{f}(\mathbf{x}) = \mathbf{0}\}.$$

This problem arises when a system of N nonlinear equations involving N unknowns and K parameters must be solved and analysed. Until recently, most parametric studies have involved only one parameter because of the unavailability of methods for handling the more general case.

These problems have been considered in several different ways. For the case $K = 1$ when $\mathscr{M}$ is a set of dimension 1, curve following methods have been useful; see for example Allgower and Georg **[3]** and Keller **[13]**. For $K \geq 2$, three distinct approaches have been used. Fink and Rheinboldt **[9]** and Rheinboldt **[14]** have used one-dimensional continuation methods to investigate $\mathscr{M}$ in a neighborhood of a solution point. Allgower and Schmidt **[5]**, **[6]**, Allgower and Gnutzmann **[4]**, and Gnutzmann **[12]** have used piecewise linear methods to approximate a connected component of $\mathscr{M}$. Rheinboldt **[15]** has developed a moving frame algorithm to triangulate a part of the manifold. It is the piecewise linear approach that will be described here.

Earlier work concentrated on the construction of $\mathscr{M}$ in the case $K = 2$ with the simple application to production of computer graphics representation of surfaces which are described by a single equation in three

---

1980 *Mathematics Subject Classification* (1985 *Revision*). Primary 65H10; Secondary 58C15.

The final version of this paper will be submitted for publication elsewhere.

variables; see Allgower and Gnutzman [**4**]. Gnutzmann [**12**] recently presented a complete study of the case where $K \geq 2$. His dissertation includes analysis of the problem, a description of an algorithm, numerical examples, and an implementation of the algorithm in PASCAL.

We wish to concentrate here on the problem where there are K distinguished parameters, i.e., when $\mathbf{f}$ can be written as $\mathbf{f}(x_1, \ldots, x_N, y_1, \ldots, y_K) \in \mathbf{R}^N$. We know by the Implicit Function Theorem that if the partial Jacobian with respect to $\mathbf{x}$ is of rank N at a solution point $(\mathbf{x}^0, \mathbf{y}^0)$, then an implicit function which determines $\mathbf{x}$ as a function of $\mathbf{y}$ is defined near $(\mathbf{x}^0, \mathbf{y}^0)$. This implicit function is of a great deal of interest in itself, specifically, as a tool in the Liapunov–Schmidt reduction procedure of bifurcation theory. We concentrate here on its numerical determination. The method which we propose is based on the piecewise linear methods [**4**], [**5**], [**6**], [**12**] combined with insights provided by the continuation approach of Fink and Rheinboldt [**9**] and Rheinboldt [**14**].

We will carry out the description of the calculation of this numerical implicit function in several stages. We begin by describing the piecewise linear tools such as triangulations, simplices, piecewise linear mapping that are used in defining a piecewise linear approximation to $\mathcal{M}$, call it $\mathcal{M}_{\mathcal{T}}$. We discuss some of the properties of $\mathcal{M}_{\mathcal{T}}$ and describe two algorithms for its construction. The first algorithm is an amalgamation of those described in Allgower and Schmidt [**6**], Allgower and Gnutzmann [**4**], and Gnutzmann [**12**]. The second is an algorithm for following curves in $\mathcal{M}_{\mathcal{T}}$ and will be useful in the determination of a piecewise linear implicit function which is related to the manifold $\mathcal{M}_{\mathcal{T}}$. We then describe the use of Newton's method to calculate the actual values of the implicit function.

PL tools. The basic tool in our construction is a piecewise linear approximation to the mapping $\mathbf{f}$. This approximation is determined by the values of $\mathbf{f}$ relative to a triangulation of $\mathbf{R}^{N+K}$ into $(N+K)$-simplices as follows.

Given a set V, the *affine hull* of V is

$$\operatorname{aff}(V) = \left\{ z = \sum_{i=0}^{N+K} \lambda_i v_i : \quad \sum_{i=0}^{N+K} \lambda_i = 1, \quad v_i \in V, \quad \lambda_i \in \mathbf{R} \right\}.$$

The *affine dimension* of a set V is the dimension of the linear space

$$\{z_1 - z_2 : z_i \in \operatorname{aff}(V)\}.$$

A finite set is *affinely independent* if its affine dimension is one less than its cardinality.

Given $N+K$ affinely independent points $\mathbf{v}_0, \ldots, \mathbf{v}_{N+K}$, an $(N+K)$-*simplex* σ is the convex hull of this set of vertices denoted as $\sigma = [\mathbf{v}_0, \ldots, \mathbf{v}_{N+K}]$.

For any $j \in \{0, \ldots, N+K\}$, a *j-face,*τ, of σ is the convex hull of some $(j+1)$ of these vertices, $\tau = [\mathbf{v}_{i_0}, \ldots, \mathbf{v}_{i_j}]$.

A *triangulation* of $\mathbf{R}^{N+K}$ is a collection of $(N+K)$-simplices, $\mathscr{T} = \{\sigma\}$, whose union covers $\mathbf{R}^{N+K}$, whose members intersect pairwise in at most a common j-face, and whose intersection with any compacta consists of a finite number of simplices.

The *diameter* of a simplex is

$$\operatorname{diam}(\sigma) = \max\{\|\mathbf{v}_1 - \mathbf{v}_2\|_\infty : \mathbf{v}_i \in \sigma\}.$$

The *meshsize* of a triangulation $\mathscr{T}$ is

$$\operatorname{mesh}(\mathscr{T}) = \sup\{\operatorname{diam}(\sigma) : \sigma \in \mathscr{T}\}.$$

In order to be useable for numerical computation, the $(N+K)$-simplices and lower-order faces of a triangulation should be easily generated, stored, and compared. A triangulation which satisfies these criteria is that due to Freudenthal [**10**]. This triangulation of $\mathbf{R}^{N+K}$ is based at a specific point $\mathbf{v}^0$ and has meshsize proportional to δ. It will be denoted as $\mathscr{F}_{N+K}[\mathbf{v}^0;\delta]$ and is defined as follows:

The vertices are lattice points in $\{\mathbf{v}^0\} + \delta Z^{N+K}$ where Z denotes the set of integers.

An $(N+K)$-simplex σ can be identified with a vertex $\mathbf{z}_0 \in Z^{N+K}$ and a permutation Π of $\{1, \ldots, N+K\}$ in the sense that $\sigma = [\mathbf{v}_0, \ldots, \mathbf{v}_{N+K}]$ where,

$$\mathbf{v}_0 = \mathbf{v}^0 + \delta \mathbf{z}_0 \qquad \text{and} \qquad \mathbf{v}_i = \mathbf{v}_{i-1} + \delta \mathbf{e}_{\Pi(i)} \quad \text{for } i = 1, \ldots, N+K;$$

here $\mathbf{e}_j$ is a standard unit vector.

Denote $\sigma \in \mathscr{F}_{N+K}[\mathbf{v}^0;\delta]$ as $\sigma = [\mathbf{z}_0;\Pi]$ when the combinatorial structure is important.

The lower-order faces $\tau = [\mathbf{v}_0, \ldots, \mathbf{v}_j]$ can similarly be identified with a vertex $\mathbf{z}_0$ and a partition Φ of $\{1, \ldots, N+K\}$ into $j+1$ subsets $\Phi_1, \ldots, \Phi_j, \Phi_{j+1}$, all nonempty except possibly Φ_{j+1} so that

$$\mathbf{v}_0 = \mathbf{v}^0 + \delta \mathbf{z}_0 \qquad \text{and} \qquad \mathbf{v}_i = \mathbf{v}_{i-1} + \delta \sum_{j \in \Phi(i)} \mathbf{e}_j \quad \text{for } i = 1, \ldots, j.$$

The diameter of each σ in $\mathscr{F}_{N+K}[\mathbf{v}^0;\delta]$ is δ so the meshsize of the Freudenthal triangulation is the same.

The *piecewise linear mapping* $\mathbf{f}_{\mathscr{T}} : \mathbf{R}^{N+K} \longrightarrow \mathbf{R}^N$ is determined by the values of $\mathbf{f}$ at vertices of simplices of the triangulation $\mathscr{T}$. It is defined for all of $\mathbf{R}^{N+K}$ by interpolation within the simplices. Each point $\mathbf{x}$ lies in some simplex σ (not necessarily unique) and thus has a barycentric representation relative to the vertices of σ:

(1)
$$\mathbf{x} = \sum_{i=0}^{N+K} \lambda_i \mathbf{v}_i, \text{ where } \sum_{i=0}^{N+K} \lambda_i = 1, \quad \text{and} \quad \lambda_i \geq 0 \quad \text{for } i = 0, \ldots, N+K.$$

The value of $\mathbf{f}_{\mathcal{T}}$ at $\mathbf{x}$ is given by

$$\text{(2)} \qquad \mathbf{f}_{\mathcal{T}}(\mathbf{x}) = \sum_{i=0}^{N+K} \lambda_i \mathbf{f}(\mathbf{v}_i), \quad \text{where } \lambda_i \text{ is determined by (1).}$$

If $\mathbf{x}$ is in the intersection of two or more simplices, then $\mathbf{x}$ lies in the relative interior of some j-face τ. The only nonzero λ_i in formula (2) correspond to the vertices of τ which are common to any simplex σ containing $\mathbf{x}$. Thus the definition of $\mathbf{f}_{\mathcal{T}}$ depends only on $\mathbf{x}$ and not on a particular simplex containing $\mathbf{x}$. $\mathbf{f}_{\mathcal{T}}$ clearly has the interpolating property and moreover is continuous provided only that $\mathbf{f}$ is defined at all vertices of simplices in the triangulation.

From the alternative representation of formulas (1) and (2) for $\mathbf{x} \in \sigma$

$$\text{(1')} \qquad \mathbf{x} = \mathbf{v}_0 + \sum_{i=1}^{N+K} \lambda_i (\mathbf{v}_i - \mathbf{v}_0), \qquad \lambda_i \geq 0,$$

and

$$\text{(2')} \qquad \mathbf{f}_{\mathcal{T}}(\mathbf{x}) = \mathbf{f}(\mathbf{v}_0) + \sum_{i=1}^{N+K} \lambda_i (\mathbf{f}(\mathbf{v}_i) - \mathbf{f}(\mathbf{v}_0)),$$

it follows that

$$\text{(3)} \qquad \mathbf{f}_{\mathcal{T}}(\mathbf{x}) = (\mathbf{f}(\mathbf{v}_0) - [\Delta\mathbf{f}(\sigma)][\Delta\sigma]^{-1}\mathbf{v}_0) + [\Delta\mathbf{f}(\sigma)][\Delta\sigma]^{-1}\mathbf{x}$$

where the matrices $[\Delta\mathbf{f}(\sigma)]$ and $[\Delta\sigma]$ are defined by columns as

$$\text{(4)} \qquad [\Delta\mathbf{f}(\sigma)] = [(\mathbf{f}(\mathbf{v}_1) - \mathbf{f}(\mathbf{v}_0)) \cdots (\mathbf{f}(\mathbf{v}_{N+K}) - \mathbf{f}(\mathbf{v}_0))]$$

and

$$\text{(5)} \qquad [\Delta\sigma] = [(\mathbf{v}_1 - \mathbf{v}_0) \cdots (\mathbf{v}_{N+K} - \mathbf{v}_0)].$$

It can also be shown that if $\mathbf{f}$ has a Jacobian $D\mathbf{f}(\mathbf{x})$ which is Lipschitz continuous in a neighborhood of the simplex σ, then there exist constants K_1 and K_2 so that

$$\|\mathbf{f}(\mathbf{x}) - \mathbf{f}_{\mathcal{T}}(\mathbf{x})\| \le K_1(\operatorname{diam}\sigma)^2 \tag{6}$$

and

$$\left\| D\mathbf{f}(\mathbf{x}) - [\Delta\mathbf{f}(\sigma)][\Delta\sigma]^{-1} \right\| \le K_2(\operatorname{diam}\sigma), \tag{7}$$

where $\mathbf{x} \in \sigma$, and the norms are the sup norm and an associated matrix norm, respectively. Thus not only can an approximation of $\mathbf{f}$ be made, but also a reasonable approximation of the Jacobian $D\mathbf{f}(\mathbf{x})$ is possible. See Allgower and Gnutzmann [**4**] and Gnutzmann [**12**] for related results.

To guarantee the local existence of the solution manifold near a point $\mathbf{x} \in \mathcal{M}$, the usual regularity assumption on $\mathbf{f}$ is that $D\mathbf{f}(\mathbf{x})$ is of full rank. The analogous condition for $\mathbf{f}_{\mathcal{T}}$ is that $[\Delta\mathbf{f}(\sigma)]$ be of full rank.

The manifold $\mathcal{M}_{\mathcal{T}}$. With this approximation, we define the piecewise linear zero set

$$\mathcal{M}_{\mathcal{T}} = \{\mathbf{x} : \mathbf{f}_{\mathcal{T}}(\mathbf{x}) = \mathbf{0}\}.$$

In light of the estimate above, this set will provide an approximation for the zero manifold $\mathcal{M} = \{\mathbf{x} : \mathbf{f}(\mathbf{x}) = \mathbf{0}\}$ in the sense that, for any compact subset C of $\mathcal{M}_{\mathcal{T}}$, there exists a constant K so that if $\mathbf{x} \in C$ then $\|\mathbf{f}(\mathbf{x})\| \le K(\operatorname{mesh}\mathcal{T})^2$.

A sufficient condition for a given simplex σ to meet $\mathcal{M}_{\mathcal{T}}$ is that there exist a solution $\Lambda = (\lambda_0, \dots, \lambda_{N+K})^{\mathrm{T}}$ to the system

$$\sum_{i=0}^{N+K} \lambda_i = 1, \qquad \sum_{i=0}^{N+K} \lambda_i \mathbf{f}(\mathbf{v}_i) = \mathbf{0}, \quad \lambda_i \ge 0. \tag{8}$$

The matrix form for this system is

$$\begin{pmatrix} 1 & \cdots & 1 \\ \mathbf{f}(\mathbf{v}_0) & \cdots & \mathbf{f}(\mathbf{v}_{N+K}) \end{pmatrix} \Lambda = \begin{pmatrix} 1 \\ \mathbf{0} \end{pmatrix}, \quad \Lambda \ge \mathbf{0}. \tag{9}$$

We refer to the $(N+1) \times (N+K+1)$ matrix on the left-hand side of this equation as the *labeling matrix* and denote it by $[\mathbf{f}_{\mathcal{T}}(\sigma)]$. If $[\Delta\mathbf{f}(\sigma)]$ is of full rank, then $[\mathbf{f}_{\mathcal{T}}(\sigma)]$ will be of full rank also.

By the fundamental theorem of linear programming (Dantzig [**8**]), the existence of such a solution in the case where $[\mathbf{f}_{\mathcal{T}}(\sigma)]$ is of full rank $N+1$ is equivalent to the existence of a solution with at most $N+1$ nonzero entries and with the property that the corresponding $N+1$ columns of the matrix above are linearly independent, a basic feasible solution in the terminology of linear programming. These columns comprise an $(N+1)\times(N+1)$ invertible submatrix of $[\mathbf{f}_{\mathcal{T}}(\sigma)]$ called a *basis matrix.* The inverse of the basis matrix contains the nonnegative λ entries in its first column.

The N-face τ with vertices corresponding to the columns of this submatrix contains a point in $\mathcal{M}_{\mathcal{T}}$ whose barycentric coordinates relative to this N-face are the nonnegative entries in the first column of the inverse of this submatrix. If all these entries are positive then the zero-point lies in the interior of the N-face. If on the other hand some entries are zero, then the zero-point lies in several N-faces and there is ambiguity in determining which N-face contains this point. For this reason, we require the stronger condition on an N-face τ: that the $(N+1)\times(N+1)$ basis matrix, which we will refer to as $[\mathbf{f}_{\mathcal{T}}(\tau)]$, have an inverse with the property that the leading (nonzero) entry in each row be positive, that is, that it be a *lexicographic positive* inverse. An N-face which satisfies this condition is said to be *completely labeled.*

The completely labeled N-face, τ, contains a zero-point of $\mathbf{f}_{\mathcal{T}}$ which need not lie in the relative interior of τ. However, it can be shown that τ is completely labeled if and only if there exists $\varepsilon_0 > 0$ such that for each ε, $0 < \varepsilon < \varepsilon_0$, the N-face τ contains a point $\mathbf{x}(\varepsilon)$ such that

$$\mathbf{f}_{\mathcal{T}}(\mathbf{x}(\varepsilon)) = \overrightarrow{\varepsilon}, \quad \text{where } \overrightarrow{\varepsilon} = (\varepsilon, \varepsilon^2, \dots, \varepsilon^N)^{\mathrm{T}}.$$

Moreover, for $0 < \varepsilon < \varepsilon_0$, the points $\mathbf{x}(\varepsilon)$ will lie in the interior of τ. While $\mathbf{x}(0)$ may lie in several N-faces there can be no ambiguity about which N-face contains the solution curve to the perturbed equation in its interior. The point $\mathbf{x}(0)$ in a completely labeled N-face τ is called the *labeling center* of τ.

If a simplex σ contains a completely labeled N-face τ, then any one of the K columns of the labeling matrix $[\mathbf{f}_{\mathcal{T}}(\sigma)]$ not in the basis $[\mathbf{f}_{\mathcal{T}}(\tau)]$ may be exchanged for some column of $[\mathbf{f}_{\mathcal{T}}(\tau)]$ to form a new basis. The exchange of basis operation is the same as that applied in linear programming modified by the Charnes–Cooper anticycling procedure (Charnes, Cooper, and Henderson [7]). This modification chooses the column to leave the basis $[\mathbf{f}_{\mathcal{T}}(\tau)]$, so that the new basis matrix has a lexicographic positive inverse.

We define a j-face, $N+1 \leq j \leq N+K$, to be *transverse* if it contains a completely labeled N-face. From the argument above, it follows that a transverse simplex contains at least $K+1$ distinct completely labeled N-faces. Furthermore, each completely labeled N-face τ is contained in several simplices each of which is transverse. These last two observations are crucial in carrying out the construction of $\mathcal{M}_{\mathcal{T}}$.

We say that a property holds *generically* for $\mathcal{M}_{\mathcal{T}}$ if for some $\varepsilon_0 > 0$ the property holds for each $\mathcal{M}_{\mathcal{T}}(\varepsilon) = \{\mathbf{x}\colon \mathbf{f}_{\mathcal{T}}(\mathbf{x}) = \vec{\varepsilon}\}$ with $0 < \varepsilon < \varepsilon_0$. The intersection of $\mathcal{M}_{\mathcal{T}}$ with a transverse $(N+K)$-simplex σ is the convex hull of the labeling centers of the completely labeled N-faces of σ. Since the rank of the labeling matrix in equation (9) is $N+1$, the solution manifold for this equation has affine dimension $K+1$, so the piece of $\mathcal{M}_{\mathcal{T}}$ lying in σ is generically a K-polytope. The sides of each K-polytope of affine dimension $j\colon 0 \le j \le K$ lie generically in a transverse $(N+j)$-face. The properties of a triangulation imply that the intersection of the pieces of $\mathcal{M}_{\mathcal{T}}$ lying in different transverse simplices are either empty or are generically sides of the corresponding K-polytopes of appropriately lower dimension. At most two K-polytopes can intersect generically in a side of affine dimension $K-1$. Finally, as can be seen from the description of the algorithm, given any two K-polytopes in $\mathcal{M}_{\mathcal{T}}$, there exists a sequence of K-polytopes with the property that two consecutive polytopes in the sequence intersect generically in a side of affine dimension $K-1$. Thus, $\mathcal{M}_{\mathcal{T}}$ is generically a K-piecewise linear manifold; see Gnutzmann **[12]**.

An algorithm for determining $\mathcal{M}_{\mathcal{T}}$. The construction of $\mathcal{M}_{\mathcal{T}}$ begins with a point $\mathbf{x}^0$ for which $\mathbf{f}(\mathbf{x}^0) = \mathbf{0}$ and $D\mathbf{f}(\mathbf{x}^0)$ is of rank N. An N-face τ is constructed with $\mathbf{x}^0$ as barycenter and with nontrivial projection onto the normal space to the manifold $\mathcal{M}$ at $\mathbf{x}^0$. It can be shown that if $\operatorname{diam}(\tau)$ is sufficiently small, τ is completely labeled.

A completely labeled N-face is contained in several transverse $(N+K)$-simplices, the determination of which is a purely combinatorial problem easily solved in the case of the Freudenthal triangulation. Allgower and Georg **[2]** show how to determine a sequence of simplices of this triangulation by the process of pivoting across facets. This procedure was exploited by Allgower and Schmidt **[5]**, **[6]**, Allgower and Gnutzmann **[4]**, and Gnutzmann **[12]** to determine a sequence of transverse simplices all of which contain a given completely labeled N-face.

Given a transverse $(N+K)$-simplex σ, determining all its completely labeled N-faces is equivalent to finding all the linear programming basis matrices for the labeling matrix $[\mathbf{f}_{\mathcal{T}}(\sigma)]$. When $K = 2$, there are only two columns not in the basis for a given completely labeled N-face, so one need only choose one of these to enter the basis and perform the exchange of basis to find a new completely labeled N-face. This process can be repeated using the column which did not just previously leave the basis until the starting basis is recovered. Gnutzmann [12] handles the case $K > 2$ by using an efficient table search procedure.

An algorithm for the construction of $\mathcal{M}_{\mathcal{T}}$. GIVEN: a compactum $\mathcal{C}$, $\delta > 0$, a triangulation $\mathcal{T}$ of meshsize δ, and a point

$$\mathbf{x}^0 \in \mathcal{M} \cap \mathcal{C} \text{ such that } D\mathbf{f}(\mathbf{x}^0) \text{ is of rank } N;$$

Find a completely labeled N-face τ containing $\mathbf{x}^0$.
Determine all (transverse) $(N+K)$-simplices σ containing τ as an N-face. Set $\Sigma := \{\sigma\}$ and $T := \{\tau\}$.
While $\Sigma \neq \emptyset$ do:
  Select $\sigma \in \Sigma$.
  Determine all completely labeled N-faces of σ; call this set $T(\sigma)$.
  For each $\tau \in T(\sigma)$,
    find and save the labeling center of τ;
    if $\tau \in T$ then delete τ from $T(\sigma)$; otherwise insert τ into T.
  For each $\tilde{\tau} \in T(\sigma)$,
    determine all $\tilde{\sigma}$ containing $\tilde{\tau}$ such that $\tilde{\sigma} \subset \mathcal{C}$;
    insert all $\tilde{\sigma}$ in Σ;
  Delete σ from Σ.

For the sake of efficiency, the list of N-faces, T, need contain only those faces whose labeling centers are on the boundary of the portion of the manifold which has been constructed, since these are the only N-faces which can be encountered in the future. See the references cited above for detailed algorithms for constructing $\mathcal{M}_{\mathcal{T}}$. The dissertation of Gnutzmann [12] also contains an implementation of his algorithm in a PASCAL program. It should be noted that the costliest procedures to implement are the determination of all completely labeled N-faces of a transverse simplex and the determination of whether or not new N-faces or $(N+K)$-simplices have been previously encountered. Gnutzmann handles these latter problems with a binary search procedure that depends heavily on the combinatorial structure of the Freudenthal triangulation. Alexander [1] has provided a measure called the *average intersection density*, or "aid", for the expected number of intersections per unit cube of an affine space of dimension K with the N-faces of the triangulation $\mathcal{F}_{N+K}[\mathbf{0}; 1]$. This number gives some idea of the work involved in constructing $\mathcal{M}_{\mathcal{T}}$. Alexander shows that, for fixed K, aid $= C_K N^{3K/2}$ asymptotically as $N \to \infty$, for some constant independent of N. This shows that the work needed to construct $\mathcal{M}_{\mathcal{T}} \cap \mathcal{C}$ for fixed mesh($\mathcal{F}$) is polynomial in N. Numerical experiments with moderate-sized $N(\leq 10)$ and $K(\leq 4)$ have been carried out using Gnutzmann's PASCAL code on a microcomputer. They bear out the work estimates resulting from Alexander's measure.

The K-piecewise linear manifold can be put to a number of uses. One of these is to use its points as starting points for Newton's method. If mesh($\mathcal{T}$) is chosen sufficiently small then, for all $\mathbf{x} \in \mathcal{M}_{\mathcal{T}}$ with σ a transverse simplex containing $\mathbf{x}$, by inequalities (6) and (7) both $\|\mathbf{f}(\mathbf{x})\|$ and $\|[\Delta\mathbf{f}(\sigma)][\Delta\sigma]^{-1} - D\mathbf{f}(\mathbf{x})\|$ can be made as small as desired. The two factors in this approximation to the Jacobian were defined in equations (4) and (5). If $D\mathbf{f}$ is Lipschitz continuous and nonsingular on an open set containing $\mathcal{M} \cap \mathcal{C}$, then Newton's method can be applied to some augmented system of equations to drive points from $\mathcal{M}_{\mathcal{T}}$ to $\mathcal{M}$. Gnutzmann **[12]** uses the LQ factorization of $[\Delta\mathbf{f}(\sigma)]$ for each transverse σ to construct modified Newton iterates which are constrained to lie in the normal space to $\mathcal{M}_{\mathcal{T}} \cap \sigma$. This process drives the labeling centers of completely labeled N-faces to $\mathcal{M}$, where the points may be endowed with the same adjacency structure as the labeling centers had in $\mathcal{M}_{\mathcal{T}}$.

The numerical implicit function. We now change notation so that $\mathbf{x} \in \mathbf{R}^{N+K}$ is replaced by $(\mathbf{x}, \mathbf{y}) \in \mathbf{R}^N \times \mathbf{R}^K$, indicating the parameters as $\mathbf{y}$. We seek to construct a numerical implicit function $\mathbf{x}(\mathbf{y})$ that solves $\mathbf{f}(\mathbf{x}, \mathbf{y}) = \mathbf{0}$ in some neighborhood of a solution point $(\mathbf{x}^0, \mathbf{y}^0)$ for which the partial Jacobian $D_{\mathbf{x}}\mathbf{f}(\mathbf{x}^0, \mathbf{y}^0)$ has rank N. A Freudenthal triangulation of a neighborhood of $\mathbf{y}^0$ in the parameter space, $\mathbf{R}^K$, is used. A different algorithm from that described above is used to give the values of the implicit function at the vertices of this triangulation. The value of the implicit function at any other point, $\tilde{\mathbf{y}}$, in the neighborhood will be calculated by applying a modified Newton's method to the equation $\mathbf{f}(\mathbf{x}, \tilde{\mathbf{y}}) = \mathbf{0}$. The iterates will be constrained by the condition that $\mathbf{y} = \tilde{\mathbf{y}}$. The starting value of the iterations is determined by linear interpolation of the implicit function values at the vertices of the K-simplex to which $\tilde{\mathbf{y}}$ belongs.

As in the discussion preceding the previous algorithm, there exists an N-face, τ, with barycenter $(\mathbf{x}^0, \mathbf{y}^0)$ which lies in the N-dimensional affine set $\mathbf{y} = \mathbf{y}^0$. This N-face exists because the affine set has a full range projection onto the normal space to $\mathcal{M}$ at $(\mathbf{x}^0, \mathbf{y}^0)$. Let $\mathbf{v}^0 = (\tilde{\mathbf{x}}, \mathbf{y}^0)$ be the smallest vertex in τ and δ be the diameter; then the underlying triangulation for the piecewise linear algorithm is $\mathcal{F}_{N+K}[\mathbf{v}^0; \delta]$. The triangulation has the property that if P is the projection of $\mathbf{R}^{N+K}$ onto $\mathbf{R}^K$ along $\mathbf{R}^N$, then $P\mathscr{F}_{N+K}[\mathbf{v}^0;\delta] = \mathscr{F}_K[\mathbf{y}^0;\delta]$. This follows since for $\sigma \in \mathscr{F}_{N+K}[\mathbf{v}^0;\delta]$, $\sigma = [\mathbf{v}_0, \Pi]$, where $P\mathbf{v}_0 = \mathbf{y}_0$ is a point in $\{\mathbf{y}^0\} + \delta Z^K$. Furthermore, since

$$P\mathbf{e}_i = \begin{cases} \mathbf{0} & \text{for } 1 \le i \le N, \\ \mathbf{e}_i & \text{for } N+1 \le i \le N+K, \end{cases}$$

then $P\sigma = [\mathbf{y}_0, \widetilde{\Pi}]$ where $\widetilde{\Pi}$ is the permutation of $\{N+1, \dots, N+K\}$ obtained by deleting any of the values of Π which are less than $N+1$.

It should be clear that not all of the piecewise linear manifold $\mathscr{M}_{\mathscr{T}}$ needs to be constructed to approximate the values of an implicit function at the vertices of $\mathscr{F}_K[\mathbf{y}^0;\delta]$. It suffices to find the labeling centers of those completely labeled N-faces lying in the N-dimensional affine sets where $\mathbf{y}$ is a vertex of $\mathscr{F}_K[\mathbf{y}^0;\delta]$. Call such a special N-face a *nodal N-face* and the corresponding special value of $\mathbf{y}$ a *node*.

The proposed method is to construct piecewise linear curves lying in the intersection of $\mathscr{M}_{\mathscr{T}}$ with the $(N+1)$-dimensional affine sets

$$\{(\mathbf{x},\mathbf{y}) \in \mathbf{R}^{N+K} : (y_1, \dots, \widehat{y_i}, \dots, y_K) = \text{constant}\},$$

where by $\widehat{y_i}$ we mean that y_i is omitted from the K-tuple. This intersection is essentially determined by the N piecewise linear equations $\mathbf{f}_{\mathscr{T}}(\mathbf{x},\mathbf{y}) = \mathbf{0}$ in only $N+1$ variables $(x_1, \dots, x_N, y_i)$, so the piecewise linear curve can be easily followed using the piecewise linear continuation method described by Allgower and Georg [3]. In order to access all nodal N-faces, a list is maintained of nodal N-faces which have been encountered along with the direction(s) through which they have been traversed. When an individual curve has been completed, then one of the nodal N-faces on the list is used as a starter for a new curve in an as-yet-unused direction through that N-face. The process continues until all nodal N-faces so encountered have been traversed in each of the K directions.

An algorithm for generating nodal N-faces. GIVEN: $(\mathbf{x}^0, \mathbf{y}^0) \in \mathscr{M}$ such that $\operatorname{rank}(D_{\mathbf{x}}\mathbf{f}(\mathbf{x}^0, \mathbf{y}^0)) = N$, $\delta > 0$ and a compactum $\mathscr{C}$.
Find τ , nodal N-face containing $(\mathbf{x}^0, \mathbf{y}^0)$.
Let $\Xi = \{\tau\}$ represent the nodal face list.
While $\Xi \neq \varnothing$ do:
  Select $\tau \in \Xi$ and an unused direction, y_i.
  Mark y_i as used for τ.
  If all $y_i, i = 1, \dots, K$, are used for τ, delete τ from Ξ.
  Apply the piecewise linear continuation algorithm as described by Allgower and Georg [3] to the mapping $\mathbf{H}(\mathbf{x}, y_i) = \mathbf{f}(\mathbf{x}, \tilde{y}_1, \dots, y_i, \dots, \tilde{y}_K)$

where $\tilde{y}_j, j \neq i$, are the constant values of the corresponding y_j 's in the nodal N-face τ. If the boundary of $\mathscr{C}$ is reached then restart from τ in the opposite direction and stop when the boundary of $\mathscr{C}$ is reached again. Otherwise, the curve will return to the starting nodal N-face τ and stop.

For each nodal N-face, $\tilde{\tau}$, on this piecewise linear curve,

Mark y_i as used for $\tilde{\tau}$.

If all $y_i, i = 1, \ldots, K$, have been used as directions for $\tilde{\tau}$ then delete $\tilde{\tau}$ from Ξ.

If $\tilde{\tau} \notin \Xi$, insert $\tilde{\tau}$ into Ξ.

Record the coordinates of the labeling center of $\tilde{\tau}$ so as to reflect the relation of the parameter vector, $P\tilde{\tau}$, to the other nodes of $\mathscr{F}_K[\mathbf{y}^0; \delta]$.

Modifications needed to construct the numerical implicit function. When a nodal N-face, τ, is encountered during continuation, the **x**-coordinates of its labeling center are recorded as a function of the **y**-coordinates of the corresponding node $P\tau \in \mathscr{F}_K[\mathbf{y}^0; \delta]$. This gives an approximation to the value of the implicit function at the node.

If a new nodal N-face $\hat{\tau}$ is encountered with the same **y**-coordinates as an earlier nodal N-face, then the **x**-coordinates are not recorded since this represents a new implicit function; indeed, $\hat{\tau}$ is deleted from Ξ. Continuation also ceases along this curve in the current direction.

The values of the implicit function at a node of $\mathscr{F}_K[\mathbf{y}^0; \delta]$ are corrected by applying the modified Newton's method to points constrained to lie in the N-dimensional affine set containing the nodal N-face $\tau = [\mathbf{v}_0, \ldots, \mathbf{v}_N]$. It can be shown that the iteration can be carried out using the approximation to the partial Jacobian provided by $[\Delta\mathbf{f}(\tau)][\Delta\tau]^+$, where the last factor is the Moore–Penrose inverse. If this iteration is applied to points written in terms of their affine coordinates relative to the vertices of τ,

$$\mathbf{x} = (I - P)\sum_{i=0}^{N} \lambda_i \mathbf{v}_i, \qquad \sum_{i=0}^{N} \lambda_i = 1,$$

then the sequence of iterates, $\{\Lambda^m\}$, of these affine coordinates is determined by solving

$$\begin{pmatrix} 1 & \cdots & 1 \\ \mathbf{f}(\mathbf{v}_0) & \cdots & \mathbf{f}(\mathbf{v}_N) \end{pmatrix} \left(\Lambda^m - \Lambda^{m-1}\right) = -\begin{pmatrix} 0 \\ \mathbf{f}(\sum_{i=0}^{N} \lambda_i^{m-1} \mathbf{v}_i) \end{pmatrix}$$

with Λ^0 the vector of barycentric coordinates of the labeling center of τ.

It can be shown that if the partial Jacobian, $D_{\mathbf{x}}\mathbf{f}(\mathbf{v})$, $\mathbf{v} \in \tau$, is bounded with a bounded inverse, and if $\mathbf{f}$ has a Lipschitz continuous derivative, then for δ sufficiently small, the labeling center for τ will lie in the domain of attraction for this iteration scheme, so the points with affine coordinates Λ^m will converge to the value of the implicit function.

When a value of the parameter vector $\tilde{\mathbf{y}}$ sufficiently close to $\mathbf{y}^0$ is given, then one can construct the corresponding value of the implicit function, $\mathbf{x}$, as follows. The simplex $\tilde{\sigma} \in \mathscr{F}_K[\mathbf{y}^0; \delta]$ to which $\tilde{\mathbf{y}}$ belongs is determined by finding the integral vector part, $\mathbf{z}_0$, and the fractional vector part, $\mathbf{F}$, of the vector

$$\frac{\tilde{\mathbf{y}} - \mathbf{y}^0}{\delta} = \mathbf{z}_0 + \mathbf{F},$$

where $\mathbf{z}_0 \in \mathbf{Z}^K$, and where each component of $\mathbf{F}$ satisfies $0 \leq F_i < 1$. The permutation, Π, associated with $\tilde{\sigma}$ is determined by ordering the components of $\mathbf{F}$ in nonincreasing order

$$F_{\Pi(1)} \geq F_{\Pi(2)} \geq \cdots \geq F_{\Pi(K)}.$$

Then $\tilde{\mathbf{y}} \in \tilde{\sigma} = [\mathbf{z}_0; \Pi]$ since it can be shown that in terms of the vertices of this simplex,

$$\mathbf{v}_0 = \mathbf{y}^0 + \delta \mathbf{z}_0, \qquad \mathbf{v}_i = \mathbf{v}_{i-1} + \delta \mathbf{e}_{\Pi(i)} \quad \text{for } i = 1, \ldots, K,$$

the given parameter vector, $\tilde{\mathbf{y}}$, has barycentric coordinates

$$\lambda_i = \begin{cases} 1 - F_{\Pi(1)} & \text{for } i = 0, \\ F_{\Pi(K)} & \text{for } i = 1, \\ F_{\Pi(K-i+1)} - F_{\Pi(K-i)} & \text{for } i = 2, \ldots, K. \end{cases}$$

Observe that these barycentric coordinates are all nonnegative and are easily computed.

If $\tilde{\mathbf{y}}$ is sufficiently close to $\mathbf{y}^0$, then the values of the implicit function at the vertices of $\tilde{\sigma}$ have been stored as the corresponding $\mathbf{x}$'s and may be easily recovered. Their affine combination with the λ's as determined above gives a value of $\tilde{\mathbf{x}}$ corresponding to $\tilde{\mathbf{y}}$. Newton's method constrained to $\mathbf{y} = \tilde{\mathbf{y}}$ yields

$$\mathbf{x}^0 = \tilde{\mathbf{x}}, \quad \mathbf{x}^m = \mathbf{x}^{m-1} - [D_{\mathbf{x}}\mathbf{f}(\mathbf{x}^{m-1}, \tilde{\mathbf{y}})]^{-1}\mathbf{f}(\mathbf{x}^{m-1}, \tilde{\mathbf{y}}), \quad m = 1, 2, \ldots.$$

If $\mathbf{f}$ is sufficiently smooth and if the Jacobian is bounded and invertible at $(\tilde{\mathbf{x}}, \tilde{\mathbf{y}})$ then, for δ sufficiently small, this point will be in the domain of attraction of Newton's method, and the value of the implicit function will be determined.

A more complete discussion of this approach to determining a numerical implicit function will appear in a future work which will elaborate on the algorithm, give illustrative examples and applications, and provide a comparison with other methods. It will also discuss the manner by which turning points can be handled and multiple branches of the implicit function can be generated.

REFERENCES

1. J. C. Alexander, *Average intersection and pivoting densities*, SIAM J. Numer. Anal. **24** (1987), 129–146.

2. E. L. Allgower and K. Georg, *Generations of triangulations by reflections*, Utilitas Math. **16** (1979), 123–129.

3. ——, *Simplicial and continuation methods for approximating fixed points and solutions to systems of equations*, SIAM Rev. **22** (1980), 28–85.

4. E. L. Allgower and S. Gnutzmann, *An algorithm for piecewise linear approximation of implicitly defined two-dimensional surfaces*, SIAM J. Numer. Anal. **24** (1987), 452–469.

5. E. L. Allgower and P. H. Schmidt, *Piecewise linear approximation of solution manifolds for nonlinear systems of equations*, Selected Topics in Operations Research, G. Hammer and D. Pallaschke (eds.), Lecture Notes in Economics and Mathematical Systems, **226**, Springer-Verlag, Berlin, Heidelberg, New York, 1984, pp. 339–347.

6. ——, *An algorithm for piecewise linear approximation of an implicitly defined manifold*, SIAM J. Numer. Anal. **22** (1985), 322–346.

7. A. Charnes, W. W. Cooper, and A. Henderson, *An Introduction to Linear Programming*, John Wiley and Sons, New York, 1953.

8. G. B. Dantzig, *Linear Programming and Extensions*, Princeton University Press, Princeton, NJ, 1963.

9. J. P. Fink and W. C. Rheinboldt, *Solution manifolds and submanifolds of parametrized equations and their discretization errors*, Numer. Math. **45** (1984), 323–343.

10. H. Freudenthal, *Simplizialzerlegungen von beschränkter Flachheit*, Annals of Math. **43** (1942), 580–582.

11. K. Georg, *PL continuation methods*, preprint available as notes of this conference.

12. S. Gnutzmann, *Stückweise lineare Approximation implizit definierter Mannigfaltigkeiten*, (Doctoral dissertation), Universität Hamburg, Hamburg, 1988.

13. H. B. Keller, *Global homotopies and Newton methods*, Recent Advances in Numerical Analysis, C. deBoor and G. H. Golub (eds.), Academic Press, New York, 1978, pp. 73–94.

14. W. C. Rheinboldt, *Numerical Analysis of Parametrized Nonlinear Equations*, John Wiley and Sons, New York, 1986.

15. ——, *On a moving frame algorithm and the triangulation of equilibrium manifolds*, Bifurcation: Analysis, Algorithms, Applications, T. Küpper, R. Seidel, and H. Troger (eds.), **79**, Birkhäuser Verlag, Basel, 1987, pp. 256–267.

THE UNIVERSITY OF AKRON

Lectures in Applied Mathematics
Volume **26** (1990)

Nonstandard Scaling Matrices in Trust Region Methods

HUBERT SCHWETLICK

1. Introduction. In recent years trust region methods have proved to be a practically reliable and theoretically satisfactory tool for solving unconstrained minimization problems

$$\min\{f(\mathbf{x}) : \mathbf{x} \in \mathbf{R}^n\} \tag{1}$$

where $f: \mathbf{R}^n \to \mathbf{R}$ is a given sufficiently smooth nonlinear function; cf. the survey of Moré [**M2**], the lecture of Moré [**M3**] at this seminar, and the corresponding chapters in the books of Dennis and Schnabel [**DS**] and Fletcher [**F1**].

In the kth iteration step of such a trust region method the function f is approximated in the neighborhood of the current iterate $\mathbf{x} = \mathbf{x}^k$ by a quadratic model function

$$m(\mathbf{s}) = f(\mathbf{x}) + \mathbf{g}^{\mathrm{T}}\mathbf{s} + \tfrac{1}{2}\mathbf{s}^{\mathrm{T}}\mathbf{B}\mathbf{s} \approx f(\mathbf{x} + \mathbf{s}), \tag{2}$$

where $\mathbf{g} = \nabla f(\mathbf{x})$ denotes the gradient of f at $\mathbf{x}$, and $\mathbf{B} = \mathbf{B}^{\mathrm{T}}$ is an appropriately chosen approximation of the Hessian $\nabla^2 f(\mathbf{x})$. Then the trust region step $\mathbf{s} = \mathbf{s}^k$ is computed as an approximate solution to the restricted problem

$$\min\{m(\mathbf{s}) : \|\mathbf{D}\mathbf{s}\| \le \Delta\}, \tag{3}$$

where $\Delta = \Delta_k > 0$ is the given trust region radius, and $\mathbf{D} = \mathbf{D}_k$ is an n-dimensional nonsingular scaling matrix that defines the elliptic norm

$$\|\mathbf{s}\|_{\mathbf{D}} = \|\mathbf{D}\mathbf{s}\| = (\mathbf{s}^{\mathrm{T}}\mathbf{W}\mathbf{s})^{1/2}, \qquad \mathbf{W} = \mathbf{D}^{\mathrm{T}}\mathbf{D}, \tag{4}$$

generated by the positive definite matrix $\mathbf{W}$. The basic norm is always the Euclidean norm $\|\mathbf{u}\| = \|\mathbf{u}\|_{\mathbf{I}} = (\mathbf{u}^{\mathrm{T}}\mathbf{u})^{1/2}$.

If the trust region step $\mathbf{s}$ leads to a sufficient decrease in the objective function in the sense of

$$f(\mathbf{x}) - f(\mathbf{x}+\mathbf{s}) \geq \delta[m(\mathbf{0}) - m(\mathbf{s})] = \delta[-\mathbf{g}^T\mathbf{s} - \tfrac{1}{2}\mathbf{s}^T\mathbf{B}\mathbf{s}] \tag{5}$$

with fixed $\delta \in (0,1)$, then $\mathbf{s}$ is accepted and gives the next iterate $\bar{\mathbf{x}} = \mathbf{x}+\mathbf{s}$. Otherwise, $\mathbf{s}$ is rejected and a new step $\mathbf{s}$ is computed using the same model $m(\mathbf{s})$ but a reduced radius Δ.

Practical trust region algorithms differ in the way the subproblem (3) is approximately solved. Note that this subproblem always has a solution $\mathbf{s}$. If $\mathbf{B}$ is positive definite (in the following, $\mathbf{B} > 0$ or $\mathbf{B} \geq 0$ means that the symmetric matrix $\mathbf{B}$ is positive definite or positive semidefinite, respectively) and if the Newton-like step $\mathbf{s}^N = -\mathbf{B}^{-1}\mathbf{g}$ as global minimizer of m satisfies $\|\mathbf{D}\mathbf{s}^N\| \leq \Delta$, then $\mathbf{s} = \mathbf{s}^N$ solves (3). Otherwise, with the exception of certain special cases (including the so-called "hard case" that can occur when $\mathbf{B}$ is indefinite), the solution of (3) is given by $\mathbf{s} = \mathbf{s}(\lambda)$, where $\mathbf{s}(\lambda)$ solves the linear system

$$(\mathbf{B} + \lambda\mathbf{W})\mathbf{s} = -\mathbf{g}, \tag{6}$$

and $\lambda > 0$ is chosen in such a way that, first,

$$\mathbf{B} + \lambda\mathbf{W} > 0 \tag{7}$$

holds so that (6) has the unique solution $\mathbf{s} = \mathbf{s}(\lambda) = -(\mathbf{B}+\lambda\mathbf{W})^{-1}\mathbf{g}$ and, second, λ satisfies the scalar equation

$$\varphi(\lambda) = \|\mathbf{D}\mathbf{s}(\lambda)\| = [\mathbf{s}(\lambda)^T\mathbf{W}\mathbf{s}(\lambda)]^{1/2} = \Delta. \tag{8}$$

For handling the exceptional cases mentioned above systems of type (6) have to be solved, too; see Gay **[G]**, Sorensen **[So]**, Moré and Sorensen **[MS]**, and Schultz, Schnabel, and Byrd **[SSB]**.

In most implementations equation (8) is approximately solved using values $\varphi(\lambda)$ and possibly

$$\dot{\varphi}(\lambda) = -[\mathbf{s}(\lambda)^T\mathbf{W}(\mathbf{B}+\lambda\mathbf{W})^{-1}\mathbf{W}\mathbf{s}(\lambda)]/\varphi(\lambda) \tag{9}$$

at trial points λ satisfying (7); see Moré **[M1]** and the literature cited above. In practice, condition (7) is checked by trying to compute the Cholesky factorization

$$\mathbf{B} + \lambda\mathbf{W} = \mathbf{R}(\lambda)^T\mathbf{R}(\lambda). \tag{10}$$

Here and in the following, Cholesky factorization is always understood as a symmetric triangular factorization with a regular triangular factor, i.e., the case of a semidefinite matrix is excluded. If the attempt succeeds, then $\mathbf{B} + \lambda\mathbf{W}$ is positive definite and $\mathbf{R}(\lambda)$ is used for computing

$\varphi(\lambda)$, $\dot{\varphi}(\lambda)$, and other quantities needed by the trust region algorithm. Otherwise, λ is increased, and a test on the occurrence of exceptional cases is made.

Obviously, the combinatorial costs for performing one iteration step depend essentially on the costs for computing the factorization (10) for several values of λ satisfying (7) until $\mathbf{s}(\lambda)$ can be accepted as a sufficiently accurate solution to (8) and, moreover, fulfills the descent condition (5). Since changing λ leads, in general, to a full-rank change of $\mathbf{B} + \lambda\mathbf{W}$, the factorization (10) known for the old λ cannot be used for the new one in order to save costs. Hence, if n is not small, there is a considerable overhead in combinatorial costs compared to classical line search methods where only one system of type (6) has to be solved per iteration step. Therefore, from the beginning of the history of trust region methods, a lot of research has been done to make the step computationally cheaper by modifying the basic trust region algorithm.

A first possibility consists of restricting the set of feasible $\mathbf{s}$ in subproblem (3) to a small part $\mathbf{S} \subset \mathbf{R}^n$ such that the modified problem

$$\min\{m(\mathbf{s}) \colon \mathbf{s} \in \mathbf{S} \text{ and } \|\mathbf{Ds}\| \leq \Delta\}$$

can be solved more easily. The oldest methods of this type are the dogleg methods of Powell [**P1**], [**P2**], where $\mathbf{S}$ is a dogleg-like polygonal arc. The choice of $\mathbf{S}$ as a p-dimensional subspace with $p \ll n$ containing at least the gradient direction $\mathbf{g}$ seems to be a more natural one. In this case, the equations to be solved have the same structure as (6) but the dimension is reduced to p; see Knoth [**K**], Bulteau and Vial [**BV**], Westin [**W**], and the remarks in [**M2**].

A second possibility consists of using iterative techniques for computing $\mathbf{s}$, especially preconditioned CG methods; see [**M2**] and the survey of Fletcher [**F2**] in these proceedings.

This paper is devoted to *a third possibility* which is again based on direct factorization of $\mathbf{B} + \lambda\mathbf{W}$, but tries to choose appropriate nonstandard scaling matrices $\mathbf{D} = \mathbf{D}_k$ such that the repeated factorization for several λ is significantly cheaper than in the standard case. Recall that the standard scaling is a diagonal one, i.e., the standard choice is

$$\mathbf{D} = \operatorname{diag}(d_1, \ldots, d_n), \quad d_i > 0 \qquad (i = 1, \ldots, n);$$

see [**M1**] and [**M2**]. The first example of using nonstandard scaling matrices seems to go back to [**So**]. It is based on symmetric factorizations of the possibly indefinite matrix $\mathbf{B}$ and can be interpreted as a standard scaling in a transformed system of coordinates where $\mathbf{B}$ has block-diagonal form. In Section 2, such transformations of variables are introduced for later use, and the approach of **[So]** is briefly discussed.

In Section 3, another type of nonstandard scaling matrices is introduced. They are based on the assumption that, possibly after simultaneous changes of rows and columns, a large leading block of $\mathbf{B}$ having dimension l near n is positive definite. Then l Cholesky steps can be performed, and the resulting truncated Cholesky factorization suggests a choice of three different versions of nonstandard scaling matrices. Based on this truncated factorization, efficient algorithms for factoring $\mathbf{B}+\lambda\mathbf{W}$ and computing further quantities such as $\mathbf{s}(\lambda)$ and $\varphi(\lambda)$ for different values of λ are given. Section 4 is devoted to the special case of nonlinear least squares problems $\mathbf{f} = \mathbf{F}^{\mathrm{T}}\mathbf{F}/2$ when $\mathbf{B} = \mathbf{J}^{\mathrm{T}}\mathbf{J}$ is taken as an approximation to $\nabla^2\mathbf{f}$ that uses only the Jacobian $\mathbf{J}$ of $\mathbf{F}$. In this Gauss–Newton case, a theoretically equivalent, but practically preferable, implementation based on numerically stable orthogonal transformations as proposed in Schwetlick and Tiller, **[ST1]**, **[ST2]** is described. In Section 5, some concluding comments are given.

2. Transformation of variables and Sorensen's approach. Let $\mathbf{G}$ be the nonsingular matrix defining the transformation of variables

$$\mathbf{s} \to \hat{\mathbf{s}} = \mathbf{G}\mathbf{s}. \tag{11}$$

In the new variables $\hat{\mathbf{s}}$, problem (3) is written as

$$\min\{\hat{m}(\hat{\mathbf{s}}) \colon \|\widehat{\mathbf{D}}\hat{\mathbf{s}}\| = (\hat{\mathbf{s}}^{\mathrm{T}}\widehat{\mathbf{W}}\hat{\mathbf{s}})^{1/2} = \Delta\},$$

where

$$\hat{m}(\hat{\mathbf{s}}) = m(\mathbf{G}^{-1}\hat{\mathbf{s}}) = f(\mathbf{x}) + \hat{\mathbf{g}}^{\mathrm{T}}\hat{\mathbf{s}} + \tfrac{1}{2}\hat{\mathbf{s}}^{\mathrm{T}}\widehat{\mathbf{B}}\hat{\mathbf{s}}$$

with transformed coefficients

$$\hat{\mathbf{g}} = \mathbf{G}^{-\mathrm{T}}\mathbf{g}, \qquad \widehat{\mathbf{B}} = \mathbf{G}^{-\mathrm{T}}\mathbf{B}\mathbf{G}^{-1} \tag{12}$$

and transformed scaling and metric matrices

$$\widehat{\mathbf{D}} = \mathbf{D}\mathbf{G}^{-1}, \qquad \widehat{\mathbf{W}} = \widehat{\mathbf{D}}^{\mathrm{T}}\widehat{\mathbf{D}} = \mathbf{G}^{-\mathrm{T}}\mathbf{W}\mathbf{G}^{-1}. \tag{13}$$

Incidentally, the choice $\mathbf{G} = \mathbf{D}$ leads to the unscaled "normal" form with $\widehat{\mathbf{D}} = \widehat{\mathbf{W}} = \mathbf{I}$.

Since the pencils $\{\mathbf{B}, \mathbf{W}\}$ and $\{\widehat{\mathbf{B}}, \widehat{\mathbf{W}}\}$ are congruent, they have the same eigenvalues. Moreover, due to Sylvester's theorem, the number of positive, vanishing, and negative eigenvalues of these pencils is equal to the number of corresponding eigenvalues of $\mathbf{B}$ itself. In the new variables $\hat{\mathbf{s}}$, the system (6) associated with (3) has the form

$$(\widehat{\mathbf{B}} + \lambda\widehat{\mathbf{W}})\hat{\mathbf{s}} = -\hat{\mathbf{g}}. \tag{14}$$

Now $\widehat{\mathbf{D}}$ is considered as not yet fixed, and $\mathbf{G}$ and $\widehat{\mathbf{W}}$ are chosen such that both transforming $\mathbf{s} \leftrightarrow \hat{\mathbf{s}}$ and solving the system (14) for several values of λ can be done efficiently. To this end, **[So]** considers the so-called Bunch–Kaufman–Parlett (=: BKP) factorization

$$\mathbf{P}_{\mathrm{BKP}}\mathbf{B}\mathbf{P}_{\mathrm{BKP}}^{\mathrm{T}} = \mathbf{L}_{\mathrm{BKP}}\mathbf{D}_{\mathrm{BKP}}\mathbf{L}_{\mathrm{BKP}}^{\mathrm{T}} \tag{15}$$

of the possibly indefinite matrix $\mathbf{B}$. Here $\mathbf{P}_{\mathrm{BKP}}$ is a permutation, $\mathbf{L}_{\mathrm{BKP}}$ is unit lower triangular, and $\mathbf{D}_{\mathrm{BKP}}$ is block diagonal with (1, 1) or (2, 2) diagonal blocks. Computing (15) requires $\sim n^3/6$ operations in the dense case; see **[BKP]**; cf. also Section 6.1 of Kiełbasiński and Schwetlick **[KS]**. Then Sorenson takes

$$\mathbf{G} = \mathbf{G}_{\mathrm{BKP}} = \mathbf{L}_{\mathrm{BKP}}^{\mathrm{T}}\mathbf{P}_{\mathrm{BKP}}, \qquad \widehat{\mathbf{W}} = \widehat{\mathbf{W}}_{\mathrm{BKP}} = \mathbf{I} \tag{16}$$

in order to obtain

$$\widehat{\mathbf{B}} = \widehat{\mathbf{B}}_{\mathrm{BKP}} = \mathbf{D}_{\mathrm{BKP}}.$$

For this choice the matrix $\widehat{\mathbf{B}}+\lambda\widehat{\mathbf{W}}$ has the same structure as $\mathbf{D}_{\mathrm{BKP}}$ so that solving the system (14) costs only $O(n)$ operations, and the transformations $\mathbf{s} \leftrightarrow \hat{\mathbf{s}}$ require $\sim n^2/2$ operations. Let us note that the choice (16) corresponds to the metric in the original variables which is generated by $\mathbf{W} = \mathbf{P}_{\mathrm{BKP}}^{\mathrm{T}}\mathbf{L}_{\mathrm{BKP}}\mathbf{L}_{\mathrm{BKP}}^{\mathrm{T}}\mathbf{P}_{\mathrm{BKP}}$. However, if the condition of $\mathbf{W}$ is large, then the norms generated by $\mathbf{W}$ may not be a good measure for the length of $\mathbf{s}$. This can cause an increase in the overall number of iteration steps; cf. the comment in **[So]**. Therefore, in order to avoid badly conditioned metric matrices, another type of nonstandard scaling matrices will be considered in the next section.

3. Nonstandard scaling matrices based on truncated Cholesky factorization. In the following it is assumed that the matrix $\mathbf{B}$ allows us to perform l Cholesky steps, possibly with diagonal pivoting. Thus, it possesses the truncated Cholesky factorization

$$\mathbf{P}\mathbf{B}\mathbf{P}^{\mathrm{T}} = \mathbf{R}^{\mathrm{T}}\widehat{\mathbf{B}}\mathbf{R}, \tag{17}$$

where $\mathbf{P}$ is the permutation that characterizes the simultaneous changes of rows and columns during the factorization, and $\mathbf{R}$ and $\widehat{\mathbf{B}}$ have the form

$$\widehat{\mathbf{B}} = \begin{pmatrix} \mathbf{I} & \mathbf{0} \\ \mathbf{0}^{\mathrm{T}} & \mathbf{M}_3 \end{pmatrix}, \qquad \mathbf{R} = \begin{pmatrix} \mathbf{R}_1 & \mathbf{R}_2 \\ \mathbf{0}^{\mathrm{T}} & \mathbf{I} \end{pmatrix} \begin{matrix} \}\, l \\ \}\, t = n - l \end{matrix}, \tag{18}$$

where $\mathbf{R}_1$ is upper triangular with positive diagonal entries and the $\{1, 1\}$-blocks have dimension l. If the permuted matrix $\widetilde{\mathbf{B}}$ is introduced and partitioned according to

$$\widetilde{\mathbf{B}} = \mathbf{P}\mathbf{B}\mathbf{P}^{\mathrm{T}} = \begin{pmatrix} \mathbf{B}_1 & \mathbf{B}_2 \\ \mathbf{B}_2^{\mathrm{T}} & \mathbf{B}_3 \end{pmatrix} \begin{matrix} \}\, l \\ \}\, t \end{matrix}, \tag{19}$$

then (17) can be written equivalently as

$$\mathbf{B}_1 = \mathbf{R}_1^{\mathrm{T}}\mathbf{R}_1, \quad \mathbf{B}_2 = \mathbf{R}_1^{\mathrm{T}}\mathbf{R}_2, \quad \mathbf{B}_3 = \mathbf{R}_2^{\mathrm{T}}\mathbf{R}_2 + \mathbf{M}_3. \tag{20}$$

Note that $\widetilde{\mathbf{B}}$ is the representation of $\mathbf{B}$ in the permuted variables

$$\widetilde{\mathbf{s}} = \mathbf{P}\mathbf{s}. \tag{21}$$

Because $\mathbf{B}_1 > 0$, the definiteness of $\mathbf{B}$ is characterized by that of the Schur complement

$$\mathbf{M}_3 = \mathbf{B}_3 - \mathbf{R}_2^{\mathrm{T}}\mathbf{R}_2 = \mathbf{B}_3 - \mathbf{B}_2^{\mathrm{T}}\mathbf{B}_1^{-1}\mathbf{B}_2. \tag{22}$$

The factorization (17), (18) exists if and only if $\mathbf{B}$ has at least l positive eigenvalues. We suppose further that

$$l \gg t = n - l$$

holds. This assumption is not very restrictive since one can expect that a reasonable approximation $\mathbf{B}$ of the Hessian at a point $\mathbf{x}$ obtained by a descent method will have, if any at all, only a few nonpositive eigenvalues.

In order to obtain the simply structured matrix $\widehat{\mathbf{B}}$ of (18) as a transformed coefficient matrix we set

$$\widehat{\mathbf{s}} = \mathbf{R}\mathbf{P}\mathbf{s} = \mathbf{R}\widetilde{\mathbf{s}}, \quad \text{i.e.,} \quad \mathbf{G} = \mathbf{R}\mathbf{P}. \tag{23}$$

Now the form of $\widehat{\mathbf{B}}$ suggests choosing a metric matrix $\widehat{\mathbf{W}}$ of the same form, namely,

$$\widehat{\mathbf{W}} = \widehat{\mathbf{W}}^{(3)} = \begin{pmatrix} \mathbf{I} & \mathbf{0} \\ \mathbf{0}^{\mathrm{T}} & \mathbf{W}_3 \end{pmatrix} \quad \text{with } \mathbf{W}_3 = \mathbf{C}_3^{\mathrm{T}}\mathbf{C}_3 > \mathbf{0}, \quad \mathbf{C}_3 \text{ nonsingular}, \tag{24}$$

where $\mathbf{C}_3$ is an arbitrary scaling matrix of dimension t, e.g., the standard diagonal scaling matrix associated with the block $\mathbf{M}_3$ of $\widehat{\mathbf{B}}$. The choice (24) is denoted as version 3 since it corresponds to version 3 of nonstandard scalings proposed by [ST2] in the Gauss-Newton case; see also Section 4. With this $\widehat{\mathbf{W}}$ the matrix of the transformed system (14) becomes

$$\widehat{\mathbf{B}} + \lambda\widehat{\mathbf{W}} = \begin{pmatrix} (1+\lambda)\mathbf{I} & \mathbf{0} \\ \mathbf{0}^{\mathrm{T}} & \mathbf{M}_3 + \lambda\mathbf{W}_3 \end{pmatrix}.$$

Obviously, $\widehat{\mathbf{B}} + \lambda\widehat{\mathbf{W}}$, and hence $\mathbf{B} + \lambda\mathbf{W}$, is positive definite if and only if $\mathbf{M}_3 + \lambda\mathbf{W}_3$ is positive definite, i.e., if the latter matrix allows the Cholesky factorization

$$\mathbf{M}_3 + \lambda\mathbf{W}_3 = \mathbf{R}_3(\lambda)^{\mathrm{T}}\mathbf{R}_3(\lambda). \tag{25}$$

If this factorization exists, then

$$\widehat{\mathbf{R}}(\lambda) = \begin{pmatrix} \sqrt{1+\lambda}\mathbf{I} & \mathbf{0} \\ \mathbf{0}^{\mathrm{T}} & \mathbf{R}_3(\lambda) \end{pmatrix} \quad \text{and} \quad \widetilde{\mathbf{R}}(\lambda) = \widehat{\mathbf{R}}(\lambda)\mathbf{R}$$
$$= \begin{pmatrix} \sqrt{1+\lambda}\mathbf{R}_1 & \sqrt{1+\lambda}\mathbf{R}_2 \\ \mathbf{0}^{\mathrm{T}} & \mathbf{R}_3(\lambda) \end{pmatrix}$$

are the Cholesky factors of $\widehat{\mathbf{B}}+\lambda\widehat{\mathbf{W}}$ and $\widetilde{\mathbf{B}}+\lambda\widetilde{\mathbf{W}}$, respectively. Remember that variables marked by a tilde correspond to the permuted variables $\tilde{\mathbf{s}}$; cf. (21). In these coordinates, $\mathbf{B}$ has the form (19), and the metric is defined by the matrix $\widetilde{\mathbf{W}} = \mathbf{R}^{\mathrm{T}}\widehat{\mathbf{W}}\mathbf{R}$, i.e., by

$$\widetilde{\mathbf{W}} = \begin{pmatrix} \mathbf{R}_1^{\mathrm{T}}\mathbf{R}_1 & \mathbf{R}_1^{\mathrm{T}}\mathbf{R}_2 \\ \mathbf{R}_2^{\mathrm{T}}\mathbf{R}_1 & \mathbf{R}_2^{\mathrm{T}}\mathbf{R}_2 + \mathbf{W}_3 \end{pmatrix} = \begin{pmatrix} \mathbf{B}_1 & \mathbf{B}_2 \\ \mathbf{B}_2^{\mathrm{T}} & \mathbf{B}_2^{\mathrm{T}}\mathbf{B}_1^{-1}\mathbf{B}_2 + \mathbf{W}_3 \end{pmatrix}. \tag{26}$$

Let us point out that computing $\widetilde{\mathbf{R}}(\lambda)$ for a new λ only requires computing the t-dimensional factorization (25) with $\sim t^3/6$ operations, whereas at the other blocks only the scalar factors $\sqrt{1+\lambda}$ have to be replaced by their new values.

Looking at formula (26) suggests two further choices of nonstandard metrics, namely, version 2, defined by

$$\widetilde{\mathbf{W}} = \begin{pmatrix} \mathbf{B}_1 & \mathbf{B}_2 \\ \mathbf{B}_2^{\mathrm{T}} & \mathbf{B}_3 + \mathbf{W}_3 \end{pmatrix}, \qquad \widehat{\mathbf{W}} = \begin{pmatrix} \mathbf{I} & \mathbf{0} \\ \mathbf{0}^{\mathrm{T}} & \mathbf{M}_3 + \mathbf{W}_3 \end{pmatrix} \tag{27}$$

and version 1 given by

$$\widetilde{\mathbf{W}} = \begin{pmatrix} \mathbf{B}_1 & \mathbf{0} \\ \mathbf{0}^{\mathrm{T}} & \mathbf{W}_3 \end{pmatrix}, \qquad \widehat{\mathbf{W}} = \begin{pmatrix} \mathbf{I} & -\mathbf{R}_2 \\ -\mathbf{R}_2^{\mathrm{T}} & \mathbf{R}_2^{\mathrm{T}}\mathbf{R}_2 + \mathbf{W}_3 \end{pmatrix} \tag{28}$$

with $\mathbf{W}_3 = \mathbf{C}_3^{\mathrm{T}}\mathbf{C}_3$ as in (24). Whereas in versions 1 and 3, any $\mathbf{W}_3 > 0$ leads to a $\mathbf{W} > 0$, in version 2, $\mathbf{W} > 0$ if and only if $\mathbf{M}_3 + \mathbf{W}_3 > 0$; cf. (27). Therefore, in case of an indefinite $\mathbf{M}_3$, a good choice of $\mathbf{W}_3$ would require some information about the negative eigenvalues of $\mathbf{M}_3$. In order to avoid such additional problems, we restrict version 2 to the case

$$\mathbf{M}_3 \geq 0, \quad \text{i.e,} \ \mathbf{B} \geq 0.$$

This assumption is automatically fulfilled, e.g., for Gauss-Newton approximations $\mathbf{B} = \mathbf{J}^{\mathrm{T}}\mathbf{J}$ (see Section 4), but not, in general, for the "optimal" choice $\mathbf{B} = \nabla^2 f(\mathbf{x})$.

In versions 1 and 2, checking $\mathbf{B}+\lambda\mathbf{W}>0$ and computing the Cholesky factorization can be done with costs comparable to those of version 3. The results are summarized as follows.

PROPOSITION 1. *Let nonstandard metrics be defined by the matrices* $\widetilde{\mathbf{W}}=\widetilde{\mathbf{W}}^{(i)}$ *of version i given in* (26), (27), *and* (28), *respectively. Then*

(29)
$$\widehat{\mathbf{B}}+\lambda\widehat{\mathbf{W}}=\mathbf{R}^{-\mathrm{T}}(\widetilde{\mathbf{B}}+\lambda\widetilde{\mathbf{W}})\mathbf{R}^{-1}=\mathbf{R}^{-\mathrm{T}}\mathbf{P}(\mathbf{B}+\lambda\mathbf{W})\mathbf{P}^{\mathrm{T}}\mathbf{R}^{-1}=\widehat{\mathbf{K}}(\lambda)^{\mathrm{T}}\widehat{\mathbf{B}}(\lambda)\widehat{\mathbf{K}}(\lambda),$$

where

$$\widehat{\mathbf{K}}(\lambda)=\begin{pmatrix}\mathbf{I} & -\delta_{i1}\frac{\lambda}{1+\lambda}\mathbf{R}_2\\ \mathbf{0}^{\mathrm{T}} & \mathbf{I}\end{pmatrix},\quad \widehat{\mathbf{B}}(\lambda)=\begin{pmatrix}(1+\lambda)\mathbf{I} & \mathbf{0}\\ \mathbf{0}^{\mathrm{T}} & \mathbf{M}_3(\lambda)\end{pmatrix}$$

and

$$\mathbf{M}_3(\lambda)=\mathbf{M}_3+\delta_{i2}\lambda\mathbf{M}_3+\delta_{i1}\frac{\lambda}{1+\lambda}\mathbf{R}_2^{\mathrm{T}}\mathbf{R}_2+\lambda\mathbf{W}_3, \tag{30}$$

δ_{ij} *being the Kronecker symbol, equal to* 1 *if* $i=j$ *and* 0 *otherwise. Moreover,* $\widehat{\mathbf{B}}(\lambda)$ *and, hence, each of the matrices* $\widehat{\mathbf{B}}+\lambda\widehat{\mathbf{W}}$, $\widetilde{\mathbf{B}}+\lambda\widetilde{\mathbf{W}}$, *and* $\mathbf{B}+\lambda\mathbf{W}$ *is positive definite if and only if* $\mathbf{M}_3(\lambda)$ *is positive definite, i.e., if it allows the Cholesky factorization*

$$\mathbf{M}_3(\lambda)=\mathbf{R}_3(\lambda)^{\mathrm{T}}\mathbf{R}_3(\lambda) \tag{31}$$

with nonsingular $\mathbf{R}_3(\lambda)$. *If the factorization* (31) *exists then*

$$\widehat{\mathbf{E}}(\lambda)=\begin{pmatrix}\sqrt{1+\lambda}\mathbf{I} & \mathbf{0}\\ \mathbf{0}^{\mathrm{T}} & \mathbf{R}_3(\lambda)\end{pmatrix}$$

is the Cholesky factor of $\widehat{\mathbf{B}}(\lambda)$ *and, hence,*

$$\widehat{\mathbf{R}}(\lambda)=\widehat{\mathbf{E}}(\lambda)\widehat{\mathbf{K}}(\lambda)\quad\text{and}\quad\widetilde{\mathbf{R}}(\lambda)=\widehat{\mathbf{E}}(\lambda)\widehat{\mathbf{K}}(\lambda)\mathbf{R} \tag{32}$$

are the Cholesky factors of $\widehat{\mathbf{B}}+\lambda\widehat{\mathbf{W}}$ *and* $\widetilde{\mathbf{B}}+\lambda\widetilde{\mathbf{W}}$, *respectively.*

Incidentally, the factorization (29) implies

$$\begin{aligned}\widetilde{\mathbf{W}}&=\lim_{\lambda\to\infty}\frac{\widetilde{\mathbf{B}}+\lambda\widetilde{\mathbf{W}}}{\lambda}=\mathbf{R}^{\mathrm{T}}\widehat{\mathbf{K}}(\infty)^{\mathrm{T}}\left(\frac{\widehat{\mathbf{B}}(\lambda)}{\lambda}\right)_{\lambda=\infty}\widehat{\mathbf{K}}(\infty)\mathbf{R}\\ &=\begin{pmatrix}\mathbf{R}_1^{\mathrm{T}} & \mathbf{0}\\ (1-\delta_{i1})\mathbf{R}_2^{\mathrm{T}} & \mathbf{I}\end{pmatrix}\begin{pmatrix}\mathbf{I} & \mathbf{0}\\ \mathbf{0}^{\mathrm{T}} & \mathbf{W}_3+\delta_{i2}\mathbf{M}_3\end{pmatrix}\begin{pmatrix}\mathbf{R}_1 & (1-\delta_{i1})\mathbf{R}_2\\ \mathbf{0}^{\mathrm{T}} & \mathbf{I}\end{pmatrix}.\end{aligned}$$

Therefore, if $\widetilde{\mathbf{D}}_3$ is any matrix satisfying

$$\widetilde{\mathbf{D}}_3^{\mathrm{T}}\widetilde{\mathbf{D}}_3=\mathbf{W}_3+\delta_{i2}\mathbf{M}_3=\mathbf{C}_3^{\mathrm{T}}\mathbf{C}_3+\delta_{i2}\mathbf{M}_3, \tag{33}$$

then

$$\widetilde{\mathbf{W}} = \widetilde{\mathbf{D}}^{\mathrm{T}}\widetilde{\mathbf{D}} \quad \text{with} \quad \widetilde{\mathbf{D}} = \begin{pmatrix} \mathbf{R}_1 & (1-\delta_{i1})\mathbf{R}_2 \\ \mathbf{0}^{\mathrm{T}} & \widetilde{\mathbf{D}}_3 \end{pmatrix}. \tag{34}$$

In versions 1 and 3, $\widetilde{\mathbf{D}}_3 = \mathbf{C}_3$ is possible, and in version 2, any $\widetilde{\mathbf{C}}_3$ with

$$\widetilde{\mathbf{C}}_3^{\mathrm{T}}\widetilde{\mathbf{C}}_3 = \mathbf{C}_3^{\mathrm{T}}\mathbf{C}_3 + \mathbf{M}_3$$

can be taken as $\widetilde{\mathbf{D}}_3$. Thus, the metrics defined by $\widetilde{\mathbf{W}}$ are generated by the nonstandard scaling matrices $\widetilde{\mathbf{D}} = \widetilde{\mathbf{D}}^{(i)}$ given by

$$\widetilde{\mathbf{D}}^{(1)} = \begin{pmatrix} \mathbf{R}_1 & \mathbf{0} \\ \mathbf{0}^{\mathrm{T}} & \mathbf{C}_3 \end{pmatrix}, \quad \widetilde{\mathbf{D}}^{(2)} = \begin{pmatrix} \mathbf{R}_1 & \mathbf{R}_2 \\ \mathbf{0}^{\mathrm{T}} & \widetilde{\mathbf{C}}_3 \end{pmatrix}, \quad \widetilde{\mathbf{D}}^{(3)} = \begin{pmatrix} \mathbf{R}_1 & \mathbf{R}_2 \\ \mathbf{0}^{\mathrm{T}} & \mathbf{C}_3 \end{pmatrix}. \tag{35}$$

The factorizations provided by Proposition 1 are now used for computing quantities such as $\mathbf{s}(\lambda)$, $\varphi(\lambda)$, ... needed by the trust region method. As an example, an algorithm for checking $\mathbf{B} + \lambda\mathbf{W} > 0$ and computing $\mathbf{s}(\lambda)$ will be derived. Based on the factorization (29), the defining equation (6) takes, in the permuted variables $\tilde{\mathbf{s}}$, the form

$$(\widetilde{\mathbf{B}} + \lambda\widetilde{\mathbf{W}})\tilde{\mathbf{s}} = \mathbf{R}^{\mathrm{T}}\widehat{\mathbf{K}}(\lambda)^{\mathrm{T}}\widehat{\mathbf{B}}(\lambda)\widehat{\mathbf{K}}(\lambda)\mathbf{R}\tilde{\mathbf{s}} = -\tilde{\mathbf{g}} = -\mathbf{P}\mathbf{g}.$$

This is equivalent to

$$\widehat{\mathbf{B}}(\lambda)\widehat{\mathbf{K}}(\lambda)\mathbf{R}\tilde{\mathbf{s}} = -\widehat{\mathbf{K}}(\lambda)^{-\mathrm{T}}\mathbf{R}^{-\mathrm{T}}\tilde{\mathbf{g}} = -\bar{\mathbf{g}};$$

in block notation,

$$\begin{pmatrix} (1+\lambda)\mathbf{R}_1 & [1+(1-\delta_{i1})\lambda]\mathbf{R}_2 \\ \mathbf{0}^{\mathrm{T}} & \mathbf{M}_3(\lambda) \end{pmatrix} \begin{pmatrix} \tilde{\mathbf{s}}_1 \\ \tilde{\mathbf{s}}_2 \end{pmatrix} = -\begin{pmatrix} \bar{\mathbf{g}}_1 \\ \bar{\mathbf{g}}_2 \end{pmatrix} \tag{36}$$

with

$$\bar{\mathbf{g}} = \begin{pmatrix} \bar{\mathbf{g}}_1 \\ \bar{\mathbf{g}}_2 \end{pmatrix} = \begin{pmatrix} \mathbf{R}_1^{-\mathrm{T}} & \mathbf{0} \\ -\frac{1+(1-\delta_{i1})\lambda}{1+\lambda}\mathbf{R}_2^{\mathrm{T}}\mathbf{R}_1^{-\mathrm{T}} & \mathbf{I} \end{pmatrix} \begin{pmatrix} \tilde{\mathbf{g}}_1 \\ \tilde{\mathbf{g}}_2 \end{pmatrix}. \tag{37}$$

This representation is the basis for the following algorithm.

ALGORITHM 1 (TEST ON $\mathbf{B}+\lambda\mathbf{W} > 0$ AND COMPUTATION OF $\mathbf{s}(\lambda)$ FOR $\lambda \geq 0$).

(S1) *Compute the truncated Cholesky factorization* (17), (18).

(S2) *Set* $\tilde{\mathbf{g}} = \mathbf{P}\mathbf{g}$.

(S3) *Solve* $\mathbf{R}_1^{\mathrm{T}}\bar{\mathbf{g}}_1 = \tilde{\mathbf{g}}_1$ *for* $\bar{\mathbf{g}}_1$; *compute* $\mathring{\mathbf{g}}_2 = \mathbf{R}_2^{\mathrm{T}}\bar{\mathbf{g}}_1$.

(S4) *Try to compute the Cholesky factorization* (31) *with regular* $\mathbf{R}_3(\lambda)$. *If the attempt does not succeed, stop* ($\mathbf{B}+\lambda\mathbf{W}$ *is not positive definite*).

(S5) *Solve*

$$\mathbf{R}_3(\lambda)^{\mathrm{T}}\mathring{\mathbf{s}}_2 = -\left\{\tilde{\mathbf{g}}_2 - \frac{1+(1-\delta_{i1})\lambda}{1+\lambda}\mathring{\mathbf{g}}_2\right\} \quad \textit{for } \mathring{\mathbf{s}}_2.$$

(S6) *Solve* $\mathbf{R}_3(\lambda)\tilde{\mathbf{s}}_2 = \mathring{\mathbf{s}}_2$ *for* $\tilde{\mathbf{s}}_2$; *compute* $\mathring{\mathbf{g}}_1 = \mathbf{R}_2\tilde{\mathbf{s}}_2$.

(S7) *Solve*

$$\mathbf{R}_1\tilde{\mathbf{s}}_1 = -\frac{1}{1+\lambda}\{\bar{\mathbf{g}}_1 + [1+(1-\delta_{i1})\lambda]\mathring{\mathbf{g}}_1\} \quad \textit{for } \tilde{\mathbf{s}}_1.$$

(S8) *Set* $\mathbf{s} = \mathbf{s}(\lambda) = \mathbf{P}^{\mathrm{T}}\tilde{\mathbf{s}}$.

Steps (S1)–(S3) have to be performed only once at the beginning of the kth trust region step. Since $l \gg t$, the costs for this fixed part are dominated by the term $\sim n^3/6$ caused by the truncated Cholesky factorization in (S1). Only the remaining steps (S4)–(S8) have to be performed repeatedly for each λ. The dominant term in the operation count is $\sim l^2/2$ caused by solving the triangular system in (S7). However, when only $\varphi(\lambda) = \|\mathbf{D}\mathbf{s}(\lambda)\|$ is needed, then these costs can be saved as follows: Because of (34),

$$\|\mathbf{D}\mathbf{s}\|^2 = \|\widetilde{\mathbf{D}}\tilde{\mathbf{s}}\|^2 = \|\mathbf{R}_1\tilde{\mathbf{s}}_1 + (1-\delta_{i1})\mathbf{R}_2\tilde{\mathbf{s}}_2\|^2 + \|\widetilde{\mathbf{D}}_3\tilde{\mathbf{s}}_2\|^2. \tag{38}$$

The term $\mathbf{R}_1\tilde{\mathbf{s}}_1$ can be replaced by the right-hand side of the equation in (S7) leading to

$$\|\mathbf{R}_1\tilde{\mathbf{s}}_1 + (1-\delta_{i1})\mathbf{R}_2\tilde{\mathbf{s}}_2\|^2 = \|\bar{\mathbf{g}}_1 + \delta_{i1}\mathring{\mathbf{g}}_1\|^2/(1+\lambda)^2,$$

and the second norm can be written as

$$\|\widetilde{\mathbf{D}}_3\tilde{\mathbf{s}}_2\|^2 = \tilde{\mathbf{s}}_2^{\mathrm{T}}\widetilde{\mathbf{D}}_3^{\mathrm{T}}\widetilde{\mathbf{D}}_3\tilde{\mathbf{s}}_2 = \|\mathbf{C}_3\tilde{\mathbf{s}}_2\|^2 + \delta_{i2}\tilde{\mathbf{s}}_2^{\mathrm{T}}\mathbf{M}_3\tilde{\mathbf{s}}_2$$

where (33) has been used. Therefore, when only $\varphi(\lambda)$ is needed, (S7) and (S8) can be replaced by

(S7′) *Set* $\varphi(\lambda) = \{\|\bar{\mathbf{g}}_1 + \delta_{i1}\mathring{\mathbf{g}}_1\|^2/(1+\lambda)^2 + \|\mathbf{C}_3\tilde{\mathbf{s}}_2\|^2 + \delta_{i2}\tilde{\mathbf{s}}_2^{\mathrm{T}}\mathbf{M}_3\tilde{s}_2\}^{1/2}$.

Then the costs of (S4)–(S7′) are of order $\sim (lt + t^3/6)$ caused by the computation of $\mathring{\mathbf{g}}_1$ and the Cholesky factor $\mathbf{R}_3(\lambda)$. The quantity $\dot{\varphi}(\lambda)$ can be computed without using $\tilde{\mathbf{s}}_1$ too; see Schwetlick and Gersonde [SG] for details. Let us point out, however, that all operation counts are valid only for dense $\mathbf{B}$. If $\mathbf{B}$ is sparse and possibly structured, then the costs depend on the sparsity pattern and the special technique used for computing the Cholesky factorization.

4. Gauss-Newton methods for nonlinear least squares problems. In this section nonlinear least squares problems

$$\min\left\{\frac{1}{2}\|\mathbf{F}(\mathbf{x})\|^2 : \mathbf{x} \in \mathbf{R}^n\right\} \tag{39}$$

are considered, where $\mathbf{F}: \mathbf{R}^n \to \mathbf{R}^m$, $m \geq n$, is given. Problem (39) is the special case of (1) with

$$f(\mathbf{x}) = \tfrac{1}{2}\|\mathbf{F}(\mathbf{x})\|^2 = \tfrac{1}{2}\mathbf{F}(\mathbf{x})^{\mathrm{T}}\mathbf{F}(\mathbf{x}).$$

In this case

$$g = \nabla f = \mathbf{J}^{\mathrm{T}}\mathbf{F},$$

where $\mathbf{J} = \mathbf{J}(\mathbf{x}) = \mathbf{F}'(\mathbf{x})$ denotes the (m, n)-Jacobian of $\mathbf{F}$. As above, the arguments $\mathbf{x}$ have been omitted.

The matrix $\mathbf{B}$ should be an approximation of the Hessian

$$\nabla^2 f = \mathbf{J}^{\mathrm{T}}\mathbf{J} + \mathbf{S}, \qquad \mathbf{S} = \sum_{i=1}^{m} \mathbf{F}_i \cdot \nabla^2 \mathbf{F}_i.$$

In the following, $\mathbf{B}$ is chosen as the so-called Gauss-Newton approximation

$$\mathbf{B} = \mathbf{J}^{\mathrm{T}}\mathbf{J} \geq 0$$

by neglecting the second-derivative term $\mathbf{S}$. The model function $m(\mathbf{s})$ is then the Gauss-Newton model

$$m(\mathbf{s}) = \tfrac{1}{2}\mathbf{F}^{\mathrm{T}}\mathbf{F} + \mathbf{F}^{\mathrm{T}}\mathbf{J}\mathbf{s} + \tfrac{1}{2}\mathbf{s}^{\mathrm{T}}\mathbf{J}^{\mathrm{T}}\mathbf{J}\mathbf{s} = \tfrac{1}{2}\|\mathbf{F} + \mathbf{J}\mathbf{s}\|^2.$$

Omitting the factor $\frac{1}{2}$, the subproblem (3) reads now as

$$\min\{\|\mathbf{F} + \mathbf{J}\mathbf{s}\|^2 : \|\mathbf{D}\mathbf{s}\| \leq \Delta\}, \tag{40}$$

and the associated linear system (6) has the special form

$$(\mathbf{J}^{\mathrm{T}}\mathbf{J} + \lambda\mathbf{D}^{\mathrm{T}}\mathbf{D})\mathbf{s} = -\mathbf{J}^{\mathrm{T}}\mathbf{F}. \tag{41}$$

These linear equations are the normal equations to the extended linear least squares problem

$$\min\left\{\left\|\begin{pmatrix}\mathbf{F}\\ \mathbf{0}\end{pmatrix} + \begin{pmatrix}\mathbf{J}\\ \sqrt{\lambda}\mathbf{D}\end{pmatrix}\mathbf{s}\right\|^2 : \mathbf{s} \in \mathbf{R}^n\right\}$$

that will briefly be written as

$$\begin{pmatrix}\mathbf{F}\\ \mathbf{0}\end{pmatrix} + \begin{pmatrix}\mathbf{J}\\ \sqrt{\lambda}\mathbf{D}\end{pmatrix}\mathbf{s} \cong \mathbf{0}. \tag{42}$$

For $\lambda > 0$, $\mathbf{B} + \lambda\mathbf{W} = \mathbf{J}^T\mathbf{J} + \lambda\mathbf{D}^T\mathbf{D} > 0$ such that (41), (42) have, for these λ, the unique solution

$$\mathbf{s} = \mathbf{s}(\lambda) = -(\mathbf{J}^T\mathbf{J} + \lambda\mathbf{D}^T\mathbf{D})^{-1}\mathbf{J}^T\mathbf{F}.$$

Moreover, the limit

$$\mathbf{s}(0) = \lim_{\lambda\to+0} \mathbf{s}(\lambda) = -\mathbf{D}^{-1}(\mathbf{J}\mathbf{D}^{-1})^{+}\mathbf{F}$$

exists, and the solutions $\mathbf{s}$ of (40) always have the form $\mathbf{s} = \mathbf{s}(\lambda)$ with an appropriate $\lambda \geq 0$, i.e., the "hard case" disappears; see **[M1]**.

For computing $\mathbf{s}(\lambda)$ and other quantities, the methods of Section 3 can immediately be applied to the normal equation matrix $\mathbf{B} = \mathbf{J}^T\mathbf{J}$ and the gradient $\mathbf{g} = \mathbf{J}^T\mathbf{F}$. However, this normal equation approach has certain disadvantages with respect to numerical stability and the class of problems that can be solved in a given floating-point arithmetic. Therefore, a different approach based on numerically stable orthogonal transformations is proposed in the following.

At first, the truncated Cholesky factorization (17), (18) as basic factorization is replaced by the **QR** factorization

$$\mathbf{Q}^T\mathbf{J}\mathbf{P}^T = \widetilde{\mathbf{R}} = \begin{pmatrix} \mathbf{R}_1 & \mathbf{R}_2 \\ \mathbf{0} & \mathbf{R}_3 \\ \mathbf{0} & \mathbf{0} \end{pmatrix} \begin{matrix} \}\, l \\ \}\, t \\ \}\, m-n \end{matrix}, \tag{43}$$

where $\mathbf{R}_1$ and $\mathbf{R}_3$ are upper triangular of dimension l and t, respectively, and $\mathbf{P}$ is a permutation describing the changes of columns. We assume further that $\mathbf{R}_1$ is regular, i.e., that the first block $\mathbf{J}_1$ of the permuted matrix

$$\widetilde{\mathbf{J}} = \mathbf{J}\mathbf{P}^T = (\underbrace{\mathbf{J}_1}_{l} \quad \underbrace{\mathbf{J}_2}_{t}) \tag{44}$$

has full rank l. Recall that $\widetilde{\mathbf{J}}$ is the representation of $\mathbf{J}$ in the permuted variables $\tilde{\mathbf{s}} = \mathbf{P}\mathbf{s}$. Now (43), (44) imply $\widetilde{\mathbf{B}} = \widetilde{\mathbf{J}}^T\widetilde{\mathbf{J}} = \widetilde{\mathbf{R}}^T\widetilde{\mathbf{R}}$, i.e.,

$$\begin{aligned}\widetilde{\mathbf{B}} &= \begin{pmatrix} \mathbf{J}_1^T\mathbf{J}_1 & \mathbf{J}_1^T\mathbf{J}_2 \\ \mathbf{J}_2^T\mathbf{J}_1 & \mathbf{J}_2^T\mathbf{J}_2 \end{pmatrix} = \begin{pmatrix} \mathbf{R}_1^T\mathbf{R}_1 & \mathbf{R}_1^T\mathbf{R}_2 \\ \mathbf{R}_2^T\mathbf{R}_1 & \mathbf{R}_2^T\mathbf{R}_2 + \mathbf{R}_3^T\mathbf{R}_3 \end{pmatrix} \\ &= \begin{pmatrix} \mathbf{R}_1^T & \mathbf{0} \\ \mathbf{R}_2^T & \mathbf{I} \end{pmatrix} \begin{pmatrix} \mathbf{I} & \mathbf{0} \\ \mathbf{0}^T & \mathbf{R}_3^T\mathbf{R}_3 \end{pmatrix} \begin{pmatrix} \mathbf{R}_1 & \mathbf{R}_2 \\ \mathbf{0}^T & \mathbf{I} \end{pmatrix}.\end{aligned}$$

Therefore, the **QR** factorization (43) leads to a factorization (17), (18) with

$$\mathbf{R} = \begin{pmatrix} \mathbf{R}_1 & \mathbf{R}_2 \\ \mathbf{0}^T & \mathbf{I} \end{pmatrix} \quad \text{and} \quad \widehat{\mathbf{B}} = \begin{pmatrix} \mathbf{I} & \mathbf{0} \\ \mathbf{0}^T & \mathbf{R}_3^T\mathbf{R}_3 \end{pmatrix},$$

i.e., $\mathbf{R}_1$, $\mathbf{R}_2$ of the **QR** factorization play the role of $\mathbf{R}_1$, $\mathbf{R}_2$ from the truncated Cholesky factorization, and $\mathbf{M}_3$ and $\mathbf{R}_3$ are related by

$$\mathbf{M}_3 = \mathbf{R}_3^{\mathrm{T}}\mathbf{R}_3. \tag{45}$$

Let us remember that when **P** in (17) and (43) is the same permutation, then in exact arithmetic, the Cholesky block $(\mathbf{R}_1|\mathbf{R}_2)$ of (17) and the **QR** block $(\mathbf{R}_1|\mathbf{R}_2)$ of (43) differ only in the sign of rows, i.e., they are essentially identical. Moreover, diagonal pivoting in Cholesky's method is equivalent to **QR** factorization with column changing as realized, e.g., in the LINPACK routines, such that these two pivoting strategies lead to the same permutation **P**.

Since the formulas of Section 3 require the diagonal entries of $\mathbf{R}_1$ to be only nonzero, but not positive, they are also valid for the factors obtained by **QR** factorization (43). In the following these formulas are equivalently evaluated in such a way that only quantities computed by orthogonal transformations are used, but not normal-equation-like terms such as $\mathbf{J}^{\mathrm{T}}\mathbf{J}$ or $\mathbf{J}^{\mathrm{T}}\mathbf{F}$. For doing this we go back to (36), (37). By using (43) and (44), we can express the permuted gradient $\tilde{\mathbf{g}}$ as

$$\tilde{\mathbf{g}} = \mathbf{P}\mathbf{g} = \widetilde{\mathbf{J}}^{\mathrm{T}}\mathbf{F} = (\mathbf{Q}^{\mathrm{T}}\widetilde{\mathbf{J}})^{\mathrm{T}}(\mathbf{Q}^{\mathrm{T}}\mathbf{F}) = \widetilde{\mathbf{R}}^{\mathrm{T}}\widetilde{\mathbf{F}} = \begin{pmatrix} \mathbf{R}_1^{\mathrm{T}}\widetilde{\mathbf{F}}_1 \\ \mathbf{R}_2^{\mathrm{T}}\widetilde{\mathbf{F}}_1 + \mathbf{R}_3^{\mathrm{T}}\widetilde{\mathbf{F}}_2 \end{pmatrix}, \tag{46}$$

where $\widetilde{\mathbf{F}}$ and its blocks are defined by

$$\widetilde{\mathbf{F}} = \mathbf{Q}^{\mathrm{T}}\mathbf{F} = \begin{pmatrix} \widetilde{\mathbf{F}}_1 \\ \widetilde{\mathbf{F}}_2 \\ \widetilde{\mathbf{F}}_3 \end{pmatrix} \begin{matrix} \}\, l \\ \}\, t \\ \}\, m-n \end{matrix} . \tag{47}$$

Inserting $\tilde{\mathbf{g}}$ from (46) into (37) yields

$$\bar{\mathbf{g}} = \begin{pmatrix} \bar{\mathbf{g}}_1 \\ \bar{\mathbf{g}}_2 \end{pmatrix} = \begin{pmatrix} \widetilde{\mathbf{F}}_1 \\ \delta_{i1}\frac{\lambda}{1+\lambda}\mathbf{R}_2^{\mathrm{T}}\widetilde{\mathbf{F}}_1 + \mathbf{R}_3^{\mathrm{T}}\widetilde{\mathbf{F}}_2 \end{pmatrix} \tag{48}$$

so that the second equation of (36) becomes

$$\mathbf{M}_3(\lambda)\tilde{\mathbf{s}}_2 = \left\{(1+\delta_{i2}\lambda)\mathbf{R}_3^{\mathrm{T}}\mathbf{R}_3 + \delta_{i1}\frac{\lambda}{1+\lambda}\mathbf{R}_2^{\mathrm{T}}\mathbf{R}_2 + \lambda\mathbf{C}_3^{\mathrm{T}}\mathbf{C}_3\right\}\tilde{\mathbf{s}}_2 = -\bar{\mathbf{g}}_2; \tag{49}$$

recall (24) and (45). If $i \neq 1$, then only the square factors $\mathbf{R}_3$ and $\mathbf{C}_3$ of dimension t occur on the left-hand side, but in case $i = 1$ there is also the factor $\mathbf{R}_2$ of size (l, t). Therefore, in case $i = 1$, a further transformation

$$\mathbf{Q}_2^{\mathrm{T}}\mathbf{R}_2 = \begin{pmatrix} \mathbf{R}_{21} \\ \mathbf{0} \end{pmatrix} \begin{matrix} \}\, t \\ \}\, l-t \end{matrix} , \qquad \mathbf{Q}_2^{\mathrm{T}}\widetilde{\mathbf{F}}_1 = \begin{pmatrix} \widetilde{\mathbf{F}}_{11} \\ \widetilde{\mathbf{F}}_{12} \end{pmatrix} \begin{matrix} \}\, t \\ \}\, l-t \end{matrix} \tag{50}$$

is performed, where $\mathbf{Q}_2$ is orthogonal and $\mathbf{R}_{21}$ is upper triangular and of dimension t. Then

$$\mathbf{R}_2^{\mathrm{T}}\mathbf{R}_2 = \mathbf{R}_{21}^{\mathrm{T}}\mathbf{R}_{21}, \qquad \mathbf{R}_2^{\mathrm{T}}\widetilde{\mathbf{F}}_1 = \mathbf{R}_{21}^{\mathrm{T}}\widetilde{\mathbf{F}}_{11}, \tag{51}$$

i.e., in (49), $\mathbf{R}_2$ and $\widetilde{\mathbf{F}}_1$ can be replaced by $\mathbf{R}_{21}$ and $\widetilde{\mathbf{F}}_{11}$, respectively. Now, we see that the equations (49) modified according to (51) are, apart from a factor $1+\delta_{i2}\lambda$, just the normal equations to

$$\begin{pmatrix} \mathbf{R}_3 \\ \delta_{i1}\beta(\lambda)\mathbf{R}_{21} \\ \gamma(\lambda)\mathbf{C}_3 \end{pmatrix} \tilde{\mathbf{s}}_2 + \begin{pmatrix} \alpha(\lambda)\widetilde{\mathbf{F}}_2 \\ \delta_{i1}\beta(\lambda)\widetilde{\mathbf{F}}_{11} \\ \mathbf{0} \end{pmatrix} \cong 0, \tag{52}$$

where

$$\alpha(\lambda) = \frac{1}{1+\delta_{i2}\lambda}, \quad \beta(\lambda) = \sqrt{\frac{\lambda}{(1+\lambda)(1+\delta_{i2}\lambda)}}, \quad \gamma(\lambda) = \sqrt{\frac{\lambda}{1+\delta_{i2}\lambda}}.$$

When $\mathbf{C}_3$ is chosen as the standard diagonal scaling matrix, then the matrix of (52) has structure

in case $i=1$ and

in case $i = 2,\ 3$.

Such matrices can be transformed to upper triangular form using, e.g., Givens rotations; see **[M1]** or Dennis, Gay, Welsch **[DGW]**, where an implementation based on fast Givens rotations is given. As a result the transformation

$$\mathbf{Q}_3^{\mathrm{T}} \begin{pmatrix} \mathbf{R}_3 \\ \delta_{i1}\beta(\lambda)\mathbf{R}_{21} \\ \gamma(\lambda)\mathbf{C}_3 \end{pmatrix} = \begin{pmatrix} \mathbf{R}_3(\lambda) \\ \mathbf{0} \\ \mathbf{0} \end{pmatrix}, \quad \mathbf{Q}_3^{\mathrm{T}} \begin{pmatrix} \alpha(\lambda)\widetilde{\mathbf{F}}_2 \\ \delta_{i1}\beta(\lambda)\widetilde{\mathbf{F}}_{11} \\ \mathbf{0} \end{pmatrix} = \begin{pmatrix} \widetilde{\mathbf{F}}_{31} \\ \widetilde{\mathbf{F}}_{32} \\ \widetilde{\mathbf{F}}_{33} \end{pmatrix} \tag{53}$$

is obtained where $\mathbf{Q}_3 = \mathbf{Q}_3(\lambda)$ is orthogonal and $\mathbf{R}_3(\lambda)$ is upper triangular and of size t. Then the solution to (52) and, hence, to (49) can be computed from the triangular system

$$\mathbf{R}_3(\lambda)\tilde{\mathbf{s}}_2 + \widetilde{\mathbf{F}}_{31} = \mathbf{0}.$$

After having obtained $\tilde{\mathbf{s}}_2$, we compute the first block $\tilde{\mathbf{s}}_1$ from the first equation of (36) as in step (S7) of Algorithm 1 using $\bar{\mathbf{g}}_1 = \widetilde{\mathbf{F}}_1$; cf. (48). Thus, we can compute $\mathbf{s}(\lambda)$ by the following algorithm.

ALGORITHM 2 (COMPUTATION OF $\mathbf{s}(\lambda)$ FOR $\lambda > 0$ IN THE GAUSS-NEWTON CASE).

(S1) *Compute the basic* **QR** *factorization* (43) *and transform* **F** *orthogonally into* $\widetilde{\mathbf{F}}$ *according to* (47).

(S2) *If* $i = 1$, *then perform the additional orthogonal transformation* (50).

(S3) *Compute* $\mathbf{R}_3(\lambda)$ *and* $\widetilde{\mathbf{F}}_{31}$ *by the orthogonal transformation* (53).

(S4) *Solve* $\mathbf{R}_3(\lambda)\tilde{\mathbf{s}}_2 = -\widetilde{\mathbf{F}}_{31}$ *for* $\tilde{\mathbf{s}}_2$; *compute* $\mathring{\mathbf{F}}_1 = \mathbf{R}_2\tilde{\mathbf{s}}_2$.

(S5) *Solve*

$$\mathbf{R}_1\tilde{\mathbf{s}}_1 = -\frac{1}{1+\lambda}\{\widetilde{\mathbf{F}}_1 + [1 + (1 - \delta_{i1})\lambda]\mathring{\mathbf{F}}_1\} \quad \textit{for } \tilde{\mathbf{s}}_1.$$

(S6) *Set* $\mathbf{s} = \mathbf{s}(\lambda) = \mathbf{P}^{\mathrm{T}}\tilde{\mathbf{s}}$.

As in Algorithm 1, the preparatory steps (S1) and (S2) have to be performed only once, and the operation count is dominated by the term $\sim (m - n/3)n^2$ caused by the **QR** factorization (17) when the latter is realized by Householder or fast Givens transformations. For each λ, only steps (S3)–(S6) have to be performed repeatedly, and the costs are determined by the term $\sim l^2/2$ resulting from the solution of the triangular system in (S5). However, when only $\varphi(\lambda) = \|\mathbf{D}\mathbf{s}(\lambda)\|$ is needed, then solving this system can be avoided analogously to the general case investigated in Section 3.

In the Gauss-Newton case step (S7′) reads as

$$\varphi(\lambda)^2 = \|\widetilde{\mathbf{F}}_1 + \delta_{i1}\mathring{\mathbf{F}}_1\|^2/(1+\lambda)^2 + \|\mathbf{C}_3\tilde{\mathbf{s}}_2\|^2 + \delta_{i2}\|\mathbf{R}_3\tilde{\mathbf{s}}_2\|^2.$$

For versions 2 and 3, the first norm reduces to $\|\widetilde{\mathbf{F}}_1\|^2$. For version 1 the computation of $\|\widetilde{\mathbf{F}}_1 + \mathring{\mathbf{F}}_1\|^2$ can be simplified by using (50) according to

$$\|\widetilde{\mathbf{F}}_1+\mathring{\mathbf{F}}_1\|^2 = \|\widetilde{\mathbf{F}}_1+\mathbf{R}_2\tilde{\mathbf{s}}_2\|^2 = \|\mathbf{Q}_2^{\mathrm{T}}(\widetilde{\mathbf{F}}_1+\mathbf{R}_2\tilde{\mathbf{s}}_2)\|^2 = \|\widetilde{\mathbf{F}}_{11}+\mathbf{R}_{21}\tilde{\mathbf{s}}_2\|^2+\|\widetilde{\mathbf{F}}_{12}\|^2.$$

Therefore, if only $\varphi(\lambda)$ is needed, (S4)–(S6) can be replaced by

(S4″) *Solve* $\mathbf{R}_3(\lambda)\tilde{\mathbf{s}}_2 = -\widetilde{\mathbf{F}}_{31}$ *for* $\tilde{\mathbf{s}}_2$. *If* $i = 1$ *then compute* $\mathring{\mathbf{F}}_{11} = \mathbf{R}_{21}\tilde{\mathbf{s}}_2$.

(S5″) *If* $i = 1$ *then compute* $\rho^2 = \|\widetilde{\mathbf{F}}_{11} + \mathring{\mathbf{F}}_{11}\|^2 + \|\widetilde{\mathbf{F}}_{12}\|^2$; *else compute* $\rho^2 = \|\widetilde{\mathbf{F}}_1\|^2$. *Set*

$$\varphi(\lambda) = \{\rho^2/(1+\lambda)^2 + \|\mathbf{C}_3\tilde{\mathbf{s}}_2\|^2 + \delta_{i2}\|\mathbf{R}_3\tilde{\mathbf{s}}_2\|^2\}^{1/2}.$$

The derivative $\dot{\varphi}(\lambda)$ can be computed in a similar manner; see **[ST2]** for details.

We finish this section with the remark that Algorithm 2 can also be derived directly from the least squares formulation (42) or its transformed form

$$\begin{pmatrix}\widetilde{\mathbf{F}}\\ \mathbf{0}\end{pmatrix} + \begin{pmatrix}\widetilde{\mathbf{R}}\\ \sqrt{\lambda}\widetilde{\mathbf{D}}\end{pmatrix}\mathbf{s} \cong \mathbf{0}$$

using the scaling matrices $\widetilde{\mathbf{D}}$ from (34) and (35). In addition, the square scaling matrix $\widetilde{\mathbf{D}}^{(2)}$ can be replaced by the rectangular matrix

$$\widetilde{\mathbf{D}}^{(2,1)} = \begin{pmatrix} \mathbf{R}_1 & \mathbf{R}_2 \\ \mathbf{0} & \mathbf{R}_3 \\ \mathbf{0} & \mathbf{C}_3 \end{pmatrix}$$

since both generate the same $\widetilde{\mathbf{W}}$. In the same equivalent sense, the scaling matrices $\widetilde{\mathbf{D}}^{(i,2)}$,

$$\widetilde{\mathbf{D}}^{(1,2)} = \begin{pmatrix} \mathbf{J}_1 & \mathbf{0} \\ \mathbf{0} & \mathbf{C}_3 \end{pmatrix}, \quad \widetilde{\mathbf{D}}^{(2,2)} = \begin{pmatrix} \mathbf{J}_1 & \mathbf{J}_2 \\ \mathbf{0} & \mathbf{C}_3 \end{pmatrix}, \quad \widetilde{\mathbf{D}}^{(3,2)} = \begin{pmatrix} \mathbf{J}_1 & \mathbf{J}_1\mathbf{J}_1^+\mathbf{J}_2 \\ \mathbf{0} & \mathbf{C}_3 \end{pmatrix},$$

of size $(m+t,\ n)$ can be used for defining $\widetilde{\mathbf{W}}^{(i)}$. In [ST2] this approach has been chosen.

5. Concluding remarks. As a first example of nonstandard scaling matrices based on truncated factorizations, version 3 has been proposed by [ST1] in the Gauss-Newton case for solving so-called orthogonal least distance regression problems. In this paper the steps $\mathbf{s}$ are exactly the steps $\mathbf{s}(\lambda)$ from version 3 of Section 4, but they are controlled there in a Marquardt-like manner by λ instead of Δ; see also Schwetlick, [S1] and [S2]. Later in [ST2] it is shown that this step can be considered to be a genuine trust region step in a nonstandard metric, and the versions 1 and 2 are introduced there. In [SG] the general case of arbitrary coefficients $\mathbf{B}$ and $\mathbf{g}$ is treated on the basis of a truncated Cholesky factorization as in Section 3 of this survey.

In the two last-mentioned papers some further problems are addressed. The first is the appropriate choice of the partition, i.e., the choice of $\mathbf{P}$ and l in the truncated factorizations. For some classes of problems such as the orthogonal regression problem mentioned above, such a partition is known a priori. If this is not the case, then l can be determined adaptively using pivoted factorization techniques and a criterion such as

$$|r_{ll}| \geq \varepsilon \quad \text{or} \quad |r_{ll}| \geq \varepsilon |r_{11}| \tag{54}$$

for defining l.

The second problem is to find conditions that guarantee the nonstandard trust region methods to have the same good global convergence behavior as the standard methods. Such conditions are discussed, e.g., by [So] and [M2] for arbitrary scalings and applied to the nonstandard scalings in the papers just mentioned.

Numerical experience available so far shows that the nonstandard methods work very well in case of large structured problems where a sufficiently definite block is known a priori as in orthogonal regression. For small and medium-sized least squares problems, the adaptive choice according to (54) leads in general to competitive results. On the other hand, in the case of general minimization problems, the situation is not so clear and requires further investigations, especially in case of large sparse problems. For such problems, the methods of Section 3 could be alternatives to modified line search methods which use so-called directions of negative curvature in case of indefinite **B**; see **[SSB]** and the references given there for such methods. The advantage of the nonstandard methods could be that they can better take into account a possible nonconvexity of $f(\mathbf{x}+\mathbf{s})$ if t is not too small.

References

[BBS] P. T. Boggs, R. H. Byrd, and R. B. Schnabel, *A stable and efficient algorithm for nonlinear orthogonal distance regression*, SIAM J. Sci. Statist. Comput. **8** (1987), 1052–1078.

[BV] J. P. Bulteau and J. P. Vial, *A restricted trust region algorithm for unconstrained optimization*, J. Optim. Theory Appl. **47** (1985), 413–435.

[BKP] J. R. Bunch, L. Kaufman, and B. N. Parlett, *Decomposition of a symmetric matrix*, Numer. Math. **31** (1976), 31–48.

[DGW] J. E. Dennis, D. M. Gay, and R. E. Welsch, *An adaptive nonlinear least squares algorithm*, ACM Trans. Math. Software **7** (1981), 348–368.

[DS] J. E. Dennis and R. B. Schnabel, *Numerical Methods for Unconstrained Optimization and Nonlinear Equations*, Prentice-Hall, Englewood Cliffs, New Jersey, 1983.

[F1] R. Fletcher, *Practical Methods of Optimization,* 2nd edition, Wiley, Chichester, 1987.

[F2] ——, *Low storage methods for unconstrained optimization*, these proceedings, 1990.

[G] D. M. Gay, *Computing optimal locally constrained steps*, SIAM J. Sci. Statist. Comput. **2** (1981), 186–197.

[KS] A. Kiełbasiński and H. Schwetlick, *Numerical linear algebra, a computer-oriented introduction* (in German), Deutscher Verlag der Wissenschaften, Berlin, and Harri Deutsch Verlag, Frankfurt, 1988.

[K] O. Knoth, *Marquardt-like methods for minimizing nonlinear functions* (in German), Dissertation A, Martin Luther University, Department of Mathematics, Halle, 1983.

[M1] J. J. Moré, *The Levenberg–Marquardt algorithm: Implementation and theory*, Lecture Notes in Mathematics **630**, G. A. Watson (ed.) Springer-Verlag, Berlin, 1978, 105–116.

[M2] ——, *Recent developments in algorithms and software for trust region methods*, Mathematical programming—the state of the art (A. Bachem, M. Grötschel, and B. Korte, eds.), Springer-Verlag, Berlin, 1983, pp. 258–287.

[M3] ——, *Trust region methods for systems of nonlinear equations*, these proceedings, 1990.

[MS] J. J. Moré and D. C. Sorensen, *Computing a trust region step*, SIAM J. Sci. Statist. Comput. **4** (1983), 553–572.

[P1] M. J. D. Powell, *A hybrid method for nonlinear equations*, Numerical methods for nonlinear algebraic equations, P. Rabinowitz (ed.), Gordon and Breach, London, 1970.

[P2] ——, *A new algorithm for unconstrained optimization*, Nonlinear programming, J. B. Rosen, O. L. Mangasarian, and K. Ritter (eds.), Academic Press, New York, 1970, 31–65.

[S1] H. Schwetlick, *On the convergence of regularized Gauss–Newton methods* (in German), Ž. Vyčisl. Mat. i Mat. Fiz. **13** (1973), 1371–1382.

[S2] ——, *Numerical solution of nonlinear equations* (in German), Deutscher Verlag der Wissenschaften, Berlin, and Oldenbourg Verlag, Munich.

[SG] H. Schwetlick and J. Gersonde, *Nonstandard scaling matrices in trust region methods for unconstrained optimization*, Manuscript (1988), Martin Luther University, Department of Mathematics, Halle.

[ST1] H. Schwetlick and V. Tiller, *Numerical methods for estimating parameters in nonlinear models with errors in the variables*, Technometrics **27** (1985), 17–24.

[ST2] ——, *Nonstandard scaling matrices for trust region Gauss–Newton methods*, SIAM J. Sci. Statist. Comput. **10** (1989).

[SSB] G. A. Shultz, R. B. Schnabel, and R. H. Byrd,, *A family of trust-region-based algorithms for unconstrained minimization with strong global convergence properties*, SIAM J. Numer. Anal. **22** (1985), 47–67.

[So] D. C. Sorensen, *Newton's method with a model trust region modification*, SIAM J. Numer. Anal. **19**, 409–426.

[W] L. Westin, *An inexpensive, conceptually simple trust region method for unconstrained optimization*, Report UMINF—135.86 (1987), University of Umeå, Institut of Information Processing, Umeå.

MARTIN LUTHER UNIVERSITY
GERMAN DEMOCRATIC REPUBLIC HALLE

Lectures in Applied Mathematics
Volume **26** (1990)

Numerical Determination of Breathers and Forced Oscillations of Nonlinear Wave Equations

MICHAEL W. SMILEY

Abstract. A numerical method for finding time-periodic solutions of nonlinear wave equations is presented. The spatial domain is assumed to be $\mathbb{R}^1$ or $\mathbb{R}^3$, and solutions are to satisfy a decay condition at infinity. Since multiple solutions exist, the problems are ill-posed. The method is based on a constructive method for establishing existence of small norm solutions, which also characterizes multiplicity.

1. Introduction. In this article we describe some work in progress concerning the numerical solution of wave equations, subjected to periodic conditions in time rather than the usual initial conditions in time. Although there is certainly an intrinsic value in being able to numerically solve these problems to aid in applications, we present our work as a numerical development that allows an interplay between theoretical and experimental work in mathematics. As a result of this interplay, progress towards a complete resolution of the problem has been made and, we feel, will continue to be made. In particular, we are interested in obtaining solutions of the nonlinear wave problem

$$u_{tt} - \Delta u + \alpha g(u) = f, \qquad (t,x) \in \mathbb{R} \times \mathbb{R}^N \qquad (N = 1 \text{ or } 3), \tag{1.1}$$

$$\lim_{|x|\to\infty} u(t,x) = 0, \qquad t \in \mathbb{R}, \tag{1.2}$$

$$u(t+T,x) = u(t,x), \qquad (t,x) \in \mathbb{R} \times \mathbb{R}^N. \tag{1.3}$$

1980 *Mathematics Subject Classification* (1985 *Revision*). Primary 35B10, 35L05, 65N99.

As a typical nonlinearity, one may consider $g(u) = m^2u + A\sin u$, which corresponds to the forced sine-Gordon equation with mass term (cf. **[5]**). We assume that $\alpha \in \mathbb{R}$, and $f : \mathbb{R} \times \mathbb{R}^N \to \mathbb{R}$ is T-periodic in t, radially symmetric in x, and square-integrable over $(0,T) \times \mathbb{R}^N$ in an exponentially weighted sense. If f is identically zero we say that a time-dependent solution u is a *breather*, a term which is suggestive of the undulations that would be observed in the spatial profile over time. We will be interested in determining radially symmetric solutions $u(t,x) = \tilde{u}(t,|x|)$. Since the radial Laplacian has the form $\Delta u = u_{rr} + \frac{N-1}{r}u_r$, we can make a standard change of variables in (1.1), $w(t,r) = r^{\frac{N-1}{2}}u(t,x)$, to obtain

$$(1.4)\quad w_{tt} - w_{rr} + q_N(r)w + \alpha r^{\frac{N-1}{2}} g\left(w/r^{\frac{N-1}{2}}\right) = h, \qquad (t,r) \in \mathbb{R} \times \mathbb{R}^+,$$

where $q_N(r) = (N-1)(N-3)/(4r^2)$ and $h(t,r) = r^{\frac{N-1}{2}} f(t,x)$. Observe that $u, f \in \mathbf{L}^2\left((0,T) \times \mathbb{R}^N\right)$ if and only if $w, h \in \mathbf{L}^2\left((0,T) \times \mathbb{R}^+\right)$. Also note that $q_N(r)$ is identically zero when $N = 1$ or 3.

In dimensions $N = 1$ or 3, weak solutions of the problem have been shown (cf. **[7]**) to exist in the following sense. We consider distributional solutions $w(t,r)$ of (1.4), which satisfy the decay and periodicity conditions

$$(1.5)\qquad w(t,\cdot) \in \mathbf{L}^2(\mathbb{R}^+), \qquad \text{(a.e.)}\quad t \in \mathbb{R},$$

$$(1.6)\qquad w(t+T,r) = w(t,r), \qquad \text{(a.e.)}\quad (t,r) \in \mathbb{R} \times \mathbb{R}^+.$$

The space of test functions used is $\mathscr{D}_T = \{\varphi \in \mathbf{C}^\infty(\mathbb{R}\times\mathbb{R}^+) : \varphi(t+T,r) = \varphi(t,r)$ for all $(t,r) \in \mathbb{R} \times \mathbb{R}^+$, and $\varphi(t,\cdot) \in \mathbf{C}_0^\infty(\mathbb{R}^+)$ for all $t \in \mathbb{R}\}$. We will need (cf. **[6]**) to work in the weighted Hilbert space $\mathbf{H}_\delta = \{h \in \mathbf{L}^2\left((0,T) \times \mathbb{R}^+\right) : \|h\|_\delta < +\infty\}$ where

$$\|h\|_\delta = \left\{\frac{2}{T}\int_0^T \int_0^{+\infty} |h(t,r)|^2 e^{2\delta r} dr dt\right\}^{1/2}.$$

Let $\mathbf{H}_\delta^1 = \{w \in \mathbf{H}_\delta : w_t, w_r \in \mathbf{H}_\delta\}$ be the corresponding Sobolev space of functions having first-order distributional derivatives (relative to $\mathscr{D}_T$) also in $\mathbf{H}_\delta$. We use the Hilbert space norm in $\mathbf{H}_\delta^1$, and denote it by $\|\cdot\|_{1,\delta}$. Consider $\alpha_1 \in \mathbb{R}$ as fixed, and α near α_1. Let $\theta_n = 2\pi n/T$ for integers $n \geq 0$, and let m denote the integer such that $\theta_m^2 < \lambda_1 = \alpha_1 g'(0) \leq \theta_{m+1}^2$. We adopt the convention $\theta_{-1}^2 = -\infty$ so that $-1 \leq m < +\infty$. Assuming $\lambda > 0$ we set $\beta_n = \sqrt{\lambda - \theta_n^2}$, $0 \leq n \leq m$; and by convention $\beta_{-1} = +\infty$. With this notation established we record the main result of **[7]**.

THEOREM 1.1. *Let $g \in \mathbf{C}^2(\mathbb{R})$ with both $g'(u)$ and $g''(u)$ bounded on $\mathbb{R}$. Let $\delta \in (0, \beta_m)$ and suppose $h \in \mathbf{H}_\delta$. If $m \geq 0$ there is a $(2m+1)$-dimensional manifold $M_{\lambda_1,\delta} \subset \mathbf{H}^1_\delta$ of solutions of (1.4)–(1.6), which is local to the origin in $\mathbf{H}^1_\delta$, for each h sufficiently small and each α near α_1. If $m = -1$ then there is a unique (local) solution.*

REMARK. We have recently extended (cf. **[8]**) the above result to all dimensions $N \geq 1$. However, the methods used require that (1.4) be considered on a domain $\mathbb{R}^+ \times (R, +\infty)$ for $R > 0$. The decay condition (1.5) is replaced by $w(t, \cdot) \in \mathbf{L}^2(R, +\infty)$. This corresponds to an exterior domain $\|x\| > R$ for (1.1)–(1.3).

A fundamental question is raised by the results presented in Theorem 1.1. What is the solution regularity and behavior at the origin? The notion of solution being used here is weaker than the usual one of (radially symmetric) distributions on all of $\mathbb{R}^N$. Each of the test functions in $\mathscr{D}_T$ is identically zero in a neighborhood of the origin. Hence $\mathscr{D}_T$ does not provide an assessment of whether (1.1) is satisfied at the origin. The definition being used only requires that solutions be square-integrable in a neighborhood of the origin. The question of whether any of the solutions shown to exist have additional regularity at the origin, so that they are solutions in the stronger sense, remains unresolved.

In the one-dimensional case, the first derivative u_x will generally fail to be continuous along the line $x = 0$ in the (x, t)-plane. However, there is a known example (cf. **[3]**) of a smooth (breather) solution. Numerically (cf. Section 5) we have verified that this solution is on the manifold of solutions $M_{\lambda_1,\delta}$ described in Theorem 1.1. Thus, in at least one instance, the solutions we have shown to exist have more regularity at the origin. It has been one of the goals of our numerical work to investigate this aspect of the nature of the solutions belonging to $M_{\lambda_1,\delta}$.

In the three-dimensional case, the function $u(t, x)$ will generally be of order $O(r^{-1})$ as $r \to 0^+$. In fact we obtain solutions of the form $u(t, x) = w(t, r)/r$, where $w(t, r)\exp(\delta r)$ is bounded on $\mathbb{R} \times \mathbb{R}^+$. Thus solutions are (exponentially) localized and integrable at the origin, but may fail to be bounded at the origin. Theoretically it may be possible to exploit the finite dimensionality of $M_{\lambda_1,\delta}$ to determine whether submanifolds of smooth solutions exist. Currently the question of regularity at the origin remains open.

Our numerical interest in the problem was originally motivated by our desire to understand the nature of this class of problems. Little was known theoretically, and yet these problems are widely applicable as physical models. Initial efforts led to knowledge of the associated linear problem, as expressed in Theorem 2.1 of the next section (cf. [6]), and subsequent work led to the nonlinear result above. Since many questions remain open, as we have indicated above, the numerical work continues to be of interest and value in guiding theoretical developments.

It is clear from the above result that there is a bifurcation in the solution set as λ_1 crosses the values θ_n^2. This is an unusual bifurcation in the sense that it is the dimension of the manifold of solutions $M_{\lambda_1,\delta}$ that changes rather than simply a finite count of solutions. Investigating the nature of this bifurcation is also of interest in our numerical work. Due to the nature of the theoretical developments appropriate to the problem, this investigation has required modifications in the usual approach to a continuation algorithm. After presenting some needed background material in Sections 2 and 3, we describe this aspect of the problem more thoroughly in Section 4. In Section 5 we make some remarks concerning implementations, and present some numerical results.

2. Manifolds and parameterizations. The analysis of [7] shows that the manifold of solutions of (1.4)–(1.6), stated to exist in Theorem 1.1, is naturally parameterized by the null space of the associated linear operator. In general suppose $\lambda \in \mathbb{R}$, and let $m \geq 0$ be such that $\theta_m^2 < \lambda \leq \theta_{m+1}^2$. Set $\beta_n = \beta_n(\lambda) = \sqrt{\lambda - \theta_n^2}$ as before. The linear span of the set of functions

$$\{\exp(-\beta_n r)\cos\theta_n t, \exp(-\beta_n r)\sin\theta_n t\}_{n=0}^m$$

forms a $(2m+1)$-dimensional linear space. Equipped with the topology inherited from $\mathbf{H}_\delta^1$ we obtain a Hilbert subspace $\mathbf{N}_{\lambda,\delta} \subset \mathbf{H}_\delta^1$. Observe that the spanning functions vary with λ, while the topology varies with δ. An easy computation shows that

$$(2.1)\qquad \begin{cases} \|\exp(-\beta_0(\cdot))\|_{1,\delta} = \left[\frac{1+\beta_0^2}{\beta_0-\delta}\right]^{1/2}, & \text{and} \\ \left\|\exp(-\beta_n(\cdot))\begin{Bmatrix}\cos\theta_n(\cdot)\\ \sin\theta_n(\cdot)\end{Bmatrix}\right\|_{1,\delta} = \left[\frac{1+\theta_n^2+\beta_n^2}{2(\beta_n-\delta)}\right]^{1/2}. \end{cases}$$

Thus we obtain an orthonormal basis, $\{u_n\}_{n=0}^m$, $\{v_n\}_{n=1}^m$, for $\mathbf{N}_{\lambda,\delta}$ by setting

$$(2.2)\qquad \begin{cases} u_0(r) = \left[\frac{\beta_0-\delta}{1+\beta_0^2}\right]^{1/2} \exp(-\beta_0 r), \\ u_n(t,r) = \left[\frac{2(\beta_n-\delta)}{1+\theta_n^2+\beta_n^2}\right]^{1/2} \exp(-\beta_n r)\cos\theta_n t, & 1 \leq n \leq m, \\ v_n(t,r) = \left[\frac{2(\beta_n-\delta)}{1+\theta_n^2+\beta_n^2}\right]^{1/2} \exp(-\beta_n r)\sin\theta_n t, & 1 \leq n \leq m. \end{cases}$$

We may define an orthogonal projection $P_{\lambda,\delta} : \mathbf{H}^1_\delta \to \mathbf{N}_{\lambda,\delta}$ in the following way. Starting with $w \in \mathbf{H}^1_\delta$, we first obtain the Fourier coefficient functions

$$\text{(2.3)} \qquad \begin{cases} a_n(r) = \frac{2}{T}\int_0^T w(t,r)\cos\theta_n t\,dt, & 0 \le n \le m, \\ b_n(r) = \frac{2}{T}\int_0^T w(t,r)\sin\theta_n t\,dt, & 1 \le n \le m. \end{cases}$$

Next we use the orthogonal projections

$$Q_{\beta_n} : \mathbf{H}^1_\delta(\mathbb{R}^+) \to \operatorname{span}\{\exp(-\beta_n r)\}.$$

Here $\mathbf{H}^1_\delta(\mathbb{R}^+)$ denotes the usual Sobolev space $\mathbf{H}^1(\mathbb{R}^+)$ equipped with the exponentially weighted norm. One finds that

$$\text{(2.4)} \quad Q_{\beta_n}h(r) = \frac{2(\beta_n-\delta)}{1+\beta_n^2}\Big[(1+2\delta\beta_n-\beta_n^2)H(\beta_n-2\delta) + \beta_n h(0)\Big]\exp(-\beta_n r),$$

with $H(s) = \mathscr{L}\{h(r)\}$ denoting the Laplace transform of $h \in H^1_\delta(\mathbb{R}^+)$. Thus we are led to define

(2.5)

$$\begin{cases} P_{\lambda,\delta}w = \frac{1}{2}c_0u_0(r) + \sum_{n=1}^{m}[c_nu_n(t,r) + d_nv_n(t,r)], \quad \text{where} \\ c_0 = 2\left[\frac{\beta_0-\delta}{1+\beta_0^2}\right]^{1/2}\Big[(1+2\delta\beta_0-\beta_0^2)A_0(\beta_0-2\delta) + \beta_0a_0(0)\Big], \\ c_n = \frac{[2(\beta_n-\delta)(1+\theta_n^2+\beta_n^2)]^{1/2}}{1+\beta_n^2}\Big[(1+2\delta\beta_n-\beta_n^2)A_n(\beta_n-2\delta) + \beta_na_n(0)\Big], \\ d_n = \frac{[2(\beta_n-\delta)(1+\theta_n^2+\beta_n^2)]^{1/2}}{1+\beta_n^2}\Big[(1+2\delta\beta_n-\beta_n^2)B_n(\beta_n-2\delta) + \beta_nb_n(0)\Big], \end{cases}$$

with $A_n(s) = \mathscr{L}\{a_n(r)\}$, $B_n(s) = \mathscr{L}\{b_n(r)\}$.

Let L_λ denote the linear operator defined by $L_\lambda w = w_{tt} - w_{rr} + \lambda w$ and having the domain specified by the boundary conditions (1.5)–(1.6). Notice that $\mathbf{N}_{\lambda,\delta}$ is the null space of L_λ in $\mathbf{H}^1_\delta$. The following result (cf. **[6]**) guarantees that L_λ has a continuous partial inverse $K_\lambda : \mathbf{H}_\delta \to \mathbf{H}^1_\delta$ when $\lambda > 0$.

THEOREM 2.1. *Let $\lambda > 0$ and m be the integer for which $\theta_m^2 < \lambda \le \theta_{m+1}^2$. Let $\delta \in (0, \beta_m)$ and $h \in \mathbf{H}_\delta$. Then there is a $(2m+1)$-dimensional affine space of solutions $w \in \mathbf{H}^1_\delta$ of $L_\lambda w = h$. All solutions have the form $w = w_0 + w_1$, where $w_0 \in \mathbf{N}_{\lambda,\delta}$ and $P_{\lambda,\delta}w_1 = 0$. Moreover there is a constant $c(\lambda,\delta)$, with the property that $c(\lambda,\delta) \to +\infty$ as $\delta \to 0^+$ or as $\delta \to \beta_m^-$, such that $\|w_1\|_{1,\delta} \le c(\lambda,\delta)\|h\|_\delta$. Thus the correspondence $h \to w_1$ defines a bounded linear map K_λ.*

REMARK. The above result also holds when $\lambda \leq 0$. In this case $K_\lambda = L_\lambda^{-1}$ is a true inverse since there are unique solutions.

3. Linear inversion. The algorithm we implement as the numerical linear solver, corresponding to the theoretically established map K_λ, is based ultimately on the classical Paley–Wiener Theorem [**4**]. Consider the generalized boundary value problem

$$\begin{cases} w'' + bw = h, & 0 < r < +\infty, \\ w \in \mathbf{L}^2(\mathbb{R}^+) \end{cases} \tag{3.1}$$

for a scalar function $w : \mathbb{R}^+ \to \mathbb{R}$. The function h is assumed to belong to the class $\mathbf{L}_\delta^2(\mathbb{R}^+)$ for some $\delta > 0$. Here $\mathbf{L}_\delta^2(\mathbb{R}^+)$ denotes the exponentially weighted $\mathbf{L}^2$ space. By considering Laplace transforms, we are led (cf. [**6**]) to the following situations which vary with b.

LEMMA 3.1. *Assume that $h \in \mathbf{L}_\delta^2(\mathbb{R}^+)$ for some $\delta > 0$, and let β denote a real number.*

(i) *If $b = \beta^2 \geq 0$ then* (3.1) *has the unique solution*

$$w^*(r) = \begin{cases} -\beta^{-1} \int_r^{+\infty} \sin \beta(r-\sigma) h(\sigma)\, d\sigma, & \beta > 0, \\ -\int_r^{+\infty} (r-\sigma) h(\sigma)\, d\sigma, & \beta = 0. \end{cases}$$

(ii) *If $b = -\beta^2 < 0$ then* (3.1) *has the* 1*-parameter family of solutions*

$$w(r) = w_0 e^{-\beta r} + \frac{1}{2\beta} \left\{ H(\beta) e^{-\beta r} - \int_0^{+\infty} e^{-\beta|r-\sigma|} h(\sigma)\, d\sigma \right\} \tag{3.2}$$

parameterized by $w_0 \in \mathbb{R}$; here $H(s) = \mathscr{L}\{h(r)\}$.

From all of the solutions provided by (3.2), we pick the unique one with the property that $Q_\beta w = 0$, where Q_β is the projection introduced earlier $Q_\beta : \mathbf{H}_\delta^1(\mathbb{R}^+) \to \operatorname{span}\{\exp(-\beta r)\}$. In the notation of (3.2) let $w_1(r) = w(r) - w_0 \exp(-\beta r)$. In [**6**] it is shown that $\mathscr{L}\{w_1(r)\} = W_1(s) = [H(s) - H(\beta)]/(s^2 - \beta^2)$. Since $w_1(0) = 0$ it follows from (2.4) that $Q_\beta w_1(r) = c_{\beta\delta} \exp(-\beta r)$, where

$$\begin{aligned} c_{\beta\delta} &= \frac{2(\beta - \delta)(1 + 2\delta\beta - \beta^2)}{1 + \beta^2} W_1(\beta - 2\delta) \\ &= \frac{1 + 2\delta\beta - \beta^2}{2\delta(1 + \beta^2)} [H(\beta) - H(\beta - 2\delta)]. \end{aligned} \tag{3.3}$$

Hence the unique solution we seek is $w^*(r) = w_1(r) - c_{\beta\delta} \exp(-\beta r)$.

In considering the problem (1.4)–(1.6), linearized at the origin, we are led to an infinite system, in which all equations have the form given in (3.1). These equations have $b = b_n = \theta_n^2 - \lambda$, and $h(r)$ equal to (the negative of) one of the Fourier coefficient functions of $h(t,r)$, as given in (2.3) with $h(t,r)$ replacing $w(t,r)$. Analysis of these component equations, including a priori estimates of the solution which are independent of b, leads to Theorem 2.1.

At this point we can more precisely describe the solution $w_1 = K_\lambda h$ of Theorem 2.1, in a way that allows for numerical approximation. We present w_1 in the form

$$w_1(t,r) = a_0(r) + \sum_{n=1}^{\infty} [a_n(r)\cos\theta_n + b_n(r)\sin\theta_n t]. \tag{3.4}$$

Let

$$c_n(r) = \frac{2}{T}\int_0^T h(t,r)\cos\theta_n t\,dt,$$

$$d_n(r) = \frac{2}{T}\int_0^T h(t,r)\sin\theta_n t\,dt,$$

and $C_n(s) = \mathscr{L}\{c_n(r)\}$, $D_n(s) = \mathscr{L}\{d_n(r)\}$. Then we have the explicit formulas

(3.5)

$$a_n(r) = \begin{cases} -\frac{1}{2\beta_n}\left\{C_n(\beta_n)e^{-\beta_n r} - \int_0^{+\infty} e^{-\beta_n|r-\sigma|}c_n(\sigma)\,d\sigma\right\} \\ \qquad + \frac{1+2\delta\beta_n-\beta_n^2}{2\delta(1+\beta_n^2)}[C_n(\beta_n) - C_n(\beta_n - 2\delta)]e^{-\beta_n r} \quad (n \le m) \\ \frac{1}{\beta_n}\int_r^{+\infty}\sin\beta_n(r-\sigma)c_n(\sigma)\,d\sigma \qquad (n > m) \end{cases}$$

for the Fourier coefficient functions $a_n(r)$. Analogous expressions hold for $b_n(r)$; we replace c_n, C_n by d_n, D_n, respectively. If $\beta_{m+1} = 0$ then the limiting form (as $\beta \to 0$) of the second formula applies in the case $n = m+1$. Some remarks on the computation of these quantities will be made in Section 5.

4. The continuation problem. Let $\alpha_1, \alpha \in \mathbb{R}$ and define $\psi(\alpha, r, w) = \alpha_1 g'(0)w - \alpha r^{\frac{N-1}{2}} g(w/r^{\frac{N-1}{2}})$, with $N = 1$ or 3. Clearly ψ also depends on α_1, but for the moment we think of α_1 as being fixed. Using ψ we rewrite (1.4) in the form

$$w_{tt} - w_{rr} + \lambda_1 w = \psi(\alpha, r, w) + h, \qquad (t,r) \in \mathbb{R}\times\mathbb{R}^+, \tag{4.1}$$

with $\lambda_1 = \alpha_1 g'(0)$. Theoretically (cf. [7]) we know that the solutions of (1.4)–(1.6) in $\mathbf{H}_\delta^1$ are characterized as fixed points of the map

$$F(w_1) = K_{\lambda_1}[\psi(\alpha, r, w_0 + w_1) + h]. \tag{4.2}$$

The corresponding solution is $w = w_0 + w_1$, and $P_{\lambda_1,\delta}w = w_0$. Furthermore, it follows from the proof of Theorem 1.1 that F is a contraction mapping on a neighborhood of the origin, when α is near α_1 and w_0, h are small. Thus numerically w_1 can be determined by using a fixed point iteration scheme. In addition, by allowing α to vary we can determine how the set of solutions changes with respect to this parameter.

In fact there are several parameters that are used in (locally) characterizing the solution set. These include all of the components of $w_0 \in \mathbf{N}_{\lambda_1,\delta}$. In general w_1 depends continuously on α, w_0, and h. That is, there is a continuous (nonlinear) map $S_{\lambda_1,\delta} : \mathbf{J} \times \mathscr{B}_0 \times \mathbf{B} \to \mathbf{N}^{\perp}_{\lambda_1,\delta}$ defined on an interval $\mathbf{J}$ containing α_1, and neighborhoods of the origin $\mathscr{B}_0 \subset \mathbf{N}_{\lambda_1,\delta}$, $\mathbf{B} \subset \mathbf{H}_\delta$. Here $\mathbf{N}^{\perp}_{\lambda_1,\delta}$ denotes the orthogonal complement of $\mathbf{N}_{\lambda_1,\delta}$ in $\mathbf{H}^1_\delta$. The map $S_{\lambda_1,\delta}$ is given implicitly by $F(w_1) = w_1$, where $w_1 = S_{\lambda_1,\delta}(\alpha, w_0, h)$.

Let $\varphi_{\lambda_1,\delta} : \mathbb{R}^{2m+1} \to \mathbf{N}_{\lambda_1,\delta}$ denote the canonical correspondence

$$\varphi_{\lambda_1,\delta}(p,q) = p_0u_0(r) + \sum_{n=1}^{m} [p_nu_n(t,r) + q_nv_n(t,r)]. \tag{4.3}$$

Clearly any path $\gamma : [a,b] \to \mathbb{R}^{2m+1}$ gives rise to a path $\gamma_{\lambda_1,\delta} : [a,b] \to \mathbf{N}_{\lambda_1,\delta}$ by the composition $\gamma_{\lambda_1,\delta} = \varphi_{\lambda_1,\delta} \circ \gamma$. Assuming h is fixed, the path $\gamma_{\lambda_1,\delta}$ then determines a path of solutions $\Gamma_{\lambda_1,\delta} = \gamma_{\lambda_1,\delta} + S_{\lambda_1,\delta} \circ \gamma_{\lambda_1,\delta}$ in $\mathbf{H}^1_\delta$. In this way we can prescribe variations in the parameters which determine w_0 and hence w_1. If α, w_0 are to vary simultaneously then we should consider paths $[a,b] \to \mathbb{R} \times \mathbf{N}_{\lambda_1,\delta}$.

One of our objectives is to investigate the nature of the bifurcation in the set of solutions, as $\lambda = \alpha g'(0)$ crosses the values θ_n^2. This requires α_1 to change. Algorithmically we need to select a sequence of values $\{\alpha_{1j}\}$, with $\lambda_{1j} = \alpha_{1j}g'(0)$ approaching θ_n^2, and successively vary α through intervals J_j centered at α_{1j}. This continuation algorithm will in general force two critical parameters in the problem to change. First of all we will be changing λ_1, which obviously changes the linear operator L_{λ_1}, and hence the null space $\mathbf{N}_{\lambda_1,\delta}$ used in the parameterization of $M_{\lambda_1,\delta}$, the manifold of solutions in $\mathbf{H}^1_\delta$. A secondary consideration is that of changing the norm parameter δ. To see why this may be necessary, suppose that λ_1 is approaching θ_m^2 from above. We must always have $0 < \delta < \beta_m = \sqrt{\lambda_1 - \theta_m^2}$, so that $K_{\lambda_1} : \mathbf{H}_\delta \to \mathbf{H}^1_\delta$ is well-defined. However $\beta_m \to 0$ as $\lambda_1 \to \theta_m^2$. If δ is held fixed eventually this inequality will be violated.

Another problem with selecting a fixed $\delta > 0$ stems from possibly intrinsic properties of smooth solutions. For example, there is a known breather (cf. [3]) $u(t,x) = 4\arctan(\beta \sin t/\cosh \beta x)$, in which $\beta = \sqrt{\alpha - 1}$, satisfying (1.1)–(1.3) when $N = 1$, $T = 2\pi$, and $g(u) = \sin u$. In the notation we have established $\beta = \beta_1(\lambda) = \sqrt{\lambda - \theta_1^2}$. It is easy to see that $u(t,x) = 8\beta \exp(-\beta|x|) \sin t + O(\exp(-3\beta|x|))$ as $|x| \to +\infty$. Moreover we have $4 \arctan \beta |\sin t| \leq |u(t,x)| \cosh \beta x \leq 4\beta |\sin t|$ for all (t,x). Thus it follows from (2.1) that $4k_\beta c_{\delta\beta} \leq \|u\|_{1,\delta} \leq 8c_{\beta\delta}$, where $k_\beta = \beta^{-1} \arctan \beta$ and $c_{\beta\delta} = \left[(2+\beta^2)/(2(\beta - \delta))\right]^{1/2}$. Clearly $c_{\delta\beta} = O(\beta^{1/2})$ as $\beta \to 0^+$ if $\delta = \varepsilon\beta$, with ε fixed in $(0,1)$. However, if δ is held fixed then $c_{\beta\delta} = O\left((\beta - \delta)^{-1/2}\right)$ as $\beta \to \delta^+$. In the latter case $u(t,x)$ fails to be in a neighborhood of the origin for β near δ.

In light of these considerations it is reasonable to allow the topological linear space $\mathbf{N}_{\lambda_1,\delta}$ to vary during a continuation algorithm. In particular the spanning functions as well as the metric should be allowed to change. However, we must take care to insure that the inherent changes in the parameterization of the solution manifold $M_{\lambda_1,\delta} \subset \mathbf{H}^1_\delta \subset \mathbf{H}^1_0$ are made in a consistent way.

Consider the diagram below in which we have used γ_1, γ_0 to denote homeomorphisms in $\mathbb{R}^{2m+1}$, and σ to denote a path in $\mathbb{R}^{2m+1}$. We assume $\delta_0 \leq \delta_1$, so that $\mathbf{H}^1_{\delta_1} \subset \mathbf{H}^1_{\delta_0}$; this inclusion is denoted by $\mathbf{i}$ in the diagram. Suppose $w \in \mathbf{H}^1_{\delta_1}$ is a solution, and $w = w_0^i + w_1^i$, where

$$w_0^i = \varphi_{\lambda_i,\delta_i} \circ \gamma_i(p,q), \qquad w_1^i = S_{\lambda_i,\delta_i}(\alpha, w_0^i, h) \qquad (i = 0, 1).$$

If $w = w_0^1 + w_1^1$ has just been determined, and we want to change parameterizations from $\mathbf{N}_{\lambda_1,\delta_1}$ to $\mathbf{N}_{\lambda_0,\delta_0}$, then γ_0 must be chosen so that the diagram commutes. That is, $\gamma_0 = \varphi^{-1}_{\lambda_0,\delta_0} \circ P_{\lambda_0,\delta_0} \circ \mathbf{i} \circ \Gamma_{\lambda_1,\delta_1} \circ \varphi_{\lambda_1,\delta_1} \circ \gamma_1$.

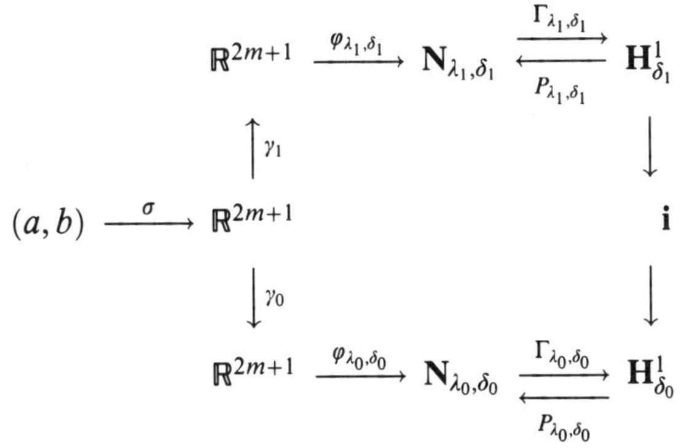

$$\begin{array}{ccccccc}
 & & \mathbb{R}^{2m+1} & \xrightarrow{\varphi_{\lambda_1,\delta_1}} & \mathbf{N}_{\lambda_1,\delta_1} & \underset{P_{\lambda_1,\delta_1}}{\overset{\Gamma_{\lambda_1,\delta_1}}{\rightleftarrows}} & \mathbf{H}^1_{\delta_1} \\
 & & \uparrow{\scriptstyle \gamma_1} & & & & \downarrow \\
(a,b) & \xrightarrow{\sigma} & \mathbb{R}^{2m+1} & & & & \mathbf{i} \\
 & & \downarrow{\scriptstyle \gamma_0} & & & & \downarrow \\
 & & \mathbb{R}^{2m+1} & \xrightarrow{\varphi_{\lambda_0,\delta_0}} & \mathbf{N}_{\lambda_0,\delta_0} & \underset{P_{\lambda_0,\delta_0}}{\overset{\Gamma_{\lambda_0,\delta_0}}{\rightleftarrows}} & \mathbf{H}^1_{\delta_0}
\end{array}$$

From a different perspective, we can think of the above diagram as implicitly determining γ_1 when γ_0 is prescribed. Thus if we have a fixed reference manifold in which measurements are to be made, and γ_0 is a prescribed path in this manifold, we can use fixed point iteration, for example, to determine the implicitly defined path γ_1.

5. Remarks on implementation. Corresponding to a function $w : \mathbb{R} \times \mathbb{R}^+ \to \mathbb{R}$, which is T-periodic in t, we consider an approximate function $W = \{W_{*,n,m}\}$ which should satisfy

$$W_{*,n,m} \simeq w(t_{*,n}, r_{*,m}), \qquad 0 \le n, m \le N_* \quad \text{and} \quad * = i, m, o.$$

In general $0 \le t_{*,n} \le T$, and $0 \le r_{i,m} \le R_1 \le r_{m,m} \le R_2 \le r_{o,m} \le R_3$. In our implementation W should be thought of as a set of three square matrices $\{W_i, W_m, W_o\}$, with W_* being $(N_*+1)\times(N_*+1)$ and corresponding to the values of w on an inner, middle, and outer rectangular grid $\mathscr{R}_*$, respectively. Since solutions decay exponentially, we can effectively use an outer grid with mesh points widely spaced and R_3 approximating $+\infty$ as determined by the decay rate. We expect little variation in the values of w on the outer grid. On the other hand we expect large variations close to the origin. To maintain accuracy in our approximations we use a very fine grid close to the origin $r = 0$. The middle grid bridges the gap between the inner and outer grid, and has intermediate mesh widths. The notation we employ here is meant to reflect our implementation of this data structure as a Pascal record (of arrays).

For each $*, m$ there is a sequence of values $\{W_{*,n,m}\}_{n=1}^{N_*-1}$ corresponding to the T-periodic function $w(t, r_{*,m})$, and hence a sequence of Fourier coefficients $\{a_{*,n,m}\}_{n=0}^{\frac{N_*}{2}}$, $\{b_{*,n,m}\}_{n=1}^{\frac{N_*}{2}-1}$ with

$$\begin{aligned} a_{*,n,m} &= \frac{2}{N_*}\sum_{j=0}^{N_*-1} W_{*,j,m}\cos(\theta_n t_{*,j}), \qquad 0 \le n \le \frac{N_*}{2}, \\ b_{*,n,m} &= \frac{2}{N_*}\sum_{j=0}^{N_*-1} W_{*,j,m}\sin(\theta_n t_{*,j}), \qquad 1 \le n \le \frac{N_*}{2} - 1. \end{aligned} \tag{5.1}$$

With the defining equations (5.1) we have (cf. [2])

$$\begin{aligned} W_{*,n,m} = \frac{1}{2}a_{*,o,m} + \sum_{j=1}^{\frac{N_*}{2}-1} [a_{*,j,m}\cos(\theta_j t_{*,n}) \\ + \beta_{*,j,m}\sin(\theta_j t_{*,n})] + \frac{1}{2}a_{*,\frac{N_*}{2},m}\cos\left(\theta_{\frac{N_*}{2}} t_{*,n}\right). \end{aligned} \tag{5.2}$$

Clearly (5.2) corresponds to the formal representation

$$w(t,r) = \frac{1}{2}a_0(r) + \sum_{j=1}^{\infty}\left[a_j(r)\cos\theta_j t + b_j(r)\sin\theta_j t\right].$$

The Fourier transforms (5.1) and (5.2) give an isomorphic correspondence between the point value representation $\{W_{*,n,m}\}$ and the Fourier coefficient function representation $\{a_{*,n,m}\}$, $\{b_{*,n,m}\}$. Numerically these are efficiently computed by fast Fourier transforms (cf. **[1]**, **[2]**). We will denote the latter representation by $\{TW_{*,n,m}\}$ and use the relations

$$TW_{*,n,m} = a_{*,n,m} \qquad \left(0 \le n \le \frac{N_*}{2}\right),$$

$$TW_{*,\frac{N_*}{2}+n,m} = b_{*,n,m} \qquad \left(1 \le n \le \frac{N_*}{2} - 1\right)$$

to define the correspondence between notations.

The numerical linear solver is based on the developments of Section 3. It requires the improper integrals appearing in (3.5) to be computed. These include the Laplace transforms which appear. Fortunately the exponential decay makes this a stable calculation. Thus after obtaining the numerical transform $\{TH_{*,n,m}\}$ corresponding to $h(t,r)$, the algorithm then computes an approximation of the corresponding function $w_1(t,r)$ according to the formulas (3.4)–(3.5). In doing this one can take advantage of the fact that the difference between $w_1(t,r_1)$ and $w_1(t,r_2)$ involves a definite integral over (r_1,r_2). Thus, working segment by segment, the improper integral over $(0,+\infty)$ need only be evaluated once to obtain the values of $w_1(t,r)$ for all values $r = r_{*,m}$.

We have tested the algorithm described above by using the known solution $u(t,x) = 4\arctan(\beta\sin t/\cosh\beta x)$ from **[3]**, which satisfies (1.1)–(1.3) when f is identically zero, $g(u) = \sin u$, $T = 2\pi$, $N = 1$, and $\beta = \sqrt{\alpha - 1}$ for $\alpha > 1$. Table 1 below contains various measures of accuracy obtained with different grid sizes and choices of α. As indicated above N_* $(* = i,m,o)$ is the number of subdivisions of the rectangle $\mathscr{R}_*$ $(* = i,m,o)$. The second column of the table gives these values, and the third column gives the values R_i $(i = 1,2,3)$ indicating the subdivisions in the x direction delineating the rectangles. The notation *FD resid* is used to denote the maximum residual in the (centered) finite difference equations approximating (1.1); the edges $x = 0, R_1, R_2, R_3$ were not checked. *EE resid* is used to denote the maximum residual in the energy equation $E(t) = \text{constant}$, where

$$E(t) = \int_0^{+\infty} \left[w_t^2 + w_x^2 + 2\alpha(1 - \cos w) \right] dr.$$

Here w denotes the approximate solution. These values seem to indicate less accuracy but are apparently a reflection of the large size of $E(t)$, which is approximately $E = 5.0, 15.7$, and 22.1 for $\alpha = 1.1, 2.0$, and 3.0 respectively. In each case w was obtained by first computing the projection component d_1 (see (2.5)) of the known solution u, and then continuing from the zero solution by incrementing the corresponding component of w from 0 to d_1. All other parameters were held fixed.

α	N_i, N_m, N_o	R_1, R_2, R_3	FD resid	EE resid	$\|w - u\|_\infty$	$\|u\|_\infty$
1.1	$8, 16, 32$	$2.2, 8.9, 37.9$	0.014	0.16	0.004	1.23
	$16, 32, 64$		0.004	0.04	0.002	
2.0	$8, 16, 32$	$0.7, 2.8, 12.0$	0.156	0.90	0.013	3.14
	$16, 32, 64$		0.047	0.23	0.009	
3.0	$8, 16, 32$	$0.5, 2.0, 8.5$	0.884	1.60	0.038	3.82
	$16, 32, 64$		0.111	0.59	0.017	

Table 1

In addition to the accuracy and convergence tests, we also computed approximations to the values $w_r(t, 0)$ in an effort to gain information on smoothness at the origin. In particular we let the component $d_1 = d_1(w)$ of the approximate solution vary near the computed value $d_1 = d_1(u)$ of the known solution, with all other parameters fixed. We were able to see that u seems to be an isolated smooth solution. The output data presented in Table 2 was obtained for $\alpha = 1.1$ and 2.0, respectively. The grids were subdivided according to $N_* = (16, 32, 64)$ and $N_* = (32, 64, 64)$, respectively, with finite difference residuals on the order of 10^{-3}, 10^{-2}, respectively.

$\alpha = 1.1$			$\alpha = 2.0$	
$d_1 = d_1(w)$	$\|w_r(t,0)\|_\infty$		$d_1 = d_1(w)$	$\|w_r(t,0)\|_\infty$
4.498	0.104		7.414	0.550
4.698	0.051		7.614	0.284
$d_1(u) = 4.898$	0.005		$d_1(u) = 7.814$	0.023
5.098	0.051		8.014	0.297
5.298	0.101		8.214	0.602

Table 2

In conclusion we feel that the numerical results show that the proposed algorithm provides an effective means of determining approximate solutions of (1.1)–(1.3). Furthermore the algorithm provides a powerful tool for the study of solution properties. By exploiting this tool we hope to gain an enhanced understanding of the currently unresolved aspects of the problem.

REFERENCES

1. J. Cooley and J. Tukey, *An algorithm for machine calculation of complex Fourier series*, Math. Comp. **19** (1965), 297–301.

2. G. Dahlquist and Å. Björck,, *Numerical Methods*, translated by N. Anderson, Prentice-Hall, 1974.

3. G. L. Lamb, Jr., *Elements of Soliton Theory*, John Wiley and Sons, New York, 1980.

4. R. Paley and N. Wiener, *Fourier Transforms in the Complex Domain*, AMS Colloq. Publ., vol. 19, Providence, R.I., 1934.

5. M. C. Reed, *Abstract Non-Linear Wave Equations*, Lecture Notes in Math. 507, Springer-Verlag, New York, 1976.

6. M. Smiley, *Time-periodic solutions of wave equations on* $\mathbf{R}^1$ *and* $\mathbf{R}^3$, Math. Methods Appl. Sci. **10**(4) (1988), 457–475.

7. ———, *Breathers and forced oscillations of nonlinear wave equations on* $\mathbf{R}^3$, Journal für die reine und angewandte Mathematik **398** (1989), 25–35.

8. ———, *Complex-valued time-periodic solutions and breathers of nonlinear wave equations*, submitted to Differential and Integral Equations.

DEPARTMENT OF MATHEMATICS, IOWA STATE UNIVERSITY

Lectures in Applied Mathematics
Volume **26** (1990)

Numerical Solutions of Singular Stochastic Control Problems in Bounded Intervals

MIN SUN

Abstract. The numerical solution of singular stochastic control problems in bounded intervals is studied. Linear programming algorithms are used for solving the discretized systems of the associated dynamic programming equations. Numerical results are presented for several examples. Extensions to numerical solutions of other nonlinear differential equations and to higher-dimensional problems are discussed.

1. Introduction. Several singular stochastic control problems in bounded intervals have been considered in Sun [27]. The problems can be described briefly as follows.

We consider optimal control of one-dimensional diffusion processes $y_x(t) = y_x^v(t)$, solutions of

$$\begin{aligned}&dy_x(t) = g(y_x(t))\,dt + \sigma CC(y_x(t))\,dw_t + dv_t,\\ &y_x(0) = x \in [0, 1],\\ &v_t, \text{ increasing adapted control processes with } v_0 = 0.\end{aligned}$$

1980 *Mathematics Subject Classification* (1985 *Revision*). Primary 49C20, 49D35, 93E20, 93E25.

Our objective is to minimize expected cost functionals $J_x^i(v)$ $(i = 1, 2, 3, 4)$ over the class of admissible control policies, where

$$J_{x,\alpha}^1(v) = \mathrm{E}\left\{\int_0^{\tau_x} f(y_x(t))e^{-\alpha t}\,dt + \int_0^{\tau_x} \phi(y_x(t))e^{-\alpha t}\,dv^c(t) + \psi(y_x(\tau_x))e^{-\alpha\tau_x} + \sum_{0<s\le\tau_x} e^{-\alpha s}\int_{y_x(s-)}^{y_x(s)} \phi(z)\,dz\right\}$$

$$J_{x,\alpha}^2(v) = \mathrm{E}\left\{\int_0^{\infty} f(y_x^r(t))e^{-\alpha t}\,dt + \int_0^{\infty} \phi(y_x^r(t))e^{-\alpha t}\,dv^c(t) + \sum_{0<s} e^{-\alpha s}\int_{y_x^r(s-)}^{y_x^r(s)} \phi(z)\,dz\right\}$$

$$J_x^3(v) = \liminf_{T\to\infty}\mathrm{E}\left\{\int_0^{T} f(y_x^r(t))\,dt + \int_0^{T} \phi(y_x^r(t))\,dv^c(t) + \sum_{0<s\le T}\int_{y_x^r(s-)}^{y_x^r(s)} \phi(z)\,dz\right\}1/T$$

$$J_x^4(v) = \liminf_{T\to\infty}\mathrm{E}\left\{\int_0^{\tau_x\Lambda T} f(y_x(t))\,dt + \int_0^{\tau_x\Lambda T} \phi(y_x(t))\,dv^c(t) + \sum_{0<s\le\tau_x\Lambda T}\int_{y_x(s-)}^{y_x(s)} \phi(z)\,dz\right\}1/\mathrm{E}(\tau_x\Lambda T)$$

$v^c(\cdot)$ is the continuous part of $v(\cdot)$
$\tau_x = \inf\{t \ge 0\colon y_x(t) \in \{0, 1\}\}$
$y_x^r(\cdot)$ is the diffusion reflected at the boundary of $[0, 1]$ with the drift term
$g(y_x^r(\cdot))\,dt + dv_t$ and the diffusion term $\sigma(y_x^r(\cdot))\,dw_t$ in $(0, 1)$.

If we denote by $u_\alpha^i(\cdot)$ the optimal cost functions, i.e.,

$$u_\alpha^i(x) = \inf\{J_{x,\alpha}^i(v)\colon v(\cdot)\},\quad i = 1, 2, 3, 4,$$

then we can show (see Sun [27] for detailed derivations) that under proper conditions the u's satisfy some appropriate differential inequalities below.

$u_\alpha^1(\cdot)$ is the maximum solution of

$$
(1.1)\qquad
\begin{aligned}
&-(\sigma^2/2)\ddot{u} - g\dot{u} + \alpha u \le f, \quad \dot{u} \ge -\phi, \quad 0 < x < 1,\\
&(-(\sigma^2/2)\ddot{u} - g\dot{u} + \alpha u - f)(\dot{u} + \phi) = 0, \quad 0 < x < 1,\\
&u(0) = \psi(0), \quad u(1) = \psi(1),\\
&u \in C^{1,1}[0,1], \quad \alpha \ge 0.
\end{aligned}
$$

Moreover, $u_\alpha^i(\cdot) \to u_0^i(\cdot)$ in $C^{1,r}[0,1]$ (C^1 functions with Hölder continuous derivatives) for all $r \in (0,1)$.

$u_\alpha^2(\cdot)$ is the maximum solution of

$$
(1.2)\qquad
\begin{aligned}
&-(\sigma^2/2)\ddot{u} - g\dot{u} + \alpha u \le f, \quad \dot{u} \ge -\phi, \quad 0 < x < 1,\\
&(-(\sigma^2/2)\ddot{u} - g\dot{u} + \alpha u - f)(\dot{u} + \phi) = 0, \quad 0 < x < 1,\\
&\dot{u}(0) = \dot{u}(1) = 0,\\
&u \in C^{1,1}[0,1], \quad \alpha > 0.
\end{aligned}
$$

$u_0^3(\cdot)$ is a constant function, denoted by λ, for which there exists $w \in C^{1,1}[0,1]$ such that (λ, w) is a solution of

$$
(1.3)\qquad
\begin{aligned}
&-(\sigma^2/2)\ddot{w} - g\dot{w} + \lambda \le f, \quad \dot{w} \ge -\phi, \quad 0 < x < 1,\\
&(-(\sigma^2/2)\ddot{w} - g\dot{w} + \lambda - f)(\dot{w} + \phi) = 0, \quad 0 < x < 1,\\
&\dot{w}(0) = \dot{w}(1) = 0,\\
&w \in C^{1,1}[0,1], \quad \alpha = 0.
\end{aligned}
$$

For a fixed $x' \in (0,1)$, let $\lambda = u_0^4(x')$. Then there exists $w \in C^{1,1}[0,1]$ such that (λ, w) is a solution of

$$
(1.4)\qquad
\begin{aligned}
&-(\sigma^2/2)\ddot{w} - g\dot{w} + \lambda \le f, \quad \dot{w} \ge -\phi, \quad 0 < x < 1,\\
&(-(\sigma^2/2)\ddot{w} - g\dot{w} + \lambda - f)(\dot{w} + \phi) = 0, \quad 0 < x < 1,\\
&w(0) = w(1) = w(x') = 0,\\
&w \in C^{1,1}[0,1], \quad \alpha = 0.
\end{aligned}
$$

We can also find explicit examples of the four control problems solved in Sun [27]. With those analytic solutions, optimal control laws have been obtained, which behave "singularly".

In this article, we solve these problems, especially the examples given in Sun [27], numerically. We will approximate the unknown value functions $u_\alpha^i(x)$ as well as optimal feedback control policies. In Section 2, we introduce a finite difference scheme to discretize the variational inequalities and the associated boundary conditions (1.1)–(1.4). Then, as the main contribution, we use the well-established linear programming (LP) algorithms to solve the discretized variational inequalities.

We can show that LP formulation is well posed and uniquely solvable. Numerical results are provided in Section 3, and generalizations and discussions follow in Section 4.

Several approaches to numerical solutions of differential equations and differential inequalities are available. Let us just mention a few references in these areas. The probabilistic approach for the partial differential equations can be found in Kushner [**18**] and Menaldi [**21**]. The probabilistic approach for directly approximating the original control problems is also widely used (cf. Bensoussan and Runggaldier [**5**]). The classical finite difference scheme for linear differential equations can be found in Twizell [**32**] or any other numerical analysis book. The iteration approach for solving discretized linear or nonlinear differential equations is also well known (cf. Gladwell and Wait [**13**]). An iteration scheme for differential inequalities is used in Chow and Menaldi [**6**] and Bancora-Imbert et al. [**1**]. The finite element approach to differential equations can be found in any book on finite element method (e.g., Oden and Reddy [**24**]). The finite element method has been used to solve differential inequalities in, e.g., Gonzalez and Rofman [**15**]. The systematic analysis of the last method is given in Glowinski et al. [**14**].

Linear programming algorithms are well known (cf. Solow [**26**]) and have been successfully used to solve linear programming problems in operations research and many other areas of application. The LP formulation in control theory was made use of at least as early as 1960 (cf. Manne [**20**]). Other applications in control could be found, e.g., in Gabasov et al. [**12**] and Gabasov and Kirillova [**11**]. In a series of their papers, linear programming problems are formulated in algorithms for solving deterministic linear optimal control problems. Their approach is based on the discretization of state equations and an application of Pontryagin's maximum principle. In this article, the linear programming algorithm is employed as an alternative approach to solving differential inequalities, which could be regarded as another application of linear programming in differential equations and in control theory. Our approach of using LP algorithms in control theory is based on the dynamic programming principle.

Among the advantages of the LP algorithm over other methods are: (a) It is easier to show the existence and uniqueness of the LP problems arising in discretization of HJB (Hamilton–Jacobi–Bellman) equations or inequalities (cf. Sun [**28**], [**29**]). Especially when the solution to a discretized HJB equation is not unique, the unique LP solution can be used to characterize the maximum solution of the discretized HJB equation, which is usually the discretized value function of control problems. (b) Algorithms for solving LP problems are well known, and readily available nowadays, including some special ones using the sparsity.

As the concluding remark of the introduction, let us point out that we can find a lot more applications of the linear programming algorithm in control theory and related areas in partial differential equations. It is well known that a class of fully nonlinear second-order PDE's called HJB equations, arise in control of diffusion processes. The discretization of such nonlinear PDE's results in "piecewise linear" equations. Our algorithms could be used to solve the discretized problems. More discussions on this issue and related ones can be found in Section 4.

2. Discretization and linear programming formulations. To be specific, let us discretize the unit interval by binary partitions

$$0 = x_1 < x_2 < \cdots < x_N = 1,$$

where

$$N = 2^{N_0} + 1 \text{ for some positive integer } N_0 \geq 3,$$
$$x_i = (i-1)\Delta x, \quad i = 1, \ldots, N, \quad \text{with } \Delta x = 1/(N-1).$$

We simply denote $u(x_i)$ by u_i.

Next, we discretize the differential operators in basically the same way as in Kushner **[18]**, Chow and Menaldi **[6]**, and Bancora-Imbert et al. **[1]**. An important feature of their discretization procedure is that it preserves the maximum principle. But it is only of first-order accuracy.

$$-g(x_i)\dot{u}(x_i)\colon \ (|g_i|/\Delta x)[h_i(u_i - u_{i+1}) + h_i'(u_i - u_{i-1})],$$
$$-(1/2)\sigma^2(x_i)\ddot{u}(x_i)\colon \ -\sigma_i^2(u_{i+1} - 2u_i + u_{i-1})/(2\Delta x^2)$$
$$\text{with } h_i = I_{(g_i \geq 0)} \text{ and } h_i' = 1 - h_i.$$

Now a discretized version of

$$-(\sigma^2/2)\ddot{u} - g\dot{u} + \alpha u \leq f, \quad 0 < x < 1,$$
$$\dot{u} \geq -\phi, \quad 0 < x < 1,$$

is given by

$$-\sigma_i^2(u_{i+1} - 2u_i + u_{i-1})/(2\Delta x^2) + (|g_i|/\Delta x)[h_i(u_i - u_{i+1}) + h_i'(u_i - u_{i-1})]$$
$$+ \alpha u_i \leq f_i, \quad (1/\Delta x)(u_i - u_{i+1}) \leq \phi_i, \quad i = 2, \ldots, N-1.$$

Equivalently, this can be written as

$$u_i - [\alpha_2(i) + h_i'\alpha_3(i)]u_{i-1} - [\alpha_2(i) + h_i\alpha_3(i)]u_{i+1} \leq \alpha_1(i),$$
$$u_i - u_{i+1} \leq \phi_i\Delta x, \quad i = 2, \ldots, N-1,$$

where the α's can be easily identified from the context.

It is clear that the original differential inequality problems (1.1)–(1.4) have their discretized versions as follows:

(2.1)

$$u_1 = \psi_1, \qquad u_N = \psi_N,$$
$$\{u_i - [\alpha_2(i) + h_i'\alpha_3(i)]u_{i-1} - [\alpha_2(i) + h_i\alpha_3(i)]u_{i+1} - \alpha_1(i)\}$$
$$\times \{u_i - u_{i+1} - \phi_i\Delta x\} = 0,$$
$$u_i - [\alpha_2(i) + h_i'\alpha_3(i)]u_{i-1} - [\alpha_2(i) + h_i\alpha_3(i)]u_{i+1} \le \alpha(i),$$
$$u_i - u_{i+1} \le \phi_i\Delta x, \quad i = 2, \ldots, N-1.$$

(2.2)

$$u_1 - u_2 = 0, \qquad u_{N-1} - u_N = 0,$$
$$\{u_i - [\alpha_2(i) + h_i'\alpha_3(i)]u_{i-1} - [\alpha_2(i) + h_i\alpha_3(i)]u_{i+1} - \alpha_1(i)\}$$
$$\times \{u_i - u_{i+1} - \phi_i\Delta x\} = 0,$$
$$u_i - [\alpha_2(i) + h_i'\alpha_3(i)]u_{i-1} - [\alpha_2'(i) + h_i\alpha_3(i)]u_{i+1} \le \alpha_1(i),$$
$$u_i - u_{i+1} \le \phi_i\Delta x, \quad i = 2, \ldots, N-1.$$

(2.3)

$$w_1 - w_2 = 0, \qquad w_{N-1} = w_N = 0,$$
$$w_i - [\alpha_2(i) + h_i'\alpha_3(i)]w_{i-1} - [\alpha_2(i) + h_i\alpha_3(i)]w_{i+1} + w_{N+1}/\alpha_0(i) \le \alpha_1(i),$$
$$w_i - w_{i+1} \le \phi_i\Delta x, \qquad i = 2, \ldots, N-1,$$
$$\{w_i - [\alpha_2(i) + h_i'\alpha_3(i)]w_{i-1} - [\alpha_2(i) + h_i\alpha_3(i)]w_{i+1} + w_{N+1}/\alpha_0(i)$$
$$- \alpha_1(i)\}\{w_i - w_{i+1} - \phi_i\Delta x\} = 0, \quad i = 2, \ldots, N-2,$$
$$-w_{N+1} \le 0$$

(2.4)

$$w_1 = w_{k_0} = w_N = 0, \qquad (k_0 - 1)\Delta x < x' \le k_0\Delta x,$$
$$w_i - [\alpha_2(i) + h_i'\alpha_3(i)]w_{i-1} - [\alpha_2(i) + h_i\alpha_3(i)]w_{i+1} + w_{N+1}/\alpha_0(i) \le \alpha_1(i),$$
$$w_i - w_{i+1} \le \phi_i\Delta x, \quad i = 2, \ldots, N-1,$$
$$\{w_i - [\alpha_2(i) + h_i'\alpha_3(i)]w_{i-1} - [\alpha_2(i) + h_i\alpha_3(i)]w_{i+1} + w_{N+1}/\alpha_0(i) - \alpha_1(i)\}$$
$$\times \{w_i - w_{i+1} - \phi_i\Delta x\} = 0, \quad i = 2, \ldots, N-1 \text{ (but } i \neq k_0),$$
$$-w_{N+1} \le 0.$$

These discretized problems can be regarded as generalized linear complementarity problems (cf. Mangasarian **[19]**, Sun **[30]**) or implicit complementarity problems (cf. Pang **[25]**). Alternatively, these problems can be reformulated as linear programming problems below.

$$(\mathrm{LP}_1)\qquad \begin{aligned}&\text{Minimize } -\sum_{i=1}^{N} u_i, \text{ subject to}\\ &u_1 = \psi_1, \qquad u_N = \psi_N,\\ &u_i - \beta_{-1}(i)u_{i-1} - \beta_1(i)u_{i+1} \le \alpha_1(i),\\ &u_i - u_{i+1} \le \phi_i \Delta x, \quad i = 2, \dots, N-1.\end{aligned}$$

$$(\mathrm{LP}_2)\qquad \begin{aligned}&\text{Minimize } -\sum_{i=1}^{N} u_i, \text{ subject to}\\ &u_1 - u_2 = 0, \qquad u_{N-1} - u_N = 0,\\ &u_i - \beta_{-1}(i)u_{i-1} - \beta_1(i)u_{i+1} \le \alpha_1(i),\\ &u_i - u_{i+1} \le \phi_i \Delta x, \quad i = 2, \dots, N-1.\end{aligned}$$

$$(\mathrm{LP}_3)\qquad \begin{aligned}&\text{Minimize } -\sum_{i=1}^{N} w_i - 2(N_0 - 2)w_{N+1}, \text{ subject to } (\lambda = w_{N+1})\\ &w_1 - w_2 = 0, \qquad w_{N-1} = w_N = 0,\\ &w_i - \beta_{-1}(i)w_{i-1} - \beta_1(i)w_{i+1} + w_{N+1}/\alpha_0(i) \le \alpha_1(i),\\ &w_i - w_{i+1} \le \phi_i \Delta x, \quad i = 2, \dots, N-1,\\ &-w_{N+1} \le 0.\end{aligned}$$

$$(\mathrm{LP}_4)\qquad \begin{aligned}&\text{Minimize } -\sum_{i=1}^{N} w_i - 2(N_0 - 2)w_{N+1}, \text{ subject to } (\lambda = w_{N+1})\\ &w_1 = w_{k_0} = w_N = 0, \qquad (k_0 - 1)\Delta x < x' \le k_0 \Delta x,\\ &w_i - \beta_{-1}(i)w_{i-1} - \beta_1(i)w_{i+1} + w_{N+1}/\alpha_0(i) \le \alpha_1(i),\\ &w_i - w_{i+1} \le \phi_i \Delta x, \quad i = 2, \dots, N-1,\\ &-w_{N+1} \le 0,\end{aligned}$$

where

$$\beta_{-1}(i) = \alpha_2(i) + h_i'\alpha_3(i) \quad \text{and} \quad \beta_1(i) = \alpha_2(i) + h_i\alpha_3(i), \quad i = 2, \dots, N-1.$$

The passage from the systems of algebraic inequalities to the corresponding LP problems is based on the maximality conditions on the value functions u of the original control problems. Further justifications will be made in the following theorems. Problems (1.1) and (1.2) (without the complementarity conditions), on the other hand, can be regarded as continuous LP analogues.

THEOREM 2.1. *The linear programming solution to problem* (LP_1) *exists, and is unique. Furthermore, it solves* (2.1), *and is the maximum subsolution to* (2.1) *in the sense that for any other subsolution to* (2.1), *say* $\{v_i\}_{i=1}^N$,

$$v_i \leq u_i \text{ for all } i = 1, \dots, N.$$

THEOREM 2.2. *Problem* (LP_2) *has a unique solution. The solution is the maximum solution to* (2.2).

One is referred to Sun **[29]** for more details related to the previous theorems. In the proofs of those two theorems, arguments are more or less based on the fact that $0 \leq \beta_{-1}(i) + \beta_1(i) \leq 1$ for all i. But for the problems (2.3) and (2.4), we have to deal with "eigenproblems" for which the standard maximum principle is no longer valid if we regard the eigenvalue as a part of the solution. Analytical results related to these two problems are presented in several theorems below.

THEOREM 2.3. *A solution to* (LP_3) *exists. Any solution of* (LP_3) *satisfies* (2.3). *Moreover, the solution to* (LP_3) *satisfying*

$$w_{N-1} - \beta_{-1}(N-1)w_{N-2} - \beta_1(N-1)w_N + w_{N+1}/\alpha_0(N-1) = \alpha_1(N-1)$$

is unique.

PROOF. Let $\{w^k = (w_1^k, \dots, w_N^k, w_{N+1}^k)\}_{k=1,2,\dots}$ be a minimizing sequence for (LP_3). It is easy to see that $(w_1^k, \dots, w_N^k)$ is bounded from above. It follows from

$$w_i^k - \beta_{-1}(i)w_{i-1}^k - \beta_1(i)w_{i+1}^k + w_{N+1}^k/\alpha_0(i) \leq \alpha_1(i) \quad \text{with } i = N-1$$

that w_{N+1}^k is bounded. Then we want to show that $(w_1^k, \dots, w_N^k)$ is also bounded from below.

Suppose that $\{w_{N-2}^k\}$ is not bounded from below, say $w_{N-2}^k \to -\infty$. Then

$$w_i^k - w_{i+1}^k \leq \phi_i \Delta x, \quad i = 2, \dots, N-3,$$

implies that w_i^k approaches $-\infty$ as $k \to \infty$ for $i = 1, \dots, N-2$. But this contradicts the fact that $\{w^k\}_{k=1,2,\dots}$ is a minimizing sequence for (LP_3). Then induction arguments show that $\{w_i^k\}$ is bounded from below for all i. Consequently, the existence of a solution to (LP_3) follows by compactness.

It is straightforward to prove that any solution to (LP_3) must satisfy (2.3).

To prove uniqueness, we need the following lemma.

LEMMA 2.1. *Let* $\{(w_1^k, \dots, w_N^k, w_{N+1}^k)\}_{k=1,2}$ *be two solutions to* (LP_3). *If* $d_{N+1} \geq 0$, $d_1 = d_2 \leq d_3 \leq \cdots \leq d_{N-1} = d_N = 0$, *where* $d_i = w_i^1 - w_i^2$.

PROOF. According to (2.3) we have

$$d_i \leq \max\{\beta_{-1}(i)d_{i-1}+\beta_1(i)d_{i+1}-d_{N+1}/\alpha_0(i), d_{i+1}\} \quad \text{for } i = 2, \dots, N-2.$$

Then the desired result can be obtained by induction.

COROLLARY 2.1. *Let* $\{(w_1^k, \dots, w_N^k, w_{N+1}^k)\}_{k=1,2}$ *be two solutions to* (LP_3). *If* $d_{N+1} \leq 0$, $d_1 = d_2 \geq d_3 \geq \cdots \geq d_{N-1} = d_N = 0$, *where* $d_i = w_i^1 - w_i^2$.

Now we return to the proof of Theorem 2.3.

Let $\{(w_1^k, \dots, w_N^k, w_{N+1}^k)\}_{k=1,2}$ be two solutions to (LP_3) satisfying

$$w_{N-1} - \beta_{-1}(N-1)w_{N-2} - \beta_1(N-1)w_N + w_{N+1}/\alpha_0(N-1) = \alpha_1(N-1).$$

With the notation in Lemma 2.1, let us assume that $d_{N+1} \geq 0$. Then we have

$$-\beta_{-1}(N-1)d_{N-2} + d_{N+1}/\alpha_0(N-1) = 0.$$

Hence $d_1 = \cdots = d_N = 0$ by Lemma 2.1 and Corollary 2.1, which proves uniqueness.

The uniqueness result stated in Theorem 2.3 is not very general. However, we are able to get a better result by introducing a new linear programming problem associated with (2.3) as follows.

$$(\mathrm{LP}_3)^* \qquad \begin{aligned} &\text{Minimize } -\sum_{i=1}^{N-1} \beta_1(i+1)w_i \\ &- w_{N+1}/\max\{\alpha_0(i)\colon 2 \leq i \leq N-2\}, \text{ subject to} \\ &w_1 - w_2 = 0, \qquad w_{N-1} = w_N = 0, \qquad -w_{N+1} \leq 0 \\ &w_i - \beta_{-1}(i)w_{i-1} - \beta_1(i)w_{i+1} + w_{N+1}/\alpha_0(i) \leq \alpha_1(i) \\ &w_i - w_{i+1} \leq \phi_i\Delta x, \quad i = 2, \dots, N-1. \end{aligned}$$

THEOREM 2.4. *A solution to* $(\mathrm{LP}_3)^*$ *exists. Any solution of* $(\mathrm{LP}_3)^*$ *satisfies* (2.3). *Moreover, the solution to* $(\mathrm{LP}_3)^*$ *is unique.*

PROOF. The proof of the first two parts is the same as that for (LP_3). To show uniqueness, let us use the same notation as in the proof of Theorem 2.3. Note that Lemma 2.1 and Corollary 2.1 remain true for $(\mathrm{LP}_3)^*$.

Suppose $d_{N+1} > 0$ and $d_1 \neq 0$. Then we show that

$$R_i \equiv d_i - \beta_{-1}(i)d_{i-1} - \beta_1(i)d_{i+1} + d_{N+1}/\alpha_0(i) > 0, \quad i = 2, \dots, N-1.$$

In fact,

$$\begin{aligned} R_i &\geq d_{i-1} - \beta_{-1}(i)d_{i-1} - \beta_1(i)d_{i+1} + d_{N+1}/\alpha_0(i) \\ &\geq \beta_1(i)d_{i-1} + d_{N+1}/\alpha_0(i) \\ &\geq -\sum_{j=2}^{i-1} \beta_1(j)d_{j-1} - \sum_{j=i+1}^{N} \beta_1(j)d_{j-1} > 0. \end{aligned}$$

Hence, in view of (2.3), we have for $(w_1^2, \dots, w_N^2, w_{N+1}^2)$ in Lemma 2.1

$$w_i^2 - \beta_{-1}(i)w_{i-1}^2 - \beta_1(i)w_{i+1}^2 + w_{N+1}^2/\alpha_0(i) < \alpha_1(i), \quad i = 2, \dots, N-1.$$

Consequently, there exists $\delta > 0$ such that $(w_1^2, \dots, w_N^2, w_{N+1}^2 + \delta)$ is still feasible for $(\mathrm{LP}_3)^*$, which is impossible.

THEOREM 2.5. *A solution to* (LP_4) *exists. Any solution of* (LP_4) *satisfies* (2.4). *Moreover, the solution to* (LP_4) *satisfying*

$$w_{k_0} - \beta_{-1}(k_0)w_{k_0-1} - \beta_1(k_0)w_{k_0+1} + w_{N+1}/\alpha_0(k_0) = \alpha_1(k_0)$$

is unique.

PROOF. The proof is close to that of Theorem 2.3, using Lemma 2.2 below. We leave the details to the reader.

LEMMA 2.2. *Let* $\{(w_1^k, \dots, w_N^k, w_{N+1}^k)\}_{k=1,2}$ *be two solutions to* (LP_4) *satisfying*

$$\{w_i - \beta_{-1}(i)w_{i-1} - \beta_1(i)w_{i+1} + w_{N+1}/\alpha_0(i) - \alpha_1(i)\}\{w_i - w_{i+1} - \phi_i \Delta x\} = 0 \tag{2.5}$$

at $i = k_0$. *If* $d_{N+1} \geq 0$, $d_i \leq 0$ *for all* i, *where* $d_i = w_i^1 - w_i^2$. *Moreover,* $0 = d_{k_0} = d_{k_0+1} = \cdots = d_{N-1} = d_N = 0$.

COROLLARY 2.2. *If* $\{\phi_i\colon i = k_0+1, \dots, N-1\}$ *is not identically zero, the solution of* (LP_4) *satisfying* (2.5) *at* $i = k_0$ *is unique.*

Again the uniqueness results stated in Theorem 2.5 and Corollary 2.2 can be improved by considering $(\mathrm{LP}_4)^*$ below. The new results are given in Theorem 2.6, the proof of which is omitted due to the analogy to that of Theorem 2.4.

$$(\mathrm{LP}_4)^* \qquad \begin{aligned} &\text{Minimize } -\sum_{i=1}^{N-1} w_i \\ &- w_{N+1}/[1+\max\{\alpha_0(i)\colon 2 \le i \le N-1\}], \text{ subject to} \\ &w_1 = w_{k_0} = w_N = 0, \qquad -w_{N+1} \le 0, \\ &w_i - \beta_{-1}(i)w_{i-1} - \beta_1(i)w_{i+1} + w_{N+1}/\alpha_0(i) \le \alpha_1(i), \\ &w_i - w_{i+1} \le \phi_i \Delta x, \quad i = 2, \dots, N-1. \end{aligned}$$

THEOREM 2.6. *A solution to* $(\mathrm{LP}_4)^*$ *exists. Any solution of* $(\mathrm{LP}_4)^*$ *satisfies* (2.4). *Moreover, the solution to* $(\mathrm{LP}_4)^*$ *is unique.*

3. Examples. With the basic analytical results above, we are now ready to solve numerically the four different examples given in Sun [27].

EXAMPLE 1.

$$\begin{aligned} &\sigma(\cdot) = 1,\ \psi(0) = 0,\ \psi(1) = 1/4,\ \alpha = 0,\ \phi(\cdot) = 0, \\ &f(x) = x^2,\ g(\cdot) = 0, \\ &\text{Minimize } J^1_{x,\alpha}(v). \end{aligned}$$

Let us solve the corresponding linear programming problem (LP_1) to get an approximate value function $\{u_i\}$ and an approximate feedback control law.

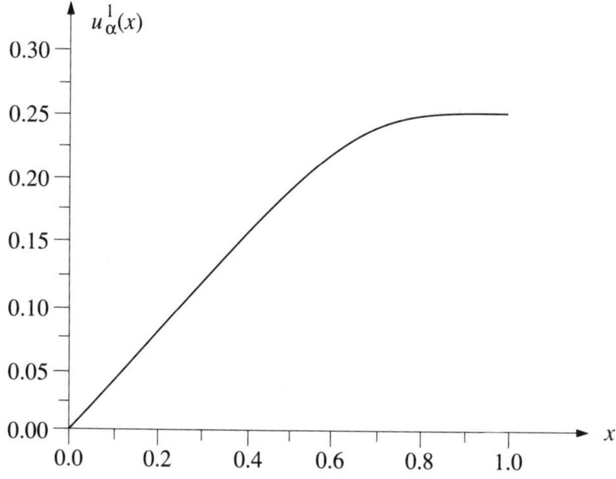

FIGURE 1. Graph of Value Function in Example 1

Table 1.1

Solution of Example 1

i	u_i
1	.000000
2	.012405
3	.024808
4	.037203
5	.049581
6	.061929
7	.074229
8	.086460
9	.098598
10	.110614
11	.122475
12	.134146
13	.145585
14	.156750
15	.167593
16	.178062
17	.188102
18	.197653
19	.206654
20	.215036
21	.222730
22	.229660
23	.235750
24	.240917
25	.245074
26	.248133
27	.250000
28	.250695
29	.250000
30	.250802
31	.250000
32	.250916
33	.250000

Table 1.2

Free-Boundary Checking for Example 1

i	$D_1(i)$	$D_2(i)$
1		
2	.0000000	−.3969252
3	.0000095	−.3967722
4	−.0000210	−.3963758
5	.0000172	−.3956130
6	.0000038	−.3943611
7	−.0000229	−.3925002
8	.0000076	−.3899066
9	.0000153	−.3864574
10	−.0000153	−.3820324
11	.0000229	−.3765085
12	−.0000381	−.3697646
13	.0000076	−.3616784
14	.0000153	−.3521256
15	−.0000076	−.3409865
16	−.0000153	−.3281393
17	.0000381	−.3134596
18	−.0000458	−.2968278
19	.0000153	−.2781215
20	.0000000	−.2572165
21	.0000000	−.2339926
22	−.0000381	−.2083285
23	.0000305	−.1801000
24	−.0000076	−.1491849
25	.0000000	−.1154633
26	.0000000	−.0788117
27	−.0602875	−.0409923
28	−.0000076	.0000007
29	−1.5322420	−.0017090
30	.0000000	−.0000007
31	−1.7587740	−.0018315
32	.0000000	.0000000

* $D_1(i) = u_i - [\alpha_2(i) + h_i'\alpha_3(i)]u_{i-1} - [\alpha_2(i) + h_i\alpha_3(i)]u_{i+1} - \alpha_1(i)$

** $D_2(i) = u_i - u_{i+1} - \phi_i \Delta x$

Numerical results to (LP_1) are given in Table 1.1. The solution is graphed in Fig. 1. With the numerical solution $\{u_i\}_{i=1}^N$, we would like to check the free-boundary condition, from which we may be able to find an approximate optimal feedback control policy. The result of free-boundary checking is presented in Table 1.2. An approximate optimal feedback control law can be stated as follows: If the initial position x is in (0, 29/32), do not exercise any control until the state process $y_x(\cdot)$ hits $\partial[0, 29/32]$. If $y_x(\cdot)$ hits 0, then the evolution stops, and so no control action has taken place. If $y_x(\cdot)$ hits 29/32, then just use enough control to let $y_x(\cdot)$ jump from 29/32 to 1 and the evolution stops there. If the initial position x is in [29/32, 1), then immediately push $y_x(\cdot)$ to 1 and the evolution stops.

EXAMPLE 2.

$$\sigma(\cdot) = 1,\ \phi(x) = 0.02\,\mathrm{Sinh}(x),\ \alpha = 1/2,$$
$$f(x) = 1 - x,\ g(\cdot) = 0,$$
$$\text{Minimize } J^2_{x,\alpha}(v).$$

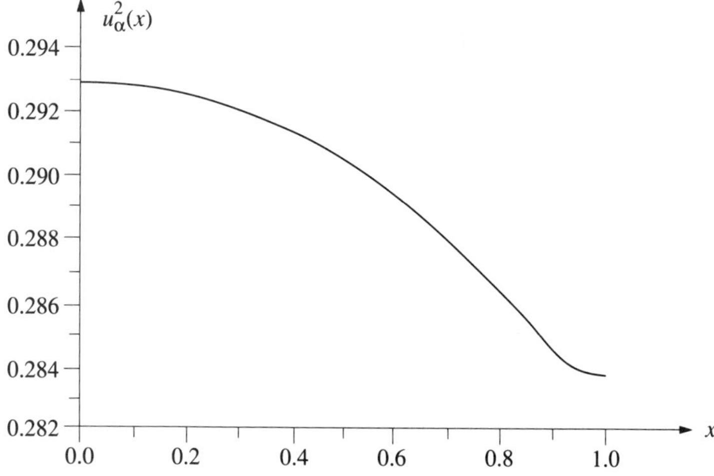

FIGURE 2. Graph of Value Function in Example 2

The numerical results to problem (LP_2) are presented in Table 2.1 and Table 2.2, with the graph in Fig. 2. An approximate feedback control law can be described as follows: If the initial position x is in (0, 27/32), just use enough control to let $y_x(\cdot)$ jump instantaneously from x to 27/32 and then keep it inside [27/32, 1] by the reflection at the

Table 2.1

Solution of Example 2

i	u_i
1	.292860
2	.292860
3	.292840
4	.292801
5	.292743
6	.292664
7	.292566
8	.292448
9	.292311
10	.292153
11	.291975
12	.291776
13	.291557
14	.291317
15	.291056
16	.290774
17	.290470
18	.290144
19	.289796
20	.289426
21	.289033
22	.288616
23	.288176
24	.287712
25	.287223
26	.286709
27	.286169
28	.285604
29	.285012
30	.284454
31	.283991
32	.283775
33	.283775

Table 2.2

Free-Boundary Checking for Example 2

i	$D_1(i)$	$D_2(i)$
1		
2	−.8123169	−.0003123
3	−.7810669	−.0003124
4	−.7498322	−.0003132
5	−.7185669	−.0003140
6	−.6873322	−.0003153
7	−.6560669	−.0003168
8	−.6248474	−.0003187
9	−.5935669	−.0003211
10	−.5623169	−.0003236
11	−.5310974	−.0003269
12	−.4998169	−.0003297
13	−.4685669	−.0003328
14	−.4373169	−.0003366
15	−.4060364	−.0003400
16	−.3748474	−.0003446
17	−.3435669	−.0003501
18	−.3123474	−.0003553
19	−.2810364	−.0003599
20	−.2498474	−.0003657
21	−.2185669	−.0003723
22	−.1873169	−.0003785
23	−.1560669	−.0003851
24	−.1248474	−.0003929
25	−.0935364	−.0004008
26	−.0623474	−.0004084
27	−.0310364	−.0004167
28	−.0000610	−.0004322
29	−.0000305	−.0014314
30	−.0000305	−.0043839
31	−.0468750	−.0107584
32	.0000305	−.0190948

* $D_1(i) = u_i - [\alpha_2(i) + h_i'\alpha_3(i)]u_{i-1} - [\alpha_2(i) + h_i\alpha_3(i)]u_{i+1} - \alpha_1(i)$

** $D_2(i) = u_i - u_{i+1} - \phi_i\Delta x$

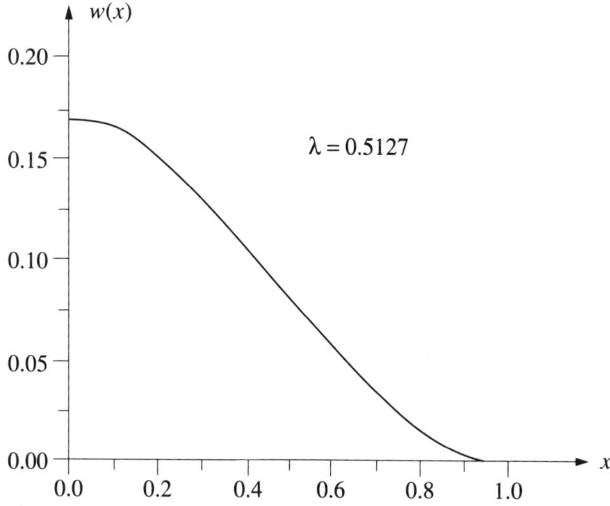

FIGURE 3. Graph of Function $w(x)$ in Example 3

endpoints. If the initial position x is in $[27/32, 1)$, just use a minimal amount of control to keep $y_x(\cdot)$ inside $[27/32, 1]$ by the same kind of reflection.

EXAMPLE 3. $\sigma(\cdot) = 1$, $\alpha = 0$, $g(\cdot) = 0$,

$$\phi(x) = \begin{cases} x & \text{if } x < 1/4, \\ 1/4 & \text{if } x \geq 1/4, \end{cases}$$

$$f(x) = \begin{cases} -2x + 2 & \text{if } x < 1/3, \\ -5x + 3 & \text{if } 1/3 \leq x < 1/2, \\ 1 - x & \text{if } x \geq 1/2, \end{cases}$$

Minimize $J_x^3(v)$.

Numerical results have been obtained (but not included here due to the page limitation). An approximate optimal control policy can also be constructed as in Example 2. Fig. 3 illustrates our numerical results.

EXAMPLE 4. $\sigma(\cdot) = 1$, $\alpha = 0$, $g(\cdot) = 0$, $x' = 1/2$,

$$\phi(x) = \begin{cases} 0 & \text{if } x < 1/2, \\ (2/3)x^3 + (3/4)x - 11/24 & \text{if } x \geq 1/2, \end{cases}$$

$$f(x) = \begin{cases} -(1/4)x + 9/64 & \text{if } x < 1/2, \\ (x - 5/8)^2 & \text{if } x \geq 1/2, \end{cases}$$

Minimize $J_x^4(v)$.

Table 3.1

Solution of Example 4

i	u_i
1	.000000
2	.000000
3	.000000
4	.000000
5	.000000
6	.000000
7	.000000
8	.000000
9	.000000
10	.000000
11	.000000
12	.000000
13	.000000
14	.000000
15	.000000
16	.000000
17	.000000
18	.000000
19	.000013
20	.000050
21	.000114
22	.000210
23	.000334
24	.000481
25	.000641
26	.000801
27	.000944
28	.001049
29	.001091
30	.001041
31	.000868
32	.000534
33	.000000
34	.015625 $(=\lambda)$

Table 3.2

Free-Boundary Checking for Example 4

i	$D_1(i)$	$D_2(i)$
1		
2	−.1171875	.0000000
3	−.1093750	.0000000
4	−.1015625	.0000000
5	−.0937500	.0000000
6	−.0859375	.0000000
7	−.0781250	.0000000
8	−.0703125	.0000000
9	−.0625000	.0000000
10	−.0546875	.0000000
11	−.0468750	.0000000
12	−.0390625	.0000000
13	−.0312500	.0000000
14	−.0234375	.0000000
15	−.0156250	.0000000
16	−.0078125	.0000000
17	−.0000000	.0000000
18	−.0000000	−.0404867
19	−.0000000	−.0833537
20	−.0000000	−.1286011
21	−.0000000	−.1762289
22	.0000000	−.2262370
23	−.0000000	−.2786255
24	.0000001	−.3333944
25	−.0000002	−.3905436
26	.0000001	−.4500733
27	.0000001	−.5119833
28	−.0000001	−.5762736
29	−.0000001	−.6429443
30	.0000001	−.7119955
31	.0000000	−.7834269
32	.0000000	−.8572388

* $D_1(i) = w_i - [\alpha_2(i) + h_i'\alpha_3(i)]w_{i-1} - [\alpha_2(i) + h_i\alpha_3(i)]w_{i+1} + w_{N+1}/\alpha_0(i) - \alpha_1(i)$

** $D_2(i) = w_i - w_{i+1} - \phi_i \Delta x$

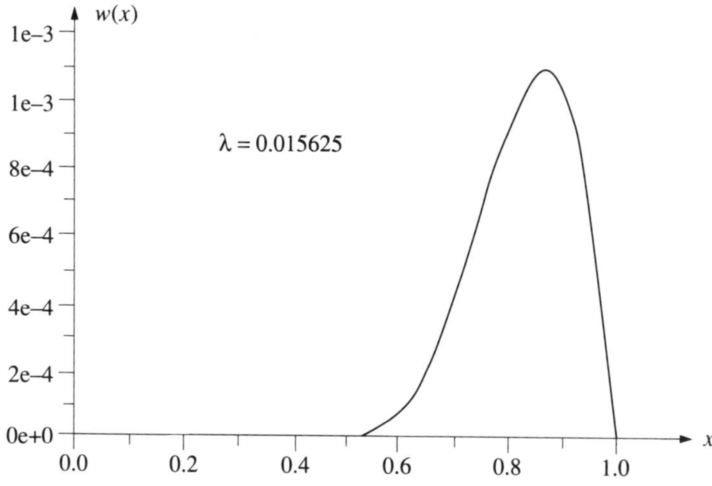

FIGURE 4. Graph of Function $w(x)$ in Example 4

Some of the numerical results for (LP_4) can be found in Table 3.1. and Table 3.2. Fig. 4 shows the graph of our solution. It is not difficult to find an approximate optimal control policy from our numerical solution.

4. Discussions.

4.1. *Comparison of numerical solutions with analytical solutions.* All the above numerical results have been obtained on a Macintosch personal computer. It takes about a minute to find the linear programming solution of any one of (LP_1), (LP_2), (LP_3), and (LP_4), when the number of grid points is about 40.

Apparently, the numerical results for Dirichlet boundary problems are more satisfactory than those for Neumann boundary problems. This simply tells us that the precise knowledge of boundary values of unknown value functions will result in a better approximation. Of course, this is what we can expect.

In Examples 3 and 4, we have attacked the differential inequalities directly. A solution to such differential inequalities consists of a constant λ and a function $w(\cdot)$. The constant λ plays the role of optimal average cost. A classical approach to find the optimal average cost λ is to find the discounted value function $u_\alpha(x)$ and then to approximate λ

Table 4.1 **Table 4.2**

Solution of Example 1 with New Data and Free-Boundary Checking

i	u_i	i	$D_1(i)$	$D_2(i)$
1	.250000	1		
2	.281669	2	−.0000610	−.9274797
3	.306014	3	.0000458	−.7321677
4	.323523	4	−.0000153	−.5524793
5	.334685	5	.0000153	−.3884168
6	.339987	6	−.0000153	−.2399793
7	.339918	7	.0000153	−.1071668
8	.334966	8	−1.0353090	−.0223322
9	.327642	9	−1.5000762	−.0000005
10	.319341	10	−1.2499240	−.0000005
11	.310064	11	−1.0000610	.0000000
12	.299810	12	−.7499542	−.0000005
13	.288579	13	−.5000305	.0000000
14	.276372	14	−.2499847	−.0000005
15	.263189	15	−.0000076	−.0000002
16	.249029	16	.0000076	−.0078127
17	.234380	17	.0000076	−.0312498
18	.219732	18	−.0001144	−.0625031
19	.205084	19	.0000305	−.0937557
20	.190435	20	.0001068	−.1250014
21	.175787	21	.0001678	−.1562533
22	.161139	22	.0001068	−.1875052
23	.146490	23	−.0000076	−.2187521
24	.131842	24	.0003815	−.2499905
25	.117193	25	.0000000	−.2812285
26	.102544	26	.0000000	−.3124785
27	.087895	27	.0000000	−.3437285
28	.073245	28	−.0000114	−.3749789
29	.058596	29	.0000000	−.4062293
30	.043947	30	.0000114	−.4374789
31	.029298	31	−.0000038	−.4687287
32	.014649	32	.0000000	−.4999788
33	.000000			

* $D_1(i) = u_i - [\alpha_2(i) + h_i'\alpha_3(i)]u_{i-1} - [\alpha_2(i) + h_i\alpha_3(i)]u_{i+1} - \alpha_1(i)$

** $D_2(i) = u_i - u_{i+1} - \phi_i \Delta x$

Table 5

Computed Values of $u_{i,j}^n$ for the 2-Dimensional Example

Given in Chow and Menaldi [6]

When $t = \Delta t$: (centered at the origin)

.78942	.51250	.31557	.19864	.16160	.20160	.32159	.52157	.80154
.74942	.47250	.27557	.15864	.12160	.16160	.28159	.48157	.76154
.70942	.43250	.23557	.11864	.08160	.12160	.24159	.44157	.72154
.66942	.39250	.19557	.07864	.04160	.08160	.20159	.40157	.68154
.64154	.36157	.16159	.04160	.00160	.04160	.16159	.36157	.64154
.68154	.40157	.20159	.08160	.04160	.07864	.19557	.39250	.66942
.72154	.44157	.24159	.12160	.08160	.11864	.23557	.43250	.70942
.76154	.48157	.28159	.16160	.12160	.15864	.27557	.47250	.74942
.80154	.52157	.32159	.20160	.16160	.19864	.31557	.51250	.78942

When $t = 2\Delta t$: (centered at the origin)

.74759	.48333	.29895	.19383	.16320	.20314	.32304	.52292	.80276
.71754	.45017	.26275	.15490	.12320	.16314	.28304	.48292	.76276
.68798	.41742	.22683	.11608	.08320	.12314	.24304	.44292	.72276
.65891	.38510	.19128	.07742	.04320	.08314	.20304	.40292	.68276
.64276	.36292	.16304	.04314	.00320	.04314	.16304	.36292	.64276
.68276	.40292	.20304	.08314	.04320	.07742	.19128	.38510	.65891
.72276	.44292	.24304	.12314	.08320	.11608	.22683	.41742	.68798
.76276	.48292	.28304	.16314	.12320	.15490	.26275	.45017	.71754
.80276	.52292	.32304	.20314	.16320	.19383	.29895	.48333	.74759

When $t = 3\Delta t$: (centered at the origin)

.70795	.45624	.28414	.18911	.16479	.20463	.32437	.52406	.80369
.68677	.42897	.25097	.15173	.12479	.16463	.28437	.48406	.76369
.66699	.40282	.21858	.11382	.08479	.12463	.24437	.44406	.72369
.64848	.37782	.18712	.07634	.04479	.08463	.20437	.40406	.68369
.64369	.36406	.16437	.04463	.00480	.04463	.16437	.36406	.64369
.68369	.40406	.20437	.08463	.04479	.07634	.18712	.37782	.64848
.72369	.44406	.24437	.12463	.08479	.11382	.21858	.40282	.66699
.76369	.48406	.28437	.16463	.12479	.15173	.25096	.42897	.68677
.80369	.52406	.32437	.20463	.16479	.18991	.28414	.45624	.70795

by $\alpha u_\alpha(x)$ for small $\alpha > 0$. But we have found approximate $(\lambda, w(\cdot))$ without going through $u_\alpha(\cdot)$.

From the numerical results reported in this article, we can see that these approximate solutions are fairly close to exact analytic solutions presented in Sun [27]. As one knows, it is impossible to find exact solutions in general. Numerical solutions are desired instead. Even for the theoretical study, numerical solutions can help a lot sometimes. As an illustration, we now redo Example 1 with

$$\sigma(\cdot) = 1,\ \psi(0) = 1/4,\ \psi(1) = 0,\ \alpha = 0,\ g(\cdot) = 0,$$

$$\phi(x) = x \quad \text{and} \quad f(x) = \begin{cases} 8(1/2 - x) & \text{if } x \leq 1/2, \\ 0 & \text{if } x > 1/2. \end{cases}$$

Then we get the numerical results in Table 4.1 and Table 4.2. An important observation should be made on the free boundary. It is clear from Table 4.2 that the so-called continuation region is no longer convex. In fact, it is not connected. However, in all the examples of singular control problems we have seen, the continuation regions are convex. The last example leads us to believe that for the singular control problems in bounded domains the convexity of continuation region can no longer be guaranteed, even if a continuation region exists. Consequently, it will be hard to search for an optimal feedback control law without numerical solutions.

4.2. *Extensions to higher-dimensional problems.* There are several references where singular stochastic control problems in $\mathbf{R}^d (d > 1)$ are treated (e.g., Chow and Menaldi **[6]**, Menaldi and Robin **[22]**, Sun and Menaldi **[31]**, and Menaldi and Taksar **[23]**). For the singular control problem considered in Menaldi and Robin **[22]**, one needs to solve

$$\begin{gathered} Lu + \alpha u \leq f, \qquad \partial u/\partial x_i \geq 0, \\ (Lu + \alpha u - f)\prod_i \partial u/\partial x_i = 0 \text{ in } \mathbf{R}^n. \end{gathered}$$

In Chow and Menaldi **[6]**, one gets

$$\begin{gathered} \partial_t u - (1/2)r^2\partial_y^2 u + (py + q^2x)\partial_y u - y\partial_x u \leq 0, \quad |\partial u/\partial y| \leq c, \\ [\partial_t u - (1/2)r^2\partial_y^2 u + (py + q^2x)\partial_y u - y\partial_x u][|\partial u/\partial y| - c] = 0, \\ u|_T = f. \end{gathered}$$

A numerical solution to the last equation was reported in Chow and Menaldi (**[6]**, Table 1). They used an iterative algorithm. For the same problem, the LP algorithm has been tested. Numerical experiments (cf. Table 5) show that the two different algorithms produce basically the same approximate solution.

From the numerical analysis point of view, the extension of the current LP algorithm to higher-dimensional problems is straightforward if one ignores the capacity of computing facilities for the moment. The difficulty lies in the theoretical study of the original higher-dimensional differential inequalities. Another related issue will be addressed in Section 4.4.

4.3. *Applications to other kinds of control problems and nonlinear PDE's.* What is important now is for us to capture the main ideas of our approach of solving singular control problems and to be able to adopt the same technique to solve other classes of control problems and nonlinear PDE's. Among other popular classes of stochastic control problems are regular control problems of continuous time (cf. Krylov **[17]**, Fleming and Rishel **[10]**), optimal stopping time problems (cf. Bensoussan and Lions **[4]**), impulsive control problems (cf. Bensoussan and Lions **[3]**), optimal switching problems (cf. Evans and Friedman **[9]**), and dynamic allocation problems (cf. Karatzas **[16]**). With the dynamic programming principle, one can derive for each control problem a dynamic programming equation which is in general a fully nonlinear second-order partial differential equation. For example, for the optimal cyclic switching problem discussed in Evans and Friedman **[9]**, one needs to find value functions $u_i(\cdot)$ $(i = 1, \dots, m)$ satisfying

$$\begin{aligned}&L^i u_i + \alpha u_i \le f^i, \qquad u_i \le k_i + \alpha u_{i+1} \quad (u_{m+1} = u_1),\\ &(L^i u_i + \alpha u_i - f^i)(u_i - k_i - \alpha u_{i+1}) = 0, \quad i = 1, \dots, m,\\ &\text{plus appropriate boundary conditions,}\end{aligned}$$

where L^i are elliptic second-order partial differential operators. In this case, one might let $u = (u_1, \dots, u_m)$. Then the discretized dynamic programming equation may read

$$A^1 u - b^1 \le 0, \qquad A^2 u - b^2 \le 0,$$
$$(A^1 u - b^1)(A^2 u - b^2) = 0.$$

In fact, it is not difficult to see that the discretized dynamic programming equations for all the control problems mentioned above have a common structure, i.e.,

$$\operatorname{Max}\{A^i u - b^i : i = 1, \dots, k\} = 0.$$

With an appropriate discretization procedure, the above equation may be equivalent to the following linear programming problem:

$$\begin{aligned}&\text{Minimize } -\sum u_j\\ &\text{Subject to } A^i u - b^i \le 0, \quad i = 1, \dots, k.\end{aligned} \tag{4.1}$$

As an explicit example of a nonlinear PDE which can be solved with our LP algorithm, we consider (in G, some domain in $\mathbf{R}^2$)

$$-\Delta u + |x_1\partial_1 u + x_2\partial_2 u| + \alpha u - f = 0,$$

plus some boundary conditions.

Then it is easy to see that the above nonlinear PDE is equivalent to

$$\operatorname{Max}\{-\Delta u - v(x_1\partial_1 u + x_2\partial_2 u) + \alpha u - f \colon v = 1, -1\} = 0,$$

with the same boundary conditions.

One must have recognized that the last equation is the dynamic programming equation of a bang-bang control problem. Therefore, our original nonlinear PDE can be approximated by an LP problem.

4.4. *Improved linear programming algorithms.* How to solve the linear programming problems obtained from the discretization of HJB equations is a practical issue of our method. Simplex algorithms are well known and readily available. Our four LP problems (LP_1)–(LP_4) have been solved quickly by such an algorithm even on a personal computer. But for the application of general control problems, one has to face a large number of linear constraints. This will be the case when higher-dimensional problems need to be solved. In those cases, a general simplex algorithm easily becomes inefficient or impractical. Sparse LP algorithms are available to overcome this problem (cf., e.g., Duff and Stewart **[8]**, Beale **[2]**, and Coleman **[7]**) to a certain extent. However, sparse LP algorithms in the LP literature generally cannot take the full advantage of special properties that an LP solution must have in our cases (e.g., Theorem 2.1). Those properties result from the stochastic control interpretation of LP solutions. A special sparse algorithm presented in Sun **[28]** greatly improves the efficiency of the regular LP algorithm for solving discretized HJB equations. For example, there are $k \times N$ constraints in LP problem (4.1). But with that algorithm we only use N constraints for each iteration. Furthermore, the working matrix preserves a fixed sparsity structure as the execution goes from one iteration to another, which allows us to use a fixed sparse procedure for solving a system of linear equations in the entire program. But greater success of our LP approach depends on further developments of more efficient sparse algorithms.

4.5. *Iterative algorithms.* Iterative algorithms are very popular in numerical analysis. As pointed out in Section 4.2, such an algorithm has been used to solve a singular control problem. Here we would like to present an iterative algorithm for the two singular control problems of time average cost criterions considered earlier in the article. Now let us solve (2.4) by the following iterative algorithm.

To describe our algorithm, let us introduce a family of nonlinear operators T_λ, defined for $w = (w_i)$ and λ by

$$(T_\lambda w)_i = \min\{w_{i+1} + \phi_i \Delta x, \\ \alpha_1(i) - \lambda/\alpha_0(i) + [\alpha_2(i) + h_i'\alpha_3(i)]w_{i-1} + [\alpha_2(i) + h_i\alpha_3(i)]w_{i+1}\}.$$

Then our algorithm can be stated as follows.

Step 1. Initialization.

$$w_i^0 = \lambda^0 = 0 \text{ for all } i.$$

Step 2. Updating from (w^k, λ^k) to (w^{k+1}, λ^{k+1}).

$$(1) \qquad w_i^{k+0.5} = w_i^k \vee (T_{\lambda^k} w^k)_i, \quad i = 2, \dots, N-1,$$

$$w_i^{k+0.5} = w_{i_0}^{k+0.5} = 0, \quad i = 1 \text{ or } N, \quad i_0 = [(N-1)/2],$$

$$(2) \qquad w_i^{k+1} = (T_{\lambda^k} w^{k+0.5})_i, \quad i = 2, \dots, N-1,$$

$$w_i^{k+1} = w_{i_0}^{k+1} = 0, \quad i = 1 \text{ or } N,$$

$$\lambda^{k+1} = (\alpha_1(i_0) + [\alpha_2(i_0) + h_i'\alpha_3(i_0)]w_{i_0-1}^{k+1} \\ + [\alpha_2(i_0) + h_i\alpha_3(i_0)]w_{i_0+1}^{k+1})\alpha_0(i_0).$$

Numerical results show that with the maximal tolerable error $\varepsilon = 10^{-7}$, we need about 90 iterations. The resulting approximate solution is very close to the LP solution obtained earlier. When we apply a similar iterative algorithm to solve (2.3), we can get a solution close to that of (LP_3), as well.

The iterative algorithm works satisfactorily, especially with Dirichlet boundary condition problems. But it is not easy to see the convergence analytically. However, with the LP approach, we have both analytic and numerical results.

References

1. M. C. Bancora-Imbert, P. L. Chow, and J. L. Menaldi, *On the numerical approximation of an optimal correction problem*, SIAM J. Sci. Statist. Comput. **9** (1988), 970–991.

2. E. M. L. Beale, *Sparseness in linear programming*, in J. K. Reid and A. E. R. E. Harwell (Editors), Large Sparse Sets of Linear Equations, Academic Press, London, 1971, 1–15.

3. A. Bensoussan and J. L. Lions, *Controle impulsionnel et inequations quasivariationnelles*, Dunod, Paris, 1982.

4. ——, *Applications des inequations variationnelles en controle stochastique*, Dunod, Paris, 1978.

5. A. Bensoussan and W. Runggaldier, *An approximation method for stochastic control problems with partial observation of the state—a method for constructing ε-optimal controls*, Acta Applicandae Mathematicae **10** (1987), 145–170.

6. P. L. Chow and J. L. Menaldi, *On the numerical solution of a stochastic optimal correction problem*, Transaction of the Third Army Conference on Applied Mathematics and Computing, Atlanta, Georgia, ARO Report 86-1 (1985), 531–558.

7. T. F. Coleman, *Large Sparse Numerical Optimization*, Springer-Verlag, Berlin, 1984.

8. I. S. Duff and G. W. Stewart, *Sparse Matrix Proceedings*, SIAM, Philadelphia, 1978.

9. L. C. Evans and A. Friedman, *Optimal stochastic switching and the Dirichlet problem for the Bellman equation*, Trans. Amer. Math. Soc. **253** (1979), 365–389.

10. W. H. Fleming and R. W. Rishel, *Deterministic and Stochastic Optimal Control*, Springer-Verlag, New York, 1975.

11. R. Gabasov and F. M. Kirillova, *Constructive methods of parametric and functional optimization*, IFAC 8th World Congress **1** (1981), 537–542.

12. R. Gabasov, F. M. Kirillova, and O. I. Kostyukova, *Algorithms to solve the linear optimal control problem*, Sov. Phys. Dokl. **29** (1984), 89–91.

13. I. Gladwell and R. Wait, *A Survey of Numerical Methods for Partial Differential Equations*, Clarendon Press, Oxford, 1979.

14. R. Glowinski, J. L. Lions, and R. Tremolieres, *Numerical Analysis of Variational Inequalities*, North-Holland, Amsterdam, 1981.

15. R. Gonzalez and E. Rofman, *On deterministic control problems: an approximation procedure for optimal cost* I. *The stationary problem,* II. *The nonstationary case*, SIAM J. Control Optim. **23** (1985), 242–285.

16. I. Karatzas, *Gittins indices in the dynamic allocation problem for diffusion processes*, The Annals of Probability **12** (1984), 173–192.

17. N. V. Krylov, *Controlled Diffusion Processes*, Springer-Verlag, New York, 1980.

18. K. Kushner, *Probabilistic Methods for Approximations in Stochastic Control and for Elliptic Equations*, Academic Press, New York, 1977.

19. O. L. Mangasarian, *Solution of symmetric linear complementarity problems by iterative methods*, J. Optim. Theory and Appl. **22** (1977), 465–485.

20. A. S. Manne, *Linear programming and sequential decisions*, Management Sci. **6** (1960), 259–267.

21. J. L. Menaldi, *Some estimates for finite difference approximations*, SIAM J. Control Optim. **27** (1989), 579–607.

22. J. L. Menaldi and M. Robin, *On some cheap control problems for diffusion processes*, Trans. Amer. Math. Soc. **278** (1983), 771–802.

23. J. L. Menaldi and M. I. Taksar, *Optimal correction problem of a multidimensional stochastic system*, preprint (1987).

24. J. T. Oden and J. N. Reddy, *An Introduction to the Mathematical Theory of Finite Elements*, John Wiley and Sons, New York, 1976.

25. J. S. Pang, *On the convergence of a basic iterative method for the implicit complementarity problem*, J. Optim. Theory and Appl. **37** (1982), 149–162.

26. D. Solow, *Linear Programming, an Introduction to Finite Improvement Algorithms*, North-Holland, New York, 1984.

27. M. Sun, *Singular control problems in bounded intervals*, Stochastics **21** (1987), 303–344.

28. ——, *Singular stochastic control problems solved by a sparse simplex method*, IMA J. Math. Control and Inf. **6** (1989), 27–38.

29. ——, *Some analytic aspects of numerical solutions of singular stochastic control problems in bounded domains*, IMA J. Math. Control and Inf. **6** (1989), 21–26.

30. ——, *Monotonicity of Mangasarian's iterative algorithm for generalized linear complementarity problems*, accepted for publication in J. Math. Anal. Appl. (1989).

31. M. Sun and J. L. Menaldi, *Monotone control of a damped oscillator under random perturbations*, IMA J. Math. Control and Inf. **5** (1988), 169–186.

32. E. H. Twizell, *Computational Method for Partial Differential Equations*, Ellis Horwood Limited, Chichester, 1984.

DEPARTMENT OF MATHEMATICS
THE UNIVERSITY OF ALABAMA, UNIVERSITY, AL 35487-0350

Lectures in Applied Mathematics
Volume **26** (1990)

Large Least-Squares Problems and the Need for Automating the Generation of Adjoint Codes

W. C. THACKER

Abstract. Some important least-squares problems that arise in oceanography and meteorology are large because they are based on systems of partial-differential equations. The function to be minimized is defined with the aid of a computer code, so the equations stating that the gradient should vanish are not available explicitly. It is possible to construct a second code that evaluates the gradient for a computational effort approximately equal to that of evaluating the function. Since the gradient code can be derived from the function code by following a well-defined set of rules, it should be possible to construct a compiler-like utility to do this task automatically. Such a compiler should find wide applicability, not only within oceanography and meteorology, but for optimization problems in general, as well as for solving systems of nonlinear equations and for checking the sensitivity of outputs of complicated codes to their inputs.

Introduction. Oceanography and meteorology are concerned with two vast fluid-dynamic systems: the world's oceans and its atmosphere. Their three-dimensional, time-varying temperature, pressure, density, and velocity fields are governed by partial-differential equations expressing the conservation of momentum, mass, and energy. In addition, conservation of salt is important in the sea, as is conservation of water vapor in the air. Due to the vast range of spatial and temporal scales of motion that are involved, numerical models simulating the behavior of the sea and the air require an immense number of discrete variables to approximate the solutions of these equations.

1980 *Mathematics Subject Classification* (1985 *Revision*). Primary 49D07, 68Q40, 93B35, 76D99.

Several nonlinear least-squares problems will be discussed here, all of which have one thing in common: the cost function that is to be minimized is defined via a large, complicated computer code centering around a numerical model for simulating atmospheric or oceanic flows. At its minimum the gradient of the cost function must vanish, but there are no closed-form expressions for the gradient, just as there is no closed-form expression for the cost function. Because of the immense number of variables involved, gradient-independent algorithms for computing the minimum can be ruled out as being far too slow, so the gradient must be evaluated numerically. Finite-difference approximations of the gradient are impractical for two reasons: first, the cost function would have to be evaluated for independent perturbations of each of the many variables, which would require an unaffordable amount of computation, and second, due to truncation error, the gradient is likely to be inaccurate unless the perturbations are appropriately small and the precision of the arithmetic is sufficiently high, further increasing the already unacceptable computational cost. Although algebraic-manipulation programs such as MACSYMA **[25]** or MATHEMATICA **[38]** might be used to derive expressions for the gradient vector for relatively simple small codes, they are not designed to handle the large complicated codes typically encountered in meteorology or oceanography. Even if expressions for the gradient vector could be constructed, although the problem of truncation error would be overcome, the problem of computational efficiency would persist; each of the many components of the gradient would have to be evaluated separately, which would entail an effort proportional to the number of variables characterizing the state of the ocean or atmosphere.

From the code defining the cost function it is possible to derive a new code that evaluates the gradient for approximately the same computational expense as a single evaluation of the cost function. Because of its relationship to the adjoints of the underlying partial-differential equations, this new code is referred to as the *adjoint code.* Its construction is a straightforward but tedious task ideally suited for a computer. A compiler that accepts function-evaluating codes as input and produces derivative-evaluating codes as output could be a highly valued utility, not only for problems arising in oceanography and meteorology, but for a wide variety of other applications as well. Some design features of such a compiler are discussed.

Fitting models to measurements: Weather prediction as an example. A fundamental problem is that of determining the state of the ocean or the atmosphere from measurements ([**11**], [3], [**26**], [**27**], [**9**], [**12**], [**4**], [**21**], [**22**], [**14**], [**2**], [**19**], [**23**], [**29**], [**18**], [**24**], [**39**], [**36**]). One approach to this problem is to fit a model to the measurements. Just as a straight line can be routinely fit to a scatter diagram, it is also possible to fit an approximate solution of a system of partial differential equations to a collection of oceanographic or meteorological data. To see how, think of the straight line as being the solution of a second-order differential equation [**34**]. For computational purposes it is necessary to replace the differential equation with some discrete counterpart, so the straight line might be represented by values at grid points. These values are determined by boundary conditions or initial conditions, which play the same role as the slope and intercept parameters in the usual two-parameter formula for the straight line. The straight-line fit is recast into the problem of determining boundary conditions or initial conditions for a differential equation.

Consider the problem of predicting the weather. The future state of the atmosphere is determined from the present state (the initial conditions) together with any addition or deletion of mass, energy, momentum, or humidity (boundary conditions plus internal sources and sinks). A numerical weather-forecast model provides a solution to the governing partial differential equations. Computing this solution can be referred to as the forward problem.

As the actual equations governing the evolution of the atmosphere are much too complicated to discuss here, a simplified system similar to that used in the early days of weather prediction [7] will be taken as an example. The vertical extent of the atmosphere is ignored, and the motion is characterized by the vertical component ζ of the vorticity vector, which is a function of the horizontal coordinates x and y and time t. For simplicity, the spherical geometry of the earth's surface will be ignored, except that the variation with latitude of the vertical component β of the earth's angular velocity will be retained as a constant; this is referred to as the β-plane approximation. The evolution of the model atmosphere is governed by the barotropic vorticity equation:

$$\frac{\partial\zeta}{\partial t}+\beta\frac{\partial\psi}{\partial x}-\frac{\partial\psi}{\partial y}\frac{\partial\zeta}{\partial x}+\frac{\partial\psi}{\partial x}\frac{\partial\zeta}{\partial y}=\varepsilon\nabla^2\zeta, \tag{1a}$$

where ψ is the stream function, which is related to the vorticity via a Poisson equation:

$$\nabla^2 \psi = \zeta. \tag{1b}$$

Nonlinearity enters via the Jacobian determinant $\partial(\psi, \zeta)/\partial(x, y)$, which accounts for advection of vorticity by the atmospheric flow. The eddy-viscosity coefficient ε is assumed to be a constant independent of time or location. For simplicity, the surface of the earth can be modelled as a rectangular domain with periodic boundary conditions. Then the forward problem is the solution of (1a) and (1b) for $\zeta(x, y, t)$ and $\psi(x, y, t)$, given the initial vorticity field $\zeta(x, y, 0)$ and values for the parameters β and ε.

Suppose that measurements d_ν, $\nu = 1, \ldots, \nu_{\max}$, are available in some space-time domain; then, given values for the necessary inputs (initial conditions, boundary conditions, parameters, etc.), the weather-forecast model can be used to predict values m_ν that the measurements should have, assuming the inputs are correct and the model is valid. The inverse problem is to determine the values of the unknown model inputs that make the predictions as close to the measurements as possible. In the vorticity-equation example the data might be measurements of u and v, the x- and y-components of the velocity, at space-time points (x_ν, y_ν, t_ν); their counterparts in the vorticity-equation model would be $-\partial\psi/\partial y$ and $\partial\psi/\partial x$, respectively, with values determined by the initial vorticity field.

Clearly, a finite number of measurements will never be sufficient to determine the initial conditions as a unique function of x and y. One solution to this dilemma is to assume that measurements are available continuously at all space-time points [**18**], but such an assumption, although mathematically convenient, is physically unreasonable. The alternative is to replace the differential-equation model with a discrete model. In the examples discussed here discretization is accomplished using finite-difference methods, but this is not essential. Carrera and Neuman [**6**] give an example within the context of groundwater hydrology in which finite-element methods are used. For spectral methods, see [**20**] for an oceanographic example and [**8**] for a meteorological example.

A finite-difference approximation for the vorticity-equation could be:

$$\begin{aligned} &\frac{1}{\Delta_t}\left(\zeta^n_{i,j} - \zeta^{n-1}_{i,j}\right) + \frac{\beta}{\Delta_x}\left(\psi^{n-1}_{i+1,j} - \psi^{n-1}_{i-1,j}\right) + A^{n-1}_{i,j}(\psi, \zeta) \\ &\quad = \frac{\varepsilon}{\Delta_x^2}\left(\zeta^{n-1}_{i+1,j} - 2\zeta^{n-1}_{i,j} + \zeta^{n-1}_{i-1,j}\right) + \frac{\varepsilon}{\Delta_y^2}\left(\zeta^{n-1}_{i,j+1} - 2\zeta^{n-1}_{i,j} + \zeta^{n-1}_{i,j-1}\right), \end{aligned} \tag{2a}$$

where

$$
\begin{aligned}
A^n_{i,j}(\psi,\zeta) = \frac{1}{12\Delta_x\Delta_y} & [(\psi^n_{i,j-1} + \psi^n_{i+1,j-1} - \psi^n_{i,j+1} - \psi^n_{i+1,j+1})(\zeta^n_{i+1,j} - \zeta^n_{i,j}) \\
& + (\psi^n_{i-1,j-1} + \psi^n_{i,j-1} - \psi^n_{i-1,j+1} - \psi^n_{i,j+1})(\zeta^n_{i,j} - \zeta^n_{i-1,j}) \\
& + (\psi^n_{i+1,j} + \psi^n_{i+1,j+1} - \psi^n_{i-1,j} - \psi^n_{i-1,j+1})(\zeta^n_{i,j+1} - \zeta^n_{i,j}) \\
& + (\psi^n_{i+1,j-1} + \psi^n_{i+1,j} - \psi^n_{i-1,j-1} - \psi^n_{i-1,j})(\zeta^n_{i,j} - \zeta^n_{i,j-1}) \\
& + (\psi^n_{i+1,j} - \psi^n_{i,j+1})(\zeta^n_{i+1,j+1} - \zeta^n_{i,j}) + (\psi^n_{i,j-1} - \psi^n_{i-1,j})(\zeta^n_{i,j} - \zeta^n_{i-1,j-1}) \\
& + (\psi^n_{i,j+1} - \psi^n_{i-1,j})(\zeta^n_{i-1,j+1} - \zeta^n_{i,j}) + (\psi^n_{i+1,j} - \psi^n_{i,j-1})(\zeta^n_{i,j} - \zeta^n_{i+1,j-1})]
\end{aligned}
\tag{2b}
$$

is a conservative approximation to the Jacobian determinant due to Arakawa [**1**]. A five-point Laplacian could be used for the Poisson equation:

$$
\frac{1}{\Delta_x^2}\left(\psi^n_{i+1,j} - 2\psi^n_{i,j} + \psi^n_{i-1,j}\right) + \frac{1}{\Delta_y^2}\left(\psi^n_{i,j+1} - 2\psi^n_{i,j} + \psi^n_{i,j-1}\right) = \zeta^n_{i,j}. \tag{2c}
$$

Given values for the initial vorticities $\zeta^0_{i,j}$, $i = 1,\ldots,i_{\max}$, $j = 1,\ldots,j_{\max}$, the initial stream-function values for $\psi^0_{i,j}$, $i = 1,\ldots,i_{\max}$, $j = 1,\ldots,j_{\max}$, can be found by solving the discrete Poisson equation (2c) using some algorithm such as successive over-relaxation. Then, after evaluating the Jacobian using (2b), the vorticities at time level $n = 1$ are given by the discrete vorticity equation (2a). By repeating this procedure it is possible to compute the discrete vorticity and stream-function fields at any time level $n > 0$ and then to compute any function that might depend on these future values.

The inverse problem can be formulated by defining a cost function J that is quadratic in the differences between the data and their model counterparts:

$$
J = \frac{1}{2}(\mathbf{m} - \mathbf{d})^{\mathrm{T}}\mathbf{W}(\mathbf{m} - \mathbf{d}) \tag{3}
$$

where $\mathbf{d}$ and $\mathbf{m}$ are column vectors containing the data and their model counterparts, respectively, and where $\mathbf{W}$ is a positive, symmetric matrix of coefficients (the inverse of the error-covariance matrix for the observations) that allows more accurate measurements to receive greater weight. The inverse problem is to find the model inputs which minimize J.

If the model counterparts of the data are linear functions of the inputs, then this is a linear least-squares problem. This would be the case if the underlying partial differential equations and boundary conditions were linear and the data were linearly related to the variables appearing in the equations. In general, however, this is a nonlinear least-squares problem. Nonlinearities enter in several ways. The first is via the Eulerian expression for the total temporal derivative following a parcel of fluid (i.e., the Jacobian term in the vorticity equation). The second is via flow-control statements (e.g., IF statement) in the model code. For example, if local conditions are correct, water vapor can condense into rain drops and release latent heat, which causes further convection and condensation. A more complicated model that simulates such processes would include flow-control statements that would introduce abrupt (logical) nonlinearities into the definition of the cost function, which are similar to the nonlinearity of the absolute-value function. Even if the partial differential equations are linear, the model counterparts of the data might be nonlinear functions. For example, it might only be possible to measure wind speed rather than the velocity vector; wind speed, which is the norm of the velocity vector, is a nonlinear function. Furthermore, the model counterparts of the data and thus the cost function generally will be nonlinear functions of the parameters like the eddy-viscosity, if such parameters are to be varied in determining the fit to the data.

A serious worry is whether the observations are sufficient to determine the sought-after model inputs. Huge numbers of measurements are needed to sample even a rather limited part of the atmosphere or the ocean, and they are both expensive and difficult to obtain. On the other hand the resolution of the model can be increased simply by using a denser computational grid, so the number of unknowns is likely to be greater than the number of observations. This is like trying to fit a straight line to a single datum; the fit is not unique. Ideally, there should be many more data than unknowns for the fit to inspire confidence. Even then, there is still the chance that the fit is not unique. For example, lots of observations over Europe won't say much about the weather in the South Pacific. To put it less dramatically, some linear combinations of the inputs might not be determined by the data.

Something must be done to guarantee that the fit is well determined. If all available data have been used and no more can be had, then it is still possible to utilize prior knowledge or prejudice [35]. Suppose that temperatures are generally known to be within a certain range; then bogus data can be used to incorporate this knowledge. Of course, the corresponding elements of the $\mathbf{W}$ matrix must reflect the reliability of these bogus data and how their errors are correlated. In the weather-prediction example, it might be possible to use the results of a previous forecast as bogus data, and their reliability should be estimated by the inverse of the covariance matrix for the forecast errors. (If new data are assimilated into the forecast as they become available, then the least-squares problem takes the form of Kalman filtering [**33**]. What is currently done operationally is similar, except that the error covariances of the forecast are modelled rather than calculated for lack of computational power.) When neither forecast nor past experience is available, it is still possible to require that the fit be smooth. This prejudice agrees with an implicit assumption that underlies the use of discrete methods for solving partial differential equations: unresolved features are assumed not to exist. Only now this assumption is also imposed with respect to the observational net. Again, bogus data can be used; in this case they correspond to hypothetical measurements indicating that certain partial derivatives of the meteorological or oceanographic fields are approximately zero. Thus, the bias toward smooth fits is like using smoothing splines to interpolate through data voids [**37**].

Even with enough data, a nonlinear least-squares problem might not have a unique solution. Rather than the cost function being flat in the vicinity of the minimum, which is the case when the data are insufficient, there might be several isolated relative minima. Hopefully, one minimum is much less than the others, so that it clearly corresponds to the physical solution. If this is the case, it might be possible to identify the inverse of the Hessian matrix as the error-covariance matrix of the variables determined by fitting the model to the data and thereby to establish confidence intervals for the recovered parameters. (Of course, computing and inverting these extremely large matrices will be quite a task!) If not, seemingly reasonable estimates for the model inputs might lie in several disjoint regions, and something might have to be done to discriminate among the possibilities. No attempt is generally made to find all of the minima; that is too difficult for such a large problem. Instead, it is necessary to rely upon physical intuition to provide a good starting guess for an iterative algorithm that finds a single local minimum. Then, computational budget allowing, a check might be made to see whether several equally reasonable guesses all lead to the same minimum.

In the vorticity-equation model suppose that observations of u and v were available at all space-time grid points (x_i, y_j, t_n), $i = 1, \ldots, i_{\max}$, $j = 1, \ldots, j_{\max}$, $n = 0, \ldots, n_{\max}$. Although these are far more data than could ever be expected from actual measurements, they are worth considering because they provide a simple example that is free from the worry of data insufficiency. Assuming all hypothetical measurements to be uncorrelated so that their error covariances vanish and to be equally accurate so that their error variances are all given by the same constant σ^2, then the weight matrix is proportional to the identity, and the least-squares cost function is:

$$J = \frac{1}{2\sigma^2} \sum_{i=1}^{i_{\max}} \sum_{j=1}^{j_{\max}} \sum_{n=0}^{n_{\max}} \left[\left(-\frac{1}{2\Delta_y} \left(\psi_{i,j+1}^n - \psi_{i,j-1}^n \right) - u_{i,j}^n \right)^2 + \left(\frac{1}{2\Delta_x} \left(\psi_{i+1,j}^n - \psi_{i-1,j}^n \right) - v_{i,j}^n \right)^2 \right], \tag{3a}$$

where periodicity requires that $i = i_{\max} + 1$ be interpreted as $i = 1$ and $i = 0$ as $i = i_{\max}$, etc. In this vorticity-equation model, the model inputs are the values of the initial vorticity field $\zeta_{i,j}^0$, the parameters β and ε, the observational error variance σ, and the grid increments Δ_x, Δ_y, and Δ_t. Because the boundary conditions are taken to be periodic, they are not to be determined from the data. For this example suppose that β, ε, σ, Δ_x, Δ_y, and Δ_t are known constants, and only the initial values of the discrete vorticities are unknowns to be determined from the data.

In computing the best least-squares fit to data, the cost function is not given by an explicit formula; it can only be evaluated using a computer code. This code can be thought of as having two parts: the first part computes the solution of the governing partial differential equations, given values for the necessary inputs; the second computes the model counterparts of the observations from the solution found by the first part and collects the model-data differences to evaluate the cost. The equations that must be solved to find the best fit to the data, which are often called the normal equations **[10]**, state that the gradient of the cost function must vanish. If the gradient were approximated using finite differences, at least as many function evaluations would be required as there are unknowns to be determined by the data; this possibility can be ruled out as being too computationally expensive and as being too inaccurate due to the truncation error of the finite-difference approximation. The alternative is to think of the cost function as being a composite function (i.e., a function of the model-data differences, which are in turn functions of the local stream-function variables, which are ultimately functions of the initial conditions) and to write a code that evaluates the gradient by repeated applications of the chain rule.

Although straightforward application of the chain rule in situations like these can be so tedious as to be practically impossible, this task can be greatly simplified through the use of Lagrange multipliers. Lagrange multipliers are generally used in constrained optimization **[31]** to avoid having to eliminate some of the arguments of the function being optimized and then solving a smaller unconstrained optimization problem. Here the essential idea is to replace a large unconstrained optimization problem by a much larger constrained optimization problem; each equation that must be solved in computing the cost function can be treated as a constraint as long as the additional variables appearing in these equations are included in the list of arguments of the cost function. At first it appears ridiculous to replace an impossibly large problem with an even larger problem, but this is only a conceptual step toward constructing code to evaluate the gradient. The point is that, just as it is easy to eliminate the additional variables to recover the unconstrained problem, it will be equally easy to eliminate the Lagrange multipliers to compute the gradient vector.

A Lagrangian function can be defined by appending to the cost function residuals of all the model equations with each multiplied by an unknown Lagrange multiplier. (Even more unknowns!) Let $\mathbf{R}$ and $\mathbf{\Lambda}$ denote vectors of residuals and multipliers, respectively; then the Lagrangian L can be written:

$$L = J - \mathbf{\Lambda}^{\mathrm{T}}\mathbf{R}, \tag{4}$$

where the choice of the minus sign is arbitrary. For example, if the functional dependence of the model counterparts of the data on the variables appearing in the governing partial differential equations is simple enough to be contained entirely in J (see equation (3a) for an example), then $\mathbf{R} = \mathbf{0}$ are just the finite-difference equations that constitute the numerical model (see (4a)); otherwise $\mathbf{R} = \mathbf{0}$ will also contain additional equations needed for defining the model counterparts of the data. The Lagrangian L is a function, not only of the model inputs that are to be determined by the data (initial conditions, possibly boundary conditions, transport parameters, etc.), but also of all intermediate values that are introduced in the model equations and of all Lagrange multipliers. The minimum of the cost function corresponds to a saddle point of the Lagrangian, so for minimum cost all derivatives of the Lagrangian must vanish. Suppose that only the derivatives with respect to the new variables (both the additional model variables and the Lagrange multipliers, but not the model inputs) vanish; this condition defines a system of equations that can be solved for the new variables in terms of the model inputs. Then the remaining derivatives of the Lagrangian (with respect to the model inputs) can be evaluated to give the sought-after gradient of the cost function.

In the vorticity-equation example the Lagrange function would be:

$$
\begin{aligned}
L = \frac{1}{2\sigma^2}\sum_{i=1}^{i_{\max}}\sum_{j=1}^{j_{\max}}\sum_{n=0}^{n_{\max}}\Bigg[&\left(-\frac{1}{2\Delta_y}\left(\psi^n_{i,j+1}-\psi^n_{i,j-1}\right)-u^n_{i,j}\right)^2 \\
&+\left(\frac{1}{2\Delta_x}\left(\psi^n_{i+1,j}-\psi^n_{i-1,j}\right)-v^n_{i,j}\right)^2\Bigg] \\
-\sum_{i=1}^{i_{\max}}\sum_{j=1}^{j_{\max}}\sum_{n=1}^{n_{\max}}\lambda^n_{i,j}\Bigg[&\frac{1}{\Delta_t}\left(\zeta^n_{i,j}-\zeta^{n-1}_{i,j}\right)+\frac{\beta}{\Delta_x}\left(\psi^{n-1}_{i+1,j}-\psi^{n-1}_{i-1,j}\right)+A^{n-1}_{i,j}(\psi,\zeta) \\
&-\frac{\varepsilon}{\Delta_x^2}\left(\zeta^{n-1}_{i+1,j}-2\zeta^{n-1}_{i,j}+\zeta^{n-1}_{i-1,j}\right)-\frac{\varepsilon}{\Delta_y^2}\left(\zeta^{n-1}_{i,j+1}-2\zeta^{n-1}_{i,j}+\zeta^{n-1}_{i,j-1}\right)\Bigg] \\
-\sum_{i=1}^{i_{\max}}\sum_{j=1}^{j_{\max}}\sum_{n=0}^{n_{\max}-1}\mu^n_{i,j}\Bigg[&A^n_{i,j}(\psi,\zeta) \\
&-\frac{1}{12\Delta_x\Delta_y}\Big(\left(\psi^n_{i,j-1}+\psi^n_{i+1,j-1}-\psi^n_{i,j+1}-\psi^n_{i+1,j+1}\right)\left(\zeta^n_{i+1,j}-\zeta^n_{i,j}\right) \\
&+\left(\psi^n_{i-1,j-1}+\psi^n_{i,j-1}-\psi^n_{i-1,j+1}-\psi^n_{i,j+1}\right)\left(\zeta^n_{i,j}-\zeta^n_{i-1,j}\right) \\
&+\left(\psi^n_{i+1,j}+\psi^n_{i+1,j+1}-\psi^n_{i-1,j}-\psi^n_{i-1,j+1}\right)\left(\zeta^n_{i,j+1}-\zeta^n_{i,j}\right) \\
&+\left(\psi^n_{i+1,j-1}+\psi^n_{i+1,j}-\psi^n_{i-1,j-1}-\psi^n_{i-1,j}\right)\left(\zeta^n_{i,j}-\zeta^n_{i,j-1}\right) \\
&+\left(\psi^n_{i+1,j}-\psi^n_{i,j+1}\right)\left(\zeta^n_{i+1,j+1}-\zeta^n_{i,j}\right)+\left(\psi^n_{i,j-1}-\psi^n_{i-1,j}\right)\left(\zeta^n_{i,j}-\zeta^n_{i-1,j-1}\right) \\
&+\left(\psi^n_{i,j+1}-\psi^n_{i-1,j}\right)\left(\zeta^n_{i-1,j+1}-\zeta^n_{i,j}\right)+\left(\psi^n_{i+1,j}-\psi^n_{i,j-1}\right)\left(\zeta^n_{i,j}-\zeta^n_{i+1,j-1}\right)\Big)\Bigg] \\
-\sum_{i=1}^{i_{\max}}\sum_{j=1}^{j_{\max}}\sum_{n=0}^{n_{\max}}\phi^n_{i,j}\Bigg[&\frac{1}{\Delta_x^2}\left(\psi^n_{i+1,j}-2\psi^n_{i,j}+\psi^n_{i-1,j}\right) \\
&+\frac{1}{\Delta_y^2}\left(\psi^n_{i,j+1}-2\psi^n_{i,j}+\psi^n_{i,j-1}\right)-\zeta^n_{i,j}\Bigg],
\end{aligned}
$$

(4a)

where $\lambda^n_{i,j}$, $\mu^n_{i,j}$, and $\phi^n_{i,j}$ are Lagrange multipliers associated with the residuals of equations (2a), (2b), and (2c), respectively. Note that, if the expression (2b) for the Jacobian determinant were included in the residual of the vorticity equation explicitly, then there would be no need for $\mu_{i,j}^n$.

The next step is to require that the derivatives of the Lagrange function with respect to all of its arguments except for the model inputs should vanish. (At the *minimum* of J, *all* derivatives of L, including those with respect to the model inputs, vanish.) This requirement provides a system of equations that can be used to solve for the Lagrange multipliers as functions of the model inputs:

$$\frac{\partial \mathbf{R}}{\partial \mathbf{q}}^{\mathrm{T}} \mathbf{\Lambda} = \frac{\partial J}{\partial \mathbf{q}}, \tag{5}$$

where $\mathbf{q}$ represents the vector of all arguments of the Lagrangian except for the model inputs. This system of linear equations is sometimes referred to as the *adjoint* model, since the equations comprise discrete counterparts to the adjoints of the partial differential equations that underlie the numerical model. (Adjoints of nonlinear differential equations are taken to be the same as the adjoints of the corresponding linearized equations. A linearized model has a parabolic cost surface that is tangent to the original cost surface at the point about which the equations are linearized, so the cost and gradient are the same for the nonlinear and linearized models at that point.) It might also be called an adjoint model to signify that it involves the transpose of the Jacobian matrix $\partial \mathbf{R}/\partial \mathbf{q}$ of the model equations. Then, differentiating with respect to the unknown model inputs provides expressions for evaluating the gradient for the original unconstrained optimization problem in terms of the Lagrange multipliers:

$$\mathbf{g} = \frac{\partial J}{\partial \mathbf{p}} - \frac{\partial \mathbf{R}}{\partial \mathbf{p}}^{\mathrm{T}} \mathbf{\Lambda}, \tag{6}$$

where $\mathbf{p}$ is the vector of all model inputs. The gradient vector $\mathbf{g}$ can be evaluated by passing once through the model and once through the adjoint model, which requires roughly the computational effort of only two evaluations of the cost function!

The system of equations for eliminating the additional variables is easily obtained for the example. Differentiating the Lagrangian (4a) with respect to the final stream-function values $\psi_{i,j}^{n_{\max}}$ at all grid points,

$i = 1, \ldots, i_{\max}$ and $j = 1, \ldots, j_{\max}$, gives

$$-\frac{1}{\Delta_x^2}\left(\phi_{i+1,j}^{n_{\max}} - 2\phi_{i,j}^{n_{\max}} + \phi_{i-1,j}^{n_{\max}}\right) - \frac{1}{\Delta_y^2}\left(\phi_{i,j+1}^{n_{\max}} - 2\phi_{i,j}^{n_{\max}} + \phi_{i,j-1}^{n_{\max}}\right)$$

$$= -\frac{1}{\sigma^2}\left[\frac{1}{4\Delta_x^2}\left(\psi_{i+2,j}^{n_{\max}} - 2\psi_{i,j}^{n_{\max}} + \psi_{i-2,j}^{n_{\max}}\right) - \frac{1}{2\Delta_x}\left(v_{i+1,j}^{n_{\max}} - v_{i-1,j}^{n_{\max}}\right)\right.$$
$$\left. + \frac{1}{4\Delta_y^2}\left(\psi_{i,j+2}^{n_{\max}} - 2\psi_{i,j}^{n_{\max}} + \psi_{i,j-2}^{n_{\max}}\right) + \frac{1}{2\Delta_y}\left(u_{i,j+1}^{n_{\max}} - u_{i,j-1}^{n_{\max}}\right)\right]. \tag{5a}$$

Differentiating with respect to the final vorticity values $\zeta_{i,j}^{n_{\max}}$, $i = 1, \ldots, i_{\max}$ and $j = 1, \ldots, j_{\max}$, gives

$$\frac{1}{\Delta_t}\left(\lambda_{i,j}^{n_{\max}}\right) = \phi_{i,j}^{n_{\max}}. \tag{5b}$$

Differentiating with respect to the Jacobian variables, $A_{i,j}^n(\psi, \zeta)$, $i = 1, \ldots, i_{\max}$, $j = 1, \ldots, j_{\max}$, and $n = 0, \ldots, n_{\max} - 1$, gives

$$\mu_{i,j}^n = -\lambda_{i,j}^{n+1}. \tag{5c}$$

Differentiating with respect to the remaining stream-function values $\psi_{i,j}^n$, $i = 1, \ldots, i_{\max}$, $j = 1, \ldots, j_{\max}$, and $n = 0, \ldots, n_{\max} - 1$, gives

$$-\frac{1}{\Delta_x^2}\left(\phi_{i+1,j}^n - 2\phi_{i,j}^n + \phi_{i-1,j}^n\right) - \frac{1}{\Delta_y^2}\left(\phi_{i,j+1}^n - 2\phi_{i,j}^n + \phi_{i,j-1}^n\right)$$
$$- A_{i,j}^n(\mu, \zeta) + \frac{\beta}{2\Delta_x}\left(\lambda_{i+1,j}^{n+1} - \lambda_{i-1,j}^{n+1}\right)$$
$$= -\frac{1}{\sigma^2}\left[\frac{1}{4\Delta_x^2}\left(\psi_{i+2,j}^n - 2\psi_{i,j}^n + \psi_{i-2,j}^n\right) - \frac{1}{2\Delta_x}\left(v_{i+1,j}^n - v_{i-1,j}^n\right)\right.$$
$$\left. + \frac{1}{4\Delta_y^2}\left(\psi_{i,j+2}^n - 2\psi_{i,j}^n + \psi_{i,j-2}^n\right) + \frac{1}{2\Delta_y}\left(u_{i,j+1}^n - u_{i,j-1}^n\right)\right], \tag{5d}$$

where the discrete Jacobian $A_{i,j}^n(\mu, \zeta)$ is defined by Arakawa's expression (2b). Finally, differentiating the Lagrangian with respect to the remaining (noninitial) vorticity variables $\zeta_{i,j}^n$, $i = 1, \ldots, i_{\max}$, $j = 1, \ldots, j_{\max}$, and $n = 1, \ldots, n_{\max} - 1$, gives

$$-\frac{1}{\Delta_t}\left(\lambda_{i,j}^n - \lambda_{i,j}^{n+1}\right) - A_{i,j}^n(\psi, \mu) + \phi_{i,j}^n$$
$$= -\frac{\varepsilon}{\Delta_x^2}\left(\lambda_{i+1,j}^{n+1} - 2\lambda_{i,j}^{n+1} + \lambda_{i-1,j}^{n+1}\right) - \frac{\varepsilon}{\Delta_y^2}\left(\lambda_{i,j+1}^{n+1} - 2\lambda_{i,j}^{n+1} + \lambda_{i,j-1}^{n+1}\right), \tag{5e}$$

where the discrete Jacobian $A_{i,j}^n(\psi, \mu)$ is also given by (2b).

The periodic boundary conditions for (2a, b, c) require the indices i and j to be interpreted cyclically, which implies periodic boundary conditions for the adjoint equations. Equations (5a) through (5e) are solved in order of decreasing temporal index n. First the discrete Poisson equation (5a) is solved for the multipliers $\phi_{i,j}^{n_{\max}}$ using the observations $u_{i,j}^{n_{\max}}$ and $v_{i,j}^{n_{\max}}$ and computed values of the stream function $\psi_{i,j}^{n_{\max}}$, which depend on the values of the model inputs $\zeta_{i,j}^{0}$. The multipliers $\lambda_{i,j}^{n_{\max}}$ are then given by (5b). Equations (5c), (5d), and (5e) are then used to step backward in time. On each step, first $\mu_{i,j}^{n}$ is given by (5c), then the discrete Poisson equation (5d) is solved for $\phi_{i,j}^{n}$, and finally $\lambda_{i,j}^{n}$ is given by (5e). On reaching $n = 0$, (5c) and (5d) are used to compute $\mu_{i,j}^{0}$ and $\phi_{i,j}^{0}$; there are no multipliers $\lambda_{i,j}^{0}$ appearing in (4a).

An expression for the gradient of the cost function (3a) is found by differentiating the Lagrangian (4a) with respect to the input variables $\zeta_{i,j}^{0}$ for $i = 1, \dots, i_{\max}$ and $j = 1, \dots, j_{\max}$:

$$
\begin{aligned}
g_{ij} = {} & \frac{1}{\Delta_t}\left(\lambda_{i,j}^{1}\right) - A_{i,j}^{0}(\psi,\mu) + \phi_{i,j}^{0} \\
& + \frac{\varepsilon}{\Delta_x^2}\left(\lambda_{i+1,j}^{1} - 2\lambda_{i,j}^{1} + \lambda_{i-1,j}^{1}\right) + \frac{\varepsilon}{\Delta_y^2}\left(\lambda_{i,j+1}^{1} - 2\lambda_{i,j}^{1} + \lambda_{i,j-1}^{1}\right).
\end{aligned}
\tag{6a}
$$

Note that the gradient would be given by $\lambda_{i,j}^{0}/\Delta_t$ as computed by (5e) for $n = 0$.

Rather than fitting a numerical model to data, it is also possible to take a functional-analytic approach and fit the underlying partial differential equations to continuous data [**32**], [**8**]. In this case the multipliers would be functions, the constraint terms in the Lagrangian would constitute a functional inner product, and the adjoint equations could be derived by integration by parts. Note that the adjoint equations should be no more difficult to solve than the original equations; still, it would be necessary to solve them numerically. So to avoid any possible difficulties associated with incompatibility between the discretizations of the original and adjoint equations and to secure the interpretation of the results as being the fit of a particular discrete model to data, it seems best to stick with the discrete-model approach. In fact, the point of this paper is that an adjoint code can in principle be derived directly from a model code without having to deal with either the discrete model or the underlying differential equations. The procedure for generating the adjoint code uses Lagrange multipliers in much the same way as they were used in the vorticity-model example, except that they are associated with definitions of variables appearing in the code. The important thing is that the procedure is systematic and thus can be automated, so all the work required to derive the adjoint equations and then to code them can be relegated to a compiler-like utility. This is discussed further in the final section of this paper.

Although integration by parts might provide a comfortable check on the form of the equations comprising the adjoint model, the equations are derived easily enough by differentiating the Lagrangian; the terms that contribute to each equation can be found by inspection. There are only a few terms in the Lagrangian that involve any particular variable, and they appear in the residuals at adjacent grid points and possibly in the cost function. So all that is needed for constructing the discrete adjoint model is the identification and differentiation of these terms. Still, this tedious task can be time-consuming and error-prone, so it would be highly desirable to have a computer do it.

The adjoint model will resemble a linearization of the original model quite closely, except for the appearance of forcing terms proportional to model-data differences. This is most easily seen in the continuous formulation, which can be obtained by letting the mesh size approach zero. Equations (5c), (5d), and (5e) can be combined in this limit to give

$$\begin{aligned}(7a)\qquad &-\frac{\partial \nabla^2 \lambda}{\partial t} - \beta\frac{\partial \lambda}{\partial x} - \nabla^2\frac{\partial(\psi,\lambda)}{\partial(x,y)} - \frac{\partial(\lambda,\nabla^2\psi)}{\partial(x,y)} \\ &= \varepsilon\nabla^4\lambda - \frac{1}{\sigma^2}\nabla^2\left(\nabla^2\psi + \frac{\partial u}{\partial y} - \frac{\partial v}{\partial x}\right),\end{aligned}$$

which, except for the data-misfit terms, can be recognized to be the adjoint of the linear equation governing a small perturbation δ of the stream function ψ obtained by combining and linearizing equations (1a) and (1b):

$$(7b)\qquad \frac{\partial \nabla^2 \delta}{\partial t} + \beta\frac{\partial \delta}{\partial x} + \nabla^2\frac{\partial(\psi,\delta)}{\partial(x,y)} + \frac{\partial(\delta,\nabla^2\psi)}{\partial(x,y)} = \varepsilon\nabla^4\delta.$$

In the adjoint equation (7a) the signs of the terms corresponding to odd-order partial derivatives are reversed; waves in the original model will appear as antiwaves carrying information from the data back to the initial time to be used to improve the initial conditions that define the best-fit trajectory. On the other hand, the signs of dissipative terms are not reversed, so there is no un-diffusion.

Note that the supplementary variables are introduced sequentially as the forecast integration proceeds by finite steps forward in time. In fact, even if the model must solve a Poisson equation for pressure at each time step, new variables are introduced at each step of the Poisson-solver algorithm. Consequently, if the adjoint code is derived directly from the model code, the Jacobian $\partial\mathbf{R}/\partial\mathbf{q}$ in (5) has only zeros above the diagonal; so the multipliers can be easily determined sequentially, starting with the last and finishing with the first. The backward integration of the adjoint model requires about the same computational effort as the forward integration of the forecast model.

Now the question of computing the minimum of the cost function must be faced. As was the case with the gradient, the Hessian matrix is not known. It might be evaluated numerically, either using finite differences to approximate derivatives of the gradient vector or using a specially constructed code. In either case this would require approximately twice as many function evaluations as there are unknowns. Furthermore, the Hessian should not be expected to be sparse. If all data were at the initial time, it would be sparse (reflecting the underlying locality of the differential equations); but in general the adjoint equations propagate the information contained in the time-distributed data backwards to the initial time, thereby filling in the zeros of the Hessian. Consequently, algorithms that exploit the Hessian matrix are likely to be too expensive to use for these large problems.

To date, only conjugate-gradient methods have been used for these problems. If these methods are to be efficient, preconditioning should be considered. To do so requires some understanding of the sources of ill conditioning. One obvious source is through the choice of dimensions of the variables that enter the calculations; this can be handled by rescaling (diagonal preconditioning). Ill conditioning can also result from the number and distribution of the observations; as mentioned above, the Hessian might even be singular if no attempt is made to use prior knowledge or prejudice. If, for example, the results of a previous forecast are taken as bogus data and if they dominate the actual observations that are being assimilated, then the error-covariance matrix of the forecast can be used as an efficient preconditioner. On the other hand, if a prior prejudice for smoothness is enforced by using bogus observations of derivatives of model fields, then the problem resembles one for computing smoothing splines, i.e., the solution of an elliptical system of equations with the order of differentiation twice that used in defining the bogus data; preconditioning methods that work for those problems might be helpful here.

So far, multigrid methods have not yet been tried for these problems, although they have been used successfully for similar parabolic boundary-control problems that are formulated in terms of adjoint equations [17]. It seems that they might also be appropriate here, especially in situations where a prejudice toward a smooth fit is being enforced. However, in the ideal case where there are many more observations than there are unknowns, and, where the observations are distributed in such a way that no bogus data are needed, it is not clear whether multigrid methods will work efficiently. There would have to be some reason to expect that the normal equations are like discretizations of elliptical partial differential equations, so that errors associated with the various spatial scales can easily be filtered out. This would indeed be the case if, whenever there is one particular type of measurement at some time, there would be an entire field of such measurements at that time; but in general the observations are scattered about, so the answer does not seem to be obvious. Of course, it is also not clear how to precondition conjugate gradients in this situation.

Steady oceanic circulation: Inverse problem. Weather prediction is basically an initial-value problem, and the previous discussion has focused on the determination of the initial conditions by fitting a numerical-forecast model to observations. The cost function was computed by integrating the model forward in time and then the gradient by a backward integration of the adjoint model. This same forward-backward procedure will also yield the gradient when boundary conditions or transport parameters are determined from the data along with the initial conditions. The formalism applies equally well to steady-state problems for which there is no integration over time. The forward-backward procedure simply becomes the solution of a boundary-value problem followed by the solution of an adjoint boundary-value problem.

As an example, consider the general circulation of the world's oceans. Over the years oceanographers have found that, when they return to any region of the ocean, measurements there are much the same as on previous visits. This leads to the concept of water-mass analysis: local water can be identified by its salinity, temperature, dissolved oxygen, etc. But how were the water masses formed and how did they get to be distributed as they are? The prevailing idea is that over the years a dynamical equilibrium has resulted from the relatively steady boundary fluxes of heat, salt, and momentum. One way to test this idea would be to fit a steady model to all available oceanographic measurements by varying the boundary fluxes and check if the optimal fluxes appear to be reasonable.

For the steady case the cost function can still be defined to be a quadratic function of the differences between the data and their model counterparts as in (3), and the Lagrangian function can still be defined by appending the model constraints as in (4); the adjoint model will be given by (5) and the gradient of the cost function by (6). The procedure for evaluating the gradient is essentially the same as before: given model inputs $\mathbf{p}$, boundary fluxes in this case, solve the boundary-value problem $\mathbf{R} = \mathbf{0}$ to evaluate the cost function and any coefficients that might be needed in the adjoint model. Then solve the adjoint boundary-value problem for the Lagrange multipliers $\mathbf{\Lambda}$ and use them to evaluate the gradient vector $\mathbf{g}$. Whereas the initial-value problem required a time integration of the model and a backward integration of the adjoint, the steady problem requires the solution of model and adjoint boundary-value problems of comparable difficulty. So, again, the evaluation of the gradient requires about the same computational effort as a second function evaluation. All that is needed is an adjoint model. A compiler that would construct it from the model code would save a lot of drudgery.

A slight modification of the barotropic vorticity equation can provide a relevant example for the steady oceanic circulation:

$$\beta\frac{\partial \psi}{\partial x} - \frac{\partial \psi}{\partial y}\frac{\partial \zeta}{\partial x} + \frac{\partial \psi}{\partial x}\frac{\partial \zeta}{\partial y} = \varepsilon\nabla^2\zeta + \chi \tag{8a}$$

and

$$\nabla^2\psi = \zeta, \tag{8b}$$

where $\chi(x, y)$ represents the curl of the wind stress, which drives the barotropic oceanic circulation. In a three-dimensional model, χ would represent the flux of angular momentum from the atmosphere into the ocean, but in a vertically integrated model such as this, it is simply an inhomogeneous forcing term. Given the wind stress, the solution is determined by the boundary conditions for ζ and ψ. At continental or island boundaries both ζ and ψ should be zero. But suppose that the domain is some rectangular region in the center of the ocean far from physical boundaries; then the boundary conditions would not in general be zero. They would depend on the flow outside the boundaries, just as the initial conditions depend on the flow before the initial time. An inverse problem would be to determine the boundary conditions from data.

The discrete model would be:

$$\frac{\beta}{\Delta_x}(\psi_{i+1,j}-\psi_{i-1,j})+A_{i,j}(\psi,\zeta) = \frac{\varepsilon}{\Delta_x^2}(\zeta_{i+1,j}-2\zeta_{i,j}+\zeta_{i-1,j})+\frac{\varepsilon}{\Delta_x^2}(\zeta_{i,j+1}-2\zeta_{i,j}+\zeta_{i,j-1})+\chi_{i,j} \tag{9a}$$

and

$$\frac{1}{\Delta_x^2}(\psi_{i+1,j}-2\psi_{i,j}+\psi_{i-1,j})+\frac{1}{\Delta_y^2}(\psi_{i,j+1}-2\psi_{i,j}+\psi_{i,j-1})=\zeta_{i,j}, \tag{9b}$$

where $A_{i,j}(\psi,\zeta)$ is Arakawa's Jacobian (2b). The forward problem is to compute $\zeta_{i,j}$ and $\psi_{i,j}$ for $i=1,\dots,i_{\max}$ and $j=1,\dots,j_{\max}$, given values for the forcing $\chi_{i,j}$ and for the parameters β, ε, Δ_x, and Δ_y, and given the boundary values $\zeta_{0,j}$, $\psi_{0,j}$, $\zeta_{i_{\max}+1,j}$, and $\psi_{i_{\max}+1,j}$ for $j=0,\dots,j_{\max}+1$ and $\zeta_{i,0}$, $\psi_{i,0}$, $\zeta_{i,j_{\max}+1}$, and $\psi_{i,j_{\max}+1}$ for $i=0,\dots,i_{\max}+1$.

Suppose, as before, that velocity observations are available for every grid point over an interval of time. A cost function for fitting the steady model to these data can be defined as in (3a), but the stream function would not have the superscript n. It could still appear on the velocity data, since the real ocean is time dependent, but information about temporal evolution would be meaningless in this model; this superscript simply provides a mechanism to represent observations at all times, so that the fit can be interpreted as a climatic mean. A Lagrangian can be defined much as in (4a), but with no temporal summation in the constraint terms. Then the adjoint model equations would be:

$$\begin{aligned}&-\frac{1}{\Delta_x^2}(\phi_{i+1,j}-2\phi_{i,j}+\phi_{i-1,j})-\frac{1}{\Delta_y^2}(\phi_{i,j+1}-2\phi_{i,j}+\phi_{i,j-1})\\ &\quad+A_{i,j}(\lambda,\zeta)+\frac{\beta}{2\Delta_x}(\lambda_{i+1,j}-\lambda_{i-1,j})\\ &\quad+\frac{1}{\sigma^2}\sum_{n=0}^{n_{\max}}\Big[\frac{1}{4\Delta_y^2}(\psi_{i,j+2}-2\psi_{i,j}+\psi_{i,j-2})+\frac{1}{2\Delta_y}\left(u_{i,j+1}^n-u_{i,j-1}^n\right)\\ &\quad+\frac{1}{4\Delta_x^2}(\psi_{i+2,j}-2\psi_{i,j}+\psi_{i-2,j})-\frac{1}{2\Delta_x}\left(v_{i+1,j}^n-v_{i-1,j}^n\right)\Big]=0\end{aligned} \tag{10a}$$

and

$$\begin{aligned}&A_{i,j}(\psi,\lambda)+\phi_{i,j}+\frac{\varepsilon}{\Delta_x^2}(\lambda_{i+1,j}-2\lambda_{i,j}+\lambda_{i-1,j})\\ &\qquad+\frac{\varepsilon}{\Delta_y^2}(\lambda_{i,j+1}-2\lambda_{i,j}+\lambda_{i,j-1})=0,\end{aligned} \tag{10b}$$

where the discrete Jacobians $A_{i,j}(\lambda,\zeta)$ and $A_{i,j}(\psi,\lambda)$ are defined as usual by Arakawa's expression (2b), and where the Lagrange multipliers $\mu_{i,j}$ for the definition of $A_{i,j}(\psi,\zeta)$ have been eliminated. These equations are obtained by differentiating the Lagrangian with respect to the streamfunction and vorticity variables at the interior grid points. As there are no Lagrange multipliers on the boundary, special equations are needed for the adjoint equations for points one grid interval from the boundary; it is simpler just to define constants $\phi_{0,j}=0$, $\phi_{i_{\max}+1,j}=0$, $\lambda_{0,j}=0$, $\lambda_{i_{\max}+1,j}=0$, etc., so that the adjoint equations (10a) and (10b) hold at *all* interior points. Thus, the adjoint model resembles a linearization of the physical model with homogeneous boundary conditions. Note that the data enter as though they were observations of vorticity, and they enter as a temporal average with an error variance $\sigma^2/(n_{\max}+1)$.

Equations for the gradient of the cost function are obtained by differentiating the Lagrangian with respect to the boundary variables, since these are the inputs to be determined from the data. Differentiating with respect to $\psi_{0,j}$, $j=1,\ldots,j_{\max}$, gives

$$\begin{aligned}\frac{\partial J}{\partial \psi_{0,j}} = &-\frac{1}{\Delta_x^2}\phi_{1,j} \\ &+\frac{1}{12\Delta_x\Delta_y}\Big[(\lambda_{1,j-1}-\lambda_{1,j+1})\,\zeta_{1,j}+\lambda_{1,j}\,(\zeta_{1,j+1}-\zeta_{1,j-1})\Big] \\ &+\frac{\beta}{2\Delta_x}\lambda_{1,j}+\frac{1}{\sigma^2}\sum_{n=0}^{n_{\max}}\left[\frac{1}{4\Delta_x^2}\psi_{2,j}-\frac{1}{2\Delta_x}v_{1,j}^n\right].\end{aligned} \tag{11a}$$

Similarly, differentiating with respect to $\zeta_{0,j}$, $j=1,\ldots,j_{\max}$, gives

$$\frac{\partial J}{\partial \zeta_{0,j}}=\frac{1}{12\Delta_x\Delta_y}\left[(\lambda_{1,j-1}-\lambda_{1,j+1})\,\psi_{1,j}+\lambda_{1,j}\,(\psi_{1,j+1}-\psi_{1,j-1})\right]+\frac{\varepsilon}{\Delta_x^2}\lambda_{1,j}. \tag{11b}$$

Rather than writing distinct formulae for every boundary segment, it is simpler to use the expressions for the derivatives that have been set equal to zero in (10a) and (10b), applying them at the boundary points, and recognizing that all undefined symbols should be interpreted as having values of zero. If the eddy-viscosity parameter is to be determined from the data, then the derivative of the Lagrangian with respect to ε is also needed:

$$\begin{aligned}\frac{\partial J}{\partial \varepsilon}=\sum_{i=1}^{i_{\max}}\sum_{j=1}^{j_{\max}}\sum_{n=1}^{n_{\max}}\lambda_{i,j}^n\Bigg[&\frac{1}{\Delta_x^2}\left(\zeta_{i+1,j}^{n-1}-2\zeta_{i,j}^{n-1}+\zeta_{i-1,j}^{n-1}\right) \\ &+\frac{1}{\Delta_y^2}\left(\zeta_{i,j+1}^{n-1}-2\zeta_{i,j}^{n-1}+\zeta_{i,j-1}^{n-1}\right)\Bigg].\end{aligned} \tag{11c}$$

If the wind-stress forcing is to be determined from the oceanographic data, the derivatives with respect to $\chi_{i,j}$, $i = 1, \ldots, i_{\max}$ and $j = 1, \ldots, j_{\max}$, are needed:

$$\frac{\partial J}{\partial \chi_{i,j}} = \lambda_{i,j}. \tag{11d}$$

It is not clear, however, whether the observations of the two components of the velocity at every grid point will contain enough information to fix the values of all boundary variables, the eddy viscosity, and the forcing at every internal grid point. More information might be needed.

The procedure for fitting the steady model to data is as follows: First, guess values for the boundary conditions and solve the forward problem (9a) and (9b). Next, solve the adjoint boundary-value problem (10a) and (10b) for the Lagrange multipliers, whose values will depend both on the guess for the boundary conditions and on the values of the data. Then evaluate the gradient of the cost function (11a), (11b), etc., (11c), and (11d), and use the gradient to compute an improved guess for the boundary values. Continue until the gradient vanishes.

There is still the question of how to solve both system (9a) and (9b) of nonlinear algebraic equations that constitute the forward boundary-value problem and the adjoint system of linear equations (10a) and (10b). Only the solution of the forward problem will be discussed here; the case for the adjoint model is similar. In fact, since the model code must define a solution to the forward boundary-value problem, an adjoint code generated directly from the model code will prescribe an algorithm for solving the adjoint boundary-value problem. There are also the questions of existence and uniqueness of a steady-state solution. Because equations (9a) and (9b) are nonlinear, the cost function may have more than one minima, and all of them might be unstable solutions to a corresponding time-dependent model.

Steady oceanic circulation: Forward problem. The equilibrium state of an oceanic circulation model is usually computed by time-stepping until transients have been damped out by friction. Unfortunately, because processes with a broad range of time scales are being simulated, an enormous amount of computer time is needed to reach a steady state [5]. This raises the question of whether there might be a more efficient algorithm.

Rather than time-stepping a transient model, perhaps it is better to try to find a solution to the steady model equations directly. In doing so it is necessary to solve a system of N coupled nonlinear equations in N unknowns, where N is an enormous number, perhaps larger than 10^6. One possibility would be to use Newton's method. The model is linearized about some guess for the solution, and a new guess is obtained by solving the linearized equations. This requires an efficient method for solving a sequence of large, sparse, stiff, nonsymmetric linear systems. The alternative considered here is to cast this into the form of a least-squares problem: The steady state corresponds to the minimum of a cost function defined as a positive quadratic function of the residuals of the steady model equations. If the cost function is taken simply as the sum of the squares of the residuals, difficulties might be anticipated; the equations for the minimum will be much more ill-conditioned than the original system. This problem can be avoided, either by introducing a conditioning operator into the definition of the cost function **[16]** or by applying preconditioning to the gradient in the usual manner **[15]**. Again, the equations that determine the minimum of the cost function are not known explicitly, since the cost function is defined by a model code. And again, this boils down to the need to evaluate the gradient of the cost function, and a gradient-evaluating code must be constructed.

The cost function for the model boundary-value problem can be written as

$$J_m = \frac{1}{2}\mathbf{R}^{\mathrm{T}}\mathbf{P}\mathbf{R}, \tag{12}$$

where $\mathbf{R}$ is the vector of residuals of the steady-model equations, and where $\mathbf{P}$ is a positive symmetric preconditioning operator. The subscript m is used to distinguish this cost function from J, which is used to fit the model to data, and from J_a, which might be used in computing the solution to the adjoint boundary-value problem. (Minima of J_m and of J_a must be found repeatedly in the computation of the minimum of J.) The gradient of J_m is

$$\mathbf{g}_m = \frac{\partial \mathbf{R}}{\partial \mathbf{q}}^{\mathrm{T}}\mathbf{P}\mathbf{R}, \tag{13}$$

where $\mathbf{q}$ represents the unknown steady-state solution vector, and where $\partial \mathbf{R}/\partial \mathbf{q}$ is the Jacobian matrix for the steady model equations. The Jacobian $\partial \mathbf{R}/\partial \mathbf{q}$ is just the linear operator associated with the linearized steady equations, and its transpose is the adjoint of the steady-model operator. Note that the Jacobian matrix need not be represented explicitly; multiplication by its transpose is the job of an adjoint code. This can be obtained without too much work by editing the adjoint code for computing the gradient of the cost function J, which must already exist if the steady model is being fit to data. The gradient $\mathbf{g}_m$ can be computed by evaluating the residuals of the model equations, preconditioning, and then plugging the results into the adjoint code. Or, if a compiler-like utility were available, a gradient-evaluating code could be obtained directly from a code for evaluating the steady-state cost function (12), without the need to edit the code for computing the gradient of the data-fitting cost function. Neglecting the cost of preconditioning by the operator $\mathbf{P}$, the gradient evaluation again requires the equivalent of approximately two evaluations of the cost function. Multigrid methods are likely to be appropriate for computing the minimum of J_m to find the steady circulation.

For the vorticity-equation example the cost function could be

$$J_m = \frac{1}{2}\sum_{i=1}^{i_{\max}}\sum_{j=1}^{j_{\max}}\left(ar_{i,j}^2 + bs_{i,j}^2\right), \tag{12a}$$

where

$$\begin{aligned} r_{i,j} = {} & \frac{\beta}{\Delta_x}(\psi_{i+1,j} - \psi_{i-1,j}) + A_{i,j}(\psi,\zeta) + \chi_{i,j} \\ & - \frac{\varepsilon}{\Delta_x^2}(\zeta_{i+1,j} - 2\zeta_{i,j} + \zeta_{i-1,j}) - \frac{\varepsilon}{\Delta_x^2}(\zeta_{i,j+1} - 2\zeta_{i,j} + \zeta_{i,j-1}) \end{aligned} \tag{12b}$$

and

$$s_{i,j} = \frac{1}{\Delta_x^2}(\psi_{i+1,j} - 2\psi_{i,j} + \psi_{i-1,j}) + \frac{1}{\Delta_y^2}(\psi_{i,j+1} - 2\psi_{i,j} + \psi_{i,j-1}) - \zeta_{i,j} \tag{12c}$$

are residuals for the vorticity equation and the Poisson equation, respectively, at grid point (i, j). The coefficients a and b can be thought of as diagonal elements of a preconditioning matrix, which insure that the cost function is dimensionally homogeneous. Given values for the boundary conditions, wind-stress forcing, and parameters, the gradient of J_m must vanish for the steady-state solution. The equations for the gradient are

$$\begin{aligned} \frac{\partial J_m}{\partial \psi_{i,j}} = {} & -a\left[A_{i,j}(r,\zeta) + \frac{\beta}{2\Delta_x}(r_{i+1,j} - r_{i-1,j})\right] \\ & + b\left[\frac{1}{\Delta_x^2}(s_{i+1,j} - 2s_{i,j} + s_{i-1,j}) + \frac{1}{\Delta_y^2}(s_{i,j+1} - 2s_{i,j} + s_{i,j-1})\right] \end{aligned} \tag{13a}$$

and

$$\text{(13b)}\qquad \frac{\partial J_m}{\partial \zeta_{i,j}} = -a\left[A_{i,j}(\psi, r) + \frac{\varepsilon}{\Delta_x^2}(r_{i+1,j} - 2r_{i,j} + r_{i-1,j}) + \frac{\varepsilon}{\Delta_y^2}(r_{i,j+1} - 2r_{i,j} + r_{i,j-1})\right] - bs_{i,j}.$$

Comparing with the adjoint-model equations (10a) and (10b), $-ar_{i,j}$ corresponds to $\lambda_{i,j}$ and $-bs_{i,j}$ to $\phi_{i,j}$.

The steady state could also be computed using only the vorticity-equation residuals ($b = 0$), if the Poisson equation is treated as a constraint. Lagrange multipliers $\phi_{i,j}$ would be needed, which would be eliminated using adjoint equations. In that case the cost function would involve fewer unknowns, but to compute the gradient of J_m two Poisson equations would have to be solved, one for the stream-function $\psi_{i,j}$ and one for the multipliers $\phi_{i,j}$. (See the example in the following section on weak constraints.)

Weak constraints. Sometimes when fitting a model to data, it is useful to make allowances for the errors in the model. This is usually done within the context of Kalman filtering by adding stochastic forcing terms with known statistics to the model equations **[13]**. Within the least-squares context, this can be accomplished by minimizing the sum of two terms, one representing the misfit between model and data and the other representing the model error:

$$\text{(14)}\qquad J_w = \frac{1}{2}(\mathbf{m} - \mathbf{d})^{\mathrm{T}}\mathbf{W}(\mathbf{m} - \mathbf{d}) + \frac{1}{2}\mathbf{R}^{\mathrm{T}}\mathbf{P}\mathbf{R}.$$

Note that the residuals enter in the same way as model counterparts of bogus zero-valued data; thus, the residuals can be identified as stochastic forcing terms with zero means and with variances determined by the inverse of $\mathbf{P}$. Because the model equations are not required to be satisfied exactly, no Lagrangian function is defined; the minimum of J_w is a compromise between the data and the dynamics. In this case the model is said to provide weak constraints, thus the subscript w; when the model equations are presumed to be error-free and must be satisfied exactly, they are referred to as strong constraints **[28]**.

Allowing for model errors results in a much larger unconstrained optimization problem than does the strong-constraint formalism. Not only the model inputs $\mathbf{p}$ but also the additional variables $\mathbf{q}$ are arguments of J_w; values for the stochastic forcing are being recovered from the data, real and bogus, along with the model inputs.

Again a code is needed to evaluate the gradient of J_w:

$$\mathbf{g}_w = \frac{\partial \mathbf{m}^{\mathrm{T}}}{\partial \mathbf{z}} \mathbf{W}(\mathbf{m} - \mathbf{d}) + \frac{\partial \mathbf{R}^{\mathrm{T}}}{\partial \mathbf{z}} \mathbf{PR}, \tag{15}$$

where $\mathbf{z} = (\mathbf{p}^{\mathrm{T}}, \mathbf{q}^{\mathrm{T}})^{\mathrm{T}}$ is the vector containing all arguments of J_w. If an adjoint code for equations (5) and (6) of the strong-constraint problem is available, not too much effort would be needed to modify it so that it would be applicable here; if not, a gradient-evaluating code must be written by hand with a great deal of labor, unless a compiler is available to do that task.

Suppose there are doubts concerning the validity of the steady vorticity-equation model. For example, the parameterization of eddy viscosity might be questioned, the wind-stress forcing might be poorly known, or the assumption that the flow should be steady might be doubted. The Poisson equation relating stream-function to vorticity can be assumed to be valid, but allowance should be made for uncertainties in the vorticity equation. To do so, add to the vorticity equation an unknown term $r_{i,j}$, which can be assumed a priori to have zero mean and variance given by $1/a$.

If temporally averaged observations of velocity are available at every grid point, it is possible to estimate the vorticity and stream function. An appropriate cost function would be

$$\begin{aligned} J_w = \frac{1}{2\sigma^2} \sum_{i=1}^{i_{\max}} \sum_{j=1}^{j_{\max}} \Bigg[& \left(-\frac{1}{2\Delta_y} (\psi_{i,j+1} - \psi_{i,j-1}) - u_{i,j} \right)^2 \\ & + \left(\frac{1}{2\Delta_x} (\psi_{i+1,j} - \psi_{i-1,j}) - v_{i,j} \right)^2 \Bigg] \\ + \frac{a}{2} \sum_{i=1}^{i_{\max}} \sum_{j=1}^{j_{\max}} & (r_{i,j})^2 . \end{aligned} \tag{14a}$$

The forward problem requires that not only the boundary conditions but also the residuals (stochastic forcing terms) be specified before stream-function and vorticity can be computed. Thus, the inverse problem requires that derivatives with respect to these variables must vanish.

Note that the residual $r_{i,j}$ of the steady model is the temporal derivative $(\zeta^n_{i,j} - \zeta^{n-1}_{i,j})/\Delta_t$ in a time-dependent model. Thus, the weak steady constraints can be thought of as strong time-varying constraints that are penalized to give preference to the steady state. By replacing $r_{i,j}$ with $(\zeta^1_{i,j} - \zeta^0_{i,j})/\Delta_t$ and using the unsteady model equations for a single time step as constraints, the Lagrangian function is:

$$
\begin{aligned}
L = \frac{1}{2\sigma^2}\sum_{i=1}^{i_{\max}}\sum_{j=1}^{j_{\max}} \Bigg[\left(-\frac{1}{2\Delta_y}(\psi_{i,j+1}-\psi_{i,j-1}) - u_{i,j} \right)^2 \\
+ \left(\frac{1}{2\Delta_x}(\psi_{i+1,j}-\psi_{i-1,j}) - v_{i,j} \right)^2 \Bigg] \\
+ \frac{a}{2\Delta_t^2}\sum_{i=1}^{i_{\max}}\sum_{j=1}^{j_{\max}} \left(\zeta^1_{i,j} - \zeta_{i,j} \right)^2 \\
- \sum_{i=1}^{i_{\max}}\sum_{j=1}^{j_{\max}} \lambda_{i,j} \Bigg[\frac{1}{\Delta_t}\left(\zeta^1_{i,j} - \zeta_{i,j} \right) + \frac{\beta}{\Delta_x}(\psi_{i+1,j}-\psi_{i-1,j}) \\
+ \frac{1}{12\Delta_x\Delta_y} \Big((\psi_{i,j-1}+\psi_{i+1,j-1}-\psi_{i,j+1}-\psi_{i+1,j+1})(\zeta_{i+1,j}-\zeta_{i,j}) \\
+ (\psi_{i-1,j-1}+\psi_{i,j-1}-\psi_{i-1,j+1}-\psi_{i,j+1})(\zeta_{i,j}-\zeta_{i-1,j}) \\
+ (\psi_{i+1,j}+\psi_{i+1,j+1}-\psi_{i-1,j}-\psi_{i-1,j+1})(\zeta_{i,j+1}-\zeta_{i,j}) \\
+ (\psi_{i+1,j-1}+\psi_{i+1,j}-\psi_{i-1,j-1}-\psi_{i-1,j})(\zeta_{i,j}-\zeta_{i,j-1}) \\
+ (\psi_{i+1,j}-\psi_{i,j+1})(\zeta_{i+1,j+1}-\zeta_{i,j}) + (\psi_{i,j-1}-\psi_{i-1,j})(\zeta_{i,j}-\zeta_{i-1,j-1}) \\
+ (\psi_{i,j+1}-\psi_{i-1,j})(\zeta_{i-1,j+1}-\zeta_{i,j}) + (\psi_{i+1,j}-\psi_{i,j-1})(\zeta_{i,j}-\zeta_{i+1,j-1}) \Big) \\
- \frac{\varepsilon}{\Delta_x^2}(\zeta_{i+1,j}-2\zeta_{i,j}+\zeta_{i-1,j}) - \frac{\varepsilon}{\Delta_y^2}(\zeta_{i,j+1}-2\zeta_{i,j}+\zeta_{i,j-1}) \Bigg] \\
- \sum_{i=1}^{i_{\max}}\sum_{j=1}^{j_{\max}} \phi_{i,j} \Bigg[\frac{1}{\Delta_x^2}(\psi_{i+1,j}-2\psi_{i,j}+\psi_{i-1,j}) \\
+ \frac{1}{\Delta_y^2}(\psi_{i,j+1}-2\psi_{i,j}+\psi_{i,j-1}) - \zeta_{i,j} \Bigg].
\end{aligned}
\tag{15a}
$$

The superscript indicating $n = 0$ has been dropped to indicate the steady-state solution. Also, Arakawa's expression for the Jacobian has been used explicitly in the vorticity constraint, so no μ multipliers are needed.

The adjoint equations are obtained, as before, by differentiating L with respect to supplementary variables. With the replacement $r_{i,j} = (\zeta^1_{i,j} - \zeta_{i,j})/\Delta_t$, there are two choices for the input (control) variables to replace $r_{i,j}$. If they are taken to be $\zeta^1_{i,j}$, with $\zeta_{i,j}$ determined simultaneously with $\psi_{i,j}$ in a forward boundary-value problem, then $\lambda_{i,j}$ and $\phi_{i,j}$ will be determined simultaneously by the adjoint boundary-value problem. Below, the initial vorticities $\zeta_{i,j}$ replace the stochastic forcing variables $r_{i,j}$ as inputs, and $\zeta^1_{i,j}$ joins $\psi_{i,j}$ as a supplemental variable. With this choice the forward and adjoint boundary-value problems involve simply Poisson equations for $\psi_{i,j}$ and $\phi_{i,j}$, with simple prognostic equations for $\zeta^1_{i,j}$ and $\lambda_{i,j}$. Differentiating with respect to $\zeta^1_{i,j}$ and $\psi_{i,j}$ gives the adjoint model equations:

$$\lambda_{i,j} = \frac{a}{\Delta_t}\left(\zeta^1_{i,j} - \zeta_{i,j}\right) \tag{16a}$$

and

$$\begin{aligned} &\frac{1}{\Delta_x^2}(\phi_{i+1,j} - 2\phi_{i,j} + \phi_{i-1,j}) + \frac{1}{\Delta_y^2}(\phi_{i,j+1} - 2\phi_{i,j} + \phi_{i,j-1}) \\ &\quad = A_{i,j}(\lambda, \zeta) + \frac{\beta}{2\Delta_x}(\lambda_{i+1,j} - \lambda_{i-1,j}) \\ &\quad + \frac{1}{\sigma^2}\left[\frac{1}{4\Delta_y^2}(\psi_{i,j+2} - 2\psi_{i,j} + \psi_{i,j-2})\right. \\ &\quad + \frac{1}{2\Delta_y}(u_{i,j+1} - u_{i,j-1}) + \frac{1}{4\Delta_x^2}(\psi_{i+2,j} - 2\psi_{i,j} + \psi_{i-2,j}) \\ &\quad \left. - \frac{1}{2\Delta_x}(v_{i+1,j} - v_{i-1,j})\right]. \end{aligned} \tag{16b}$$

All λ's and ϕ's that don't appear in the definition of the Lagrangian (15a) should be set to zero, which gives homogeneous boundary conditions for (16a) and (16b).

Differentiating with respect to the initial (steady) vorticity variables $\zeta_{i,j}$ gives

$$\begin{aligned} \frac{\partial J_w}{\partial \zeta_{i,j}} &= A_{i,j}(\psi, \lambda) + \phi_{i,j} \\ &\quad + \frac{\varepsilon}{\Delta_x^2}(\lambda_{i+1,j} - 2\lambda_{i,j} + \lambda_{i-1,j}) + \frac{\varepsilon}{\Delta_y^2}(\lambda_{i,j+1} - 2\lambda_{i,j} + \lambda_{i,j-1}). \end{aligned} \tag{17a}$$

By setting all undefined quantities to zero, (17a) provides expressions for the derivative with respect to either interior or boundary variables. The parts of the gradient for stream-function variables on the boundary are given by

$$
\begin{aligned}
\frac{\partial J_w}{\partial \psi_{i,j}} = &- \frac{1}{\Delta_x^2}(\phi_{i+1,j} - 2\phi_{i,j} + \phi_{i-1,j}) - \frac{1}{\Delta_y^2}(\phi_{i,j+1} - 2\phi_{i,j} + \phi_{i,j-1}) \\
&+ A_{i,j}(\lambda, \zeta) + \frac{\beta}{2\Delta_x}(\lambda_{i+1,j} - \lambda_{i-1,j}) \\
&+ \frac{1}{\sigma^2}\left[\frac{1}{4\Delta_y^2}(\psi_{i,j+2} - 2\psi_{i,j} + \psi_{i,j-2}) + \frac{1}{2\Delta_y}(u_{i,j+1} - u_{i,j-1}) \right. \\
&\qquad \left. + \frac{1}{4\Delta_x^2}(\psi_{i+2,j} - 2\psi_{i,j} + \psi_{i-2,j}) - \frac{1}{2\Delta_x}(v_{i+1,j} - v_{i-1,j})\right],
\end{aligned}
\tag{17b}
$$

where i and j refer to boundary points, and all undefined quantities are set to zero.

The solution procedure would be as follows: Guess the initial (steady) vorticity and boundary conditions, step the model forward once to compute initial stream function and vorticity at time level $n = 1$, use (16a) and (16b) to compute the Lagrange multipliers, and use (17a) and (17b) to compute the gradient. Use the gradient to improve the guess and iterate until convergence. The solution will be a compromise between the steady state and the data. When there are no data, i.e., when $\sigma^2 = \infty$, this reduces to a procedure for computing the steady-state solution.

Automating adjoints. It is easy to imagine a compiler that accepts the function-evaluation code as input and produces a gradient-evaluation code as output. In fact, an example of such a compiler already exists: JAKE, written by Speelpenning [**30**]. Unfortunately, it is somewhat limited, so it is not likely to be able to handle an arbitrary, off-the-shelf, function-evaluating code as input. Suitably generalized, such a tool would be useful for the myriad of optimization problems in all realms of applied mathematics. But its applications would not be limited to optimization; it could also be used to compute the sensitivity of the outputs of large codes to variations in their inputs. For an extreme example, imagine a "Star Wars" code; the output of such a utility would be a code that could identify those situations that might trigger a momentous decision.

The model code consists of two types of statements: control statements and assignment statements. For now, disregard the control statements; as they are routinely handled every time the model code is compiled for execution, they should not present substantially greater difficulties here. (There is a minor problem with nondifferentiability, analogous to that of the absolute-value function where its argument vanishes, associated with branching paths caused, for example, by an IF statement.) The assignment statements represent the constraints that are to be incorporated into the definition of the Lagrange function, each having its own Lagrange multiplier. Each assignment statement can be thought of as defining a new model variable; so to avoid confusion, suppose that all variables are given distinct names in the code. This restriction can later be dropped. Differentiation with respect to variables appearing on the left-hand sides of the assignment statements yields equations that determine the values of the multipliers, and differentiation with respect to the model's input variables yields expressions for the gradient. Thus, the strategy is to solve for the multipliers and then to use them to evaluate the gradient.

Suppose that the model were rewritten so that all control statements are removed and so that no two variables have the same name. Then code would be a simple sequence of assignment statements:

$$v_i = E_i, \qquad i = 1, \ldots, I, \tag{18}$$

where v_i is the variable defined by the expression E_i in the ith assignment statement. Note that E_i can be formed from any of the previously defined variables v_j, $j < i$, and from any of the model inputs x_k, $k = 1, \ldots, K$. If the function to be differentiated is denoted by J, then the Lagrange function can be written

$$L = J - \sum_{i=1}^{I} a_i \, (v_i - E_i) \,, \tag{19}$$

where a_i denotes the ith Lagrange multiplier. (The choice of sign in front of the summation is arbitrary.) As the last task of the code is to evaluate the function J, J can be replaced by v_I.

The requirement that the derivative of L with respect to v_i should vanish is

$$a_i = \sum_{j=i+1}^{I} a_j \frac{\partial E_j}{\partial v_i}, \qquad i = 1, \ldots, I-1, \tag{20}$$

with $a_I = 1$. (The other choice for the sign in front of the summation in the definition of the Lagrangian would have resulted in $a_I = -1$.) Thus, the Lagrange multipliers can be evaluated in the reversed order, starting with a_I and ending with a_1. Then the gradient can be evaluated using equations obtained by differentiating with respect to the model's inputs x_k:

$$g_k = \sum_{j=1}^{I} a_j \frac{\partial E_j}{\partial x_k}, \qquad k = 1, \ldots, K, \tag{21}$$

where g_k is the kth entry in the gradient vector. Thus, the adjoint code would consist of a sequence of assignment statements for a_i, with i decreasing, followed by those for g_k.

The expressions for the derivatives $\partial E_j / \partial v_i$ and $\partial E_j / \partial x_k$ would be relatively simple, but still the compiler must be able to construct them. To do so it might simplify each assignment statement into a sequence of simpler assignments, much like what is done in compiling an executable code. If Lagrange multipliers are introduced for each substep, only derivatives of elementary expressions would be needed, and these could be looked up in a table. And if the adjoint code is to be able to evaluate these expressions, some values input to or computed by the model code must be passed to the adjoint code. As this would require some modification of the model code, the output of the compiler should be thought of as a new code that computes both the function and its gradient.

The following example illustrates how the compiler should work. Suppose the input code evaluates a function of two variables:

$$J = \left(x_1{}^2 + x_2{}^2\right)^{\frac{1}{2}} \exp\left[-\left(x_1{}^2 + x_2{}^2\right)\right]. \tag{22}$$

(More likely, something like this would appear as a single assignment statement in an actual code, in which case the example code would represent an expansion into a sequence of simpler assignment statements.) The code might be written in FORTRAN, with **X1** and **X2** representing x_1 and x_2:

(23)

```
V1 = X1**2
V2 = X2**2
V3 = V1 + V2
V4 = SQRT(V3)
V5 = EXP(-V3)
V6 = V4 * V5 ,
```

where **V6** is the computed value of the function J. Then the output of the compiler might be the code:

(24)
```
V1 = X1**2
V2 = X2**2
V3 = V1 + V2
V4 = SQRT(V3)
V5 = EXP(-V3)
V6 = V4 * V5
A6 = 1
A5 = A6 * V4
A4 = A6 * V5
A3 = A4 * (0.5/V4) + A5 * (-V5)
A2 = A3
A1 = A3
G1 = A1 * (2*X1)
G2 = A2 * (2*X2),
```

where **V6**, **G1**, and **G2** represent the values returned for J, $\partial J/\partial x_1$, and $\partial J/\partial x_2$, respectively, given values for x_1 and x_2.

It is interesting to compare this approach with that of JAKE. Speelpenning **[30]** never talks about Lagrange multipliers. He thinks of the assignment statements as defining a series of transformations the product of which maps the input variables into the space of all variables that are used: input variables, intermediate variables introduced in the process of evaluating the function that is being differentiated, and finally the function itself. To produce the gradient vector, the product of the Jacobian matrices for these transformations must be multiplied by a vector. Speelpenning noted that, since less work is required to multiply a matrix by a vector than by a second matrix, the final product could be computed more efficiently by first multiplying the vector and the adjacent matrix factor, then the resulting vector with the next matrix factor, and so on until the gradient is evaluated. This process amounts to exactly the same thing as eliminating Lagrange multipliers.

The code produced by JAKE does not look like that in the above example. Instead, lines of code would be inserted after the evaluation of each of the variables **V1** through **V6**, which would save the derivatives of these variables on a stack. Then, at the end of the code, a subroutine would be called to take these coefficients off the stack, in effect, evaluating the Lagrange multipliers in the same order as in the example code, and ending up with the gradient.

The idea of using a stack is quite clever. First, it solves the problem of what to do about flow-control statements in the input code. They are simply identified and then ignored; whatever the execution path through the code, the stack will contain the necessary information for evaluating the gradient, and this information will appear in exactly the order in which it will be needed. Second, it also solves the problem of how to save the coefficients (**V4**, **V5**, **0.5/V4**, −**V5**, **2∗X1**, **2∗X2**) that are needed in computing the multipliers. On the other hand, a stack might not be efficient when the function-evaluation code involves the numerical solution of differential equations. Consider, for example, the case where the differential equations are linear. All of the coefficients of the model code, which are already available for solving the differential equations, would have to be saved a second time on the stack, possibly with some signs reversed. Even if the equations are nonlinear, it should not be necessary to save the derivatives of *all* expressions appearing in assignment statements; for the above example, **V4** might be saved and **0.5/V4** computed. And if the model is vectorizable, the adjoint model should also be vectorizable; but a stack is not suited to vectorization. A compiler is need, which produces code containing DO loops and a more efficient use of memory.

Some other useful features are also missing from JAKE. The most important improvement would be to allow the input code to call subroutines. A subroutine with p input variables and q output variables can be thought of as vector-valued function-evaluating code, and its adjoint would have q inputs and p outputs. In the example above the adjoint portion of the code started out with the single input **A6** being given a value of unity; for each subroutine in the model code, there would be an adjoint subroutine in the adjoint code having input variables passed to it when it is called. The input to the compiler could be a vector-valued function, and by allowing each of the inputs in turn to have a value of unity while the others all are zero, the output code would compute the Jacobian matrix for the subroutine. And if a gradient-evaluating code constructed by the compiler is given to the compiler as input, the output would be a Hessian evaluating code. To do this, COMMON and EQUIVALENCE statements should be supported. And for large problems the compiler should be able to recognize when, due to insufficient fast memory, information is saved on disc to be recalled for later use. These are only technical problems that should not be too difficult to overcome. Once such a compiler is written, a most valuable tool will be available that should expedite the solution of many interesting problems.

References

1. A. Arakawa, *Computational design for long-term numerical integration of the equations of fluid motion: Two-dimensional incompressible flow. Part* 1, J. Comp. Phys. **1** (1966), 119–143.
2. A. F. Bennett and P. C. McIntosh, *Open ocean modeling as an inverse problem: Tidal theory*, J. Phys. Oceanogr. **12** (1982), 1004–1018.
3. P. Bergthorsson and B. R. Döös, *Numerical weather map analysing*, Tellus **7** (1955), 329–340.
4. F. E. Bretherton, R. E. Davis, and C. B. Fandry, *A technique for objective analysis and design of oceanographic experiments applied to MODE-73*, Deep Sea Research **23** (1976), 559–582.
5. K. Bryan, *Accelerating convergence to equilibrium of ocean-climate models*, Journal of Physical Oceanography **14** (1984), 666–673.
6. J. Carrera and S. P. Neuman, *Estimation of aquifer parameters under transient and steady conditions: 2. Uniqueness, stability, and solution algorithms*, Water Resources Research **22** (1986), 211–227.
7. J. G. Charney, R. Fjörtoft, and J. von Neumann, *Numerical integration of the barotropic vorticity equation*, Tellus **2** (1950), 237–254.
8. P. Courtier and O. Talagrand, *Variational assimilation of meteorological observations with the adjoint vorticity equation. I: Numerical results*, Q. J. R. Meteorol. Soc. **113** (1987), 1329–1347.
9. G. P. Cressman, *An operational objective analysis system*, Mon. Wea. Rev. **87** (1959), 367–374.
10. N. A. Draper and H. Smith, *Applied Regression Analysis*, Wiley, New York, 1966, 407 pp.
11. A. Eliassen, *A provisional report on calculation of spatial covariance and autocorrelation of the pressure field*, Rapp 5, Vidensk.-Akad. Inst. for Vaer-Og Klmaforsk., Oslo. (Reprinted in *Dynamic Meteorology: Data Assimilation Methods*, Springer-Verlag, New York, 1954, pp. 319–330.)
12. L. S. Gandin, *Objective Analysis of Meteorological Fields*, translated from the Russian, Israel Program for Scientific Translations, Jerusalem, 1963, 242 pp.
13. A. Gelb, *Applied Optimal Estimation*, MIT Press, Cambridge, MA, 1974, 374 pp.
14. M. Ghil, S. Cohn, J. Tavantzis, K. Bube, and E. Isaacson, *Applications of estimation theory to numerical weather prediction*, Dynamic Meteorology: Data Assimilation Methods, Springer-Verlag, New York, 1981, pp. 139–224.
15. P. E. Gill, W. Murray, and M. H. Wright, *Practical Optimization*, Academic Press, Orlando, FL, 1981, 401 pp.
16. R. Glowinski and J. Periaux, *Finite element, least squares and domain decomposition methods for the numerical solution of nonlinear problems in fluid dynamics*, Numerical Methods in Fluid Dynamics, F. Brezzi, ed., Springer-Verlag, Berlin, 1983, pp. 1–114.
17. W. Hackbusch, *On the fast solving of parabolic boundary control problems*, SIAM J. Control and Optimization **17** (1979), 231–244.
18. F. X. Le Dimet and O. Talagrand, *Variational algorithms for analysis and assimilation of meteorological data: Theoretical aspects.*, Tellus **A38** (1986), 97–110.

19. J. M. Lewis and J. C. Derber, *The use of adjoint equations to solve a variational adjustment problem with advective constraints*. Tellus **37** (1985), 309–322.

20. R. B. Long and W. C. Thacker, *Data assimilation into a numerical equatorial ocean model. Part 1: The model and the assimilation algorithm*, Dynamics of Atmospheres and Oceans (1989), 379–412.

21. A. C. Lorenc, *A global three-dimensional multivariate statistical interpolation scheme*, Mon. Weather. Rev. **109** (1981), 701–721.

22. ——, *Optimal non-linear objective analysis*, Q. J. R. Meteorol. Soc. **114** (1988), 205–240.

23. C. Potier and C. Verken, *Numerical variational methods for data filtering and interpolation*, J. Atmos. Oceanic Technology **2** (1985), 528–538.

24. C. Provost and R. Salmon, *A variational method for inverting hydrographic data*, J. Mar. Res. **44** (1986), 1–34.

25. R. H. Rand, *Computer Algebra in Applied Mathematics: An Introduction to MACSYMA*, Pitman, Boston, 1984, 181 pp.

26. Y. Sasaki, *An objective analysis based on the variational method*, Journ. Met. Soc. Japan **36** (1958), 77–85.

27. ——, *Proposed inclusion of time variation terms, observational and theoretical, in numerical variational objective analysis*, Journ. Met. Soc. Japan **47** (1969), 115–124.

28. ——, *Some basic formalisms in numerical variational analysis*, Monthly Weather Review **98** (1970), 875–883.

29. J. Schröter and C. Wunsch, *Solution of nonlinear finite difference ocean models by optimization methods with sensitivity and observational strategy analysis*, J. Phys. Oceanogr. **16** (1986), 1855–1874.

30. B. Speelpenning, *Compiling fast partial derivatives of functions given by algorithms*, 1980, Ph.D. Thesis, Department of Computer Science, University of Illinois at Urbana-Champaign, UIUCDCS-R-80-1002, 74 pp.

31. G. Strang, *Introduction to Applied Mathematics*, Wellesley-Cambridge Press, Wellesley, MA, 1986, 758 pp.

32. O. Talagrand and P. Courtier, *Variational assimilation of meteorological observations with the adjoint vorticity equation. I: Theory.*, Q. J. R. Meteorol. Soc. **113** (1987), 1311–1328.

33. W. C. Thacker, *Relationships between statistical and deterministic methods of data assimilation*, Variational Methods in Geosciences, Y. K. Sasaki, ed., Elsevier, New York, 1986, pp. 173–179.

34. ——, *Three lectures on fitting numerical models to observations*, GKSS Forschungszentrum Geesthacht GmbH, Geesthacht, Federal Republic of Germany, External Report GKSS87/E/65, 1987, 64 pp.

35. ——, *Fitting models to inadequate data by enforcing spatial and temporal smoothness*, Journal of Geophysical Research **93** (1988), 10655–10665.

36. W. C. Thacker and R. B. Long, *Fitting dynamics to data*, Journal of Geophysical Research **93** (1988), 1227–1240.

37. G. Wahba and J. Wendelberger, *Some new mathematical methods for variational objective analysis using splines and cross validation*, Monthly Weather Review **108** (1980), 1122–1143.

38. S. Wolfram, *Mathematica*, Addison-Wesley, New York, 1988, 749 pp.

39. C. Wunsch, *Transient tracers as a problem in control theory*, J. Geophys. Res. **93** (1988), 8099–8110.

Atlantic Oceanographic and Meteorological Laboratory, 4301 Rickenbacker Causeway, Miami FL 33149 USA

Lectures in Applied Mathematics
Volume **26** (1990)

Newton-like Methods for Underdetermined Systems

HOMER F. WALKER

Abstract. Underdetermined systems of nonlinear equations arise in a variety of contexts. Here we consider analogues for underdetermined systems of Newton's method and quasi-Newton methods which have proved very successful for systems with an equal number of equations and unknowns.

1. Introduction. The main purpose of this note is to survey recent work in **[5]** and **[34]** on Newton-like methods for underdetermined nonlinear systems. For the benefit of the nonspecialist, we also provide a little background on methods for systems with an equal number of equations and unknowns, which we refer to as *well-determined* systems even though they may not be well-posed in the classical sense of having unique solutions which depend continuously on problem parameters. Many technical details are not given here; for these, the reader should consult **[5]**, **[34]**, and other referenced works.

Our notational conventions, which are not strictly observed but are intended to serve as helpful guidelines for remembering what is what, are the following: unless otherwise indicated, lowercase letters denote vectors and scalars, and capital letters denote matrices and operators. Boldface uppercase letters denote vector spaces, subspaces, and affine subspaces. For positive integers p and q, $\mathbf{R}^p$ denotes p-dimensional real Euclidean space and $\mathbf{R}^{p\times q}$ denotes the space of real $p \times q$ matrices. We refer particularly to $\mathbf{R}^n$ and $\mathbf{R}^{\bar{n}}$ for $\bar{n} \geq n$, and for convenience, we

1980 *Mathematics Subject Classification* (1985 *Revision*). Primary 65H10.
This work was supported by U. S. Department of Energy Grant DE-FG02-86ER25018, Department of Defense/Army Grant DAAL03-88-K, and NSF Grant DMS-0088995, all with Utah State University.

set $\bar{n} = n + m$ for $m \geq 0$. Vectors with bars are in $\mathbf{R}^{\bar{n}}$; without bars, they are in $\mathbf{R}^n$ or $\mathbf{R}^m$ unless otherwise indicated. We often partition vectors, e.g., we write $\bar{x} \in \mathbf{R}^{\bar{n}}$ as $\bar{x} = (x, \lambda)$ for $x \in \mathbf{R}^n$ and $\lambda \in \mathbf{R}^m$, and we do not distinguish between (x, λ) and $\binom{x}{\lambda}$. We also often partition matrices, e.g., we write $B \in \mathbf{R}^{n \times \bar{n}}$ as $B = [\mathscr{B}, \mathscr{C}]$ for $\mathscr{B} \in \mathbf{R}^{n \times n}$ and $\mathscr{C} \in \mathbf{R}^{n \times m}$. The dimensions of vector and matrix partitions are made clear in each case, usually by the context. We use "Jacobian" to mean "Jacobian matrix," and we denote the full Jacobian of a function F by F'. If F is a function of $\bar{x} = (x, \lambda) \in \mathbf{R}^{\bar{n}}$, then we denote partial Jacobians $\partial F / \partial x$ by F_x, $\partial F / \partial \lambda$ by F_λ, etc. We assume throughout the following that there are given but unspecified vector norms on $\mathbf{R}^n$, $\mathbf{R}^m$, and $\mathbf{R}^{\bar{n}}$, together with their associated induced matrix norms, and we denote all of these norms by $|\cdot|$. Similarly, we assume there is a given but unspecified matrix norm on $\mathbf{R}^{n \times \bar{n}}$ associated with a matrix inner product, and we denote this norm by $\|\cdot\|$. A projection onto a subspace or affine subspace which is orthogonal with respect to $\|\cdot\|$ is denoted by P with the subspace or affine subspace appearing as a subscript.

The problem of interest is that of solving a (possibly) underdetermined system of nonlinear equations, which we write as

PROBLEM 1.1. Given $F : \mathbf{R}^{\bar{n}} \to \mathbf{R}^n$ with $\bar{n} \geq n$, find $\bar{x}_* \in \mathbf{R}^{\bar{n}}$ such that

$$F(\bar{x}_*) = 0. \tag{1.1}$$

We are primarily interested in the strictly underdetermined case $\bar{n} > n$, but we explicitly allow the well-determined case $\bar{n} = n$ as well.

Instances of Problem 1.1 arise in a variety of contexts, sometimes naturally and sometimes as artifacts of mathematical methods or problem formulations. A particularly important source is parameter-dependent problems. In these, $\bar{x}$ is typically regarded as $\bar{x} = (x, \lambda)$, where $x \in \mathbf{R}^n$ is an "independent" variable and $\lambda \in \mathbf{R}^m$ is a parameter. For example, λ is the fluid Reynolds number in the driven cavity problem described by Glowinski, Keller, and Reinhart **[20]**, and it is a load parameter in the study of snap-through buckling of a shallow elastic arch outlined by Kamat, Watson, and Vanden Brink **[24]**. As in **[20]** and **[24]**, one is often interested in tracing a curve of solutions of (1.1) over some range of parameter values using homotopy or continuation methods. These methods are often of predictor-corrector form, and one must solve a problem of the form (1.1) in the corrector steps. These methods may also be applied to solving difficult nonparametric problems by embedding them in parametric families and following solution curves from easy-to-solve problems to the problems of interest. See the survey of

Allgower and Georg **[2]** as a general reference on continuation and homotopy methods.

Examples of Problem 1.1 also arise in constrained problems, in which (1.1) determines a constraint surface, and in the solution of time-dependent nonlinear differential equations by implicit methods. In the latter, $\bar{n} = n + 1$ and $\bar{x} = (x, \lambda)$, where $\lambda \in \mathbf{R}^1$ is the "time" variable. Here, one actually considers a sequence of problems of the form (1.1), one for each time step. Since λ is fixed in each of these problems, they are, strictly speaking, a sequence of well-determined problems. However, they are usually similar in essential ways; in particular, it is typically necessary for efficiency to carry over Jacobian and other information from one problem to the next. Consequently, it is usually more appropriate to view them in the context of parameter-dependent underdetermined systems.

In the following, we consider iterative methods for solving Problem 1.1 which can be regarded as extensions of Newton's method and quasi-Newton methods which have proved very successful for solving well-determined systems. In Section 2 we discuss extensions of Newton's method; these use exact Jacobians at each iteration. In Section 3 we consider extensions of quasi-Newton methods in which approximate Jacobians are maintained by least-change secant updating. In Section 4 we conclude with some simple examples which demonstrate interesting properties of these updating methods.

2. Extensions of Newton's method. We first consider Problem 1.1 in the well-determined case, which we write as

PROBLEM 2.1. Given $F : \mathbf{R}^n \to \mathbf{R}^n$, find x_* such that $F(x_*) = 0$.

For a given approximate solution x, the classical *Newton's method* for determining a next approximate solution x_+ is

$$x_+ = x - F'(x)^{-1}F(x). \tag{2.1}$$

It is easy to show that under mild assumptions there is a constant β such that if x is sufficiently near x_*, then one has $|x_+ - x_*| \leq \beta|x - x_*|^2$, which implies that sequences of iterates produced by Newton's method enjoy *q-quadratic* local convergence; see, e.g., **[14]** or **[26]**. (See those references also for precise definitions of various types of convergence referred to here.) This very fast local convergence is the major strength of Newton's method and is the principal reason why it has been viewed as a model method for the derivation of many other successful methods.

A possible difficulty is that if the starting point is not near a solution, then a sequence of Newton iterates may not converge. Consequently, in order to have a trustworthy practical algorithm, one must augment the basic Newton iteration (2.1) with safeguards which modify the Newton step if necessary to ensure progress toward a solution. A discussion of such safeguards is beyond the scope of this note (see Dennis and Schnabel **[14]** for a full treatment), and we consider only (2.1) and other basic iterations here. There are two other potential difficulties associated with Newton's method which are likely to be especially serious if n is large. The first is that the up to n^2 scalar functions which determine $F'(\bar{x})$ must be evaluated at each iteration, and this may be undesirably expensive or even practically impossible. The second is that solving an order-n linear system for the step $-F'(x)^{-1}F(x)$ requires up to $O(n^3)$ arithmetic operations per iteration.

For general $\bar{n} \geq n$, an extension of the Newton iteration (2.1) requires

$$F(\bar{x}) + F'(\bar{x})(\bar{x}_+ - \bar{x}) = 0. \tag{2.2}$$

If $\bar{n} > n$, then (2.2) does not determine a unique $\bar{x}_+$, and particular algorithms are obtained by specifying particular choices of $\bar{x}_+$ satisfying (2.2). We consider two widely applicable algorithms which are used in homotopy or continuation methods and which are treated in **[34]**.

The first algorithm is the *normal flow algorithm*: Given an approximate solution $\bar{x}$, determine $\bar{x}_+$ by

$$\bar{x}_+ = \bar{x} - F'(\bar{x})^+F(\bar{x}). \tag{2.3}$$

In (2.3), the superscript "+" indicates pseudo-inverse. That is, the step $-F'(\bar{x})^+F(\bar{x})$ is the solution of $F'(\bar{x})\bar{s} = -F(\bar{x})$ having minimal norm, i.e., the solution which is orthogonal to the null-space of $F'(\bar{x})$, i.e., the solution which is in the span of the columns of $F'(\bar{x})^{\mathrm{T}}$. The algorithm takes its name from the $\bar{n} = n + 1$ case, in which the iteration steps are asymptotically normal to the *Davidenko flow*; see **[11]** and **[35]**. For any $\bar{n}$, it is clear that the step $-F'(\bar{x})^+F(\bar{x})$ is normal to the manifold $\{\bar{y} \in \mathbf{R}^{\bar{n}} : F(\bar{y}) = F(\bar{x})\}$. The iteration (2.3) is considered in the $\bar{n} = n + 1$ case by Georg **[18]**, who credits Haselgrove **[23]** with originating it. Ben-Israel **[4]** treats a related *chord method*, viz., an iteration $\bar{x}_+ = \bar{x} - F'(\bar{x}_0)^+F(\bar{x})$, where $\bar{x}_0$ is the initial approximate Jacobian.

The second algorithm is the *augmented Jacobian algorithm*. We write this algorithm as follows for a specified $V \in \mathbf{R}^{m\times\bar{n}}$: given an approximate solution $\bar{x}$, determine $\bar{x}_+$ by

$$\bar{x}_+ = \bar{x} + \bar{s}, \text{ where } F'(\bar{x})\bar{s} = -F(\bar{x}) \text{ and } V\bar{s} = 0. \tag{2.4}$$

There are various practically effective ways of choosing V. For example, in the $\bar{n} = n+1$ case Georg [**18**] and Rheinboldt [**30**] take V to be the transpose of a well-chosen unit basis vector in $\mathbf{R}^{\bar{n}}$, while Watson, Billups, and Morgan [**35**] take V to be the transpose of an approximate tangent vector to the solution curve.

A local convergence analysis for the normal flow and augmented Jacobian algorithms is given in [**34**] which extends the usual local convergence analysis for Newton's method, and we outline this analysis here. This analysis is a *Kantorovich-type* analysis (see, e.g., [**26**]) in that no assumptions are made about existence of or nearness to solutions of (1.1). Such an analysis is necessary for $\bar{n} > n$, since in general solutions are not isolated and therefore one cannot hope to single out a priori particular solutions to which iteration sequences might converge. This analysis is of course valid for $\bar{n} = n$ and so provides a Kantorovich-type analysis in the well-determined case. We believe that this analysis is simpler and more straightforward than the usual Kantorovich analysis for well-determined systems, which uses the "method of majorization" (see [**26**]).

We first give the analysis for the normal flow iteration. An appropriate hypothesis on F is the following, which is analogous to the standard hypothesis made in the usual local convergence analysis for Newton's method:

NORMAL FLOW HYPOTHESIS. *F is differentiable and F' is of full rank n in an open convex set Ω, and the following hold:*

(i) *there exist $\gamma \geq 0$ and $p \in (0,1]$ such that $|F'(\bar{y}) - F'(\bar{x})| \leq \gamma|\bar{y} - \bar{x}|^p$ for all $\bar{x}, \bar{y} \in \Omega$;*
(ii) *there is a constant μ for which $|F'(\bar{x})^+| \leq \mu$ for all $\bar{x} \in \Omega$.*

For our local convergence theorems, we also define for $\eta > 0$

$$\Omega_\eta = \{\bar{x} \in \Omega : |\bar{y} - \bar{x}| < \eta \Rightarrow \bar{y} \in \Omega\}. \tag{2.5}$$

THEOREM. *Let F satisfy the Normal Flow Hypothesis and suppose Ω_η is given by* (2.5) *for some $\eta > 0$. Then there is an $\varepsilon > 0$ depending only on γ, p, μ, and η such that if $\bar{x}_0 \in \Omega_\eta$ and $|F(\bar{x}_0)| < \varepsilon$, then the iterates $\{\bar{x}_k\}_{k=0,1,\ldots}$ determined by the normal flow algorithm* (2.3) *are well-defined and converge to a point $\bar{x}_* \in \Omega$ such that $F(\bar{x}_*) = 0$. Furthermore, there is a constant β for which*

$$|\bar{x}_{k+1} - \bar{x}_*| \leq \beta|\bar{x}_k - \bar{x}_*|^{1+p}, \qquad k = 0, 1, \ldots. \tag{2.6}$$

PROOF SKETCH. If $\bar{x} \in \Omega$ and $\bar{s} = -F'(\bar{x})^+ F(\bar{x})$, then $|\bar{s}| \le \mu |F(\bar{x})|$. If also $\bar{x}_+ = \bar{x} + \bar{s} \in \Omega$, then a standard argument gives $|F(\bar{x}_+)| = |F(\bar{x}_+) - F(\bar{x}) - F'(\bar{x})\bar{s}| \le \frac{\gamma}{1+p}|\bar{s}|^{1+p}$. If $\bar{s}_+ = -F'(\bar{x}_+)^+ F(\bar{x}_+)$, then these inequalities give $|\bar{s}_+| \le \frac{\gamma\mu}{1+p}|\bar{s}|^{1+p}$. It follows from the first and third inequalities by an easy induction that there is an $\varepsilon > 0$ depending only on γ, p, μ, and η such that if $\bar{x}_0 \in \Omega_\eta$ and $|F(\bar{x}_0)| < \varepsilon$, then the normal flow iterates are well-defined and constitute a Cauchy sequence with limit $x_* \in \Omega$. The second inequality implies $F(\bar{x}_*) = 0$. A closer look reveals that (2.6) holds as well; see **[34]** for details.

A slight variation of the Normal Flow Hypothesis is appropriate for the augmented Jacobian algorithm.

AUGMENTED JACOBIAN HYPOTHESIS. *F is differentiable and $\begin{bmatrix} F'(\bar{x}) \\ V \end{bmatrix}$ is nonsingular in an open convex set Ω, and the following hold:*

(i) *there exist $\gamma \ge 0$ and $p \in (0,1]$ such that $|F'(\bar{y}) - F'(\bar{x})| \le \gamma|\bar{y} - \bar{x}|^p$ for all $\bar{x}$, $\bar{y} \in \Omega$;*

(ii) *there is a constant $\bar{\mu}$ for which $\left|\begin{bmatrix} F'(\bar{x}) \\ V \end{bmatrix}^{-1}\right| \le \bar{\mu}$ for all $\bar{x} \in \Omega$.*

Our local convergence theorem for the augmented Jacobian algorithm is the following:

THEOREM. *Let F and V satisfy the Augmented Jacobian Hypothesis and suppose Ω_η is given by* (2.5) *for some $\eta > 0$. Then there is an $\varepsilon > 0$ depending only on γ, p, $\bar{\mu}$, and η such that if $\bar{x}_0 \in \Omega_\eta$ and $|F(\bar{x}_0)| < \varepsilon$, then the iterates $\{\bar{x}_k\}_{k=0,1,\ldots}$ determined by the augmented Jacobian algorithm are well-defined and converge to a point $\bar{x}_* \in \Omega$ such that $F(\bar{x}_*) = 0$. Furthermore, there is a constant β for which* (2.6) *holds.*

PROOF SKETCH. For given $\bar{x}_0 \in \Omega$ we define

$$\overline{F}(\bar{x}) \equiv \begin{pmatrix} F(\bar{x}) \\ V(\bar{x} - \bar{x}_0) \end{pmatrix}$$

for $\bar{x} \in \Omega$. It is easy to see that starting at $\bar{x}_0$, the augmented Jacobian algorithm applied to F is equivalent to the normal flow algorithm applied to $\overline{F}$ (which is just Newton's method). Furthermore, $\overline{F}$ satisfies the Normal Flow Hypothesis with constants which are independent of $\bar{x}_0$ since $\overline{F}'$ is independent of $\bar{x}_0$. The theorem then follows from the preceding theorem applied to $\overline{F}$.

3. Extensions of least-change secant update methods. We have noted that there are two potential difficulties associated with Newton's method for the well-determined Problem 2.1 which are likely to be especially serious for large n, viz., the need at each iteration to evaluate $F'(x)$, which requires the evaluation of up to n^2 scalar functions, and the need to solve for the step $-F'(x)^{-1}F(x)$, which requires up to $O(n^3)$ arithmetic operations. It is clear that there are similar difficulties associated with the normal flow and augmented Jacobian algorithms. In this section we outline methods which offer hope of avoiding these difficulties while retaining much of the effectiveness of the algorithms on which they are modeled.

3.1. *A review.* We begin by reviewing appropriate methods for the well-determined Problem 2.1. For a given approximate solution x, a *quasi-Newton method* for determining a next approximate solution x_+ is a method of the general form

$$x_+ = x - B^{-1}F(x). \tag{3.1}$$

In (3.1), B is clearly meant to be regarded as an approximation of $F'(x)$, and particular quasi-Newton methods are determined by specifying ways of obtaining the next approximate Jacobian B_+ for the next iteration. Our interest here is in determining B_+ as an *update* of B in ways which are simple enough to avoid much of the Jacobian evaluations and arithmetic of Newton's method, yet effective enough to yield methods with adequately fast local convergence.

To obtain guidelines for such updating, we recall the familiar *secant method* for the case $n = 1$: If x and x_- are current and past approximate solutions, then determine x_+ by

$$x_+ = x - \left[\frac{F(x) - F(x_-)}{x - x_-}\right]^{-1} F(x).$$

The secant method avoids the evaluation of $F'(x)$ and, under hypotheses that give local q-quadratic convergence for Newton's method, enjoys local *q-superlinear* convergence, i.e., if $\{x_k\}_{k=0,1,\ldots}$ is a sequence of secant iterates which converges to a solution x_*, then

$$|x_{k+1} - x_*| \le \alpha_k |x_k - x_*|, \qquad k = 0, 1, \ldots, \tag{3.2}$$

for a sequence $\{\alpha_k\}_{k=0,1,\ldots}$ which converges to zero. (Actually, if F is twice continuously differentiable, then one has the delightful result that

$$\lim_{k\to\infty} \frac{|x_{k+1} - x_*|}{|x_k - x_*|^{(\sqrt{5}+1)/2}} = \left|\frac{F''(x_*)}{2F'(x_*)}\right|^{(\sqrt{5}-1)/2};$$

see Ostrowski [27].)

The secant method suggests requiring that B_+ satisfy the *secant equation*

$$B_+s = y, \text{ where } s = x_+ - x, \text{ and } y = F(x_+) - F(x), \tag{3.3}$$

i.e., that B_+ be a *secant update.* Clearly, (3.3) determines B_+ uniquely only if $n = 1$, in which case the secant method results. If $n > 1$, then an accepted principle for specifying B_+ is the following: *Make the least possible change in B* to obtain B_+ satisfying (3.3), together with any auxiliary conditions. The auxiliary conditions referred to here are typically structural conditions such as symmetry, sparsity, etc., that reflect structural properties of F'. The rationale underlying this principle is that B presumably has useful information about F', and this should be corrupted as little as possible in incorporating the current information expressed in (3.3). The phrase "least possible change" can be interpreted in several ways. The earliest secant updates were of the lowest possible *rank* of all updates satisfying (3.3) and perhaps auxiliary conditions such as symmetry; however, this minimum-rank interpretation cannot be used to obtain full-rank updates such as the sparsity-preserving updates developed in the 1970s.

An interpretation by which all successful updates can be derived is the following: make the least possible change in B *as measured by a suitable matrix norm* to obtain B_+ satisfying (3.3) and any auxiliary conditions. We illustrate this minimum-norm interpretation by deriving the simplest, most generally applicable update—the (*first*) *Broyden update* [7]. We take the norm on $\mathbf{R}^{n\times n}$ to be the Frobenius norm $\|\cdot\|_{\mathrm{Fr}}$, given by $\|M\|_{\mathrm{Fr}} \equiv \sqrt{\operatorname{trace}\{MM^{\mathrm{T}}\}}$ for $M \in \mathbf{R}^{p\times q}$ for any p and q, and require that B_+ satisfy both (3.3) and

$$\|B_+ - B\|_{\mathrm{Fr}} = \min_{\overline{B}s=y} \|\overline{B} - B\|_{\mathrm{Fr}}. \tag{3.4}$$

It is easy to see that (3.3) and (3.4) uniquely determine

$$B_+ = B + \frac{(y - Bs)s^{\mathrm{T}}}{s^{\mathrm{T}}s}, \tag{3.5}$$

which is the (first) Broyden update. The iteration (3.1) with the next approximate Jacobian B_+ given by (3.5) is *Broyden's method,* which is the most successful generally applicable secant-updating alternative to Newton's method. Under the same hypotheses that yield local q-quadratic convergence of Newton's method, Broyden's method enjoys local q-superlinear convergence, i.e., an inequality (3.2) holds for a solution x_* and $\{x_k\}_{k=0,1,\dots}$ generated by Broyden's method provided x_0

and the initial approximate Jacobian B_0 are sufficiently close to x_* and $F'(x_*)$, respectively. Also, it is clearly unnecessary to evaluate $F'(x)$ at each iteration, although it is considered advisable in practice to take B_0 equal to either $F'(x_0)$ or a finite-difference approximation of it. Since an order-n linear system must be solved at each iteration, it may not be clear that Broyden's method can be implemented with fewer than $O(n^3)$ arithmetic operations per iteration. However, because B_+ is a simple rank-one update of B, this can be done in at least two ways: (1) using the Sherman–Morrison formula [**26**, p. 50], and (2) updating the QR factorization of B [**19**]. Both ways require $O(n^2)$ arithmetic operations per iteration. The second way requires a little more arithmetic than the first, but it is somewhat more stable and is usually preferred.

The minimum-norm change interpretation of the least-change principle has been formalized in the notion of a *least-change secant update* by Dennis and Schnabel [**13**], which we now outline. Suppose one is given the following ingredients:

(i) $B \in \mathbf{R}^{n\times n}$ and $s, y \in \mathbf{R}^n$;
(ii) an inner product norm $\|\cdot\|$ on $\mathbf{R}^{n\times n}$;
(iii) an affine subspace $\mathbf{A} \subseteq \mathbf{R}^{n\times n}$ reflecting some structure of F'.

Then set $\mathbf{Q}(y,s) = \{M \in \mathbf{R}^{n\times n} : Ms = y\}$, the affine subspace of matrices in $\mathbf{R}^{n\times n}$ which satisfy the secant equation determined by s and y, and define the least-change secant update of B in $\mathbf{A}$ with respect to s, y, and $\|\cdot\|$ to be

$$B_+ = P_{\mathbf{A}\cap\mathbf{Q}(y,s)}(B). \tag{3.6}$$

It may happen that $\mathbf{A} \cap \mathbf{Q}(y,s) = \emptyset$, in which case take B_+ to be the orthogonal projection of B onto the affine subspace of elements of $\mathbf{A}$ which are closest to $\mathbf{Q}(y,s)$ in the norm $\|\cdot\|$ (cf. Dennis and Walker [**15**]). However, (3.6) conveys the sense of the definition and nearly always suffices, and so we use this and other analogous definitions below without further clarification.

Particular least-change secant updates are obtained by making particular choices of $\mathbf{A}$ and $\|\cdot\|$. For perspective, we review the choices that lead to various known updates, omitting the formulas for the more complicated or less well-known updates. For these the reader is referred to [**13**] or to specific references given below.

We first take $\|\cdot\| = \|\cdot\|_{\mathrm{Fr}}$. Then as seen above, choosing $\mathbf{A} = \mathbf{R}^{n\times n}$ gives the (first) Broyden update (3.5). Taking $\mathbf{A}$ to be the subspace of symmetric matrices gives the *Powell symmetric Broyden* (*PSB*) *update*

[29] for symmetric B,

$$B_+ = B + \frac{(y - Bs)s^T + s(y - Bs)^T}{s^T s} - \frac{s^T(y - Bs)}{s^T s}\frac{ss^T}{s^T s}. \tag{3.7}$$

Taking $\mathbf{A}$ to be the subspace of matrices having a particular pattern of sparsity gives the *sparse Broyden* or *Schubert update* ([10], [31], [13]), and taking $\mathbf{A}$ to be the subspace of matrices having symmetry as well as a particular pattern of sparsity gives the *sparse symmetric update* of Marwil [25] and Toint [33] (see also [13]).

We now take the norm to be a *weighted Frobenius norm* $\|\cdot\|_W$, defined as follows: Assuming $y^T s > 0$, let $W \in \mathbf{R}^{n\times n}$ be a positive-definite symmetric matrix such that $Ws = y$ and define

$$\|M\|_W \equiv \sqrt{\text{trace}\,\{W^{-1}MW^{-1}M^T\}}$$

for $M \in \mathbf{R}^{n\times n}$. The idea is that such a W, by taking s to $y \approx F'(x)s$, represents an ideal but unknown scaling determined by $F'(x)$ when $F'(x)$ is positive-definite symmetric, as is likely to be the case, e.g., in a minimization problem. The reader may be troubled by the fact that if $y^T s > 0$, then there are many W such that $Ws = y$. However, for the most immediate choices of $\mathbf{A}$, the resulting updates do not depend on W. If $\mathbf{A} = \mathbf{R}^{n\times n}$, then the resulting update is a rank-one update of Pearson [28], which is a weighted-norm analogue of the (first) Broyden update. Taking $\mathbf{A}$ to be the subspace of symmetric matrices gives the well-known *Davidon–Fletcher–Powell* (*DFP*) *update* ([12], [17]), given for symmetric B by

$$B_+ = B + \frac{(y - Bs)y^T + y(y - Bs)^T}{y^T s} - \frac{s^T(y - Bs)}{y^T s}\frac{yy^T}{y^T s}. \tag{3.8}$$

This update has the very desirable property that if B is positive-definite symmetric and $y^T s > 0$, then B_+ is also positive-definite symmetric.

There are also least-change *inverse* secant updates obtained by applying the minimum-norm interpretation of the least-change principle to the *inverse* of the matrix being updated. That is, given the ingredients (i)–(iii) above with invertible B, define the least-change inverse secant update of B^{-1} in $\mathbf{A}$ with respect to s, y, and $\|\cdot\|$ by

$$(B_+)^{-1} = P_{\mathbf{A}\cap\mathbf{Q}(s,y)}(B^{-1}). \tag{3.9}$$

As with the direct least-change secant updates above, particular least-change inverse secant updates are obtained from (3.9) by making particular choices of $\mathbf{A}$ and $\|\cdot\|$. We give some of these here, again omitting the less well-known formulas. The formulas given express the update in terms of B rather than B^{-1}, because that is usually preferred for implementation. With $\|\cdot\| = \|\cdot\|_{\mathrm{Fr}}$ and $\mathbf{A} = \mathbf{R}^{n\times n}$, one obtains the *second Broyden update*, given by

$$B_+ = B + \frac{(y - Bs)y^{\mathrm{T}}B}{y^{\mathrm{T}}Bs}. \tag{3.10}$$

With this norm and with $\mathbf{A}$ the subspace of symmetric matrices, one has the *Greenstadt update* **[22]**, a rank-two symmetry-preserving inverse analogue of the PSB update. Now suppose $y^{\mathrm{T}}s > 0$ and that W is a positive-definite symmetric matrix such that $Wy = s$. Taking the norm to be the weighted Frobenius norm $\|\cdot\|_W$ and $\mathbf{A} = \mathbf{R}^{n\times n}$ gives a rank-one update of G. McCormick (see Pearson **[28]**), which is an inverse-update analogue of the Pearson update. With this norm, taking $\mathbf{A}$ to be the subspace of symmetric matrices gives the *Broyden–Fletcher–Goldfarb–Shanno* (*BFGS*) *update* (**[8]**, **[9]**, **[16]**, **[21]**, **[32]**), given by

$$B_+ = B + \frac{yy^{\mathrm{T}}}{y^{\mathrm{T}}s} - \frac{Bss^{\mathrm{T}}B}{s^{\mathrm{T}}Bs}, \tag{3.11}$$

which is an inverse-update analogue of the DFP update.

A general local convergence analysis has been given by Dennis and Walker **[15]** for least-change secant update methods, i.e., quasi-Newton methods which use least-change secant and inverse secant updates. This analysis is quite extensive in scope and allows for the possibilities of (1) computing part of the Jacobian or inverse Jacobian while approximating the rest through updating, (2) relaxing the choice of y in the secant equation (3.3), and (3) not requiring that $F'(x)$ or $F'(x)^{-1}$ be in $\mathbf{A}$ for x in the region of interest. This analysis shows that under reasonable conditions, least-change secant update methods exhibit local q-superlinear convergence if at all possible and otherwise local q-linear convergence which is optimal in a certain sense. Local here means not only that the initial approximate solution should be near a solution x_*, but also that the initial approximate Jacobian should be near $F'(x_*)$. While the full scope of this analysis is necessary for important applications such as nonlinear least-squares problems, we can summarize the main point for most applications as follows: *If F satisfies standard hypotheses and if $F'(x) \in \mathbf{A}$ (or $F'(x)^{-1} \in \mathbf{A}$) for all x near a solution x_*, then least-change secant update methods enjoy local q-superlinear convergence to x_*.*

Although the analysis of **[15]** gives the same local convergence results for all methods using least-change secant and inverse secant updates and in particular for methods using all of the specific updates mentioned here, it is interesting to note that in practice least-change inverse secant updates are generally less effective than their direct update counterparts, with one very notable exception: The BFGS update has shown itself to be somewhat more robust than the DFP update and is generally preferred. We also note that although the first Broyden update is generally more effective than the second, there is some evidence that the second may be competitive on stiff ODE applications ([**1**], [**6**]).

3.2. *Least-change secant update methods for underdetermined systems.* In view of the success of least-change secant update methods for well-determined systems, it is natural to consider analogues of these methods for underdetermined systems. Here we outline extensions of the notions of least-change secant and inverse secant updates which apply to nonsquare matrices, and then consider analogues of the normal flow and augmented Jacobian algorithms which use these updates. Except where noted otherwise, the material in this section is taken from [**5**] and [**34**], and the reader is referred to those papers for more details.

In the following, it is often necessary to assume that an $n \times n$ submatrix of an $n \times \bar{n}$ matrix exhibits a particular property such as symmetry or invertibility. Both for convenience and because it is natural in many applications such as those arising from parameter-dependent systems, we assume that the submatrix consists of the first n columns. We stress, though, that any other subset of n columns will do just as well and that all updates below are obtained by making a minimum-norm change in the entire matrix. Also, for convenience we write $\bar{s} = (s, t)$ for $s \in \mathbf{R}^n$ and $t \in \mathbf{R}^m$ throughout the following.

The extension of the notion of a least-change secant update is quite straightforward. We assume that we are given analogues of the ingredients for the square-matrix case, viz.,

(i) $B \in \mathbf{R}^{n\times\bar{n}}$, $\bar{s} \in \mathbf{R}^{\bar{n}}$, and $y \in \mathbf{R}^n$;
(ii) an inner product norm $\|\cdot\|$ on $\mathbf{R}^{n\times\bar{n}}$;
(iii) an affine subspace $\mathbf{A} \subseteq \mathbf{R}^{n\times\bar{n}}$ reflecting some structure of F'.

We set $\mathbf{Q}(y, \bar{s}) = \{M \in \mathbf{R}^{n\times\bar{n}} : M\bar{s} = y\}$ and define the least-change secant update of B in $\mathbf{A}$ with respect to $\bar{s}$, y, and $\|\cdot\|$ to be

$$B_+ = P_{\mathbf{A}\cap\mathbf{Q}(y,\bar{s})}(B). \tag{3.12}$$

As in the square-matrix case, particular updates are obtained by particular choices of $\mathbf{A}$ and $\|\cdot\|$. With $\|\cdot\| = \|\cdot\|_{\mathrm{Fr}}$ and $\mathbf{A} = \mathbf{R}^{n\times\bar{n}}$, one obtains an extension of the (first) Broyden update (3.5), viz.,

$$B_+ = B + \frac{(y - B\bar{s})\bar{s}^T}{\bar{s}^T\bar{s}}. \tag{3.13}$$

An extension of the PSB update (3.7) is obtained with $\|\cdot\| = \|\cdot\|_{Fr}$ and

$$\mathbf{A} = \left\{ M = [\mathscr{M}, \mathscr{N}] \in \mathbf{R}^{n\times\bar{n}} : \mathscr{M} = \mathscr{M}^T \in \mathbf{R}^{n\times n} \right\}. \tag{3.14}$$

For $B \in \mathbf{A}$, this update is

$$\begin{aligned} B_+ = B &+ \frac{(y - B\bar{s})\bar{s}^T + \left[s(y - B\bar{s})^T, (y - B\bar{s})t^T\right]}{\bar{s}^T\bar{s} + t^Tt} \\ &- \frac{s^T(y - B\bar{s})}{\bar{s}^T\bar{s} + t^Tt}\,\frac{s\bar{s}^T}{\bar{s}^T\bar{s}}. \end{aligned} \tag{3.15}$$

Note that while (3.13) is pleasingly like its square-matrix analogue (3.5), (3.15) has "extra" denominator terms t^Tt which have no counterparts in (3.7). Except for these terms, though, the update of the first n columns of B in (3.15) looks very much like (3.7). With $\|\cdot\| = \|\cdot\|_{Fr}$, there are also extensions of the sparse Broyden (Schubert) update, obtained by taking $\mathbf{A}$ to be the subspace of matrices having a particular pattern of sparsity, and of the sparse symmetric update (due to Beattie and Weaver-Smith **[3]**), obtained by taking $\mathbf{A}$ to be the subspace of matrices having a particular pattern of sparsity and such that the first n columns constitute a symmetric matrix.

To extend the DFP update (3.8), we take $\mathbf{A}$ to be given by (3.14) and take $\|\cdot\|$ to be a weighted Frobenius norm $\|\cdot\|_W$ defined as follows: We write $B = [\mathscr{B}, \mathscr{C}]$ for $\mathscr{B} \in \mathbf{R}^{n\times n}$ and $\mathscr{C} \in \mathbf{R}^{n\times m}$, and we assume that $\hat{y} \equiv y - \mathscr{C}t$ satisfies $\hat{y}^Ts > 0$. The rationale is that an extension of the DFP update is presumably of interest when $F_x(\bar{x})$ is positive-definite symmetric, where $\bar{x} = (x, \lambda)$ for $x \in \mathbf{R}^n$ and $\lambda \in \mathbf{R}^m$. If $\bar{s} = \bar{x}_+ - \bar{x}$ for nearby $\bar{x}_+$ and $\bar{x}$, and if $\mathscr{C}$ is a good approximation of $F_\lambda(\bar{x})$, then $\hat{y} \approx F_x(\bar{x})s$, and so one can expect $\hat{y}^Ts > 0$. For $M = [\mathscr{M}, \mathscr{N}] \in \mathbf{R}^{n\times\bar{n}}$, we then define

$$\|M\|_W \equiv \sqrt{\operatorname{trace}\{W^{-1}\mathscr{M}W^{-1}\mathscr{M}^T + W^{-1}\mathscr{N}\mathscr{N}^T\}} \tag{3.16}$$

for a positive-definite symmetric $W \in \mathbf{R}^{n\times n}$ satisfying $Ws = \hat{y}$. This norm gives the update

$$\begin{aligned} B_+ = B &+ \frac{(y - B\bar{s})\bar{\hat{y}}^T + \left[\hat{y}(y - B\bar{s})^T, (y - B\bar{s})t^T\right]}{\bar{s}^T\bar{\hat{y}} + t^Tt} \\ &- \frac{s^T(y - B\bar{s})}{\bar{s}^T\bar{\hat{y}} + t^Tt}\,\frac{\hat{y}\bar{\hat{y}}^T}{\bar{s}^T\bar{\hat{y}}}, \end{aligned} \tag{3.17}$$

in which $\bar{\hat{y}} = (\hat{y}, t)$. In principle, we might have defined $\|\cdot\|_W$ in any of several ways which extend the square-matrix weighted Frobenius norm used to obtain the DFP update (3.8). The definition used here has the crucial property that the resulting update (3.17) does not depend on W. Like (3.15), (3.17) has "extra" denominator terms $t^T t$, but otherwise the update of the first n columns is much like the DFP update (3.8).

The extension of the notion of a least-change inverse secant update may not seem straightforward but can be achieved in a reasonable way. Our approach is to assume that B is of full rank n, then to select n columns of B which constitute an invertible submatrix (assumed to be the first n columns for convenience here), and finally to apply the least-change principle to the inverse of this submatrix together with another m columns derived from B. There are several ways in which this might be done; we do the following: given the ingredients (i)–(iii) above with $B = [\mathscr{B}, \mathscr{C}]$ such that $\mathscr{B} \in \mathbf{R}^{n\times n}$ is invertible, we set $K = [\mathscr{K}, \mathscr{L}]$, where $\mathscr{K} = \mathscr{B}^{-1}$ and $\mathscr{L} = -\mathscr{B}^{-1}\mathscr{C}$; then we define the least-change inverse secant update of K in $\mathbf{A}$ with respect to $\bar{s}$, y, and $\|\cdot\|$ to be

$$K_+ = P_{\mathbf{A}\cap\mathbf{Q}(s,\bar{y})}(K), \tag{3.18}$$

where we set $\bar{y} = (y, t)$. If desired, one can recover B_+ from K_+ by writing $K_+ = [\mathscr{K}_+, \mathscr{L}_+]$ for $\mathscr{K}_+ \in \mathbf{R}^{n\times n}$ and $\mathscr{L}_+ \in \mathbf{R}^{n\times m}$ and setting $B_+ = [\mathscr{B}_+, \mathscr{C}_+]$, where $\mathscr{B}_+ = \mathscr{K}_+^{-1}$ and $\mathscr{C}_+ = -\mathscr{K}_+^{-1}\mathscr{L}_+$. Note that $K_+\bar{y} = s$ if and only if $B_+\bar{s} = y$.

The definition (3.18) may in some ways seem arbitrary, but methods using updates given in this way are supported by the local convergence analysis outlined below and given in detail in **[34]** and by the local convergence analysis in **[5]**. To further make (3.18) seem reasonable, we note that T. Ypma **[36]** has observed that

$$\begin{bmatrix} \mathscr{B} & \mathscr{C} \\ 0 & I \end{bmatrix}^{-1} = \begin{bmatrix} \mathscr{B}^{-1} & -\mathscr{B}^{-1}\mathscr{C} \\ 0 & I \end{bmatrix} = \begin{bmatrix} \mathscr{K} & \mathscr{L} \\ 0 & I \end{bmatrix},$$

and so (3.18) can be obtained via a conventional least-change inverse secant update of a square matrix. Also, it might seem that a natural alternative to (3.18) is to make a minimum-norm change in the *pseudo-inverse* of B, resulting in a *least-change pseudo-inverse secant update* of B. However, it is observed in **[34]** that such updates may have certain inherent instabilities.

We give some particular least-change inverse secant updates obtained from (3.18) with various choices of $\|\cdot\|$ and $\mathbf{A}$. As in the square-matrix case, the update formulas given here are in terms of B rather than

K. This is both for easy comparison with the updates previously given and because this is how these updates are likely to be implemented in practice. With $\|\cdot\| = \|\cdot\|_{\mathrm{Fr}}$ and $\mathbf{A} = \mathbf{R}^{n\times\bar{n}}$, (3.18) gives an extension of the second Broyden update (3.10), viz.,

$$B_+ = B + \frac{(y - B\bar{s})\left[y^{\mathrm{T}}B + \begin{pmatrix}0\\ t\end{pmatrix}^{\mathrm{T}}\right]}{y^{\mathrm{T}}B\bar{s} + t^{\mathrm{T}}t}. \tag{3.19}$$

With $\|\cdot\| = \|\cdot\|_{\mathrm{Fr}}$ and $\mathbf{A}$ given by (3.14) as in the extensions of the PSB and DFP updates, the resulting update is an extension of the Greenstadt update. To obtain an extension of the BFGS update (3.11), we again use $\mathbf{A}$ given by (3.14) and define a weighted Frobenius norm as follows: We set $\hat{s} = s - \mathscr{L}t \approx F_x(\bar{x})^{-1}y$, assume $y^{\mathrm{T}}\hat{s} > 0$, and define $\|\cdot\|_W$ by (3.16) for a positive-definite symmetric $W \in \mathbf{R}^{n\times n}$ satisfying $Wy = \hat{s}$. The resulting update is $B_+ = [\mathscr{B}_+, \mathscr{C}_+]$, where $\mathscr{B}_+ \in \mathbf{R}^{n\times n}$ and $\mathscr{C}_+ \in \mathbf{R}^{n\times m}$ are given by

$$\begin{aligned}\mathscr{B}_+ &= \mathscr{B} + \frac{(B\bar{s})^{\mathrm{T}}\mathscr{B}^{-1}(B\bar{s})}{d}yy^{\mathrm{T}} - \frac{c}{d}(B\bar{s})(B\bar{s})^{\mathrm{T}}\\ &\quad + \frac{2t^{\mathrm{T}}t}{d}\left[y(B\bar{s})^{\mathrm{T}} + (B\bar{s})y^{\mathrm{T}}\right],\\ \mathscr{C}_+ &= \mathscr{B}_+\mathscr{B}^{-1}\left[\mathscr{C} + \frac{2}{d_2}yt^{\mathrm{T}} - \frac{d_2 + y^{\mathrm{T}}\mathscr{B}^{-1}y}{d_1 d_2}(B\bar{s})t^{\mathrm{T}}\right],\end{aligned} \tag{3.20}$$

where

$$\begin{aligned}d_1 &= y^{\mathrm{T}}\mathscr{B}^{-1}B\bar{s} + t^{\mathrm{T}}t,\\ d_2 &= y^{\mathrm{T}}\mathscr{B}^{-1}B\bar{s} + 2t^{\mathrm{T}}t,\\ c &= d_2 + \frac{\left(y^{\mathrm{T}}\mathscr{B}^{-1}y + t^{\mathrm{T}}t\right)d_2}{d_1} - y^{\mathrm{T}}\mathscr{B}^{-1}y,\\ d &= 4\left(t^{\mathrm{T}}t\right)^2 + (B\bar{s})^{\mathrm{T}}\mathscr{B}^{-1}(B\bar{s})c.\end{aligned}$$

These expressions are perhaps not very attractive, but we have been unable to find more appealing ones. In fact, the implementation of this update is not as difficult as it might appear; see **[5]** for details and for possibly more tractable alternative expressions.

We now outline analogues of the normal flow algorithm (2.3) and the augmented Jacobian algorithm (2.4) which use least-change secant and inverse secant updates. If an approximate solution $\bar{x}$ and an approximate Jacobian B are given, then the normal flow iteration is

$$\begin{aligned}\bar{x}_+ &= \bar{x} - B^+F(\bar{x}),\\ B_+ &= \text{least-change (inverse) secant update of } B,\end{aligned} \tag{3.21}$$

and the augmented Jacobian iteration is

$$\begin{aligned} \bar{x}_+ &= \bar{x} + \bar{s}, \text{ where } B\bar{s} = -F(\bar{x}) \text{ and } V\bar{s} = 0, \\ B_+ &= \text{least-change (inverse) secant update of } B. \end{aligned} \tag{3.22}$$

The updates are determined by (3.12) or (3.18) for specified $\|\cdot\|$ and $\mathbf{A}$ and for $\bar{s} = \bar{x}_+ - \bar{x}$ and $y = F(\bar{x}_+) - F(\bar{x})$.

A local convergence analysis is given in **[34]** which applies to the iterations (3.21) and (3.22). This analysis is very broad and allows, among other things, for alternative choices for y and for not requiring that $F'(\bar{x})$ actually have the structure reflected in $\mathbf{A}$. (In saying here and below that a matrix $M \in \mathbf{R}^{n\times\bar{n}}$ has the structure reflected in $\mathbf{A}$, we mean either that $M \in \mathbf{A}$ or, writing $M = [\mathscr{M}, \mathscr{N}]$ for $\mathscr{M} \in \mathbf{R}^{n\times n}$ and $\mathscr{N} \in \mathbf{R}^{n\times m}$, that $[\mathscr{M}^{-1}, -\mathscr{M}^{-1}\mathscr{N}] \in \mathbf{A}$, according to whether least-change secant or inverse secant updates are of interest.) This analysis can also be easily extended to include situations in which part of the Jacobian is computed while the rest is approximated by updating. Like the analysis in **[15]**, the analysis in **[34]** shows that under reasonable hypotheses the iterations (3.21) and (3.22) exhibit local q-superlinear convergence if possible and otherwise local q-linear convergence which is optimal in a certain sense. Like the analysis in Section 2 for the normal flow and augmented Jacobian algorithms, it is a Kantorovich-type analysis. We summarize the main point of this analysis for most applications as follows:

THEOREM. *Under the Normal Flow Hypothesis (or the Augmented Jacobian Hypothesis), if $F'(\bar{x})$ has the structure reflected in $\mathbf{A}$ for all $\bar{x} \in \Omega$, and if Ω_η is given by* (2.5) *for some $\eta > 0$, then there are $\varepsilon > 0$ and $\delta > 0$ such that if $\bar{x}_0 \in \Omega_\eta$ and B_0 having the structure reflected in $\mathbf{A}$ satisfy $|F(\bar{x}_0)| < \varepsilon$ and $|B_0 - F'(\bar{x}_0)| < \delta$, then the iterates generated by* (3.21) (*or* (3.22)) *converge q-superlinearly to $\bar{x}_* \in \Omega$ such that $F(\bar{x}_*) = 0$.*

There is an important limitation to this analysis: It does not apply to an iteration (3.21) or (3.22) which uses the extension of either the DFP update (3.17) or the BFGS update (3.20) because these updates are least-change with respect to norms which depend on current approximate Jacobians. However, it applies to an iteration (3.21) or (3.22) whenever the update is least-change with respect to a norm which remains the same throughout all iterations (a "fixed-scale" update in the terminology of **[15]**), e.g., the extensions of the first and second Broyden updates, the PSB and Greenstadt updates, and the sparse Broyden (Schubert) and sparse-symmetric updates, all of which are least-change with respect to the Frobenius norm. We also note that other iterations using least-change secant and inverse secant updates are analyzed in **[5]**. These are closely related to (3.21) and (3.22) in that they satisfy an analogue of (2.2), viz., $F(\bar{x}) + B(\bar{x}_+ - \bar{x}) = 0$, but they allow the specification of some variable values at each iteration and so may be more useful than (3.21) and (3.22) for some parameter-dependent problems.

4. A short tale of two updates. We conclude this note with a quick look at some simple numerical experiments reported in **[34]** which involve the extensions of the first and second Broyden updates, given by (3.13) and (3.19), respectively. Our purpose is to give some indication of the performance of the iterations in Section 3 and to point out some issues associated with them by looking at the simplest cases.

The experiments involved the normal flow algorithm (2.3) and iteration (3.21) with the updates (3.13) and (3.19) applied to simple scalar-valued functions of two variables. For perspective we also included a chord method of the form $\bar{x}_+ = \bar{x} - B^+F(\bar{x})$ in which the approximate Jacobian B is not updated but held fixed at its initial value. For each $F : \mathbf{R}^2 \to \mathbf{R}^1$ and each initial approximate solution $\bar{x}_0$ used in the experiments, we took $F'(\bar{x}_0)$ as the initial approximate Jacobian in iteration (3.21) and in the chord method. Computing was done in double precision, and in each trial all algorithms ran until the residual norm was no greater than 10^{-12}.

We first took $F(\bar{x}) = x_1 - 2x_2^3 + 9x_2^2 - 12x_2$ for $\bar{x} = (x_1, x_2) \in \mathbf{R}^2$ and considered starting points $\bar{x}_0 = (5, 0)$ near the solution curve and $\bar{x}_0 = (0, 5)$ somewhat farther away. The results are given in Tables 1 and 2. In these and in Table 3 below, "Normal Flow" refers to iteration (2.3), and "First Broyden" and "Second Broyden" refer to iteration (3.21) with the updates (3.13) and (3.19), respectively. The results in Table 1 are about what one might expect; perhaps the only noteworthy feature is the large number of iterations required by the chord method to meet the very small residual tolerance. In Table 2, however, it is seen that the second Broyden update outperformed the first, in contrast with experience with these updates in the well-determined case. Actually, inspection of the iterates suggests that neither update did very well in this trial; both gave rise to iterates which wandered far from the ultimate limit. However, in a number of other trials from other starting points, the second Broyden update consistently beat the first.

It is also seen clearly in Table 2 that the first Broyden update and the chord method yielded the same limit, and indeed this is the case in

Method	Number of Iterations	Final Iterate
Normal Flow	7	(4.864, .7997)
First Broyden	10	(4.929, .8531)
Second Broyden	10	(4.927, .8516)
Chord Method	273	(4.929, .8531)

Table 1. Results for $F(\bar{x}) = x_1 - 2x_2^3 + 9x_2^2 - 12x_2$ with $\bar{x}_0 = (5, 0)$.

Method	Number of Iterations	Final Iterate
Normal Flow	9	(1.226, .1112)
First Broyden	30	(.06936, .005806)
Second Broyden	17	(4.711, 1.355)
Chord Method	208	(.06936, .005806)

Table 2. Results for $F(\bar{x}) = x_1 - 2x_2^3 + 9x_2^2 - 12x_2$ with $\bar{x}_0 = (0, 5)$.

Table 1 as well. *This is no accident*: One has in general that the step $-B^+F(\bar{x})$ in (3.21) is in the range of B^{T}, and so it follows from (3.13) that the range of B_+^{T} is contained in the range of B^{T} and, by induction, in the range of B_0^{T}, where B_0 is the initial approximate Jacobian. This is to say that in general *the iterates produced by* (3.21) *with the first Broyden update* (3.13) *must lie in the affine subspace* $\bar{x}_0 + \mathit{range}\{B_0^{\mathrm{T}}\}$. This also implies that in general *the iteration* (3.21) *with the first Broyden update* (3.13) *can be viewed as a special case of iteration* (3.22) in which V is any matrix such that the range of V^{T} is the null-space of B_0. Related observations have been made by Georg **[18]** in the $\bar{n} = n+1$ case. Here, one sees that the iterates produced by the first Broyden update (3.13) and by the chord method must lie in a single line which has a unique intersection with the solution curve; therefore the limits of these iterates must be the same, viz., this point of intersection.

It is apparent from this that *iterates produced by* (3.21) *with the first Broyden update* (3.13) *cannot converge to a solution if* $\bar{x}_0 + \mathrm{range}\{B_0^{\mathrm{T}}\}$ *does not intersect the solution set.* The same holds for the chord method, and indeed a similar warning is in order for the augmented Jacobian algorithm (2.4) and for the iteration (3.22): the iterates cannot converge to a solution if $\bar{x}_0 + \text{null-space}\{V\}$ does not intersect the solution set. On the other hand, (3.21) with the second Broyden update (3.19) does not seem to share this limitation. For illustration we took $F(\bar{x}) = x_1^2 - x_2$ for $\bar{x} = (x_1, x_2) \in \mathbf{R}^2$ and considered the starting point $\bar{x}_0 = (1, -1)$. For $B_0 = F'(\bar{x}_0)$ one has $\bar{x}_0 + \text{range}\{B_0^{\mathrm{T}}\} = \{(1, -1) + t(2, -1) : -\infty < t < \infty\}$, which does not intersect the solution curve. The results for all algorithms are summarized in Table 3.

Method	Number of Iterations	Final Iterate
Normal Flow	14	(-.01868, .0003489)
First Broyden	*	*
Second Broyden	16	(.1985, .03942)
Chord Method	*	*

* unable to converge

Table 3. Results for $F(\bar{x}) = x_1^2 - x_2$ with $\bar{x}_0 = (1, -1)$.

This discussion seems to indicate that the second Broyden update (3.19) has advantages over the first Broyden update (3.13) in some circumstances. However, there are possibly unattractive features of the second Broyden update as well. For one thing, this update distinguishes particular subsets of columns of matrices being updated, and this may be unappealing. For another, it may encounter difficulty in predictor-corrector algorithms for homotopy or continuation methods. In these algorithms the predictor step $\bar{s}$ is typically in an approximate tangent direction to the solution curve, so that $B\bar{s} = 0$. If (3.19) is used for updating on such a step near a turning point at which $t = 0$, then trouble is likely to ensue. The first Broyden update does not have this problem and indeed seems particularly well-suited to updating following a tangent step (Georg [**18**]), but as we have seen it cannot incorporate tangent information on the corrector steps. We feel that what is needed are updating methods which can effectively incorporate tangent information on both predictor and corrector steps and which do not distinguish particular subsets of columns of matrices being updated.

BIBLIOGRAPHY

1. P. Alfeld, *Two devices for improving the efficiency of stiff ODE solvers*, Proc. 1979 SIGNUM Meeting on Numerical Ordinary Differential Equations, Univ. Illinois Dept. Computer Science, Report 79-1710, Urbana, pp. 24-1 to 24-3.

2. E. Allgower and K. Georg, *Simplicial and continuation methods for approximating fixed points and solutions to systems of equations*, SIAM Rev. **22** (1980), 28–85.

3. C. A. Beattie and S. Weaver-Smith, *Secant methods for structural model identification*, Virginia Polytechnic Institute and State University Interdisciplinary Center for Appl. Math. Rep. ICAM-TR-88-0601, Blacksburg, VA, 1988.

4. A. Ben-Israel, *A modified Newton-Raphson method for the solution of systems of equations*, Israel J. Math. **3** (1965), 94–98.

5. S. K. Bourji and H. F. Walker, *Least-change secant updates of nonsquare matrices*, Utah State University Math. Stat. Dept. Res. Rep. December, 1987/38, Logan, UT, 1987, SIAM J. Numer. Anal. (to appear).

6. P. N. Brown, A. C. Hindmarsh, and H. F. Walker, *Experiments with quasi-Newton methods in solving stiff ODE systems*, SIAM J. Sci. Stat. Comput. **6** (1985), 297–313.

7. C. G. Broyden, *A class of methods for solving nonlinear simultaneous equations*, Math. Comp. **19** (1965), 577–593.

8. ——, *A new double-rank minimization algorithm*, AMS Notices **16** (1969), 670.

9. ——, *The convergence of a class of double-rank minimization algorithms, Parts* I *and* II, J. Inst. Math. Appl. **6** (1971), 76–90, 222–236.

10. ——, *The convergence of an algorithm for solving sparse nonlinear systems*, Math. Comp. **25** (1971), 285–294.

11. D. F. Davidenko, *On the approximate solution of systems of nonlinear equations*, Ukrain. Mat. Ž. **5** (1953), 196–206.

12. W. C. Davidon, *Variable metric methods for minimization*, Argonne Nat. Labs. Report ANL-5990 Rev.

13. J. E. Dennis, Jr. and R. B. Schnabel, *Least change secant updates for quasi-Newton methods*, SIAM Rev. **21** (1979), 443–459.

14. ——, *Numerical Methods for Unconstrained Optimization and Nonlinear Equations*, Prentice-Hall Series in Computational Mathematics, Englewood Cliffs, NJ, 1983.

15. J. E. Dennis, Jr. and H. F. Walker, *Convergence theorems for least change secant update methods*, SIAM J. Numer. Anal. **18** (1981), 949–987.

16. R. Fletcher, *A new approach to variable metric algorithms*, Comput. J. **13** (1970), 317–322.

17. R. Fletcher and M. J. D. Powell, *A rapidly convergent descent method for minimization*, Comput. J. **6** (1963), 163–168.

18. K. Georg, *Numerical integration of the Davidenko equation*, Numerical Solution of Nonlinear Equations: Proceedings, Bremen 1980, E. L. Allgower, K. Glashoff, and H.-O. Peitgen, eds., Lecture Notes in Mathematics 878, Springer-Verlag, Berlin, 1980, pp. 128–161.

19. P. E. Gill, G. H. Golub, W. Murray, and M. A. Saunders, *Methods for modifying matrix factorizations*, Math. Comp. **28** (1974), 505–536.

20. R. Glowinski, H. B. Keller, and L. Reinhart, *Continuation-conjugate gradient methods for the least squares solution of nonlinear boundary value problems*, SIAM J. Sci. Stat. Comput. **6** (1985), 793–832.

21. D. Goldfarb, *A family of variable-metric methods derived by variational means*, Math. Comp. **24** (1970), 23–26.

22. J. Greenstadt, *Variations on variable metric methods*, Math. Comp. **24** (1970), 1–18.

23. C. B. Haselgrove, *The solution of nonlinear equations and of differential equations with two-point boundary conditions*, Computing J. **4** (1961), 255–259.

24. M. P. Kamat, L. T. Watson, and D. J. Vanden Brink, *An assessment of quasi-Newton sparse update techniques for nonlinear structural analysis*, Comp. Meth. Appl. Mech. Eng. **26** (1981), 363–375.

25. E. Marwil, *Convergence results for Schubert's method for solving sparse nonlinear equations*, SIAM J. Numer. Anal. **16** (1979), 588–604.

26. J. M. Ortega and W. C. Rheinboldt, *Iterative Solution of Nonlinear Equations in Several Variables*, Academic Press, New York, 1970.

27. A. M. Ostrowski, *Solution of Equations in Euclidean and Banach Spaces*, Academic Press, New York, 1973.

28. J. D. Pearson, *Variable metric methods of minimization*, Comput. J. **12** (1969), 171–178.

29. M. J. D. Powell, *A new algorithm for unconstrained optimization*, Nonlinear Programming, J. B. Rosen, O. L. Mangasarian, and K. Ritter, eds., Academic Press, New York, 1970.

30. W. C. Rheinboldt, *Numerical Analysis of Parametrized Nonlinear Equations*, University of Arkansas Lecture Notes in the Mathematical Sciences 7, John Wiley and Sons, New York, NY, 1986.

31. L. K. Schubert, *Modification of a quasi-Newton method for nonlinear equations with a sparse Jacobian*, Math. Comp. **24** (1978), 27–30.

32. D. F. Shanno, *Conditioning of quasi-Newton methods for function minimization*, Math. Comp. **24** (1970), 647–656.

33. Ph. L. Toint, *On sparse and symmetric matrix updating subject to a linear equation*, Math. Comp. **31** (1977), 954–961.

34. H. F. Walker and L. T. Watson, *Least-change secant update methods for underdetermined systems*, Utah State University Math. Stat. Dept. Res. Rep. August/88/41, Logan, UT, 1988, Virginia Polytechnic Institute and State University Comp. Sci. Dept. Rep. 88-28 and Interdisciplinary Center for Appl. Math. Rep. 88-09-03, Blacksburg, VA, 1988, SIAM J. Numer. Anal. (to appear).

35. L. T. Watson, S. C. Billups, and A. P. Morgan, *Algorithm 652: HOMPACK: A suite of codes for globally convergent homotopy algorithms*, ACM Trans. Math. Software **13** (1987), 281–310.

36. T. Ypma, private communication, 1988.

UTAH STATE UNIVERSITY

Lectures in Applied Mathematics
Volume **26** (1990)

Sard's Theorem and Its Improved Versions in Numerical Analysis

Y. YOMDIN

1. Introduction. Virtually any algorithm in nonlinear analysis requires for its proper work that some Jacobians be nonzero. Sard's theorem (see, e.g., [**5**]) essentially claims that this is the case for randomly chosen data. Some numerical algorithms, such as continuation methods for solution of nonlinear systems of equations, explicitly involve a random choice of parameters, appealing to Sard's theorem as a justification (see, e.g., [**1**], [**2**], [**3**], [**4**]).

However, the role of this theorem is restricted to an a posteriori confirmation of the efficiency (experimentally well known) of the above type algorithms. It gives no specific recommendations on how to organize such algorithms or how to optimize the choice of the parameters involved. This is due to an inherent qualitativeness of the information provided by Sard's theorem.

The purpose of this paper is to discuss much stronger, quantitative results now available ([**6**], [**7**], [**8**]) in the direction of the classical Sard's theorem. We show that in principle these results may produce practical recipes for organization of algorithms in nonlinear problems.

Let us recall the statement of Sard's theorem. To simplify the presentation we consider below only the case of functions $f\colon \mathbf{R}^n \to \mathbf{R}$, although all the results can be extended to the mappings of $\mathbf{R}^n$ to $\mathbf{R}^m$ or smooth manifolds.

1980 *Mathematics Subject Classification* (1985 *Revision*). Primary 58C27, 65H05.
Supported by NSF Grant No. DMS-8610730.

A point $x \in \mathbf{R}^n$ is called a critical point of $f: \mathbf{R}^n \to \mathbf{R}$ if $\operatorname{grad} f(x) = 0$. A value of f at its critical point is called a critical value.

Then Sard's (or, more precisely, M. Morse's–A. P. Morse's–A. Sard's ...) theorem claims that *for f k-times continuously differentiable, with $k \geq n$, the Lebesgue measure of the set $\Delta(f)$ of all the critical values of f is zero.* Usually this result is interpreted in the following way: A randomly picked value $c \in \mathbf{R}$ is regular for f.

Notice that (at least formally) Sard's theorem is not applicable at all to computations with finite accuracy. Indeed, if we cannot check whether $\operatorname{grad} f(x)$ is zero *exactly*, the notion of a critical point disappears. The same happens with the conclusion of this theorem. Since in numerical computations we work only with rational numbers, Sard's theorem does not exclude the possibility that *any* value we pick is critical. One can say that computers understand neither the assumption nor the conclusion of this theorem.

This reasoning is, of course, too formal. In practice, $m(\Delta(f)) = 0$ usually does mean that a randomly picked rational number is not in $\Delta(f)$. However, the following point (which essentially concerns the same weakness of Sard's theorem—its qualitative character) is of extreme importance in numerical computations. One needs *the Jacobians there to be well separated from zero*, and not merely to be nonzero. Sard's theorem provides no information of this type.

It turns out that a much better description of the behavior of smooth mappings near their critical set is possible. Theorem 1.1 of **[6]** gives quantitative restrictions on the geometry of not only critical but also "near-critical" values of C^k-mappings. These restrictions imply the usual Sard's theorem, but they are much stronger and, most important, they are meaningful for finite-accuracy data. In Section 2 we state the simplest version of this result and give some of its consequences. Thus, the above-mentioned difficulties with an explanation of well-known effects disappear.

However, the results of **[6]**–**[8]** allow us to do much more: A knowledge of the distribution of values with respect to their regularity allows one to optimize the choice of parameters involved in an algorithm, such as the stepsize of the curve-following procedure, etc. In Section 3 we give an example (oversimplified, of course) of such an optimization.

2. Main results. To simplify the presentation, we consider only polynomial functions. The results to their full extent (including some infinite-dimensional situations) can be found in **[6]**–**[8]**.

DEFINITION 2.1. Let $f\colon B^n \to \mathbf{R}$ be a C^1 function, $\gamma \geq 0$, where B^n denotes the unit ball in $\mathbf{R}^n$. The set of γ-critical points of f is defined as

$$\Sigma(f, \gamma) = \{x \in B^n \mid \|\operatorname{grad} f(x)\| \leq \gamma\}.$$

The set of γ-critical values is defined as

$$\Delta(f, \gamma) = f(\Sigma(f, \gamma)).$$

Now let $f\colon B^n \to \mathbf{R}$ be a polynomial of degree d.

THEOREM 2.2. *The set $\Delta(f, \gamma)$ can be covered by $(2d)^n$ intervals of length γ.*

Theorem 2.2 is the main result on near-critical values that we use in this paper. In the following corollaries we give some of its consequences.

COROLLARY 2.3. *The measure of γ-critical values of f does not exceed $(2d)^n\gamma$.*

DEFINITION 2.4. A value $\xi \in \mathbf{R}$ is called a γ-regular value of f if $\xi \notin \Delta(f, \gamma)$.

Thus, ξ is γ-regular if at any $x \in f^{-1}(\xi)$, $\|\operatorname{grad} f(x)\| \geq \gamma$. This is exactly the information we need, e.g., to organize effectively the numerical "tracing" of the hypersurface $f = \xi$.

COROLLARY 2.5. *Let ξ be picked randomly in the interval $[a, b]$. Then for any $\gamma \geq 0$ the probability that ξ is a γ-regular value for f is at least $1 - (2d)^n\gamma/(b-a)$.*

In Section 3 we use the distribution of values ξ with respect to their regularity, given by Corollary 2.5, to find an optimal value of the stepsize in a curve-following algorithm.

The next corollary shows that regular values of f can be found in any sufficiently dense grid. Notice that one cannot get this type of conclusion using the Lebesgue measure, as in the usual Sard's theorem.

COROLLARY 2.6. *Let $Z_h = \{x_i\}$ be a grid in $[a, b]$, $x_i = x_0 + ih$, $h > 0$. Then for any $\gamma \geq 0$ at least $(b-a)/h - (2d)^n(\gamma/h + 1)$ points in this grid are γ-regular.*

Indeed, $(2d)^n$ intervals of length γ can cover at most $(2d)^n(\gamma/h + 1)$ points in the grid.

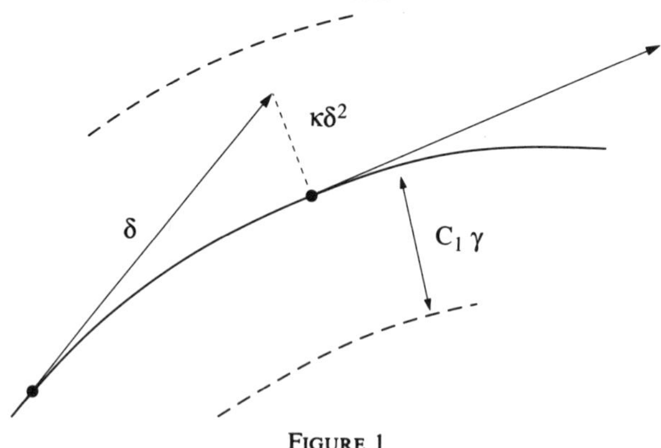

FIGURE 1

COROLLARY 2.7. *For any* $h \leq h_0 = (b-a)/(2d)^n$ *and* $\gamma(h) = h_0 - h$, *there are* $\gamma(h)$*-regular values in the grid* Z_h.

Let us mention an additional application of the results above: computer-assisted proofs. More precisely, using these results, one can find the size of the grid and the accuracy of computations, required to *guarantee* the correctness of the final result (e.g., to guarantee that we have found *all* the solutions of a given system).

3. Example. Consider the following problem: For a given polynomial $F(x,y)$ of degree d we want to follow numerically the curve $F(x,y) = \xi$ inside the unit disk $B^2 \subseteq \mathbf{R}^2$. The point ξ is assumed to be picked randomly in $[0,1]$.

We use one of the standard predictor-corrector methods, with the only "free" parameter, the stepsize δ (see Figure 1).

We assume that the execution time for one step is T, independently of δ.

The only requirement for the method we consider below to work is that the result of the predictor on each step must be within the "attraction basin" of the corrector. The size of this basin at a point (x,y) on the curve is well known to be of order $C_1\|\operatorname{grad} F(x,y)\|$.

On the other hand, the distance of the predictor result from the curve is $\kappa\delta^2$, where κ is the curvature of this curve at (x,y). An easy computation shows that $\kappa = |F_{tt}|/\|\operatorname{grad} F\| \leq C_2/\|\operatorname{grad} F\|$, where t is the tangent direction to the curve.

Thus, the condition of applicability of the algorithm with the stepsize δ takes the form

$$(*) \qquad \frac{C_2}{\|\operatorname{grad} F(x,y)\|}\delta^2 \leq C_1\|\operatorname{grad} F(x,y)\|,$$

for each point (x, y) on the curve $F = \xi$.

Now if the value ξ is γ-regular for F, then, by definition, at each point (x, y) on the curve $F = \xi$, $\|\operatorname{grad} F(x, y)\| \geq \gamma$, and we may replace $(*)$ by

$$(**)\qquad \frac{C_2}{\gamma}\delta^2 \leq C_1\gamma \quad \text{or} \quad \delta \leq C_3\gamma, \quad \text{where} \quad C_3 = \left(\frac{C_1}{C_2}\right)^{1/2}.$$

Now assume that we apply our algorithm as follows. We fix some value δ of the stepsize, and start the algorithm. If it breaks down, i.e., if the condition $(**)$ is not satisfied at some point on the curve, we pass to a very small value α of the stepsize (given from the very beginning) and start the algorithm once more.

Now the optimization problem is the following: *to find the value δ, for which the average execution time of the algorithm will be minimal.* (Average here is taken with respect to ξ, uniformly distributed in [0, 1].)

For δ fixed and for ξ a γ-regular value, the execution time is of order $(C_4/\delta)T$, if $\gamma \geq 1/C_s\delta$, and it is $(C_4/\alpha)T$, if $\gamma < 1/C_s\delta$, by the condition $(**)$. (Here C_4 is the length of the curve $F = \xi$.)

Using the distribution given by Corollary 2.5, we find that the average time is

$$\left(1 - (2d)^2 \cdot \tfrac{1}{C_3}\delta\right) \tfrac{C_4}{\delta} T + (2d)^2 \tfrac{1}{C_3}\delta \tfrac{C_4}{\alpha} T$$
$$= C_4 T \left(\tfrac{1}{\delta} + (2d)^2 \tfrac{\delta}{C_3\alpha} - (2d)^n \cdot \tfrac{1}{C_3}\right).$$

Differentiating, we find that this expression attains its minimum for

$$\delta = C_5\alpha^{1/2}, \quad \text{where} \quad C_5 = \tfrac{C_3^{1/2}}{2d}.$$

REMARK. Exactly the same result would be obtained if, instead of following one curve $F = \xi$ for a random ξ, we followed curves $F = \xi_i$, $i = 1, 2, \ldots$, for a dense grid (ξ_i) in [0, 1].

Although the above computations in no way form a rigorous proof, we expect the main conclusion to be true: *If the strategy as above is adopted, then the starting value δ of the stepsize should be of order of the square root of the minimal value α.*

Taking a bigger risk, we may estimate the coefficients in the expressions above. For a normalized polynomial F of degree d, the worst-case estimates are the following: $C_1 = 1/d^4$, $C_2 = d^4$, and hence $C_3 = 1/d^4$, $C_4 = \pi d$, and we obtain $C_5 = 1/(2d^3)$.

However, typically the values of these constants are much better, and we expect, say, for F of degree 5, C_5 to be of order 1/20.

In conclusion, notice that the constants, given in the results of [**6**]–[**8**], are the worst-case ones. To make these results practically applicable one should find instead *typical* values of the above constants, which are, of course, much better.

Some results in this direction, as well as a more detailed exposition of the results in the present paper, will appear separately.

References

1. E. Allgower and K. Georg, *Simplicial and continuation methods for approximating fixed points and solutions to systems of equations*, SIAM Review **22** (1980), 28–85.

2. H. B. Keller, test problem, this volume.

3. T. Y. Li, T. Sauer, and J. A. Yorke, *Numerically determining solutions of systems of polynomial equations*, Bull. Amer. Math. Soc. **18** (1988), 173–177.

4. A. P. Morgan and A. Sommese, *Computing all solutions to polynomial systems using homotopy continuation*, Applied Math. Comp. **24** (1987), 115–138.

5. A. Sard, *The measure of the critical values of differentiable maps*, Bull. Amer. Math. Soc. **48** (1942), 883–890.

6. Y. Yomdin, *The geometry of critical and near critical values of differentiable mappings*, Math. Ann. **264** (1983), 495–515.

7. ——, *Approximative complexity of functions*, LMN **1317** (1988), 21–43.

8. ——, *Metric semialgebraic geometry with applications in smooth analysis*, submitted for publication.

The Institute for Advanced Study and

The Weizmann Institute of Science

Lectures in Applied Mathematics
Volume **26** (1990)

Finite Difference Approximation of Sparse Jacobian Matrices in Newton-like Methods

TJALLING J. YPMA

Abstract. The initial expository sections of this paper aim to acquaint the reader with some basic ideas and results concerning techniques for approximating sparse Jacobian matrices by finite differences. In the subsequent sections more recent developments in the use of such techniques in the context of Newton-like iterative methods are described. In particular, we show that for convergence to occur in either the damped or undamped versions of the resulting iterative methods, it is essential to maintain a sufficiently recent approximation to the Jacobian matrix. Monitoring devices for this purpose are outlined.

1. Introduction. Consider solving the nonlinear equation

$$\mathbf{F}(\mathbf{x}) = \mathbf{0} \tag{1}$$

where $\mathbf{F}\colon \mathbf{R}^N \to \mathbf{R}^N$ is assumed to be continuously differentiable on some appropriate subspace of $\mathbf{R}^N$. We write (1) componentwise as

$$\begin{aligned} f_1(x_1, x_2, \dots, x_N) &= 0 \\ f_2(x_1, x_2, \dots, x_N) &= 0 \\ \vdots \quad &= \vdots \\ f_N(x_1, x_2, \dots, x_N) &= 0. \end{aligned} \tag{2}$$

The standard locally convergent Newton iteration for solving (1) generates successive estimates $\mathbf{x}^k \in \mathbf{R}^N$ of a solution $\mathbf{x}^* \in \mathbf{R}^N$ of (1) as

$$\mathbf{x}^{k+1} = \mathbf{x}^k - [\mathbf{F}'(\mathbf{x}^k)]^{-1}\mathbf{F}(\mathbf{x}^k), \qquad k = 0, 1, \dots, \tag{3}$$

1980 *Mathematics Subject Classification* (1985 *Revision*). Primary 65H10.
Research partially funded by the Foundation for Research Development, Council for Scientific and Industrial Research, Pretoria, South Africa.

and our concern is with the case in which the Jacobian matrix $\mathbf{F}'(\mathbf{x}^k)$ is replaced by an approximation $\mathbf{B}_k$ calculated using finite differences. We also consider the case in which a damping parameter $t_k \in \mathbf{R}$, $t_k > 0$, is introduced in order to increase the extent of the region of convergence of the algorithm. Thus, we examine iterative methods of the form

$$\mathbf{x}^{k+1} = \mathbf{x}^k - t_k \mathbf{B}_k^{-1}\mathbf{F}(\mathbf{x}^k), \qquad t_k > 0, \quad k = 0, 1, \ldots . \tag{4}$$

Our interest is mainly in techniques for efficiently calculating $\mathbf{B}_k$, and the convergence properties of (4) when using these approximations to $\mathbf{F}'(\mathbf{x}^k)$. Iteration (4) is an instance of the more general iteration

$$\mathbf{x}^{k+1} = \mathbf{x}^k + t_k \mathbf{d}^k, \qquad t_k > 0, \quad k = 0, 1, \ldots , \tag{5}$$

where $\mathbf{d}^k \in \mathbf{R}^N$, and several of our results are derived in this more general setting.

The initial sections of this paper are expository. We introduce the reader to the fundamental ideas concerning the efficient approximation of sparse Jacobian matrices by finite differences, and provide the basic references. In subsequent sections we examine first the possibilities of calculating $\mathbf{B}_k$ in the context of iteration (4) using few evaluations of $\mathbf{F}$, and then examine the convergence properties of iteration (4) using these approximations. In particular, we show that for both local and wider convergence to be guaranteed it is essential to maintain fairly recent approximations to the Jacobian matrix. Some devices for ensuring that this is done are outlined.

2. Finite difference approximations. In this section we briefly recall some fundamental ideas and work concerning finite difference approximation of Jacobian matrices in Newton-like methods without any particular reference to sparsity.

The element in the ith row and jth column of $\mathbf{F}'(x)$ is $\partial f_i(x)/\partial x_j$, and it is natural to approximate this by a forward difference, that is, to use

$$\frac{\partial f_i(x)}{\partial x_j} \approx h_j^{-1}[f_i(\mathbf{x} + h_j\mathbf{e}_j) - f_i(\mathbf{x})] \tag{6}$$

where $h_j \in \mathbf{R}$ is some suitably small steplength and $\mathbf{e}_j \in \mathbf{R}^N$ is the jth column of the $N \times N$ identity matrix. The size of h_j determines the accuracy of the approximation; guidelines for the selection of h_j may be found, for example, in **[9]**, **[17]**.

Combining (6) and (2), we note immediately that the entire jth column of $\mathbf{F}'(x)$ can be approximated by the single difference

$$h_j^{-1}[\mathbf{F}(\mathbf{x} + h_j\mathbf{e}_j) - \mathbf{F}(\mathbf{x})] \tag{7}$$

provided that one is willing to use the same h_j for every component function f_i, $i = 1, 2, \ldots, N$ (as opposed to using, say, $h_{i,j}$ when approximating $\partial f_i/\partial x_j$). The form (7) may be more convenient than (6) in situations in which different component functions have common subexpressions, or when individual component functions are not readily accessible. In certain situations one has no access to the individual component functions at all, so that (7) is the only feasible option. The methods described in this paper concentrate on the form (7) and a generalization of that form.

Using (7), we see that $N + 1$ evaluations of $\mathbf{F}$ suffice to approximate the entire matrix $\mathbf{F}'(\mathbf{x})$. For the Newton-like iteration (4) an approximation to the matrix $\mathbf{F}'(\mathbf{x}^k)$ is needed, and in that case the value of $\mathbf{F}(\mathbf{x}^k)$ required by (7) is already known. This shows that by examining finite difference approximations within the context of iteration (4) it is possible to reduce the number of evaluations of $\mathbf{F}$ required to calculate an approximation $\mathbf{B}_k$ of $\mathbf{F}'(\mathbf{x}^k)$. In particular, it is frequently possible to exploit known $\mathbf{F}$ values from previous iterations to obtain a cheap approximation. Differences other than the obvious (6) are often useful for such purposes. Numerous such schemes, in which the number of evaluations of $\mathbf{F}$ per iteration may be considerably less than N, but which take no special account of any sparsity that may be present, exist and have been described in the references **[15]**, **[17]**, where reasons why not all such schemes are useful are also given. A very general local convergence analysis for such schemes appears in **[18]**, **[19]**. We point out in Sections 3 and 8 of this paper that in some senses all these schemes may be regarded as special cases of the general technique for approximating sparse Jacobian matrices to be described in the next section; consequently, the cautionary statements in Sections 6 and 7 concerning the need for sufficiently recent Jacobian approximations to ensure convergence of both the damped and undamped Newton-like iterations are relevant not only to the sparse case but to these finite difference schemes in general.

3. Sparse approximation. We assume from now on that the Jacobian matrix is sparse and that the elements in the Jacobian matrix that we wish to approximate, which we refer to as *entries*, are in known positions. These positions define the *pattern* of the matrix. Note that the entries are not necessarily *all* the (possibly) nonzero entries of the matrix, although for simplicity we shall henceforth assume this to be the case—thus ignoring cases in which certain nonzero entries in the matrix are known a priori or are otherwise easily available and hence do not require approximation [20]. With this assumption we observe that the pattern corresponds to the functionality of the individual component functions. Frequently the pattern is regular and easily identifiable; this is typically the case when the nonlinear system of equations arises from some discretization of a differential equation. We shall call patterns with such regular structures *structured patterns.* The tridiagonal form

$$
\begin{pmatrix}
* & * & & & \\
* & * & * & & \\
 & * & * & * & \\
 & & * & * & * \\
 & & & \ddots & \ddots & \ddots
\end{pmatrix}, \tag{8}
$$

where every $*$ represents a matrix entry, is a classic example of a structured pattern.

If one has access to the individual component functions as in (2), then one can use (6) for each i and j in the pattern to generate all the individual Jacobian entries efficiently. However, when such access is not convenient, one might choose to use (7). Consider applying (7) with $j = 1$ in the case of a problem with pattern (8). In this case, recalling that the pattern corresponds to the functionality of the individual component functions, the calculated difference (7) produces a vector with only two nonzero elements corresponding to the two entries in the first column of (8), and the remaining cost of the $\mathbf{F}$ evaluations produces only unwanted zeroes. This waste can be avoided by a simple observation first reported by [6], who noted that by carefully selecting the points where $\mathbf{F}$ is evaluated, one can generate approximations to all the entries in *several* columns of the Jacobian matrix simultaneously from a *single* difference.

The principle of the method of [6] is to allocate every column of the Jacobian matrix to one of a number of groups, in such a way that for every $i = 1, 2, \ldots, N$, in each group there is at most one column with an entry in row i. Such an allocation of columns is called a *consistent partition* of the columns of the matrix, and we shall denote it by a collection of sets, each set corresponding to a particular group and containing the indices of the columns in the corresponding group. For example, in the pattern

$$\begin{pmatrix} * & * & & \\ & * & * & \\ * & & & * \\ & & * & \end{pmatrix}$$

a possible consistent partition consists of $\{1, 3\}$ and $\{2, 4\}$, but the second column may not be in the same group as either the first or the third column. Similarly, for the tridiagonal case (8) a consistent partition is $\{1, 4, 7, \ldots\}$, $\{2, 5, 8, \ldots\}$, $\{3, 6, 9, \ldots\}$. Notice immediately that in general there is no unique consistent partition; for example, in the tridiagonal case another possible consistent partition is $\{1\}$, $\{2\}$, $\{3\}, \ldots$. In particular, the number of groups may vary from one consistent partition to another; one consistent partition that is always available is to allocate every column to a different group, as above. Many of the techniques for difference approximations that take no account of sparsity may be regarded as sparse techniques which use the latter partition—though we do not recommend that this approach be taken in practice.

Given a consistent partition, it is possible to approximate all the entries in all the columns of one particular group by one difference of the form (7), as follows. Let $j_1, j_2, \ldots, j_m$ be the indices of the columns in a particular group, say group l; then calculate the difference $\boldsymbol{\delta}^l \in \mathbf{R}^N$ as

$$\boldsymbol{\delta}^l = \mathbf{F}(\mathbf{x} + h_{j_1}\mathbf{e}_{j_1} + \mathbf{h}_{j_2}\mathbf{e}_{j_2} + \cdots + h_{j_m}\mathbf{e}_{j_m}) - \mathbf{F}(\mathbf{x}) \tag{9}$$

from which the entries in the j_kth column of the approximate Jacobian matrix $\mathbf{B}$ can be found as the corresponding elements of $h_{j_k}^{-1}\boldsymbol{\delta}^l$, for every j_k, $k = 1, 2, \ldots, m$. This can be readily appreciated in the case of the pattern (8), in which for the group $\{1, 4, 7, \ldots\}$ the difference

$$\mathbf{F}(\mathbf{x} + h_1\mathbf{e}_1 + h_4\mathbf{e}_4 + h_7\mathbf{e}_7 + \cdots) - \mathbf{F}(\mathbf{x}),$$

by virtue of the functionality identified by the pattern (8), has elements

$$\begin{gathered} f_1(\mathbf{x} + h_1\mathbf{e}_1) - f_1(\mathbf{x}) \\ f_2(\mathbf{x} + h_1\mathbf{e}_1) - f_2(\mathbf{x}) \\ f_3(\mathbf{x} + h_4\mathbf{e}_4) - f_3(\mathbf{x}) \\ f_4(\mathbf{x} + h_4\mathbf{e}_4) - f_4(\mathbf{x}) \\ f_5(\mathbf{x} + h_4\mathbf{e}_4) - f_5(\mathbf{x}) \\ f_6(\mathbf{x} + h_7\mathbf{e}_7) - f_6(\mathbf{x}) \\ \vdots \end{gathered}$$

which by comparison with (6) readily produces all the desired entries in columns $1, 4, 7, \ldots$ of **B**. We shall see below that differences of a form more general than (9) may also be used.

This idea first appeared in **[6]** and was extended to the approximation of sparse Hessian matrices in the context of optimization in **[16]** and a number of subsequent papers.

We note that if (9) is used then the number of evaluations of **F** required to approximate all the matrix entries is one more than the number of groups in the partition. For example, in the tridiagonal case with the partition consisting of three groups, only four evaluations of **F** suffice, irrespective of the value of N. We shall discuss later the possibility of reducing this number of evaluations of **F** when the Jacobian approximation is done in the context of a Newton-like iteration; but observe that generally it is valuable to have a partition with as few groups as possible. Thus, it is relevant to ask what the minimum number of groups in a partition may be, and how such a minimal partition may be found. Evidently the definition of a consistent partition ensures that the number of groups cannot be less than the greatest number of entries in any one row of the matrix. It was shown in **[14]** that the entire matrix can indeed by approximated using this minimal number of evaluations of **F**, though using a technique which differs from that presented here and in other respects seems likely to be more expensive on the majority of problems. In general the problem of finding the minimal partition in our sense is an NP-complete problem **[14]**, **[4]**. The latter result is established by showing that the problem of finding a minimal partition is equivalent to solving the graph-coloring problem. It follows that heuristics for the graph-coloring problem may be applied to find near-minimal partitions efficiently. This approach has been adopted in **[4]**, in which a number of techniques are proposed; the corresponding software is described in **[5]**. Given a matrix pattern, this software will efficiently generate a consistent partition which in most cases is near-minimal; the software can also be used to generate the corresponding Jacobian approximation effectively. Recent developments in graph-coloring algorithms (e.g., **[3]**) may lead to further improvements in these techniques.

It should be prominently noted that the techniques for finding near-minimal partitions mentioned above can often be avoided, particularly when the pattern is structured. For example we have seen that in the tridiagonal case the minimal partition of three groups is easily identified. More generally, it may be fairly easy to recognize the minimal partitions obtainable in such highly structured patterns as arise from the standard discretizations of differential equations. A number of results of this sort are given in [11] and [14], and the insights presented there may be extended to other cases.

4. A generalization. At this point it is convenient for subsequent discussion and analysis to introduce some additional notation and a useful generalization of (9) given in [20], motivated by the relationship between this work and [18]. Noting that (9) can be written in the form

$$\boldsymbol{\delta}^l = \mathbf{F}(\mathbf{x}+\mathbf{h}^l) - \mathbf{F}(\mathbf{x}), \qquad \mathbf{h}^l \in \mathbf{R}^N, \tag{10}$$

and assuming that there are M groups in the partition, we define $\mathbf{H}$ to be an $N \times M$ matrix with columns $\mathbf{h}^l$, $l = 1, 2, \ldots, M$, whose entry pattern corresponds to the pattern of the perturbation of $\mathbf{x}$ used when calculating the difference (9) for the lth group. Thus, for the minimal partition of the tridiagonal matrix, $\mathbf{H}$ has the pattern

$$\begin{pmatrix} * & & \\ & * & \\ & & * \\ * & & \\ & * & \\ & & * \\ \vdots & \vdots & \vdots \end{pmatrix}.$$

Let $\mathbf{P}$, $\mathbf{Q}$ be two further $N \times M$ matrices with columns $\mathbf{p}^l$, $\mathbf{q}^l$, $l = 1, 2, \ldots, M$, whose difference $\mathbf{p}^l - \mathbf{q}^l$ has the same pattern of entries as $\mathbf{h}^l$. We are not for the moment concerned with the values of these entries, just the patterns. Then an alternative to (10) and hence also to (9) is to use the difference

$$\boldsymbol{\delta}^l = \mathbf{F}(\mathbf{x}+\mathbf{p}^l) - \mathbf{F}(\mathbf{x}+\mathbf{q}^l) \tag{11}$$

to construct the entries in all the columns in group l of the matrix, obtaining entries in the j_kth column as the corresponding entries of $(\mathbf{p}^l_{j_k} - \mathbf{q}^l_{j_k})^{-1}\boldsymbol{\delta}^l$. Clearly (11) generalizes (10), and it is easy to show [20] that the resulting approximation $\mathbf{B}$ to the Jacobian matrix $\mathbf{F}'(\mathbf{x})$ satisfies

$$\mathbf{B}\mathbf{h}^l = \mathbf{F}(\mathbf{x}+\mathbf{p}^l) - \mathbf{F}(\mathbf{x}+\mathbf{q}^l) = \left(\int_0^1 \mathbf{F}'(\mathbf{x}+t\mathbf{h}^l)\,dt\right)\mathbf{h}^l$$

so that $\mathbf{B}$ satisfies a secant condition [9] and can be regarded as an average Jacobian along the line $\mathbf{h}^l$.

To illustrate the utility of this generalization, suppose $\mathbf{H}$ is given, and let

$$\mathbf{P} = (\mathbf{h}^1, \mathbf{h}^1 + \mathbf{h}^2,\ \mathbf{h}^1 + \mathbf{h}^2 + \mathbf{h}^3, \ldots, \mathbf{h}^1 + \mathbf{h}^2 + \mathbf{h}^3 + \cdots + \mathbf{h}^M),$$
$$\mathbf{Q} = (\mathbf{0}, \mathbf{h}^1, \mathbf{h}^1 + \mathbf{h}^2, \ldots, \mathbf{h}^1 + \mathbf{h}^2 + \mathbf{h}^3 + \cdots + \mathbf{h}^{M-1});$$

then clearly $\mathbf{P} - \mathbf{Q} = \mathbf{H}$, while the differences calculated in producing the approximation $\mathbf{B}$ take the form

$$\begin{gathered}\mathbf{F}(\mathbf{x} + \mathbf{h}^1) - \mathbf{F}(\mathbf{x}) \\ \mathbf{F}(\mathbf{x} + \mathbf{h}^1 + \mathbf{h}^2) - \mathbf{F}(\mathbf{x} + \mathbf{h}^1) \\ \mathbf{F}(\mathbf{x} + \mathbf{h}^1 + \mathbf{h}^2 + \mathbf{h}^3) - \mathbf{F}(\mathbf{x} + \mathbf{h}^1 + \mathbf{h}^2) \\ \vdots \\ \mathbf{F}(\mathbf{x} + \mathbf{h}^1 + \mathbf{h}^2 + \cdots + \mathbf{h}^M) - \mathbf{F}(\mathbf{x} + \mathbf{h}^1 + \mathbf{h}^2 + \cdots + \mathbf{h}^{M-1})\end{gathered}$$

so that again only $M+1$ evaluations of $\mathbf{F}$ are required to construct $\mathbf{B}$, as opposed to the $2M$ which (11) appears to require at first sight. In the context of iteration (4) we set $\mathbf{x} = \mathbf{x}^k$ and it is then possible to choose the values of the entries in $\mathbf{H}$ to be such that $\mathbf{x}^k + \mathbf{h}^1 + \mathbf{h}^2 + \cdots + \mathbf{h}^M = \mathbf{x}^{k-1}$, and since $\mathbf{F}(\mathbf{x}^{k-1})$ is available from the previous iteration this reduces the number of required evaluations of $\mathbf{F}$ by one. This technique enables one, for example, to generate successive approximations to tridiagonal Jacobians using only two additional evaluations of $\mathbf{F}$ per iteration. This idea appears in **[8]**, **[20]** and illustrates the value of examining these Jacobian approximation techniques in the context of the Newton-like iterations (4) in which they are used.

5. Updating Jacobian approximations. The major costs in iteration (4) are the construction of $\mathbf{B}_k$ and the linear algebra involved in calculating $\mathbf{d}^k = \mathbf{B}_k^{-1}\mathbf{F}(\mathbf{x}^k)$. These can be reduced by efficiently updating $\mathbf{B}_{k-1}$ to obtain $\mathbf{B}_k$, and updating the factors of the matrix $\mathbf{B}_{k-1}$ correspondingly if a direct solution technique is used to solve the linear equation $\mathbf{B}_k\mathbf{d}^k = \mathbf{F}(\mathbf{x}^k)$. The latter problem appears to be relatively unexplored in the current context at the time of this writing. We discuss below and in Sections 6 and 7 only the question of efficiently updating $\mathbf{B}_{k-1}$.

The most obvious approach is simply to calculate an entirely new approximation to the Jacobian matrix at every iteration, but this makes each iteration relatively expensive in terms of the number of evaluations of $\mathbf{F}$ required per iteration. An obvious cheap possibility is to set $\mathbf{B}_k = \mathbf{B}_{k-1}$. It is well known that this option of not updating the matrix approximation for a number of successive iterations may be successful, at some cost in the rate of convergence, but we shall show later that it can lead to failure of the iteration to converge unless the Jacobian approximation is at least periodically reevaluated so that at all times a sufficiently recent approximation is in use.

A related compromise proposal in **[20]** is to produce $\mathbf{B}_k$ from $\mathbf{B}_{k-1}$ by replacing only the entries in a few of the groups of columns by new differences of the form (11), cycling through all the groups in some systematic way as the iteration proceeds. Clearly this reduces the number of evaluations of **F** required per iteration. We proved in **[20]** that this technique does produce a locally convergent method. We investigate below some other aspects of this process, particularly the global convergence properties, but note immediately that this process embraces also the possibility of updating either all or none of the groups of columns at any iteration.

We now discuss a suggestion which initially appears to be useful in the above context but which in our experience failed to measure up to its promise. It is based on the observation that in a consistent partition it is often possible to move columns from one group to another without violating the conditions for maintaining a consistent partition. This suggests the possibility that when in the course of an iteration a particular group of columns is being updated, some other columns may *temporarily* be appended to that group and have their corresponding entries updated at no additional cost in **F** evaluations. This idea of *local expansion* has been used with some success in the nonlinear least squares problem **[10]** and was suggested by Moré for nonlinear equations **[13]**. We have conducted numerous computational experiments with this idea, in the context of updating a few groups of columns in the Jacobian approximation at every iteration in a cyclic fashion. Our experience was that, although it was frequently possible to increase the number of columns in any one group rather significantly, the resulting apparently improved approximation to the Jacobian matrix had essentially no beneficial effect on either the convergence behavior or the efficiency (in terms of the number of evaluations of **F** and the number of iterations required to achieve convergence) of the Newton-like method. This was true in all of the many cases we tested—irrespective of the degree of sparsity, whether we updated only a few or several groups of columns per iteration, started near or far from the solution, required high or low accuracy, and for all forms of nonlinearity we tested. Our conclusion is that in general the additional cost of identifying which columns may be appended to which group in a local expansion is not offset by any substantial benefits, and hence that this additional cost is not warranted. This contrasts somewhat with the reported experience in the nonlinear least squares case [10].

6. Convergence. We now investigate the local and global convergence properties of the iteration (4). In particular, we show that if groups of columns are not updated sufficiently frequently then iteration (4) may fail to converge at some stage. Most techniques for obtaining a large domain of convergence for such methods are based on replacing the problem of solving $\mathbf{F}(\mathbf{x}) = \mathbf{0}$ by solving

$$\min_x f(x) \quad \text{where} \quad f(x) = \frac{1}{2}\|\mathbf{F}(\mathbf{x})\|_2^2.$$

We recall some fundamental ideas from the solution of the optimization problem $\min_x f(x)$ for a general objective function f by iterations of the form (5) [9]. The standard steplength (or damped) techniques involve selecting $\mathbf{d}^k$ to be a *direction of descent* for f at $\mathbf{x}^k$, that is, to ensure that $\mathbf{d}^k$ satisfies

$$\nabla f(\mathbf{x}^k)^{\mathrm{T}}\mathbf{d}^k < 0, \tag{12}$$

which guarantees the existence of $t_k > 0$ such that

$$f(\mathbf{x}^k + t_k\mathbf{d}^k) < f(\mathbf{x}^k). \tag{13}$$

In order to be able to apply standard local convergence theorems it is necessary to strengthen inequality (13) [9], but we shall investigate only this weak form. In our case (13) is equivalent to

$$\|\mathbf{F}(\mathbf{x}^k + t_k\mathbf{d}^k)\|_2 < \|\mathbf{F}(\mathbf{x}^k)\|_2$$

and the question is whether $\mathbf{d}^k = -\mathbf{B}_k^{-1}\mathbf{F}(\mathbf{x}^k)$ satisfies (12). In the present context $\nabla f(x) = \mathbf{F}'(\mathbf{x})^{\mathrm{T}}\mathbf{F}(\mathbf{x})$ so (12) takes the form

$$\mathbf{F}(\mathbf{x}^k)^{\mathrm{T}}\mathbf{F}'(\mathbf{x}^k)\mathbf{d}^k < 0, \tag{14}$$

which may be rewritten

$$\mathbf{F}(\mathbf{x}^k)^{\mathrm{T}}(\mathbf{F}'(\mathbf{x}^k)\,\mathbf{d}^k + \mathbf{F}(\mathbf{x}^k) - \mathbf{F}(\mathbf{x}^k)) < 0$$

and hence as

$$\mathbf{F}(\mathbf{x}^k)^{\mathrm{T}}(\mathbf{F}'(\mathbf{x}^k)\,\mathbf{d}^k + \mathbf{F}(\mathbf{x}^k)) < \mathbf{F}(\mathbf{x}^k)^{\mathrm{T}}\mathbf{F}(\mathbf{x}^k).$$

Applying the Cauchy–Schwarz inequality, we see that (14) is satisfied if $\mathbf{d}^k$ satisfies

$$\frac{\|\mathbf{F}'(\mathbf{x}^k)\mathbf{d}^k + \mathbf{F}(\mathbf{x}^k)\|_2}{\|\mathbf{F}(\mathbf{x}^k)\|_2} < 1, \tag{15}$$

though this is clearly a stronger requirement than (14).

Condition (15) is essentially the inexact Newton condition of [7] for the particular norm $\|\cdot\|_2$, and hence in a more precise form is known to be sufficient for local convergence of the iteration (5) with $t_k = 1$ as proved in [7]. Condition (15) has previously been shown to be a sufficient condition for $\mathbf{d}^k$ to be a descent direction [1], [2], where essentially the following more direct argument is used to show that if $\mathbf{d}^k$ satisfies

$$\frac{\|\mathbf{F}'(\mathbf{x}^k)\mathbf{d}^k + \mathbf{F}(\mathbf{x}^k)\|}{\|\mathbf{F}(\mathbf{x}^k)\|} < 1 \tag{16}$$

for any norm $\|\cdot\|$ then there is some $t_k > 0$ such that

$$\|\mathbf{F}(\mathbf{x}^k + t_k\mathbf{d}^k)\| < \|\mathbf{F}(\mathbf{x}^k)\|. \tag{17}$$

They argue as follows. Writing

$$\begin{aligned}\mathbf{F}(\mathbf{x}+t\mathbf{d}) &= [\mathbf{F}(\mathbf{x}) + t\mathbf{F}'(\mathbf{x})\mathbf{d}] + [\mathbf{F}(\mathbf{x}+t\mathbf{d}) - \mathbf{F}(\mathbf{x}) - t\mathbf{F}'(\mathbf{x})\mathbf{d}]\\ &= [(1-t)\mathbf{F}(\mathbf{x}) + t(\mathbf{F}(\mathbf{x}) + \mathbf{F}'(\mathbf{x})\mathbf{d})]\\ &\quad + \left[\int_0^t \mathbf{F}'(\mathbf{x}+s\mathbf{d}) - \mathbf{F}'(\mathbf{x})\,ds\right]\mathbf{d},\end{aligned}$$

the triangle inequality enables us to deduce that for any $\lambda > 0$

$$\|\mathbf{F}(\mathbf{x}+t\mathbf{d})\| \le \left[(1-t) + t\frac{\|\mathbf{F}(\mathbf{x}) + \mathbf{F}'(\mathbf{x})\mathbf{d}\|}{\|\mathbf{F}(\mathbf{x})\|}\right]\|\mathbf{F}(\mathbf{x})\| + \frac{\gamma(t\|\mathbf{d}\|)^{1+\lambda}}{1+\lambda} \tag{18}$$

where

$$\gamma = \max_{s\in[0,t]} \frac{\|\mathbf{F}'(\mathbf{x}+s\mathbf{d}) - \mathbf{F}'(\mathbf{x})\|}{s^\lambda\|\mathbf{d}\|^\lambda}.$$

Note that γ may be assumed to be bounded if λ is such that $\mathbf{F}'$ satisfies the Hölder continuity condition

$$\|\mathbf{F}'(\mathbf{y}) - \mathbf{F}'(\mathbf{z})\| \le L\|\mathbf{y}-\mathbf{z}\|^\lambda \tag{19}$$

for some $L, \lambda > 0$ and all $\mathbf{y}, \mathbf{z}$ in some suitable subspace of $\mathbf{R}^N$ containing $\mathbf{x}^*$. With this assumption we observe that for t sufficiently small we may ignore the final (higher-order) term of (18) and hence satisfy (17) if (16) holds.

For the case $\mathbf{d}^k = -\mathbf{B}_k^{-1}\mathbf{F}(\mathbf{x}^k)$, condition (15) is

$$\frac{\|-\mathbf{F}'(\mathbf{x}^k)\mathbf{B}_k^{-1}\mathbf{F}(\mathbf{x}^k) + \mathbf{F}(\mathbf{x}^k)\|_2}{\|\mathbf{F}(\mathbf{x}^k)\|_2} < 1,$$

which we rewrite as

$$\frac{\|(\mathbf{B}_k - \mathbf{F}'(\mathbf{x}^k))\mathbf{B}_k^{-1}\mathbf{F}(\mathbf{x}^k)\|_2}{\|\mathbf{F}(\mathbf{x}^k)\|_2} < 1. \tag{20}$$

The crucial observation is that (20) can be satisfied provided that $\mathbf{B}_k$ is a sufficiently good approximation of $\mathbf{F}'(\mathbf{x}^k)$; by the Neumann lemma **[15]** the latter condition also guarantees that $\mathbf{B}_k$ is invertible provided that $\mathbf{F}'(\mathbf{x}^k)$ is invertible. Subject to (19) and some technical assumptions which need not concern us here, it was shown in **[20]** that

$$(21) \qquad \|\mathbf{B}_k - \mathbf{F}'(\mathbf{x}^k)\|_2 \leq LN^{1/2}\left(\sum_{j=1}^{M} \max\{\|\mathbf{p}_k^j\|_2, \|\mathbf{q}_k^j\|_2\}\right)^{1/2},$$

where $\mathbf{p}_k^l$ and $\mathbf{q}_k^l$ are the columns of the matrices $\mathbf{P}_k$, $\mathbf{Q}_k$ used in the construction of $\mathbf{B}_k$ via (11) and the matrix subscripts indicate the use of possibly different matrices $\mathbf{P}_k$ and $\mathbf{Q}_k$ as the iteration progresses.

From (21) it is apparent that the direction $\mathbf{d}^k$ satisfies (15) provided that $\|\mathbf{p}_k^l\|_2$ and $\|\mathbf{q}_k^l\|_2$ are sufficiently small; hence these are the conditions which ensure that (4) will enjoy not only local convergence when $t_k = 1$ but also wider convergence when the steplength parameter $t_k > 0$ is used to ensure that (17) holds. On the other hand, convergence might fail to occur if $\|\mathbf{p}_k^l\|_2$ or $\|\mathbf{q}_k^l\|_2$ are allowed to grow with k. Unfortunately, this can easily happen when certain columns of the Jacobian matrix approximation are not updated from one iteration to the next, as we now show. Suppose that at the kth iteration we approximate the columns in the lth group of the partition using

$$\boldsymbol{\delta}_k^l = \mathbf{F}(\mathbf{x}^k + \mathbf{p}_k^l) - \mathbf{F}(\mathbf{x}^k + \mathbf{q}_k^l)$$

and do not update these columns at the next iteration; then the perturbations $\mathbf{p}_{k+1}^l$ and $\mathbf{q}_{k+1}^l$ *implicitly* being used at the next iteration satisfy

$$\boldsymbol{\delta}_{k+1}^l = \mathbf{F}(\mathbf{x}^{k+1} + \mathbf{p}_{k+1}^l) - \mathbf{F}(\mathbf{x}^{k+1} + \mathbf{q}_{k+1}^l) = \mathbf{F}(\mathbf{x}^k + \mathbf{p}_k^l) - \mathbf{F}(\mathbf{x}^k + \mathbf{q}_k^l) = \boldsymbol{\delta}_k^l,$$

which we may interpret as

$$(22) \qquad \mathbf{p}_{k+1}^l = \mathbf{p}_k^l + (\mathbf{x}^k - \mathbf{x}^{k+1}) \quad \text{and} \quad \mathbf{q}_{k+1}^l = \mathbf{q}_k^l + (\mathbf{x}^k - \mathbf{x}^{k+1}).$$

Similar expressions hold in subsequent iterations if the columns of group l are not updated; at iteration $k+n$ we find

$$(23) \qquad \mathbf{p}_{k+n}^l = \mathbf{p}_k^l + (\mathbf{x}^k - \mathbf{x}^{k+n}) \quad \text{and} \quad \mathbf{q}_{k+n}^l = \mathbf{q}_k^l + (\mathbf{x}^k - \mathbf{x}^{k+1}).$$

Equation (23) reveals the potential for growth of $\|\mathbf{p}_k^l\|_2$ and $\|\mathbf{q}_k^l\|_2$ with increasing k and the consequent possibility that through failure to satisfy (15) the iteration could fail either due to failure to find a descent direction or through failure of local convergence to occur.

7. Computational aspects. The phenomenon described in Section 6 is not apparent in computation when starting from a point $\mathbf{x}^0$ near the solution $\mathbf{x}^*$, since then successive estimates $\mathbf{x}^k$ are relatively close to one another so little growth is possible. However, we have commonly experienced failure of the iteration due to this phenomenon when starting far from $\mathbf{x}^*$. In extensive computational testing of (4), in which $\mathbf{B}_k$ was calculated by updating several groups of columns at every iteration, we found that virtually every failure of the iteration (other than those caused by encountering a singular matrix) was due to failure to obtain a good descent direction, which was caused by the use of a poor Jacobian approximation, which in turn was caused by our failure to update the groups of columns sufficiently frequently. That this was indeed the cause of the difficulties was confirmed by simply increasing the number of groups of columns being updated at every iteration and noting that invariably when a sufficient number of groups were updated simultaneously no failure of this sort occurred.

On a positive note we observed that when the technique of updating only a few groups of columns at every iteration resulted in a convergent iteration we were usually able to reduce the number of evaluations of $\mathbf{F}$ required to obtain convergence well below the number required when updating all the groups of columns at every iteration. The exceptions were the cases in which we updated only very few groups of columns at every iteration; in these cases the Jacobian approximation deteriorated to such an extent that although convergence still occurred, it was very slow. This suggests that there is some "optimal" number of groups to update at every iteration.

We also compared this technique with the alternative of simply updating the entire Jacobian approximation every few iterations, leaving it unchanged during the intermediate steps. In this case we found that provided that we updated an optimal number of groups of columns per iteration (in the sense of the preceding paragraph) we could save a significant number of evaluations of $\mathbf{F}$ compared to the number required when updating the entire matrix a fixed optimal number of iterations apart. On the other hand, the latter technique requires fewer matrix factorizations, so in practice it is important to weigh the relative costs of evaluations of $\mathbf{F}$ against the costs of the linear algebra involved. In particular, we caution that even if only one group of columns of a sparse matrix is updated then in general a high-rank update of the matrix has occurred and hence the corresponding updating of the matrix factors seems likely to be expensive. In this sense there are situations in which it pays to update a significant number (rather than just a few) groups of columns in one iteration—requiring one expensive matrix factorization at this stage—and then perform no further update for a number of subsequent iterations and hence no further factorizations. Provided that these updates are done sufficiently often the iteration will converge, as shown by the analysis outlined above.

The empirical observations described above suggest the need for a scheme to automatically monitor the accuracy of Jacobian approximations and hence trigger a new update of either the entire matrix or a few groups of columns when the need becomes apparent. A key condition for convergence to occur is (15), so an obvious approach is to construct an approximation to the left-hand term in (15) and trigger a complete update of all the columns of the Jacobian approximation when the value of this estimate approaches 1. Noting that both $\mathbf{F}(\mathbf{x}^k)$ and $\mathbf{F}(\mathbf{x}^k + t_k\mathbf{d}^k)$ are at hand after a successful iteration of (4), and that

$$\mathbf{F}(\mathbf{x}^k + t_k\mathbf{d}^k) - \mathbf{F}(\mathbf{x}^k) \approx t_k\mathbf{F}'(\mathbf{x}^k)\,\mathbf{d}^k, \tag{24}$$

we are led to the cheap estimate

$$\frac{\|\mathbf{F}'(\mathbf{x}^k)\mathbf{d}^k + \mathbf{F}(\mathbf{x}^k)\|_2}{\|\mathbf{F}(\mathbf{x}^k)\|_2} \approx \frac{\|\mathbf{F}(\mathbf{x}^k + t_k\mathbf{d}^k) - (1 - t_k)\mathbf{F}(\mathbf{x}^k)\|_2}{t_k\|\mathbf{F}(\mathbf{x}^k)\|_2}.$$

Of course, this only provides an a posteriori estimate, available only after a successful iteration of (4) has occurred, but nevertheless a value of the right-hand side near 1 could be interpreted as indicating the need for a substantial update of the Jacobian approximation. We have conducted several experiments in which this estimate was used to trigger a completely new Jacobian approximation when the value of the right-hand side exceeded 0.75, with a resulting significant reduction in the incidence of failure due to lack of a descent direction. However, this proposal is as yet tentative and we are investigating other means, based on (14) and (24), for detecting whether an effective descent direction is available. An almost equally obvious alternative motivated by (22) is simply to monitor $\|\mathbf{x}^{k+1} - \mathbf{x}^k\|_2$ and update the entire Jacobian matrix approximation when this value is "large" in some sense, while (23) suggests that if the columns in group l were last updated at iteration k using (11) with $\mathbf{p}_k^l = \mathbf{h}^l$ and $\mathbf{q}_k^l = 0$ (that is, essentially using (10)), then after iteration j one might calculate

$$\|\mathbf{p}_k^l + (\mathbf{x}^j - \mathbf{x}^k)\|_2 = \|\mathbf{p}_k^l + (\mathbf{x}^j - \mathbf{x}^{j-1}) + \cdots + (\mathbf{x}^{k+1} - \mathbf{x}^k)\|_2$$

and update group l when this value becomes "large". The latter two proposals coincide with the traditional rule-of-thumb that if you have taken a large step then you should calculate a new Jacobian approximation, with the standard justification that if a large step has been taken then the Jacobian is likely to have changed significantly, an idea supported by our argument above.

8. Concluding remarks. We offer two concluding remarks. First, we note that our convergence results apply also to the dense case, where in the latter we regard every column as an individual group in the consistent partition. Our notation here has deliberately been consistent with that in [**20**] in order to ease such comparisons. In particular, in the dense case, too, it is essential to maintain a sufficiently recently calculated approximation to the Jacobian matrix if either local or global convergence is to occur. The monitoring devices we have proposed above can be used in this case too. Finally, we point out that these sparse finite difference techniques can be extended to the practically important case of block-structured Jacobians; an example of this can be found in [**12**].

References

1. O. Axelsson, *On global convergence of iterative methods*, Iterative Solution of Nonlinear Systems of Equations, Springer Lecture Notes in Mathematics 953, Springer-Verlag, Berlin, 1982.

2. R. E. Bank and D. J. Rose, *Global approximate Newton methods*, Numer. Math. **37** (1981), 279–295.

3. S. Bokhari, *Assignment Problems in Parallel and Distributed Computing*, Kluwer Academic Publ., Boston, 1987.

4. T. F. Coleman and J. J. Moré, *Estimation of sparse Jacobian matrices and graph-coloring problems*, SIAM J. Numer. Anal. **20** (1983), 187–209.

5. T. F. Coleman, B. S. Garbow, and J. J. Moré, *Software for estimating sparse Jacobian matrices: Algorithm* 618, ACM Trans. Math. Software **10** (1984), 329–347.

6. A. R. Curtis, M. J. D. Powell, and J. K. Reid, *On the estimation of sparse Jacobian matrices*, J. Inst. Math. Applic. **13** (1974), 117–119.

7. R. S. Dembo, S. C. Eisenstat, and T. Steihaug, *Inexact Newton methods*, SIAM J. Numer. Anal. **19** (1982), 400–408.

8. J. E. Dennis, Jr. and G. Li, *A hybrid algorithm for solving sparse nonlinear systems of nonlinear equations*, Math. Comput. **50** (1988), 155–166.

9. J. E. Dennis, Jr. and R. B. Schnabel, *Numerical Methods for Unconstrained Optimization and Nonlinear Equations*, Prentice-Hill, Englewood Cliffs, New Jersey, 1983.

10. J. E. Dennis, Jr. and T. Steihaug, *On the successive projections approach to solving least-squares problems*, Report 83–18, Math. Sciences, Rice University, Houston, Texas, 1983.

11. D. Goldfarb and Ph. Toint, *Optimal estimation of Jacobian and Hessian matrices that arise in finite difference calculations*, Math. Comput. **43** (1984), 69–88.

12. W. Mönch, *Secant methods for sparse systems of nonlinear equations with a special structure*, Computing **30** (1983), 213–223.

13. J. J. Moré, private communication (1985).

14. G. N. Newsam and J. D. Ramsdell, *Estimation of sparse Jacobian matrices*, SIAM J. Alg. Disc. Meth. **4** (1983), 404–418.

15. J. M. Ortega and W. C. Rheinboldt, *Iterative Solution of Nonlinear Equations in Several Variables*, Academic Press, New York, 1970.

16. M. J. D. Powell and Ph. Toint, *On the estimation of sparse Hessian matrices*, SIAM J. Numer. Anal. **16** (1979), 1060–1074.

17. H. Schwetlick, *Numerische Lösung nichtlinearer Gleichungen*, Oldenbourg Verlag, Munich, 1979.

18. T. J. Ypma, *Local convergence of difference Newton-like methods*, Math. Comput. **41** (1983), 527–536.

19. T. J. Ypma, *Difference Newton-like methods under weak continuity conditions*, Computing **33** (1984), 51–64.

20. T. J. Ypma, *Efficient estimation of sparse Jacobian matrices by differences*, J. Comp. Appl. Math. **18** (1987), 17–28.

Western Washington University

Lectures in Applied Mathematics
Volume **26** (1990)

A Collection of Nonlinear Model Problems

JORGE J. MORÉ

Abstract. This paper presents a collection of nonlinear problems. The aim of this collection is to provide model problems of scientific interest to researchers interested in algorithm development. The problems and their origin are described, and references are given to other papers that have dealt with these problems.

Model problems of scientific interest are an important tool for researchers interested in algorithm development, and thus it was natural to use the AMS-SIAM Summer Seminar on the *Computational Solution of Nonlinear Systems of Equations* as the occasion to solicit suitable problems for a collection of nonlinear problems. This paper describes the problems and the origin of the problems that were submitted. References to other papers that have dealt with these problems are included.

There are other collections of test problems in the numerical analysis community. Two of the best known collections are the Harwell-Boeing sparse matrix test problem collection [**1**], and the netlib collection of linear programming test problems [**2**]. Both of these collections are for linear problems. Among the collections of nonlinear test problems we note the collections of Moré, Garbow, and Hillstrom [**4**] and that of Hock and Schittkowski [**3**].

The usefulness of these collections derives from their availability as software in a convenient format. At present this is certainly not the case for the collection described in this paper. We intend, however,

---

1980 *Mathematics Subject Classification* (1985 *Revision*). Primary 65H10, 65K05, 65K10, 90C30.

Work supported in part by the Applied Mathematical Sciences subprogram of the Office of Energy Research of the U.S. Department of Energy under Contract W-31-109-Eng-38.

to develop a software collection of large-scale nonlinear test problems which would include, for example, many of the problems in this paper.

The problems and contributors to this collection are as follows:

G. Auchmuty: Unconstrained Variational Principles for Eigenvalue Problems.

R. Fletcher: Distillation Column Test Problem.

R. Glowinski and H. B. Keller: The Bratu Problem.

C. T. Kelley: The Chandrasekhar H-equation.

M. Kostreva: The Elastohydrodynamic Lubrication Problem.

R. Mejia: Model Problems for Renal Network Flows.

H. D. Mittelmann: The Obstacle Bratu Problem.

A. B. Poore: A Tubular Chemical Reactor Model.

W. C. Rheinboldt: An Aircraft Stability Test Problem.

W. C. Rheinboldt: A Semi-Conductor Problem.

W. C. Rheinboldt: A Shallow Arch.

W. C. Rheinboldt: A Trigger Circuit.

H. F. Walker: The Inverse Elastic Rod Problem.

H. F. Walker: The Driven Cavity Problem.

One of the important features of this collection is that all of the problems arise in an application area. There are no toy problems in this collection; the solution of problems from this collection is an important challenge to algorithm developers. Most of the problems are phrased in terms of systems of nonlinear equations, but there are also problems in variational inequalities, nonlinear complementarity, and minimization.

Additional information on the problems in this collection, or suggestions for additional problems, will be appreciated.

ACKNOWLEDGMENTS. The idea behind this paper first arose during my lecture in the form of a casual request for test problems of scientific interest. I am sure that this request would have suffered the same fate as similar requests if Giles Auchmuty had not proposed that this casual request be made into a formal request, or if Gene Allgower had not seconded the proposition of Giles.

This paper was also influenced by related efforts in the mathematical programming community to collect large-scale problems. See, for example, Toint [5].

Finally, Rosemary Barrera deserves a special thanks for her help in transforming the contributions into LATEX.

REFERENCES

1. I. S. Duff, R. G. Grimes, and J. G. Lewis, *Sparse matrix test problems*, Harwell Laboratory, Computer Science and Systems Division report CSS 191, Harwell, England, 1987.

2. D. M. Gay, *Electronic mail distribution of linear programming test problems*, COAL Newsletter No. 13 (1985), 10–12.

3. W. Hock and K. Schittkowski, *Test examples for nonlinear programming codes*, Lecture Notes in Economics and Mathematical Systems **187**, Springer-Verlag, New York, 1981.

4. J. J. Moré, B. S. Garbow, and K. E. Hillstrom, *Testing unconstrained optimization software*, ACM Trans. Math. Software **7** (1981), 17–41.

5. Ph. L. Toint, *Call for test problems in large-scale nonlinear optimization*, COAL Newsletter No. 16 (1987), 5–10.

1. Unconstrained variational principles for eigenvalue problems (contributed by G. Auchmuty). The following problems provide some interesting examples for testing unconstrained optimization methods. They all involve nonconvex functions which, in general, have many critical points of various different types. The size of these problems may range from small to as large as desired.

Eigenvalues of a real symmetric matrix. Let $\mathbf{A}$ be a real, symmetric $n \times n$ matrix with eigenvalues $\lambda_n \leq \lambda_{n-1} \leq \cdots \leq \lambda_1$. When μ is not an eigenvalue of $\mathbf{A}$, consider the problem of minimizing the function

$$F(\mathbf{x}, \mu) = \frac{1}{2}\|(\mathbf{A} - \mu\mathbf{I})\mathbf{x}\|^2 - \frac{1}{q}\|\mathbf{x}\|^q, \tag{1.1}$$

where $\mathbf{x}$ is in $\mathbf{R}^n$, $\|\cdot\|$ represents the Euclidean norm, and $1 \leq q < 2$.

This function is bounded below on $\mathbf{R}^n$, and its critical points occur at certain eigenvectors of $(\mathbf{A} - \mu\mathbf{I})^2$ whose norm is related to the corresponding eigenvalue. Using functional calculus, one can relate the eigenvalues and eigenvectors of $(\mathbf{A} - \mu\mathbf{I})^2$ to those of $\mathbf{A}$. This problem was analyzed in Section 5 of **[1]**, for the case $q = 1$, and in Example 7.1 of **[2]** for the case $1 < q < 2$.

The critical points of $F(\cdot, \mu)$ occur at $|\lambda_j - \mu|^r \mathbf{e}^{(j)}$ where $r = 2/(q-2)$ and $\mathbf{e}^{(j)}$ is a unit vector obeying

$$(\mathbf{A} - \mu\mathbf{I})^2\mathbf{x} = (\lambda_j - \mu)^2\mathbf{x}. \tag{1.2}$$

The corresponding critical values are $-[\gamma|\lambda_j - \mu|^\gamma]^{-1}$ with $\gamma = -qr$. The minimizers occur at those critical points corresponding to the eigenvalues of $\mathbf{A}$ closest to μ.

The function $F(\cdot, \mu)$ defined by (1.1) is infinitely often continuously differentiable except at the origin. In particular, its Hessian is explicitly computable and, at each nonzero critical point, the eigenvectors of this Hessian are the same as those of $(\mathbf{A}-\mu\mathbf{I})^2$. Hence, the eigenvalues of the Hessian can be found, so one can describe the type of the critical point and determine when it is degenerate. By varying $\mathbf{A}$, q, or μ, one can choose functions that have various different features for testing various optimization algorithms.

Weighted eigenvalue problems. Let $\mathbf{A}$ and $\mathbf{C}$ be two real, symmetric, positive definite matrices and consider the problem of minimizing

$$F(\mathbf{x}) = \frac{1}{2}\langle \mathbf{Ax}, \mathbf{x}\rangle - \frac{1}{q}\langle \mathbf{Cx}, \mathbf{x}\rangle^{q/2} \tag{1.3}$$

with $1 \leq q < 2$. When $\mathbf{A}$ is symmetric, but not positive definite, one can replace $\mathbf{A}$ by $\mathbf{B} = \mathbf{A} + \mu\mathbf{C}$ for some μ large enough, so that $\mathbf{B}$ is positive definite and the following analysis generalizes.

The critical points of this function are eigenvectors of the weighted eigenvalue problem

$$\mathbf{Ax} = \nu\mathbf{Cx} \tag{1.4}$$

with $\langle \mathbf{Cx}, \mathbf{x}\rangle = \nu^{-r}$ and $r = 2/(2-q)$. The minimal value is $-[2\gamma\nu_n^{\gamma}]^{-1}$, where ν_n is the least eigenvalue of (1.4) and $\gamma = qr/2$.

This problem is analyzed in Example 7.2 of **[2]**, where it is shown that this is minimized at a (weighted) eigenvector of $(\mathbf{A}, \mathbf{C})$ corresponding to the smallest eigenvalue. Again one can derive explicit formulas for the Hessian and analyze the type of each critical point of this function.

Positive weighted eigenproblems. Let $\mathbf{A}$, $\mathbf{C}$ be real, symmetric $n \times n$ matrices with $\mathbf{C}$ positive definite. Consider the problem of minimizing

$$G(\mathbf{x}, \mu) = \frac{1}{p}\langle \mathbf{Cx}, \mathbf{x}\rangle^{p/2} - \frac{1}{2}\langle (\mu\mathbf{C} - A)\mathbf{x}, \mathbf{x}\rangle \tag{1.5}$$

where $2 < p < \infty$. When $\mu = 0$, this problem is analyzed in Example 7.3 of **[2]**, and if $\mathbf{C} = \mathbf{I}$, $\mu = 0$, it is treated in Section 6 of **[1]**. When μ is sufficiently negative this function is strictly convex and 0 is the only critical point and is a global minimizer of $G(\cdot, \mu)$ on $\mathbf{R}^n$.

As μ increases, new critical points bifurcate from the trivial critical point at 0 when $\mu = \nu_n, \nu_{n-1}, \ldots, \nu_1$. These critical points are eigenvectors of $(\mathbf{A}, \mathbf{C})$ with $\langle \mathbf{Cx}, \mathbf{x}\rangle$ being related to a function of $|\nu_j - \mu|$. These new branches either consist of isolated points or they are diffeomorphic to spheres of dimension d depending on whether the corresponding eigenvalue is simple or has multiplicity $d + 1$. Again for this problem, the Hessian and its eigenvalues are explicitly computable, so one can describe the type of any critical point.

Eigenproblems of the above type often arise in stability theory as well as in various science and engineering applications. Many examples of finding the natural frequencies and mode shapes of various engineering systems are described in **[3]**. Problems of finding the energy levels and the quantum states of various atoms and molecules are described in **[4]**.

REFERENCES

1. G. Auchmuty, *Unconstrained variational principles for eigenvalues of real symmetric matrices*, SIAM J. Math. Anal. **20** (1989), 1186–1207.

2. G. Auchmuty, *Duality algorithms for nonconvex variational principles*, Numer. Funct. Anal. and Optim. **10** (1989), 211–264.

3. R. D. Blevins, *Formulas for Natural Frequency and Mode Shape*, Van Nostrand, New York, 1979.

4. I. Levine, *Quantum Chemistry*, 3rd edition, Allyn & Bacon, New York, 1983.

2. Distillation column test problem (contributed by R. Fletcher). An n-stage distillation column has a reboiler, $n-2$ plates, and a condenser, which we index respectively (see Figure 1.1.1 in **[1]**) by $i = 0, 1, 2, \ldots, n-1$. A steady-state solution is sought in which a feed stream supplies material between stages k and $k+1$, and products are removed from the top and bottom of the column. The material on each stage is a mixture of m components indexed by $j = 1, 2, \ldots, m$, which may be in either liquid or vapor phase. The amount of vapor passing from stage i to $i+1$ in given time is denoted by v_i and liquid from stage i to $i-1$ by l_i. The proportion of component j in the liquid l_i is denoted by x_{ij} and in the vapor v_i by y_{ij}. We denote the temperature of stage i by t_i, the heat content per unit of liquid by h_i, and per unit of vapor by H_i. The amount of material removed from the top and bottom of the plant in unit time is denoted by d and b, respectively. The following equations determine the equilibrium conditions.

Material balance on each stage.

STAGE 0.

$$l_1 x_{1j} = v_0 y_{0j} + b x_{0j}, \qquad j = 1, \ldots, m. \tag{2.1}$$

STAGES $i = 1, \ldots, n-2$.

$$v_{i-1} y_{i-1,j} + l_{i+1} x_{i+1,j} = v_i y_{ij} + l_i x_{ij}, \qquad j = 1, \ldots, m. \tag{2.2}$$

In addition, if $i = k$ then a constant f_j^l occurs on the left-hand side (l.h.s.) and if $i = k + 1$ a constant f_j^v occurs on the l.h.s.; these account for the feed input.

STAGE $n - 1$.

$$y_{n-2,j} = x_{n-1,j}, \qquad j = 1, \ldots, m. \tag{2.3}$$

Material balance over the column.

$$l_i = v_{i-1} + b, \qquad i = 1, \ldots, k, \tag{2.4}$$

$$l_i = v_{i-1} - d, \qquad i = k + 1, \ldots, n - 1. \tag{2.5}$$

It is also assumed that input and output to the column balance, i.e., $\sum_j (f_j^v + f_j^l) = b + d$.

Equilibrium conditions.

$$y_{ij} = k_{ij} x_{ij}, \qquad j = 1, \ldots, m, \quad i = 0, \ldots, n - 1. \tag{2.6}$$

Here the k_{ij} are certain functions of t_i defined below.

Summation equations.

$$\sum_{j=1}^{m} y_{ij} = 1, \qquad i = 0, \ldots, n - 1. \tag{2.7}$$

It is also a consequence of (2.1)–(2.7) that $\sum_j x_{ij} = 1$, but it seems better to enforce (2.7) explicitly.

Heat balance equations.

STAGE 0.

$$l_1 h_1 + q = v_0 H_0 + b h_0. \tag{2.8}$$

Here q is the heat rate supplied to the reboiler.

STAGES $i = 1, \ldots, n - 2$.

$$v_{i-1} H_{i-1} + l_{i+1} h_{i+1} = v_i H_i + l_i h_i. \tag{2.9}$$

In addition, if $i = k$ then a constant h_f occurs on the l.h.s., and if $i = k + 1$, a constant H_f occurs on the l.h.s.; these account for the heat content of the feed (see below).

Other relationships. The k_{ij} are functions of t_i defined by the relationship

$$k_{ij} = \frac{1}{\pi_i} \exp\left(a_j + \frac{b_j}{c_j + t_i}\right), \tag{2.10}$$

where a_j, b_j, c_j for $j = 1, \ldots, m$ (Antoine constants) and π_i for $i = 0, \ldots, n-1$ (stage pressures) are data values.

The h_i are functions of x_{ij} and t_i defined by

$$h_i = \sum_{j=1}^{m} x_{ij}(\alpha_j + \alpha'_j t_i + \alpha''_j t_i^2), \tag{2.11}$$

where α_j, α'_j, α''_j for $j = 1, \ldots, m$ (liquid enthalpy constants) are data. Similarly, the H_i are functions of y_{ij} and t_i defined by

$$H_i = \sum_{j=1}^{m} y_{ij}(\beta_j + \beta'_j t_i + \beta''_j t_i^2), \tag{2.12}$$

where β_j, β'_j, β''_j for $j = 1, \ldots, m$ (vapor enthalpy constants) are data.

The quantities h_f and H_f in (2.9) are defined by

$$h_f = \sum_{j=1}^{m} f_j^l(\alpha_j + \alpha'_j t_f + \alpha''_j t_f^2) \tag{2.13}$$

and

$$H_f = \sum_{j=1}^{m} f_j^v(\beta_j + \beta'_j t_f + \beta''_j t_f^2), \tag{2.14}$$

where t_f is the feed temperature (data).

A system of equations. We define the variables to be the x_{ij} for $i = 0, \ldots, n-1$ and $j = 1, \ldots, m$; the t_i for $i = 0, \ldots, n-1$; and the v_i for $i = 0, \ldots, n-2$. Equations (2.4)–(2.6) and (2.10)–(2.12) are used to define the quantities l_i, y_{ij}, k_{ij}, h_i, and H_i, respectively, as functions of these variables. The equations that still must be satisfied are those in (2.1)–(2.3) and (2.7)–(2.9). These form the equations of the problem. Thus, there are $n(m+2)-1$ equations and $n(m+2)-1$ variables. To scale the equations, I multiply equations (2.1) and (2.2) by 0.01, and equations (2.8) and (2.9) by 10^{-6}.

Data. The data required to define the problem consist of the Antoine constants a_i, b_i, c_i in (2.10); the liquid enthalpy constants α_i, α'_i, α''_i in (2.11); the vapor enthalpy constants β_i, β'_i, β''_i in (2.12); the constants

f_i^l and f_i^v in (2.2); the constant t_f in (2.13) and (2.14); the constants b, d, q in (2.2), (2.3)–(2.5), and (2.8); and the constants π_i in (2.10):

$$n, m, k;$$
$$a_1, b_1, c_1, \ldots, a_m, b_m, c_m;$$
$$\alpha_1, \alpha'_1, \alpha''_1, \ldots, \alpha_m, \alpha'_m, \alpha''_m;$$
$$\beta_1, \beta'_1, \beta''_1, \ldots, \beta_m, \beta'_m, \beta''_m;$$
$$f_1^l, \ldots, f_m^l, f_1^v, \ldots, f_m^v;$$
$$t_f, b, d, q;$$
$$\pi_0, \ldots, \pi_{n-1}.$$

In addition, initial values for the variables

$$x_{01}, \ldots, x_{0m}, x_{11}, \ldots, x_{1m}, \ldots, x_{n-1,1}, \ldots, x_{n-1,m};$$
$$t_0, \ldots, t_{n-1};$$
$$v_0, \ldots, v_{n-2};$$

are required.

Numerical data for three problems, set out in the above format, are given below.

Numerical experience. To get Newton's method with a line search on the sum of squares, I used a standard Gauss–Newton subroutine (setting # of variables = # of residuals). This solved the hydrocarbon-6 and hydrocarbon-20 problems in 5 and 13 iterations, respectively (the problems have 29 and 99 variables, respectively). However, the routine does not solve the methanol-8 problem (31 variables).

In an attempt to solve the latter, I included equations $\sum_j x_{ij} = 1$ for $i = 0, \ldots, n-1$ among the residuals and solved an overdetermined least squares problem. This appeared to find a local minimum with sum of squares $\simeq 10^{-2}$. I still do not know if there exists a solution to the methanol-8 problem. I tried this idea on the hydrocarbon problems and it also failed to solve those!

References

1. R. Fletcher, *Practical methods of optimization*, 2nd edition, Wiley, New York, 1987.

Appendix: Numerical data.

6	3	2						
9.647	−2998.00	230.66	9.953	−3448.10	235.88	9.466	−3347.25	215.31
0.	37.6	0.	0.	48.2	0.	0.	45.4	0.
8425.	24.2	0.	9395.	35.6	0.	10466.	31.9	0.
30.	30.	40.	0.	0.	0.			
100.	40.	60.	2500000.					
6*1.								

Table 2.1. Hydrocarbon-6 data

20	3	9						
9.647	−2998.00	230.66	9.953	−3448.10	235.88	9.466	−3347.25	215.31
0.	37.6	0	0.	48.2	0.	0.	45.4	0.
8425.	24.2	0	9395.	35.6	0.	10466.	31.9	0.
30.	30.	40.	0.	0.	0.			
100.	40.	60.	2500000.					
20*1.								

Table 2.2. Hydrocarbon-20 data

8	2	2					
18.5751	−3632.649	239.2	18.3443	−3841.2203	228.		
0.	15.97	.0422	0.	18.1	0.		
9566.67	−1.59	.0422	10834.67	8.74	0.		
451.25	684.25	0.	0.				
89.	693.37	442.13	8386200.				
1210.	1200.	1190.	1180.	1170.	1160.	1150.	1140.

Table 2.3. Methanol-8 data

0.	.2	.9	0.	.2	.8	.05	.3	.8
.1	.3	.6	.3	.5	.3	.6	.6	0.
6*100.								
5*300.								

Table 2.4. Hydrocarbon-6 initial values

0.	.3	1.	0.	.3	.9	.01	.3	.9
.02	.4	.8	.05	.4	.8	.07	.45	.8
.09	.5	.7	.1	.5	.7	.15	.5	.6
.2	.5	.6	.25	.6	.5	.3	.6	.5
.35	.6	.5	.4	.6	.4	.4	.7	.4
.42	.7	.3	.45	.75	.3	.45	.75	.2
.5	.8	.1	.5	.8	0.			
20*100.								
19*300.								

Table 2.5. Hydrocarbon-20 initial values

.09203	.908	.1819	.8181	.284	.716	.3051	.6949
.3566	.6434	.468	.532	.6579	.3421	.8763	.1237
120.	110.	100.	88.	86.	84.	80.	76.
886.37	910.01	922.52	926.45	935.56	952.83	975.73	

Table 2.6. Methanol-8 initial values

107.47	102.4	97.44	96.3	93.99	89.72	83.71	78.31

Table 2.7. Alternative t_i values for Table 2.6

5.47225e-3	.129537	.864991			
2.00767e-2	.222556	.757368			
6.45143e-2	.330253	.605232			
.117201	.442833	.439966			
.249447	.501568	.248985			
.496352	.413642	9.00063e-2			
132.683	127.942	120.569	113.650	103.977	93.1902
288.902	288.558	288.669	293.339	303.021	

Table 2.8. Hydrocarbon-6 solution

2.52926e-7	1.94816e-3	.998052
1.04680e-6	3.81029e-3	.996189
4.22935e-6	7.21394e-3	.992782
1.69425e-5	1.34022e-3	.986581
6.76008e-5	2.45442e-2	.975388
2.65354e-4	4.42547e-2	.955480
1.02604e-3	7.80309e-2	.920943
3.84530e-3	.132701	.863454
1.36661e-2	.212539	.773795
4.44913e-2	.308528	.646981
4.95711e-2	.418733	.531696
5.56062e-2	.549980	.394414
6.17424e-2	.677091	.261166
6.76299e-2	.776392	.155978
7.44920e-2	.839436	8.60723e-2
8.61773e-2	.868949	4.48732e-2
.111427	.866286	2.22877e-2
.168534	.821085	1.03809e-2
.288999	.706735	4.26616e-3
.500000	.498701	1.29894e-3

138.218	138.142	138.004	137.753	137.300	136.498	135.106
132.762	128.946	122.975	119.604	115.963	112.742	110.347
108.710	107.442	105.958	103.419	98.833	92.079	

290.694	290.658	290.594	290.484	290.303	290.023	289.633
289.158	288.756	288.483	291.408	295.179	298.905	301.799
303.640	304.613	305.095	306.010	309.615		

Table 2.9. Hydrocarbon-20 solution

3. The Bratu problem (contributed by R. Glowinski and H. B. Keller).

Formulation of the problem. Let Ω be a bounded domain of $\mathbf{R}^N$, and denote by Γ the boundary of Ω. With $\lambda \in \mathbf{R}$, we are interested in finding a function u solving

$$-\Delta u = \lambda e^u \text{ in } \Omega, \qquad u = 0 \text{ on } \Gamma, \tag{3.1}$$

where, in (3.1), $\Delta = \nabla^2 = \sum_{i=1}^n \partial^2/\partial x_i^2$ is the Laplace operator. Problem (3.1) is frequently known as the *Bratu problem.*

Physical relevance of the problem. Problem (3.1) is a simplified model for nonlinear diffusion phenomena taking place, for example, in combustion and semiconductors. In the context of combustion, problem (3.1) is related to the Arrhenius law modeling exothermic phenomena; refer to, e.g., Aris [1] for details.

Mathematical results concerning problem (3.1). If one has $\lambda \leq 0$, problem (3.1) has a unique solution and since the operator

$$v \to -\Delta v - \lambda e^{v}$$

is monotone increasing a large number of numerical methods will work quite well for the finite-dimensional problems approximating (3.1).

From now on, we shall consider only the case where $\lambda > 0$. In this case problem (3.1) may have one, several, or no solutions. A typical family of solutions is depicted in Figure 3.1. Pairs such as u_c, λ_c are called *fold points* (or *limit points*). For more information on the mathematical aspects of problems such as (3.1) see, e.g., Crandall and Rabinowitz [**2**], or Keller and Cohen [**9**], and the references therein.

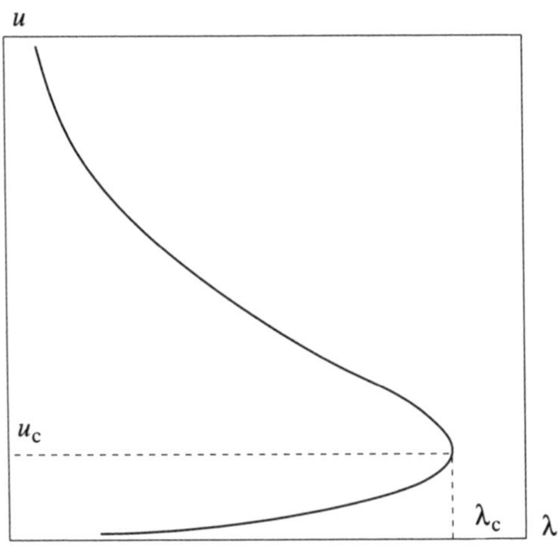

FIGURE 3.1. A family of solutions for the Bratu problem

Variational formulation of problem (3.1). Formulation (3.1) is well suited to *finite-difference discretizations.* However, in the case of domains Ω of complicated shape a *finite-element* treatment of (3.1) may be preferable. Such treatment may be based on the following variational formulation: Find $u \in H_0^1(\Omega)$ such that

$$\int_\Omega \nabla u \cdot \nabla v \, dx = \lambda \int_\Omega e^{u} v \, dx, \quad \forall \; v \text{ in } H_0^1(\Omega). \tag{3.2}$$

In (3.2) we have used

$$\nabla = \left\{\frac{\partial}{\partial x_i}\right\}_{i=1}^{n}, \quad \nabla u \cdot \nabla v = \sum_{i=1}^{n} \frac{\partial u}{\partial x_i}\frac{\partial v}{\partial x_i}, \quad dx = dx_1 \cdots dx_n,$$

$$H_0^1(\Omega) = \left\{v|\ v \in \mathscr{L}^2(\Omega),\ \frac{\partial v}{\partial x_i} \in \mathscr{L}^2(\Omega),\ \forall\ i = 1, \dots, N, v = 0 \text{ on } \Gamma\right\}$$

with $\partial v/\partial x_i$ being taken in the *distribution* sense.

Finite-difference discretization of the Bratu problem. In this section we consider the case where $\Omega = (0, 1) \times (0, 1)$; for the Bratu problem on unbounded domains see, for example, Hagstrom and Keller [**3**].

With J a positive integer we define $h = 1/(J+1)$ and then

$$M_{ij} = \{ih, jh\}, \qquad 0 \leq i,\ j \leq J+1;$$

for the interior grid points M_{ij} we have $1 \leq i,\ j \leq J$. To approximate problem (3.1) we shall use the following finite-difference scheme:

(3.3)

$$-\frac{u_{i+1j} + u_{i-1j} + u_{ij+1} + u_{ij-1} - 4u_{ij}}{h^2} = \lambda e^{u_{ij}}, \qquad 1 \leq i,\ j \leq J,$$

$$u_{kl} = 0 \quad \text{if } M_{kl} \in \Gamma. \tag{3.4}$$

In (3.3), u_{ij} is an approximation to $u(M_{ij})$. The generalization to three-dimensional domains, and to domains more complicated than the unit square $(0, 1) \times (0, 1)$ is left to the reader.

Finite-element approximation of the Bratu problem. Suppose for simplicity that Ω is a polygonal domain of $\mathbf{R}^2$. Let $\mathscr{T}_h$ be a triangulation of Ω such that

$$\bigcup_{T \in \mathscr{T}_h} = \overline{\Omega}. \tag{3.5}$$

Here the triangles T of $\mathscr{T}_h$ are supposed to be *closed* and h denotes the length of the largest edge(s) of the $T \in \mathscr{T}_h$. We approximate $H_0^1(\Omega)$ by

$$V_{0h} = \{v_h|\ v_h \in \mathbf{C}^0(\overline{\Omega}),\ v_h|_T \in P_1,\ \forall\ \mathscr{T} \in \mathscr{T}_h,\ v_h = 0 \text{ on } \Gamma\}. \tag{3.6}$$

In (3.6) $P_1 \equiv$ space of polynomials in two variables of degree ≤ 1. We approximate (3.1) via (3.2) by

$$\begin{cases} u_h \in V_{0h}, \\ \int_\Omega \nabla u_h \cdot \nabla v_h \, dx = \lambda \int_\Omega e^{u_h} v_h \, dx, \quad \forall\ v_h \in V_{0h}. \end{cases} \tag{3.7}$$

In practice one uses the *trapezoidal rule* to evaluate the integrals in the right-hand side of (3.7). For more details on the numerical treatment of the Bratu problem by finite-element methods see Glowinski [**4**, Chapter 7], and Glowinski, Keller, and Reinhart [**5**].

Comments on the numerical solutions of the approximate Bratu problems. The solution of the Bratu problem by a combination of *finite differences, multigrid* and *arc length continuation* methods is discussed with details in Chan and Keller [**6**], Bolstad and Keller [**7**] (see also the references therein). The solution by a combination of *preconditioned conjugate gradient, arc length continuation,* and *finite-element methods* is discussed in, e.g., [**4**], [**5**].

Further comments. (A) If Ω is the unit square $(0,1)\times(0,1)$, the branch of solutions has a *limit point* at $\lambda = \lambda_c = 6.80812\cdots$ and $u = u_c$, with $u_c(.5,.5) = 1.39166\cdots$.

(B) It is most interesting to do calculations with $\Omega \equiv$ unit ball in $\mathbf{R}^N$ for $N = 1, 2, 3, \ldots$. Seek radially symmetric solutions $u = u(r)$ and impose the regularity condition

$$\left.\frac{du(r)}{dr}\right|_{r=0} = 0.$$

The solution paths for $N = 1$ and $N = 2$ are as in Figure 3.1. For $N = 3, 4, \ldots, 10$, they oscillate about a value $\lambda^{\infty}(N)$ for which value there are infinitely many positive, smooth solutions. For

$$N = 1, \quad \lambda_c(1) = 3.5138\cdots; \qquad \text{for } 2 < N < 10, \quad \lambda^{\infty}(N) = 2(N-2).$$

The solution for $N = 1$ is given in [**3**].

(C) Another interesting calculation is to take λ real-valued and u complex-valued. Taking $u = u_1 + iu_2$, the *complex Bratu problem* is then equivalent to the following system:

$$\begin{cases} -\Delta u_1 = \lambda e^{u_1}\cos u_2 & \text{in } \Omega, \\ -\Delta u_2 = \lambda e^{u_1}\sin u_2 & \text{in } \Omega, \\ u_1 = 0 \text{ and } u_2 = 0 & \text{on } \Gamma. \end{cases} \tag{3.8}$$

Solving (3.8) is an almost immediate generalization of what has been seen before. The complexification approach is an elegant way to take λ beyond the critical value λ_c since, if we allow complex-valued solutions, the limit point $\{u_c, \lambda_c\}$ of the real-valued Bratu problem is just a bifurcation point for the complex-valued Bratu problem; see Henderson and Keller [**8**]. Solving the complex-valued Bratu problem is also a test problem relevant to the nonlinear library presently in construction.

REFERENCES

1. R. Aris, *The Mathematical Theory of Diffusion and Reaction in Permeable Catalysts, Vol. I*, Clarendon Press, Oxford, 1975.

2. M. G. Crandall and P. Rabinowitz, *Some continuation and variational methods for positive solutions of nonlinear elliptic eigenvalue problems*, Arch. Rat. Mech. Anal. **58** (1975), 207–218.

3. T. Hagstrom and H. G. Keller, *Asymptotic boundary conditions and numerical methods for nonlinear elliptic problems on unbounded domains*, Math. of Comp. **48** (1987), 449–470.

4. R. Glowinski, *Numerical Methods for Nonlinear Variational Problems*, Springer-Verlag, New York, 1984.

5. R. Glowinski, H. B. Keller, and L. Reinhart, *Continuation-conjugate gradient methods for the least squares solution of nonlinear boundary value problems*, SIAM J. Sci. Statist. Comput. **6** (1985), 793–832.

6. T. F. C. Chan and H. B. Keller, *Arc-length continuation and multigrid techniques for nonlinear elliptic eigenvalue problems*, SIAM J. Sci. Statist. Comput. **3** (1982), 173–194.

7. J. Bolstad and H. B. Keller, *A multigrid continuation method for elliptic problems with folds*, SIAM J. Sci. Statist. Comput. **7** (1986), 1081–1104.

8. M. Henderson and H. B. Keller, *Complex bifurcation*, SIAM J. Appl. Math. (to appear).

9. H. B. Keller and D. S. Cohen, *Some positive problems suggested by nonlinear heat generation*, J. Math. Mech. **16** (1967), 1361–1376.

4. The Chandrasekhar H-equation (contributed by C. T. Kelley). This test problem is a quadratic nonlinear equation

$$\mathcal{F}(H)(x) = H(x) - 1 - \frac{c}{2}\int_0^1 \frac{xH(x)H(y)}{x+y}\,dy = 0. \tag{4.1}$$

In (4.1) the unknown is the function $H \in C[0,1]$ and c is a parameter. The equation was introduced by Chandrasekhar **[1]**, in the context of radiative transfer problems. Chandrasekhar also introduced generalizations of (4.1) for matrix-valued functions. Equation (4.1) has all the essential features of the more general problems. Physically reasonable values of the parameters are $0 < c \leq 1$. As a test problem, any real or complex c can be used. Bifurcation properties of the equation were analyzed for the first time in **[4]**. Nonsingularity of the Fréchet derivative for $c \neq 1$ is a result in **[2]**.

Here is a list of some known facts about the equation and its solution:

(1) For $c = 0$ and $c = 1$ there is a unique solution. For other values of $c > 0$ there are two solutions. The solution of physical interest is characterized by analyticity in the parameter c in the complex c-plane for $\mathrm{Re}(c) > 0$.

(2) The Fréchet derivative of $\mathcal{F}$ is nonsingular for all complex $c \neq 1$. At $c = 1$ complex bifurcation takes place. For real $c > 1$ there are two complex conjugate solutions.

(3) At $c = 1$ the Fréchet derivative has a one-dimensional null space spanned by $\varphi(x) = xH(x)$. The branch point at $c = 1$ is a simple fold.

(4) Facts (1)–(3) remain true if the integral is replaced by a quadrature rule that integrates constants exactly and does not include the origin as a node.

The fact that there is a simple fold at $c = 1$ allows one to observe the behavior of Newton's method, the chord method, and Broyden's method at such points. One may also try to track the nonphysical solution from $c = 1$ to $c = 0$ where it becomes unbounded. I have not done this; tell me how it went. Aside from the singularity, this problem is not difficult. The generalizations for matrix-valued H-functions present difficulty only in that they have a large number of unknowns. It is easy to increase the dimension of approximations to (4.1) and thereby observe (or fail to observe) mesh independence. It is also worth noting that the Fréchet derivative is not known up to a compact operator. This fact affects the superlinear convergence of quasi-Newton methods [**3**].

For $c < 1$ there seems to be a unique continuous solution and a second solution that is discontinuous. I cannot prove any of this, however. To observe the second discontinuous solution, begin a Newton iteration with the initial iterate $H_0 = 10^6$.

To compare results with published tables, such as those in [**1**], one may use Nyström interpolation. Equation (4.1) may be rewritten as

$$H(x) = \left(1 - \frac{c}{2}\int_0^1 \frac{xH(y)}{x+y}\,dy\right)^{-1}. \tag{4.2}$$

Thus, given a quadrature rule with nodes $\{x_i\}_{i=1}^N$ and weights $\{w_i\}_{i=1}^N$ and discrete solution $\{H_i\}_{i=1}^N$ to

$$H_i = 1 + \frac{c}{2}\sum_{j=1}^{N} \frac{x_i H_i H_j}{x_i + x_j} w_j, \tag{4.3}$$

one may approximate H for any x by

$$H(x) \approx \left(1 - \frac{c}{2}\sum_{j=1}^{N} \frac{xH_j}{x+x_j} w_j\right)^{-1}.$$

REFERENCES

1. S. Chandrasekhar, *Radiative Transfer*, Dover, New York, 1960.

2. C. T. Kelley, *Solution of the Chandrasekhar H-equation by Newton's method*, J. Math. Phys. **21** (1980), 1625–1628.

3. C. T. Kelley and J. I. Northrup, *A pointwise quasi-Newton method for integral equations*, SIAM J. Numer. Anal. **25** (1988), 1138–1155.

4. T. W. Mullikin, *Some probability distributions for neutron transport in a half space*, J. Appl. Prob. **5** (1968), 357–374.

5. The elastohydrodynamic lubrication problem (contributed by M. Kostreva). The elastohydrodynamic lubrication problem has received much attention in the mechanical engineering literature. A recent computer search of the literature turned up over 700 papers published on the subject. Recent applications of the direct algorithm of complementarity (Habetler and Kostrave [**2**], Murty [**6**]) to revised formulations of the problem have produced some very encouraging results. (See Kostreva [**3**] and [**4**], Kostreva and Hatton [**5**], Oh [**7**], Oh and Goenka [**9**], Oh [**8**], and Oh, Li, and Goenka [**10**].) This accomplishment is mainly due to the application of direct algorithms to models of interest to researchers. Many times these models were not solvable using previously developed techniques such as successive overrelaxation, Newton–Raphson iteration, or inverse iteration. In fact, sometimes incorrect solutions were obtained and published by researchers using the older techniques (see Kostreva [**3**] for an example). Once the direct algorithm of complementarity was tried, many of the earlier difficulties disappeared, and many new questions could be answered. Prospects for further applications and extensions seem extremely good, given the results obtained to date. However, it should be noted that success on this problem is highly dependent on the ability to solve large systems of nonlinear equations.

For ease of presentation, only the simplest model will be presented here: the line contact model of the elastohydrodynamic lubrication of cylinders. Such a model would be considered in the design of roller bearings which are to be lubricated by oil.

The standard mathematical model consists of three equations and three boundary conditions: Reynolds' differential equation for the pressure in the lubricant, a linear integral equation for the deformation of the cylinders, and another linear integral equation representing the balance of load. In dimensionless form, given the dimensionless parameters α and λ and an inlet point x_a, find $p(x)$, $h(x)$, the free boundary

x_b, and the aggregate variable k satisfying

$$\frac{d}{dx}\left(\frac{h(x)^3}{e^{\alpha p}}\frac{dp}{dx}\right) = \lambda\frac{dh}{dx},$$

$$h(x) = x^2 + k - \frac{2}{\pi}\int_a^b p(s)\ln|x-s|\,ds,$$

$$\frac{2}{\pi}\int_a^b p(s)\,ds = 1,$$

with the boundary conditions $p(x_a) = 0$ and $p(x_b) = p'(x_b) = 0$.

To apply the direct algorithm of complementarity, finite-difference approximations are applied to the above equations. Let x_F denote a point far downstream of x_a. Then let $x_F = x_a + n\Delta x$. Then for $i = 1, 2, \ldots, n$, let $p_i = p(x_a + i\Delta x)$ and for $j = i \pm \frac{1}{2}$, let $h_j = h(x_a + j\Delta x)$. For convenience, take $p_0 = k$. To reduce the dimensions, make the direct substitution of h_j into the finite-difference approximation to Reynolds' equation. Then for $i = 1, 2, \ldots, n$,

$$\begin{aligned} R_i(p_0, p_1, \ldots, p_n) = -\frac{1}{(\Delta x)^2}&\left[\frac{(h_{i+\frac{1}{2}})^3}{\exp(\alpha p_{i+\frac{1}{2}})}(p_{i+1} - p_i)\right. \\ &\left. - \frac{(h_{i-\frac{1}{2}})^3}{\exp(\alpha p_{i-\frac{1}{2}})}(p_i - p_{i-1})\right] + \frac{\lambda}{\Delta x}(h_{i+\frac{1}{2}} - h_{i-\frac{1}{2}}). \end{aligned}$$

The remaining equality for the load balance becomes

$$R_0 = 1 - (2/\pi)\sum w_i p_i \Delta x,$$

so that k and R_0 form a complementary pair of variables. The values of p at half points are obtained by averaging neighboring points, while the h_j and R_0 values are obtained by applying the trapezoid rule. See Kostreva [**3**] for more details.

The final outcome is: Find $p = (p_0, p_1, \ldots, p_n)$, so that

$$\begin{aligned} &R_0(p) = 0, \\ &R_i(p) \geq 0, \quad p_i \geq 0, \quad p_i R_i(p) = 0, \qquad i = 1, \ldots, n. \end{aligned}$$

The strategy of compiling solutions into a database for later use has been implemented in ROLLUBE (Kostreva and Hatton [**5**]). With this approach the previously computed solutions are readily available for use as starting points in new computations.

As indicated above, the model shown here is the simplest elastohydrodynamic lubrication model. Oh [**7**], Oh and Goenka [**9**], and Oh, Li, and Goenka [**10**] develop models in which the ordinary differential equation (Reynolds' equation) is replaced by a two-dimensional partial differential equation, the integral equations require multiple integrals, and the free boundary is a curve rather than a point. These papers also feature dynamic loading. Such models are much more complex to solve, but the same direct complementarity approach has worked.

The above differential-integral equations model, with the uniform grid approximations, has recently been revised to accommodate variable grid approximations (Bissett and Glander [**1**]) in an effort to solve some particularly difficult cases. For example, when the parameter $\alpha > 5.0$ and the parameter $\lambda > 2.0$, the pressure function can develop a large spike, which is both interesting to engineers and quite difficult to compute. While the solutions in Kostreva [**3**] and Kostreva [**4**] considered uniform mesh solutions leading to 100 and 100–400 equations, Bissett and Glander have considered as many as 1,000 equations in their variable mesh approximations. Such sets of equations test the limits of *any* of today's computers and would also benefit from further algorithmic studies.

REFERENCES

1. E. J. Bissett and D. W. Glander, *A highly accurate approach that resolves the pressure spike of elastohydrodynamic lubrication*, Trans. ASME Journal of Tribology **110** (1988), 241–246.

2. G. J. Habetler and M. M. Kostreva, *Nonlinear complementarity problems*, SIAM Journal on Control and Optimization **16** (1978), 504–511.

3. M. M. Kostreva, *Elastohydrodynamic lubrication: A nonlinear complementarity problem*, Internat. J. Numer. Methods Fluids **4** (1984), 377–397.

4. ——, *Pressure spikes and stability consideration in elastohydrodynamic lubrication models*, Trans. ASME, Journal of Tribology **106** (1984), 386–395.

5. M. M. Kostreva and M. B. Hatton, *ROLLUBE—A computer package for modeling roller bearing lubrication and elasticity*, Proc. ASME International Computers in Engineering Conference and Exhibit (Las Vegas, 1984), 614–620.

6. K. G. Murty, *Note on a Bard-type scheme for solving the complementarity problem*, Opsearch **11** (1974), 123–130.

7. K. P. Oh, *The numerical solution of dynamically loaded elastohydrodynamic contacts as a nonlinear complementarity problem*, Trans. ASME, Journal of Tribology **106** (1984), 88–95.

8. ——, *The formulation of the mixed lubrication problem as a generalized nonlinear complementarity problem*, Trans. ASME, Journal of Tribology **106** (1986), 598–603.

9. K. P. Oh and P. K. Goenka, *The elastohydrodynamic lubrication of a journal bearing under dynamic loading*, Trans. ASME, Journal of Tribology **107** (1985), 107, 389.

10. K. P. Oh, C. H. Li, and P. K. Goenka, *Elastohydrodynamic lubrication of piston skirts*, Trans. ASME, Journal of Tribology **109** (1987), 356–362.

6. Model problems for renal network flows (contributed by R. Mejia).

Six-tube model. Consider a prototypical model of the renal concentrating mechanism, described by Stephenson, Tewarson, and Mejia [12], that consists of six leaky tubes with fluid and solute flow in one space-dimension, as illustrated in Figure 6.1. The model is described by the following differential equations:

$$\frac{dP_i}{dx} = -R_i F_{iv}, \quad \frac{dF_{iv}}{dx} = -J_{iv}, \quad \frac{dF_{ik}}{dx} = -J_{ik}, \tag{6.1}$$

where

$$F_{ik} = F_{iv} C_{ik}, \tag{6.2}$$

for $0 \le i \le 6$; k solutes, $k = 1, 2$; and position x, $0 \le x \le 1$. P_i is the hydrostatic pressure in the ith structure; R_i is the resistance to flow; F_{iv} is the volume flow in the ith tube; J_{iv} is the transmural volume flux out of the ith tube; J_{ik} is the transmural flux of the kth solute out of the ith tube; and C_{ik} is the concentration of the kth solute in the ith structure. In the medulla, $x = 0$ at the border of the cortex and the outer medulla, and $x = 1$ at the extreme inner medulla. In the cortex, flow in tube 3 is from $x = 0$ at the junction with tube 2 to $x = 1$ at the junction with tube 4 (see Figure 6.1).

The transmural fluxes are defined as

$$\begin{aligned} J_{iv} &= h_{iv}\left[\sum_{k=1}^{2}(C_{qk} - C_{ik})\sigma_{ik} + P_i - P_q\right], \\ J_{ik} &= h_{ik}(C_{ik} - C_{qk}) + (1 - \sigma_{ik})J_{iv}\left(\frac{C_{ik} + C_{qk}}{2}\right) + \frac{a_{ik}C_{ik}}{b_{ik} + C_{ik}}, \end{aligned} \tag{6.3}$$

where $q = 6$ for $i = 1, 2, 4, 5$, and $q = 0$ for $i = 3$. h_{iv} is the hydraulic permeability of the ith tube; σ_{ik} is the Staverman reflection coefficient of the kth solute in the ith tube, $0 \le \sigma_{ik} \le 1$; h_{ik} is the solute permeability of the kth solute in the ith tube; a_{ik} is the maximum rate of active transport of the kth solute from the ith tube; and b_{ik} is the half-maximal concentration for active transport. Discontinuities in the transport coefficients result in jumps in (6.3), hence, in discontinuous right-hand sides in (6.1).

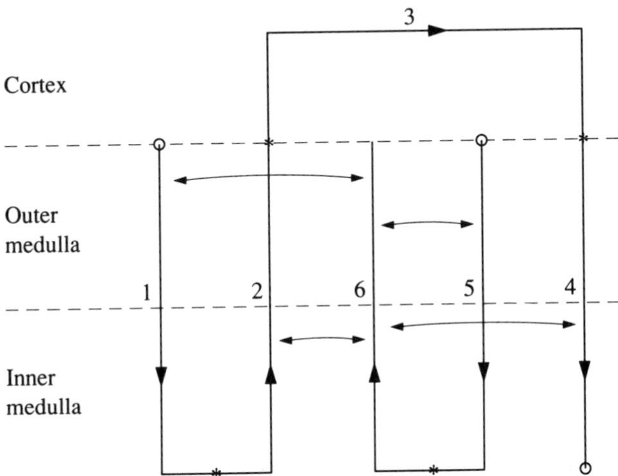

FIGURE 6.1. Schematic diagram of a six-tube model of the mammalian kidney. Arrows show the direction of flow in each tube and the transport of water and solutes between tubes. Boundary data are prescribed at ⊙; matching boundary conditions are indicated by $*$.

In tube 6,

$$J_{6v} = -\sum_{\substack{1\le i\le 5\\ i\ne 3}} J_{iv}, \qquad J_{6k} = -\sum_{\substack{1\le i\le 5\\ i\ne 3}} J_{ik} \quad \forall k. \tag{6.4}$$

In the cortex

$$P_0 = P_0^0 \quad \text{and} \quad C_{0k} = C_{0k}^0 \quad \forall k. \tag{6.5}$$

Boundary conditions are as follows:

$$\begin{aligned} &F_{1v}(0) = F_{1v}^0, && C_{1k}(0) = C_{1k}^0 \quad \forall k, \\ &P_2(1) = P_1(1), && F_{2v}(1) = -F_{1v}(1), && C_{2k}(1) = C_{1k}(1) \quad \forall k, \\ &P_3(0) = P_2(0), && F_{3v}(0) = F_{2v}(0), && C_{3k}(0) = C_{2k}(0) \quad \forall k, \\ &P_4(0) = P_3(1), && F_{4v}(0) = -F_{3v}(1), && C_{4k}(0) = C_{3k}(1) \quad \forall k, \\ &P_4(1) = P_4^1, \end{aligned} \tag{6.6}$$

and

$$\begin{aligned} &P_5(0) = P_5^0, && F_{5v}(0) = F_{5v}^0, && C_{5k}(0) = C_{5k}^0 \quad \forall k, \\ &P_6(1) = P_5(1), && F_{6v}(1) = -F_{5v}(1), && C_{6k}(1) = C_{5k}(1) \quad \forall k, \end{aligned} \tag{6.7}$$

where the following data are specified:

$$\begin{aligned}&P_0^0, \quad P_4^1, \quad P_5^0,\\ &F_{1v}^0, \quad F_{5v}^0,\\ &C_{0k}^0, \quad C_{1k}^0, \quad C_{5k}^0 \quad \forall k.\end{aligned}$$

A second-order, centered trapezoidal differencing scheme was first used **[12]** to discretize equations (6.1), and Newton's method was used to solve the nonlinear algebraic equations. Many discretization schemes have since been applied to this problem. Midpoint, cubic overhang, corrected trapezoidal, and fifth-degree overhang methods were considered in **[14]**. Multiple shooting **[5]**, adaptive finite differences with deferred corrections, and Richardson extrapolation were compared in **[10]**. Simpson's rule was used in **[13]**, and a sixth-order formula was derived for this problem in **[4]**.

Various attempts to exploit the connectivity of the system have also been made. The Jacobian matrix has been permuted to a bordered block triangular form **[15]** and then used to compute the correction to the solution vector of the nonlinear equations. A sparse matrix approach is described in **[1]**, and partitioning of the multipoint boundary value problem into a series of small initial value problems and a reduced boundary value problem is described in **[9]**. Quasi-Newton methods have been used to solve the global nonlinear system, while symbolic manipulation has been used to pre-solve for the correction to the solution vector of the small initial value problems described in **[9]**.

Model parameters that have commonly been used in the literature are given below with γ chosen so that $0.4 \le \gamma \le 0.5$, $R_i = 0$, $a_{i2} = b_{i2} = 0$ $\forall i$, and $\sigma_{ik} = 1$ $\forall i, k$ in (6.1)–(6.3).

i	h_{iv}	h_{i1}	h_{i2}	a_{i1}	b_{i1}	
1	10	0	0	0	0	
2	0	0	0	1.8	0	$0 \le x < \gamma$
	0	10	0	0	0	$\gamma \le x \le 1$
3	1	0	0	0.75	1	
4	10	0	0	0	0	$0 \le x < \gamma$
	10	0	0.01	0	0	$\gamma \le x \le 1$
5	0	1,000	1,000	0	0	

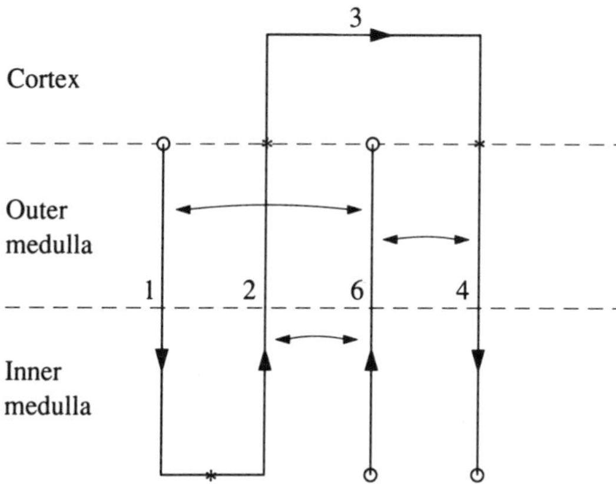

FIGURE 6.2. Schematic diagram of a central core model of the mammalian kidney. Symbols are as in Figure 6.1. The central core, tube 6, has a blind end at $x = 1$ in the inner medulla.

Boundary data used are as follows:

$$P_0^0 = P_5^0 = P_4^1 = 0,$$
$$F_{1v}^0 = F_{5v}^0/5 = 1,$$
$$C_{01}^0 = C_{11}^0 = C_{51}^0 = 1,$$
$$C_{02}^0 = C_{12}^0 = C_{52}^0 = 0.05.$$

Central core model. A problem with particular physiological relevance is a central core model of the mammalian kidney, where tube 5 in Figure 6.1 is deleted, and tube 6 is closed at one end ($x = 1$) to form a central core (see Figure 6.2) with boundary conditions

$$P_6(0) = P_0^0, \tag{6.8}$$

$$F_{6v}(1) = F_{6k}(1) = 0, \tag{6.9}$$

$$C_{6k}(1) = \frac{J_{6k}(1)}{J_{6v}(1)}, \quad \forall k, \tag{6.10}$$

substituted for those in (6.7). The bulk flow equations (6.2) for the core are rewritten to include diffusive flow as follows:

$$F_{6k} = F_{6v}C_{6k} - D_{6k}\frac{C_{6k}}{dx} \tag{6.11}$$

$\forall k$, and D_{6k} is the diffusion coefficient of the kth solute.

Equations (6.1), (6.3)–(6.6), and (6.8)–(6.11) describe the central core model first proposed by Stephenson **[11]**. This problem was first solved using both a fourth-order predictor-corrector and a Runge-Kutta method [7], and using a second-order approximation for the derivatives in [12]. Multiple stable solutions have been identified and studied numerically [8], [3], and conditions for existence and uniqueness of solutions are considered in [3], [2]. Note that Garner et al. in [3], [2] use the boundary condition

$$C_{6k}(0) = C_{0k}^{0} \tag{6.12}$$

$\forall k$ instead of (6.10). However, condition (6.10) has often been used because it is satisfied naturally when $D_{6k} = 0$.

An extension of this central core model to five sets of tubes, similarly linked, is treated in [6] in these proceedings.

REFERENCES

1. P. Farahzad and R. P. Tewarson, *An efficient numerical method for solving the differential equation of renal counterflow systems*, Comput. Biol. Med. **8** (1978), 57–64.
2. J. B. Garner and R. B. Kellogg, *Existence and uniqueness of solutions in general multisolute renal flow problems*, J. Math. Biol. **26** (1988), 455–464.
3. J. B. Garner, R. B. Kellogg, and J. L. Stephenson, *Mathematical analysis of a model for the renal concentrating mechanism*, Math. Biosci. **65** (1983), 125–150.
4. S. Gupta and R. P. Tewarson, *An accurate solution of renal models*, Math. Biosci. **65** (1983), 199–207.
5. P. Lory, *Numerical solution of a kidney model by multiple shooting*, Math. Biosci. **50** (1980), 117–128.
6. R. Mejia, *Interactive program for continuation of solutions of large systems of nonlinear equations*, this volume.
7. R. Mejia and J. L. Stephenson, *Numerical solution of a central core model of the renal medulla*, Proc. of Summer Computer Simulation Conference (Montreal, Canada, 1973), Simulations Councils, Inc., La Jolla, California.
8. ——, *Multiple solutions of convection–diffusion equations describing the renal concentrating mechanism*, Numerical Methods for Engineering, 2nd International Congress, E. Absi, R. Glowinski, P. Lascaux, and H. Veysseyre (eds.), GAMNI, Dunod, Paris, France, 1980.
9. ——, *Numerical solution of multinephron kidney equations*, J. Comput. Phys. **32** (1979), 235–246.
10. R. Mejia, J. L. Stephenson, and R. J. LeVeque, *A test problem for kidney models*, Math. Biosci. **50** (1980), 129–131.
11. J. L. Stephenson, *Concentrating engines and the kidney*: I. *Central core model of the renal medulla*, Biophys. J. **13** (1973), 512–545; II. *Multisolute central core system*, Biophys. J. **13** (1973), 546–567.
12. J. L. Stephenson, R. P. Tewarson, and R. Mejia, *Quantitative analysis of mass and energy balance in non-ideal models of the renal counterflow system*, Proc. Nat. Acad. Sci. USA **71** (1974), 1618–1622.
13. R. P. Tewarson, *On the use of Simpson's rule in renal models*, Math. Biosci. **55** (1981), 1–5.
14. R. P. Tewarson and P. Farahzad, *On the numerical solution of differential equations for renal counterflow systems*, Comp. Biomed. Res. **11** (1978), 381–391.

15. R. P. Tewarson, A. Kydes, J. L. Stephenson, and R. Mejia, *Use of sparse matrix techniques in numerical solution of differential equations for renal counterflow systems*, Comp. Biomed. Res. **9** (1976), 507–520.

7. The obstacle Bratu problem (contributed by H. D. Mittelmann). This problem is considered in [**1**], [**2**], [**5**], [**6**]; see also the references in these papers. Let

$$\begin{gathered} -\Delta u \leq \lambda e^u \text{ in } \Omega, \quad u = 0 \text{ on } \partial\Omega, \\ u \leq \psi \text{ and } (-\Delta u - \lambda e^u)(u - \psi) = 0 \text{ in } \Omega, \end{gathered} \tag{7.1}$$

where Ω is (say) a two-dimensional domain and $\psi \geq 0$ is an obstacle function. Problem (7.1) may equivalently be written as a constrained minimization problem or as a variational inequality. It may be discretized by finite elements or finite differences. A simple finite-difference discretization in the case $\Omega = [0, 1]^2$ is

$$\begin{gathered} \mathbf{Ax} \leq \lambda \mathbf{e}(\mathbf{x}), \quad \mathbf{x} \leq \psi, \\ (\mathbf{Ax} - \lambda \mathbf{e}(\mathbf{x}))(\mathbf{x} - \psi) = 0, \end{gathered} \tag{7.2}$$

where $\mathbf{A}$ denotes the standard five-point difference matrix including the zero boundary conditions and the components x_{ij} of the vector $\mathbf{x}$ are approximations to $u(ih, jh)$ for $i, j = 1, \ldots, n$. In this formulation $h = 1/(n+1)$ is the discretization parameter, the vector $\mathbf{e}(\mathbf{x})$ has components $e^{x_{ij}}$, and ψ has components $\psi(x_{ij})$.

Problem (7.2) serves only as a model problem for testing numerical methods to be applied to similar problems in the applications. See, for example, [**4**] for an obstacle problem for the nonlinear beam.

Numerically problem (7.2) will generally be solved by a continuation technique. It is interesting to see how different methods perform in the computation of the solution curves $(\lambda, \mathbf{x})$ of (7.2). These curves are only piecewise smooth (see Figure 7.1) and exhibit for not too fine discretizations—h not too small, say $h = \frac{1}{8}$ or $\frac{1}{16}$—many turning (fold) points and points of nonsmoothness. In addition to this scalar parameter h that may be changed, the problem depends on the obstacle function ψ and the nonlinearity, here exponential. While only constant ψ was considered in [**6**], [**7**], nonconstant obstacles were used in [**1**], [**2**], [**5**], and different nonlinearities in [**1**], [**7**]. Finite-difference discretizations were dealt with in [**5**], [**6**], [**7**], finite elements in [**1**], while a finite-difference multigrid method was presented in [**2**].

For theoretical results on a general class of problems of type (7.1) see [**1**], [**4**], [**5**] and the literature cited there. For results on a general class of finite-dimensional constrained optimization problems of type

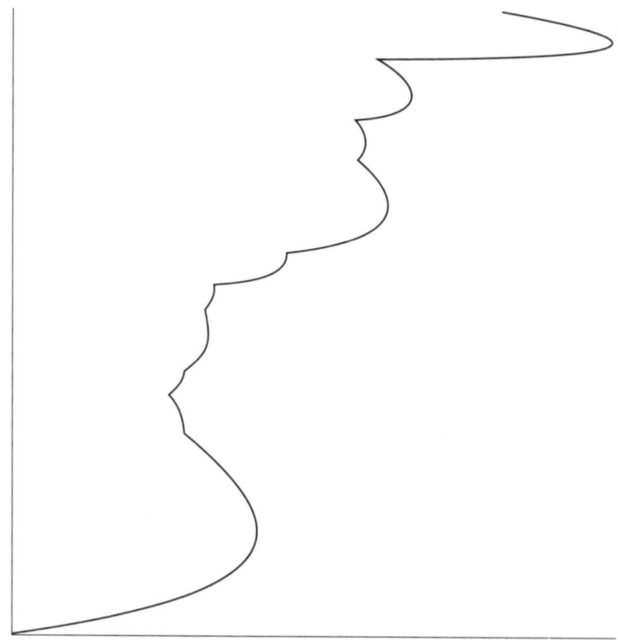

FIGURE 7.1. Typical solution curves (λ, r) with $r = \|\mathbf{x}\|$ for problem (7.2)

(7.2) addressing in particular the piecewise smoothness of the solution curves, see [3] and the paper by Jongen in this volume.

REFERENCES

1. F. Conrad, R. Herbin, and H. D. Mittelmann, *Approximation of obstacle problems by continuation methods*, SIAM J. Numer. Anal. **25** (1988), 1409–1431.

2. R. H. W. Hoppe and H. D. Mittelmann, *A multigrid continuation strategy for parameter-dependent variational inequalities*, J. Comput. Appl. Math. **26** (1989), 35–46.

3. H. Th. Jongen, P. Jonker, and F. Twilt, *Critical sets in parametric optimization*, Math. Programming **34** (1986), 333–353.

4. E. Miersemann and H. D. Mittelmann, *On the continuation for variational inequalities depending on an eigenvalue parameter*, Math. Meth. in the Appl. Sci. **11** (1989), 95–104.

5. ———, *Continuation for parametrized nonlinear variational inequalities*, J. Comput. Appl. Math. **26** (1989), 23–34.

6. H. D. Mittelmann, *On continuation for variational inequalities*, SIAM J. Numer. Anal. **24** (1987), 1374–1381.

7. ———, *Nonlinear parametrized equations: New results for variational problems and inequalities*, this volume.

8. A tubular chemical reactor model (contributed by Aubrey B. Poore). One of the much studied models in chemical reactor theory is the nonadiabatic chemical reactor in which there is processed a simple, first-order, exothermic chemical reaction $A \to B$. The mathematical model can be described by a set of nonlinear parabolic partial differential equations which in dimensionless form can be written as [3], [4], [9]

$$\frac{\partial u}{\partial \tau} = \frac{1}{Pe_m}\frac{\partial^2 u}{\partial s^2} - \frac{\partial u}{\partial s} - Df(u,v;\gamma) \qquad (0 < s < 1,\ \tau > 0), \tag{8.1}$$

$$\frac{\partial v}{\partial \tau} = \frac{1}{Pe_h}\frac{\partial^2 v}{\partial s^2} = \frac{\partial v}{\partial s} - \beta(v - v_0) + BDf(u,v;\gamma) \quad (0 < s < 1,\ \tau > 0), \tag{8.2}$$

with boundary and initial conditions

$$\frac{\partial u}{\partial s} = Pe_m(u-1) \quad \text{and} \quad \frac{\partial v}{\partial s} = Pe_h(v-1) \qquad (s = 0,\ \tau > 0), \tag{8.3}$$

$$\frac{\partial u}{\partial s} = \frac{\partial v}{\partial s} = 0 \qquad (s = 1,\ \tau > 0), \tag{8.4}$$

$$u = u_{\text{in}}(s) \quad \text{and} \quad v = v_{\text{in}}(s) \qquad (0 < s < 1,\ \tau = 0). \tag{8.5}$$

The assumptions made in the derivation of these equations are that the velocity profile is flat with constant velocity; the dimensionless chemical concentration u and temperature v depend only on the axial coordinate s and time τ; the diffusion of reactant A is governed by Fick's law; heat conduction is described by Fourier's law; heat loss is due to a cooling jacket on the reactor which is kept at a constant temperature v_0 and is proportional to $v - v_0$; and the reaction rate $f(u,v;\gamma) = u\exp(\gamma - \gamma/v)$ is derived from Arrhenius kinetics. The parameters Pe_m and Pe_h are the mass and heat Peclet numbers; D, the Damkohler number; B, the heat of reaction; β, the heat transfer coefficient; and γ, the dimensionless activation energy. Each of these parameters is assumed positive except for β, which can also be zero. Due to the use of different normalizations, these equations are sometimes written in slightly different form in the literature, and the results can be confusing. For example, the normalized concentration of the reactant A, u, is sometimes replaced by $w = 1 - u$, which is a measure of the concentration of the product chemical B.

For various parameter ranges these equations exhibit a variety of phenomena and solution characteristics ranging from multiple steady states (time-independent solutions), multiple time periodic solutions, and bifurcations to steady states with internal and boundary layers. Many of these phenomena and the formidable literature are reviewed by Varma and Aris [9] and the references therein. In particular, in the asymptotic limit as the Peclet numbers Pe_h and $Pe_m \to 0$ and except for an initial diffusive layer, the solutions are governed by the continuous stirred tank reactor [1], [9]. In the large activation limit $\gamma \to +\infty$, the steady-state behavior has been classified for all remaining system parameters by Kapila and Poore [5].

The steady-state solution $u = u(s)$ and $v = v(s)$ of the parabolic system (8.1)–(8.5) are solutions of the system of two-point boundary value problems

$$\frac{1}{Pe_m}u'' - u' - Df(u,v;\gamma) = 0 \qquad (0 < s < 1), \tag{8.6}$$

$$\frac{1}{Pe_h}v'' - v' - \beta(v - v_0) + BDf(u,v;\gamma) = 0 \qquad (0 < s < 1), \tag{8.7}$$

$$u' = Pe_m(u-1) \quad \text{and} \quad v' = Pe_h(v-1) \qquad (s = 0), \tag{8.8}$$

$$u' = v' = 0 \qquad (s = 1). \tag{8.9}$$

Two special cases of these equations deserve further comment. First, if the Peclet numbers are equal, say to Pe, and the reactor is adiabatic, i.e., $\beta = 0$, then $u = (B + 1 - v)/B$, where v solves

$$\frac{1}{Pe}v'' - v' + BDf((B+1-v)/B, v;\gamma) = 0 \qquad (0 < s < 1), \tag{8.10}$$

$$v' = Pe(v-1) \text{ at } s = 0 \quad \text{and} \quad v' = 0 \text{ at } s = 1. \tag{8.11}$$

For $0 < Pe_h \ll 1$ and $0 < Pe_m \ll 1$, the solutions are given by $v(s) = C + O(Pe_h)$ and $u(s) = T + O(Pe_m)$ as Pe_h and $Pe_m \to 0$, where the constants C and T satisfy the two algebraic equations

$$C = Df(C,T;\gamma) \quad \text{and} \quad T = (BC + \beta v_o)/(1+\beta).$$

These equations govern the time-independent solutions of the continuous stirred tank reactor [2], [6], [7], [8], and admit one, two, or three solutions depending on the parameter ranges of $\beta \geq 0$, $B > 0$, $D > 0$, and $\gamma > 0$. In the large activation energy limit ($\gamma \to +\infty$), the solutions are far more complex with both internal and boundary layers and are described in the work of Heinemann, Kapila, and Poore [3], [4], [5].

The numerical solution of the boundary value problem (8.6)–(8.9) ranges from being quite easy to next-to-impossible, depending on the parameter values. The difficulty in solving these equations is almost always due to the nonlinearity $f(u,v;\gamma)$ arising from Arrhenius kinetics, which can cause the equations to be extremely stiff in small regions of the solution profile. These regions are characterized by sharp thin spikes which occur in the temperature profile v, and can be missed by a coarse discretization.

Thus, discrete versions of these equations should provide an excellent source of test problems for the nonlinear equation solvers. My impression from the chemical engineering literature is that simple finite differences are not as efficient as collocations in computing the solutions. (Someone will surely show me otherwise.)

Two examples are presented in Figures 8.1 and 8.2, where the maximum temperature ($\max\{v(s)\colon 0 \leq s \leq 1\}$) versus the Damkohler number D is sketched to illustrate the existence of one to seven solutions. The temperature v_0 is taken to be zero and the Peclet numbers are equal ($Pe_h = Pe_m = Pe$). The solid lines represent stable states while the dashed ones correspond to the unstable states relative to the parabolic system (8.1)–(8.5). In both figures the solutions are computed using collocation and continuation methods **[3]**, **[4]**. The solutions along the lower branch are approximated quite well by the solution of (8.6)–(8.9) with $D = 0$, and in fact the continuation procedure is started from this value of the Damkohler number. At larger activation energies γ, the solutions on the upper branches are more difficult to compute. Thus, the solution profiles in Figure 8.2 are generally more difficult to compute than those in Figure 8.1.

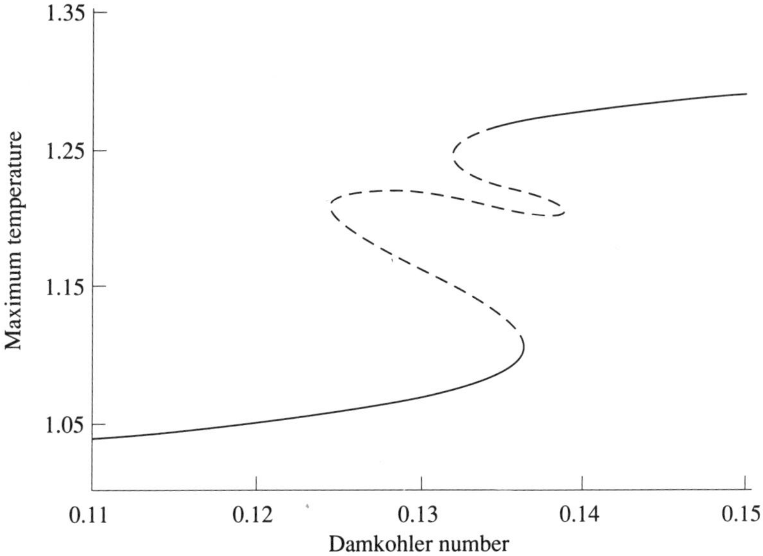

FIGURE 8.1. Existence of five solutions ($Pe = 5$, $B = 0.5$, $\gamma = 25$, and $\beta = 2$)

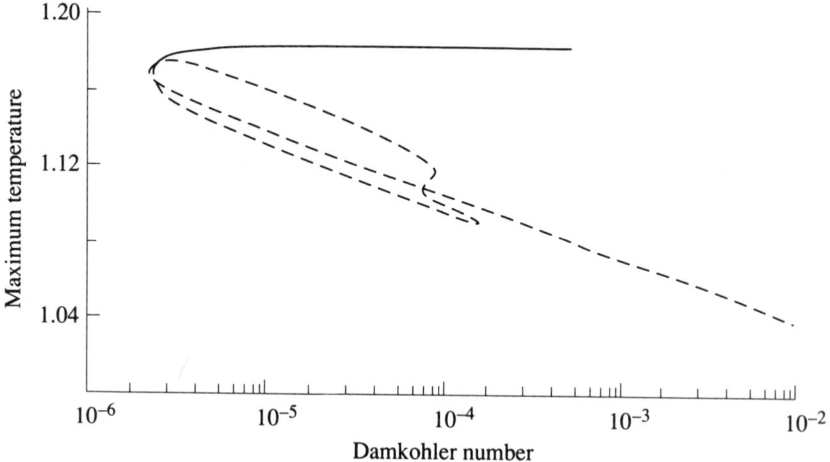

FIGURE 8.2. Existence of seven solutions ($Pe = 5$, $B = 0.5$, $\gamma = 125$, and $\beta = 4$)

REFERENCES

1. D. S. Cohen and A. B. Poore, *Tubular chemical reactors: The "lumping approximation" and bifurcation of oscillatory states*, SIAM J. Appl. Math. **27** (1974), 416–429.
2. M. Golubitsky and B. L. Keyfitz, *A qualitative study of the steady-state solutions for a continuous flow stirred tank chemical reactor*, SIAM J. Math. Anal. **11** (1980), 316–339.
3. R. F. Heinemann and A. B. Poore, *Multiplicity, stability and oscillatory dynamics of the tubular reactor*, Chem. Eng. Sci. **36** (1981), 1411–1419.
4. ———, *The effect of activation energy on tubular reactor multiplicity*, Chem. Eng. Sci. **37** (1982), 128–131.
5. A. Kapila and A. B. Poore, *The steady response of a nonadiabatic tubular reactor: New multiplicities*, Chem. Eng. Sci. **37** (1982), 57–68.
6. A. B. Poore, *A model equation arising from chemical reactor theory*, Arch. Rat. Mech. Anal. **52** (1973), 358–388.
7. A. Uppal, W. H. Ray, and A. B. Poore, *On the dynamic behavior of continuous stirred tank reactors*, Chem. Eng. Sci. **29** (1974), 967–985.
8. ———, *Dynamic behavior of the CSTR—Influence of reactor residence time*, Chem. Eng. Sci. **31** (1976), 205–214.
9. A. Varma and R. Aris, *Stirred pots and empty tubes*, Chemical Reactor Theory: A Review, L. Lapidus and N. R. Amundson (eds.), Prentice-Hall, Englewood Cliffs, New Jersey, 1977.

9. An aircraft stability test problem (contributed by W. C. Rheinboldt). Maneuvering airplanes sometimes undergo sudden changes in motion in response to the pilot's control inputs. In particular, the combined effects of rotational coupling and nonlinear aerodynamics may cause sudden jumps in the roll rate when a certain critical roll rate is exceeded. The basic theory explaining this nonlinear response was established by Phillips [**1**]. Further studies may be found, for example, in [**2**], [**3**], [**4**]. The model used here is a simplified version of a model given in [**4**]. It constitutes a system of force-balance equations in which gravity terms have been neglected and only linear and quadratic terms have been retained.

These equilibrium equations have the (dimensionless) form

$$F: \mathbf{R}^8 \to \mathbf{R}^5, \qquad F\mathbf{x} \equiv \mathbf{A}\mathbf{x} + \varphi(\mathbf{x}) = \mathbf{0} \tag{9.1}$$

where $\mathbf{A} \in L(\mathbf{R}^8, \mathbf{R}^5)$ is a constant matrix and $\varphi: \mathbf{R}^8 \to \mathbf{R}^5$ a quadratic nonlinear map. The variables x_1, x_2, x_3 represent the roll rate, pitch rate, and yaw rate, respectively; x_4 is the incremental angle of attack; and x_5 the side-slip angle. The last three variables x_6, x_7, x_8 are the controls—that is, the elevator, aileron, and rudder deflections, respectively.

For a particular aircraft the matrix $\mathbf{A}$ is defined by

$$\mathbf{A} = \begin{pmatrix} -3.933 & 0.107 & 0.126 & 0 & -9.99 & 0 & -45.83 & -7.64 \\ 0 & -0.987 & 0 & -22.95 & 0 & -28.37 & 0 & 0 \\ 0.002 & 0 & -0.235 & 0 & 5.67 & 0 & -0.921 & -6.51 \\ 0 & 1.0 & 0 & -1.0 & 0 & -0.168 & 0 & 0 \\ 0 & 0 & -1.0 & 0 & -0.196 & 0 & -0.0071 & 0 \end{pmatrix}.$$

Moreover, the nonlinear part has the form

$$\varphi(\mathbf{x}) = \begin{pmatrix} -0.727x_2x_3 + 8.39x_3x_4 - 684.4x_4x_5 + 63.5x_4x_2 \\ 0.949x_1x_3 + 0.173x_1x_5 \\ -0.716x_1x_2 - 1.578x_1x_4 + 1.132x_4x_2 \\ -x_1x_5 \\ x_1x_4 \end{pmatrix}.$$

As discussed, for instance, in [**5**], the set of solutions $\mathbf{x}$ of (9.1) constitutes a three-dimensional differentiable manifold in $\mathbf{R}^8$. A typical continuation problem might be to trace the curves defined for fixed settings of x_6 and x_8. Some paths of this type for elevator settings $x_6 = 0.1$, 0.0, -0.008, -0.05 and a zero-rudder deflection $x_8 = 0$ were computed with the continuation code PITCON and are shown in [**5**], [**6**]. These paths contain a number of limit points with respect to x_7 and, in addition, for $x_8 = 0$ there are several bifurcation points, some of which are listed in [**7**].

The computation of the indicated paths represents a good test problem for continuation processes. The detection and exact determination of the limit points was used as a test problem in [**8**] and there also several numerical data are given. Some pictures showing the results of a computation of the two-dimensional solution submanifold defined by the condition $x_8 = 0$ can be found in [**9**].

References

1. W. H. Phillips, *Effect of steady rolling on longitudinal stability*, NASA TN 1627 (1948).

2. A. A. Schy and M. E. Hannah, *Prediction of jump phenomena in roll-coupled maneuvers of airplanes*, J. of Aircraft **14** (1977), 375–382.

3. J. W. Young, A. A. Schy, and K. G. Johnson, *Prediction of jump phenomena in aircraft maneuvers, including nonlinear aerodynamic effects*, J. Guidance and Control **1** (1978), 26–31.

4. R. K. Mehra, W. C. Kessel, and J. V. Carroll, *Global stability and control analysis of aircraft at high angles of attack*, ONR Report CR-215-248-1, 2, 3 (1977).

5. W. C. Rheinboldt, *Numerical Analysis of Parametrized Nonlinear Equations*, Wiley-Interscience, New York, 1986.

6. ——, *Computation of critical boundaries on equilibrium manifolds*, SIAM J. Numer. Anal. **19** (1982), 653–669.

7. A. Jepson and A. Spence, *Folds in solutions of two-parameter systems and their calculations, Part* I, SIAM J. Numer. Anal. **22** (1985), 347–368.

8. R. G. Melhem and W. C. Rheinboldt, *A comparison of methods for determining turning points of nonlinear equations*, Computing **29** (1982), 201–226.

9. W. C. Rheinboldt, *On a moving frame algorithm and the triangulation of equilibrium manifolds*, Bifurcation: Analysis, Algorithms, Applications, *T. Kuepper, R. Seydel, and H. Troger* (*eds.*), Birkhauser Verlag, Basel, 1987.

10. A semiconductor problem (contributed by W. C. Rheinboldt). Two-point boundary value problems

$$-u''(t) = f(u,t), \quad a < t < b, \qquad u(a) = u_a, \quad u(b) = u_b,$$

arise in many applications and provide excellent test cases for nonlinear equation solvers. For

$$f(u,t) = c_a e^{\beta(u_a - u)} - c_b e^{\beta(u - u_b)} + d(t),$$
$$d(t) = \begin{cases} -c_a & \text{for } t \le 0 \\ c_b & \text{for } t > 0 \end{cases}$$

the resulting problem occurs in the analysis of the electrostatic potential of semiconductors (see, e.g., [**1**], [**2**]).

For an application of continuation methods, it is useful to embed the problem into the one-parameter family

$$-u''(t) = f(u,t,\lambda), \quad a < t < b, \qquad u(a) = \lambda u_a, \quad u(b) = \lambda u_b,$$
$$f(u,t,\lambda) = \lambda c_a e^{\lambda\beta(\lambda u_a - u)} - \lambda c_b e^{\lambda\beta(u-\lambda u_b)} + d(t,\lambda),$$
$$d(t,\lambda) = \begin{cases} -\lambda c_a & \text{for } t \le 0 \\ \lambda c_b & \text{for } t > 0. \end{cases}$$

Then for $\lambda = 0$ the solution is trivial.

On the mesh

$$a = t_0 < t_1 < \cdots < t_{n+1} = b$$

we use piecewise linear elements

$$u(t) = \lambda u_a \left(\frac{b-t}{b-a}\right) + \lambda u_b \left(\frac{t-a}{b-a}\right) + \sum_{i=1}^{n} x_i \varphi_i(t)$$

where, as usual,

$$\varphi_i(t) = \begin{cases} \dfrac{t - t_{i-1}}{t_i - t_{i-1}} & \text{for } t_{i-1} \le t \le t_i \\ \dfrac{t_{i+1} - t}{t_{i+1} - t_i} & \text{for } t_i \le t \le t_{i+1} \\ 0 & \text{elsewhere.} \end{cases}$$

Then the unknown vector $\mathbf{x}$ is determined by

$$F\mathbf{x} \equiv \mathbf{A}\mathbf{x} - G(\mathbf{x},\lambda) = \mathbf{0}$$

where $\mathbf{A} = (a_{ij})$ with

$$a_{ij} = \int_a^b \varphi_i'(t)\varphi_j'(t)\,dt,$$

and $G(\mathbf{x},\lambda) = (g_i(\mathbf{x},\lambda))$ with

$$g_j(\mathbf{x},\lambda) = \int_a^b f\left(\lambda u_a \left(\frac{b-t}{b-a}\right) + \lambda u_b \left(\frac{t-a}{b-a}\right) + \sum_{i=1}^{n} x_i\varphi_i(t), t_j, \lambda\right) \varphi_j(t)\,dt + \lambda\left(\frac{u_a - u_b}{b-a}\right)\int_a^b \varphi_j'(t)\,dt.$$

Particularly interesting and computationally challenging values of the parameters are

$$a = -9 \times 10^{-5}, \quad b = 10^{-5}, \quad u_a = 0, \quad u_b = 700,$$
$$\beta = 40, \quad c_a = 10^{12}, \quad c_b = 10^{13}.$$

The solution process then requires double-precision. The following data can be used for single-precision calculations:

$$a = -9 \times 10^{-3}, \quad b = 10^{-3}, \quad u_a = 0, \quad u_b = 25,$$
$$\beta = 20, \quad c_a = 10^r, \quad r = 6 \text{ or } 7.$$

In either case the mesh is defined in the table.

i	0	1	2	3	4	5	6	7	8	9	10
$10^3 t_i$	−9	−7	−5	−4.5	−4.25	−4	−3.75	−3.5	−3	−1	−1

References

1. S. J. Polak, et al., *A continuation method for the calculation of electrostatic potential in semiconductors*, N. V. Philips Gloeilampen-fabrieken, Eindhoven, The Netherlands, Tech. Rep. ISA-TIS/CARD, 1979.

2. C. den Heijer and W. C. Rheinboldt, *On steplength algorithms for a class of continuation methods*, SIAM J. Num. Anal. **18** (1981), 925–948.

11. A shallow arch (contributed by W. C. Rheinboldt). The shallow arch represents one of the simplest mechanical structures illustrating many of the elastic instabilities; in particular, it may "snap through" or alternately buckle asymmetrically. The following finite-element model of a thin, shallow, circular arch has been used frequently as a test case; see, in particular, [1], [2], [3], as well as [4] and [5]. In dimensionless form the total potential energy is given by

$$\pi = \int_{-\theta_0}^{\theta_0} \left\{ \left[\left(\frac{dw}{d\theta} - u \right) + \frac{1}{2}\alpha_0 \left(\frac{du}{d\theta} \right)^2 \right]^2 + \alpha_1 \left(\frac{d^2u}{d\theta^2} \right)^2 - \alpha_2 p u \right\} d\theta,$$

where u, w are the radial and axial displacements, p is the radial load, θ_0 is the opening angle of the arc, and α_0, α_1, α_2 are constants. For pinned ends the boundary conditions are

$$(11.1) \qquad u(\pm\theta_0) = 0, \qquad w(\pm\theta_0) = 0, \qquad \frac{d^2u}{d\theta^2}(\pm\theta_0) = 0.$$

Consider the uniform mesh $\theta_i = -\theta_0 + i\Delta\theta_0$, $\Delta\theta_0 = \theta_0/(2m)$, $i = 0, \ldots, m$. In [1] Walker introduced the displacement functions

$$u_i(s) = \sum_{j=1}^{6} u_{ij}\varphi_j(s), \quad w_i(s) = \sum_{j=1}^{4} w_{ij}\psi_j(s), \qquad i = 0, \ldots, m,$$

where s is defined by $\theta = \theta_i + s\Delta\theta_0$, $0 \le s \le 1$, and the shape functions are given in the table.

j	φ_j	ψ_j
1	$1 - \varphi_4(s)$	$1 - \psi_3(s)$
2	$s - 6s^3 + 8s^4 - 3s^5$	$s(1-s)^2$
3	$\frac{1}{2}s^2(1-s)^3$	$s^2(3-2s)$
4	$10s^3 - 15s^4 + 6s^5$	$s^2(s-1)$
5	$-s^3(s-1)(3s-4)$	–
6	$\frac{1}{2}s^3(1-s)^2$	–

From the definition of the variables we have

$$\left.\begin{aligned} u_{i,j+3} &= u_{i+1,j}, \quad j = 1,2,3 \\ w_{i,j+2} &= w_{i+1,j}, \quad j = 1,2 \end{aligned}\right\} \quad i = 1,2,\ldots,m-1$$

and the boundary conditions (11.1) are equivalent with

$$u_{1,1} = u_{m,4} = 0, \quad w_{1,1} = w_{m,3} = 0, \quad u_{1,3} = u_{m,6} = 0.$$

Thus, a standard application of the finite-element procedure results in a system of $5m - 1$ equations in as many unknowns.

For the data given in [3] the dimensionless constants are $\alpha_0 = 0.0684$, $\alpha_1 = 3.8716 \times 10^{-6}$, $\alpha_2 = 1.6550 \times 10^{-1}$, $\theta_0 = 15°$, and $m = 8$. For the constant radial load

$$p(\theta) = \theta, \qquad -\theta_0 \le \theta \le \theta_0,$$

a symmetry breaking bifurcation point occurs before a limit point is reached; that is, the structure will lose stability at the bifurcation point and buckle asymmetrically. It is interesting to study the changes of these points when an asymmetric load, such as

$$p(\theta) = \begin{cases} 0.9\theta, & \text{for } -\theta_0 \le \theta < 0 \\ 1.1\theta, & \text{for } 0 \le \theta \le \theta_0, \end{cases}$$

is used. A more challenging computational problem is to determine the location of a worst load, such as

$$p(\theta) = \begin{cases} \left(1 - \dfrac{4}{\theta_0}(\nu - \theta)\right)\mu, & \text{for } \max\left(-\theta_0,\ \nu - \dfrac{\theta_0}{4}\right) \le \theta \le \nu \\ \left(1 - \dfrac{4}{\theta_0}(\theta - \nu)\right)\mu, & \text{for } \nu \le \theta \le \min\left(\theta_0,\ \nu + \dfrac{\theta_0}{4}\right) \\ 0, & \text{elsewhere.} \end{cases}$$

Thus, these loads are piecewise linear "hat" functions which have the value μ at $\theta = \nu$ and are zero outside the interval of width $\theta_0/2$ centered at ν. For each ν we are interested in determining the smallest value of μ where stability is lost. In other words, we wish to compute the corresponding foldline on the two-dimensional equilibrium manifold of the problem. Its projection onto the (ν, μ)-plane is a W-shaped curve which shows that the worst locations of the load correspond to small nonzero angles ν. A figure of this curve is given in [5] and simplicial approximation of a relevant part of the manifold is shown in [6].

REFERENCES

1. A. C. Walker, *A nonlinear finite-element analysis of shallow circular arches*, Int. J. Solids Struct. **5** (1969), 97–107.

2. A. D. Kerr and M. T. Soifer, *The linearization of the prebuckling state and its effect on the determining instability load*, Trans. ASME J. of Appl. Mech. **36** (1969), 775–783.

3. S. T. Mau and R. H. Gallagher, *A finite-element procedure for nonlinear prebuckling and initial postbuckling analysis*, NASA Report, NASA-CR 1936, 1972.

4. R. G. Melhem and W. C. Rheinboldt, *A comparison of methods for determining turning points of nonlinear equations*, Computing **29** (1982), 201–226.

5. I. Babuska and W. C. Rheinboldt, *Adaptive finite-element processes in structural mechanics*, Elliptic Problem Solvers II, G. Birkhoff and A. Schoenstadt (eds.), Academic Press, New York, 1984, pp. 345–378.

6. W. C. Rheinboldt, *On the computation of multidimensional solution manifolds of parametrized equations*, Num. Math. **53** (1988), 165–181.

12. A trigger circuit (contributed by W. C. Rheinboldt). The operation of a particular trigger circuit was described in [1] (see also [2]). The nodal equations of the circuit may be written in the form

$$F\mathbf{x} \equiv \mathbf{A}\mathbf{x} + \varphi(\mathbf{x}) = \mathbf{0}$$

where

$$F: \mathbf{R}^7 \to \mathbf{R}^6, \qquad \mathbf{A} \in L(\mathbf{R}^7, \mathbf{R}^6), \qquad \varphi: \mathbf{R}^7 \to \mathbf{R}^6,$$

and all variables denote voltages and, in particular, x_6 is the output voltage and x_7 the input voltage. Accordingly, we are especially interested in the dependence between x_6 and x_7. All coefficients of the matrix A are zero except those given in the following list:

$$a_{11} = R_0^{-1} + R_1^{-1} + R_2^{-1}, \quad a_{12} = -R_1^{-1}, \quad a_{13} = R_0^{-1}, \quad a_{17} = R_2^{-1}$$
$$a_{21} = -R_1^{-1}, \quad a_{22} = R_1^{-1} + R_3^{-1}, \quad a_{16} = -R_3^{-1}$$
$$a_{31} = -R_0^{-1}, \quad a_{33} = R_0^{-1} + R_4^{-1}, \quad a_{34} = -R_4^{-1}$$
$$a_{43} = -R_4^{-1}, \quad a_{44} = R_4^{-1} + R_5^{-1} + R_6^{-1}, \quad a_{45} = -R_5^{-1}$$
$$a_{54} = -R_5^{-1}, \quad a_{55} = R_5^{-1} + R_7^{-1}, \quad a_{56} = -R_7^{-1}$$
$$a_{62} = -R_3^{-1}, \quad a_{65} = -R_7^{-1}, \quad a_{66} = R_3^{-1} + R_7^{-1} + R_8^{-1}$$

while

$$\varphi(\mathbf{x}) = (0, g_1(x_2), 0, 0, g_1(x_5), -R_8^{-1}, g_2(x_3 - x_1))^{\mathrm{T}}$$

where

$$g_1(u) = (5.6 \times 10^{-8})\exp(25u - 1)$$

models a diode and

$$g_2(u) = 7.65 \arctan(1,962u)$$

an amplifier.

For the resistance values

$$R_0 = 10^4, \quad R_1 = 39.0, \quad R_2 = 51.0, \quad R_3 = 10.0, \quad R_4 = 25.5,$$
$$R_5 = 1.0, \quad R_6 = 0.62, \quad R_7 = 13.0, \quad R_8 = 0.201$$

there are the following two limit points:

x^1	x^2
0.049366971	0.235777668
0.547358409	0.662968764
0.049447207	0.237597699
0.049447411	0.237602341
0.129201309	0.620832106
1.166019152	9.608996879
0.601853012	0.322866124

REFERENCES

1. G. Poenisch and H. Schwetlick, *Computing turning points of curves implicitly defined by nonlinear equations depending on a parameter*, Computing **26** (1981), 107–121.

2. R. G. Melhem and W. C. Rheinboldt, *A comparison of methods for determining turning points of nonlinear equations*, Computing **29** (1982), 201–226.

13. The inverse elastic rod problem (contributed by H. F. Walker). This problem is considered in [2], which gives a number of other references going back to Euler. A large planar deflection of a thin rod, or *elastica*, is modeled by

$$EI\frac{d\theta}{ds} = Qx - Py + M,$$

where EI is the flexural rigidity, θ is the local angle of inclination, and s is the arclength along the elastica. See Figure 13.1. For simplicity, we take $EI = 1$ and assume that the elastica is of unit length with the left endpoint clamped horizontally at the origin. This gives the initial-value problem

$$\frac{d\theta}{ds} = Qx - Py + M, \quad \frac{dx}{ds} = \cos\theta, \quad \frac{dy}{ds} = \sin\theta,$$
$$x(0) = y(0) = \theta(0) = 0.$$

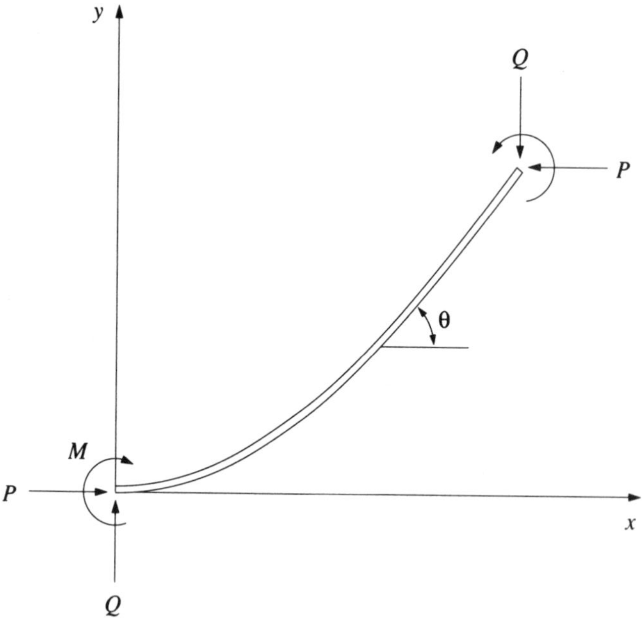

FIGURE 13.1. The coordinate system

The problem is to determine Q, P, and M so that $x(1)$, $y(1)$, and $\theta(1)$ have specified values, that is, to solve

$$F(Q,P,M) = \begin{pmatrix} x(1) - a \\ y(1) - b \\ \theta(1) - c \end{pmatrix} = \mathbf{0}$$

for given a, b, and c. Note that the evaluation of $F(Q,P,M)$ requires the numerical integration of the initial-value problem, and suitably controlling the error in this can be an interesting issue.

The problem can be treated as a simple nonlinear system problem or as a homotopy or continuation problem. As a simple nonlinear system problem, it can be quite challenging for Newton-like methods because the elastica is very sensitive to end conditions, especially for more complicated shapes which require large forces and torques to achieve. Unadorned Newton's method can require an extremely good initial guess for convergence, and even accepted global convergence safeguards are of limited utility. Homotopy methods seem to fare better. An approach suggested in [2] is first to choose initial Q_0, P_0, M_0 and a_0, b_0, c_0 such that $(x(1), y(1), \theta(1)) = (a_0, b_0, c_0)$ when $(Q, P, M) = (Q_0, P_0, M_0)$ and then to track the zero curve of

$$\overline{F}(Q, P, M, \lambda) \equiv \begin{pmatrix} x(1) - [\lambda a + (1-\lambda)a_0] \\ y(1) - [\lambda b + (1-\lambda)b_0] \\ \theta(1) - [\lambda c + (1-\lambda)c_0] \end{pmatrix}$$

from $\lambda = 0$ to $\lambda = 1$.

REMARKS. (1) Jacobian matrices can be evaluated either by finite differences or by integrating appropriate auxiliary initial-value problems. (2) For $a = 0$, $b = 2/\pi$, $c = \pi$, the exact solution is $Q = 0$, $P = 0$, $M = \pi$, i.e., the elastica is a semicircle.

See [2] for more details, including some interesting instances of the problem and some test results for various algorithms applied to it. See also [1] for similar problems, including some that are more difficult than this one.

REFERENCES

1. L. T. Watson, *Engineering applications of the Chow-Yorke algorithm*, Appl. Math. Comput. **9** (1981), 111–133.

2. L. T. Watson and C. Y. Wang, *A homotopy method applied to elastic problems*, Int. J. Solids Struct. **17** (1981), 29–37.

14. The driven cavity problem (contributed by H. F. Walker). This is a classical problem from incompressible fluid flow which is considered in [1] and [2]. The steady-state flow of an incompressible and viscous Newtonian fluid in a domain $\Omega \subset \mathbf{R}^2$ is modeled by the Navier–Stokes equations. If ψ is a *stream function*, the level curves of which represent the streamlines of the velocity field, then an appropriate boundary-value problem for ψ arising from the Navier–Stokes equations is a particular case of the general nonlinear biharmonic problem

$$\nu\Delta^2\psi + \frac{\partial\psi}{\partial x_1}\frac{\partial}{\partial x_2}\Delta\psi - \frac{\partial\psi}{\partial x_2}\frac{\partial}{\partial x_1}\Delta\psi = f \quad \text{in } \Omega, \tag{14.1}$$

with boundary conditions

$$\psi = g_1 \quad \text{and} \quad \frac{\partial\psi}{\partial n} = g_2 \quad \text{on } \partial\Omega. \tag{14.2}$$

The *viscosity* ν is the reciprocal of the Reynolds number Re. An equivalent formulation is

$$(14.3)\qquad \begin{aligned} -\nu\Delta\omega + \frac{\partial\psi}{\partial x_2}\frac{\partial\omega}{\partial x_1} - \frac{\partial\psi}{\partial x_1}\frac{\partial\omega}{\partial x_2} &= f \quad \text{in } \Omega, \\ -\Delta\psi = \omega \quad \text{in } \Omega,& \end{aligned}$$

with boundary conditions (14.2) and where ω is the *vorticity* of the flow. See **[2]** for more background and details.

In **[2]**, problem (14.3) with boundary conditions (14.2) is used as a test problem for an arclength continuation method with $\Omega = (0,1) \times (0,1)$, $f \equiv 0$, $g_1 \equiv 0$, and $g_2(x_1, x_2) = 1$ if $x_2 = 1$ and $g_2(x_1, x_2) = 0$ if $0 \leq x_2 < 1$. The method is applied to determine solutions for $100 \leq Re = 1/\nu \leq 3,000$ of a discrete problem obtained from a variational formulation of (14.2) by piecewise-quadratic finite-element approximation.

In **[1]**, problem (14.1) with boundary conditions (14.2) and with the same Ω, f, g_1, and g_2 as in **[2]** is used as a test problem for Newton iterative methods. The methods are applied to a discretized problem obtained by piecewise-linear finite-element approximation with $Re = 1/\nu = 500$, 1,500, 2,000, 3,000, 5,000.

References

1. P. N. Brown and Y. Saad, *Hybrid Krylov methods for nonlinear systems of equations*, Lawrence Livermore National Laboratory Report UCRL-97645, 1987, to appear in SIAM J. Sci. Stat. Comp.

2. R. Glowinski, H. B. Keller, and L. Rheinhart, *Continuation-conjugate gradient methods for the least-squares solution of nonlinear boundary value problems*, SIAM J. Sci. Stat. Comp. **6** (1985), 793–832.

ARGONNE NATIONAL LABORATORY